Methods in Enzymology

Volume 65
NUCLEIC ACIDS
Part I

METHODS IN ENZYMOLOGY

EDITORS-IN-CHIEF

Sidney P. Colowick   Nathan O. Kaplan

Methods in Enzymology

Volume 65

Nucleic Acids

Part I

EDITED BY

Lawrence Grossman

DEPARTMENT OF BIOCHEMISTRY
THE JOHNS HOPKINS UNIVERSITY
SCHOOL OF HYGIENE AND PUBLIC HEALTH
BALTIMORE, MARYLAND

Kivie Moldave

DEPARTMENT OF BIOLOGICAL CHEMISTRY
CALIFORNIA COLLEGE OF MEDICINE
UNIVERSITY OF CALIFORNIA
IRVINE, CALIFORNIA

1980

ACADEMIC PRESS

A Subsidiary of Harcourt Brace Jovanovich, Publishers

New York   London   Toronto   Sydney   San Francisco

ACADEMIC PRESS, INC.
111 Fifth Avenue, New York, New York 10003

United Kingdom Edition published by
ACADEMIC PRESS, INC. (LONDON) LTD.
24/28 Oval Road, London NW1 7DX

Library of Congress Cataloging in Publication Data
Main entry under title:

Nucleic acids.

(Methods in enzymology, v. 12, 20–21, 29–30, 65)
Pts. C, E–F have title: Nucleic acids and protein
synthesis: with editor's names in reverse order on t. p.
Includes bibliographical references.
1. Nucleic acids. 2. Protein biosynthesis.
I. Grossman, Lawrence, Date ed. II. Moldave,
Kivie, Date ed. III. Title: Nucleic acids and
protein synthesis. IV. Series: Methods in enzymology.
v. 12 [etc.] [DNLM: 1. Nucleic acids––Biosynthesis.
2. Proteins––Biosynthesis. W 1 Me9615K v. 30 1974
QU 55 N964 1974]
QP601.M49 vol. 12, etc. 574.1'925'08s [QP620]
ISBN 0–12–181965–5 (v. 65) [574.8'732] 74–26909

PRINTED IN THE UNITED STATES OF AMERICA

81 82 83   9 8 7 6 5 4 3 2

Table of Contents

Section I. Assays for [Class II Restriction] Endonucleases

Section II. Techniques for Labeling Termini

Section III. Purification of Restriction Enzymes

Section IV. Purification and Properties of Enzymes Acting at Sites with Altered Bases

Section V. Endonuclease Cleavage Mapping Techniques

Section VI. Determination of DNA Fragment Sizes

Section VII. Determination of Fragment Ordering

Section VIII. Nucleotide Sequencing Techniques

Section IX. Localization of Functional Sites on Chromosomes

Contributors to Volume 65

Article numbers are in parentheses following the names of contributors.
Affiliations listed are current.

KAN L. AGARWAL (19), *Department of Biochemistry, University of Chicago, Chicago, Illinois 60637*

CHANDER P. BAHL (78), *Cetus Corporation, Berkeley, California 94710*

SUSAN M. BERGET (69), *Department of Biochemistry, Rice University, Houston, Texas 77001*

ARNOLD J. BERK (69), *Department of Microbiology, University of California, Los Angeles, California 90024*

KATHLEEN L. BERKNER (5), *Department of Biology, Massachusetts Institute of Technology, Cambridge, Massachusetts 02139*

THOMAS A. BICKLE (11, 16), *Department of Microbiology, Biozentrum, University of Basel, CH-4056 Basel, Switzerland*

M. L. BIRNSTIEL (55), *Institut für Molekularbiologie II der Universität Zürich, Hönggerberg, 8093 Zürich, Switzerland*

ROBERT BLAKESLEY (23), *Bethesda Research Laboratory, Rockville, Maryland 20850*

P. G. BOSELEY (55), *Department of Biological Sciences, University of Warwick, Coventry CV4 7A1, United Kingdom*

SIERD BRON (15), *Department of Genetics, Centre of Biological Sciences, University of Groningen, 9751 NN Haren (Gn), Kerklaan 30, The Netherlands*

R. BROUSSEAU (61), *Division of Biological Sciences, National Research Council of Canada, Ottawa, Ontario K1A OR6, Canada*

NIGEL L. BROWN (48), *Department of Biochemistry, University of Bristol, Bristol Medical School, Bristol BS8 1TD, England*

A. G. BRUCE (9), *Department of Biochemistry, University of Illinois, Urbana, Illinois 61801*

A. I. BUKHARI (52), *Cold Spring Harbor Laboratory, Cold Spring Harbor, New York 11724*

JOHN CAMERON (49), *Department of Biochemistry, Stanford University School of Medicine, Stanford, California 94305*

GORDON G. CARMICHAEL (47), *Department of Pathology, Harvard Medical School, Boston, Massachusetts 02115*

JAMES F. CATTERALL (21), *Department of Cell Biology, Baylor College of Medicine, Houston, Texas 77030*

GEORGE CHACONAS (10), *Cold Spring Harbor Laboratory, Cold Spring Harbor, New York 11724*

MARK D. CHALLBERG (6), *Department of Microbiology, The Johns Hopkins University School of Medicine, Baltimore, Maryland 21205*

THOMAS R. CHAUNCEY (19), *Department of Biochemistry, University of Chicago, Chicago, Illinois 60637*

JACK G. CHIRIKJIAN (23), *Department of Biochemistry, Georgetown University Medical Center, Washington, D.C. 20007*

ALAN D. D'ANDREA (31), *Sidney Farber Cancer Institute, Harvard Medical School, Boston, Massachusetts 02115*

KATHLEEN J. DANNA (53), *Department of Molecular, Cellular, and Developmental Biology, University of Colorado, Boulder, Colorado 80309*

SANTANU DASGUPTA (51), *Biophysics Laboratory and Department of Biochemistry, University of Wisconsin, Madison, Wisconsin 53706*

RONALD W. DAVIS (49), *Department of Biochemistry, Stanford University School of Medicine, Stanford, California 94305*

BRUCE DEMPLE (29), *Department of Biochemistry, University of California, Berkeley, California 94720*

ASHLEY R. DUNN (54), *Cold Spring Harbor Laboratory, Cold Spring Harbor, New York 11724*

MARSHALL H. EDGELL (40), *Division of Health Affairs, Department of Bacteriology and Immunology, University of North Carolina, Chapel Hill, North Carolina 27514*

ARGIRIS EFSTRATIADIS (38), *Department of Biological Chemistry, Harvard Medical School, Boston, Massachusetts 02115*

T. E. ENGLAND (9), *Department of Biochemistry, University of Illinois, Urbana, Illinois 61801*

PAUL T. ENGLUND (6), *Department of Physiological Chemistry, The Johns Hopkins University School of Medicine, Baltimore, Maryland 21205*

GEORGE C. FAREED (67), *Department of Microbiology and Immunology, and Molecular Biology Institute, University of California, Los Angeles, California 90024*

JENNIFER FAVALORO (68), *Imperial Cancer Research Fund Laboratories, Lincoln's Inn Fields, London WC2A 3PX, England*

WILLIAM R. FOLK (5), *Department of Biological Chemistry, University of Michigan, Ann Arbor, Michigan 48109*

M. J. FRASER (33), *Department of Biochemistry, McGill University, McIntyre Medical Sciences Building, Montreal, Quebec, Canada H3G 1Y6*

ERROL C. FRIEDBERG (25), *Department of Pathology, Stanford University School of Medicine, Stanford, California 94305*

ANN K. GANESAN (25), *Department of Biological Sciences, Stanford University, Stanford, California 94305*

FREDERICK T. GATES III (29), *Department of Biochemistry, University of California, Berkeley, California 94720*

P. K. GHOSH (59), *Department of Internal Medicine, Yale University School of Medicine, New Haven, Connecticut 06510*

WALTER GILBERT (57), *Department of Biochemistry and Molecular Biology, Harvard University, Cambridge, Massachusetts 02138*

SHIRLEY GILLAM (65), *Department of Biochemistry, University of British Columbia,*

Vancouver, British Columbia V6T 1W5, Canada

HOWARD M. GOODMAN (8), *The Howard Hughes Medical Institute, Department of Biochemistry and Biophysics, University of California, San Francisco, California 94143*

A. GRAESSMANN (74), *Institut für Molekularbiologie und Biochemie, Freien Universität Berlin, D-1000 Berlin 33, Federal Republic of Germany*

M. GRAESSMANN (74), *Institut für Molekularbiologie und Biochemie, Freien Universität Berlin, D-1000 Berlin 33, Federal Republic of Germany*

F. L. GRAHAM (75), *Departments of Biology and Pathology, McMaster University, Hamilton, Ontario L8S 4K1, Canada*

LAWRENCE GROSSMAN (28), *Department of Biochemistry, The Johns Hopkins University School of Hygiene and Public Health, Baltimore, Maryland 21205*

RAMESH C. GUPTA (63), *Department of Pharmacology, Baylor College of Medicine, Texas Medical Center, Houston, Texas 77030*

STEPHEN C. HARDIES (41), *Department of Bacteriology and Immunology, University of North Carolina, Chapel Hill, North Carolina 27514*

WILLIAM A. HASELTINE (31, 64), *Sidney Farber Cancer Institute, Harvard Medical School, Boston, Massachusetts 02115*

JEROME L. HINES (19), *Department of Biochemistry, University of Chicago, Chicago, Illinois 60637*

JAY HIRSH (79), *Department of Biological Chemistry, Harvard Medical School, Boston, Massachusetts 02115*

WOLFRAM HÖRZ (15), *Institut für Physiologische Chemie, Physikalische Biochemie und Zellbiologie, Universität München, Goethestrasse 33, D-8000 München 2, Federal Republic of Germany*

GLENN T. HORN (41), *Department of Biochemistry, University of Wisconsin, Madison, Wisconsin 53706*

H. M. HSIUNG (61), *Division of Biological*

Sciences, National Research Council of Canada, Ottawa, Ontario K1A OR6, Canada

ROLAND IMBER (16), Department of Microbiology, Biozentrum, University of Basel, CH-4056 Basel, Switzerland

ROSS B. INMAN (51), Biophysics Laboratory and Department of Biochemistry, University of Wisconsin, Madison, Wisconsin 53706

PETER G. N. JEPPESEN (39), MRC Clinical and Population Cytogenetics Unit, Western General Hospital, Edinburgh EH4 2XU, United Kingdom

ALEXANDER D. JOHNSON (76), Department of Biochemistry and Molecular Biology, The Biological Laboratories, Harvard University, Cambridge, Massachusetts 02138

LORRAINE JOHNSRUD (31), Biological Laboratories, Harvard University, Cambridge, Massachusetts 02138

ROBERT KAMEN (68), Imperial Cancer Research Fund Laboratories, Lincoln's Inn Fields, London WC2A 3PX, England

D. KAMP (52), Cold Spring Harbor Laboratory, Cold Spring Harbor, New York 11724

HARUMI KASAMATSU (67), Department of Biology and Molecular Biology Institute, University of California, Los Angeles, California 90024

DENNIS G. KLEID (20), Division of Molecular Biology, Genentech, Inc., South San Francisco, California 94080

BARBARA KLEIN (41), Department of Chemistry, University of Wisconsin, Madison, Wisconsin 53706

RUUD N. H. KONINGS (72), Laboratory of Molecular Biology, University of Nijmegen, Nijmegen, The Netherlands

LAURENCE JAY KORN (60), Department of Embryology, Carnegie Institution of Washington, Baltimore, Maryland 21210

DAVID LACKEY (4), Department of Biochemistry, University of California, Berkeley, California 94720

SANFORD A. LACKS (17), Department of

Biology, Brookhaven National Laboratory, Upton, New York 11973

CHING-JUH LAI (66, 73), Laboratory of Infectious Diseases, National Institute of Allergy and Infectious Diseases, National Institutes of Health, Bethesda, Maryland 20014

JACQUELYNN E. LARSON (41), Department of Biochemistry, University of Wisconsin, Madison, Wisconsin 53706

RONALD A. LASKEY (45), Medical Research Council Laboratory of Molecular Biology, Cambridge CB2 2QH, England

M. LASKOWSKI, SR. (34, 35), Laboratory of Enzymology, Roswell Park Memorial Institute, Buffalo, New York 14263

P. LEBOWITZ (59), Department of Internal Medicine, Yale University School of Medicine, New Haven, Connecticut 06510

YAN HWA LEE (23), Department of Biochemistry, Georgetown University Medical Center, Washington, D.C. 20007

TOMAS LINDAHL (36), Department of Medical Chemistry, University of Gothenburg, 400 33 Gothenburg, Sweden

CHRISTINA P. LINDAN (31), Sidney Farber Cancer Institute, Harvard Medical School, Boston, Massachusetts 02115

STUART LINN (4, 29), Department of Biochemistry, University of California, Berkeley, California 94720

JOHN T. LIS (42), Department of Biochemistry, Molecular and Cell Biology, Cornell University, Ithaca, New York 14853

SIV LJUNGQUIST (27), Department of Medical Cell Genetics, Medical Nobel Institute, Karolinska Institutet, S-104 01 Stockholm, Sweden

GARY K. MCMASTER (47), Swiss Institute for Experimental Cancer Research, CH-1066 Epalinges, Switzerland

TOM MANIATIS (38), Division of Biology, California Institute of Technology, Pasadena, California 91125

GARRY M. MARLEY (13), Department of Microbiology, The Johns Hopkins University School of Medicine, Baltimore, Maryland 21205

ALLAN M. MAXAM (57), Department of Biochemistry and Molecular Biology, Harvard University, Cambridge, Massachusetts 02138

J. J. MICHNIEWICZ (61), Division of Biological Sciences, National Research Council of Canada, Ottawa, Ontario K1A OR6, Canada

PAUL MODRICH (12), Department of Biochemistry, Duke University Medical Center, Durham, North Carolina 27710

T. MOSS (55), Institut für Molekularbiologie II der Universität Zürich, Hönggerberg, 8093 Zürich, Switzerland

C. MUELLER (74), Institut für Molekularbiologie und Biochemie, Freien Universität Berlin, D-1000 Berlin 33, Federal Republic of Germany

S. A. NARANG (61, 78), Division of Biological Sciences, National Research Council of Canada, Ottawa, Ontario K1A OR6, Canada

SANDRA K. NEUENDORF (41), Department of Biochemistry, University of Wisconsin, Madison, Wisconsin 53706

JOSEPH R. NEVINS (70), The Rockefeller University, New York, New York 10021

CARL O. PABO (76), Department of Biochemistry and Molecular Biology, The Biological Laboratories, Harvard University, Cambridge, Massachusetts 02138

RICHARD A. PADGETT (49), Department of Biochemistry, Stanford University School of Medicine, Stanford, California 94305

NIKOS PANAYOTATOS (41), Department of Biochemistry, University of Wisconsin, Madison, Wisconsin 53706

RICHARD C. PARKER (44, 50), Department of Microbiology and Immunology, University of California, San Francisco, California 94143

ROGER K. PATIENT (41), Imperial Cancer Research Fund, Mill Hill Laboratories, London NW7 1AD, England

FINN SKOU PEDERSEN (64), Sidney Farber Cancer Institute, Harvard Medical School, Boston, Massachusetts 02115

M. PIATAK (59), Department of Human Genetics, Yale University School of Medicine, New Haven, Connecticut 06510

VINCENZO PIRROTTA (11, 16), European Molecular Biology Laboratory, D-6900 Heidelberg, Federal Republic of Germany

FRED I. POLSKY (40), Division of Health Affairs, University of North Carolina, Chapel Hill, North Carolina 27514

ARIEL PRUNELL (43), Institut de Recherches en Biologie Moléculaire, Université Paris VII, 75005 Paris, France

CARY L. QUEEN (60), Laboratory of Biochemistry, National Cancer Institute, National Institutes of Health, Bethesda, Maryland 20205

ALAIN RAMBACH (22), Institut Pasteur, 75724 Paris Cédex 15, France

ERIKA RANDERATH (63), Department of Pharmacology, Baylor College of Medicine, Texas Medical Center, Houston, Texas 77030

KURT RANDERATH (63), Department of Pharmacology, Baylor College of Medicine, Texas Medical Center, Houston, Texas 77030

V. B. REDDY (59), Department of Human Genetics, Yale University School of Medicine, New Haven, Connecticut 06510

SHEIKH RIAZUDDIN (24, 30, 37), Nuclear Institute for Agriculture and Biology, Faisalabad, Pakistan

RICHARD J. ROBERTS (1), Cold Spring Harbor Laboratory, Cold Spring Harbor, New York 11724

J. D. ROCHAIX (71), Department de Biologie Moléculaire, Université de Genève, CH-1211 Genève 4, Switzerland

STEPHEN G. ROGERS (26), Department of Microbiology and Immunology, Indiana University School of Medicine, Indianapolis, Indiana 46202

RANAJIT ROYCHOUDHURY (7), *Abbott Laboratories, Department 474, R1-B, North Chicago, Illinois 60064*

ROBERT A. RUBIN (12), *Department of Biochemistry, Duke University Medical Center, Durham, North Carolina 27710*

THOMAS P. ST. JOHN (49), *Department of Biochemistry, Stanford University School of Medicine, Stanford, California 94305*

JOSEPH SAMBROOK (54), *Cold Spring Harbor Laboratory, Cold Spring Harbor, New York 11724*

ROBERT T. SAUER (76), *Department of Biology, Massachusetts Institute of Technology, Cambridge, Massachusetts 02139*

STEWART SCHERER (49), *Department of Biochemistry, Stanford University School of Medicine, Stanford, California 94305*

ROBERT SCHLEIF (2, 79), *Department of Biochemistry, Brandeis University, Waltham, Massachusetts 02254*

PATRICIA C. SEAWELL (25), *Department of Biological Sciences, Stanford University, Stanford, California 94305*

BRIAN SEED (44), *Division of Biology, California Institute of Technology, Pasadena, California 91125*

ERIK SELSING (41), *Department of Microbiology and Immunology, University of Washington, Seattle, Washington 98195*

NANCY L. SHAPER (28), *Department of Biochemistry, The Johns Hopkins University School of Hygiene and Public Health, Baltimore, Maryland 21205*

PHILLIP A. SHARP (69), *Center for Cancer Research and Department of Biology, Massachusetts Institute of Technology, Cambridge, Massachusetts 02139*

ANDREW J. H. SMITH (58), *MRC Laboratory of Molecular Biology, Hills Road, Cambridge CB2 2QH, England*

HAMILTON O. SMITH (13, 46), *Department of Microbiology, The Johns Hopkins University School of Medicine, Baltimore, Maryland 21205*

LEONARD A. SMITH (23), *Laboratory of Biochemistry, National Cancer Institute, Bethesda, Maryland 20205*

MICHAEL SMITH (48, 65), *Department of Biochemistry, University of British Columbia, Vancouver, British Columbia V6T 1W5, Canada*

M. TAKANAMI (3, 56), *Institute for Chemical Research, Kyoto University, Uji, Kyoto-Fu, Japan*

MARJORIE THOMAS (49), *Department of Biochemistry, Stanford University School of Medicine, Stanford, California 94305*

RICHARD TREISMAN (68), *Imperial Cancer Research Fund Laboratories, Lincoln's Inn Fields, London WC2A 3PX, England*

CHEN-PEI D. TU (62), *Department of Medicine, Stanford University School of Medicine, Stanford, California 94305*

O. C. UHLENBECK (9), *Department of Biochemistry, University of Illinois, Urbana, Illinois 61801*

A. J. VAN DER EB (75), *Department of Medical Biochemistry, Sylvius Laboratories, University of Leiden, 2333 AL Leiden, The Netherlands*

JOHAN H. VAN DE SANDE (10), *Division of Medical Biochemistry, University of Calgary, Alberta, Canada T2N 1N4*

PÁL VENETIANER (14), *Institute of Biochemistry, Biological Research Center, Hungarian Academy of Sciences, Szeged, 6701 Hungary*

VOLKER M. VOGT (32), *Department of Biochemistry, Molecular and Cell Biology, Cornell University, Ithaca, New York 14853*

BERNARD WEISS (26), *Department of Microbiology, The Johns Hopkins University School of Medicine, Baltimore, Maryland 21205*

S. M. WEISSMAN (59), *Department of Human Genetics, Yale University School of Medicine, New Haven, Connecticut 06510*

N. E. WELKER (21), *Department of Biochemistry and Molecular Biology, Northwestern University, Evanston, Illinois 60201*

ROBERT D. WELLS (41), *Department of Biochemistry, University of Wisconsin, Madison, Wisconsin 53706*

GARY A. WILSON (18), *Department of Microbiology, University of Rochester Medical Center, Rochester, New York 14642*

RAY WU (7, 62, 78), *Department of Biochemistry, Molecular and Cell Biology, Cornell University, Ithaca, New York 14853*

FRANK E. YOUNG (18), *Department of Microbiology, University of Rochester Medical Center, Rochester, New York 14642*

GEOFFREY ZUBAY (77), *Department of Biological Sciences, Columbia University, New York, New York 10027*

Preface

During the interim following publication of Volumes XXIX, Part E and XXX, Part F of "Nucleic Acids and Protein Synthesis" there has been a remarkable transformation of DNA methodology. With the availability of restriction endonucleases which can recognize unique sequences, newer techniques which facilitate the sequencing of DNA, and DNA cloning methodology, the molecular biologist has at his disposal the most precise tools for isolating and characterizing gene structures and for closer examination of gene expression. This volume is dedicated to these newer methods and to Daniel Nathans and Hamilton O. Smith whose discoveries led to this explosion of new information. Their personal assistance facilitated the organization of this volume.

Recent advances in DNA methodology applied to gene expression and cloning, elaborated upon here, are presented in greater depth in Volume 68, "Recombinant DNA," edited by Ray Wu.

LAWRENCE GROSSMAN
KIVIE MOLDAVE

METHODS IN ENZYMOLOGY

EDITED BY

Sidney P. Colowick and Nathan O. Kaplan

VANDERBILT UNIVERSITY
SCHOOL OF MEDICINE
NASHVILLE, TENNESSEE

DEPARTMENT OF CHEMISTRY
UNIVERSITY OF CALIFORNIA
AT SAN DIEGO
LA JOLLA, CALIFORNIA

METHODS IN ENZYMOLOGY

EDITORS-IN-CHIEF

Sidney P. Colowick   Nathan O. Kaplan

VOLUME XXXV. Lipids (Part B)
Edited by JOHN M. LOWENSTEIN

VOLUME XXXVI. Hormone Action (Part A: Steroid Hormones)
Edited by BERT W. O'MALLEY AND JOEL G. HARDMAN

VOLUME XXXVII. Hormone Action (Part B: Peptide Hormones)
Edited by BERT W. O'MALLEY AND JOEL G. HARDMAN

VOLUME XXXVIII. Hormone Action (Part C: Cyclic Nucleotides)
Edited by JOEL G. HARDMAN AND BERT W. O'MALLEY

VOLUME XXXIX. Hormone Action (Part D: Isolated Cells, Tissues, and Organ Systems)
Edited by JOEL G. HARDMAN AND BERT W. O'MALLEY

VOLUME XL. Hormone Action (Part E: Nuclear Structure and Function)
Edited by BERT W. O'MALLEY AND JOEL G. HARDMAN

VOLUME XLI. Carbohydrate Metabolism (Part B)
Edited by W. A. WOOD

VOLUME XLII. Carbohydrate Metabolism (Part C)
Edited by W. A. WOOD

VOLUME XLIII. Antibiotics
Edited by JOHN H. HASH

VOLUME XLIV. Immobilized Enzymes
Edited by KLAUS MOSBACH

VOLUME XLV. Proteolytic Enzymes (Part B)
Edited by LASZLO LORAND

VOLUME XLVI. Affinity Labeling
Edited by WILLIAM B. JAKOBY AND MEIR WILCHEK

VOLUME XLVII. Enzyme Structure (Part E)
Edited by C. H. W. HIRS AND SERGE N. TIMASHEFF

VOLUME XLVIII. Enzyme Structure (Part F)
Edited by C. H. W. HIRS AND SERGE N. TIMASHEFF

[1] Directory of Restriction Endonucleases

By RICHARD J. ROBERTS

This article is intended to serve as a directory to the restriction endonucleases which have not been characterized. All endonucleases which cleave DNA at a specific sequence have been considered to be restriction enzymes, although in most cases there is no direct genetic evidence for the presence of a host-controlled restriction-modification system. Certain strains are omitted from the table to save space.

Thus the many different *Staphylococcus aureus* isolates which contain an isoschizomer of *Sau*3A[1] are not listed individually. Similarly the many strains of gliding bacteria (orders: Myxobacterales and Cytophagales) which showed evidence of specific endonucleases during a large-scale screening[2] are still rather poorly characterized.

Within the table the source of each microorganism is given either as an individual or a National Culture Collection. The enzymes are named in accordance with the proposal of Smith and Nathans.[3] When two enzymes recognize the same sequence (i.e., are isoschizomers), the prototype (i.e., the first example isolated) is indicated in parentheses in column 3 of the table. The recognition sequences (column 4 of the table) are abbreviated so that only one strand, reading $5' \rightarrow 3'$, is indicated and the point of cleavage, when known, is indicated by an arrow ($\downarrow$). When two bases appear in parentheses, either one may appear at that position within the recognition sequence. Where known, the base modified by the corresponding methylase is indicated by an asterisk. (It should be noted that the base preceding the asterisk is the one methylated.) A* is N^6-methyl-adenosine; C* is 5-methylcytosine. The frequency of cleavage (columns five to eight) is experimentally determined for bacteriophage lambda (λ) and adenovirus-2 (Ad2) DNAs, but represents the computer-derived values from the published sequences of SV40[4] and ϕX174[5] DNAs. When more than one reference appears (column 9 of the table), the first contains the purification procedure for the restriction enzyme, the second concerns its recognition sequence, the third contains the purification procedure for the methylase, and the fourth describes its recognition sequence. In some cases two references appear in one of these categories when two independent groups have reached similar conclusions.

RESTRICTION ENDONUCLEASES

Microorganism	Source	Enzyme	Sequence	Number of cleavage sites				References
				λ	Ad2	SV40	ϕX174	
Achromobacter immobilis	ATCC 15934	AimI	?	?	?	?	?	6
Acinetobacter calcoaceticus	R. J. Roberts	AccI	GT↓$\binom{A}{C}\binom{G}{T}$AC	7	8	1	3	7
		AccII (FnuDII)	CGCG	>50	>50	0	14	7
Agrobacterium tumefaciens	ATCC 15955	AtuAI	?	>30	>30	?	?	8
Agrobacterium tumefaciens B6806	E. Nester	AtuBI (EcoRII)	CC$\binom{A}{T}$GG	>35	>35	16	2	9
Agrobacterium tumefaciens ID 135	C. Kado	AtuII (EcoRII)	CC$\binom{A}{T}$GG	>35	>35	16	2	10
Agrobacterium tumefaciens C58	E. Nester	AtuCI (BclI)	TGATCA	7	5	1	0	8
Anabaena catanula	CCAP 1403/1	AcaI	?	?	?	?	?	11
Anabaena cylindrica	A. deWaard	AcyI	GPu↓CGPyC	>14	>14	0	7	12
Anabaena subcylindrica	K. Murray	AsuI	G↓GNCC	>30	>30	11	2	11
Anabaena variabilis	K. Murray	AvaI	C↓PyCGPuG	8	?	0	1	13
		AvaII	G↓G$\binom{A}{T}$CC	>17	>30	6	1	13, 14, and 15
		AvaIII	ATGCAT	?	?	3	0	16, 17, and 18
Anabaena variabilis[uw]	E. C. Rosenvold	AvrI (AvaI)	CPyCGPuG	8	?	0	1	19
		AvrII	CCTAGG	1	2	2	0	19

Arthrobacter luteus	ATCC 21606	*Alu*I	AG↓CT	>50	>50	35	24	20
Arthrobacter pyridinolis	R. DiLauro	*Apy*I	CC$\binom{A}{T}$GG	>35	>35	16	2	21
Bacillus amyloliquefaciens F	ATCC 23350	*Bam*FI (*Bam*HI)	GGATCC	5	3	1	0	22
Bacillus amyloliquefaciens H	F. E. Young	*Bam*HI	G↓GATCC	5	3	1	0	23, 24
Bacillus amyloliquefaciens K	T. Kaneko	*Bam*KI (*Bam*HI)	GGATCC	5	3	1	0	22
Bacillus amyloliquefaciens N	T. Ando	*Bam*NI (*Bam*HI)	GGATCC	5	3	1	0	25
		*Bam*N$_x$	?	?	?	?	?	25 and 26
Bacillus brevis S	A. P. Zarubina	*Bbv*SI	GC*$\binom{T}{A}$GC	specific methylase				27
Bacillus brevis	ATCC 9999	*Bbv*I	GC$\binom{T}{A}$GC	>30	>30	23	14	28
Bacillus caldolyticus	A. Atkinson	*Bcl*I	T↓GATCA	7	5	1	0	29
Bacillus cereus	ATCC 14579	*Bce*14579	?	>10	?	?	?	22
Bacillus cereus	1AM 1229	*Bce*1229	?	>10	?	?	?	22
Bacillus cereus	T. Ando	*Bce*170 (*Pst*I)	CTGCAG	18	25	2	1	22
Bacillus cereus Rf sm st	T. Ando	*Bce*R (*Fnu*DII)	CGCG	>50	>50	0	14	22
Bacillus globigii	G. A. Wilson	*Bgl*I	GCCNNNN↓NGGC	22	12	1	0	30 and 31, 32
		*Bgl*II	A↓GATCT	>6	12	0	0	30 and 31, 33
Bacillus megaterium 899	B899	*Bme*899	?	>5	?	?	?	22
Bacillus megaterium B205-3	T. Kaneko	*Bme*205	?	>10	?	?	?	22
Bacillus megaterium	J. Upcroft	*Bme*I	?	>10	>20	4	?	34
Bacillus pumilus AHU1387	T. Ando	*Bpu*I	?	6	>30	2	?	35
Bacillus sphaericus	1AM 1286	*Bsp*1286	?	?	?	?	?	22

(continued)

RESTRICTION ENDONUCLEASES (continued)

Microorganism	Source	Enzyme	Sequence	Number of cleavage sites				References
				λ	Ad2	SV40	φX174	
Bacillus sphaericus R	P. Venetianer	BspRI (HaeIII)	GGCC	>50	>50	19	11	36
Bacillus stearothermophilus 1503-4R	N. Welker	BstI (BamHI)	GGATCC	5	3	1	0	37
Bacillus stearothermophilus 240	T. Atkinson	BstAI	?	?	?	?	?	38
Bacillus stearothermophilus ET	N. Welker	BstEI	?	?	?	?	?	39
		BstEII	?	11	8	0	?	39
		BstEIII	?	>7	?	?	?	39
Bacillus subtilis strain X5	T. Trautner	BsuRI (HaeIII)	GG↓C*C	>50	>50	19	11	40, 41, 42
Bacillus subtilis Marburg 168	T. Ando	BsuM	?	>10	?	?	?	22
Bacillus subtilis	ATCC 6633	Bsu6633	?	>20	?	?	?	22
Bacillus subtilis	IAM 1076	Bsu1076 (HaeIII)	GGCC	>50	>50	19	11	22
Bacillus subtilis	IAM 1114	Bsu1114 (HaeIII)	GGCC	>50	>50	19	11	22
Bacillus subtilis	IAM 1247	Bsu1247 (PstI)	CTGCAG	18	25	2	1	22, 43
Bacillus subtilis	ATCC 14593	Bsu1145	?	>20	?	?	?	22
Bacillus subtilis	IAM 1192	Bsu1192	?	>10	?	?	?	22
Bacillus subtilis	IAM 1193	Bsu1193	?	>30	?	?	?	22
Bacillus subtilis	IAM 1231	Bsu1231	?	>20	?	?	?	22
Bacillus subtilis	IAM 1259	Bsu1259	?	>8	?	?	?	22

Organism	Source	Enzyme	Sequence					Reference
Bordetella bronchiseptica	ATCC 19395	BbrI (HindIII)	AAGCTT	6	11	6	0	44
Brevibacterium albidum	ATCC 15831	BalI	TGG↓CCA	15	17	0	0	45
Brevibacterium luteum	ATCC 15830	BluI (XhoI)	C↓TCGAG	1	5	0	1	46
		BluII (HaeIII)	GGCC	>50	>50	19	11	47
Caryophanon latum L	H. Mayer	ClaI	AT↓CGAT	12	?	0	0	48
Chloroflexus aurantiacus	A. Bingham	CauI (AvaII)	GG(A/T)CC	>30	>30	6	1	49
		CauII	?	>30	>30	0	?	49
Chromobacterium violaceum	ATCC 12472	CviI	?	?	?	?	?	6
Corynebacterium humiferum	ATCC 21108	ChuI (HindIII)	AAGCTT	6	11	6	0	6
		ChuII (HindII)	GTPyPuAC	34	>20	7	13	6
Corynebacterium petrophilum	ATCC 19080	CpeI (BclI)	TGATCA	7	5	1	0	50
Diplococcus pneumoniae	S. Lacks	DpnI	GA*↓TC	?	?	?	0	51, 52 and 53
Diplococcus pneumoniae	S. Lacks	DpnII (MboI)	GATC	>50	>50	7	0	51, 52
Enterobacter cloacae	H. Hartmann	EclI	?	15	?	?	?	54
		EclII (EcoRII)	CC(A/T)GG	>35	>35	16	2	54
Enterobacter cloacae	DSM 30056	EcaI	G↓GTNACC	12	?	0	0	55
Escherichia coli RY13	R. N. Yoshimori	EcoRI	G↓AA*TTC	5	5	1	0	56, 57, 56, 58
		EcoRI'	PuPuA↓TPyPy	>10	>10	24	16	59
Escherichia coli R245	R. N. Yoshimori	EcoRII	↓CC*(A/T)GG	>35	>35	16	2	60, 61 and 62, 60
Escherichia coli B	W. Arber	EcoB	TGA*(N)$_8$TGCT	?	?	?	?	63, 64 and 65, 66

(continued)

RESTRICTION ENDONUCLEASES (continued)

Microorganism	Source	Enzyme	Sequence	Number of cleavage sites				References
				λ	Ad2	SV40	φX174	
Escherichia coli K	M. Meselson	EcoK	AAC(N)$_6$GTGC	?	?	?	?	67, 68, 69
Escherichia coli (PI)	K. Murray	EcoPI	AGACC	?	?	?	?	70, 71, 72 and 73, 74
Escherichia coli P15	W. Arber	EcoP15	?	?	?	?	?	75
Fusobacterium nucleatum A	M. Smith	FnuAI (HinfI)	G↓ANTC	>50	>50	10	21	76
		FnuAII (MboI)	GATC	>50	>50	7	0	44
Fusobacterium nucleatum C	M. Smith	FnuCI (MboI)	↓GATC	>50	>50	7	0	76
Fusobacterium nucleatum D	M. Smith	FnuDI (HaeIII)	GG↓CC	>50	>50	19	11	76
		FnuDII	CG↓CG	>50	>50	0	14	76
		FnuDIII (HhaI)	GCG↓C	>50	>50	2	18	76
Fusobacterium nucleatum E	M. Smith	FnuEI (Sau3A)	↓GATC	>50	>50	7	0	76
Fusobacterium nucleatum 48	M. Smith	Fnu48I	?	>50	?	?	?	76
Haemophilus aegyptius	ATCC 11116	HaeI	$\binom{A}{T}$GG↓CC$\binom{T}{A}$	?	?	11	6	77
		HaeII	PuGCGC↓Py	>30	>30	1	8	78, 79
		HaeIII	GG↓C*C	>50	>50	19	11	80, 41, 81
Haemophilus aphrophilus	ATCC 19415	HapI	?	>30	?	?	?	44
		HapII (HpaII)	C↓CGG	>50	>50	1	5	82, 83

Microorganism	Source	Enzyme (isoschizomer)	Sequence					References
Haemophilus gallinarum	ATCC 14385	*Hga*I	GACGC	>50	>50	0	14	82, 84 and 85
Haemophilus haemoglobinophilus	ATCC 19416	*Hhg*I (*Hae*III)	GGCC	>50	>50	19	11	44
Haemophilus haemolyticus	ATCC 10014	*Hha*I	GC*G↓C	>50	>50	2	18	86, 86, 87
		*Hha*II (*Hin*fI)	GANTC	>50	>50	10	21	88
Haemophilus influenzae 1056	J. Stuy	*Hin*1056I (*Fnu*DII)	CGCG	>50	>50	0	14	89
		*Hin*1056II	?	>30	>30	0	4	89
Haemophilus influenzae serotype b, 1076	J. Stuy	*Hin*bIII (*Hin*dIII)	AAGCTT	6	11	6	0	89
Haemophilus influenzae R$_b$	C. A. Hutchison	*Hin*bIII (*Hin*dIII)	AAGCTT	6	11	6	0	90 and 42
Haemophilus influenzae serotype c, 1160	J. Stuy	*Hin*cII (*Hin*dII)	GTPyPuAC	34	>20	7	13	89
Haemophilus influenzae serotype c, 1161	J. Stuy	*Hin*cII (*Hin*dII)	GTPyPuAC	34	>20	7	13	89
Haemophilus influenzae R$_c$	A. Landy, G. Leidy	*Hin*cII (*Hin*dII)	GTPyPuAC	34	>20	7	13	91
Haemophilus influenzae R$_d$ (exo mutant)	S. H. Goodgal	*Hin*dI	CA*C	specific methylase				92, 93
		*Hin*dII	GTPy↓PuA*C	34	>20	7	13	94, 95, 92, 93
		*Hin*dIII	A*↓AGCTT	6	11	6	0	96, 96, 92, 93
		*Hin*dIV	GA*C	specific methylase				92, 93
Haemophilus influenzae R$_d$123	V. Tanyashin	*Hin*dGLU	?	?	?	?	?	97
Haemophilus influenzae R$_f$	C. A. Hutchison	*Hin*fI	G↓ANTC	>50	>50	10	21	90, 98 and 99
		*Hin*fII (*Hin*dIII)	AAGCTT	6	11	6	0	87

(*continued*)

RESTRICTION ENDONUCLEASES (continued)

Microorganism	Source	Enzyme	Sequence	Number of cleavage sites				References
				λ	Ad2	SV40	φX174	
Haemophilus influenzae H-1	M. Takanami	HinHI (HaeII)	PuGCGCPy	>30	>30	1	8	82
Haemophilus parahaemolyticus	C. A. Hutchison	HphI	GGTGA	>50	>50	4	9	90, 100
Haemophilus parainfluenzae	J. Setlow	HpaI	GTT↓AAC	11	6	4	3	101, 102
		HpaII	C↓C*GG	>50	>50	1	5	101, 102, 81
Haemophilus suis	ATCC 19417	HsuI (HindIII)	AAGCTT	6	11	6	0	44
Herpetosiphon giganteus HP1023	J. H. Parish	HgiAI	$G\binom{T}{A}GC\binom{T}{A}{\downarrow}C$	20	?	0	3	103
Klebsiella pneumoniae OK8	J. Davies	KpnI	GGTAC↓C	2	8	1	0	104, 105
Micrococoleus species	D. Comb	MstI	TGCGCA	>10	>15	0	1	106, 106a
Moraxella bovis	ATCC 10900	MboI	↓GATC	>50	>50	7	0	107
		MboII	GAAGA	>50	>50	15	11	107, 108 and 109
Moraxella glueidi LG1	J. Davies	MglI	?	?	?	?	?	104
Moraxella glueidi LG2	J. Davies	MglII	?	?	?	?	?	104
Moraxella nonliquefaciens	ATCC 19975	MnoI (HpaII)	C↓CGG	>50	>50	1	5	44, 110
		MnoII	?	>10	>6	2	?	44
	ATCC 17953	MnlI	CCTC	>100	>100	52	35	111
Moraxella nonliquefaciens	ATCC 17954	MnnI	GTPyPuAC	>34	>20	7	13	112
		MnnII (HaeIII)	GGCC	>50	>50	19	11	112
		MnnIII	?	>50	>50	?	?	112
		MnnIV (HhaI)	GCGC	>50	>50	2	18	112

Organism	Source	Enzyme	Sequence					Ref.
Moraxella osloensis	ATCC 19976	*Mos*I (*Mbo*I)	GATC	>50	>50	7	0	107
Moraxella species	R. J. Roberts	*Msp*I (*Hpa*II)	CCGG	>50	>50	1	5	113
Myxococcus virescens	H. Reichenbach	*Mvi*I	?	1	?	?	?	114
		*Mvi*II	?	?	?	?	?	114
Neisseria gonorrhoeae	G. Wilson	*Ngo*I (*Hae*II)	PuGCGCPy	>30	>30	1	8	115
Neisseria gonorrhoeae	CDC 66	*Ngo*II (*Hae*III)	GGCC	>50	>50	19	11	116
Oerskovia xanthineolytica	R. Shekman	*Oxa*I (*Alu*I)	AGCT	>50	>50	35	24	117
		*Oxa*II	?	?	?	?	?	117
Proteus vulgaris	ATCC 13315	*Pvu*I	CGATCG	4	7	0	0	28
		*Pvu*II	CAG↓CTG	15	22	3	0	28
Providencia alcalifaciens	ATCC 9886	*Pal*I (*Hae*III)	GGCC	>50	>50	19	11	34
Providencia stuartii 164	J. Davies	*Pst*I	CTGCA↓G	18	25	2	1	104, 118
Pseudomonas facilis	M. VanMontagu	*Pfa*I	?	>30	>30	?	?	47, 89
Rhodopseudomonas sphaeroides	R. Lascelles	*Rsp*I	?	3	12	0	?	119
Rhodopseudomonas sphaeroides	S. Kaplan	*Rsh*I (*Pvu*I)	CGATCG	4	7	0	0	120
Serratia marcescens S_b	C. Mulder	*Sma*I	CCC↓GGG	3	12	0	0	121, 122
Serratia species SAI	B. Torheim	*Ssp*I	?	?	?	?	?	123
Staphylococcus aureus 3A	E. E. Stobberingh	*Sau*3A (*Mbo*I)	GATC	>50	>50	7	0	124

(continued)

RESTRICTION ENDONUCLEASES (continued)

Microorganism	Source	Enzyme	Sequence	Number of cleavage sites				References
				λ	Ad2	SV40	φX174	
Staphylococcus aureus	E. E. Stobberingh	*Sau*96I (*Asu*I)	G↓GNCC	>30	>30	11	2	125
Streptococcus faecalis var. *zymogenes*	R. Wu	*Sfa*I (*Hae*III)	GG↓CC	>50	>50	19	11	126
Streptococcus faecalis ND547	D. Clewell	*Sfa*NI	GATGC	>50	>30	6	12	8
Streptomyces achromogenes	ATCC 12767	*Sac*I	GAGCT↓C	2	7	0	0	127
		*Sac*II	CCGC↓GG	3	>25	0	1	127
		*Sac*III	?	>30	>30	?	?	127
Streptomyces albus	CMI 52766	*Sal*PI (*Pst*I)	CTGCAG	18	25	2	1	128
Streptomyces albus subspecies *pathociclicus*	KCC S0166	*Spa*I (*Xho*I)	CTCGAG	1	5	0	1	129
Streptomyces albus G	J. M. Ghysen	*Sal*I	G↓TCGAC	2	3	0	0	130
		*Sal*II	?	>30	?	?	?	130
Streptomyces bobiliae	ATCC 3310	*Sbo*I	?	?	?	?	?	131
Streptomyces bradiae	ATCC 3535	*Sbr*I	?	?	?	?	?	131
Streptomyces cupidosporus	KCC S0316	*Scu*I (*Xho*I)	CTCGAG	1	5	0	1	131
Streptomyces exfoliatus	H. Takahashi	*Sex*I (*Xho*I)	CTCGAG	1	6	0	1	129
Streptomyces goshikiensis	H. Takahashi	*Sgo*I (*Xho*I)	CTCGAG	1	6	0	1	129
Streptomyces griseus	ATCC 23345	*Sgr*I	?	0	7	0	?	127
Streptomyces hygroscopicus	?	*Shy*I	?	2	?	?	?	132
Streptomyces lavendulae	ATCC 8644	*Sla*I (*Xho*I)	C↓TCGAG	1	6	0	1	131

Streptomyces luteoreticuli	H. Takahashi	SluI (XhoI)	CTCGAG	1	6	0	1	129
Streptomyces stanford	S. Goff, A. Rambach	SstI(SacI)	GAGCT↓C	2	7	0	0	133, 134
		SstII (SacII)	CCGC↓GG	3	>25	0	1	133
		SstIII (SacIII)	?	>30	>30	?	?	133
Thermoplasma acidophilum	D. Searcy	ThaI (FnuDII)	CG↓CG	>50	>50	0	14	135
Thermopolyspora glauca	ATCC 15345	TglI (SacII)	CCGCGG	3	>25	0	1	28
Thermus aquaticus YTI	J. I. Harris	TaqI	T↓CGA	>50	>50	1	10	136
		TaqII	?	>30	>30	4	6	44
Xanthomonas amaranthicola	ATCC 11645	XamI (SalI)	GTCGAC	2	3	0	0	130
Xanthomonas badrii	ATCC 11672	XbaI	T↓CTAGA	1	4	0	0	137
Xanthomonas holcicola	ATCC 13461	XhoI	C↓TCGAG	1	6	0	1	46
		XhoII	?	>20	>20	4	?	89
Xanthomonas malvacearum	ATCC 9924	XmaI	C↓CCGGG	3	12	0	0	122
		XmaII (PstI)	CTGCAG	18	25	2	1	122
Xanthomonas nigromaculans	ATCC 23390	XniI (PvuI)	CGATCG	4	7	0	0	112
Xanthomonas oryzae	M. Ehrlich	XorI (PstI)	CTGCAG	18	25	2	1	138
		XorI (PvuI)	CGATCG	4	7	0	0	138
Xanthomonas papavericola	ATCC 14180	XpaI (XhoI)	C↓TCGAG	1	6	0	1	138

References

[1] E. E. Stobberingh, R. Schiphof, and J. S. Sussenbach, *J. Bacteriol.* **131**, 645 (1977).
[2] H. Mayer and H. Reichenbach, *J. Bacteriol.* **136**, 708 (1978).
[3] H. O. Smith and D. Nathans, *J. Mol. Biol.* **81**, 419 (1973).
[4] V. B. Reddy, B. Thimmappaya, R. Dhar, K. N. Subramanian, B. S. Zain, J. Pan, P. K. Ghosh, M. L. Celma, and S. M. Weissman, *Science* **200**, 494 (1978).
[5] F. Sanger, G. M. Air, B. G. Barrell, N. L. Brown, A. R. Coulson, J. C. Fiddes, C. A. Hutchison III, P. M. Slocombe, and M. Smith, *Nature (London)* **265**, 687 (1977).
[6] S. A. Endow and R. J. Roberts, unpublished observations.
[7] M. Zabeau and R. J. Roberts, unpublished observations.
[8] D. Sciaky and R. J. Roberts, unpublished observations.
[9] G. Roizes, M. Patillon, and A. Kovoor, *FEBS Lett.* **82**, 69 (1977).
[10] J. M. LeBon, C. Kado, L. J. Rosenthal, and J. Chirikjian, *Proc. Natl. Acad. Sci. U.S.A.* **75**, 4097 (1978).
[11] S. G. Hughes, T. Bruce, and K. Murray, unpublished observations.
[12] A. DeWaard, J. Korsuize, C. P. van Beveren, and J. Maat, *FEBS Lett.* **96**, 106 (1978).
[13] K. Murray, S. G. Hughes, J. S. Brown, and S. Bruce, *Biochem. J.* **159**, 317 (1976).
[14] G. Sutcliffe and G. Church, unpublished observations.
[15] C. Fuchs, E. C. Rosenvold, A. Honigman, and W. Szybalski, *Gene* **4**, 1 (1978).
[16] G. Roizes, P.-C. Nardeux, and R. Monier, *FEBS Lett.* **104**, 39 (1979).
[17] H. Shimatake and M. Rosenberg, unpublished observations.
[18] K. Denniston-Thompson, D. D. Moore, K. E. Kruger, M. E. Furth, and F. R. Blattner, *Science* **198**, 1051 (1978).
[19] E. C. Rosenvold and W. Szybalski, unpublished observations.
[20] R. J. Roberts, P. A. Myers, A. Morrison, and K. Murray, *J. Mol. Biol.* **102**, 157 (1976).
[21] R. DiLauro, unpublished observations.
[22] T. Shibata, S. Ikawa, C. Kim, and T. Ando, *J. Bacteriol.* **128**, 473 (1976).
[23] G. A. Wilson and F. E. Young, *J. Mol. Biol.* **97**, 123 (1975).
[24] R. J. Roberts, G. A. Wilson, and F. E. Young, *Nature (London)* **265**, 82 (1977).
[25] T. Shibata and T. Ando, *Biochim. Biophys Acta* **442**, 184 (1976).
[26] T. Shibata and T. Ando, *Mol. Gen. Genet.* **138**, 269 (1975).
[27] B. F. Vanyushin and A. P. Dobritsa, *Biochim. Biophys. Acta* **407**, 61 (1975).
[28] T. R. Gingeras and R. J. Roberts, unpublished observations.
[29] A. H. A. Bingham, T. Atkinson, D. Sciaky, and R. J. Roberts, *Nucleic Acids Res.* **5**, 3457 (1978).
[30] G. A. Wilson and F. E. Young, *in* "Microbiology 1976" (D. Schlessinger, ed.), p. 350. Am. Soc. Microbiol., Washington, D. C., 1976.
[31] C. H. Duncan, G. A. Wilson, and F. E. Young, *J. Bacteriol.* **134**, 338 (1978).
[32] T. Bickle, unpublished observations.
[33] V. Pirrotta, *Nucleic Acids Res.* **3**, 1747 (1976).
[34] R. E. Gelinas, P. A. Myers, and R. J. Roberts, unpublished observations.
[35] S. Ikawa, T. Shibata, and T. Ando, *J. Biochem. (Tokyo)* **80**, 1457 (1976).
[36] A. Kiss, B. Sain, É. Csordás-Tóth, and P. Venetianer, *Gene* **1**, 323 (1977).
[37] J. Catterall and N. Welker, *J. Bacteriol.* **129**, 1110 (1977).
[38] A. H. A. Bingham, R. J. Sharp, and T. Atkinson, unpublished observations.
[39] R. B. Meagher, unpublished observations.
[40] S. Bron, K. Murray, and T. A. Trautner, *Mol. Gen. Genet.* **143**, 13 (1975).
[41] S. Bron and K. Murray, *Mol. Gen. Genet.* **143**, 25 (1975).
[42] U. Gunthert, M. Freund, and T. A. Trautner, *Abst., FEBS Symp., 12th, 1978.*
[43] T. Hoshino, T. Uozumi, S. Horinouchi, A. Ozaki, T. Beppu, and K. Arima, *Biochim. Biophys. Acta* **479**, 367 (1977).

[44] R. J. Roberts and P. A. Myers, unpublished observations.
[45] R. E. Gelinas, P. A. Myers, G. A. Weiss, R. J. Roberts, and K. Murray, *J. Mol. Biol.* **114**, 433 (1977).
[46] T. R. Gingeras, P. A. Myers, J. A. Olson, F. A. Hanberg, and R. J. Roberts, *J. Mol. Biol.* **118**, 113 (1978).
[47] M. Van Montagu, unpublished observations.
[48] H. Mayer, R. Grosschedl, H. Schutte, and G. Hobom, unpublished observations.
[49] A. H. A. Bingham and J. Darbyshire, unpublished observations.
[50] J. Fisherman, T. R. Gingeras, and R. J. Roberts, unpublished observations.
[51] S. Lacks and B. Greenberg, *J. Biol. Chem.* **250**, 4060 (1975).
[52] S. Lacks and B. Greenberg, *J. Mol. Biol.* **114**, 153 (1977).
[53] G. E. Geier and P. Modrich, *J. Biol. Chem.* **254**, 1408 (1979).
[54] H. Hartmann and W. Goebel, *FEBS Lett.* **80**, 285 (1977).
[55] H. Mayer, E. Schwarz, M. Melzer, and G. Hobom, unpublished observations.
[56] P. J. Greene, M. C. Betlach, H. M. Goodman, and H. W. Boyer, *Methods Mol. Biol.* **7**, 87 (1974).
[57] J. Hedgpeth, H. M. Goodman, and H. W. Boyer, *Proc. Natl. Acad. Sci. U.S.A.* **69**, 3448 (1972).
[58] A. Dugaiczyk, J. Hedgpeth, H. W. Boyer, and H. M. Goodman, *Biochemistry* **13**, 503 (1974).
[59] K. Murray, J. S. Brown, and S. A. Bruce, unpublished observations.
[60] R. N. Yoshimori, Ph.D. Thesis, Univ. California, San Francisco (1971).
[61] C. H. Bigger, K. Murray, and N. E. Murray, *Nature (London), New Biol.* **244**, 7 (1973).
[62] H. W. Boyer, L. T. Chow, A. Dugaiczyk, J. Hedgpeth, and H. M. Goodman, *Nature (London), New Biol.* **244**, 40 (1973).
[63] B. Eskin and S. Linn, *J. Biol. Chem.* **247**, 6183 (1972).
[64] J. A. Lautenberger, N. C. Kan, D. Lackey, S. Linn, M. H. Edgell, and C. A. Hutchison III, *Proc. Natl. Acad. Sci. U.S.A.* **75**, 2271 (1978).
[65] J. V. Ravetch, K. Horiuchi, and N. D. Zinder, *Proc. Natl. Acad. Sci. U.S.A.* **75**, 2266 (1978).
[66] J. A. Lautenberger and S. Linn, *J. Biol. Chem.* **247**, 6176 (1972).
[67] M. Meselson and R. Yuan, *Nature (London)* **217**, 1110 (1968).
[68] N. C. Kan, J. A. Lautenberger, M. H. Edgell, and C. A. Hutchison III, *Fed. Proc., Fed. Am. Soc. Exp. Biol.* **37**, 1499 (1978), and unpublished observations.
[69] A. Haberman, J. Heywood, and M. Meselson, *Proc. Natl. Acad. Sci. U.S.A.* **69**, 3138 (1972).
[70] A. Haberman, *J. Mol. Biol.* **89**, 545 (1974).
[71] B. Bachi and V. Pirrotta, unpublished observations.
[72] J. P. Brockes, *Biochem. J.* **133**, 629 (1973).
[73] J. P. Brockes, P. R. Brown, and K. Murray, *Biochem. J.* **127**, 1 (1972).
[74] J. P. Brockes, P. R. Brown, and K. Murray, *J. Mol. Biol.* **88**, 437 (1974).
[75] J. Reiser and R. Yuan, *J. Biol. Chem.* **252**, 451 (1977).
[76] A. Lui, B. C. McBride, and M. Smith, unpublished observations.
[77] K. Murray, A. Morrison, H. W. Cooke, and R. J. Roberts, unpublished observations.
[78] R. J. Roberts, J. B. Breitmeyer, N. F. Tabachnik, and P. A. Myers, *J. Mol. Biol.* **91**, 121 (1975).
[79] C.- P. D., Tu, R. Roychoudhury, and R. Wu, *Biochem. Biophys. Res. Commun.* **72**, 355 (1976).
[80] J. H. Middleton, M. H. Edgell, and C. A. Hutchison III, *J. Virol.* **10**, 42 (1972).
[81] M. B. Mann and H. O. Smith, *Nucleic Acids Res.* **4**, 4211 (1977).
[82] M. Takanami, *Methods Mol. Biol.* **7**, 113 (1974).
[83] H. Sugisaki and K. Takanami, *Nature (London), New Biol.* **246**, 138 (1973).

[84] N. L. Brown and M. Smith, *Proc. Natl. Acad. Sci. U.S.A.* **74**, 3213 (1977). *Hga*I cleaves as indicated:

$$5'\text{-GACGCNNNNN} \downarrow \qquad 3'$$
$$3'\text{-CTGCGNNNNNNNNNN} \downarrow \text{-}5'$$

[85] H. Sugisaki, *Gene* **3**, 17 (1978).

[86] R. J. Roberts, P. A. Myers, A. Morrison, and K. Murray, *J. Mol. Biol.* **103**, 199 (1976).

[87] M. B. Mann and H. O. Smith, unpublished observations.

[88] M. B. Mann, R. N. Rao, and H. O. Smith, *Gene* **3**, 97 (1978).

[89] J. A. Olson, P. A. Myers, and R. J. Roberts, unpublished observations.

[90] J. H. Middleton, P. V. Stankus, M. H. Edgell, and C. A. Hutchison III, unpublished observations.

[91] A. Landy, E. Ruedisueli, L. Robinson, C. Foeller, and W. Ross, *Biochemistry* **13**, 2134 (1974).

[92] P. H. Roy and H. O. Smith, *J. Mol. Biol.* **81**, 427 (1973).

[93] P. H. Roy and H. O. Smith, *J. Mol. Biol.* **81**, 445 (1973).

[94] H. O. Smith and K. W. Wilcox, *J. Mol. Biol.* **51**, 379 (1970).

[95] T. J. Kelly, Jr. and H. O. Smith, *J. Mol. Biol.* **51**, 393 (1970).

[96] R. Old, K. Murray, and G. Roizes, *J. Mol. Biol.* **92**, 331 (1975).

[97] V. I. Tanyashin, L. I. Li, I. O. Muizhnieks, and A. A. Baev, *Dokl. Akad. Nauk SSSR* **231**, 226 (1976).

[98] C. A. Hutchison III and B. G. Barrell, unpublished observations.

[99] K. Murray and A. Morrison, unpublished observations.

[100] D. Kleid, Z. Humayun, A. Jeffrey, and M. Ptashne, *Proc. Natl. Acad. Sci. U.S.A.* **73**, 293 (1976). *Hph*I cleaves as indicated:

$$5'\text{-GGTGANNNNNNNN} \downarrow \text{-}3'$$
$$3'\text{-CCACTNNNNNNN} \uparrow \text{-}5'$$

[101] P. A. Sharp, B. Sugden, and J. Sambrook, *Biochemistry* **12**, 3055 (1973).

[102] D. E. Garfin and H. M. Goodman, *Biochem. Biophys. Res. Commun.* **59**, 108 (1974).

[103] N. L. Brown, M. McClelland, and P. R. Whitehead, unpublished observations.

[104] D. L. Smith, F. R. Blattner, and J. Davies, *Nucleic Acids Res.* **3**, 343 (1976).

[105] J. Tomassini, R. Roychoudhury, R. Wu, and R. J. Roberts, *Nucleic Acids Res.* **5**, 4055 (1978).

[106] D. Comb, I. Schildkraut, and R. J. Roberts, unpublished observations.

[106a] T. R. Gingeras, J. P. Milazzo, and R. J. Roberts, *Nucleic Acids Res.* **5**, 4105 (1978).

[107] R. E. Gelinas, P. A. Myers, and R. J. Roberts, *J. Mol. Biol.* **114**, 169 (1977).

[108] N. L. Brown, C. A. Hutchison III, and M. Smith, *J. Mol. Biol.* (in press). *Mbo*I cleaves as indicated:

$$5'\text{-GAAGANNNNNNNN} \downarrow \text{-}3'$$
$$3'\text{-CTTCTNNNNNNN} \uparrow \text{-}5'$$

[109] S. A. Endow, *J. Mol. Biol.* **114**, 441 (1977).

[110] U. L. RajBhandary and B. Baumstark, unpublished observations.

[111] M. Zabeau, R. Greene, P. A. Myers, and R. J. Roberts, unpublished observations. *Mnl*I cleaves 5 to 10 bases from the recognition sequence.

[112] F. Hanberg, P. A. Myers, and R. J. Roberts, unpublished observations.

[113] M. Van Montagu, P. A. Myers, and R. J. Roberts, unpublished observations.

[114] D. W. Morris and J. H. Parish, *Arch. Microbiol.* **108**, 227 (1976).

[115] G. A. Wilson and F. E. Young, unpublished observations.

[116] D. J. Clanton, J. M. Woodward, and R. V. Miller, *J. Bacteriol.* **135**, 270 (1978).

[117] A. Stotz and P. Philippson, unpublished observations.

[118] N. L. Brown and M. Smith, *FEBS Lett.* **65**, 284 (1976).

[119] A. H. A. Bingham, A. Atkinson, and J. Darbyshire, unpublished observations.

[120] J. Gardner and S. Kaplan, unpublished observations.

[121] R. Greene and C. Mulder, unpublished observations.

[122] S. A. Endow and R. J. Roberts, *J. Mol. Biol.* **112**, 521 (1977).

[123] B. Torheim, personal communication.

[124] J. S. Sussenbach, C. H. Monfoort, R. Schiphof, and E. E. Stobberingh, *Nucleic Acids Res.* **3**, 3193 (1976).

[125] J. S. Sussenbach, P. H. Steenbergh, J. A. Rost, W. J. Van Leeuwen, and J. D. A. van Embden, *Nucleic Acids Res.* **5**, 1153 (1978).

[126] R. Wu, C. T. King, and E. Jay, *Gene* **4**, 329 (1978).

[127] J. R. Arrand, P. A. Myers, and R. J. Roberts, unpublished observations.

[128] K. Chater, *Nucleic Acids Res.* **4**, 1989 (1977).

[129] H. Shinotsu, H. Takahashi, and H. Saito, unpublished observations.

[130] J. R. Arrand, P. A. Myers, and R. J. Roberts, *J. Mol. Biol.* **118**, 127 (1978).

[131] H. Takahashi, M. Shimizu, H. Saito, Y. Ikeda, and H. Sugisaki, *Gene* **5**, 9 (1979).

[132] F. Walter, M. Hartmann, and M. Roth, *Abstr., FEBS Symp., 12th, 1978.*

[133] S. Goff and A. Rambach, *Gene* **3**, 347 (1978), and unpublished observations.

[134] F. Muller, S. Stoffel, and S. G. Clarkson, unpublished observations.

[135] D. McConnell, D. Searcy, and G. Sutcliffe, *Nucleic Acids Res.* **5**, 1729 (1978).

[136] S. Sato, C. A. Hutchison III, and J. I. Harris *Proc. Natl. Acad. Sci. U.S.A.* **74**, 542 (1977).

[137] B. S. Zain and R. J. Roberts, *J. Mol. Biol.* **115**, 249 (1977).

[138] J. Shedlarski, M. Farber, and M. Ehrlich, unpublished observations.

Section I

Assays for [Class II Restriction] Endonucleases

[2] Assaying of Organisms for the Presence of Restriction Endonucleases

By ROBERT SCHLEIF

The utility of sequence-specific endonucleases recommends their use in a wide variety of applications. Typically a potential user is faced with the problem of determining which of the large number of enzymes already known will cleave in acceptable locations. Thus an easy, uniform, and rapid scheme for partial purification of these enzymes would greatly assist the initial screening. A purification step fulfilling these requirements could also be of value as a first step when pure enzyme is required or of use in searching for new varieties of endonuclease. Dextran–polyethylene glycol phase partition meets the requirements of speed and convenience.[1] Here it is shown that adjustment of the salt concentration during the phase partition step allows separation of most of the interfering nuclease activities in the crude extracts of many strains containing site-specific nucleases.

Materials

Bacterial Strains. The following strains have been used: *Arthrobacter luteus, Bacillus amyloliquefaciens* H, *Brevibacterium umbra, Echerichia coli, Haemophilus aegyptius, Haemophilus aphrophilus, Haemophilus haemolyticus, Haemophilus influenzae* Rd, *Haemophilus parahaemolyticus, Haemophilus parainfluenzae, Klebsiella pneumonia, Providencia stuarti, Serratia marcescens.* The *Arthrobacter,* and *Bacillus* cells were grown on nutrient broth, Difco, 8 g/liter, and the other cells in brain heart infusion, Difco, 37 g/liter. Medium for the *Haemophilus* strains contained 2 μg/ml NAD and 10 μg/ml hemin. All cells were harvested between 1 and 6 hr after exponential growth had ceased, and were centrifuged and frozen at $-10°C$ until use.

Polymer Concentrate. Polymer concentrate was made by heating 500 ml of H_2O to near boiling, dissolving 64 g Dextran T500 and then adding 256 g polyethylene glycol 6000. Water was then added to bring the final mass to 900 g. This polymer concentrate was stored at 4°, but thoroughly mixed by heating to 65° and then cooling to 20° just before use.

---

[1] P. A. Albertson, "Partition of Cell Particles and Macromolecules." Wiley, New York, 1960.

Procedures

Assay of Endonuclease. Typically 2 μl of extract are allowed to digest 1 μg of λ phage DNA for 1 hr at 37° in 0.006 M Tris-HCl, pH 7.4, 0.006 M MgCl$_2$, 0.006 M β-mercaptoethanol. The samples are heated to 65° for 5 min, glycerol added to 5%, and bromphenol blue is added to a concentration of 0.01%. The samples are applied to 4.5 × 3 mm slots 8 mm deep in a 17.5 cm high by 14.2 cm wide by 3 mm thick agarose (Seakem, Marine Coloids, Rockland Maine) gel held between Lucite plates. Buffer in the reservoirs and gel is 0.04 M Tris base, 5 mM Na acetate, 1 mM EDTA, adjusted with acetic acid to pH 7.9. For restriction enzymes producing few cleavages, 0.7% agarose is used with a bottom retaining layer 3 cm wide of 1.4% agarose, otherwise 1.4% agarose is used through the whole gel. Electrophoresis is terminated when the bromphenol dye is about 2 cm from the bottom. Resolution of DNA bands is appreciably improved by curing the gels overnight at 4° before use. After electrophoresis one or both of the Lucite plates are removed, and the gel is stained 20 min in a solution of 1 μg/ml ethidium bromide. The DNA bands are then visualized and recorded by placing the gel directly on the Ultraviolet Products Inc. C-50 short wave (257 nm) transilluminator. The fluorescent DNA bands are photographed with a Polaroid camera loaded with Type 107 film (ASA 3000). A red Corning glass filter, 2 mm thick, CS 2-73, enhances the signal to noise level. Bleaching occasionally occurs after several minutes exposure to the uv, so it is best to photograph promptly after turning on the uv light and to minimize exposure to the uv light.

Protocol for Testing Crude Extracts for Endonuclease. Cells (5 g) are mixed with 10 ml 0.01 M Tris-HCl, pH 7.9 and 0.01 M β-mercaptoethanol and opened by sonication. Care should be taken that their temperature remains below 10°. The following additions are made per milliliter of sonicated cells: 0.5 ml H$_2$O or 0.5 ml single-stranded calf thymus DNA at 3 mg/ml in 0.01 M Tris-HCl, pH 7.9, 0.001 M EDTA; 0.5 to 0.6 g polymer concentrate; and the following volumes of 4 M NaCl: 0.002, 0.005, 0.01, 0.02, 0.04, 0.08, 0.17, 0.32, 0.64, and 1.2 ml. After all the additions, the mixtures are thoroughly stirred for 5 min and then centrifuged 10 min at 5000 g. The resulting clear supernatants are then assayed for endonuclease activity.

Figure 1 shows typical results obtained from an extract of *Haemophilus aphrophilus*. The addition of 0.64 ml of 4 M salt yields the best enzyme. In this experiment denatured calf thymus DNA was added. It was found in early experiments when cells were opened by grinding with alumina to reduce the exonuclease contamination. However, later tests with the *Hin* and *Pst* extracts have shown this addition to be unnecessary if cells are opened by sonication. Presumably the sonicated cellular

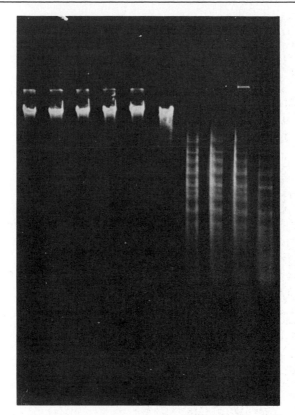

FIG. 1. Digestion patterns of λ*para107*[2] DNA by extracts from *Haemophilus parainfluenzae*. Increasing concentrations, left to right, of NaCl were used in the PEG phase partitions as described in the text. The first fraction with substantial endonucleolytic activity had 0.16 ml of 4 M NaCl added per milliliter of sonicated cells.

DNA provides sufficient sites to bind the exonuclease and carry it into the dextran phase during the phase partition.

Figure 2 shows the DNA resulting by digestion with the extract from *Hind* cells. At low concentrations of salt, only *Hind*II is released to the supernatant, whereas at higher salt concentrations both *Hind*II and *Hind*III are released to the supernatant.

The enzymes which were undetectable in phase-partitioned extracts were *Alu, Bum, Eco*RII, and *Sma*. If the phase-partition method were to be used as a first step in their purification, an NaCl concentration of 2 M or above would seem advisable during the phase partition step. Table I

[2] J. T. Lis and R. Schleif, *J. Mol. Biol.* **95,** 395 (1975).

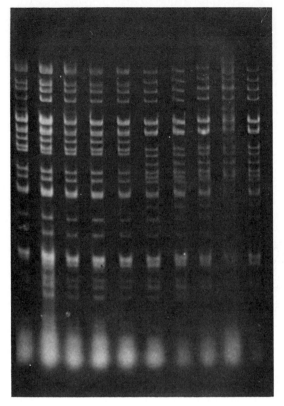

FIG. 2. Digestion pattern of λ*para107* DNA by extracts prepared from *Haemophilus influenzae* Rd.

shows the enzymes which are detectable in the phase-partitioned crude extracts and the NaCl needed for optimum purity.

Storage or Further Purification of an Extract. When an enzyme-containing extract is to be stored for extended periods it is dialyzed against $0.01\,M$ Tris-HCl, pH 7.9, $0.01\,M$ MgSO$_4$ $0.0001\,M$ EDTA, $0.01\,M$ β-mercaptoethanol, and 50% glycerol. The slight precipitate which forms during dialysis is removed by 5-min centrifugation at $5000\,g$. If the extract is to be subjected to additional purification steps, it is dialyzed into the appropriate buffer and the resulting precipitate is removed by centrifugation and discarded. Residual polyethylene glycol flows through most columns.

TABLE I

ENZYMES FOR WHICH THE PHASE PARTITION METHOD YIELDS
DETECTABLE ACTIVITY IN THE PHASE-PARTITIONED CRUDE EXTRACT

Enzyme	Volume (ml) 4 M NaCl per milliliter extract for optimal purity
Enzymes yielded with activity sufficient for many purposes	
Bam HI	0.16
Hap I, II	0.64
Hha	0.01
Hin dII	0.002–0.04
Hin dII + III	0.08 –1.2
Pst	0.32
Hae III	0.05
Enzymes yielded but probably in need of further purification	
Eco RI	0.02
Hae II	0.05
Hpa II, III	0.16
Hph	0.16
Kpn	0.8

[3] Use of Infectious DNA Assays

By M. TAKANAMI

Number of infectious centers scored by transfection of infective bacteriophage DNA are proportional to the amount of phage DNA used under appropriate conditions, and endonucleolytic cleavage of intact DNA molecules destroys the plaque-forming ability. This transfection method is therefore applicable to the assay of restriction endonucleases as well as other endonucleases and has been used for preparation of enzymes from *Haemophilus* strains.[1,2] However, this assay does not provide information on the specificity of enzyme, so that the application is limited to a preliminary survey of enzyme activity in fractions in the course of enzyme purification. The specificity of enzyme should be subsequently assay by electrophoretic analysis of reaction products on either agarose or polyacrylamide gel.

[1] M. Takanami and H. Kojo, *FEBS Lett.* **29,** 267 (1973).
[2] M. Takanami, *FEBS Lett.* **34,** 318 (1973).

METHODS IN ENZYMOLOGY, VOL. 65

For transfection, Ca-treated cells[3,4] are more commonly used because of its simplicity of use and high efficiency of transfection (number of plaques resulting per phage DNA molecules added). However, the transfection efficiency of Ca-treated cells is extremely sensitive to environmental conditions especially to salts. In contrast, transfection to spheroplasts made by lysozyme–EDTA treatment is less influenced by salts and other factors, so that the reaction mixture used for DNA digestion can assay directly. A general treatment of the use of spheroplasts is given in Volume XII, Part B, Section XII of this series. The optimal conditions for transfection of various phage DNA's have been studied in more detail.[5-7] The protocol presented in this section is of the *E. coli* K38-phage fd replicative form I (RFI) DNA system used for preparation of restriction endonucleases *Hga*I, *Hap*II, and *Hin*Hl.[1,2] The overall assay procedure has been simplified for application. Some variation of course is necessary for other phage DNA-host cell systems.

Preparation of Spheroplasts

Escherichia coli K38 is grown by shaking in 300 ml of tryptone broth (10 g tryptone, 1 g yeast extract, 5 g NaCl, 1 g glucose in 1 liter, pH 7.4) to a cell density of about 10^9 cells/ml and harvested by centrifugation. Cells are immediately suspended in 15 ml of 0.6 M sucrose, 0.04 M Tris-HCl (pH 8.5) and transferred to a water bath of 20°. After 15 min, 1.5 ml of 10% bovine serum albumin (Armour Fraction V, sterilized by filtration), 0.25 ml of 0.2 M EDTA (adjusted pH to about 8); and 0.2 ml of 0.1% lysozyme (Sigma Grade I, freshly dissolved in 0.04 M Tris-HCl, pH 8.5) are added in this order by stirring. The suspension is held for 2 min at 20° and immediately chilled by transferring to an ice bath, followed by addition of 0.2 ml of 1 M MgSO$_4$. The spheroplasts are stored in a refrigerator (about 4°). The efficiency of transfection usually drops somewhat by as much as 50% in the first day and is then maintained at this level at least for 1 week. It is suggested that the spheroplast stock be used at this level. The transfection efficiency of resulting spheroplasts depends on the period and temperature of the lysozyme treatment and also on lots of lysozyme and bovine serum albumin used. The optimal conditions for preparation of spheroplasts may be different with bacterial strains. With the identical conditions, however, reproducible results are usually obtained.

[3] M. Mandel and A. Higa, *J. Mol. Biol.* **53**, 159 (1970).
[4] S. N. Cohen, A. C. Y. Chang, and L. Hsu, *Proc. Natl. Acad. Sci. U.S.A.* **69**, 2110 (1972).
[5] R. Benzinger, I. Kleber, and R. Huskey, *J. Virol.* **7**, 646 (1971).
[6] L. Lawhorne, I. Kleber, C. Mitchell, and R. Benzinger, *J. Virol.* **12**, 733 (1973).
[7] W. D. Henner, I. Kleber, and R. Benzinger, *J. Virol.* **12**, 741 (1973).

Assay Procedure

Digestion of DNA with enzyme is carried out in the usual conditions; e.g., the reaction mixture for most of *Haemophilus* enzymes contains 10 mM Tris-HCl (pH 7.6), 7 mM MgCl$_2$, 7 mM mercaptoethanol, 0.01 A_{260} unit (about 0.5 μg) fd RFI DNA; and enzyme fraction in a final volume of 0.1 ml. After incubation for appropriate periods, 0.4 ml of dilution buffer (0.4 M sucrose, 5 mM MgSO$_4$+, 1 mM CaCl$_2$+, 0.5% albumin) and 0.1 ml of spheroplasts are added. After holding for 10 min at 37°, 0.1 ml of fresh *E. coli* K38 cells (about 10^8) and 2 ml of soft agar containing 0.4 M sucrose, 5 mM MgSO$_4$, and 1 mM CaCl$_2$ in tryptone broth (warmed to 45°) are added, and the mixture is plated in the usual manner. Under these conditions, about 500 plaques are usually formed with 0.01 A_{260} unit of fd RFI DNA. As mentioned above, the transfection efficiency differs with spheroplast preparations used. With the same preparation, however, the relative number of plaques formed are proportional to the amount of intact phage DNA presented in the transfection medium at least up to 0.05 A_{260} units. When the fractions obtained by column chromatography are submitted to assay, the enzyme activity is identified as a peak of depression in plaque number.[1,2] The efficiency of transfection is not affected by salts at least up to 0.1 M KCl or 0.1 M NaCl and also by proteins. The efficiency is however significantly depressed by the presence of other DNA (see Volume XII, Part B, p. 856 in this series). If such fractions are submitted to assay, it is necessary to take the no digestion control. The fd RFI DNA is prepared from *E. coli* 38 infected with fd as described[8] or by the general procedure for preparation of bacterial plasmids.[9] With spheroplasts, the efficiency of transfection is about ten times greater for single-stranded (SS) phage DNA than for double (RF) DNA. If the contamination of SS phage DNA is not negligible, the SS phage DNA is selectively inactivated by treatment with NH$_2$OH. Usually a concentrated DNA solution is treated for 30 min at 37° with 0.2 M NH$_2$OH, 0.1 M phosphate buffer (pH 6.2) and used to assay after dialysis.[10]

[8] M. Takanami, T. Okamoto, K. Sugimoto, and H. Sugisaki, *J. Mol. Biol.* **95**, 21 (1975).
[9] D. Vapnek and W. D. Rupp, *J. Mol. Biol.* **60**, 413 (1971).
[10] E. S. Tessman, *J. Mol. Biol.* **17**, 218 (1966).

[4] Assay for Type II Restriction Endonucleases Using the *Escherichia coli* recBC DNase and Duplex Circular DNA

By DAVID LACKEY and STUART LINN

Principle

Duplex, circular DNA is not a substrate for the recBC DNase of *Escherichia coli* (exonuclease V); however, linear DNA is digested by the enzyme to acid-soluble nucleotides.[1] Therefore, the conversion of duplex, circular DNA to linear DNA by a restriction endonuclease in the presence of an excess of recBC nuclease results in the production of acid-soluble nucleotide in amounts proportional to the endonuclease activity. We have previously described such an assay for type I restriction enzymes,[2] whereas Modrich and Zabel have applied the principle to type II restriction enzymes.[3]

Reagents

1. The reaction mixture (total volume, 0.05 ml) contains
   50 mM Tris-HCl, pH 7.4
   10 mM MgCl$_2$
   0.6 mM dithiothreitol
   1.0 mg/ml acetylated bovine serum albumin[4]
   0.33 mM ATP
   ^{3}H-Labeled colicin El DNA, 60 μM (3 nmole total DNA nucleotide, ~3000–10,000 cpm/nmole)[5]
   Fraction IX recBC enzyme, 2.4 exonuclease units[6]
2. Bovine serum albumin, 10 mg/ml

---

[1] P. J. Goldmark and S. Linn, *J. Biol. Chem.* **247**, 1849 (1972).

[2] B. Eskin and S. Linn, *J. Biol. Chem.* **247**, 6183 (1972).

[3] P. Modrich and D. Zabel, *J. Biol. Chem.* **251**, 5866 (1976).

[4] In order to eliminate contaminating nuclease activity, the albumin is acetylated with acetic anhydride as described by C. J. Epstein and R. F. Goldberger [*J. Biol. Chem.* **239**, 1087 (1964)].

[5] Colicin El DNA prepared by the method of D. B. Clewell [*J. Bacteriol.* **110**, 667 (1972)] was used to assay the *Eco*R1 and *Alu*1 restriction endonucleases. Other duplex circular DNAs which contain one or more unmodified restriction sites may also be used (e.g., fd or ϕX174 replicative forms or SV40 DNA), although phage PM2 DNA is less suitable because its highly supertwisted structure makes it slightly susceptible to the recBC DNase.

[6] One exonuclease unit produces 1.0 nmole of acid-soluble nucleotide in 30 min at 37° under standard conditions. Enzyme was purified by the procedure of Goldmark and Linn.[1]

3. Trichloroacetic acid, 7% (w/v)
4. Aqueous scintillation fluor: 9.1 g of 2,5-diphenyloxazole (PPO) and 0.61 g of 1,4-bis[2-(5-phenyloxazolyl)]benzene (POPOP) are mixed for 1 hr with 2140 ml toluene. Then 1250 ml Triton X-100 are added and mixing is continued for 5 hr.

Procedure

Enzymatic Reaction. An appropriate amount of restriction enzyme is added to the reaction mixture. The mixture is incubated for 30 min at 37°, then chilled to 0°.

Acid Precipitation and Determination of Radioactivity. Bovine serum albumin (0.05 ml of a 10 mg/ml solution) and 0.25 ml of 7% (w/v) trichloroacetic acid are added successively to the reaction mixture, with mixing following each addition. After 5 min at 0°, the samples are centrifuged for 5 min at 8000 g. An aliquot of the supernatant fluid is added to a miniature scintillation vial, brought to 0.3 ml with water (if necessary), and mixed with 3.3 ml aqueous scintillation fluor. Radioactivity is then determined in a scintillation counter. If a standard 20-ml vial is used, the aliquot should be brought to 0.9 ml and 10 ml of the fluor should be utilized.

Calculations. One unit of restriction activity is defined as the amount which produces 1 nmole of acid-soluble nucleotide during the 30-min incubation. The assay is linear in the range of 0.1–1.0 unit. A blank value of 0.3 nmole or less of acid-soluble nucleotide is typically produced in the absence of restriction enzyme due to broken DNA and/or contaminating endonuclease in the *rec*BC enzyme preparation. This value should be determined and used to correct the assay values.

If the above unit is divided by the number of nucleotides in the circular DNA substrate used, a direct estimate of the number of double-strand cleavages can be obtained. For example, colicin El DNA contains 12,800 nucleotides. Therefore, with this substrate, one unit of restriction endonuclease would produce 0.078 pmole of double-strand cleavages (i.e., convert 0.078 pmole of circular duplex DNA molecules to a linear form) during an assay incubation.

Properties of the Assay

An estimation of endonuclease activity by this assay gave excellent quantitative agreement with results obtained by agarose gel electrophoresis assays when either *Eco*R1 or *Alu*1 restriction endonucleases were utilized. *Eco*R1 breaks at one site on colicin El DNA and makes cohesive ends, whereas the *Alu*1 enzyme makes roughly 50 nonstaggered breaks with this DNA. These results indicate that the exonuclease is present in excess so that a DNA molecule is essentially digested completely to acid-soluble fragments immediately upon receiving one double-stranded restriction cleavage.

The exonuclease-coupled assay described here is well suited to the purification and quantitation of restriction enzymes, since it is simpler to perform and quantitate than an agarose gel electrophoresis assay. Both types of assay are comparable in terms of sensitivity, but the coupled assay is somewhat less specific in that gapped duplex circles are indistinguishable from restricted DNA and the coupled assay would not distinguish between random and sequence-specific DNA breakage.

In principle any exonuclease that is purified free from interfering endonuclease activity could be utilized for a coupled assay. However, our experience is that enzymes that can digest DNA from a single-strand nick (such as *E. coli* exonuclease III or the exonucleases associated with bacterial DNA polymerases) are less suitable because of the difficulty of maintaining DNA in an unnicked (form I) state. The *rec* BC DNase does not act upon nicked circular (form II) DNAs and is therefore insensitive to spurious nicking of the DNA during storage or by contaminating enzymes present in an incubation. Phage λ exonuclease does not act from nicks in DNA; however, we have not been successful in obtaining a linear relationship between restriction endonuclease and acid-soluble nucleotide in this case.

With a slight modification in procedure the coupled assay was successfully used with the type I *Eco*B restriction endonuclease.[2] The modified assay should be readily adaptable to other type I restriction enzyme assays as well.

[5] Polynucleotide Kinase Exchange as an Assay for Class II Restriction Endonucleases

By Kathleen L. Berkner and William R. Folk

Phage T4 polynucleotide kinase[1] can be used to label 5′-hydroxyl or 5′-phosphoryl termini in DNAs cleaved by type II restriction endonucleases. As 5′-phosphoryl termini are generated by most known site-specific endonucleases, phosphorylation with [γ-^{32}P]ATP may be performed, but this requires dephosphorylation of the termini with a phosphatase. Alternatively, since the polynucleotide kinase phosphorylation

---

[1] C. C. Richardson, *Proc. Natl. Acad. Sci. U.S.A.* **54**, 158 (1965).

METHODS IN ENZYMOLOGY, VOL. 65

reaction is reversible,[2] termini can be conveniently labeled in a one-step exchange reaction with [γ-^{32}P]ATP.

$$(5')\text{P-DNA} + \text{ADP} \rightleftharpoons (5')\text{HO-DNA} + \text{ADP}$$

The exchange reaction is quantitative and rapid, permitting many samples to be processed. It is not limited by the genome size or number of cleavage sites per DNA molecule and is therefore applicable to a wide variety of DNAs. However, to achieve the same extent of labeling, more polynucleotide kinase is required by the exchange reaction than by the phosphorylation reaction.[3,4] The procedures used for phosphorylation of (5')HO-DNA have been described by others.[1,5-9] In the following sections, we describe procedures for labeling endonuclease-generated 5'-phosphoryl termini using the polynucleotide kinase exchange reaction.

The Polynucleotide Kinase Exchange Assay

After cleavage of DNA with a site-specific endonuclease, the newly formed DNA 5'-phosphates are exchanged with ^{32}P from [γ-^{32}P]ATP. An example is presented for polynucleotide kinase exchange of EcoRI-digested λ DNA, where the 5'-phosphoryl groups are on protruding single-stranded (external) termini. Where applicable, differences in procedures used for 5'-phosphoryl groups on internal or blunt termini generated by other endonucleases will be included.

Reagents

Buffer A: 10 mM Tris-HCl; pH 7.4, 10 mM MgCl$_2$, 10 mM NaCl

Bacteriophage λ DNA (prepared and purified as described elsewhere[3])

Endo EcoRI (prepared and purified as described elsewhere[3]); a unit of endonuclease activity is roughly defined as that amount which can cleave 1 μg λDNA in 10 μl in 1 hr at 37°C

Buffer B: 0.22 mg/ml autoclaved gelatin, 36 mM MgCl$_2$, 9 mM dithiothreitol, 100 mM imidazole-HCl; pH 6.6, 654 μM ADP; and

[2] J. H. van de Sande, K. Kleppe, and H. G. Khorana, *Biochemistry* **12**, 5050 (1973).

[3] K. L. Berkner and W. R. Folk, *J. Biol. Chem.* **252**, 3176 (1977).

[4] G. Chaconas, J. H. van de Sande, and R. B. Church, *Biochem. Biophys. Res. Commun.* **66**, 962 (1975).

[5] C. C. Richardson, *Proced. Nucleic Acid Res.* **2** (1972).

[6] J. R. Lillehaug and K. Kleppe, *Biochemistry* **14**, 1221 (1975).

[7] J. R. Lillehaug and K. Kleppe, *Biochemistry* **14**, 1225 (1975).

[8] J. R. Lillehaug, R. K. Kleppe, and K. Kleppe, *Biochemistry* **15**, 1858 (1976).

[9] W. D. Kroeker and M. Laskowski, Sr., *Anal. Biochem.* **79**, 63 (1977).

20–30 μM [γ-^{32}P]ATP(2–10 $\times$ 10^{10} cpm/μmole); prepared according to the procedure of Glynn and Chappell[10] and purified by quaternary aminoethyl Sephadex and Sephadex G-10 chromatography.[3,11] The [γ-^{32}P]ATP is stored separately from the remaining ingredients

Phage T4 polynucleotide kinase: There are a number of procedures available for the purification of this enzyme.[5,12] The use of phage T4 mutants defective in DNA synthesis, grown in a restrictive host, results in high yields of polynucleotide kinase.[13] A unit of polynucleotide kinase is as defined by Richardson.[5] Relatively high concentrations of polynucleotide kinase are used during the exchange reaction. Therefore it is important that the preparation of polynucleotide kinase be free of interfering activities, including exonuclease, endonuclease, ligase, ATPase, and other kinase activities.

0.4 M trichloroacetic acid, 20 mM Na$_4$P$_2$O$_7$;

0.1 M Na$_4$P$_2$O$_7$

For analysis of the samples by gel electrophoresis:

2.5% Ficoll, 100 mM EDTA, 1% bromphenol blue;

4% polyacrylamide (0.15% N,N'-methylene bisacrylamide) slab gel or 1% agarose slab gel

Buffer C: 40 mM Tris-HCl, pH 7.9, 5 mM sodium acetate, 1 mM EDTA

Digestion with Endo EcoRI. Bacteriophage λ DNA, at an approximate concentration of 0.1 μg/μl, is incubated with endo *Eco*RI at 37°. The length of incubation and quantity of enzyme are determined by the desired extent of cleavage. For a limit digest, 1 μg λ DNA in 10 μl buffer A is incubated with 2 units endo *Eco*RI for 30 min at 37°. Shorter incubation times and/or less enzyme are used where only partial digestion is desired. These conditions can be determined in preliminary experiments, using agarose or polyacrylamide slab gel electrophoresis to examine the products of digestion.

To terminate endonuclease digestion, heating the reaction mixtures to 78° for less than 1 min results in complete loss of activity for many endonucleases. Quenching the digestion with EDTA cannot be used, since Mg^{2+} is required for the polynucleotide kinase exchange reaction.

[10] I. M. Glynn and J. B. Chappell, *Biochem. J.* **90,** 147 (1964).
[11] R. Symons, personal communication.
[12] A. Panet, J. H. van de Sande, P. C. Loewen, H. G. Khorana, A. J. Raae, J. R. Lillehaug, and K. Kleppe, *Biochemistry* **12,** 5045 (1973).
[13] S. G. Hughes and P. R. Brown, *Biochem. J.* **131,** 583 (1973).

Endonuclease-digested DNAs may be phenol extracted and purified prior to polynucleotide kinase labeling. However, this step has been found to be unnecessary, and by simple alteration of the reaction conditions digested DNAs can be immediately labeled with ^{32}P.

Labeling with Polynucleotide Kinase. *Eco*RI-cleaved λ DNA (1 μg in 10 μl buffer A) is mixed with an equal volume of buffer B together with 2–3 units of polynucleotide kinase. After incubation at 37° for 10–15 min, the exchange reaction is completed. A number of techniques can be used to terminate the polynucleotide kinase exchange reaction, depending on the subsequent method to be used for detecting ^{32}P incorporation. It is rapidly inactivated by EDTA (to 30 mM), heating (complete loss of activity within 30 sec at 78°), or acid (0.4 M trichloroacetic acid, 20 mM $Na_4P_2O_7$).

Detection of ^{32}*P Incorporation.* To quantitate ^{32}P incorporation into *Eco*RI-digested λ DNA, the reaction mixture (1 μg λ DNA in 20 μl buffers A + B) is precipitated by the addition of 500 μl of 0.1 M $Na_4P_2O_7$, 20 μg calf thymus DNA and 2.5 ml of 0.4 M trichloroacetic acid, 20 mM $Na_4P_2O_7$. After 30 min on ice, the DNAs are collected on Whatman GF-C filters presoaked in 0.4 M trichloroacetic acid and 20 mM $Na_4P_2O_7$. Each filter is rinsed with five 10-ml aliquots of 0.4 M trichloroacetic acid, 20 mM $Na_4P_2O_7$ and two 2-ml rinses of ethanol. The filters are dried and counted with a liquid scintillant.

To visualize ^{32}P incorporation into individual restriction fragments, the *Eco*RI digested polynucleotide kinase exchange labeled λ DNA (1 μg in 20 μl) is mixed with 2 μl 2.5% Ficoll, 100 mM EDTA and 1% bromphenol blue and subjected to electrophoresis in buffer C with 10 mM $Na_4P_2O_7$ for 10–18 hr at 35 mA/gel in a vertical slab gel (0.15 × 14 × 30 cm, Hoefer Scientific) composed of agarose or polyacrylamide. The [γ-^{32}P]ATP migrates slightly faster than the bromphenol blue in both gel matrices. For the analysis of low molecular weight DNA fragments, it is necessary to first dialyze the DNA (against 10 mM Tris-HCl, pH 7.5, 1 mM EDTA, 2 M KCl), followed by ethanol precipitation. After electrophoresis the slab gels are dried onto Whatman 3 MM or DE-81 cellulose papers and autoradiographed.

Optimal Conditions for Exchange. The ^{32}P incorporation into *Eco*RI-digested λ DNA has been found to be maximal at pH 6.6 with 300 μM ADP and 10 μM ATP. The optimal nucleotide concentrations for exchange of blunt or internal 5'-phosphates are similar: 12 μM ATP for both type of termini, and 100 μM ADP for internal termini or 200 μM ADP for phosphates at blunt termini. Exchange into all three types of termini is highly sensitive to ionic strength, with a 50% reduction in the extent of exchange observed at a KCl concentration of 35–60 mM.

The concentration of DNA (measured over the range 4–100 μg/ml of

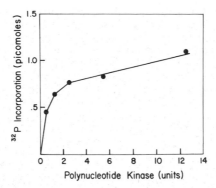

FIG. 1. The effect of polynucleotide kinase concentration upon the extent of exchange. EcoRI-digested λ DNA (3.65 μg) was incubated in 50 μl of 0.11 mg/ml autoclaved gelatin, 18 mM $MgCl_2$, 4.5 mM dithiothreitol, 45 mM KCl, 50 mM imidazole-HCl, pH 6.6, 327 μM ADP, and 12.5 μM [γ-[32] P]ATP with from 0.4 to 12.5 units of polynucleotide kinase. After 10 min at 37°, the reaction was stopped by the addition of 500 μl of 0.1 M $Na_4P_2O_7$ and the DNA precipitated in 0.4 M trichloroacetic acid and 20 mM $Na_4P_2O_7$, as described in the text.

EcoRI-digested λ DNA) does not have any effect upon the extent of ^{32}P incorporation into endonuclease-digested DNA. However, the extent of exchange is dependent upon the concentration of polynucleotide kinase used in the exchange reaction (Fig. 1). With polynucleotide kinase concentrations of 0.2–0.3 unit/μl, the extent of exchange with external 5′-phosphoryl termini is routinely 50–70% of the predicted theoretical amount. This has been observed with a number of different DNAs (λ DNA, P22 DNA, polyoma DNA, SV40 DNA) digested with different endonucleases (EcoRI, HindIII, BamHI, HpaII), where the numbers of termini are well defined. With blunt or internal termini, as will be discussed later, the amount of ^{32}P incorporation into 5′-phosphoryl termini is even lower (15–20% of the theoretical value). Incomplete exchange may be a consequence of the large quantities of enzyme used, where the number of polynucleotide kinase molecules approximates the number of 5′-phosphoryl termini.[3,12]

Use of Polynucleotide Kinase Exchange in Quantitating 5′-Phosphoryl Termini

As polynucleotide kinase exchange of ^{32}P with 5′-phosphoryl termini is incomplete, its usefulness as a quantitative assay demands that the extent of exchange be both reproducible and proportional to the number of DNA termini. The reproducibility is excellent, as the variance in extent of exchange from one sample to another is only 2–3%.[3] Greater variability appears from one experiment to another, however, and it is therefore necessary to include a sample (e.g., EcoRI-digested λ DNA) with a known number of 5'-phosphoryl termini in each experiment, so that the extent of exchange achieved in that experiment can be precisely determined.

To test the dependence of the extent of exchange upon the concentration of 5'-phosphoryl termini present, aliquots containing a constant amount of DNA, but with varying concentrations of endonuclease-generated 5'-phosphoryl termini are reacted in the polynucleotide kinase exchange assay (Fig. 2). For internal (*Hha*I-digested λ DNA), blunt (*Hin*dII-digested λ DNA), or external (*Eco*RI-digested λ DNA) 5'-phosphoryl termini, it can be seen (Fig. 2) that there is a linear relationship between the number of 5'-phosphoryl termini and the extent of ^{32}P exchange with those termini. Thus the extent of polynucleotide kinase exchange is proportional to the number of 5'-phosphoryl termini present, for all three types of termini generated by type II restriction endonucleases.

Specificity of the Polynucleotide Kinase Exchange Reaction

Structure of the 5'-Phosphoryl Termini. As mentioned earlier, the extent of exchange is dependent upon the type of DNA terminus. At a given concentration of polynucleotide kinase, the extent of ^{32}P incorporation per

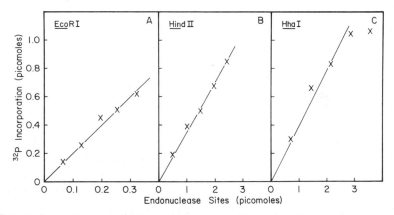

FIG. 2. Dependence of ^{32}P incorporation upon the concentration of endonuclease-generated 5'-phosphoryl termini. (A) Samples containing a constant amount of DNA (2 μg) but composed of mixtures of *Eco*RI-digested λ DNA (0.42–2 μg) and λ DNA (0.41–2 μg) in a total volume of 50 μl were incubated in 0.11 mg/ml of autoclaved gelatin, 18 mM MgCl$_2$, 4.5 mM dithiothreitol, 45 mM KCl, 50 mM imidazole-HCl, pH 6.6, 327 μM ADP, and 11.4 μM [γ-^{32}P]ATP with 1.5 units polynucleotide kinase. After 10 min at 37°, the reaction was stopped by the addition of (500 μl) 0.1 M Na$_4$P$_2$O$_7$, 20 μg calf thymus DNA and (2.5 ml) 0.4 M trichloroacetic acid, and 200 mM Na$_4$P$_2$O$_7$, and the precipitated DNA was collected on filters. Corrections were made for background ^{32}P incorporation into undigested linear λ DNA. Incorporation of ^{32}P into *Hin*dII-digested λ DNA (B) and *Hha*I-digested λ DNA (C) was measured as with *Eco*RI-digested λ DNA, except that the total amount of DNA per sample was 2.5 μg, no KCl was included during polynucleotide kinase exchange, and 5 units of polynucleotide kinase were used to label the DNA in each sample.

site for internal or blunt 5'-phosphoryl termini is less than for external 5'-phosphoryl termini (Fig. 2). The results of similar experiments with several other site-specific endonucleases are summarized in Table I. In general, polynucleotide kinase exchange of 5'-phosphates at internal or blunt positions occurs at approximately 15–25% the extent of external termini.

The extent of exchange of [γ-^{32}]ATP with the 5'-phosphoryl termini of heat denatured single-stranded DNA is the same as that observed for external 5'-phosphoryl termini of duplex DNA. The extent of exchange with single-stranded nicks, however, is approximately one-thirtieth as efficient as with 5'-phosphates located at external termini. Thus polynucleotide kinase clearly has a preference for single-stranded substrates in the exchange reaction.

Effect of DNA Fragment Size and Nucleotide Composition. DNA fragments ranging in size from 10^4 to 13.7×10^6 daltons have been labeled using the polynucleotide kinase exchange reaction. Over this 1000-fold size range, no effect of the fragment size upon extent of ^{32}P incorporation has been observed. Thus polynucleotide kinase exchange of ^{32}P with DNA fragments is proportional to the number of termini and not to the amount of DNA in each fragment. The sensitivity of detection for fragments with molecular weights as low as 10^4 illustrates the power of this technique. With alternative methods of detection which rely upon the number of nucleotides in each fragment of DNA, those fragments with molecular weights of 10^5 or less are often difficult to detect.

TABLE I

RELATIVE EXTENT OF POLYNUCLEOTIDE KINASE EXCHANGE WITH 5'-PHOSPHORYL TERMINI

Endonuclease	Cleavage site	Type of terminal 5'-phosphate	Extent of exchange relative to *Eco*RI terminus
*Bam*HI	(5')G↓GATCC	External	86
*Eco*RI	(5')G↓AATTC	External	100
*Hin*dIII	(5')A↓AGCTT	External	124
*Hpa*II	(5')C↓CGG	External	157
*Hae*II	(5')PuGCGC↓Py	Internal	24
*Hha*I	(5')GCG↓C	Internal	21
*Hin*dII	(5')GTPy↓PuAC	Blunt	16
*Hpa*I	(5')GTT↓AAC	Blunt	19

Polynucleotide kinase exchange has been used to label DNAs containing modified bases. The extent of polynucleotide kinase exchange does not appear to differ between DNAs containing uracil or hydroxymethyluracil in place of thymine or glucosylated hydroxymethylcytosine instead of cytosine.[14] DNA containing 5-bromouracil instead of thymine can also be labeled by polynucleotide kinase exchange. With restriction endonucleases that generate external 5'-phosphoryl termini (e.g., endonucleases EcoRI, HindIII, HpaII, BamHI), the extent of exchange is the same with 5-bromouracil as with thymine-containing DNAs. 5-Bromouracil-containing DNA's cleaved by restriction endonucleases that generate internal (e.g., endonucleases HaeII, HhaI) or blunt (e.g., endonucleases HpaI, HindII) 5'-phosphoryl termini, however, incorporate approximately twice as much ^{32}P as similarly cleaved thymine-containing DNAs.

Utility of the Polynucleotide Kinase Exchange Reaction

To demonstrate the utility of polynucleotide kinase exchange as an assay for detecting restriction endonuclease cleavage, it has been compared with an already established assay, which employs the electrophoretic separation and subsequent quantitation of the EcoRI digestion products of polyoma DNA (3×10^6 daltons), which contains one EcoRI site.

Polyoma [^{3}H]DNA (29 μg, 1.25×10^4 cpm/μg) was incubated in 250 μl buffer A with 5 units of endo EcoRI. Aliquots (25 μl) were withdrawn at timed intervals (from 0 to 15 min) and the activity quenched by heating at 78° for 1 min, followed by chilling on ice. Each aliquot was made 50 μl in 0.11 mg/ml of autoclaved gelatin, 18 mM MgCl$_2$, 4.5 mM dithiothreitol, 45 mM KCl, 25 mM imidazole-HCl, pH 6.6, 327 μM ADP, and 11.4 μM [γ-^{32}P]ATP (2×10^{10} cpm/μmole), and 2 units of polynucleotide kinase were added. After 10 min at 37°, the reaction was terminated and the samples were subjected to electrophoresis through a 1% agarose slab gel ($0.15 \times 14 \times 20$ cm) in buffer C at 20 mA for 7 hr. The different forms of polyoma DNA were detected after staining the gel in 1 μg/ml ethidium bromide (30 min) and exposure to long wavelength ultraviolet light. All polyoma DNA bands were cut out of the gel, dissolved in 2 ml of saturated KI, and precipitated with 100 μg calf thymus DNA in 5 ml of 0.4 M trichloroacetic acid and 20 mM Na$_4$P$_2$O$_7$. After 30 min on ice, the samples were collected on GF-C filters and rinsed with three 10-ml aliquots of 0.4 M trichloroacetic acid and 20 mM Na$_4$P$_2$O$_7$ and two 2-ml aliquots of ethanol.

Quantitation of endo EcoRI digestion of polyoma DNA by polynu-

[14] K. L. Berkner and W. R. Folk, J. Biol. Chem. 252, 3185 (1977).

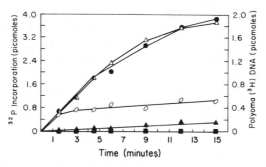

FIG. 3. Quantitation of endo *Eco*RI digestion of polyoma DNA by polynucleotide kinase exchange and by agarose slab gel electrophoresis. Polyoma [3H]-DNA was digested with endo *Eco*RI, labeled by polynucleotide kinase exchange and subjected to agarose slab gel electrophoresis, as described in the text. The polyoma DNA forms were then isolated and quantitated. Incorporation of [32P] into polyoma DNA has been corrected for the efficiency of labeling of 5'-phosphoryl termini achieved in this experiment (67% of the maximum). Form I [32P]-DNA (■——■), form II [32P]-DNA (▲——▲), form III [32P]-DNA (●——●), form II [3H]-DNA (○——○), form III [3H]-DNA (△——△).

cleotide kinase exchange (measurement of incorporated [32P]), or by quantitation of the amount of form III polyoma [3H]DNA, give identical results (Fig. 3). The small amount of label in Form II DNA reflects the inefficiency of polynucleotide kinase labeling of nicks.

Acknowledgments

Support from the National Institutes of Health, American Cancer Society, and the National Foundation is gratefully acknowledged.

Section II

Techniques for Labeling Termini

[6] Specific Labeling of 3′ Termini with T4 DNA Polymerase

By MARK D. CHALLBERG and PAUL T. ENGLUND

T4 DNA polymerase is both a 5′ → 3′ polymerase and a 3′ → 5′ exonuclease.[1] Since its rate of polymerization is greater than its rate of exonucleolytic hydrolysis, the enzyme has proved useful in analyzing and manipulating terminal DNA sequences.[2,3]

Consider this arbitrary terminal sequence of a duplex DNA:

—AGGTAAGCCAGCGC   5′
—TCCATTCGGT   3′

Depending on the composition of the deoxynucleoside triphosphates in the reaction mixture, it is possible to use the T4 polymerase to convert this structure into one of several related structures. For example, if the DNA is incubated with T4 DNA polymerase and [α-^{32}P]dTTP as the sole triphosphate, the exonuclease activity will remove the 3′-terminal dTMP. However, before the enzyme can remove more nucleotides from the strand, the polymerase activity will replace the terminal dTMP by transfer from the [α-^{32}P]dTTP. Subsequent reaction consists of alternating removal and replacement of this dTMP residue. Since the rate of incorporation is much greater than the rate of removal, the steady state product will have an unaltered structure, but the terminal dTMP residue will be labeled with ^{32}P. If the DNA is incubated with the enzyme and dATP as the sole deoxynucleoside triphosphate, the enzyme will sequentially remove seven nucleotides from the 3′ terminus and then immediately replace the dAMP by transfer from the triphosphate. It will continue to remove and replace this nucleotide. The net result will be a strand shortened by six nucleotides with dAMP as the 3′-terminal residue. If the DNA is incubated with the enzyme and dCTP, the 3′ terminus will be extended by a single residue, and if it is incubated with both dCTP and dGTP, the DNA will be fully repaired.

Applications

The combined polymerase and exonuclease activities of the T4 DNA polymerase have suggested several applications.

---

[1] M. Goulian, Z. J. Lucas, and A. Kornberg, *J. Biol. Chem.* **243**, 627 (1968).
[2] P. T. Englund, *J. Biol. Chem.* **246**, 3269 (1971).
[3] P. T. Englund, *J. Mol. Biol.* **66**, 209 (1972).

1. Radioactivity can be introduced at the 3′ terminus of a duplex without altering the structure of the terminus.[3,4]

2. One or more nucleotides can be removed from a 3′ terminus at either a nick or at the end of a duplex molecule.[2,5] The extent of removal can be precisely predicted if previous sequence data exists. Similarly, one or more nucleotides can be added to a 3′ terminus which is base-paired to an appropriate template.

3. Any linear duplex DNA molecule can be converted to a blunt-ended structure if it is incubated with polymerase and all four deoxynucleoside triphosphates. If the molecule has an unpaired 5′ terminus, the enzyme will use this sequence as a template and extend the 3′ terminus until repair is complete. Conversely, if the molecule has an unpaired 3′ terminus, the enzyme will degrade it. Degradation effectively stops when the enzyme reaches the double-stranded region of the molecule because the enzyme will replace any base-paired nucleotide which is removed.[5a]

4. A nucleotide at a 3′ terminus can be exchanged for a nucleotide analog. Modrich and Rubin have used the T4 DNA polymerase to exchange a dIMP residue for a dGMP residue in the *Eco*RI recognition site for the purpose of studying the *Eco*RI methylase specificity.[6]

5. The T4 polymerase reaction can be used in nucleotide sequence analysis. Originally, short 3′-terminal sequences were analyzed by nearest neighbor or oligonucleotide analysis of 3′-terminally labeled DNA.[3,4,5,7,8] Subsequently, Sanger and Coulson used this reaction in the "plus" technique of the "plus and minus" method for analyzing long nucleotide sequences.[9] Federoff and Brown have also used the T4 DNA polymerase in their modification of the "plus and minus" method.[10]

Reaction Conditions

Reaction conditions were chosen to maximize the ratio of polymerase to exonuclease activity. In particular, high triphosphate concentration and low temperature were used. Possible variations in these conditions are described below. A typical reaction mixture contains 63 mM Tris-HCl (pH

[4] S. S. Price, J. M. Schwing, and P. T. Englund, *J. Biol. Chem.* **248**, 7001 (1973).

[5] P. T. Englund, S. S. Price, J. M. Schwing, and P. H. Weigel, in "DNA Synthesis *In Vitro*" (R. D. Wells and R. B. Inman, eds.), p. 35. Univ. Park Press, Baltimore, Maryland, 1973.

[5a] J. F. Burd and R. D. Wells, *J. Biol. Chem.* **249**, 7094 (1974).

[6] P. Modrich and R. A. Rubin, *J. Biol. Chem.* **252**, 7273 (1977).

[7] P. H. Weigel, P. T. Englund, K. Murray, and R. W. Old, *Proc. Natl. Acad. Sci. U.S.A.* **70**, 1151 (1973).

[8] K. Murray, A. G. Isaksson-Forsen, M. Challberg, and P. T. Englund, *J. Mol. Biol.* **112**, 471 (1977).

[9] F. Sanger and A. R. Coulson, *J. Mol. Biol.* **94**, 441 (1975).

[10] N. V. Federoff and D. D. Brown, *Cell* **13**, 701 (1978).

8.0), 6.3 mM MgCl$_2$, and 6.3 mM mercaptoethanol. Each deoxynucleoside triphosphate is about 50 to 100 μM, and they may be labeled with ^{3}H or ^{32}P. The DNA concentration is about 35 pmoles of terminus per milliliter, and polymerase and the DNA termini are present in roughly equimolar quantities. (Polymerase molarity is calculated assuming a molecular weight of 112,000 and a specific activity of about 30,000 units/mg.[1]) The reaction mixture is incubated at 11° until incorporation of radioactivity into an acid-insoluble form reaches a limit. Usually only a single nucleotide is incorporated per strand.

Acid-insoluble radioactivity is measured by a filter assay. Since only a very small fraction of the radioactivity is incorporated into the DNA, it is essential that the filters be washed very efficiently. A small aliquot of the reaction mixture is diluted with 0.1 ml of ice-cold water and the following additions are made: 0.1 ml of 0.1 M EDTA; 0.025 ml of sonicated salmon sperm DNA (2.0 mg/ml); 0.1 ml of 1.0 M HCl containing 0.1 M sodium pyrophosphate; and 0.5 ml of 7% perchloric acid. After at least 5 min in an ice bath, the contents of the tube are filtered through a 2.4-cm Whatman GF/C disk (presoaked in 0.1 M sodium pyrophosphate) which is then washed with about 100 ml of cold 1.0 M HCl containing 0.1 M sodium pyrophosphate. After a rinse with ethanol, the filters are dried and counted in a scintillation counter.

Effect of Triphosphate Concentration

High concentrations of deoxynucleoside triphosphates (50–100 μM) were originally chosen to maximize the ratio of polymerase to nuclease activity, and most of our experiments have used that concentration. However, Steenbergh and co-workers used only 0.5 μM triphosphates in labeling adenovirus DNA for the purpose of sequencing the termini.[11] The sequences which they obtained by this method are fully confirmed in later studies.[12] In limited experiments we have found that the rate of incorporation of ^{32}P-nucleotides is reduced when the triphosphate concentration is in the range of 1 to 5 μM.

Effect of Temperature

It is essential to maintain the reaction temperature at 11°. At higher temperatures, the incorporation is less specific as the exonuclease appears to degrade beyond a nucleotide which the polymerase could replace. Pre-

[11] P. H. Steenbergh, J. S. Sussenbach, R. J. Roberts, and H. S. Jansz, *J. Virol.* **15**, 268 (1975).
[12] J. R. Arrand and R. J. Roberts, personal communication (1978).

sumably at elevated temperatures the ends of the duplex are locally denatured, and the presence of a triphosphate would not limit the degradation of a single-stranded 3' terminus. In the "plus and minus" method Sanger and Coulson incubate the DNA with T4 polymerase at 37°.[9] However, in that method it is not essential to stop degradation at one specific nucleotide, but only to favor stopping at any nucleotide which has a base the same as the one on the single triphosphate which is present.

Effect of DNA and Enzyme Concentration

These parameters appear to be relatively unimportant. DNA concentration has been lowered to about 4 pmoles termini/ml,[11] and could probably be lowered even further. No significant change in the maximal level of incorporation could be detected when the level of enzyme was varied from 0.3 to 1.8 enzyme molecules per DNA strand.

Effect of Reaction Time

The time required to reach a steady state depends on the number of nucleotides which have to be removed. The rate of removal undoubtedly depends on the base composition and nucleotide sequence. If the reaction involves only the exchange of the 3'-terminal nucleotide without net degradation, the steady state is reached within 10 min and possibly much sooner. Removal of 1 to 6 nucleotides probably occurs within 1 hr. Labeling the l-strand of T7 DNA with [32P]dAMP[3] required prior removal of 14 nucleotides[13] and this reaction took about 3 hr. Steenbergh and co-workers found that it also took 3 hr to label the termini of adenovirus-2 DNA with [32P]dCMP. This reaction required the removal of 24 nucleotides from each terminus.[11,12] In contrast, the T4 polymerase was unable to incorporate a [32P]dTMP into the l-strand of λ DNA even during a 165-min incubation,[7] and subsequent sequence analysis has revealed that a successful incorporation would have required a removal of 15 nucleotides from the natural terminus.[14] The inefficiency of the T4 polymerase in degrading the terminus of the λ DNA l-strand may be related to the fact that it contains about 60% G+C base pairs, whereas the more susceptible adenovirus DNA termini contain only 20% G+C base pairs. It is important to note that despite the extensive degradation required to label the adenovirus termini with [32P]dCMP and the T7 l-strand with [32P]dAMP, in both cases the nearest neighbor of these 32P-nucleotides was confirmed by subsequent sequence analysis.

[13] P. C. Loewen, *Nucleic Acids Res.* 2, 839 (1975).
[14] B. P. Nichols and J. E. Donelson, *J. Virol.* 26, 429 (1978).

Acknowledgments

This work was supported by the National Institutes of Health (Grant No. CA13602). M.D.C. was supported by NIH Training Grant No. 5T01GM00184.

[7] Terminal Transferase-Catalyzed Addition of Nucleotides to the 3' Termini of DNA

By RANAJIT ROYCHOUDHURY and RAY WU

Terminal deoxynucleotidyltransferase, which usually requires a single-stranded DNA primer,[1] can accept double-stranded DNA as primer when Co^{2+} ion is used in place of Mg^{2+} ion. In the presence of Co^{2+} ion, all forms of duplex DNA molecule can be labeled at their 3' ends regardless of whether such ends are single stranded or base paired.[2]

In this article, we describe a procedure for specific labeling of 3' termini with ribonucleotide[3-4a] for DNA sequence analysis,[4a,4b] and for adding homopolymer tracts of deoxynucleotides for recombinant DNA research.[5-7] This procedure allows the addition of nucleotides to the 3' ends of any DNA molecule without need for prior treatment of the DNA with λ exonuclease to expose the 3' terminus as single-stranded primer.[5,6]

Materials

All glassware, capillary pipettes, Pasteur pipettes, and small columns are siliconized before use. Test tubes, capillary tubes, and pipettes are conveniently siliconized by using dichlorodimethyl silane: CCl_4 (25 : 475, v/v). This is done by placing about 5 ml of the above siliconizing solution in a large petri dish under the perforated platform of a 12-inch vacuum desiccator. Glassware is then placed over the platform. A mild vacuum is applied for about 3 min by using a water pump (aspirator). After closing the stopcock, the glassware is allowed to remain under partial vacuum overnight. The chemicals evaporate and coat the glass

[1] F. J. Bollum, *in* "The Enzymes" (P. D. Boyer, ed.), 3rd ed., Vol. 10, p. 145. Academic Press, New York, 1974.
[2] R. Roychoudhury, E. Jay, and R. Wu, *Nucleic Acids Res.* **3**, 863 (1976).
[3] H. Kössel and R. Roychoudhury, *Eur. J. Biochem.* **22**, 271 (1971).
[3a] R. Roychoudhury and H. Kössel, *Eur. J. Biochem.* **22**, 310 (1971).
[4] R. Roychoudhury, *J. Biol. Chem.* **247**, 3910 (1972).
[4a] H. Kössel, R. Roychoudhury, D. Fischer, and A. Otto, Vol. 29, p. 322.
[4b] R. Roychoudhury, D. Fischer, and H. Kössel, *Biochem. Biophys. Res. Commun.* **45**, 430 (1971).
[5] D. A. Jackson, R. H. Symons, and P. Berg, *Proc. Natl. Acad. Sci. U.S.A.* **69**, 2904 (1972).
[6] P. E. Lobban and A. D. Kaiser, *J. Mol. Biol.* **78**, 453 (1973).
[7] D. Brutlag, K. Fry, T. Nelson, and P. Hung, *Cell* **10**, 509 (1977).

uniformly and efficiently. The glassware is heated in an oven at 120° for 24 hr and rinsed in deionized water. This procedure (introduced by Dr. Ernest Jay in our laboratory) saves considerable time and labor.

Sephadex G-50 is obtained from Pharmacia. The column is made from siliconized glass tube (0.8 × 80 cm) by tapering one end in a flame and then connecting the tapered end with a small Tygon tubing. A light plastic stopcock from Bel-Art Products is used for closing the column. Small glass tubes (13 × 55 mm) for fraction collection and glass insert counting vials (20 × 46 mm) are obtained from Rochester Scientific Company, 15 Jet View Drive, Rochester, New York.

Reagents and Enzymes

Cacodylic acid (Fischer Scientific Company, purified grade)
Solid KOH pellets (analytical grade)
Tris(hydroxymethyl)aminomethane (TRIZMA base from Sigma)
0.1 M $CoCl_2$ (Fischer Scientific, analytical grade)
0.1 M dithiothreitol (Calbiochem or BioRad)
Dichlorodimethyl silane (Accurate Chemical and Scientific Company, Hicksville, New York)
$[\alpha\text{-}^{32}P]rNTP$, specific activity 100–600 Ci/mmole from New England Nuclear Corporation or Amersham
Bacterial alkaline phosphatase (BAPF), electrophoretically pure grade, from Worthington Biochemical Company (Freehold, New Jersey) or RNase-free grade from Bethesda Research Laboratories (441 N. Stonestreet Avenue, Rockville, Maryland)
Micrococcal nuclease (Worthington), further purified[8] to remove traces of phosphatase
Pancreatic DNase (DNaseI), highest grade from Worthington. *Hinc*II and *Hind*III restriction endonucleases from New England Biolabs, Inc. (283 Cabot St., Beverly, Massachusetts)

Terminal transferase is purified according to Chang and Bollum.[9] The enzyme preparation should be tested for contaminating endonuclease and exonuclease activities. The most sensitive tests for detection of exonuclease and endonuclease have been described earlier.[8] The enzyme should be completely free of 3' → 5' exonuclease activity and very low in contaminating endonuclease activity. The latter is checked by determining the conversion of supercoiled DNA to relaxed DNA,[8] or by measuring homopolymer formation using covalently closed circular DNA as primer.[10] Exonuclease activity is checked by release of counts from 3'

---

[8] R. Wu, E. Jay, and R. Roychoudhury, *Methods Cancer Res.* **12**, 87 (1976).
[9] L. M. S. Chang and F. J. Bollum, *J. Biol. Chem.* **246**, 909 (1971).
[10] P. Humphries, R. Old, L. W. Coggins, T. McShane, C. Watson, and J. Paul, *Nucleic Acids Res.* **5**, 905 (1977).

terminally labeled double-stranded DNA.[8] Terminal transferase from several commercial sources, especially those from Bethesda Research Laboratories (lot 827) and P. L. Biochemicals (lot 730-9), have been adequate.

Preparation of Concentrated Buffer Mixture for the Terminal Transferase Reaction

A stock solution of 10× buffer mixture is prepared as follows. Cacodylic acid powder (13.8 g) is suspended in glass distilled water (35 ml), mixed with solid Tris base (3.0 g), and adjusted to pH 7.6 by slow addition of solid KOH with constant stirring by a magnetic stirrer. Glass-distilled water is added to a final volume of 88 ml. The solution is chilled to 0° and thoroughly mixed with 2 ml of 0.1 M dithiothreitol. The final addition of 10 ml of 0.1 M $CoCl_2$ is carried out dropwise with constant stirring by a magnetic stirrer in order to avoid precipitation. This clear solution (1 M potassium cacodylate, 250 mM Tris base, 10 mM $CoCl_2$, 2 mM dithiothreitol) may be kept at 4° for more than 1 year. If the solution is turbid, warm it up to 37° and cool to 4°. After tenfold dilution the pH of this solution becomes 6.9–7.0.

Preparation of DNA

The isolation of DNA preparation from mitochondria,[11] yeast,[12] chloroplast,[13] bacteria,[14] phage-infected bacteria,[15] plants,[16] mammals and insects[17] have been described in this series. For isolation of bacterial and animal virus DNA, more recent procedures have been reviewed.[8] The DNA samples should be checked for contamination with proteins. Histones and glycoproteins as well as polysaccharides inhibit the transferase reaction. These contaminants often can be removed by phenol extraction of the DNA samples. Chelating agents, such as EDTA or citrate, should be avoided. Salts and other contaminants in DNA preparations can usually be removed by precipitating the DNA with ethanol. The DNA solution (in polystyrene tubes, 12 × 55 mm, Walter Sarstedt) is made 0.2 M with KOAc, to which two volumes of ethanol are added. After standing at −20° for 2 hr, the tubes are centrifuged at 10,000 g in swinging bucket rotors with rubber adapters (Sorvall HB-4 or Beckman J-21 rotor) for 30 min. The supernatant is discarded and the tubes are kept inverted over tissue papers until all the ethanol has drained off. Alternatively, DNA

---

[11] G. Kalf and M. A. Grece, this series, Vol. 12, Part A, p. 533.
[12] D. Smith and H. O. Halvorson, this series, Vol. 12, Part A, p. 538.
[13] M. Eisenstadt and G. Brawerman, this series, Vol. 12, Part A, p. 541.
[14] K. Miura, this series, Vol. 12, Part A, p. 543.
[15] M. G. Smith, this series, Vol. 12, Part A, p. 545.
[16] H. Stern, this series, Vol. 12, Part B, p. 108.
[17] K. S. Kirby, this series, Vol. 12, Part B, p. 87.

solution in 2 volumes of ethanol can be chilled in a Dry Ice–ethanol bath for 15 min and centrifuged in an Eppendorf Microcentrifuge (Brinkman Company) for 10 min.[18] The traces of ethanol at the rim of the tubes are wiped off with tissue papers. The DNA sample is dissolved in a minimum volume of buffer appropriate for a desired restriction endonuclease.

Digestion of DNA with a Restriction Endonuclease

As an example, HincII endonuclease digestion of SV40 DNA was chosen for cleaving the DNA before labeling and sequencing the DNA from the 3' ends. The reaction mixture contained 10 mM Tris-HCl (pH 8.0), 10 mM MgCl$_2$, 5 mM 2-mercaptoethanol, 50 mM NaCl, 3 A_{260} unit (150 μg) of covalently closed circular SV40 DNA,[8] and 60 to 160 units of HincII in a final volume of 300 μl. After 6–12 hr of incubation at 37°, the completeness of digestion was monitored by agarose gel electrophoresis of an aliquot (2 μl) of the reaction mixture.

Restriction endonuclease preparations sometimes contain RNA, gelatin, and other trace contaminants. Some of these may inhibit the transferase reaction. Therefore, a minimum amount of the restriction endonuclease and a prolonged incubation time (4–12 hr) should be used. This will also conserve the endonuclease. We have tested most of the restriction endonucleases from New England Biolabs and from Bethesda Research Laboratory; these are usually of good quality. It is advisable to predetermine the amount of restriction endonuclease needed to digest a given amount of DNA. It is also important to carry out the restriction endonuclease digestion in a minimum volume. After the digestion, an aliquot of the reaction mixture is diluted 4 to 20 times, so that the final DNA concentration is approximately 2 A_{260} unit/ml. The reaction mixture is then supplemented with the transferase buffer and the enzyme. We found that terminal transferase can catalyze addition of nucleotides at the 3'-OH ends of duplex DNA in the presence of all the restriction endonucleases tested so far. This simple procedure minimizes the handling and the loss of DNA, which is especially important when minute quantities of DNA (such as 0.1 μg of cDNA) are used.[2]

Labeling DNA with Ribonucleotides

Terminal transferase catalyzes a limited polymerization of ribonucleotides. Originally, a maximum of two ribonucleotide additions was detected by paper chromatography using Mg^{2+} ion in the reaction mixture.[3,4] In 1973, using an improved separation procedure (homochromatography),

---

[18] A. Maxam and W. Gilbert, *Proc. Natl. Acad. Sci. U.S.A.* **74,** 560 (1977).

we noticed that replacement of Mg^{2+} by Co^{2+} ion not only leads to an efficient conversion of primer into ribo addition products but also causes the additions of more than two ribonucleotides.[19] This increased priming activity in the presence of Co^{2+} ion prompted an application of this reaction for direct labeling of duplex DNA fragments generated by restriction endonucleolytic cleavages. It proved to be successful as we noticed that all forms of duplex DNA fragments could be labeled with either ribonucleotides or deoxynucleotides.[2]

Labeling the 3' Ends of Hinc*II-Digested DNA Fragments with Ribonucleotides.* The reaction mixture contains 100 μl of *Hinc*II digested SV40 DNA fragments (1 A_{260} unit = 210 pmoles of 3'-OH ends), 20 μl of 10× buffer mixture for transferase, 1980 pmoles of [α-^{32}P]rCTP, and 2 μg (40 units) of terminal transferase in a final volume of 200 μl. Incorporation of [^{32}P]rCMP residues into DNA at 37° is monitored by sampling 2-μl aliquots at intervals (e.g., 10, 30, 60, and 120 min). The samples are spotted on 3 MM paper disks, acid-washed,[20] and counted for ^{32}P in the acid-precipitated DNA. When the incorporation approaches a plateau (usually after 1 to 8 hr), the enzyme is inactivated by adding 20 μl of neutralized phenol, 20 μl of 0.5 M EDTA and 20 μl of 1% SDS. The contents of the tube is loaded onto a column of Sephadex G-50, preequilibrated with 50 mM Tris-HCl (pH 8.0), 100 mM NaCl, and 1 mM EDTA and eluted with the same solution. Fractions of approximately 400 μl (20 drops) are collected in siliconized glass tubes (13 × 50 mm, Rochester Scientific Company) using a Gilson model F80 fraction collector. The tubes are then placed directly inside plastic scintillation vials and counted for Cerenkov radiation.

The elution profile and the distribution of radioactivity is shown in Fig. 1. The fractions in the labeled DNA peak are pooled into a siliconized graduated tube (10-ml capacity). Aliquots (1 ml) are transferred with a Pasteur pipette to clear polystyrene tubes (12 × 55 mm, catalogue No. 484, Walter Sarstedt, Princeton, New Jersey). To each tube is added 2 ml of 95% ethanol. The tubes are covered with Parafilm and the contents mixed thoroughly. After standing at −20° for 4 hr or more the tubes are centrifuged at 12,000 g in a Sorvall HB-4 rotor or a Beckman J-21 rotor with swinging bucket for 1 hr. Under these conditions, duplex DNA fragments as short as 20 nucleotide long are almost quantitatively pelleted. The ethanol supernatant is poured into tubes of the same size for counting. The original tube containing the DNA pellet is kept inverted over tissue paper until the alcohol is drained off. It is then held horizontally while 50 μl of buffer (10 mM Tris-HCl, pH 8.0, and 0.5 mM EDTA) is

[19] R. Roychoudhury and R. Wu, unpublished results (1973).
[20] F. J. Bollum, this series, Vol. 12, p. 169.

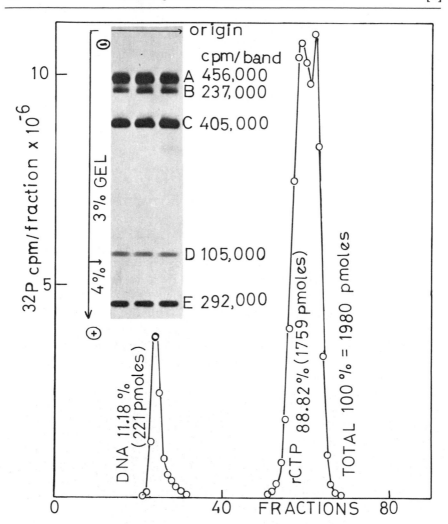

FIG. 1. Isolation of [³²P]rCMP labeled *Hinc*II fragments of SV40 DNA by gel filtration and their separation by gel electrophoresis.

applied near the rim. The tube is held at an angle of 10°–15° and gently rotated so that the drop of liquid washes all the sides of the tube before it reaches the pellet at the bottom. This tube is directly counted for ³²P.

The resuspended labeled DNA from the tubes are pooled, mixed with 10% sucrose, 1% bromphenol blue, 5 mM EDTA, and 0.1% SDS and loaded into slots of a vertical slab gel[8]. The insert in Fig. 1 shows an autoradiogram of the part of the gel containing five major *Hinc*II fragments. DNA from each band was cut out and eluted according to Wu *et al.*[8] (p. 136 and Fig. 19C). It is evident from the counts (Fig. 1) that all the bands are not labeled with equal efficiency. Because *Hin*cII recognizes four sequences

$$5'\text{GTT}\!\downarrow\!\text{AAC} \qquad 5'\text{GTT}\!\downarrow\!\text{GAC} \qquad 5'\text{GTC}\!\downarrow\!\text{AAC} \qquad 5'\text{GTC}\!\downarrow\!\text{GAC}$$

some of the duplex blunt ends will have A : T base pairs while others will have G : C base pairs. We noticed that 3' ends terminated with A : T base pairs are generally more efficiently labeled.

Labeling of Hin*dIII-Digested DNA.* The reaction mixture (200 μl) contains 1 A_{260} unit of *Hin*dIII digested SV40 DNA (180 pmoles of 3'-OH ends), 20 μl of 10× buffer mixture (with Co^{2+}), 1600 pmoles of [α-^{32}P]rGTP and 40 units of terminal transferase. After 6 hr at 37°, the labeled DNA fragments are isolated by gel filtration. Of the total radioactivity applied to the column, 22.84% (365 pmoles) are recovered as labeled DNA. This accounts for an average of two labeled residues added per DNA ends. Under similar conditions an average of one labeled residue was added per DNA end using *Hin*cII digested DNA (Fig. 1). Since all *Hin*dIII cuts have A : T base pairs at the exposed end, it appears that terminal transferase has a preference for such ends in its priming activity.

Labeling of Hin*dII + III-Digested DNA.* The reaction mixture (400 μl) contains 1.5 A_{260} unit of *Hin*dII + III digested DNA (585 pmoles of 3'-OH ends), 40 μl of 10× buffer mixture, 4520 pmoles of [α-^{32}P]rCTP, and 160 units of transferase. After 8 hr at 37° the reaction mixture is processed as described earlier, and a total of 587 pmoles of rCMP residues are incorporated into DNA fragments. When the bands from the polyacrylamide gel are cut out and counted it appears that the bands A, E, and K are most heavily labeled. This coincides with the fact that these three DNA fragments have *Hin*dIII cuts at both ends (see Fig. 3A).

Primer Conversion. The 3'-terminal labeling of duplex DNA primers is more sensitive to variables during incubation than that of single-stranded primers. These include polynucleotide concentration, salts, and the amount of terminal transferase per DNA end. Most of our earlier incubations were carried out using high concentrations of DNA (4–6 A_{260}/ml). However, we noticed that reducing the concentration of DNA to 1.5–2 A_{260}/ml improves the labeling efficiency (Table I). We recommend 1 A_{260}/ml as the optimum concentration. The amount of enzyme per DNA end (Table I, experiments 3–6) also influence the labeling. However, we usually use suboptimal enzyme levels and longer times to conserve the enzyme. The ratio of rCTP to primer also affects the rate of incorporation. In most experiments, we use a ratio of approximately 10. The extent of incorporation is lower if this ratio is lowered to 2 or 4. Higher salt concentration (300

TABLE I
FACTORS INFLUENCING LABELING REACTION OF DUPLEX DNA FRAGMENTS

Expt. No.	Restriction enzyme used in cleaving DNA	DNA concentration (A_{260} unit/ml)	Enzyme concentration (units/pmole of 3-OH ends)	rNTP/primer	Residues of rCMP added per strand
1	HindII + III 300 mM salt	2.5	0.82	11	0.67
2	HindII + III 150 mM salt	2.5	0.82	11	1.0
3	HindIII	1.5	1.77	31	3.1
4	HindIII	5.0	0.22	11	1.1
5	HindIII	5.7	0.46	9	1.7
6	HindIII	7.5	0.23	4	0.4
7	HincII	1.5	0.25	11	1.8
8	HincII	4.0	0.22	11	1.1
9	HincII	5.0	0.15	11	0.9

mM, experiment 1) inhibits the reaction. The standard buffer containing 100 mM potassium cacodylate (pH 7.0) and 1 mM CoCl$_2$ seems optimal for the ribo addition reaction.

We have reported[2] that the utilization of primers using different rNTP follow the order rGTP > rCTP = rATP > rUTP. However, since rGTP results in multiple (1–8) additions, it gives little information about the number of primer molecules participating in the reaction. Since rCTP yields products with predominantly two ribo additions and permits a fair estimate of the number of primer molecules participating in the reaction, it is the preferred substrate.

Elution of DNA from Gel

The gel band is cut out with a razor blade and placed in a siliconized glass insert counting vial (20 × 46 mm). The gel is then ground against the sides of the vial with a siliconized glass rod (6–8 mm in diameter). The gel particles are suspended after addition of 3–5 ml (about twice the volume of the gel) of 0.2 M triethylammonium bicarbonate (TEAB, pH 8.0). The vial is covered with Parafilm, shaken, and allowed to stand at room temperature for 12 hr. The vial is centrifuged at 2000 g in a clinical centrifuge, and the clear supernatant is withdrawn carefully without disturbing the gel pellet. The process is repeated twice. The combined supernatant (about 10 ml) containing DNA labeled with [³²P]rNMP is adjusted to 0.3 M NaOH by adding 10 N NaOH. The sample is incubated at room temperature for

20 hr and neutralized with concentrated HCl. The pH is adjusted to 8.0 by addition of Tris-HCl buffer (pH 8.0) and the solution is treated with 0.5–1.0 unit of phosphatase at 45° for 1 hr. The solution is then diluted threefold with water and passed through a small DEAE-cellulose column (DE 52, bed volume 50–100 μl), packed over a small wad of sterile cotton in a 1-ml plastic cone (Pipettenspitzen, Walter Sarstedt). The radioactive materials adsorbed onto DE 52 are washed twice with 1-ml portions of 0.25 M TEAB. This treatment removes the phosphatase. The plastic cone is then placed over a thick-wall (10 × 75 mm) siliconized tube (see Wu *et al.*,[8] Fig. 19C on p. 144), and the labeled DNA is eluted by adding 50 μl of 2 M NaCl in 10 mM Tris-HCl (pH 8.0) to the top of the column. After 10 min, the tube and cone are centrifuged together at low speed. The process is repeated twice. We notice that about 90% of the radioactivity is eluted (recovery is poor if DNA fragment is over 1000 base pairs in length) in a total volume of 150 μl. The content from the glass tube is transferred to a polystyrene tube (12 × 55 mm) with a siliconized Pasteur pipette. The glass tube is rinsed twice with 150-μl aliquots of glass distilled water and the rinse transferred to the polystyrene tube. After addition of 1 ml of ethanol, the DNA fragments are chilled and centrifuged as described earlier.

Separation of Two Labeled Ends of a DNA Fragment

Before carrying out sequence analysis of a DNA segment from each 3' terminus, the two labeled ends are separated. In one method, the DNA is subjected to digestion with a suitable restriction endonuclease and the resulting single-end labeled fragments are separated by gel electrophoresis. For example, the *Hinc*II fragments A, B and C (Fig. 2) obtained as ethanol precipitate are each suspended in 50 μl of restriction endonuclease buffer and digested with 30 units of *Hin*dIII at 37° for 12 hr. The fragment D is digested with 30 units of *Bam*HI and the fragment E with 4 units of *Hae*III. The digestion is stopped by adding 10 μl of a solution containing 50% sucrose, 1% SDS, 100 mM EDTA, 1% bromphenol blue, and 1% xylene cyanol. The mixture is loaded onto three slots of a polyacrylamide step gel with 3% gel at the top and 4% gel at the bottom. As shown in Fig. 2A, *Hin*dII A is separated into segments A_1 and A_2 (each containing a single labeled end) suitable for sequence analysis.

The *Hin*dIII fragments were similarly digested with *Hae*III (or *Hinc*II for fragment D) for the separation of two ends (Fig. 2B).

Nearest Neighbor Analysis. In the presence of Co^{2+} ion, the labeling at the nicks in DNA molecules is also fairly efficient. Therefore, before sequence analysis of isolated fragments, it is advisable to check the cor-

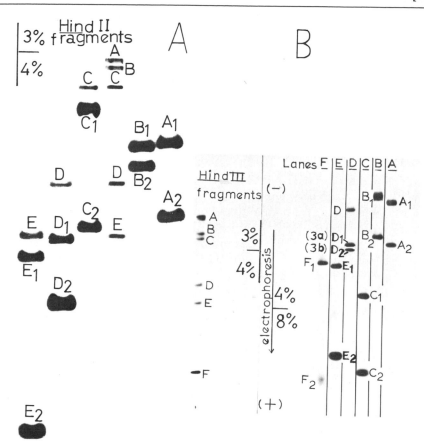

FIG. 2. Separation of unique labeled ends by electrophoresis after restriction endonuclease digestion of isolated labeled DNA fragments.

rectness of 3'-terminal labeling by nearest neighbor analysis.[21] In this analysis only one nucleotide should be made radioactive. If other nucleotides are significantly labeled (indicating labeling at the nicks) then the sequence analysis by two-dimensional fingerprinting[22,23] or by chemical degradation[18] will give "ghost-spots" or "ghost-bands."

Since *Hin*dIII endonuclease recognizes the hexanucleotide sequence[24]

$$5'\text{-A}{\downarrow}\text{AGCTT-3'}$$
$$\text{-TTCGA}{\uparrow}\text{A-}$$

[21] R. Wu, R. Padmanabhan and R. Bambara, this series, Vol. 29, p. 231.
[22] G. G. Brownlee and F. Sanger, *Eur. J. Biochem.* **11**, 395 (1969).
[23] C. D. Tu, E. Jay, C. P. Bahl, and R. Wu, *Anal. Biochem.* **74**, 73 (1976).
[24] R. Old, K. Murray, and G. Roizes, *J. Mol. Biol.* **92**, 331 (1975).

with arrows indicating the site of cleavage, the nearest neighbor analysis after the addition of a [^{32}P]rC at the 3' end should yield dAp as the radioactive nucleotide. The isolated single-end labeled HindIII fragments are (Fig. 2B) treated with alkali and phosphatase[4a] and purified by DEAE-cellulose adsorption as described earlier. This treatment gives a single [^{32}P]rCMP addition at the ends of each DNA fragment. When these are subjected to nearest neighbor analysis,[21] all of the fragments show dAp as the predominant (>80%) 3'-terminal nucleotide. Thus, the 3' ends resulting from HindIII endonuclease digestion, instead of the random nicks, have been labeled with [^{32}P]CMP.

Effect of Nucleotide Sequence at the Duplex End on Labeling Efficiency. For a clear understanding of the differences in the labeling efficiency of different duplex DNA fragments by the terminal transferase it is necessary to determine the nucleotide sequence adjacent to the site of labeling. As an illustration, Hind C and D fragments of SV40 DNA are taken into consideration. The cleavage map[25] of SV40 DNA shows the restriction endonuclease HhaI cleaves SV40 DNA at two sites that are located on Hind C and D fragments. Therefore, by using this restriction endonuclease all four labeled ends of the two Hind fragments can be separated from one another by gel electrophoresis.

Hind C and D fragments (22 pmoles each) are dissolved in 10 mM Tris-HCl (pH 8.0), 10 mM MgCl$_2$, and 5 mM mercaptoethanol and incubated with 30 units of HhaI at 37° for 11 hr. The reaction mixture is then loaded onto gel and subjected to electrophoresis. The separation of four bands in relation to original HindII + III marker fragments is shown in Fig. 3A. The long and short ends of C are designated C$_1$ and C$_2$. Similarly, Hind D fragment yielded D$_1$ and D$_2$. It is clear from the figure that radioactivity in C$_1$ end in greater han C$_2$ end, and D$_2$ end is greater than D$_1$ end. Among the four bands, the D$_1$ end is most poorly labeled. The numbers adjacent to the DNA ends indicate Cerenkov counts.

DNA from the four gel bands is extracted and treated with alkali and phosphatase in order to isolate monoaddition products.[4a] The single-end labeled DNA is dissolved in 20 μl of a solution containing 50 mM TEAB (pH 8.0), 500 μg/ml of calf thymus DNA, 5 mM MnCl$_2$, 3 mM CoCl$_2$, and 1 mM CaCl$_2$. We noticed that this combination of metal ions reduces the specificity of cleavage by pancreatic DNase so that oligonucleotides of all sizes are produced.[27] The partial digestion is carried out with 1 μl (2 μg) of pancreatic DNase at 37°. At intervals of 2, 5, and 20 min aliquots (3 μl) are transferred to a small siliconized tube (6 × 50 mm) maintained at 90° on a temperature block. The remaining solution is digested for 2 hr and combined in he tube maintained at 90°. After drying, the oligonucleotides are

[25] M. Fried and B. Griffin, *Adv. Cancer Res.,* **24,** 67 (1976).

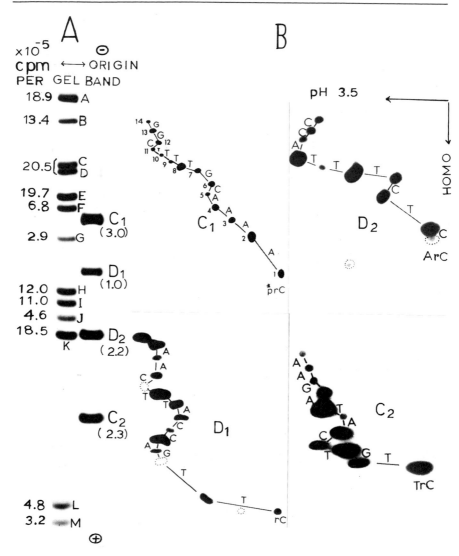

FIG. 3. Isolation of labeled DNA fragments and the separation of four labeled ends after digestion with *Hha*I restriction endonuclease. (A) Relative mobility of four labeled ends in *Hin*d(II + III) C and D fragment in relation to all the *Hin*d(II + III) fragments of SV40 DNA as markers. The marker DNA fragments and the recut DNA fragments originate from two different batches. The radioactivity, thus, bears no relationship with one another. (B) Two-dimensional map of four separated ends.

dissolved in 5 μl of water containing [^{14}C]pT marker (about 10,000 cpm), mixed in a Vortex mixer and centrifuged to bring the droplets down. From this solution a 1- to 2-μl sample is applied to a cellulose acetate strip and subjected to two-dimensional fractionation.[22,23] The maps of the isolated fragments are shown in Fig. 3. The most highly labeled fragment C_1 contains a HindIII cut at the exposed end. The 3'-terminal nucleotide sequence obtained at this end is 3' d(A–A–A–A–C–G–T–T–T–T–C–G–G$\cdots$). Therefore the duplex end consists of the following sequence.

5'pA—G—C—T— T—T— T— T—G—C—A—A—A—A—G—C—C————3'
3'   A—A—A—A—C—G— T— T— T— T—C—G—G————5'

Although the 3' end of the lower strand is overlapped by the protruding 5' end four nucleotides in length, there is a stretch of four A : T base pairs at the 3' end. This feature probably imparts considerable flexibility for the "terminal breathing" of the 3' end of the fragment C_1. As a result this end is most efficiently labeled by terminal transferase which prefers single-stranded DNA as primer.

The sequence around the C_2-end (HindII cut) is

5' pA—A—C—A—G— T—A— T—C— T—T————3'
3'   T— T—G— T—C—A— T—A—G—A—A————

There is a two-nucleotide stretch of A : T base pairs at the exposed end of the duplex. A comparison of the C_2 end and the D_1 end, both of which contain a stretch of two A : T base pairs, showed that the radioactivity in the D_1 end is about half that in the C_2 end. It appears therefore that the sequences in the vicinity of the exposed end considerably influence the labeling reaction. Thus, if we consider the first six nucleotides from the end, we notice that in the fragment D_1 (3' T—T—G—A—C—C$\cdots$), there are three G : C base pairs that furnish it with the most stable duplex configuration among these four fragments. Consequently, this is the most poorly labeled among these four duplex ends. Comparison of nucleotide sequence of the first six nucleotides in C_2 (3' T—T—G—T—C—A————) and D_2 (3' A—C—T—C—T—T————) indicates that these ends, each containing two G : C base pairs within the first six nucleotides, should be labeled with equal efficiency. This is found to be the case.

Determination of the Recognition Sequence of Restriction Endonucleases Which Produce Protruding 3' Ends. Terminal transferase provides a unique means of labeling the protruding 3' ends produced after cleavage by certain restriction endonucleases. Polynucleotide kinase cannot be used since it is specific for the 5'-OH terminus. The enzymes DNA polymerase or reverse transcriptase also cannot be used since these enzymes require a template for the 3'-terminal labeling reaction. In a circular DNA molecule with multiple cleavage sites, the protruding 3' ends contain the same sequence. Therefore, a two-dimensional map of the partial pancreatic DNase digest[23] of the mixture of DNA fragments labeled with terminal transferase yields the recognition sequence of such restriction endonucleases. This is illustrated by digesting plasmid RSF 2124[26] using the restriction endonuclease Hae II.[27]

The plasmid RSF 2124 DNA (50 μg) is digested with 20 units of Hae II, 6 mM Tris-HCl (pH 7.8), 6 mM MgCl$_2$, and 1 mM dithiothreitol at 37° for 22 hr in a final volume of 300 μl. An aliquot (10 μl) of the digested DNA is examined by agarose gel electrophoresis in the presence of ethidium bromide. More than ten fluorescent DNA bands are detectable. The remaining reaction mixture (290 μl) is transferred to a siliconized tube containing 1 nmole of recently dried [α-^{32}P]rCTP. To the tube is then added 30 μl of 10× cacodylate buffer containing Co^{2+} and 4 μl (32 units) of terminal transferase. After 60 min at 37° the reaction is terminated by adding 30 μl of 4 M NaOH. The tube is sealed with Parafilm and allowed to stand at room temperature for 20 hr. The solution is adjusted to pH 8.0 by addition of dry powder of HEPES (Calbiochem). The pH is monitored using a very thin strip of plastic-strip pH indicator (E. Merck). After neutralization, the solution is incubated at 45° with 0.6 unit of alkaline phosphatase for 1 hr.

The reaction mixture is then treated with 10 μl of 0.5 M EDTA, 30 μl of 1% SDS, and 50 μl of neutralized phenol. After thorough mixing in a Vortex mixer, the tube is allowed to stand at room temperature for 10 min. This treatment completely destroys the phosphatase activity. The contents of the tube are loaded onto a Sephadex G-50 column and eluted with 50 mM TEAB (pH 8.0). The labeled DNA fragments are eluted in the void volume in four tubes (1.6 ml total volume). An aliquot (50 μl) of this solution is subjected to electrophoresis in 4% polyacrylamide gel. As shown in Fig. 4A, a total of 13 labeled bands are detectable. When a mixture of these fragments are partially digested with pancreatic DNase[27] and subjected to fingerprinting, the sequence 3' rC—C—G—C—G—A/G————— 5' is obtained (Fig. 4B). By aligning two such ends in an antiparallel orientation, we obtain the duplex sequence

$$5'\ \ A/G—G—C—G—C\overset{\downarrow}{—}T/C$$
$$3'\ \ T/C\underset{\uparrow}{—}C—G—C—G—A/G$$

directly as the recognition sequence for Hae II endonuclease. In this reduced recognition sequence the ribonucleotide is not written since it is

[26] M. So, R. Gill, and S. Falkow, *Mol. Gen. Genet.* **142**, 239 (1976).
[27] C. P. D. Tu, R. Roychoudhury, and R. Wu. *Biochem. Biophys. Res. Commun.* **72**, 355 (1976).

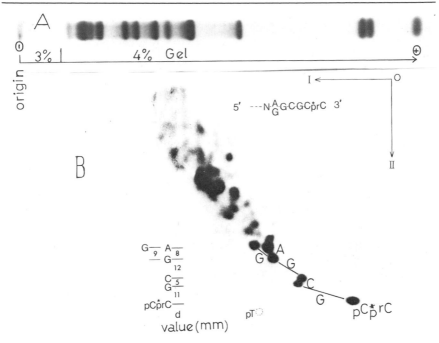

FIG. 4. Recognition sequence of *Hae*II endonuclease. (A) Labeled DNA fragments obtained after terminal labeling of *Hae*II cleaved plasmid RSF 2124 DNA. (B) Two-dimensional map of the partial digest of the mixture of DNA fragments shown in (A).

introduced by terminal transferase. There are four nucleotides 3′ C—G—C—G—— common to all fragments. The diversion comes in the fifth nucleotide which can be either an A or a G residue.

Use of Transferase-Labeled DNA Fragments for Sequence Analysis by Partial Chemical Degradation. Two types of transferase labeled DNA fragments can be used for sequence analysis using the chemical degradation procedure.[18] In one method the labeling reaction with terminal transferase is performed on native DNA. In this case the ribonucleotide addition takes place mainly at the exposed 3′ ends. The nicks in DNA are not significantly labeled. After a second restriction endonuclease digestion the two labeled ends can be separated by gel electrophoresis, treated with alkali and phosphatase, and sequenced.

In the second method, the DNA fragment is heat denatured before the transferase reaction.[18] This procedure ensures a much higher priming activity and hence better incorporation of ribonucleotides. Although in this procedure the nicks in the DNA molecule are extensively labeled, such labeled smaller fragments are removed by using the strand separation technique.[18] By using [α-³²P]rATP or [α-³²P]rGTP at about 50- to 100-fold excess over DNA ends, two to six ribonucleotides can be added at the 3′

ends. When such labeled fragments are treated with alkali each of the labeled strands contains two ^{32}P label as shown:

$$5' \; \underset{}{\overset{\text{DNA}}{\rule{2em}{0.4pt}}} \; \overset{*\;\;*}{prAp} \; 3' \quad \text{or} \quad 5' \; \underset{}{\overset{\text{DNA}}{\rule{2em}{0.4pt}}} \; \overset{*\;\;*}{prGp}$$

Since in this method virtually all the primer molecules have several ribonucleotides added, after alkali treatment each strand will be terminated with two labeled phosphates. Thus, under the above condition the phosphatase step can be eliminated and twice the counts remained with the DNA molecule. This end labeled DNA can then be sequenced by the partial chemical degradation procedure of Maxam and Gilbert.[18] Figure 5 shows a part of the sequence of the *cro* gene of phage λ labeled with rAMP by terminal transferase[28] at the *Mbo*II site.

Addition of Homopolymer Tracts of Deoxynucleotides

Labeling of duplex DNA molecules with deoxynucleotide homopolymer tracts provides a powerful tool for *in vitro* construction of recombinant DNA.[5,6,29] Two types of biochemical events take place during the process of formation of homopolymer tracts at the 3' ends of DNA strands. The first reaction, "the priming event," is determined by the number of primer molecules accepting the first nucleotide at the start of the reaction. The second reaction, the "polymerization event," consists of the successive addition of the same nucleotide to form the homopolymer tract. Some confusion exists in the literature regarding the efficacy of different metal ions for addition of homopolymer tracts to DNA. The effect of different metal ions on the polymerization event has been studied in detail but not the effect on the priming event.

More than a decade ago, Kato *et al.*[30] discovered that the polymerization event (number of nucleotides polymerized per unit time) proceeds efficiently with Mg^{2+} for the purine nucleotides and with Co^{2+} for the pyrimidine nucleotides.

In 1973, during our work with ribonucleotide addition to oligonucleotide primer, we discovered[19] that the priming event (the number of primer molecules accepting a given ribonucleotide) was more efficient in the presence of Co^{2+} than that in the presence of Mg^{2+}, regardless of the type (purine or pyrimidine) of nucleoside triphosphates used. The number of primer molecules participating in the reaction was determined by nearest neighbor analysis. This prompted us to examine this reaction with

[28] E. Schwarz, G. Scherer, G. Hobom, and H. Kössel, *Nature (London)* **272**, 410 (1978).
[29] J. F. Morrow, S. N. Cohen, A. C. Y. Chang, H. W. Boyer, H. M. Goodman, and R. Helling, *Proc. Natl. Acad. Sci. U.S.A.* **71**, 1743 (1974).
[30] K. Kato, J. M. Goncalves, G. E. Houts, and F. J. Bollum, *J. Biol. Chem.* **242**, 2780 (1967).

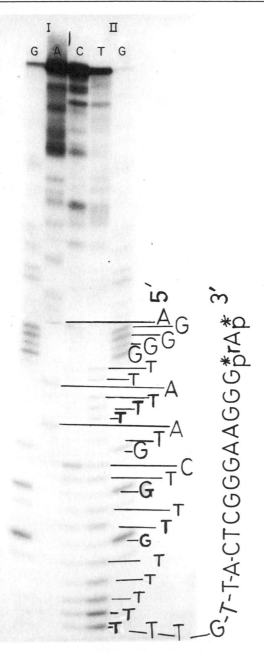

FIG. 5. Nucleotide sequence adjacent to *Mbo*II and *Ava*I cleavage sites of λdvh 93 plasmid DNA labeled at the *Mbo*II end with terminal transferase. The 17 nucleotide sequence at the labeled 3' end was derived from a separate experiment (data not shown).[28]

duplex DNA fragments produced by restriction endonuclease cleavages. We noticed that a much higher number of DNA molecules take part in the reaction in the presence of Co^{2+} than that in the presence of Mg^{2+}, regardless of the type of triphosphate used.

Procedure for Adding Deoxyribonucleotides to the 3' Ends of DNA Fragments. Most of the work on recombinant DNA is carried out with small amounts of DNA. Therefore, a small scale incubation is described. The reaction mixture in a small siliconized tube (6 × 50 mm) contains 1–10 μg (0.02–0.2 A_{260} unit) of intact DNA or DNA fragments (restriction endonuclease digested[30a] or mechanically sheared), 100 mM potassium cacodylate (pH 7.0), 1 mM $CoCl_2$, 200 μM dithiothreitol, 10–100 μM [α-^{32}P]dNTP (A, G, C, or T specific activities 50–100 Ci/mmole) and 60 units/ pmole of 3'-OH ends of terminal transferase in a final volume of 200 μl. At intervals of 0, 1, 2, 3, 4, and 5 min at 37°, an aliquot (2 μl) of the reaction mixture is monitored for acid-insoluble radioactivity.[20] During the monitoring period the reaction mixture is kept at 0°. From the above result the expected time for addition of 50–200 residues is determined. The reaction is then resumed at 37° for the addition of desired nucleotide residues. For addition of G residues, a 5–10 min incubation will usually add about 20–30 residues. Further additions cannot be made. In order to reduce the number of G residues in the poly(dG) tract a lower incubation temperature (15°) may be used.[31] The reaction is then terminated by adding 20 μl of 0.5 M EDTA and 100 μl of neutralized phenol. After standing at 0° for 10 min the tube is centrifuged at 2000–3000 g in a clinical centrifuge and the aqueous layer is withdrawn with a siliconized micropipette (50-μl capacity) in several installments without disturbing the phenol layer. The employment of this narrow (6-mm diameter) tube for phenol extraction is highly desirable because it allows a minimum of surface area at the phenol–water interphase. The phenol layer is again extracted in the same way with 100 μl of 50 mM Tris-HCl, pH 8.0, 100 mM NaCl, and 1 mM EDTA. The combined aqueous layer is ether extracted and ethanol precipitated.

Duplex DNA molecules are always less efficient primers than single-stranded DNA molecules. The most inefficient type of duplex consists of molecules with protruding 5' ends such as that found at the cohesive ends of λ DNA.[32] It is evident from Fig. 6 that even with this DNA as primer,

[30a] The restriction endonuclease reaction mixture (carried out in 10–20 μl volume) can be directly included into the reaction mixture (final volume 200 μl) for the addition of homopolymer tracts. Because of 10- to 20-fold dilution, the components of the restriction endonuclease digestion mixture do not interfere with the transferase reaction. This procedure eliminates labor and prevents the loss of DNA.

[31] L. Villa-Komaroff, A. Efstratiadis, S. Broome, P. Lomedico, R. Tizard, S. P. Naber, W. L. Chick, and W. Gilbert, *Proc. Natl. Acad. Sci. U.S.A.* **75**, 3727 (1978).

[32] R. Wu and E. Taylor, *J. Mol. Biol.* **57**, 491 (1971).

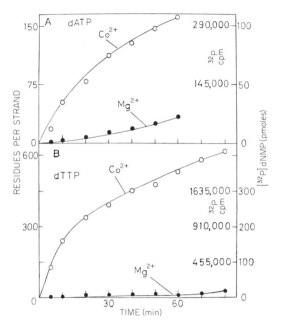

Fɪɢ. 6. Addition of homopolymer tracts to the cohesive ends of λ DNA. For experiments using Mg^{2+}, 10 mM MgCl$_2$ were used in place of 1 mM CoCl$_2$.

terminal transferase can catalyze the addition of homopolymer tracts much more efficiently in the presence of cobalt ion than that in the presence of magnesium ion, with either dATP or dTTP as substrates.[2]

In this way, proinsulin gene,[31] silk fibroin gene,[33] immunoglobulin gene,[34] human globin gene,[35] yeast leucine gene,[36] yeast nuclear DNA fragments,[37] bovine corticotropin gene,[38] rat prolactin gene,[39] and histone gene[40] were cloned successfully.

[33] Y. Oshihama and Y. Suzuki, *Proc. Natl. Acad. Sci. U.S.A.* **74**, 5363 (1977).

[34] J. G. Seidman, M. H. Edgell, and P. Leder, *Nature (London)* **271**, 582 (1978).

[35] J. T. Wilson, L. B. Wilson, J. K. deRiel, L. Villa-komaroff, A. E. Efstratiadis, B. G. Forget, and S. M. Weissman, *Nucleic Acids Res.* **5**, 563 (1978).

[36] B. Ratzkin, R. Roychoudhury, and P. P. Hung, unpublished results (1978).

[37] T. D. Petes, J. R. Broach, P. C. Wensink, L. M. Hereford, G. R. Fink, and D. Botstein, *Gene* **4**, 37 (1978).

[38] S. Nakanishi, A. Inoue, T. Kita, S. Numa, A. C. Y. Chang, S. N. Cohen, J. Nunberg, and R. T. Schimke, *Proc. Natl. Acad. Sci. U.S.A.* **75**, 6021 (1978).

[39] E. J. Gubbins, R. A. Maurer, J. L. Hartley, and J. E. Donelson, *Nucleic Acids Res.* **6**, 915 (1979).

[40] W. Schaffner, W. Topp, and M. Botchan, *in* "Specific Eucaryotic Gene," Alfred Benzon Symposium 13, Munksgaard, Copenhagen, Denmark, 1978 (in press).

Addition of homopolymer tracts to duplex DNA was first reported by Jensen et al.[41] Since then, various reaction conditions have been worked out for specific objectives. When available, the use of λ exonuclease to digest the 5' ends and thereby convert the 3' ends to single-stranded primer,[6] is the most efficient method for addition of homopolymer tracts to duplex DNA. However, the need for λ exonuclease can be eliminated by using Co^{2+} ion in the reaction mixture.

Brutlag et al.[7] observed that by using Mg^{2+} ion, the addition of poly(dA) tracts to mechanically sheared satellite DNA or to EcoR1 cleaved colE1 DNA can be successfully carried out in the presence of low salt (20 mM potassium phosphate). Humphries et al.[10] used 12.5 mM HEPES buffer (pH 7.1) in the presence of Mg^{2+} ion for a limited addition of dGMP residues at the 3' ends of duplex DNA. In our experiments we notice that all four deoxynucleotides can be added using Co^{2+} ion whereby a greater utilization of primer (in the form of duplex DNA) is achieved.

The amount of enzyme relative to DNA ends as well as the enzyme concentration are important. In our experiments, we have used 50 units of enzyme per pmole of 3'-OH ends and obtained satisfactory results. Brutlag et al.[7] have noticed that a four- to fivefold higher enzyme concentration is required for duplex DNA as compared to single-stranded DNA. They have used 680 units of enzyme for 1–10 pmoles of DNA ends. Humphries et al.[10] have used 350 units/pmole of 3' ends together with a short incubation period (5 min). Using lower levels of enzyme they obtained fewer transformants. Clearly, the use of such high levels of enzyme improves the priming event. However, the enzyme preparations often contain traces of endonuclease activity which become concentrated at a higher level of enzyme. Unless the enzyme is very pure there is the possibility of extensive nicking of DNA and the formation of multiple tails.[42]

Brutlag et al.[7] noticed that the affinity of the enzyme for poly(dA) tract is lower than that for the poly(dT) tract. Therefore, for the addition of poly(dA) tract it is necessary to use higher levels of enzyme or longer incubation periods. The polymerization event is also strongly influenced by enzyme concentration. The higher the level of enzyme, the greater the number of residues polymerized per minute.[7]

Acknowledgments

This work was supported by Research Grants GM24904 from the National Institutes of Health and 77-20313 from the National Science Foundation. We are grateful to Elisabeth Schwarz and Hans Kössel for providing us with the result of cro gene sequence obtained with transferase-labeled DNA fragment and to Ernest Jay and C. P. D. Tu for their advice and help in two-dimensional mapping and gel electrophoresis.

[41] R. H. Jensen, R. J. Wodzinski, and M. H. Rogoff, Biochem. Biophys. Res. Commun. 43, 384 (1971).
[42] W. Bender and N. Davidson, Cell 7, 595 (1976).

[8] Repair of Overlapping DNA Termini

By Howard M. Goodman

The natural cohesive termini of bacteriophage DNA[1] such as λ and the staggered phosphodiester bond cleavages produced by various restriction endonucleases[2] could have several possible arrangements.[3] If the arrangement is such that a protruding 5′ single-stranded end and an internal 3′-hydroxyl end are formed, then the termini will serve both as primer and template for repair synthesis by a DNA polymerase.[1,3]

Method

Principle. Overlapping DNA termini can be repaired using the RNA-directed DNA polymerase ("reverse transcriptase") from RNA tumor viruses and radioactive nucleoside triphosphates. The incorporation into the termini can be quantified and the sequence analyzed from nearest neighbor data.

Labeling Procedure. The DNA to be analyzed is digested with the appropriate restriction endonuclease[2] (if required), extracted with an equal volume of phenol–chloroform [DNA:phenol:$CHCl_3$(1 : 1 : 1)] and the DNA precipitated with 2 volumes of 95% ethanol after making the aqueous phase 0.3 M in sodium acetate. After chilling ($-70°$ for 30 min), the precipitate is collected by centrifugation and washed once with 95% ethanol. The DNA is then dissolved in a small volume of water (see below). These operations are easily performed in 1.5 ml Eppendorf Micro test tubes which can be centrifuged in an Eppendorf Microcentrifuge.

The overlapping termini are radioactively labeled in a reaction which contains the following ingredients in a total volume of 25 μl: 0.1 to 2 pmole of termini, 0.1 M Tris-HCl (pH 8.0), 2 mM 2-mercaptoethanol, 10 mM $MgCl_2$, 5–10 μM dNTP (various combinations of nonradioactive and [α-^{32}P]-labeled nucleoside triphosphates, specific activities 100–1000 Ci/mM), and 0.2 munits of reverse transcriptase per microgram of DNA (specific activity about 0.1 unit/μg).[4] The reaction mixture is incubated at 37° for 2 hr. The reaction is stopped by addition of 0.4 ml of 0.05 M sodium pyrophosphate (NaPP$_i$) plus 0.05 M Na$_2$ EDTA and 0.2 ml calf thymus

[1] R. Wu and E. Taylor, *J. Mol. Biol.* **57,** 491 (1971).
[2] R. J. Roberts, *Crit. Rev. Biochem.* **3,** 123 (1976).
[3] J. Hedgpeth, H. M. Goodman, and H. W. Boyer, *Proc. Natl. Acad. Sci. U.S.A.* **69,** 3448 (1972).
[4] A. J. Faras, J. M. Taylor, J. P. McDonnell, W. E. Levinson, and J. M. Bishop, *Biochemistry* **11,** 2334 (1972).

DNA (400 μg/ml) as carrier. Ten percent (60 μl) of the reaction is diluted into 0.2 ml NaPP$_i$ plus 0.2 ml calf thymus DNA (400 μg/ml) for the determination of total incorporation. After precipitation with 1 ml of 7% perchloric acid containing 0.1 M NaPP$_i$ and standing on ice for 30 min, the precipitate is collected on glass fiber filters and counted.

Nearest Neighbor Analysis. The remainder of the reaction (90%; 565 μl) is extracted with an equal volume (600 μl) of phenol and the aqueous phase containing the labeled DNA dialyzed for 2 days against at least three changes of 4 liters of 0.3 M sodium acetate and 10 mM Tris-HCl (pH 7.5). (Any other method, such as gel filtration, which removes unincorporated nucleoside triphosphates can be substituted for the dialysis step.) The DNA is then precipitated with 2 volumes of 95% ethanol and resuspended in a minimal volume of water. The DNA is then digested to completion by first digesting with 40 units of micrococcal nuclease in 10 mM sodium borate buffer (pH 8.6) plus 20 mM CaCl$_2$ for 30 min at 37° and then with 0.2 units of spleen phosphodiesterase in 50 mM Tris-HCl (pH 7.5), 10 mM MgCl$_2$, and 5 mM rAMP. The 3'-mononucleotides produced are separated by high-voltage electrophoresis (3–5 kV, 1–3 hr) on Whatman 3 MM paper in pyridine-acetate buffer, pH 3.5. Appropriate absorbance markers can be added where necessary, the pyridine removed in NH$_3$ vapor, and after drying the paper the makers are visible under a short-wave uv lamp. The incorporated radioactivity can be quantitated by cutting-out and counting the paper in an end-window Geiger counter or scintillation counter and/or visualized by autoradiography using X-ray film.

Comments

It is useful to optimize the incorporation reactions prior to performing quantitative studies by varying the time of incubation, the enzyme–template ratio, and the concentrations of nucleoside triphosphates. The method has been successfully used in determining the cleavage site for the *Eco*RI and *Eco*RII restriction endonucleases.[3,5] The reader may also want to consult a previous article in this series by Wu *et al.*[6]

Acknowledgment

This work was supported in part by a grant from the United States Public Health Service (CA 14026).

[5] H. W. Boyer, L. T. Chow, A. Dugaiczyk, J. Hedgpeth, and H. M. Goodman, *Nature (London)* **244**, 40 (1973).
[6] R. Wu, R. Padmanabhan, and R. Bambara, this series, Vol. 29, p. 231.

[9] Specific Labeling of 3' Termini of RNA with T4 RNA Ligase

By T. E. ENGLAND, A. G. BRUCE, and O. C. UHLENBECK

Introduction

T4 RNA ligase (EC = 6.5.1.3) catalyzes the formation of an internucleotide phosphodiester bond between a 5' terminal phosphate and a 3' terminal hydroxyl of oligo- or polyribonucleotides with the accompanying hydrolysis of ATP.[1-3] If the two termini are present on the same molecule, a cyclic RNA product is formed. However, if the oligoribonucleotide with the 5'-phosphate (the *donor*) is too short to cyclize or has a blocked 3' terminus, it will join with another oligoribonucleotide with a free 3'-hydroxyl (the *acceptor*) to form an intermolecular ligation product.[3] A wide variety of sequences, including donors as short as nucleoside 3',5' bisphosphates and acceptors as short as trinucleoside diphosphates, are active in the intermolecular reaction.[4] This makes RNA ligase a valuable tool for the synthesis of oligoribonucleotides.

One important application of the RNA ligase reaction is the use of an RNA molecule as an acceptor and a [5'-^{32}P]nucleoside 3',5'-bisphosphate as a donor to form a product RNA one nucleotide longer with a 3' terminal phosphate and a ^{32}P-phosphate in the last internucleotide linkage.[5] This reaction can be written as follows.

$$\text{RNA—OH} + \text{*pNp} + \text{pppA} \rightarrow \text{RNA—*pNp} + \text{pA} + \text{PP}_i$$

Since this reaction can be carried out at very low RNA concentrations and often gives very good yields, it is a convenient method for introducing a radiolabel into RNA for a variety of applications. In addition, since the radioactivity is introduced at the 3' terminus of the RNA molecule, the sequence of the RNA can be determined in the neighborhood of the label by one of the recently developed procedures.[6-8] In this sense the RNA ligase

[1] R. Silber, V. G. Malathi, and J. Hurwitz, *Proc. Natl. Acad. Sci. U.S.A.* **69**, 3009 (1972).

[2] J. W. Cranston, R. Silber, V. G. Malathi, and J. Hurwitz, *J. Biol. Chem.* **249**, 7447 (1974).

[3] G. C. Walker, O. C. Uhlenbeck, E. Bedows, and R. I. Gumport, *Proc. Natl. Acad. Sci. U.S.A.* **74**, 122 (1975).

[4] T. E. England and O. C. Uhlenbeck, *Biochemistry* **17**, 2069 (1978).

[5] T. E. England and O. C. Uhlenbeck, *Nature (London)* **275**, 560 (1978).

[6] M. Silberklang, A. Prochiantz, A. L. Haenni, and U. L. RajBhandary, *Eur. J. Biochem.* **72**, 465 (1977).

[7] A. Simoncsits, G. G. Brownlee, R. S. Brown, J. R. Rubin, and H. Guilley, *Nature (London)* **269**, 833 (1977).

[8] H. Donis-Keller, A. M. Maxam, and W. Gilbert, *Nucleic Acids Res.* **4**, 2527 (1977).

METHODS IN ENZYMOLOGY, VOL. 65

reaction complements the use of polynucleotide kinase to label the 5' termini of RNA molecules.

Materials

T4 RNA Ligase. Several satisfactory procedures for the purification of this enzyme from *E. coli* infected with bacteriophage T4 have been published.[1,3,9,10] Since RNA ligase is produced early in T4 infection as the product of gene 63,[11] a convenient source for the enzyme is *Su⁻ E. coli* infected with a bacteriophage T4 which has an amber mutation in one of the genes involved in T4 DNA replication. Large amounts of the early proteins accumulate in these infected cells. The mutant T4 bacteriophage needed for infection are grown in a *Su⁺ E. coli* strain. If the *RegA* mutation[12] is included in T4 as well, even greater amounts of RNA ligase are produced.[10] Since as much as 0.5 mg of RNA ligase is present in each gram of infected cells, purification procedures can emphasize the necessary complete removal of ribonuclease and phosphatase activities from the protein. In this work we have used RNA ligase purified by the method of McCoy *et al.*[13] which modifies and extends the procedure of Walker *et al.*[3] by using a G-150 column in place of the G-75 column and including a hydroxylapatite column followed by a DEAE-cellulose column. Although a variety of substrates and reaction conditions have been used, RNA ligase activity has generally been based upon the cyclization of an oligoadenylylate. One unit of enzyme activity corresponds to 1 nmole of $(pA)_{12}$ cyclized in 30 min at 37°.[1,13] Homogeneous enzyme preparations range from 1000 to 2500 units/mg of protein, depending on the assay conditions. Since assays vary considerably, it is best to test the optimal enzyme concentration necessary for 3' terminal labeling of RNA using tRNA as a substrate (see below). RNA ligase should be stored at quite high concentrations (2000 units/ml) for 3' end labeling reactions.

T4 Polynucleotide Kinase. This is required to prepare [5'-³²P] nucleoside 3',5'-bisphosphates from the corresponding 3'-monophosphate and [γ-³²P]ATP. This enzyme can be prepared from the same preparation of T4-infected cells used as a source for RNA ligase by a slight modification[14] of the Richardson procedure.[15]

[9] J. A. Last, and W. F. Anderson, *Arch. Biochem. Biophys.* **174,** 167 (1976).

[10] N. P. Higgens, A. P. Geballe, T. J. Snopek, A. Sugino, and N. R. Cozzarelli, *Nucleic Acids Res.* **4,** 3175 (1977).

[11] T. J. Snopek, W. B. Wood, M. P. Conley, P. Chen, and N. R. Cozzarelli, *Proc. Natl. Acad. Sci. U.S.A.* **74,** 3355 (1977).

[12] J. S. Wiberg, S. Mendelson, V. Warner, K. Hercules, C. Aldrich, and J. L. Munro, *J. Virol.* **12,** 775 (1973).

[13] M. McCoy, T. Lubben, and R. I. Gumport, *Biochem. Biophys. Acta* **562,** 149 (1979).

[14] V. Cameron and O. C. Uhlenbeck, *Biochemistry* **16,** 5120 (1977).

[15] C. C. Richardson, *Proced. Nucleic Acid Res.* **2,** 815 (1972).

[γ-^{32}P]ATP of high specific activity can be prepared from ^{32}P-orthophosphate by the method of Glynn and Chapell[16] or Schendel and Wells.[17]

Preparation of [5′-^{32}P]Nucleoside 3′,5′-Bisphosphates

Although any one of the four common deoxy- or ribonucleoside 3′,5′-bisphosphates could be used as donors in the RNA ligase reaction, studies with a common oligonucleotide acceptor[4] revealed that deoxycytidine 3′,5′-bisphosphate (pdCp) or cytidine 3′,5′-bisphosphate (pCp) were the two best donors. Since the 3′-phosphatase activity inherent in polynucleotide kinase[14] will cause more rapid degradation of pdCp during synthesis, pCp is the best available donor for the 3′ terminal labeling reaction.

The preparation of [5′-^{32}P]pCp is carried out under the conditions shown in the tabulation:

1–5 μM [γ-^{32}P]ATP (500–5000 Ci/mmole)
1 mM 3′-CMP
25 mM potassium-CHES, pH 9.5
5 mM MgCl$_2$
3 mM dithiothreitol
50 μg/ml serum albumin
60 units/ml T4 polynucleotide kinase

The reaction (50 μl) is incubated at 37° for 90 min. When this reaction is analyzed by polyethyleneimine (PEI) thin layer chromatography using 0.8 M (NH$_4$)$_2$SO$_4$ as solvent, all the radioactivity should be converted from [γ-^{32}P]ATP ($R_f \sim 0.3$) to [5′-^{32}P]pCp ($R_f \sim 0.6$). When large amounts of radioactivity are used (>1 mCi), the reaction mixture can be heated to 100° for 1 min and an equal volume of ethanol added for storage ($-20°$). Aliquots can then be used in 3′ terminal labeling reactions. Alternately, the [5′-^{32}P]pCp can be purified from unreacted [γ-^{32}P]ATP and the excess 3′-CMP by descending paper chromatography using equal volumes of 1 M ammonium acetate and 95% ethanol as a solvent.[4]

Nonradioactive pCp can be prepared in moderate amounts by increasing the ATP concentration in the above reaction to 1 mM and purifying the pCp by paper chromatography. Larger amounts of a 60 : 40 mixture of the 3′,5′ and 2′,5′ isomers of pCp can be easily prepared from cytidine and

---

[16] I. M. Glynn and J. B. Chappell, *Biochem. J.* **90**, 147 (1964).
[17] P. F. Schendel and R. D. Wells, *J. Biol. Chem.* **248**, 8319 (1973).

pyrophosphoryl chloride.[18] The 2′,5′ isomer in this mixture is not a substrate for RNA ligase.[4]

Although the procedures given here are primarily designed using ^{32}P-phosphate as the radioisotope, other isotopes can be used in the 3′ labeling reaction as well. 5-Iodocytidine 3′,5′-bisphosphate (pICp) is a donor in the RNA ligase reaction, and a procedure for the preparation of high specific activity [^{125}I]pICp is available.[19] [^{3}H]- or [^{14}C]pCp could be prepared from [^{3}H]- or [^{14}C]3′-CMP with polynucleotide kinase or from [^{3}H]- or [^{14}C]cytidine using pyrophosphoryl chloride.

3′ Terminal Labeling Reaction

In order to test the RNA ligase and pCp preparations for 3′ terminal labeling, a trial reaction with tRNA should be carried out. Although it is best to use a homogenous tRNA species, such as yeast tRNA^{Phe}, unfractionated tRNA can be used if all low molecular weight impurities have been removed. The trial reaction (30 μl) with tRNA acceptor should contain the reactants listed in the tabulation:[19]

1.2 μM tRNA (in moles of 3′ termini)
2.3 μM [5′-^{32}P]pCp (~200 Ci/mmole)
6.0 μM ATP
50 mM HEPES, pH 7.5
20 mM MgCl$_2$
3.3 mM dithiothreitol
10 μg/ml bovine serum albumin
10% (v/v) dimethyl sulfoxide
100 units/ml (50 μg/ml) T4 RNA ligase

The reaction is incubated at 5° and aliquots are withdrawn at intervals. The end-labeled tRNA is precipitated with cold 5% trichloroacetic acid and collected by Millipore filtration. If the yield is expressed in terms of moles of tRNA labeled, the reaction should reach completion in 6 hr with 100 units/ml RNA ligase. Lesser amounts of RNA ligase will give proportionately slower rates but disproportionately lower final yields (Fig. 1). The absence of a 3′-exonuclease in the RNA ligase preparation can be demonstrated by performing a base hydrolysis of the RNA product and confirming that the ^{32}P-phosphate is transferred only to 3′-AMP, the nucleotide at the 3′ terminus of tRNA.

While the reaction conditions given above optimize the 3′ labeling reaction for tRNA, some alterations in the conditions will still give effi-

[18] J. R. Barrio, M. C. G. Barrio, N. J. Leonard, T. E. England, and O. C. Uhlenbeck, *Biochemistry* **17**, 2077 (1978).
[19] A. G. Bruce and O. C. Uhlenbeck, *Nucleic Acids Res.* **10**, 3665 (1978).

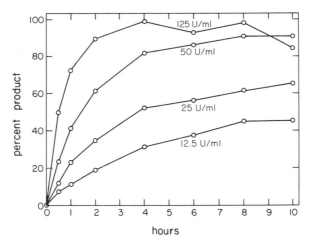

FIG. 1. Kinetics of the addition of [5′-^{32}P]pCp to yeast tRNA^{Phe} at several enzyme concentrations. The reaction conditions are given in the text. Taken with permission from Bruce and Uhlenbeck.[19]

cient labeling. The tRNA, pCp, and ATP concentrations can be proportionately increased or decreased by as much as a factor of ten without a great difference in the yield. Since the magnesium concentration and pH optima are quite broad, small alterations of these conditions are possible. However, the omission of dimethyl sulfoxide will result in a 2.5-fold decrease in yield. Finally, the temperature of the reaction should be kept at or below 15° since higher temperatures give much poorer yields.[19]

The reaction conditions used for tRNA are usually altered slightly for reactions labeling the 3′ terminus of higher molecular weight RNAs. The concentration of 3′ termini are generally reduced so as to use less of the RNA sample. The specific activity of the [5′-^{32}P]pCp is increased and its concentration decreased in order to make a more radioactive product and to make more efficient use of the label. The ATP concentration is not significantly altered even though it is well below the K_m of ATP of 12 μM.[2] Higher concentrations of ATP are not advantageous, since adenylylated enzyme which accumulates after all the available donor has been adenylylated cannot participate in the ligation step.[20] The buffer conditions optimized for tRNA are similar for high molecular weight RNA, but conceivably a different magnesium ion or dimethyl sulfoxide concentration could give better yields for individual RNAs. Finally, the enzyme concentrations are often increased to as high as 500 units/ml in order to offset the decrease in substrate concentration. It should be noted that 500 units/ml RNA ligase corresponds to 6 μM protein, which is at least fifty times the

[20] O. C. Uhlenbeck and V. Cameron, *Nucleic Acids Res.* **4**, 85 (1977).

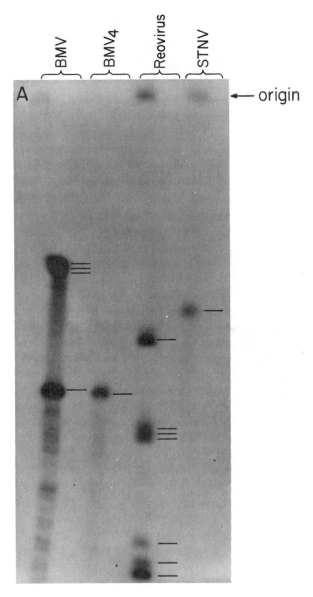

FIG. 2. Analysis of RNA ligase reactions by denaturing polyacrylamide gel electrophoresis. The labeled products are detected by autoradiography. The thin solid lines indicate the position of stained RNA molecules run in adjacent wells.

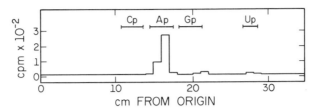

FIG. 3. Base composition analysis of 3' end labeled RNA demonstrating transfer of the ^{32}P from pCp to the 3' terminal adenosine of BMV-4 RNA.[20] Taken with permission from England and Uhlenbeck.[5]

RNA concentration. Thus, most protein molecules are present in the reaction to overcome the poor affinity for substrates and do not turn over during the incubation period. This unusual situation is a consequence of the extremely low substrate concentrations; RNA ligase acts catalytically with reasonably high turnover numbers when it is used for oligonucleotide synthesis at much higher substrate concentrations.[4,21] A typical 3' terminal labeling reaction (15 μl total volume) for BMV-4 contains the reactants listed in the tabulation:[5]

0.22 μM Brome mosaic virus (BMW) RNA (1 μg)
1.0 μM [5'-^{32}P]pCp (500–5000 Ci/mmole)
5.0 μM ATP
50 mM HEPES, pH 8.3
10 mM MgCl$_2$
3.3 mM dithiothreitol
10 μg/ml bovine serum albumin
10% (v/v) dimethyl sulfoxide
15% (v/v) glycerol (present in RNA ligase)
400 units/ml (200 μg/ml) T4 RNA ligase

Incubation is carried out at 5° and the course of the reaction can again be followed by withdrawing aliquots for trichloroacetic acid precipitation and Millipore filtration. Generally incorporation of ^{32}P into acid-precipitable RNA is complete by 6–12 hr. In this case it is especially important to analyze the reaction products by denaturing polyacrylamide gel electrophoresis. The radioactive end-labeled RNA should comigrate with unreacted RNA (Fig. 2). If significant amounts of low molecular weight impurities with free 3'-hydroxyls are present in the reaction mixture, the incorporated yield may not reflect addition to the high molecular weight RNA. A high concentration of free terminal 3'-hydroxyls could effectively compete for the available pCp and result in poor yields of the desired product.

[21] R. Dasgupta and P. Kaesberg, *Proc. Natl. Acad. Sci. U.S.A.* **74**, 4900 (1977).

If no low molecular weight impurities are detected, the end-labeled RNA can be purified from the unreacted ATP, pCp, and proteins by Sephadex G-100 or G-150 column chromatography. If impurities are present, a preparative procedure for size fractionation of RNA, such as sucrose gradient centrifugation or gel electrophoresis, should be carried out.

A wide variety of RNA molecules have been successfully labeled by RNA ligase. These include *E. coli* and eukaryotic ribosomal RNAs, several viral RNA genomes, and several messenger RNAs (Table I). Although the reaction yields vary, no case has been found where an RNA with a free terminal 3'-hydroxyl did not label. Thus, similar to 5' terminal labeling with polynucleotide kinase, the procedure is very general.

The large variation in the reaction yield, which can range from a few percent (reovirus RNA) to completion (Sendai virus RNA) can be attributed to several sources. First, it is likely that the presence of secondary structure in the RNA near the 3' terminus will inhibit the RNA ligase

TABLE I

RNAs Labeled Successfully at the 3' Terminus with T4 RNA Ligase[a]

Excellent substrates
   Yeast tRNA^{Phe}
   Unfractionated *E. coli* tRNA
   Unfractionated Bakers yeast tRNA
   Sendai viral RNA
   Newcastle disease viral RNA
   Ovalbumin mRNA
Intermediate substrates
   E. coli tRNA$_f^{Met}$
   E. coli 16 S rRNA
   E. coli 23 S rRNA
   Human 28 S rRNA
   Rabbit 18 S rRNA
   Brome mosaic viral RNA 1-4
   QB viral RNA
   QB midivariant 1 RNA (+ and − strands)
   Bovine parathyroid hormone mRNA
   Rabbit globin mRNA
Poor substrates
   Human 5 S RNA
   Satellite tobacco necrosis viral RNA
   Reovirus RNA

[a] Reactions were carried out under conditions given in the text and analyzed by acrylamide gel electrophoresis. Excellent substrates gave greater than 90% addition of pCp to the 3' terminus of the RNA. Poor substrates gave less than 10% addition and intermediate substrates ranged in between. Data taken from England and Uhlenbeck;[5] Bruce and Uhlenbeck,[19] and T. E. England, A. G. Bruce, and O. C. Uhlenbeck, unpublished data.

reaction. In experiments with model compounds, RNA ligase shows a strong preference for single-stranded RNA.[1,2] The poor reactivity of reovirus RNA is undoubtedly due to its double-stranded structure. On the other hand, the excellent reactivity of polyadenylylated messenger RNAs is partly due to the absence of 3' terminal secondary structure. Second, the sequence of nucleotides at the 3' terminus of RNA is also expected to affect the reaction yield. A comparison of a variety of oligonucleotide acceptors with a common donor revealed that oligomers with uridine in one of the three 3' terminal positions are considerably poorer acceptors than oligomers which did not contain uridine.[4] Thus, RNA molecules terminating with one or more uridines may be poorer acceptors. Finally, those RNA molecules which have a free 5'-phosphate in close proximity to the 3' terminus will also show low reaction yields, since the intramolecular cyclization will compete effectively with the intermolecular 3' labeling reaction. This situation occurs with $E.$ $coli$ tRNA$_f^{Met}$.[19] If the 5' terminal phosphate is removed with alkaline phosphatase prior to reaction, no cyclic RNA product can form, and efficient 3' end labeling occurs.

Several procedures can be followed if a low yield of 3' end-labeled RNA is obtained. First, the possibility that an inhibitor of RNA ligase, such as ammonium ion, is present in the RNA sample can be tested by adding a trace amount of a purified tRNA to the reaction mixture and determining whether it reacts to the same extent that it would in the absence of the RNA sample. Second, a poor reaction yield can often be improved by increasing either the enzyme or substrate concentrations. Higher enzyme concentrations will often result in disproportionately greater yields as seen in Fig. 1 for tRNA as an acceptor. Since the enzyme is already in considerable molar excess over the substrates in the reaction, its increased concentration presumably improves the reaction yield by increasing the proportion of substrate molecules bound to the enzyme. Similarly, an increase in the RNA concentration (with a proportional increase in the pCp concentration) will also result in a larger fraction of the RNA molecules labeled with pCp. This situation only occurs at low substrate concentrations (below ~ 1 μM). At intermediate substrate concentrations (1–100 μM), the percent reaction yield is usually independent of substrate concentration. Saturation of RNA ligase is generally not reached until above 1 mM.

If the source of low reaction yields is suspected to be due to extensive 3' terminal secondary structure, it is possible that denaturing the RNA in a low ionic strength buffer containing EDTA prior to adding it to the reaction could increase the availability of the 3' terminal nucleotides. However, if the 3' terminal secondary structure is simply a hairpin loop, it is likely to reform rapidly under the conditions of high magnesium ion concentration and low temperature necessary for efficient RNA ligase reaction. Although RNA ligase is active in modest concentrations of solvents (dimethyl sulfoxide, formamide, urea) known to disrupt RNA secondary structure, it is unlikely that high enough concentrations can be reached to be of much use. The increase in reaction yield upon addition of 10–20% dimethyl sulfoxide for several RNAs is not likely to be a secondary structure effect since the increase is also seen with oligouridylates which have no secondary structure under the conditions used.[4]

The 3' terminal *in vitro* labeled RNA can be used for a variety of applications in which radioactive RNA is required. If high specific activity [5'-^{32}P]pCp or [^{125}I]pICp are used, the radioactive RNA can be used in DNA–RNA hybridization experiments. Since the ^{32}P-phosphate is located in the last internucleotide linkage in the labeled RNA, base hydrolysis of the RNA and separation of the nucleoside 3'-monophosphates by electrophoresis will identify the 3' terminal nucleotide (Fig. 2B). Similarly, random partial hydrolysis of 3' end-labeled RNA and subsequent two-dimensional separation of the end-labeled oligonucleotides can be used to deduce the sequence of the RNA in the vicinity of the 3' terminus.[6] Finally, partial digestion of the 3' terminal labeled RNA with sequence-specific endonucleases and subsequent separation of the end-labeled oligonucleotides by acrylamide gel electrophoresis can give extensive sequence information in the neighborhood of the 3' terminus of the RNA.[7,8]

In all of the above applications, the *in vitro* labeling of the 3' terminus with RNA ligase is quite analogous to labeling of the 5' terminus with polynucleotide kinase. Since the two procedures label opposite ends of the polynucleotide chain, they complement each other and can be used to corroborate each other when used for RNA sequencing studies. However, if the application simply requires *in vitro* labeling of RNA, several advantages of the RNA ligase method can be identified. First, unlike the 5' ends of RNA, which must usually be freed of terminal phosphates or "cap" structures prior to reaction with polynucleotide kinase, the 3' ends of RNA are normally free hydroxyls and thus are reactive as acceptors with RNA ligase without modification. Second, a variety of different isotopes can be used for *in vitro* labeling with RNA ligase. Finally, the RNA ligase reaction is carried out at low temperatures under extremely mild conditions, thus minimizing degradation and allowing recovery of the labeled RNA in high yield. Thus, this procedure for labeling the 3' terminus should be a useful addition to the available techniques applicable to RNA.

[10] 5'-³²P Labeling of RNA and DNA Restriction Fragments[1]

By GEORGE CHACONAS and JOHAN H. VAN DE SANDE

Bacteriophage T4-induced polynucleotide kinase catalyzes the transfer of the γ-phosphate from ATP to the 5'-hydroxyl terminus of DNA, RNA, and 3'-mononucleotides.[2] The enzyme has been used extensively in the study of nucleic acid structure and function and is particularly useful in recently developed methods for rapid DNA and RNA sequence analysis.[3-5] Because of the requirement for a 5'-hydroxyl terminus, most substrates must be dephosphorylated with alkaline phosphatase prior to the labeling step, and the phosphomonoesterase activity must then be eliminated to avoid degradation of ATP and loss of label from polynucleotide substrates. The most effective method for inactivation of alkaline phosphatase is repeated extraction with phenol followed by dialysis or ethanol precipitation. This method is tedious, results in losses of polynucleotide, and is inappropriate for processing large numbers of samples for labeling. Other methods of inactivation which have been reported are treatment with NaOH,[6] HCl,[7] or boiling in the presence of NTA to chelate the zinc from the alkaline phosphatase metalloenzyme.[8] Treatment with acid or base requires precise titrations, adds salt which inhibits the labeling of double stranded substrates,[9] is not suitable for use with RNA, and often results in only temporary inactivation of the phosphatase which easily renatures to an active form. Similarly, treatment with NTA does not completely eliminate the phosphomonoesterase activity. An alternative is to label 5'-phosphoryl terminated substrates by phosphate exchange taking

[1] Supported by research grants from the Medical Research Council of Canada to J.H.v.d.S. and from the National Research Council of Canada to Dr. R. B. Church of the University of Calgary. The authors would like to thank Dr. R. B. Church for support and use of laboratory space.

[2] C. C. Richardson, *Proc. Natl. Acad. Sci. U.S.A.* **54,** 158 (1965).
[3] A. M. Maxam and W. Gilbert, *Proc. Natl. Acad. Sci. U.S.A.* **74,** 560 (1977).
[4] H. Donis-Keller, A. M. Maxam, and W. Gilbert, *Nucleic Acids Res.* **4,** 2527 (1977).
[5] A. Simoncsits, G. G. Brownlee, R. S. Brown, J. R. Rubin, and H. Guilley, *Nature (London)* **269,** 833 (1977).
[6] A. Jacquemin-Sablon and C. C. Richardson, *J. Mol. Biol.* **47,** 477 (1970).
[7] H. O. Smith and M. L. Birnstiel, *Nucleic Acids Res.* **3,** 2387 (1976).
[8] M. Simsek, J. Ziegenmeyer, J. Heckman, and U. L. RajBhandary, *Proc. Natl. Acad. Sci. U.S.A.* **70,** 1041 (1973).
[9] J. R. Lillehaug, R. K. Kleppe, and K. Kleppe, *Biochemistry* **15,** 1858 (1976).

advantage of the reversibility of T4 polynucleotide kinase activity.[10-12] This method has the disadvantage of requiring a large excess of [γ-^{32}P]ATP over the substrate to be labeled.

As an alternative to the methodology just described, the use of inorganic phosphate to inhibit alkaline phosphatase activity in polynucleotide kinase catalyzed phosphorylation reactions has been investigated. Conditions are described for end group labeling of a variety of substrates to between 30 and 50% of the theoretical maximum. The method is particularly useful for the phosphorylation of restriction endonuclease digests, since the endonucleolytic cleavage, dephosphorylation, and subsequent end group labeling can be performed sequentially in one reaction mixture.

Inorganic phosphate (KP$_i$) is a well-known inhibitor of alkaline phosphatase[13] and has been used previously to inhibit dephosphorylation in polynucleotide kinase reaction mixtures.[14] This inhibition as well as the effect of the inorganic phosphate upon the kinase-catalyzed phosphorylation was not well characterized. Using [γ-^{32}P]ATP as a substrate for phosphomonoesterase activity, the ability of potassium phosphate (KP$_i$) to inhibit calf intestinal alkaline phosphatase at various ATP concentrations in standard kinase buffer was investigated. Figure 1 shows that increasing concentrations of potassium phosphate progressively inhibited the conversion of [γ-^{32}P]ATP to ^{32}P$_i$ and that the percent inhibition for a given phosphate concentration did not vary over a range of ATP concentrations from 1 to 66 μM. Because the percent inhibition was invariant throughout the entire spectrum of ATP concentrations used for polynucleotide kinase labeling, this method seemed quite promising for establishing conditions for generalized use. Since the dephosphorylation reaction was found to be linear over time in the presence of potassium phosphate (data not shown), conditions for end group labeling were established using high concentrations of polynucleotide kinase in order to complete the phosphorylation in a short period of time. Figure 2 shows that in the presence of 10 μM[γ-^{32}P]ATP and 1 mM KP$_i$, the kinase catalyzed phosphorylation of yeast tRNA^{Phe} and an Hha I digest of PM2 DNA had reached a plateau by about 10 min in the presence of 4 units of polynucleotide kinase. Using a KP$_i$ concentration of 1 mM, only about 18% of the [γ-^{32}P]ATP in the reaction was degraded by residual phosphatase activity in this time under these conditions.

[10] J. H. van de Sande, K. Kleppe, and H. G. Khorana, *Biochemistry* 12, 5050 (1973).
[11] G. Chaconas, J. H. van de Sande, and R. B. Church, *Biochem. Biophys. Res. Commun.* 66, 962 (1975).
[12] K. L. Berkner and W. R. Folk, *J. Biol. Chem.* 252, 3176 (1977).
[13] T. W. Reid and I. B. Wilson, *in* "The Enzymes" (P. D. Boyer, ed.), 3rd ed., Vol. 4, p. 373. Academic Press, New York, 1971.
[14] B. Weiss, T. R. Live, and C. C. Richardson, *J. Biol. Chem.* 243, 4530 (1968).

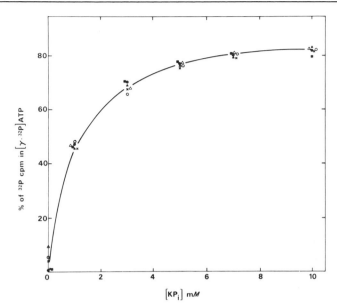

FIG. 1. Inhibition of alkaline phosphatase activity by inorganic phosphate. Reactions were run in 10 μl of 50 mM glycine buffer, pH 9.2, containing 10 mM DTT, 5 mM MgCl₂, 0.1 unit alkaline phosphatase, and increasing amounts of potassium phosphate. The reactions were incubated at 37° for 1 hr at which time 2 μl aliquots were removed for high voltage paper electrophoresis to separate [γ-³²P]ATP and ³²Pᵢ. The concentrations of ATP used in the reactions were 1 μM (●——●), 5 μM (■——■), 10 μM (▲——▲), 20 μM (×——×), 30 μM (○——○), and 66 μM (△——△).

Since inorganic phosphate is also known to inhibit polynucleotide kinase activity,[15] the effect on the phosphorylation of single- and double-stranded nucleic acids was investigated. Figure 3 shows the effect of KPᵢ on the polynucleotide kinase catalyzed labeling of yeast tRNA^Phe and a single-stranded tRNA fragment resulting from continual freezing and thawing of the RNA sample. While the phosphorylation of the single-stranded RNA was not adversely affected by the inorganic phosphate, the labeling of the highly structured tRNA was greatly reduced by increasing concentrations of KPᵢ. It is also noteworthy that under the conditions established for rapid kinase labeling, a high, suboptimal labeling of both the tRNA and the single-stranded fragment was observed in the absence of any inorganic phosphate to inhibit the calf intestinal alkaline phosphatase.

As a result of the studies reported above and other preliminary labeling experiments, a protocol was developed for polynucleotide kinase

[15] J. R. Lillehaug and K. Kleppe, *Biochemistry* **14**, 1225 (1975).

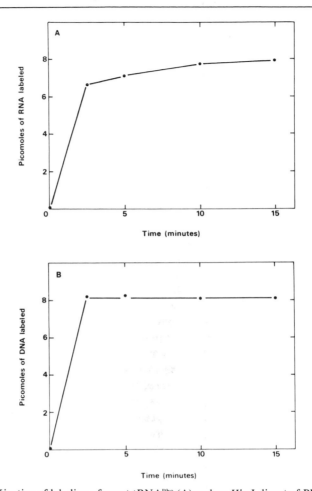

Fig. 2. Kinetics of labeling of yeast tRNA^{Phe} (A) and an *Hha*I digest of PM2 DNA (B). The phosphorylations were performed in 10 μl reactions of the standard kinase buffer containing 10 μM [γ-^{32}P]ATP, 1 mM KP$_i$ and 4 units of polynucleotide kinase. Aliquots (1 μl) were removed at various times and analyzed by DE-81 paper chromatography. The tRNA reaction (A) contained 0.16 μg of dephosphorylated RNA and 0.1 unit of phosphatase, and the *Hha*I-treated reaction (B) contained 0.3 μg of dephosphorylated PM2 DNA and 0.02 units of phosphatase.

catalyzed phosphorylations using inorganic phosphate as an inhibitor of alkaline phosphatase.

Labeling Procedure

The polynucleotide substrate is dephosphorylated with calf intestinal alkaline phosphatase. The commercially available enzyme (Grade 1,

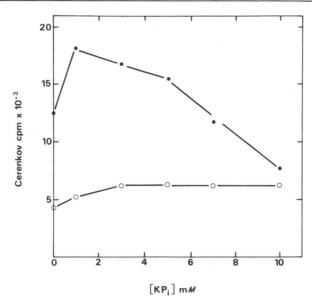

FIG. 3. The effect of inorganic phosphate on the phosphorylation of yeast tRNA^Phe (●———●) and a single-stranded tRNA fragment (○———○). Labeling reactions were performed as described in the legend to Fig. 2 except that the KP_i concentrations were varied. The phosphorylated RNA was analyzed by electrophoresis on a 6% 7 M urea acrylamide slab gel. The radioactively labeled RNA bands were visualized by autoradiography, excised from the gel and counted by measuring Cerenkov radiation.

Boehringer Mannheim Corp.) can be easily purified to a form free of any detectable endo- or exo-DNase and RNase activity by chromatography on Sephadex G-75 as previously described.[16] Furthermore, the calf intestinal enzyme has a 10- to 20-fold higher activity per milligram of protein than the bacterial enzyme as determined by a standard assay method.[17] Dephosphorylations are usually performed in 10–50 μl of Tris-HCl (pH 8) containing 0.1 mM EDTA, but may also be performed in other buffers. Restriction nuclease cleavage products may be dephosphorylated by addition of alkaline phosphatase directly to the enzyme digest. In general, as little phosphomonoesterase as possible should be used while still effecting the quantitative removal of 5'-phosphates. The amount of enzyme required may vary depending upon the polynucleotide substrate and is best determined empirically. About 0.1 unit of calf intestinal alkaline phosphatase can easily remove 50 pmoles of 5'-phosphomonoesters from double-stranded DNA or RNA. Dephosphorylation of single-stranded

[16] A. Efstratiadis, J. N. Vournakis, H. Donis-Keller, G. Chaconas, D. K. Dougall, and F. C. Kafatos Nucleic Acids Res. 4, 4165 (1977).
[17] "Worthington Enzyme Manual," p. 73. Worthington Biochemical Corp., Freehold, N.J., 1972.

substrates is performed by incubation with the enzyme at 37° for 30–45 min, while double-stranded DNA or RNA is treated at 45° for the same length of time.

The dephosphorylated reaction mixture or an aliquot thereof is labeled by addition of glycine buffer (pH 9.2), dithiothreitol, and $MgCl_2$ to give final concentrations of 50, 10, and 5 mM, respectively, in the polynucleotide kinase reaction mixture. The appropriate amount of KP$_i$ (pH 9.2) is added followed by the [γ-^{32}P]ATP and polynucleotide kinase. Phosphorylation reactions are carried out in a volume of 10–50 μl with 4 units of polynucleotide kinase at 37° for 10–15 min. Labeling reactions should contain a final alkaline phosphatase concentration of less than 10 units/ml. T4 polynucleotide kinase can be purified free of any detectable endo- or exo-DNAse and RNAse activity by a minor modification of the procedure of Panet et al.[18] as previously described.[16] [γ-^{32}P]ATP at 1000 Ci/mmole may be prepared as described by Maxam and Gilbert[3] or may be purchased from one of several commercial suppliers. If desired, the specific activity of the ATP may be determined by a published procedure.[19] The ATP and inorganic phosphate concentrations to be used are as follows.

1. For double-stranded polynucleotides, including restriction fragments with flush ends or 3' sticky ends an ATP concentration of at least 10 μM, and preferably 30 μM should be used[9] with an ATP to 5'-hydroxyl end ratio of at least 10 : 1, and a KP$_i$ concentration of 1 mM.

2. For restriction fragments with 5' sticky ends an ATP concentration of 10 μM is used with a 10 : 1 ratio of ATP to 5'-hydroxyl ends, and a KP$_i$ concentration of 5 mM.

3. For single-stranded substrates an ATP concentration of as low as 1 μM may be used with fivefold excess of ATP over 5'-hydroxyl ends and a KP$_i$ concentration of 5 mM.

Following the incubation of 37° for 10–15 min, KP$_i$ is added to a final concentration of 30 mM to inhibit dephosphorylation of the end-labeled polynucleotide. It should be noted that all subsequent manipulations of the phosphorylated nucleic acids, such as cleavage by a restriction nuclease or digestion to 5'-mononucleotides, should be performed in 20 mM KP$_i$. Furthermore, the polynucleotide kinase should be inactivated by heating to 60° for 5 min before any other enzymatic procedures are performed.

[18] A. Panet, J. H. van de Sande, P. C. Loewen, H. G. Khorana, A. J. Raae, J. R. Lillehaug, and K. Kleppe, *Biochemistry* **12,** 5045 (1973).
[19] G. Chaconas, J. H. van de Sande, and R. B. Church, *Anal. Biochem.* **69,** 312 (1975).

Fig. 4. Polyacrylamide gel electrophoresis of polynucleotide kinase labeled yeast tRNA^Phe. The tRNA (0.15 μg) was dephosphorylated in 5 μl of 5 mM Tris-HCl (pH 7.8) containing 0.1 mM EDTA and 0.03 units calf intestinal alkaline phosphatase at 45° for 20 min. The phosphorylation was performed in a volume of 10 μl containing 15 μM [γ-³²P]ATP using the conditions described for labeling of double-stranded polynucleotides. The reaction was terminated by freezing in Dry Ice–acetone and subsequently lyophilized and resuspended in 100% deionized formamide containing 0.03% xylene cyanol FF. The reaction was then applied onto a 6% 7 M urea polyacrylamide gel in 50 mM Tris-borate (pH 8.3) containing 1 mM EDTA (30 cm in length) which was run until the dye marker had migrated 15 cm from the origin. The labeled tRNA was visualized by autoradiography.

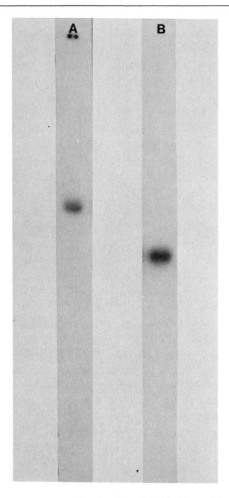

FIG. 5. Agarose gel electrophoresis of polynucleotide kinase labeled PM2 (A) and SV40 (B) linear DNAs generated by cleavage of superhelical molecules with restriction endonuclease *Mno*I. The 15 cm, 1% agarose gel in 50 m*M* Tris acetate (pH 7.8) was run until the bromphenol blue marker had migrated to the bottom of the gel. The labeled DNA was visualized by autoradiography. (A) PM2 DNA (45 μg) was incubated in 45 μl of 10 m*M* Tris-HCl (pH 7.8) containing 5 m*M* $MgCl_2$, 5 m*M* DTT and 17 units of restriction endonuclease *Mno*I at 37° for 1 hr. Calf intestinal alkaline phosphatase (0.1 unit) was added and the reaction incubated at 45° for 45 min. An aliquot of the reaction (2.5 μl) was labeled with polynucleotide kinase and [γ-^{32}P]ATP in 10 μl using the conditions for restriction fragments with 5′ sticky ends. An aliquot (1 μl) was removed from the phosphorylation reaction and added to 10 μl of 20 m*M* KP_i (pH 7.8), 10 m*M* EDTA, 10% sucrose, and 0.025% bromphenol blue for application on the gel. (B) SV40 DNA (10 μg) was incubated in 25 μl of 10 m*M* Tris-HCl (pH 7.8) containing 5 m*M* $MgCl_2$, 5 m*M* DTT, and 12 units of *Mno*I at 37° for 1 hr. Alkaline phosphatase (0.5 units) was added and the reaction incubated at 45° for 45 min at which time it was lyophilized to dryness. The DNA was dissolved and phosphorylated in 10 μl and an aliquot prepared for gel electrophoresis as described above.

FIG. 6. Agarose gel electrophoresis of polynucleotide kinase labeled SV40 DNA which had been digested with restriction endonuclease *Hha*I. Alkaline phosphatase (0.1 unit) was added to the restriction digest containing 1.5 μg of DNA in 5 μl. Following dephosphorylation at 45° for 45 min the DNA was labeled in a 20 μl reaction containing 30 μM [γ-³²P]ATP using the conditions described for the phosphorylation of double-stranded DNA. An aliquot (1 μl) was removed and analyzed on a 1% agarose gel using the conditions described in the legend to Fig. 5.

Analysis and Isolation of End-Labeled Polynucleotides

DNA or RNA phosphorylated as described above may be analyzed or isolated by analytical or preparative gel electrophoresis. Samples may be prepared for loading as follows:

Denaturing Gels in 7 M Urea. The labeling reactions are lyophilized or ethanol-precipitated after addition of carrier RNA. The reactions are dis-

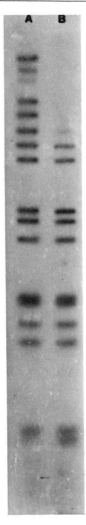

FIG. 7. Polyacrylamide gel electrophoresis of polynucleotide kinase labeled PM2 DNA which had been digested with restriction endonuclease *Hae*III. Alkaline phosphatase (0.1 unit) was added to a 10-μl digest containing 3 μg of DNA. Following dephosphorylation at 45° for 45 min, a 2-μl aliquot was removed and labeled in a 10-μl reaction containing 30 μM [γ-^{32}P]ATP using the conditions described for phosphorylation of double-stranded DNA, except that sodium phosphate was used instead of potassium phosphate to inhibit the alkaline phosphatase. Following the labeling reaction, NaP$_i$ (pH 7.8), EDTA, bromphenol blue, and sucrose were added to a final concentration of 20 and 10 mM, 0.025 and 10%, respectively. The reaction was divided into two equal portions (A) and (B) which were treated as follows: (A) The reaction was loaded directly onto the gel. (B) SDS was added to 1.2% and the reaction was heated to 70° for 5 min before loading onto the gel. The bromphenol blue marker was run to the bottom of the 30 cm, 8% acrylamide (1 : 40, bisacrylamide : acrylamide) gel in 50 mM Tris-glycine (pH 8.3). The labeled DNA was visualized by autoradiography.

solved in deionized formamide containing 0.025% bromphenol blue or xylene cyanol FF and loaded onto the gels.

Native Gels. The labeling reactions are ethanol-precipitated after addition of carrier, and resuspended in 20 mM KP$_i$ (pH 7.8),10 mM EDTA, 10% sucrose, and 0.025% tracking dye. Alternatively, KP$_i$, EDTA, sucrose, and tracking dye can be added directly to the labeling reaction before loading onto the gels.

Figures 4–7 show gel profiles of end-labeled tRNA and restriction nuclease digests phosphorylated using the conditions described here. It is noteworthy that the high molecular weight DNA fragments on the gel in Figure 7 were dephosphorylated by the alkaline phosphatase during the electrophoretic run (slot B). This problem has been observed only on native acrylamide gels and can be greatly reduced by ethanol precipitation of the labeled DNA, or by treatment of the sample with sodium dodecyl sulfate (SDS) at 70°–75° for 5 min prior to loading on the gel (slot A). If SDS is used to inactivate the alkaline phosphatase, NaP$_i$ rather than KP$_i$ should be used in the labeling reaction since potassium dodecyl sulfate will precipitate. An alternative method for eliminating the dephosphorylation problem is to ethanol precipitate the end-labeled polynucleotides and run them on gels prepared in 50 mM Tris–phosphate and 1 mM EDTA. This buffer is prepared by adjusting the pH of the Tris buffer to 7.6 with H$_3$PO$_4$. The phosphate in the gel and reservoir buffers inhibits the alkaline phosphatase activity and results in DNA restriction fragments labeled to an equal extent following acrylamide gel electrophoresis. Although the dephosphorylation problem has not been observed in agarose gels we now routinely use the Tris–phosphate–EDTA buffer system as a precautionary measure in all native agarose and acrylamide gels.

Based on the amount of radioactive DNA and RNA found in acrylamide and agarose gel bands, the efficiency of labeling using inorganic phosphate to inhibit calf intestinal alkaline phosphatase in polynucleotide kinase reaction mixtures is generally 30–50%. Complete phosphorylation has not been observed even with single-stranded oligonucleotides. The method is particularly useful for labeling restriction endonuclease digests; however, it should be noted that many commercially available restriction nucleases are supplied in phosphate buffer. The use of such enzymes will result in poor end group labeling because of the inability of alkaline phosphatase to remove 5' terminal phosphates in the presence of inorganic phosphate. Inorganic phosphate is also an inhibitor of tobacco acid pyrophosphatase[20] and can be used to inhibit this enzyme as well as alkaline phosphatase in reactions designed to enzymatically decap and end-label eukaryotic mRNA.[16]

[20] H. Shinshi, M. Miwa, K. Kato, M. Noguchi, T. Matsushima, and T. Sugimura, *Biochemistry* 15, 2185 (1976).

Section III

Purification of Restriction Enzymes

[11] General Purification Schemes for Restriction Endonucleases[1]

By VINCENZO PIRROTTA and THOMAS A. BICKLE

The range of organisms used for the production of restriction enzymes and the various properties of these enzymes is such that it is difficult to devise a purification scheme of general application. The combination of polyethyleneimine precipitation and chromatography on heparin-agarose discussed in this article has a sufficiently wide applicability and a number of advantages to recommend it as the backbone of a general purification scheme or as a first approach in the isolation of new restriction endonuclease activities. In individual cases, however, additions or modifications of this procedure are necessary to optimize the results. In this article we will first present the basic procedure and then discuss alternatives or modifications suitable to certain particular cases.

Growth and Storage of Cells

The requirements and conditions of growth vary of course with the wide variety of organisms which produce restriction enzymes. In most cases the amount of enzyme per cell does not seem to vary substantially during the growth cycle, and it is therefore possible to allow the cells to grow to stationary phase before harvesting. This results in substantial saving in media and means that the cultures do not have to be monitored carefully. Most enzymes are stable in cell pastes stored at −70° for long periods of time, so that it is feasible to grow more cells than are actually needed for a preparation.

Opening Cells

Many different methods have been used to break open cells for restriction enzyme preparations. These include osmotic shock of lysozyme-prepared spheroplasts, explosive decompression in the French press, grinding with glass beads or alumina, and sonication. This last method is probably the most generally satisfactory and universally applicable so long as the quantity of cells to be opened does not exceed 20–300 g (wet weight). The general procedure most often used in our laboratory is described below.

[1] This work was supported in part by grants from the Swiss National Foundation for Scientific Research.

Method. Thaw and suspend a frozen cell paste in an equal volume of buffer containing 10 mM Tris-HCl, pH 7.5, 1 mM EDTA, and 7 mM 2-mercaptoethanol. Sonicate at maximum power of the sonicator until most of the cells are broken, taking care that the temperature does not rise above 10°. In practice this is best done by performing the sonication with the sample immersed in an ice-water bath and sonicating for periods of 30–60 sec allowing 2–3 min between bursts for the sample to cool down. It is very often difficult to decide when the sample has been sufficiently sonicated. For some bacteria the suspension will clear as the cells break, but this is far from being a general rule. The most generally reliable method of monitoring cell breakage is by observation in the light microscope. In general, a total sonication time of 5–10 min is sufficient.

Removal of Cell Debris and Ribosomes

Once the cells have been opened, it is necessary to remove unbroken cells and cell debris. Ribosomes may constitute as much as 30% of the total protein in a crude lysate. Their removal represents a considerable purification and improves the resolution of subsequent fractionations. This is achieved by centrifuging the crude lysate at 100,000 g for 1 hr. The presence of Mg^{2+} facilitates the pelleting of ribosomes which otherwise tend to form a loose layer which is easily disturbed by decanting. Fine cell debris which often packs poorly even after ultracentrifugation can sometimes be made to aggregate into larger, more firmly packing material by a single cycle of freezing and thawing of the cell extract. For large scale preparations, the high speed centrifugation is facilitated by a previous low speed centrifugation to remove the bulk of the coarse cell debris.

Removal of Nucleic Acids

It is essential to remove as much of the nucleic acids as possible from the cell extracts before chromatography on ion exchangers, particularly phosphocellulose, since the presence of nucleic acids changes the chromatographic properties of many enzymes, probably because the enzymes prefer to bind to the nucleic acid rather than to the ion exchanger.

Several methods have been used for this: gel filtration chromatography, fractional precipitation with ammonium sulfate, and specific precipitation with agents that precipitate mainly nucleic acids but not proteins. The first of these methods, gel filtration chromatography at high ionic strength has been widely used in restriction enzyme purifications, particularly by Roberts.[2] It suffers from the disadvantage that it is only

[2] R. J. Roberts, *Crit. Rev. Biochem.* **4**, 123 (1976).

feasible with fairly small scale preparations; the amount of column material needed for a large scale preparation becomes prohibitive, and the volume of the eluate is large and contains a high concentration of salt which must be removed by dialysis before subsequent purification steps. The advantages of the method are that it is fairly rapid and that the same column can be regenerated and used many times.

The most efficient way of removing nucleic acids is by precipitation. Traditionally, this has been done with streptomycin sulfate and more recently with polyethyleneimine (PEI).[3-5]

Method. A 10% (v/v) solution of PEI (Serva, practical grade) in water is prepared, and the pH is adjusted to 7.5 with HCl. The solution has a turbid appearance. This solution is added slowly and with stirring in the cold to the high speed supernatant to a final PEI concentration of 1%. After stirring for at least 30 min (and sometimes for as long as 16 hr) the precipitated nucleic acids are removed by low speed centrifugation. The ionic conditions are important for the success of the PEI precipitation. In low salt conditions (the buffer used for sonicating the cells) some proteins including many restriction enzymes, coprecipitate with the nucleic acids. In some cases the enzyme can be recovered from the PEI pellet by extraction with buffers containing 0.1–0.2 M NaCl. When this is possible the enzymes are often pure enough to use for some purposes without further purification[4,5] (see below discussion or criteria for purity). Other enzymes cannot be isolated in an active form from the PEI pellet. For these, the PEI precipitation must be performed in relatively high (0.1–0.2 M) concentrations of salt. Under these conditions the nucleic acids still precipitate but most of the proteins remain in solution. At present, we routinely do the PEI step at 0.2 M NaCl even for those enzymes that can be recovered from the pellet when the precipitation is done in low salt. This is because we have found that the recovery of some enzymes from the PEI pellet is variable, even though the degree of purification is high.

Polyethyleneimine itself would also interfere with chromatography. It can be removed, and the proteins concentrated, by precipitation with ammonium sulfate. Solid ammonium sulfate is added to the PEI supernatant slowly and with stirring to a final concentration of 70% of saturation. After all of the ammonium sulfate has dissolved, stirring is continued for a further 30 min, and the precipitated proteins are harvested by low speed centrifugation.

[3] A. H. A. Bingham, A. F. Sharman, and T. Atkinson, *FEBS Lett.* **76,** 250 (1977).
[4] J. Sümegi, D. Breedveld, P. Hossenlopp, and P. Chambon, *Biochem. Biophys. Res. Commun.* **76,** 78 (1977).
[5] T. A. Bickle, V. Pirrotta, and R. Imber, *Nucleic Acids Res.* **4,** 2561 (1977).

Chromatographic Procedures

For most restriction enzymes further purification will be obtained by one or more steps of column chromatography. The most popular, and until recently generally useful, column material is phosphocellulose, which has the advantages of selectivity for nucleic acid binding proteins and high capacity. It suffers from the disadvantages that it is chemically unstable and difficult to equilibrate. Recently, we have used heparin covalently linked to agarose as an alternative to phosphocellulose. Proteins which bind to nucleic acids apparently recognize heparin as a nucleic acid analogue and bind to the heparin-agarose column because of this affinity rather than by ion exchange. Heparin-agarose has the advantages of very high capacity, high selectivity for nucleic acid binding proteins, chemical stability, and ease of equilibration. Restriction enzymes usually bind to heparin more tightly than nonspecific nucleases and consequently eluted at high concentrations. Because of the very high capacity, it is possible to use small bed volumes, which means that the enzymes usually elute as sharp peaks in relatively small volumes. A procedure for the preparation of heparin agarose follows:

Preparation of Heparin-Agarose. Five hundred milliliters of settled bed volume of Biogel A $1.5 M$ or Sepharose 2B are washed twice with distilled water and suspended in 1.5 liters of distilled water. The suspension is placed on a magnetic stirrer in an ice bath, and a thermometer and a pH electrode are immersed in it. Twenty-five grams of CNBr are dissolved in 50 ml of cold dimethyl formamide, and the solution is added with stirring to the suspension of agarose. The pH is then adjusted to 10.5–11.5 with 5 N NaOH. As the reaction proceeds, HBr is produced and the pH drops. The pH should be maintained within the limits given above by the addition of 5 N NaOH until the rate of change of pH with time becomes negligible (30–60 min). The temperature should be maintained below 15° by the occasional addition of crushed ice. The suspension is collected on a fritted glass filter and washed with 5 liters of distilled water. All of the above operations should be carried out in a well-ventilated hood since CNBr is volatile and extremely toxic. The CNBr-activated agarose is suspended in 1 liter of 0.1 M NaHCO$_3$ containing 1 g of heparin. We have used Sigma grade 1 heparin (Catalogue No. H-3125), other grades may also be satisfactory. The suspension is stirred at 4° overnight, 50 ml of triethanolamine is added, and the suspension is stirred for a further 4 hr. The agarose is collected on a fritted glass filter and washed with 2 liters of 1 M NaCl and 5 liters of distilled water. The agarose is then suspended in an equal volume of the buffer that will be used for chromatography, in our case, 20 mM Tris-HCl, pH 7.5, 0.5 mM EDTA. Heparin-agarose made in this way and stored at 4° is stable for at least 2 years.

Use of Heparin-Agarose. Heparin agarose is used in a very similar way to ion exchange chromatography media with the only difference being that lengthy equilibration with starting chromatography buffers is unnecessary. The required volume of heparin-agarose is made into a 30% (v/v) slurry with the storage buffer supplemented with 7 mM 2-mercaptoethanol, degassed under vacuum and packed into the column. We normally use a bed volume of 20–25 ml per 100 g wet weight of cells and a bed height 5–10 times the diameter of the column. The columns are operated at a flow rate of 0.5–1 bed volumes per hour although higher flow rates have been used without apparent loss of binding capacity. They are washed with at least 2 bed volumes of chromatography buffer (20 mM Tris-HCl, 7 mM 2-mercaptoethanol, 0.5 mM EDTA). The sample, dialyzed against the chromatography buffer, is pumped onto the column which is then washed with a further 2 bed volumes of chromatography buffer. Restriction enzymes are eluted with a gradient of NaCl generally between 0 and 0.8 M NaCl made up in 10 bed volumes of chromatography buffer.

Other Column Materials. Depending on the enzyme and on the degree of purity required, it may be necessary to employ a second chromatographic step. Phosphocellulose and DEAE-cellulose are the most convenient available materials. Chromatography on phosphocellulose will not in general be a repeat of the heparin step. The two materials separate according to different principles, and the order of elution of two enzymes from the two different columns is frequently inverted. DEAE-cellulose has also been of use in some cases. The column can be equilibrated with the same buffer used for the heparin-agarose. No restriction enzyme so far examined binds to DEAE-cellulose at higher than 0.4 M salt. Recently, we have had good results with DEAE-Sephacel (Pharmacia) which has a cellulose matrix in spherical bead form. This material is supplied preswollen and equilibrated with Tris buffer and so needs very little preparation. In our hands, it gives sharper separations than the fibrous cellulose derivatives.

Other types of chromatography systems have also been employed for restriction enzyme purifications, including DNA cellulose or agarose[6] and hydrophobic interaction chromatography.[7] This last system might be particularly useful for those enzymes that have a tendency to precipitate in low salt, since the hydrophobic matrices are loaded at relatively high salt concentrations.

In cases where more than one restriction enzyme are present in the

---

[6] H. Schaller, C. Nüsslein, F. J. Bonohoeffer, C. Kurz, and I. Nietzschann, *Eur. J. Biochem.* **26**, 474 (1972).
[7] R. E. Gelinas, P. A. Myers, G. H. Weiss, K. Murray, and R. J. Roberts, *J. Mol. Biol.* **114**, 433 (1977).

same cell extract, heparin-agarose is often not a good choice as a first chromatographic step. Since heparin separates on the basis of affinity, two restriction enzymes are frequently poorly resolved. In the case of *Haemophilus aegyptius,* for example, the extract contains vast quantities of *Hae* III and moderate quantities of *Hae* II. On heparin-agarose these two enzymes elute very close to one another, and the *Hae* II activity is over-shadowed by the *Hae* III which elutes slightly earlier. Instead the separation of the two activities is nearly complete on phosphocellulose where *Hae* II elutes first, followed by *Hae* III which is pure enough for sequencing use. If necessary, the *Hae* II fractions can be further purified, to remove traces of *Hae* III, with a small heparin-agarose column which has the additional effect of concentrating the enzyme. Another example is *Bgl* I– *Bgl* II. These two enzymes are frequently poorly resolved on heparin agarose or on phosphocellulose. In this case the best separation is obtained with DEAE-cellulose.

Problem Cases

In some instances the PEI precipitation is not sufficient to remove substances which interfere with subsequent steps. We have encountered this problem with *Mbo* and to a larger extent with *Mnl* and with *Taq.* In the case of *Mbo* and *Mnl,* the contaminating material which interfered with ammonium sulfate precipitation and with column chromatography could be removed together with the nucleic acids by precipitation with strep-tomycin at a final concentration of 5%. The supernatant was then sub-jected to ammonium sulfate precipitation before loading on heparin-agarose. In the case of *Taq,* we found that the extracts contained an acidic "slime" which interfered with binding of the enzyme to columns even after PEI precipitation. The slime could be removed by applying the 35–75% ammonium sulfate cut to a DEAE-cellulose column. A relatively large bed volume is necessary since the slime, as well as the enzymes, binds to the column. Although the column gives poor resolution, the active fractions now bind well to heparin-agarose affording good separation of *Taq* I from *Taq* II.

Assay Procedures

The most useful assay procedure for chromatographic separations of restriction enzymes is electrophoresis of digests of DNA on slab gels of polyacrylamide or agarose. This is described in detail elsewhere in this work (see Section V). Here, we would just like to emphasize that the conditions of incubation may be important for the interpretation of the results. It is very easy to overdigest the DNA by incubating the samples too long with too much enzyme. This may have two results. First, a complete digest is obtained through a large part of the gradient making it difficult to decide where the enzyme peak is located. Second, if two different enzymes are separated by the gradient the more abundant may "swamp out" the activity that is present in lesser amounts.

Criteria for Purity

Enzymes may be purified to different degrees, depending on the use that will be made of them. For mapping or comparative studies, an enzyme need only be purified to the point where digests of DNA give sharp bands on agarose or polyacrylamide gels. For DNA sequencing work this is not enough; enzymes have to be completely free from contaminating 3'- or 5'-exonucleases, single-stranded-specific nucleases, phosphatases, etc. This is best tested directly with the aid of end-labeled substrates.

Relatively few attempts have been made to purify restriction enzymes to the point where they are suitable for enzymatic and protein chemical studies, that is, to homogeneity. This is probably because, until now, interest has centered on the application of the enzymes to nucleic acid studies rather than on the enzymes themselves.

Storage of Restriction Enzymes

Purified restriction enzymes are generally stored at $-20°$ in buffers containing 50% glycerol. The glycerol can be added directly to the enzyme-containing fractions from the last step of the purification, or the fractions can be dialyzed against buffer containing 50% glycerol. In this last case, a considerable concentration of the enzyme is obtained. Highly purified enzyme preparations can lose activity upon storage. This can often be prevented by adding autoclaved gelatin or bovine serum albumin to a final concentration of 50–100 μg/ml. Some enzymes, (*Eco*RI, *Pst*I, etc.) also require a neutral detergent, such as 0.2% Triton X-100, to stabilize them.

[12] Purification and Properties of *Eco*RI Endonuclease

By ROBERT A. RUBIN and PAUL MODRICH

The *Escherichia coli* RI (*Eco*RI)[1] DNA restriction and modification enzymes recognize a common twofold symmetrical hexanucleotide sequence in duplex DNA.[3]

$$d(pG \overset{\downarrow}{}p\overset{*}{A}pApTpTpCp)$$
$$d(pCpTpTpApA\underset{*}{p}\underset{\uparrow}{}Gp)$$

*Eco*RI restriction endonuclease cleaves the DNA duplex[4] within this sequence (see arrows in above sequence), while the modification enzyme methylates the two adenine residues adjacent to the axis of symmetry (asterisks) to yield 6-methylaminopurine.[3] The presence of one 6-methylaminopurine residue within this sequence is sufficient to block single- or double-stranded cleavage by the endonuclease.[5]

*Eco*RI endonuclease has been extensively used as a reagent for the preparation of recombinant molecules.[6] In addition, the *Eco*RI enzymes are biochemically simple and hence provide an ideal system for study of sequence-specific DNA–protein interaction. We describe here a convenient method for isolation of large quantities of the endonuclease, a rapid assay procedure for quantitation of specific endonucleolytic activity, as well as physical and catalytic properties of the homogeneous protein.

Assay Methods

Principle. The assay scores conversion of covalently closed circular colicin E1 (ColE1) DNA, which contains a single *Eco*RI site,[7] to the linear

---

[1] In the nomenclature proposed by Smith and Nathans,[1a] *Eco*RI designates a plasmid-specified DNA restriction and modification system initially identified in *Escherichia coli* strains carrying the drug resistance transfer factor RTF-1.[2]

[1a] H. O. Smith and D. Nathans, *J. Mol. Biol.* **81**, 419 (1973).

[2] M. Betlach, V. Hershfield, L. Chow, W. Brown, H. M. Goodman, and H. W. Boyer, *Fed. Proc., Fed. Am. Soc. Exp. Biol.* **35**, 2037 (1976).

[3] A. Dugaiczyk, J. Hedgpeth, H. W. Boyer, and H. M. Goodman, *Biochemistry* **13**, 503 (1974).

[4] J. Hedgpeth, H. M. Goodman, and H. W. Boyer, *Proc. Natl. Acad. Sci. U.S.A.* **69**, 3448 (1972).

[5] R. A. Rubin and P. Modrich, *J. Biol. Chem.* **252**, 7265 (1977).

[6] R. F. Beers, Jr. and E. G. Bassett, eds., "Recombinant Molecules: Impact on Science and Society." Raven, New York, 1977.

[7] M. A. Lovett, D. G. Guiney, and D. R. Helinski, *Proc. Natl. Acad. Sci. U.S.A.* **69**, 3448 (1974).

METHODS IN ENZYMOLOGY, VOL. 65

form which is sensitive to hydrolysis by *rec*BC DNase.[8,9] Other assay methods which have been employed include separation of reaction products by gel electrophoresis,[10] and labeling of newly produced 5' termini by polynucleotide kinase (see Bernker and Folk [5], this volume).

Reagents

Tris (hydroxymethyl)aminomethane-HCl buffer, 1 M in water, pH 7.6 (all pH values are measured at 0.05 M in water at room temperature, unless stated otherwise)

KPO$_4$, 1 M in water, pH 7.4

NaEDTA, 0.5 M in water, pH 8.0, at 0.5 M

MgCl$_2$, 0.1 M in water

NaCl, 2 M in water

Dithiothreitol, 0.5 M in water, stored at $-20°$

Na$_3$ATP, 0.01 M in water, stored at $-20°$

Bovine serum albumin (Calbiochem, A grade), 10 mg/ml in water, stored at $-20°$ (an $E_{1\%}^{1cm}$ of 6.6 is assumed). Batches of BSA are screened for presence of contaminating endonuclease prior to use

*rec*BC nuclease, purified through the phosphocellulose step of Eichler and Lehman[11]

[^{3}H]Thymidine-labeled (7–10 cpm/pmol nucleotide) covalently closed circular ColE1 DNA, 1 mM nucleotide phosphorus (assuming $A_{260} = 6.7$ for 1 mM), in 20 mM Tris-HCl (pH 7.6), 50 mM NaCl, 1 mM EDTA. ColE1 DNA is isolated from *E. coli* JC411 *thy* (ColE1) after chloramphenicol amplification of the plasmid[12]

3 mM salmon sperm DNA (Sigma) in 0.02 M Tris-HCl (pH 7.6), 1 mM EDTA

30% trichloroacetic acid (w/v) in water

Procedure.[12] Reactions (200 μl) contain 0.1 M Tris-HCl (pH 7.6), 0.05 M NaCl, 5 mM MgCl$_2$, 0.2 mM EDTA, 0.5 mM dithiothreitol, 0.1 mM Na$_3$ATP, 0.02 mM ColE1 [^{3}H]DNA, 6.8 units of *rec*BC DNase, and *Eco*RI endonuclease. After incubation at 37° for 10 min, 0.05 ml each of 3 mM salmon sperm DNA and 30% trichloroacetic acid are added to terminate the reaction. After 10 min at 0°, precipitates are removed by centrifugation at 23,000 g for 10 min, and acid-soluble nucleotide is determined by counting 200 μl of supernatant in 10 ml Aquasol-2 scintillation fluid (New En-

[8] M. Wright, G. Buttin, and J. Hurwitz, *J. Biol. Chem.* **246**, 6543 (1971).

[9] A. E. Karu, V. MacKay, P. J. Goldmark, and S. Linn, *J. Biol. Chem.* **248**, 4874 (1973).

[10] P. J. Greene, M. C. Betlach, H. M. Goodman, and H. W. Boyer, *Methods Mol. Biol.* **7**, 87 (1974).

[11] D. C. Eichler and I. R. Lehman, *J. Biol. Chem.* **252**, 499 (1977).

[12] P. Modrich and D. Zabel, *J. Biol. Chem.* **251**, 5866 (1976).

gland Nuclear). Reaction is linear with time, and control experiments demonstrate that rec BC DNase is present in excess. The enzyme is diluted as necessary into 0.02 M KPO$_4$ (pH 7.4), 0.2 M NaCl, 0.5 mM dithiothreitol, 0.2 mM EDTA, 0.2 mg/ml bovine serum albumin, and 10% (w/v) glycerol.

Unit of Activity. One unit of endonuclease is defined as that which converts 1 pmole of ColE1 molecules (12.7 nmole nucleotide) to a rec BC DNase-sensitive form in 1 min under assay conditions.

Enzyme Purification

Several satisfactory procedures for isolation of EcoRI endonuclease are available.[10,12] The one employed in our laboratory will be described here.[12] For large scale enzyme isolations, the endonuclease I-deficient *E. coli* strain RY13 *gal end* (RI) is grown in a 200-liter culture of L broth, containing (per liter) 10 g Bactotryptone, 5 g yeast extract, 10 g NaCl, 5 g glucose, and 5 mM KPO$_4$ (pH 7.0), in a New Brunswick fermentor at maximum aeration. Culture pH is maintained at 7.0 with NaOH. When the culture reaches early stationary phase as determined by no further increase in A_{590} in 20 min, it is chilled to 5° and the cells are harvested with a refrigerated Sharples centrifuge. Cell paste is stored at $-20°$ and shows no loss of EcoRI endonuclease activity over a period of at least 4 months.

A purification from 4.73 kg *E. coli* RY13 is summarized in Table I. All steps were carried out at 0°–4°, and centrifugation was at 12,000 g for 20 to 30 min in Sorvall GSA or GS3 rotors. Unless noted otherwise, protein was determined by the method of Lowry *et al.*[13] using bovine serum albumin as standard.

Preparation of Crude Extract. After overnight thawing at 4°, 4.73 kg of cell paste were suspended in 14.1 liters of 0.02 M KPO$_4$ (pH 7.4), 15 mM 2-mercaptoethanol, and 1 mM EDTA with the aid of a Waring Blendor. Cells were disrupted by two passages through a Manton-Gaulin homogenizer at 9000 lb/in.^{2} Effluent from the homogenizer had a temperature of about 15° and was quickly chilled to 4° in an ice–salt bath (sonication can be employed for smaller quantities of cell paste). The crude extract was centrifuged and the supernatant (13.9 liters) was diluted with the above buffer to yield an A_{260} of 200.

Streptomycin Fractionation. The clarified crude extract (19.4 liters) was treated with 3.9 liters of fresh 25% (w/v) streptomycin sulfate dissolved in H$_2$O. After stirring for 45 min the precipitate was removed by centrifugation. The supernatant (fraction I, 21 liters) had a ratio of A_{280} to A_{260} of 0.83 and an A_{280} of 25.

[13] O. H. Lowry, N. J. Rosebrough, A. L. Farr, and R. J. Randall, *J. Biol. Chem.* **193,** 265 (1951).

TABLE I
PURIFICATION OF _Eco_RI ENDONUCLEASE FROM 4.73 kg OF _E. coli_ RY13

Fraction	Step	Protein (mg)	Specific activity (units/mg protein)	Recovery (%)
I	Streptomycin supernatant	330,000	14	(100)
II	Ammonium sulfate	56,000	59	70
III	Phosphocellulose	780	2,200	36
IV	Hydroxylapatite	54	23,500	27
V	DNA cellulose	53.7[a]	24,500	28

[a] Protein determined by amino acid analysis.

Ammonium Sulfate Fractionation. Solid $(NH_4)_2SO_4$ (8.2 kg) was added to fraction I (21 liters) over a period of 20 min to yield 60% saturation. After stirring for 60 min at 0°, the precipitate was collected by centrifugation. The precipitate (which is stable at 0° for at least 12 hr) was then extracted successively with solutions of 45, 40, 35, 30, and 25% saturation in $(NH_4)_2SO_4$, which were prepared by dissolving 277, 242, 208, 176, and 144 g/liter, respectively, of 0.02 M KPO$_4$ (pH 7.4), 15 mM 2-mercaptoethanol, and 1 mM EDTA. Extractions were performed by suspending the precipitate in 6.5 liters of the appropriate $(NH_4)_2SO_4$ solution, stirring for 45 min, followed by centrifugation to remove insoluble material. The 25, 30, and 35% $(NH_4)_2SO_4$ washes together contained 80% of the recovered _Eco_RI endonuclease activity. (_Eco_RI methylase was recovered primarily in the 40 and 45% $(NH_4)_2SO_4$ washes.[5]) These fractions were pooled (18.9 liters) and concentrated by precipitation with $(NH_4)_2SO_4$ (3.8 kg). After centrifugation, the pellet was suspended in 1200 ml 0.02 M KPO$_4$ (pH 7.4), 5 mM 2-mercaptoethanol, 0.5 mM EDTA, and 0.1 M KCl, and dialyzed against two 40-liter portions (3 hr per change) of 0.02 M KPO$_4$ (pH 7.4), 5 mM 2-mercaptoethanol, and 10% (w/v) glycerol (buffer A) containing 0.2 M KCl and 0.5 mM EDTA to yield fraction II.

Phosphocellulose Chromatography. Whatman P-11 phosphocellulose was precycled as described,[14] and equilibrated with buffer A containing 0.1 M KCl and 0.5 mM EDTA. A 3.5-liter column (44.5 cm × 79 cm^2) was prepared and washed with 7 liters of this buffer. Fraction II (1.56 liter) was diluted with an equal volume of buffer A containing 0.5 mM EDTA and with 3.12 liters of buffer A containing 0.1 M KCl and 0.5 mM EDTA to yield to a final KCl concentration of 0.1 M and a protein concentration of

[14] P. Modrich, Y. Anraku, and I. R. Lehman, _J. Biol. Chem._ **248**, 7495 (1973).

about 5 mg/ml. The solution was immediately applied to the column at a flow rate of 900 ml/hr. After washing with 7 liters of buffer A containing 0.2 M KCl and 0.1 mM EDTA, the column was eluted with a 36-liter linear gradient of KCl (0.2–1.0 M) in buffer A containing 0.1 mM EDTA. Fractions of about 600 ml were collected, and those containing EcoRI endonuclease activity, which eluted at 0.66 M, were pooled (fraction III).

Hydroxylapatite Chromatography. Hydroxylapatite (Bio-Rad, Bio-Gel HTP) was equilibrated with 0.02 M KPO$_4$ (pH 7.4), 0.2 M KCl, 5 mM 2-mercaptoethanol, and 10% (w/v) glycerol. A 480-ml column (19.5 cm × 24.6 cm^2) was prepared and washed with 1 liter of this buffer. Fraction III (5.6 liters) was applied to the column at a flow rate of 200 ml/hr. After washing with 500 ml of 0.15 M KPO$_4$ (pH 7.4), 5 mM 2-mercaptoethanol, and 10% (w/v) glycerol (Eastman spectral grade glycerol was used in this and subsequent steps), the column was eluted at 140 ml/hr with a 4.8-liter linear gradient of KPO$_4$ (pH 7.4, 0.15–0.85 M) containing 5 mM 2-mercaptoethanol and 10% (w/v) glycerol. Fractions containing activity, which eluted at about 0.44 M, were pooled and made 1 mM in EDTA (fraction IV).

DNA–Cellulose Chromatography. DNA–cellulose containing 8 mg heat-denatured salmon sperm DNA per gram of cellulose was prepared by the ultraviolet irradiation procedure of Litman,[15] and equilibrated with buffer A containing 0.1 M NaCl and 1 mM EDTA. A 44-ml column (14 cm × 3.1 cm^2) was prepared and washed with 90 ml of equilibration buffer. Fraction IV (280 ml) was dialyzed against two 4-liter portions of buffer A containing 0.25 M NaCl and 1 mM EDTA (2 hr per change). The dialyzed material was diluted with buffer A (0.7 volume) to yield a conductivity equivalent to buffer A containing 0.2 M NaCl. This dilution was performed in a batchwise manner on 30-ml portions which were immediately loaded onto the column at a flow rate of 130 ml/hr. After washing with 50 ml of the equilibrating buffer, the column was step-eluted with 90 ml portions of buffer A and 1 mM EDTA containing 0.25, 0.50, and 0.75 M NaCl, and fractions of 5.4 ml were taken. Activity eluted with the 0.50 M NaCl step, and pooled fractions (21.6 ml) were dialyzed overnight against 1 liter of 0.02 M KPO$_4$ (pH 7.4), 0.3 M NaCl, 2.5 mM 2-mercaptoethanol, 1 mM EDTA, and 50% (v/v) glycerol to yield fraction V (9.1 ml). When stored at −20°, this fraction has not lost significant activity (<10%) over a period of 2 years.

Properties of EcoRI Endonuclease

Purity.[12] Fraction V is physically homogeneous as judged by analytical sedimentation velocity and sedimentation equilibrium centrifugation as

[15] R. M. Litman, *J. Biol. Chem.* **243,** 6222 (1968).

well as by polyacrylamide gel electrophoresis under native or denaturing conditions. The enzyme is free of *Eco*RI methylase activity and contains no detectable double-strand endonuclease activity on T7 DNA, which is devoid of *Eco*RI sites. In addition, since products of cleavage can be quantitatively converted to covalently closed circles by DNA ligase, the preparation is free of significant exonuclease activity. However, fraction V does contain low levels of an activity which introduces single-strand breaks into ColE1 DNA methylated at the *Eco*RI site (50 pmole/min-mg) or into PM2 DNA (400 pmole/min-mg) which is devoid of *Eco*RI sites. It is not clear if this reflects a contaminating enzyme or activity of the *Eco*RI enzyme at secondary *Eco*RI* sites (below).

Physical and Chemical Properties. Denatured and reduced endonuclease behaves as a single protein species of MW 28,500 during electrophoresis on polyacrylamide gels in the presence of sodium dodecyl sulfate.[12] In the native state, however, the enzyme exists as an equilibrium mixture of dimers and tetramers of the 28,500 subunit.[12] The equilibrium constant for tetramer dissociation is on the order of 4×10^{-7} M as judged by analytical velocity sedimentation.[16] This value for the equilibrium constant together with results of kinetic analysis[12] indicate that the dimer is catalytically active. Although it has been suggested that the tetramer will comprise a significant fraction of the endonuclease within the cell,[12] it is not clear whether this aggregation state possesses endonucleolytic activity.

The ultraviolet spectra of the endonuclease under native or alkaline conditions are those of a typical protein.[12] In the case of the native enzyme, a $E_{1cm}^{1\%}$ of 8.30 has been determined at the absorption maximum of 278 nm; furthermore, the ratio of A_{280} to A_{260} of 1.86–1.90 rules out significant nucleotide content or contamination of fraction V. The amino acid composition of the enzyme is available.[12]

Catalytic Requirements. *In vitro* restriction by *Eco*RI endonuclease requires only unmodified DNA and Mg^{2+}. The pH optimum in 0.1 *M* Tris-HCl is about 7.1 to 7.5,[17] and 0.1 *M* Tris-HCl (pH 7.6 at 25°) has been employed by most investigators. It should be noted, however, that the endonuclease is unstable in this buffer at low temperatures, a presumed consequence of the elevated pH of Tris-HCl under such conditions.[12] It is, therefore, inadvisable to employ this buffer for enzyme dilution or low temperature incubation. We have found that the use of phosphate buffer (pH 7.4 at 25°) circumvents this problem.[12] A broad range of optimal Mg^{2+} (1–15 m*M*) has been reported,[10] with a concentration of 5 m*M* being routinely employed. Sodium chloride stimulates enzyme activity, and is

---

[16] P. Modrich and H. K. Schachman, unpublished experiments.
[17] R. N. Yoshimori, Ph.D. Thesis, University of California, San Francisco, Dissertation Abstracts Order No. 73-3662 (1971).

generally present at 0.05 M with concentrations above 0.1 M being inhibitory.[10]

Mechanism of Double-Strand Cleavage. The endonuclease obeys Michaelis–Menten kinetics.[12,18] At 37° the K_m for the *Eco*RI site of ColE1 DNA is 8 nM and the catalytic constant is four double-strand scissions per minute per dimer.[12] The results of steady state and transient kinetic experiments with ColE1 and SV40 DNA substrates have led to the following model for the mechanism of double strand cleavage.[12,19]

$$E + I \underset{k_{-1}}{\overset{k_1}{\rightleftharpoons}} E \cdot I \xrightarrow{k_2} E \cdot II \xrightarrow{k_3} E \cdot III \xrightarrow{k_4} E + III$$

$$k_5 \Big\updownarrow k_{-5}$$

$$E + II$$

where I, II, and III represent, respectively, form I or superhelical DNA, form II DNA containing a single-strand scission within the *Eco*RI sequence, and form III molecules which result from double-strand cleavage. Thus, double-strand cleavage by the endonuclease proceeds via an intermediate species containing a single-strand break in the *Eco*RI sequence.[12,19] In the case of ColE1 DNA it has been demonstrated that $k_2 > k_3 > k_4$.[12] Hence, the rate-limiting step in double-strand cleavage of this molecule is dissociation of the enzyme from the form III product.

Within the framework of this mechanism, the path followed by the enzyme is dependent on both the nature of the substrate and the reaction temperature, with differences being observed in the fate of the $E \cdot II$ intermediate. With ColE1 DNA at 37°, double-strand cleavage occurs without detectable dissociation of the form II intermediate from the enzyme surface.[12,19] Under identical conditions, the $E \cdot II$ intermediate formed with SV40 DNA dissociates to a significant extent,[19] although the majority of double-strand events proceed without dissociation.[19] In contrast to the behavior of the $E \cdot II$ intermediate at 37°, extensive dissociation of this complex occurs under steady state condition at 0°.[12] As shown in Table II, the major reaction product formed at low temperature is the form II species, and with an SV40 substrate this species is formed almost exclusively. Thus, the endonuclease can be employed at 0° to introduce site-specific single-strand breaks into duplex DNA.

The conclusions outlined above are based on studies with substrates containing a single *Eco*RI site. We have also examined the mode of

[18] P. J. Greene, M. S. Poonian, A. L. Nussbaum, L. Tobias, D. E. Garfin, H. W. Boyer, and H. M. Goodman, *J. Mol. Biol.* **99,** 237 (1975).
[19] R. A. Rubin and P. Modrich, *Nucleic Acids Res.* **5,** 2991 (1978).

TABLE II
SITE-SPECIFIC SINGLE STRAND CLEAVAGE BY *Eco*RI ENDONUCLEASE AT 0°[a]

Substrate	Form II	Form III
	(mole % in product)	
ColE1 DNA	58	28
SV40 DNA	72	9

[a] Reactions contained 0.02 M KPO$_4$ (pH 7.4), 0.05 M NaCl, 5 mM MgCl$_2$, 100 μg/ml bovine serum albumin, 100 μg/ml DNA, and 5 units/ml (ColE1) or 4.2 units/ml (SV40) *Eco*RI endonuclease. After 30 min (ColE1) or 50 min (SV40) at 0°, reactions were terminated by addition of EDTA to 0.02 M and extraction with redistilled phenol. Products were separated by electrophoresis on 1% agarose gels,[12] and quantitated as described.[5,19] Values shown are corrected for trace amounts of form II molecules present before digestion. ColE1 and SV40 DNA methylated *in vitro*[5] at these *Eco*RI sites were not subject to single- or double-strand scission under these conditions, indicating that cleavage observed occurred within the *Eco*RI sites.

double-strand cleavage of pVH153 DNA circles, which contain two *Eco*RI sites,[20] to determine if cleavage at one site increases the probability of cleavage at the other. As shown in Fig. 1, under steady state conditions at 37°, molecules generated by a single cleavage event accumulate to a concentration ten times that of the endonuclease. Furthermore, products resulting from a double cleavage event appear only after a lag of 1–2 min. These results indicate that most, if not all, of the second double-strand cleavage events are independent of the first and thus that the endonuclease exhibits little if any processive behavior. Therefore, if one wishes to clone a DNA fragment containing an internal *Eco*RI sequence into an *Eco*RI site of a vehicle, it should be possible to obtain the desired fragment simply by limiting the extent of the endonuclease digestion.

Reduction of Sequence Specificity. Polisky *et al.*[21] have demonstrated that under conditions of elevated pH and low ionic strength the cleavage specificity of the endonuclease is reduced to

$$d(^{\downarrow}pApApTpT) \qquad (Eco\text{RI* sequence})$$

Optimal conditions for this reaction are 0.025 M Tris-HCl (pH 8.5) and 2 mM MgCl$_2$. Although extracts and partially purified fractions derived from mutants deficient in *Eco*RI endonuclease are also deficient in *Eco*RI* activity, it has not been demonstrated that these two activities are mediated by the same enzymatic species.[21]

[20] K. A. Armstrong, V. Hershfield, and D. R. Helinski, *Science* **196**, 172 (1977).
[21] B. Polisky, P. Greene, D. E. Garfin, B. J. McCarthy, H. M. Goodman, and H. W. Boyer, *Proc. Natl. Acad. Sci. U.S.A.* **72**, 3310 (1975).

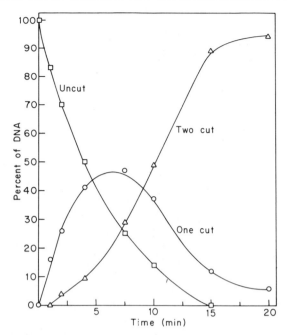

Fig. 1. Independent cleavage of 2 *Eco*RI sites of pVH153 DNA. The reaction (0.5 ml) contained 0.1 M Tris-HCl (pH 7.6), 0.05 M NaCl, 5 mM MgCl$_2$, 0.2 mM EDTA, 50 μg/ml bovine serum albumin, 2.74 nM pVH153 DNA (as molecules), and 0.14 nM *Eco*RI endonuclease as dimer. Incubation was at 37°. Samples (0.04 ml) were removed as indicated, and reaction terminated by addition of 0.02 ml of 0.03 M Na$_3$EDTA, 0.8% SDS, 30% sucrose, and 0.005% bromphenol blue. Reactants and products were separated by electrophoresis on agarose gels and quantitated as described.[5,19] Uncut, 7.3×10^6 dalton circles; one cut, 7.3×10^6 dalton linear; two cut, sum of 2.3×10^6 and 5.0×10^6 dalton linear fragments.

Acknowledgments

Work performed in the authors' laboratory was supported by grants from the National Institutes of Health and the National Science Foundation.

[13] Purification and Properties of *Hind*II and *Hind*III Endonucleases from *Haemophilus influenzae* Rd

By Hamilton O. Smith and Garry M. Marley

Haemophilus influenzae Rd is the source for two restriction endonucleases, *Hind*II and *Hind*III, that may be separated and purified using conventional ion exchange chromatography media.[1,2] The basic steps in the procedure are removal of nucleic acids from the cell extract, separa-

[1] H. O. Smith and K. W. Wilcox, *J. Mol. Biol.* **51**, 379 (1970).
[2] H. O. Smith, *Methods Mol. Biol.* **9**, 71 (1975).

tion of the two activities by DEAE–cellulose chromatography, and chromatography of each activity on phosphocellulose as a combined purification and concentration step.

Assay Method

*Hin*dII. Reaction mixtures (20 μl) contain 1 μg of phage λ DNA, 10 mM Tris-HCl (pH 7.5), 10 mM MgCl$_2$, 7 mM 2-mercaptoethanol, and 1–5 μl of enzyme fraction. Incubation is for 60 min at 37°.

*Hin*dIII. Reaction mixtures (20 μl) contain 1 μg of phage λ DNA, 10 mM Tris-HCl (pH 8.5), 10 mM MgCl$_2$, 60 mM NaCl, 7 mM 2-mercaptoethanol, and 1–5 μl of enzyme fraction. Incubation is for 60 min at 37°.

Reactions are stopped by addition of 5 μl of dye mixture containing 25% glycerol, 0.1 M Na$_2$EDTA (pH 8), 0.05% bromphenol blue. Samples are analyzed for DNA cleavage by electrophoresis on a 1% agarose gel. One unit of endonuclease activity is just sufficient to produce complete cleavage of 1 μg of phage λ DNA in 60 min.

Procedure

Materials

Growth Medium. Brain heart infusion (Difco), 18.5 g; tryptone, 5 g; yeast extract, 2.5 g; NaCl, 2.5 g; H$_2$O, 1000 ml; autoclave to sterilize. Store at 4°.

Hemin Stock Solution. Hemin (Eastman 2203), 0.1 g; histidine, 0.1 gm; triethanolamine, 4 ml; H$_2$O, 96 ml. Incubate at 70° for 15 min to sterilize. Store at 4°.

NAD Stock Solution. β-Diphosphopyridine nucleotide (β-DPN; β-NAD; Sigma No. D-5755) 10 mg/ml. Sterilize by filtration through a 0.45 micron Millipore filter. Store at −20°.

Buffer A: 30 mM Tris-HCl (pH 8.4), 14 mM 2-mercaptoethanol, 5% glycerol

Buffer B: 20 mM Tris-HCl (pH 8.4); 14 mM 2-mercaptoethanol; 0.2 M EDTA, 5% glycerol.

Buffer C: 10 mM potassium phosphate (pH 7.4), 14 mM 2-mer-captoethanol, 5% glycerol.

Purification Procedure

Growth of Cells. Six 2-liter flasks, each with 1 liter of growth medium containing 10 ml of hemin stock solution and 0.2 ml of NAD stock solution, are inoculated with 50 ml of frozen log phase *H. influenzae* Rd *com* 10⁻ (exonuclease III⁻) cells and incubated at 37° with vigorous shaking until about 3 hr beyond the beginning of stationary phase. Cells are harvested in a Sorvall GSA rotor at 8000 rpm. The pellets are quick frozen in the plastic centrifuge bottles, knocked loose as frozen chips, pooled together, weighed, and then stored at −70° until use. Yield is about 3 gm of cell pellet per liter.

Preparation of Extract. Cells (18 gm) are suspended in 120 ml of buffer A in a 250 ml stainless steel beaker and disrupted at maximum intensity on a Branson Sonifier. Temperature is maintained at around 5° by cooling in an ice–salt water bath. Sonication is continued for about 20 min until the turbidity of 50-fold diluted samples show a decrease of 70 to 80%. Cell debris is removed by centrifugation at 20,000 g for 10 min. The supernatant is recovered and adjusted to 4 mM MgCl$_2$ and 0.3 M NaCl in a final volume of about 140 ml. This is centrifuged at 100,000 g for 5 hr at 4° to remove ribosomes and fine debris. The supernatant (130 ml) is recovered (fraction I).

Ammonium Sulfate Precipitation. Ammonium sulfate is added slowly with stirring to achieve 70% of saturation (0.472 g/ml of fraction I). After standing 10 min on ice, the precipitate is collected by centrifugation at 10,000 g for 10 min. The precipitate is then dissolved in buffer B and adjusted to a final volume that gives a conductivity equal to 0.16 M ammonium sulfate in buffer B (about 15.5 mmhos at 0°).

First DEAE–Cellulose Chromatography. Nucleic acids are removed by passage through a DE52 column. The column (2.5 × 30 cm) is equilibrated with buffer B. The ammonium sulfate fraction is loaded at approximately 150 ml/hr and washed through with one column volume of buffer B containing 0.16 M ammonium sulfate. Amber colored breakthrough and early wash fractions containing the bulk of the unadsorbed protein are identified by inspection (or A_{280} measurements) and pooled (fraction II, approximately 200 ml).

Ammonium Sulfate Precipitation. Fraction II is adjusted to 0.3 M NaCl by addition of 5 M NaCl. Ammonium sulfate (0.313 g/ml of fraction II) is added slowly with stirring to give 50% of saturation. The precipitate is removed by centrifugation at 10,000 g for 10 min. The supernatant is adjusted to 70% saturation by addition of 0.159 g of ammonium sulfate per milliliter of fraction II, and the precipitate is collected by centrifugation. The precipitate representing the 50–70% ammonium sulfate cut is dissolved in a minimum volume of buffer C (about 50 ml usually).

Sephadex G-25 Chromatography. A Sephadex G-25 (medium grade) column of approximately 300 ml bed volume is equilibrated with buffer C. The 50–70% ammonium sulfate fraction is loaded and eluted with buffer

C. The void volume fractions, containing the bulk of the protein, are easily identified by their yellow color. These are pooled.

Separation of Hin*dII and* Hin*dIII Activities by DEAE-Cellulose Chromatography.* A DE52 column (1.75 × 8 cm) is equilibrated with buffer C. The Sephadex G-25 pooled fractions are loaded at a flow rate of about 1 bed volume per hour and the column is washed with 5 bed volumes of buffer C. Fractions of approximately 10 ml are collected. The breakthrough and wash fractions that contain restriction enzyme (*Hin*dIII) activity are pooled (fraction III, about 100 ml). *Hin*dII restriction endonuclease should be quantitatively retained on the column.

Phosphocellulose Chromatography of Hin*dIII.* A phosphocellulose (Whatman P11) column (1 × 8 cm) is equilibrated with buffer C. Fraction III is then loaded at about 1 bed volume/hr. This is conveniently done overnight using an inlet tubing loop that drops below the column outlet so that the bed does not run dry. After loading, the column is washed with 3 to 4 bed volumes of buffer C. The *Hin*dIII endonuclease is then eluted with buffer C containing 0.5 *M* NaCl. Fractions (2 ml) are collected and assayed for activity. Activity begins to emerge after about 1 bed volume and is mostly eluted by 2.5 bed volumes. The pooled active fractions (about 11 ml) are mixed with an equal volume of glycerol and stored at −20° (fraction IV).

Comment. Fraction IV should contain around 10,000 units/ml of *Hin*dIII endonuclease activity with a specific activity of 20,000 units/mg protein and should be free of detectable exonucleases. A DNA binding activity is usually present which interferes with DNA cleavage and gel electrophoresis when large excesses of enzyme (greater than 10 units/μg of DNA) are used. Stepwise or gradient elution may be used to obtain slightly better purity. *Hin*dIII endonuclease elutes at approximately 0.3 *M* NaCl.

Elution of Hin*dII Endonuclease from DE-52.* A 200-ml linear gradient from 0 to 0.3 *M* NaCl in buffer C is applied to the *Hin*dII activity retained on the 1.75 × 8 cm DE-52 column. *Hin*dII activity elutes in a broad peak centering at 0.1 *M* NaCl. Active fractions are pooled (fraction V, about 80 ml).

Phosphocellulose Chromatography of Hin*dII.* A 1 × 10 cm phosphocellulose column, equilibrated with buffer C, is loaded with *Hin*dII fraction V at 10 ml/hr. Elution is stepwise with 30 ml of 0.1 *M* NaCl, 30 ml of 0.2 *M* NaCl, and 50 ml of 0.3 *M* NaCl in buffer C. *Hin*dII activity elutes between 1 and 4 column bed volumes during the 0.3 *M* NaCl elution. Fractions containing activity are pooled (25 to 30 ml), adjusted to 50 m*M* Tris-HCl (pH 8.5) with a 1 *M* stock solution, and placed in a dialysis bag. The enzyme is osmotically concentrated about fivefold against dry polyethylene glycol (Carbowax 6000, Eastman 15415). An additional threefold concentration is achieved by dialysis for 18 hr against 500 ml of 50% glycerol, 50 mM NaCl, 20 mM Tris-HCl (pH 8.5), and 14 mM 2-mercaptoethanol. Enzyme is stored at $-20°$ (fraction VI).

*Comment. Hin*dII endonuclease fraction VI is free of detectable exonuclease but may contain traces of *Hin*dIII activity. Effects from the latter can be minimized by controlling reaction conditions and avoiding excessive digestion. Total yield is about 10,000 units and specific activity is $>$ 10,000 units/mg. *Haemophilus influenzae* Rc contains the isoschizomer *Hin*cII as its only restriction activity and is a better source of the *Hin*dII cleavage specificity than the Rd strain.[3]

Properties

*Enzyme Stability. Hin*dIII is stable for at least 2 years at $-20°$. Under reaction conditions it retains activity for hours at 37° and for several minutes at 65°. Activity is $>$ 20% after 10 min at 65° in the absence or presence of DNA substrate. Some loss of *Hin*dII activity has been noted in preparations stored for 6 months at $-20°$. *Hin*dII is relatively stable in 37° reactions but is $>$ 95% inactivated after 1 min at 65°.

Substrate Specificity. Both enzymes require double-stranded DNA for cleavage. *Hin*dIII cleaves the sequence (5′) A-A-G-C-T-T,[4] and *Hin*dII cleaves the sequence (5′) G-T-Py-Pu-A-C.[5] Cleavage produces 5′-phosphoryl and 3-hydroxyl termini.

*Divalent Cation Requirements. Hin*dII endonuclease requires 5–10 mM Mg^{2+} for activity; Mn^{2+}, Ca^{2+}, and Co^{2+} cannot substitute for Mg^{2+}. *Hin*dIII endonuclease requires 5–10 mM Mg^{2+} or Mn^{2+} for activity; Ca^{2+} and Co^{2+} do not confer activity. Neither enzyme shows any activity in the presence of excess EDTA. *Hin*dIII is reported to have an altered specificity in the presence of Mn^{2+}.[6]

Ionic Requirements. Ionic and pH requirements have been evaluated semiquantitatively by gel assays. *Hin*dIII is optimally active in 30–60 mM NaCl, retains some activity at 0.125 M NaCl, but is strongly inhibited in 0.25 M NaCl or in the absence of NaCl. On the other hand, *Hin*dII is optimally active in the absence of added NaCl, is noticeably inhibited in 30 mM NaCl, and is strongly inhibited in 0.125 M NaCl.

*pH Optimum. Hin*dII is optimally active in the range of pH 7.5–8.0. *Hin*dIII is optimally active in the range of pH 8.0–9.0. *Hin*dIII activity is still appreciable at pH 10.0, but is strongly inhibited below pH 7.0.

[3] A. Landy, E. Ruedisueli, L. Robinson, C. Foeller, and W. Ross, *Biochemistry* **13**, 2134 (1974).
[4] R. Old, K. Murray, and G. Roizes, *J. Mol. Biol.* **92**, 331 (1975).
[5] T. J. Kelly, Jr. and H. O. Smith, *J. Mol. Biol.* **51**, 393 (1970).
[6] M. Hsu and P. Berg, *J. Biol. Chem.* **17**, 131 (1978).

[14] Purification and Properties of the *Bsp* Endonuclease

By PÁL VENETIANER

Assay of the Enzyme

The most convenient assay is the digestion of λ phage DNA and the analysis of the products by agarose gel electrophoresis as described by Dunn and Sambrock [55] in this volume. One unit of enzyme is able to cleave completely 1 μg of λ phage DNA in 1 hr at 37°.

Reagents

Tris-HCl, pH 8, 125 mM, containing 250 mM NaCl and 100 mM MgCl
Enzyme, diluted in 0.05% bovine serum albumin
λ phage DNA, 200 μg/ml
50% sucrose, 0.2 M EDTA, 0.1% bromphenol blue

Procedure. Five microliters of λ phage DNA, 5 μl assay buffer, and the appropriately diluted enzyme are incubated at 37° for 1 hr in 25 μl final volume. Digestion is stopped by the addition of 5 μl EDTA–sucrose–bromphenol blue, and the mixture is subsequently electrophoresed. During the purification procedure much shorter incubation times can be used (3–5 min), and any other well-defined DNA molecule can serve as substrate. Owing to the very high content of the enzyme in *Bacillus sphaericus,* the assay can be carried out even in crude extracts (several thousandfold dilution and short incubation).

Preparation of the Enzyme

The restriction endonuclease of a laboratory isolate of *Bacillus sphaericus* has been discovered and partially characterized in the author's laboratory.[1] It recognizes and cleaves the sequence GGCC, thus it is an isoschisomer of the *Hae*III, *Bsu*I, *Blu*II, *Hhg*I, *Sfa*I, and *Pal*I enzymes. The enzyme had been purified to electrophoretic homogeneity.[2] The procedure described here is a modification of the published method which by applying PEI precipitation (Pirrotta and Bickle [11] this volume) instead of BioGel chromatography allows scaling up of the purification.

Growth Conditions. Bacillus spaericus R cells are grown in nutrient

---

[1] A. Kiss, B. Sain, É. Csordás-Tóth, and P. Venetianer, *Gene* **1**, 323 (1977).
[2] C. Koncz, A. Kiss, and P. Venetianer, *Eur. J. Biochem.* **89**, 523 (1978).

broth (5 g Bacto-yeast extract, 10 g Bacto-tryptone, 5 g NaCl, and 5 g glucose per liter) at 30° to early stationary phase. The strain does not grow in minimal medium. Although at 37° growth is faster, the final yield is much higher at lower temperature, therefore, 30° is recommended. After centrifugation the cell paste can be stored at −20°.

Crude Extract. Cells are suspended in 2 volumes of 0.02 *M* Tris-HCl, pH 8, 0.1 m*M* EDTA, and 0.01 *M* 2-mercaptoethanol and sonicated in 20 g portions (20 × 30 sec) under careful cooling. The extract is then centrifuged for 1 hr at 105,000 *g*.

Polyethyleneimine Precipitation. Solid NaCl is added to the supernatant to 0.1 *M*. A 10% (w/v) solution of polyethyleneimine (PEI) (Polymin P) adjusted to pH 8 is added dropwise to the extract with constant stirring for 30 min to a final concentration of 0.7%. It is advisable to titrate the required amount of PEI on a small sample. Optimal yields are obtained when approximately 60% of the total protein precipitates with the nucleic acids. In our hands this required 0.7%, but it may vary with the batch of PEI used and the protein content of the extract. After centrifugation (10 min, 10,000 *g*) the precipitate is discarded. Solid ammonium sulfate is added to the supernatant to 0.7 saturation with stirring for 30 min. After centrifugation the supernatant is discarded, and the precipitate is dissolved in a minimal amount of PC buffer (0.01 *M* potassium phosphate, pH 7.5, 0.01 *M* 2-mercaptoethanol, 0.1 m*M* EDTA). It is then equilibrated with the same buffer by extensive dialysis, or preferably by passing through a Sephadex G-25 column.

Phosphocellulose Chromatography. The dialyzed enzyme is applied to a 2 × 25 cm Whatman P 11 phosphocellulose column equilibrated with PC buffer. This column size is sufficient for preparations starting from up to 100 g cell. After washing with PC buffer, elution is carried out with a 600-ml linear gradient of 0–0.7 *M* KCl in PC buffer. Activity appears in a broad peak between 0.3 and 0.45 *M* KCl.

Hydroxylapatite Chromatography. Active fractions pooled from the previous step are dialyzed against HA buffer (0.02 *M* potassium phosphate, pH 7.5, 0.1 *M* KCl, 0.1 m*M* EDTA, 0.01 *M* 2-mercaptoethanol) and applied to a 1.5 × 15 cm hydroxylapatite column[3] equilibrated with the same buffer. After washing, a linear 300 ml gradient is applied (0.02–0.3 *M* potassium phosphate in 0.1 *M* KCl, 0.1 m*M* EDTA, 0.01 *M* 2-mercaptoethanol) and the peak of activity elutes between 0.08 and 0.12 *M* potassium phosphate.

DNA-Agarose Chromatography. Pooled active fractions are dialyzed against PC buffer and applied to a 1.5 × 15 cm DNA–agarose column (prepared from commercial chicken blood DNA according to Schaller *et*

[3] W. W. Siegelman, G. A. Wieczorek, and B. C. Turner, *Anal. Biochem.* **13**, 402 (1965).

al.[4]). After extensive washing (at least 10 column volumes) elution is carried out with a linear 300 ml gradient of 0–0.5 M KCl in PC buffer. Active enzyme elutes in a broad peak between 0.24–0.35 M. Pooled fractions are concentrated by dialysis against 55% glycerol, containing 0.01 M potassium phosphate, pH 7.5, 0.1 mM EDTA, and 0.01 M 2-mercaptoethanol.

Properties of the Enzyme

Molecular Weight. The enzyme purified as just described appeared to be homogenous on both nondenaturing and SDS-polyacrylamide gel electrophoresis (95% pure) with a mobility corresponding to 35,000 MW. As on gel chromatography on Sephadex G-150, active enzyme was eluted in a position corresponding to the same molecular weight, it seems likely that the enzyme does not have a subunit structure.

Reaction Conditions. The pH optimum of the enzyme is between pH 8 and 8.5 but some activity can be detected even at pH 5 and pH 10.

The presence of Mn^{2+} (optimally at 0.02 M) is absolutely required; Mn^{2+} can partially replace Mg^{2+}. The optimal Mn^{2+} concentration is 0.01 M. At Mn^{2+} concentrations higher than 0.02 M the enzyme appears to loose its specificity. Monovalent cations are not required and do not inhibit up to 0.1 M; Zn^{2+} is a strong inhibitor of the enzyme.

Stability. The enzyme is remarkably stable throughout the purification procedure. When stored in 50% glycerel at $-20°$ no loss of activity was observed for 1 year. In dilute solutions the enzyme is rapidly inactivated at 37° even in the presence of substrate. This loss of activity can be prevented by the presence of bovine serum albumin (0.01%). In the presence of serum albumin 20–40% of the activity survives lyophilization.

Yield. As seen in Table I, *Bsp* can be prepared in an extremely high yield. One gram of cells contains approximately 2×10^6 units of enzyme, this is 30-fold higher than reported for *Eco*RI. It can be calculated that this does not mean a much higher amount of enzyme but a higher specific activity. The estimated turnover number of *Bsp* is 200 cleavages per minute at 37° (versus 4 cleavages per minute for *Eco*RI).[5]

Substrate Specificity. *Bsp* recognizes and cleaves the sequence GGCC in double-stranded DNA. The 5′ terminus of the cleavage products is C, but it has not been established whether the terminus is "flush" or "staggered" (in the latter case the single-stranded end could be only two nucleotide long). Similarly to *Hae*III and several other restriction en-

[4] H. Schaller, C. Nüsslein, F. J. Bonohoeffer, C. Kurtz, and I. Nietzschann, *Eur. J. Biochem.* **26**, 474 (1972).
[5] P. Modrich and D. Zabel, *J. Biol. Chem.* **251**, 5866 (1976).

TABLE I

PURIFICATION OF *Bsp* ENDONUCLEASE[a]

Fraction	Volume (ml)	Enzyme (units × 10⁻⁶)	Protein (mg)	Specific activity (unit/mg)	Purification	Yield (%)
1. Crude extract	110	110	3630	3×10^4	1	100
2. PEI supernatant	43	83	1350	6×10^4	2	75
3. Phosphocellulose	120	72	24	3×10^6	100	65
4. Hydroxylapatite	37	60	7.3	8×10^6	270	54
5. DNA–agarose	48	20	0.4	4.6×10^7	1600	18

[a] Starting with 50 g of cell paste.

donucleases,[6] *Bsp* can cleave also single-stranded DNA at GGCC sequences but cleavage of single-stranded DNA requires approximately 100-fold higher amounts of enzyme. It is interesting to note that an unmodified DNA strand can be cleaved even if it is in a heteroduplex with a specifically modified (thus in itself resistant) complementary strand.[2]

[6] R. W. Blakesley and R. D. Wells, *Nature* (*London*) **257**, 421 (1975).

[15] Purification and Properties of the *Bsu* Endonuclease

By SIERD BRON and WOLFRAM HÖRZ

Introduction

For various reasons *Bacillus subtilis* is an attractive organism to study restriction and modification. It is easy to grow and to manipulate, has a fairly detailed genetic map, and, most important, possesses a simple transformation/transfection system. This enables the introduction of biologically active DNA into cells of defined R-M[1] background, and the analysis of its fate. In addition, the effects of *in vitro* treatments of biolog-

[1] Abbreviations: ATP, adenosine triphosphate; DTT, dithiothreitol; 2-ME, 2-mercaptoethanol; moi, multiplicity of infection; PFU, plaque-forming units; SAM, *S*-adenosylmethionine; SDS, sodium dodecyl sulfate; SSC, 150 mM NaCl + 15 mM trisodium citrate; R-M, restriction and modification system; r⁺m⁺, restricting and modifying phenotype; r⁻m⁻, nonrestricting and nonmodifying phenotype.

METHODS IN ENZYMOLOGY, VOL. 65

ically active DNA with restriction and modification enzymes can conveniently be studied by means of transformation and transfection.[2,3]

Restriction and modification in *B. subtilis* were demonstrated some years ago.[2,3] A site-specific type II restriction endonuclease, *Bsu*,[4] has been purified and characterized[3,5,6] from *B. subtilis* strain R.[7] The enzyme cleaves susceptible DNAs in the middle of the tetranucleotide sequence 5' GGCC 3'.

Resistance to *Bsu* is conferred to DNA by the modification methylase from the same strain,[8] which methylates the internal C residues of the recognition sequence.[9] We will describe the purification and properties of *Bsu*.

Assays

Two different methods can be conveniently used to assay *Bsu* during the successive purification steps. One involves the inactivation of transfecting bacteriophage DNA, the other, gel electrophoresis of DNA fragments (see also this volume, Section I, articles [2] and [3]).

TRANSFECTION ASSAY. The assay is based on the observation that *Bsu* rapidly destroys the transfecting activity of nonmodified SPP1.0 DNA, but not that of modified SPP1.R DNA.[3] The inactivation follows single-hit kinetics, indicating that one cut per transfecting molecule suffices to inactivate its biological activity. The assay is, therefore, very sensitive and, also because of its specificity toward nonmodified DNA, allows the de-

[2] T. A. Trautner, B. Pawlek, S. Bron, and C. A. Anagnostopoulos, *Mol. Gen. Genet.* **131**, 181 (1974).

[3] S. Bron, K. Murray, and T. A. Trautner, *Mol. Gen. Genet.* **143**, 13 (1975).

[4] According to the nomenclature proposed by H. O. Smith and D. Nathans [*J. Mol. Biol.* **81**, 419 (1973)] the full designation should be "*Bsu*RI." We will use the shorthand "*Bsu*."

[5] S. Bron and K. Murray, *Mol. Gen. Genet.* **143**, 25 (1975).

[6] K. Heininger, W. Hörz, and H. G. Zachau, *Gene* **1**, 291 (1977).

[7] *Bacillus subtilis* strain R was obtained from Dr. Delaporte, Institut Pasteur, Paris, who isolated it as an air contaminant. The strain has been referred to as X5 by K. Murray and R. W. Old [*Prog. Nucleic Acid Res. Mol. Biol.* **14**, 117 (1974)]. *Bacillus subtilis* R is neither transformable nor transfectable. However, its DNA transforms the commonly used competent *B. subtilis* 168-derived strains as efficiently as homologous 168-type DNA, indicating that the two strains are very closely related. This enabled the transfer of the r⁺m⁺ phenotype from R to 168 cells,[2] which are normally r⁻m⁻. These transformed 168 r⁺m⁺ cells maintain their transforming and transfecting properties. *Bsu* can equally well be purified from the original as from the transformed 168 cells; no enzyme is detectable in nontransformed 168 cells.

[8] U. Günthert, B. Pawlek, and T. A. Trautner, *in* "Modern Trends in Bacterial Transformation and Transfection" (A. Portolés, R. López, and M. Espinoza, eds.), p. 249. Elsevier, Amsterdam 1977.

[9] U. Günthert, K. Storm, and R. Bald, *Eur. J. Biochem.* **90**, 581 (1978).

tection of enzyme in crude cell extracts. The single-hit kinetics of inactivation provides a basis for the definition of unit of enzyme activity.

One unit of activity is defined as the amount of enzyme required to inactivate 1 μg of SPP1.0 DNA, at a concentration of 10 μg/ml in enzyme buffer, to 37% residual transfecting activity in 5 min at 37°. We will refer to this unit as "transfection unit of activity."

Two types of assay are described: one for the quantitative determination of enzyme concentration, and a more simple one for the qualitative detection of *Bsu*.

Reagents and other Requirements. All chemicals used are analytical grade. Solutions are prepared in sterile distilled water. Test tubes used for incubations with enzyme are silicone-coated.

Tris (hydroxymethyl)methylamine-HCl buffer, 1 M, pH 7.4

EDTA (disodium salt), 0.2 M, pH 7.4

TY medium: trypton (Difco), 10 g/liter; yeast extract (Difco), 5 g/liter; 0.1 mM MnCl$_2$; and 20 mM MgCl$_2$

TY agar: TY medium with 1.5% (w/v) Bacto-agar (Difco)

Minimal medium (per liter): (NH$_4$)$_2$SO$_4$, 2 g; K$_2$HPO$_4$, 14 g; KH$_2$PO$_4$, 6 g; trisodium citrate, 1 g; MgSO$_4$ · 7H$_2$O, 0.2 g; casamino acids (Difco), 0.2 g; and glucose, 5 g. The medium is supplemented with required growth factors

Enzyme buffer: 10 mM Tris-HCl, pH 7.4, 10 mM MgCl$_2$, 150 mM NaCl, 1 mM DTT, stored at $-20°$. This is the buffer used in standard conditions (see discussion of standard conditions for enzyme activity under "Properties" below)

TBT buffer: 100 mM Tris-HCl, pH 7.4, 10 mM MgCl$_2$, 100 mM NaCl

Competent r$^-$m$^-$ *B. subtilis* cells, e.g., strain 1G-20 [trpC$_2$, r$^-$m$^-$ (Bron and Venema)[10]]

Indicator cells for phage SPP1, e.g., strain 1G-20 (trpC$_2$, r$^-$m$^-$)

SPP1.0 DNA, 10 μg/ml, in enzyme buffer

SPP1.R DNA, 10 μg/ml, in enzyme buffer

Pancreatic DNase (BDH), 40 μg/ml, in 0.1 M MgCl$_2$

Preparation of Reagents and Requirements. The following regimen is routinely used to prepare competent cells.[11] Overnight cultures of strain 1G-20 (trpC$_2$, r$^-$m$^-$)[12] in minimal medium are diluted sevenfold with the same medium and grown for 3 hr at 37° with moderate aeration. The culture is then diluted twofold with minimal medium from which growth factors and casamino acids are omitted. Maximal competence is reached

[10] S. Bron and G. Venema, *Mutat. Res.* 15, 1 (1972).

[11] Any other competence regimen for *B. subtilis* will also be adequate.

[12] Any other transformable/transfectable r$^-$m$^-$ *B. subtilis* strain is adequate.

after about 2 hr of further incubation at 37° with vigorous aeration. Cells can be kept competent by adding glycerol (10%, v/v) to the cultures, which are divided in small portions and subsequently frozen and stored at −80°. Freezing and subsequent thawing usually results in a two- to tenfold reduction of transfection efficiency.

Indicator cells for SPP1 are prepared by diluting overnight cultures of sensitive strains, e.g., 1G-20 (*trp* C_2, r^-m^-), in TY medium fivefold prior to use. Stocks of overnight cultures can be stored in small portions at −80° in 10% (v/v) glycerol.

Lysates of phage SPP1.0 and SPP1.R are routinely prepared in TY medium by infecting exponentially growing cultures (approximately 2×10^7 ceils/ml) of strains 1G-20 (r^-m^-) and R (r^+m^+), respectively, at a moi of about 0.01. Lysates obtained after 4 to 6 hr at 37° with vigorous aeration, contain approximately 10^{10} PFU/ml. Phage is purified and concentrated by differential and isopycnic centrifugation.[3] DNA is subsequently extracted twice with freshly tetraborate-washed (or distilled) phenol at neutral pH and dialysed extensively against SSC. Starting from about 10^{13} PFU, 1 ml stock solution, containing 500–1000 μg DNA is usually obtained, which is stored at 4°.

Procedure Qualitative Assay. This assay is routinely used to detect *Bsu* in enzyme fractions, and is particularly useful to monitor the enzyme during the successive purification steps. Many assays can be carried out simultaneously.

About 2 hr before use, stock solutions of SPP1.0 and SPP1.R DNA are diluted to 10 μg/ml in enzyme buffer. In small test tubes 50-μl portions of each DNA are mixed at 0° with 10-μl portions of suitably diluted enzyme fractions (dilutions in enzyme buffer). The mixtures are incubated for 15 min at 37° in a water bath shaker.[13] Controls in which enzyme buffer replaces the enzyme fractions are treated identically. Then 0.5 ml of competent r^-m^- cells are added and the mixtures are incubated at 37° with shaking for another 40 min. Transfectants are subsequently scored on TY plates using 1G-20 (r^-m^-) indicator cells. Plates are routinely incubated overnight at 37°. If desired, the outcome of the assay can be judged after 5 to 6 hr, if the plates are incubated at 42°. Residual transfecting activities of both types of DNA are determined. Those fractions reducing the transfecting activity of SPP1.0 DNA, but not, or to lesser extent, that of SPP1.R DNA, contain the enzyme.

Quantitative Assay. About 2 hr before use, SPP1.0 and SPP1.R DNA are diluted to 10 μg/ml in enzyme buffer. Portions of 450 μl of each DNA in small test tubes are put for 5 min in a water bath shaker at 37° for temperature equilibration. Then 50-μl enzyme fractions, appropriately di-

---

[13] There is no need to stop the restriction reaction in this particular assay.

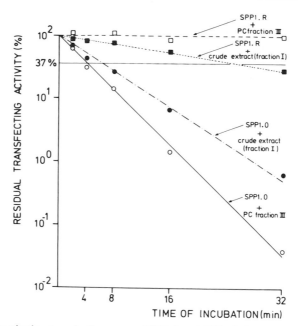

FIG. 1. Quantitative transfection assay. SPP1.0 and SPP1.R DNA were incubated for various times with 1000-fold diluted crude extract (fraction I) or 2000-fold diluted phosphocellulose fraction III. Details are described in the text.

luted with enzyme buffer, are added to both DNAs to start the reaction (zero time). Samples of 50 μl are withdrawn after 1, 2, 4, 8, 16 and 32 min, respectively, and immediately put at 68° for 10 min to inactivate the enzyme. Zero time control samples are prepared in separate test tubes as follows: 50 μl of diluted enzyme fraction is first heated for 10 min at 68° and then mixed with 450-μl DNA solution at 0°. A 50-μl sample is subsequently heated at 68° for another 10 min. All heated samples are chilled in ice and 0.5 ml of competent 1G-20 (r^-m^-) B. subtilis cells are added to each tube. Transfection is allowed for 40 min at 37° in a water bath shaker. Further uptake of DNA by the cells is then prevented by the addition of DNase to 8 μg/ml (2 min, 37°). Transfectants are scored as indicated in the qualitative assay. Residual transfecting activities of both DNAs are plotted semilogarithmically as a function of time of incubation. A typical example is given in Fig. 1. Straight inactivation curves are obtained, from which the time required to reach the 37% level of inactivation can easily be read. After correcting for the effects of nonspecific nucleases, as measured with SPP1.R DNA, the net activity of Bsu in the enzyme fraction follows from the definition of unit of enzyme activity.

A schematic presentation of both assays is given below.

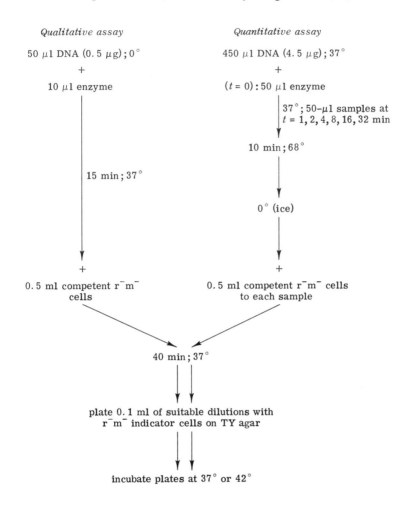

GEL ELECTROPHORESIS ASSAY

In this assay the cleavage pattern of phage or plasmid DNA obtained with *Bsu* is analyzed on agarose or polyacrylamide gels. During purification, the enzyme is conveniently monitored using SPP1 DNA: non-modified SPP1.0 DNA is cleaved into approximately 80 fragments,[3] whereas SPP1.R DNA remains intact. For quantitative determinations of enzyme concentration, DNA substrates which are cleaved into fewer fragments are to be preferred. We have chosen plasmid λdvl DNA for this purpose, which is cleaved into 17 fragments.[6] Examples of standard limit digests of λdvl and SPP1.0 DNA are shown in Fig. 6 (track a) and Fig. 7 (track a), respectively.

Reagents

10 × concentrated enzyme buffer: 0.1 M Tris-HCl, pH 7.4, 0.1 M MgCl$_2$, 1.5 M NaCl, 10 mM DTT, stored at 4°

10 × concentrated electrophoresis buffer: 0.4 M Tris(hydroxymethyl)methylamine, 0.2 M sodium acetate, 20 mM EDTA, pH 7.8, stored at 4°

EDTA (disodium salt), 0.2 M, pH 7.4

Agarose, 2% (w/v) in electrophoresis buffer

Polyacrylamide, 4% (w/v) in electrophoresis buffer

Ethidium bromide, 1 mg/ml

Bromphenol blue–glycerol–SDS mixture; 0.05% (w/v) bromphenol blue, 50% (v/v) glycerol, 2% (w/v) SDS (specially pure)

SPP1.0 and SPP1.R DNA, 300 μg/ml in 1 mM Tris-HCl, pH 7.4, 0.1 mM EDTA

λdvl DNA (Boehringer, Mannheim), 300 μg/ml in 1 mM Tris-HCl, pH 7.4, 0.1 mM EDTA

Procedure. A series of assay mixtures, 50 μl each, is prepared by mixing 5 μl 10 × concentrated enzyme buffer, 10 μl DNA solution (3 μg of DNA), 5 μl enzyme diluted to different degrees in enzyme buffer, and 30 μl water. After incubation at 37° for 1 hr in a water bath shaker, 5 μl EDTA is added (final concentration 20 mM). Five microliters of the bromphenol blue–glycerol–SDS mixture are then added and the samples are analyzed by electrophoresis in 4% (w/v) polyacrylamide, or 2% (w/v) agarose gels. Gels are immersed for 1 hr in ethidium bromide solution (2 μg/ml) after electrophoresis, and then transilluminated with long-wave ultraviolet light and photographed on Ilford FP4 films using a B + W 4x red filter.

One unit of activity ("gel unit") is defined as the amount of enzyme required to completely digest 1 μg of λdvl DNA in 1 hr at 37° under the standard conditions described above.

To convert "gel units to "transfection units" of activity, the following relationship can be used: 1 "gel unit" is approximately 200 "transfection units."

Purification Procedure

Overnight cultures of *B. subtilis* R in TY medium are diluted 100-fold with 50 liter fresh TY medium and grown for 4 to 6 hr at 37° with vigorous aeration to late log phase (approximately 5×10^8 to 10^9 cells/ml). Cells are harvested by centrifugation and stored at $-20°$. Yields are about 2 to 3 g wet cell paste per liter culture.

All subsequent operations are carried out at approximately $0°$ to $4°$. Enzyme fractions are collected in silicone-coated tubes.

Step 1. Preparation of Crude Extract. A 70-g portion of frozen cell paste is thawed in 70 ml buffer A (50 mM Tris-HCl, pH 8.4, 2 mM 2-ME, 0.1 mM EDTA). Acid-washed glass beads (0.18–0.25 mm diameter), 100 g, are added and the mixture is homogenized in an ice–water-cooled high-speed blender (Virtis "45," or Sorval Omnimixer) five times for 1 min at maximal speed, with 1-min cooling intervals. Further disruption is achieved with an ultrasonic disintegrator (MSE) four times for 30 sec at maximal output and cooling in ice water. Remaining cells,[14] debris, and glass beads are removed by centrifugation (30 min, 10,000 g). The supernatant is adjusted to 1 M KCl and subjected to high-speed centrifugation (Beckman, rotor Ti-50, 3 hr at 45,000 rpm) to sediment small cellular particles and the majority of chromosomal DNA. The supernatant, 79 ml, is collected (fraction I, crude extract).

Step 2. DEAE–Cellulose Chromatography. This step is mainly aimed to remove nucleic acids from the crude extract.[15] In the presence of 0.3 M salt, nucleic acids bind to the column, whereas the majority of proteins flow through.

Fraction I is dialyzed for a short period (4 hr) against four changes of 1 liter buffer B [20 mM Tris-HCl, pH 8.4, 0.1 mM EDTA, 14 mM 2-ME, 0.3 M KCl, and 5% (v/v) glycerol]. A column, 30×2.6 cm, is packed with preequilibrated DEAE–cellulose (Pharmacia, DEAE-Sephacel) and washed with buffer B (600 ml, 30 ml/hr) until pH and conductivity of effluent and buffer are identical. Dialyzed fraction I, 77 ml with 7400 A_{280} units, is applied to the column (30 ml/hr), which is subsequently washed (30 ml/hr) with buffer B. Flowthrough fractions,[16] the $A_{280/260}$ ratio of which should not be less than 1, are collected and pooled (DEAE–cellulose fraction II, 106 ml).

Step 3. Phosphocellulose Chromatography. The pooled fraction II (106 ml; 1500 A_{280} units) is dialyzed for a short period (3 hr) against three changes of 1 liter buffer C [16 mM Na$_2$HPO$_4$ · 7H$_2$O, 4 mM KH$_2$PO$_4$, pH 7.4, 14 mM 2-ME, 0.1 mM EDTA, 0.3 M KCl, 5%(v/v) glycerol] and

---

[14] If desired, further breakage of cells can be achieved by incubating the mixture with egg white lysozyme (200 μg/ml) for 30 min at 37° and sonicating the viscous lysate.

[15] We have occasionally included an ammonium sulfate fractionation step prior to DEAE–cellulose chromatography (*Bsu* precipitates at 50–70% saturation). However, the subsequent purification steps are equally effective without this step.

[16] Preferably the column is started at the end of the first day, the flowthrough being collected overnight.

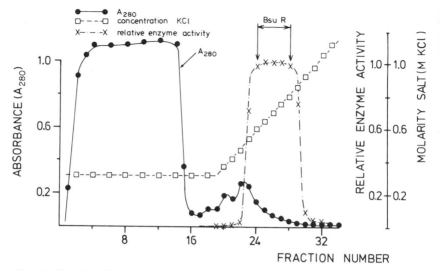

FIG. 2. Phosphocellulose chromatography. Salt concentrations were determined from conductivity measurements. Absorbance (280 nm) was measured in tenfold diluted samples and enzyme activity in 40-fold diluted samples (transfection assay). Relative enzyme activity is expressed as: 1 − residual transfecting activity, in which the residual transfecting activity is the number of transfectants per sample relative to that of the control (no enzyme). Pooled fractions with *Bsu* activity are indicated by arrows.

fractionated on phosphocellulose (Whatman cellulose phosphate P11). A column, 20 × 1.5 cm, is packed with preequilibrated phosphocellulose and washed with buffer C (500 ml, 25 ml/hr) until pH and conductivity of effluent and buffer are identical. The column is loaded with fraction II (20 ml/hr),[17] which is then washed with 50 ml buffer C (20 ml/hr), and eluted with a 200-ml linear gradient of 0.3 to 1.3 M KCl in buffer C (15 ml/hr). Fractions of 10 ml are collected. The elution profile is shown in Fig. 2. The bulk of absorbing (280 nm) material, including nonspecific nuclease activity, flows through in these conditions (0.3 M salt). Active *Bsu* fractions, eluted at 0.6–0.8 M KCl, are pooled (58.5 ml) and concentrated by ultrafiltration (Millipore, Immersible Molecular Separator Kit),[18] followed by dialysis for 6 hr against 20 mM Tris-HCl, pH 7.4, 1 mM DTT, 0.1 mM EDTA, 50%(v/v) glycerol. The concentrated phosphocellulose preparation (fraction III, 1.6 ml), which can be obtained at the end of the third day, is sufficiently pure for many purposes, such as the production of

[17] Loading of the column is possible at the end of the second day, with elution overnight.
[18] Other ultrafiltration systems are probably also adequate. Precipitation with ammonium sulfate (80% saturation) is successful only if carrier protein (such as serum albumin) is added. We have also successfully used small (1 × 1.5 cm) DEAE columns, to which the enzyme binds in buffer without salt and elutes in a small volume in buffer with 0.7 M salt.

DNA digests for gel electrophoresis. In that case the purification is stopped at this stage and the enzyme preparation, in 50% (v/v) glycerol, is stored at $-20°$ or $-80°$.

Step 4. Affinity Chromatography on DNA–Polyacrylamide Columns. If desired, fraction III can be further purified on DNA–polyacrylamide columns.[19] Highly polymerized calf thymus DNA (BDH),[20] 60 mg, is dissolved in 20 ml of 50 mM Tris-HCl, pH 7.4. Then 2 g acrylamide, 0.1 g bisacrylamide, and 0.16 ml TEMED are added at room temperature. This solution is briefly mixed with 20 ml 3% (w/v) agarose in 50 mM Tris-HCl, pH 7.4, at a temperature of 70°. Then 25 mg ammonium persulfate, dissolved in 1 ml water, is added, and the gel material is thoroughly mixed and allowed to set. Solidification occurs within 1 min. After 30 min, the gel is cut into small pieces and squeezed through a 70-mesh stainless steel screen (e.g., Retsch, ASTM 0.212 mm). The resulting slurry is suspended in 200 ml buffer D [20 mM Tris-HCl, pH 7.4, 1 mM DTT, 0.1 mM EDTA, 5% (v/v) glycerol] and left for 30 min to allow settling of the gel by gravity. The supernatant is decanted and the process of suspending and decanting is repeated four times. About 70 ml column material, sufficient for four to six columns, is obtained, which can be stored at 4° in the presence of 0.02% sodium azide.

A column, 10 × 1 cm, is poured from a thick gel slurry and washed with buffer D under a hydrostatic pressure of 20 to 40 cm. The same pressure is used in all subsequent operations, allowing flow rates up to 10 ml/hr. Higher pressures, or the use of peristaltic pumps, may cause too tight packing, resulting in decreased flow rates. The column is stripped of free DNA and possible protein with 50 ml buffer D + 1.0 M KCl (8 ml/hr), followed by 100 ml buffer D + 0.15 M KCl (8 ml/hr). Approximately 10 A_{260} units may be washed out; the final washes with 0.15 M KCl buffer do not contain absorbing (260 nm) material.

About one-third of the phosphocellulose fraction III (0.5 ml, 2.5 A_{280} units) is diluted tenfold with buffer D and 0.15 M KCl and applied to the column at 5 ml/hr. After washing with 20 ml buffer D and 0.15 M KCl (8 ml/hr), the column is eluted at 5 ml/hr with a 80-ml linear gradient of 0.15 to 1.2 M KCl in buffer D and fractions of 5 ml are collected. Figure 3 shows the elution profile; *Bsu* elutes between 0.30 and 0.50 M KCl, after the major absorption (280 nm) peak. Active fractions are pooled (12.5 ml), concentrated by ultrafiltration and dialyzed against buffer D and 50% (v/v) glycerol as indicated for the phosphocellulose fraction. The concentrated

---

[19] The procedure is a modification of the one described by L. F. Cavalieri and E. Caroll, *Proc. Natl. Acad. Sci. U.S.A.* **67**, 807 (1970).

[20] The source of the DNA is not critical. We have also successfully used *B. subtilis* DNA. Even modified DNA is perfect.

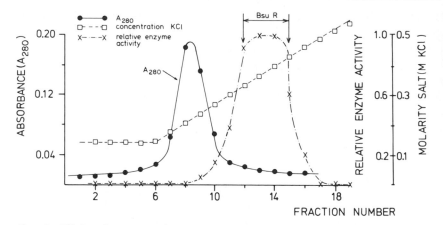

FIG. 3. Affinity chromatography on DNA-polyacrylamide. Absorbance (280 nm) was measured in undiluted samples. Salt concentration and enzyme activity were determined as described in Fig. 2. Pooled fractions with *Bsu* activity are indicated by arrows. See text for details.

preparation (fraction IV, DNA–column enzyme, 0.55 ml) is stored at $-20°$ or $-80°$.

A summary of the purification is presented in Table I, which shows that the phosphocellulose fraction III is purified approximately 225-fold, and the DNA–column fraction IV 7000-fold.

In addition to the standard procedure, an alternative purification procedure has been described.[3,6] It involves precipitation of *Bsu* from crude extracts with streptomycin sulfate, fractionation with ammonium sulfate, and chromatography on DEAE–cellulose.[3] Enzyme preparations thus obtained are approximately 400-fold purified and almost free from contaminating nucleases. However, they are occasionally slightly contaminated with nucleic acids. Further purification, with removal of nucleic acids, can be achieved by chromatography on hydroxyapatite.[6]

Properties

Purity

Proteins. Figure 4 shows a SDS-polyacrylamide gel of proteins in the various enzyme fractions. *Bsu* constitutes a minor band in the mass of proteins in the crude extract and DEAE-cellulose fraction. About 20 proteins remain in the phosphocellulose fraction, the majority being considerably smaller than *Bsu*. The DNA-column fraction IV is highly purified and virtually homogeneous.

TABLE I
PURIFICATION OF *Bsu*[a]

Fraction	Volume (ml)	Total A_{280} units	Total protein[b] (mg)	Total activity[c] (units)	Specific activity (units/mg)	Recovery (% of fraction I)	Purification[d] factor
I crude extract	79	7400	2940	8.64×10^6	2.93×10^3	100	1.0
II DEAE-cellulose	106	1500	2010	7.99×10^6	3.98×10^3	92.4	1.4
III phospho-cellulose	1.72	7.5	7.9	5.20×10^6	6.63×10^5	60.2	226
IV DNA-column	1.75	<0.1	0.077	1.60×10^6	2.08×10^7	18.5	7100

[a] Entries are calculated assuming that the total of each fraction was carried through all successive steps.
[b] Protein is determined according to O. H. Lowry, N. J. Rosebrough, A. L. Farr, and R. J. Randall, *J. Biol. Chem.* **193**, 265 (1951).
[c] Transfection units of activity.
[d] Purification is given as the relative increase of specific activity.

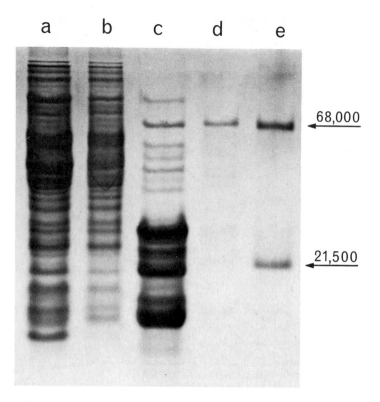

FIG. 4. SDS-polyacrylamide gel electrophoresis of protein fractions. Electrophoresis was as described by U. K. Laemmli [*Nature* (*London*) **227**, 680 (1970)], with a 10% separating and a 3% stacking gel. (a) Crude extract (fraction I), 150 μg total protein. (b) DEAE-cellulose fraction II, 100 μg total protein. (c) Phosphocellulose fraction III, 100 μg protein. (d) DNA-column fraction IV, 1 μg protein. (e) Mixture of reference proteins bovine serum albumin (3 μg) and trypsin inhibitor (5 μg). Reference positions are indicated by arrows.

Partially purified preparations may contain protein contaminants interfering with gel electrophoresis of DNA fragments. Trailing of DNA bands is then observed, which is abolished by the addition of 0.2% SDS to digested samples, or by phenol extraction or treatment with proteinase K of the digests prior to electrophoresis.

Contaminating Nonspecific Nuclease Activities. The phosphocellulose fraction III causes almost no inactivation of SPP1.R DNA under conditions in which SPP1.0 DNA is inactivated to 0.1–0.01% residual transfecting activity. This enzyme fraction also does not degrade SPP1.R DNA detectably in the gel electrophoresis assay. This indicates that under the standard assay conditions for *Bsu* (see discussion of standard conditions of enzyme activity below) the phosphocellulose fraction III is essentially free from nonspecific endonucleases acting on double-stranded DNA. Exonuclease assays are also negative: less than 0.01% of the radioactivity in labeled DNA is rendered acid-soluble under conditions in which the phosphocellulose fraction inactivates the transfecting activity of SPP1.0 DNA to less than 0.01%.

Some phosphocellulose enzyme preparations, particularly those including fractions eluted below 0.6 *M* KC1, are slightly contaminated with a nuclease specific for uracil-containing DNA, such as from phage PBS1.

The phosphocellulose fraction also contains a nuclease activity toward denatured DNA, both modified and nonmodified.

The DNA-column fraction IV is free from any contaminating nuclease activity: none of the above-mentioned activities is detectable in this enzyme preparation under standard conditions.

It should be mentioned that at exceptionally high concentration (20- to 40-fold the amount required to produce a standard limit digest) *Bsu* may lose part of its specificity and may start cleaving at sites different from the normal recognition site (see discussion of relaxation of specificity below). This precludes testing the enzyme at great excess.

Storage and Stability

Enzyme fractions stored at $-20°$ or $-80°$ in the presence of 50% (v/v) glycerol show no significant losses of activity for several years. Storage at $0°$ or $4°$ usually results in a gradual inactivation, particularly in low-salt conditions. At $30°-37°$ the enzyme remains active for at least 1 hr.

Particularly in diluted fractions, the enzyme sticks to glass, resulting in severe losses of activity. Sticking is prevented by using silicone-coated glass. Heating for 10 min at $68°$ inactivates the enzyme irreversibly.

Standard Conditions for Enzyme Activity

Only Mg^{2+} is an essential cofactor (optimal concentration 10 to 20 m*M*).[3] Activity is hardly detectable at concentrations below 0.5 m*M* Mg^{2+}. Neither Ca^{2+} nor Mn^{2+} can replace Mg^{2+} in the reaction. ATP and SAM have no effect.

Bsu is stimulated two- to fivefold by NaCl at concentrations between 0.1 and 0.2 *M*. At higher salt concentrations stimulation decreases, and restriction is prevented at concentrations exceeding 0.5 *M* NaCl. The enzyme is active in the pH range 6.5 to 8.8, with a broad maximum between pH 7.4 and 8.0. In the presence of salt (0.1 to 0.3 M) the reaction is specific for nonmodified DNA over the whole pH range. A change of specificity may occur, however, under special conditions, such as described in the discussion of relaxation of specificity below.

The above results have led to the following standard conditions: 10 mM Tris-HCl, pH 7.4, 10 mM MgCl$_2$, 1 mM DTT and 150 mM NaCl.

Physicochemical Properties

In SDS-polyacrylamide gels (Fig. 4), *Bsu* migrates as a single band to the same position as bovine serum albumin, indicating that the enzyme contains only one type of subunit of molecular weight 68,000. Upon gelfiltration (Sephadex G-150) in the presence of 0.7M NaCl, active enzyme also elutes together with monomeric bovine serum albumin.[21] These results seem to indicate that *Bsu* is a monomeric enzyme with a molecular weight of 68,000. However, upon gelfiltration under low-salt conditions (less than 0.1 M NaCl), a minority of the input enzyme activity elutes at the position of dimeric bovine serum albumin,[21,22] indicating that also dimeric enzyme may exist.

Specificity

Nucleotide Sequence Recognized by Bsu. Under standard conditions *Bsu* recognizes and cleaves the tetranucleotide sequence[5]

$$5' \text{ —G—G—}^{\downarrow}\text{C—C— } 3'$$
$$3' \text{ —C—C—}_{\uparrow}\text{G—G— } 5'$$

Any of the four possible nucleotides can occur in the positions flanking the recognition site. As indicated by arrows, points of DNA strand scission are in the middle of this sequence, so that "even" double-strand breaks are produced. Several other restriction endonucleases, e.g., *Hae*III, *Hhg*I, *Blu*II, *Pal*I, and *Sfa*I, recognize the same sequence (see article [1], this volume).

Under certain conditions, *Bsu* loses its stringent requirement for the tetranucleotide GGCC and, instead, starts cleaving at the GC dinucleotide (see discussion of relaxation of specificity below).

Substrate Specificities. Bsu specifically cleaves nonmodified DNA substrates under standard conditions; modified DNAs are protected (see discussion of relaxation of specificity below for exceptions).

---

[21] S. Bron and E. Luxen, unpublished results (1976).

[22] The majority of the enzyme activity is lost under these conditions. We have no explanation for this phenomenon.

BACTERIOPHAGE DNAs. Nonmodified DNAs from *B. subtilis* phages SPP1, SP02 and φ105 are cleaved into about 80 fragments.[3,5] A limit digest of SPP1.0 DNA is shown in Fig. 7 (track a). Modified DNAs from these phages are resistant to the enzyme.

DNA from several other *B. subtilis* phages, such as φ29, SP01, H1, PBS1, SP50, and SP8, are not cleaved by *Bsu*.[2,3] Methylation of *Bsu* recognition sites cannot be held responsible for this phenomenon.[23] The sensitivity or resistance of phage DNAs *in vitro* is paralleled *in vivo*: only SPP1, SP02 and φ105 are restricted upon infection or transfection of r^+m^+ *B. subtilis*.

In addition to DNAs from several *B. subtilis* phages, DNAs from other phages, such as λ and T7, are also sensitive. About 100 to 200 fragments are produced from λ DNA. Plasmid λdvl DNA, yielding 17 fragments,[6] is useful in enzyme assays and studies on properties of the enzyme. The fragmentation pattern of λdvl DNA is shown in Fig. 6 (track a).

BACTERIAL TRANSFORMING DNA. Transforming *B. subtilis* r^-m^- DNA is inactivated by *Bsu,* the rate of inactivation being about 5 to 10 times lower than of transfecting DNA. Cotransfer of linked markers is more sensitive than single-marker transfer. DNA extracted from r^+m^+ cells is resistant. In contrast to transfecting DNA, nonmodified transforming DNA is *not* inactivated *in vivo* by restricting competent cells.[2,3]

MODIFIED/NONMODIFIED HETERODUPLEX DNA. Artificially produced heteroduplex recombinant molecules consisting of one modified and one nonmodified strand of SPP1 DNA are inactivated both by restricting cells *in vivo*,[2] and by *Bsu in vitro*.[24] Whether the inactivation really involves restriction of heteroduplex molecules is doubtful at present. Recent models for transfection[25] predict the formation of segments of nonmodified homoduplex DNA, also from heteroduplex origin.[26] These nonmodified homoduplices might be the real substrate for restriction *in vivo*.

SINGLE-STRANDED DNA. Single-stranded transfecting SPP1 DNA is restricted by *Bsu in vitro,* and by competent r^+m^+ cells *in vivo*.[27] Figure 5 shows the kinetics of inactivation of native and denatured SPP1 DNA with *Bsu.* Down to levels of 0.1 to 1.0% residual transfecting activity, the efficiency of inactivation of denatured DNA is two- to fivefold lower than of native DNA. Under standard conditions, restriction of single-stranded DNA is specific for nonmodified DNA. Modified single-stranded DNA is

[23] U. Günthert, J. Stutz, and G. Klotz, *Mol. Gen. Genet.* **142,** 185 (1975).
[24] S. Bron and T. A. Trautner, unpublished results (1974).
[25] K. S. Loveday and M. S. Fox, *Virology* **85,** 387 (1978).
[26] The implications of this model for restriction of transfecting homo- and heteroduplex DNA are discussed by S. Bron, E. Luxen, and G. Venema (submitted for publication).
[27] S. Bron, E. Luxen, and G. Venema (submitted for publication).

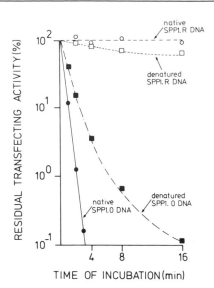

FIG. 5. Kinetics of inactivation of native and denatured DNA. SPP1.0 and SPP1.R DNA were denatured in 0.1 M NaOH and subsequently neutralized with 1 M Na_2HPO_4 solution. Native and denatured DNAs were incubated with 300-fold diluted phosphocellulose enzyme fraction III as described in the text for the quantitative transfection assay of enzyme activity. Transfections with denatured DNA were carried out in the presence of 1 mM EGTA [ethyleneglycolbis(aminoethyl ether)tetraacetic acid].

inactivated to only a very slight extent, albeit somewhat more than modified native DNA. Restriction of denatured SPP1 DNA does not require the presence of both complementary strands: each of the purified heavy (H) or light (L) strands of nonmodified SPP1 DNA is inactivated to the same extent as the mixture of the two strands.[27] Limit digests are not obtained with single-stranded DNA. These results most likely indicate that the mechanism of restriction of single-stranded SPP1 DNA involves locally intrastrand base-paired regions.

EUKARYOTIC DNA. *Bsu* has been used to generate characteristic fragmentation patterns from mouse satellite DNA,[28] the α satellite from African green monkey cells,[29] and guinea pig satellite DNAs.[30] *Bsu* has also proved capable of digesting in a specific manner DNA in mouse and rat liver nuclei.[31]

[28] W. Hörz and H. G. Zachau, *Eur. J. Biochem.* **73**, 383 (1977).
[29] F. Fittler, *Eur. J. Biochem.* **74**, 343 (1977).
[30] W. Altenburger, W. Hörz, and H. G. Zachau, *Eur. J. Biochem.* **73**, 393 (1977).
[31] W. Hörz, T. Igo-Kemenes, W. Pfeiffer, and H. G. Zachau, *Nucleic Acids Res.* **3**, 3213 (1976).

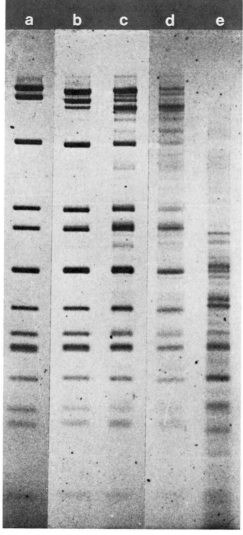

FIG. 6. Effect of *Bsu* concentration and glycerol content on λdvl DNA digests. Three micrograms of λdvl DNA were incubated for 1 hr in 10 mM Tris-HCl, pH 7.4, 10 mM MgCl$_2$, 5 mM 2-mercaptoethanol in a total volume of 20 μl with *Bsu*. (a) 6 U ("gel units"). (b) 12 U. (c) 12 U plus 12% glycerol. (d) 40 U plus 12% glycerol. (e) 80 U plus 12% glycerol. Three units of *Bsu* are required to give a limit digest under these conditions. The fragmentation patterns in a 4% polyacrylamide gel are shown.[6] The resolution is equivalent to that obtained in a 2% agarose gel. Reproduced with permission from Heininger *et al.*[6]

Relaxation of Specificity

The specificities described in the foregoing paragraph are partially lost under certain conditions. When phage or plasmid DNAs are digested with high concentrations of *Bsu,* additional bands may appear upon gel electrophoresis that are not present in standard limit digests.[6] This is shown for λdvl DNA in Fig. 6. Just how much *Bsu* is required to show this additional activity is strongly dependent on the reaction medium. Glycerol stimulates the formation of the extra bands in the *Bsu* digests; NaCl, on the other hand, is inhibitory. The pH exerts a strong influence: the number of additional bands increases up to pH 8.5. At even higher pH values all degradation is strongly suppressed. Taken together, conditions most conducive to this additional *Bsu* activity are 25 mM Tris-HCl, pH 8.5, 10 mM MgCl$_2$, 5 mM 2-ME, and 25% (w/v) glycerol. At these conditions as little as a twofold excess of *Bsu* suffices to generate additional bands, while at standard conditions the first signs of further degradation are observed only at a 20- to 40-fold excess of enzyme. The additional fragments are never produced in stoichiometric amounts; with further increases in enzyme concentration more and more fragments of smaller size are produced, indicating that there is never a plateau value obtained with the characteristics of a new limit digest.

The additional *Bsu* activity is not limited to nonmodified DNAs. It can be conveniently assayed with modified SPP1.R DNA, as is shown in Fig. 7. This DNA is degraded to a heterogeneous mixture of cleavage products, with some indication of discrete bands. The isoschizomeric *Hae*III enzyme does not display a similar activity under the same conditions.

In addition to modified SPP1.R DNA, a number of DNAs from normally resistant bacteriophages, notably hydroxymethyluracil-containing phages, such as H1, φe, SPO1, and SP8, are cleaved under special conditions. Two types of reaction are observed. In the buffer described above these DNAs are, like SPP1 and λdvl DNA, cleaved into a heterogeneous mixture of numerous fragments.[32] However, in standard buffer from which salt is omitted (10 mM Tris-HCl, pH 7.4, 10 mM MgCl$_2$, 1 mM DTT), characteristic limit digests are produced. H1 and φe DNA are cleaved into four fragments[32] and SPO1 DNA into five or six fragments.[33] In this buffer also *Hae*III, which under standard conditions has no effect, cleaves H1 and φe into the same four fragments.

The nucleotide sequences cleaved by the additional *Bsu* activities have only been analyzed for the high pH, high glycerol conditions described above (25 mM Tris-HCl, pH 8.5, 10 mM MgCl$_2$, 5 mM 2-ME, and 25% glycerol).[6] Under these conditions C is still the predominant 5′-terminal

[32] E. C. P. Heisterkamp, J. H. C. Groffen, and S. Bron, unpublished results (1978).
[33] K. Scholtz and T. A. Trautner, personal communication (1977).

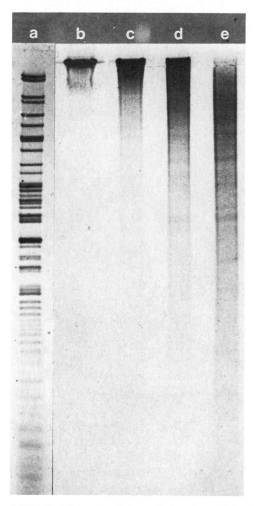

FIG. 7. Digestion of SPP1.R DNA at conditions of relaxed specificity. Two micrograms of SPP1.R DNA were digested for 1 hr in 25 mM Tris-HCl, pH 8.5, 10 mM MgCl$_2$, 5 mM 2-mercaptoethanol, and 25% glycerol in a total volume of 30 μl with *Bsu*. (c) 4 U ("gel units"). (d) 8 U. (e) 16 U. For comparison, the *Bsu* digestion pattern obtained at standard conditions with SPP1.0 DNA is shown in (a), and with SPP1.R DNA in (b). Electrophoresis was carried out on a 4% polyacrylamide gel. Reproduced with permission from Heininger *et al.*[6]

nucleotide of the cleavage products. However, all four possible nucleotides occur in the 5′ penultimate position. It is clear, therefore, that the additional fragments occur as a result of cleavage at the dinucleotide sequence GC rather than of the tetranucleotide sequence GGCC. A comparable reduction in specificity is known for the restriction endonuclease *Eco*R1.[34] The analogy with *Bsu* holds even with respect to the conditions of lower specificity.

One of the implications of the properties of *Bsu* described here is that, when using the enzyme, reaction conditions should be carefully monitored in order to avoid inadvertent loss of specificity. On the other hand, it is difficult to reach a limit digest under conditions of reduced specificity. This complicates, but may not preclude, the use of *Bsu* as a GC-specific nuclease.

[34] B. Polisky, P. Greene, D. E. Garfin, B. J. McCarthy, H. M. Goodman, and H. W. Boyer, *Proc. Natl. Acad. Sci. U.S.A.* **72**, 3310 (1975).

[16] Purification and Properties of the *Bgl*I and II Endonucleases[1]

By Thomas A. Bickle, Vincenzo Pirrotta, and Roland Imber

The presence of two distinct restriction endonucleases in *Bacillus globigii, Bgl*I and II, was first reported by Wilson and Young.[2]

Growth and Storage of Cells

Bacillus globigii can be maintained and propagated on simple media.

The strain is maintained at room temperature in the dark in stab cultures containing (per liter) 7.6 g Bacto agar and 8 g nutrient broth, pH 7.2.

Although the strain will grow in simpler media, we prefer a yeast extract-containing medium for the higher cell densities that it affords. The medium contains (per liter) 10 g Bacto tryptone, 5 g yeast extract (Difco), and 10 g NaCl, pH 7.2. After sterilization, 10 ml of sterile 10% glucose is added per liter. Cultures are grown under aeration at 37°, and cells are harvested in the stationary phase by centrifugation. The cell paste is stored at −70°. The enzymatic activities in cells prepared and stored in this way are stable for at least 1 year.

[1] This work was supported in part by grants from the Swiss National Science Foundation.
[2] G. A. Wilson and F. E. Young, *in* "Microbiology, 1976" (D. Schlessinger, ed.), p. 350. Am. Soc. Microbiol., Washington, D.C., 1976.

Purification of *Bgl*I and II

The purification of the enzymes from 100 g of frozen cell paste will be described. The procedure is applicable, with minor modifications, from 10 g to several kilograms of cells. All operations are performed at 0°–4°.

Cell Breakage. One hundred grams of frozen cells are thawed and suspended in 100 ml of a buffer containing 20 mM Tris-HCl, pH 7.5, 10 mM MgCl$_2$, 0.5 mM EDTA, and 7 mM 2-mercaptoethanol. The suspension is sonicated in an ice-water bath with a Branson Sonifier at maximum power for twenty 30-sec bursts, allowing 2–3 min between bursts for cooling.

The extract is clarified by centrifugation at 12,000 rpm for 20 min in the Sorvall GSA rotor, and then ribosomes and fine cell debris are removed by ultracentrifugation for 2 hr at 40,000 rpm in the IEC A140 rotor.

The supernatant from the high speed centrifugation is made 0.2 M in NaCl and one-tenth volume of a neutralized 10% (v/v) solution of polyethyleneimine is added slowly and with stirring. After 30 min the nucleic acid precipitate is removed by centrifugation at 5000 rpm for 10 min in the Sorvall GSA rotor. Solid ammonium sulfate is added to the supernatant to a final concentration of 70% of saturation (46.3 g ammonium sulfate per 100 ml of supernatant), and the solution is stirred for 30 min after the crystals have dissolved. The protein precipitate is recovered by centrifugation for 30 min at 10,000 rpm.

Heparin–Agarose Chromatography.[3,4] The ammonium sulfate precipitate is dissolved in about 30 ml of chromatography buffer (20 mM Tris-HCl, pH 7.5; 0.5 mM EDTA; and 7 mM 2-mercaptoethanol) and dialyzed for 3–4 hr against this buffer. A small amount of insoluble material is removed by centrifugation and the supernatant is applied to a column (1.6 × 12 cm) of heparin–agarose equilibrated with chromatography buffer at a flow rate of 15 ml/hr. The column is washed with 50 ml of chromatography buffer containing 0.2 M NaCl and eluted with a linear gradient of 250 ml between 0.2 and 0.8 M NaCl. Fractions are collected every 20 min.

Samples of 1 μl from each fraction are assayed by incubating with 1 μg λ DNA for 30 min at 30° followed by electrophoresis through an agarose gel. Digests by the purified enzymes are shown in Fig. 1. The *Bgl*I activity elutes first from the column at about 0.4 M NaCl. It is contaminated with both nonspecific nucleases and *Bgl*II and requires further purification. *Bgl*II activity elutes just after the *Bgl*I, at about 0.5 M NaCl. A substantial part of the *Bgl*II activity is sufficiently free of *Bgl*I and nonspecific nu-

[3] H. Sternbach, R. Engelhardt, and A. G. Lezius, *Eur. J. Biochem.* **60**, 51 (1975).
[4] T. A. Bickle, V. Pirrotta, and R. Imber, *Nucleic Acids Res.* **4**, 2561 (1977).

FIG. 1. Agarose gel electrophoresis of λ DNA digests. (A) *Bgl*II. (B) *Bgl*I. One microgram of phage λ DNA was digested with the purified restriction endonucleases and the reactions were analyzed by electrophoresis on 1% agarose gels.

cleases to be used directly for most purposes, including DNA sequencing, without further purification.

Further Purification of BglI. The *Bgl*I-containing fractions are dialysed against a buffer containing 10% glycerol, 10 mM Tris-HCl, pH 7.5, 100 mM NaCl, 0.5 mM EDTA, and 7 mM 2-mercaptoethanol for 4 hr and loaded on a column of DEAE–Sephacel equilibrated with the same buffer.

The column has the same dimensions as the heparin column and is run in the same buffer under similar conditions except that the NaCl gradient runs between 0.1 and 0.4 *M* salt. *Bgl*I elutes at 0.2–0.25 *M* NaCl. The enzyme obtained by this procedure is free from nonspecific nucleases and *Bgl*II (which elutes earlier from the DEAE column). It is not purified to homogeneity; SDS-polyacrylamide gel electrophoresis of the preparation reveals the presence of three major protein bands of approximate molecular weights 35,000, 45,000, and 60,000.

*Further Purification of Bgl*II. Although the *Bgl*II purified by heparin–agarose chromatography is enzymatically pure, it still contains protein species other than the enzyme. It can be purified to homogeneity by DEAE–Sephacel chromatography exactly as described for *Bgl*I except that the fractions from the heparin–agarose column are dialyzed against a buffer that contains only 50 m*M* NaCl.

Storage of the Purified Enzymes

The fractions containing enzyme are dialyzed for 3–4 hr against a buffer containing 50% glycerol; 10 m*M* Tris-HCl, pH 7.5; 100 m*M* NaCl; 0.5 m*M* EDTA; and 7 m*M* 2-mercaptoethanol and stored at −20°. Both of the enzymes are stable when stored in this way, *Bgl*I for at least 1 year, and *Bgl*II for at least 2 years. For protein chemical studies with highly purified *Bgl*II preparations we avoid the use of glycerol and store the enzyme at 4° in chromatography buffer supplemented with 100 m*M* NaCl. These preparations have retained at least 80% of their activity after 1 year.

Comments on the Purification Scheme

One major problem with the purification of the *Bgl* enzymes is the tendency of both of them to precipitate at low ionic strength. This is much more marked for *Bgl*I than for *Bgl*II. After heparin–agarose chromatography, *Bgl*I will precipitate quantitatively if it is dialyzed against a buffer containing 75 m*M* NaCl or less even in the presence of Triton X-100 at a concentration of 0.2%. Unfortunately, this property of the enzyme cannot be exploited in the purification since the contaminating nonspecific nucleases also precipitate under these conditions. To avoid precipitation we use a relatively high ionic strength (100 m*M* NaCl) for loading the DEAE–Sephacel column. This problem might possibly be avoided by replacing the DEAE step by hydrophobic interaction chromatography in high salt. This possibility is currently being investigated. *Bgl*II will also precipitate after extensive dialysis against low ionic strength buffers, although in this case the precipitation is not quantitative. This is the reason for including 50 mM NaCl in the dialysis buffer. Under these conditions *Bgl*II is fully soluble and many of the contaminating proteins precipitate.

Recently, mutants of *B. globigii* have been described that lack one or the other of the two restriction activities.[5] These strains should simplify the preparation of the two enzymes, perhaps allowing both of them to be purified by one-column procedures.

Properties of *Bgl*I

Until recently all that was known about the sequence cleaved by *Bgl*I was that it occurred mainly in very G + C rich regions and that cleavage resulted in a relatively long 3′ single-stranded protrusion. It was known also that the single site in SV40 DNA was somewhere within the long palindromic sequence that contains the origin of replication and that cleavage produced 5′ CGGC . . . ends (R. J. Roberts personal communication). Three other sites are contained within the DNA of the plasmid pBR322, which has now been completely sequenced.[6] We have mapped these three sites to within 20–50 base pair stretches of pBR322 and have looked for sequences common to these stretches and to the SV40 replication origin and absent from the rest of the genomes. The only such sequence that we find is

$$5′ \text{ GCCNNNN}\overset{\downarrow}{\text{N}}\text{NGGC } 3′$$
$$3′ \text{ CGGN}\underset{\uparrow}{\text{N}}\text{NNNCCG } 5′$$

where N can be any nucleotide. We feel sure that this is the *Bgl*I recognition sequence because it is also found in the position of two *Bgl*I sites in IS2 DNA (D. Ghosal and Sommer, personal communication) and one from adenovirus DNA (R. J. Roberts, personal communication). The same sequence has been independently deduced by W. Fiers and by R. Portmann (personal communications). The sequence consists of a symmetric hexamer with an unusually large (five base pair) nonspecific interruption at the center of symmetry. The enzyme cleaves the sequence at the positions shown by the arrows.

The optimum reaction conditions for digestion are unusual (Table I). The enzyme works best at the relatively high pH of 9.5 and requires an exceptionally high concentration of Mg^{2+} for maximum activity. The Mg^{2+} optimum is lowered by the addition of NaCl (condition B of Table I). We recommend that DNA digestions with this enzyme be carried out in the following buffer: 20 mM glycine-OH, pH 9.5; 20 mM MgCl$_2$; 150 mM

[5] C. H. Duncan, G. A. Wilson, and F. E. Young, *J. Bacteriol.* **134**, 338 (1978).
[6] G. Sutcliffe, *Cold Spring Harbor Symp. Quant. Biol.,* in press.

TABLE I
OPTIMUM REACTION CONDITIONS FOR *Bgl*I AND II

Condition	*Bgl*I (A)	*Bgl*I (B)	*Bgl*II
pH (glycine buffer)	9.5	9.5	9.5
MgCl$_2$ (mM)	40	20	10
NaCl (mM)	-	150	200
Temperature	30°–40°	30°–40°	30°–40°

NaCl; and 7 mM 2-mercaptoethanol at a temperature of 30°. *Bgl*I does not seem to be genetically related to *Bgl*II, since antibodies prepared against the purified *Bgl*II enzyme neither inhibit nor precipitate *Bgl*I.

Properties of *Bgl*II

*Bgl*II recognizes and cleaves the sequence[7]

$$5' \quad \underset{\downarrow}{} \quad 3'$$
$$\text{AGATCT}$$

$$\text{TCTAGA}$$
$$3' \quad \uparrow \quad 5'$$

at the positions shown by the arrows to leave single-stranded 4 base long 5′ protrusions with 5′-phosphates and 3′-OH groups. Fragments produced by this enzyme can, therefore, be ligated together by either *E. coli* or phage T4 DNA ligase. The sequence of the sticky ends produced by *Bgl*II is identical to that of the sticky ends produced by the enzymes *Bam*HI or *Mbo*I and so fragments produced by any of these three enzymes can be ligated together. Moreover, once a *Bam*HI end has become ligated to a *Bgl*II end, the resulting joint is resistant to cleavage by either *Bam*HI or *Bgl*II (although it will still be sensitive to *Mbo*I).

The central tetramer of the *Bgl*II recognition sequence, GATC, is also recognized by the *E. coli dam* DNA methylase which methylates the only A residue in the sequence.[8] *Bgl*II cleaves normally DNA from *E. coli* K12 in which this sequence is methylated. *Bgl*II-specific modification must therefore be on some other base in the sequence.

The purified *Bgl*II protein has a monomer molecular weight of 27,000 by SDS-polyacrylamide gel electrophoresis and behaves as a dimer during equilibrium analytical ultracentrifugation. Antibodies prepared against the purified *Bgl*II protein both precipitate *Bgl*II and inhibit its enzymatic activ-

---

[7] V. Pirrotta, *Nucleic Acids Res.* **3**, 1747 (1976).

[8] S. Hattman, personal communication.

ity. They neither inhibit nor precipitate any other restriction endonuclease that we have tested. These include $BglI$, carried by the same strain, and $BamHI$ which comes from a closely related $Bacillus$ species and which cleaves a related sequence.

The optimum reaction conditions for $BglII$ are shown in Table I. The enzyme has the same high pH optimum as $BglI$ but has a magnesium optimum more typical of other restriction endonucleases. Magnesium can be replaced by manganese (with an optimum at 10 mM) and partial activity can be obtained with a number of other divalent cations, including Ni^{2+} and Zn^{2+}. This last fact may prove useful for preparing partial digests for mapping studies. We have not succeeded in changing the cleavage specificity of the enzyme by changing the reaction conditions as has proved possible for at least two other restriction endonucleases.[9-11] The enzyme is stimulated by NaCl with an optimum at 200 mM. The stimulation is, however, at most twofold and for most purposes we prefer to omit salt from the reaction mixtures.

[9] K. Heininger, W. Hörz, and H. G. Zachau, $Gene$ 1, 291 (1977).
[10] M. Hsu and P. Berg, $Biochemistry$ 17, 131 (1978).
[11] H. Mayer, $FEBS Lett.$ 90, 341 (1978).

[17] Purification and Properties of the Complementary Endonucleases $DpnI$ and $DpnII$

By SANFORD A. LACKS

Strains of $Streptococcus$ ($Diplococcus$) pneumoniae produce one or the other of two sequence-specific endodeoxyribonucleases called endo R·$DpnI$ and endo R·$DpnII$. These enzymes fall into the category of restriction endonucleases with simple cofactor requirements. Although both enzymes recognize and cleave at the palindromic sequence 5' GATC 3' in duplex DNA, the enzymes are complementary inasmuch as $DpnI$ acts only when the adenine residues are methylated and $DpnII$ only when they are not.[1] Thus $DpnI$ is unique among restriction enzymes in that it requires a methylated sequence for action.[2] All other restriction enzymes so far isolated are inhibited by methylation of particular bases in the recognition sequence. As expected, DNA in strains of $S.$ $pneumoniae$ that produce $DpnII$ is appropriately methylated, and DNA in strains that produce $DpnI$ is not methylated in the critical sequence.

[1] S. Lacks and B. Greenberg, $J. Mol. Biol.$ 114, 153 (1977).
[2] S. Lacks and B. Greenberg, $J. Biol. Chem.$ 250, 4060 (1975).

Purification

Nuclease-Deficient Strains. The strains of *S. pneumoniae* that are used as sources of the enzymes lack polysaccharide capsules and are non-pathogenic. To reduce contamination by nonspecific nucleases, the enzymes are prepared from mutant strains deficient in the major pneumococcal deoxyribonucleases.[3,4] *Dpn*I is prepared from strain 641, which carries the mutations *end-1, noz-19,* and *exo-3.* The first two mutations reduce the major endonuclease activity to <1% of the wild type, and the latter reduces the major exonuclease activity to 3%. *Dpn*II is prepared either from strain 679,[5] which contains the *end-14* mutation that reduces endonuclease activity to <1%, or from strain 649, which contains ~10% of the endonuclease activity and ~30% of the exonuclease activity of a typical pneumococcal strain. Strain 649 differs from typical strains of *S. pneumoniae* in its failure to be lysed by detergents and in its resistance to optochin.

Growth of Cultures. The growth medium contains 10 g each of casamino acids, tryptone, yeast extract, brain heart infusion (all Difco products) and 2 g K_2HPO_4 in 1.5 liters of H_2O. It is brought to pH 7.5 with 2.5 ml of $4 N$ NaOH and autoclaved. Before use, 30 ml each of 20% (w/v) sucrose and 0.5 M K_2HPO_4, sterilized separately, are added. A flask of this medium is inoculated with 10^7 cells and incubated at 37°C *without* shaking or aeration for ~12 hr, until the culture reaches an OD_{650} ~1.2. The pH is maintained at ~7.5 by addition of $4 N$ NaOH. After growth the culture is chilled and subsequent operations are carried out in the cold.

Preparation of Extract. The cells are centrifuged and washed by suspension in 150 ml of saline buffer (0.5 M NaCl, 10 mM Tris-HCl, pH 7.5). They are centrifuged again and suspended with 6 ml saline buffer containing 0.1 mM EDTA. The cells are then disrupted by passage through a French pressure cell. Brij-35 is added to 0.1%, and after holding the crude extract for 30 min at 0°, cellular debris is removed by centrifugation. Treatment with Brij-35 helps extract some pneumococcal enzymes, but it has not been determined whether it aids the recovery of endo R·*Dpn*I or II. The volume of the clarified extract obtained by this procedure is ~8 ml. It contains ~60 mg protein per milliliter.

Gel Filtration. Approximately 6 ml of extract are applied to an agarose column (BioGel A-0.5 m), 3.9 × 140 cm, and eluted with saline buffer containing 0.1 mM EDTA. Fractions of 16 ml are collected. The high salt concentration (0.5 M) separates the enzymes from DNA, and this single

[3] S. Lacks, *J. Bacteriol.* **101**, 373 (1970).

[4] S. Lacks, B. Greenberg, and M. Neuberger, *J. Bacteriol.* **123**, 222 (1975).

[5] Dr. H. Bernheimer generously provided several strains of *S. pneumoniae* including HB264, the progenitor of 679 and 697.

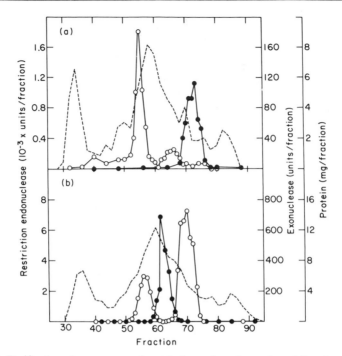

FIG. 1. Purification of pneumococcal restriction endonucleases by gel filtration on BioGel A-0.5 m. (a) DpnI. (b) DpnII. (a) Crude extract of strain 641 containing 322 mg protein applied to column 3.9 × 134 cm. (b) Crude extract of strain 679 containing 526 mg protein applied to column 3.9 × 140 cm. ●, Restriction endonuclease; ○, exonuclease; ---, protein.

fractionation step generally suffices to separate DpnI and DpnII from contaminating nucleases. Fractions are assayed for deoxyribonucleases that yield acid-soluble products as previously indicated[6] and for the restriction enzymes as indicated below.

Figure 1a and b show typical results obtained for the fractionation of DpnI and DpnII, respectively. With respect to the nucleases that release acid-soluble products, the first peak eluted from the column represents exonuclease activity of a DNA polymerase.[3] The second peak is the major exonuclease of the wild type; it is much reduced in the exo-3 mutant (Fig. 1a). When the major endonuclease is present (as in strain 649), it elutes in two peaks: one precedes the polymerase-exonuclease, and the other coincides with the major exonuclease.[4] DpnI is eluted after the second exonuclease peak. DpnII, however, is eluted between the two exonuclease peaks. Fractions sufficiently free from contaminating exonuclease activity

[6] S. Lacks, B. Greenberg, and M. Neuberger, Proc. Natl. Acad. Sci. U.S.A. 71, 2305 (1974).

TABLE I
PURIFICATION OF PNEUMOCOCCAL RESTRICTION ENDONUCLEASES

		Exonuclease	Restriction endonuclease	
Material	Protein (mg)	activity (units)	Activity (units)	Specific activity (units/mg)
*Dpn*I				
Crude extract	322	804	7,300	23
BioGel peak, Fr. 69–75	18.1	39	5,400	298
Peak fraction, Fr. 73	1.9	4	1,100	580
*Dpn*II				
Crude extract	526	5680	45,000	85
BioGel peak, Fr. 60–65	60	68	20,000	333
Peak fraction, Fr. 62	10	3	6,900	690

can usually be obtained. Table I summarizes typical results of purification at this stage.

DEAE–Cellulose Fractionation. Removal of contaminating exonuclease activity from *Dpn*II can be accomplished with DEAE–cellulose. Peak fractions from the agarose column are pooled and dialyzed against 0.01 M Tris-HCl buffer, pH 7.5. It is important to exclude fractions containing polymerase–exonuclease because that enzyme is not separated from *Dpn*II by DEAE–cellulose chromatography. The sample is applied to a 1.0 × 10 cm column of DEAE–cellulose and eluted with a linear gradient of NaCl in buffer. The major exonuclease is eluted at 0.15 M, while *Dpn*II is eluted at 0.20 M, so the peaks are well separated.

*Dpn*I can be similarly purified by DEAE–cellulose chromatography. It, too, elutes at 0.20 M NaCl. However, this enzyme when purified appears to be quite unstable, possibly because of the low protein concentration in the fractions.

Stability. *Dpn*I can be conserved frozen at −20° in the presence of bovine serum albumin at 1 mg/ml for up to 2 years. It loses activity rapidly when stored at 5° or when frozen in the absence of albumin. *Dpn*II is quite stable. It can be kept at 5° for 2 weeks or at −20° for 2 years with little loss of activity.

Assay Methods

Principle. *Dpn*I and *Dpn*II make double-strand breaks in DNA at an average frequency of less than one per 256 nucleotide pairs. Their assay depends on the demonstration or measurement of this fragmentation. Two assay methods will be discussed in detail. In one, gel electrophoresis, the extent of fragment formation is determined by separation of fragments on the basis of size. In the other, transformation assay, the inactivation of biological transforming activity as a result of fragmentation[7] is measured. Additional methods, based on sedimentation and viscosity properties of DNA, have also been used.[2]

Reagents

Buffer: Tris-HCl, 10 mM, pH 7.5

Saline buffer: NaCl, 0.5 M in buffer

Heated albumin, 4%: Dissolve 4 g bovine serum albumin (Armour, fraction V) in 80 ml of water. Bring pH to 3.5 by addition of 4 N HCl. Heat 15 min in a boiling water bath. Cool. Neutralize with 4 N NaOH. Adjust volume to 100 ml. Filter to sterilize

Assay mix: Tris-HCl, 67 mM, pH 7.6; MgCl$_2$, 6.7 mM; heated albumin, 133 μg/ml

Sterile assay mix: As above but with heated albumin, 533 μg/ml, and 2-mercaptoethanol, 4 mM. Filter to sterilize

Electrophoresis sample mix: EDTA, 2 mM; bromphenol blue, 0.75 mg/ml; sucrose, 0.3 g/ml

Gel Electrophoresis. This widely used method for the analysis of restriction enzyme activity has been described in detail.[8-10] It is based on the more rapid electrophoretic migration in polyacrylamide or agarose gels of the fragments of DNA formed by endonucleolytic action. The location of DNA in the gel after electrophoresis is revealed by fluorescence of ethidium bromide bound to the DNA. The method is applicable to the assay of *Dpn*I and *Dpn*II provided that susceptible DNA is used as substrate.[1,2] DNAs from several bacterial species, including *Escherichia coli,* as well as from the *Dpn*II producing strains of *S. pneumoniae,* are susceptible to *Dpn*I (see also section on distribution below). Many bacterial species and mammals contain DNA susceptible to *Dpn*II. Although cellular DNA can be used in this assay, the diversity of fragments produced results in a broad distribution of DNA products in the electrophoregram, which renders quantitation of enzyme activity difficult. Viral DNA substrates, which give rise to a limited number of bands, confer certain advantages: (1) smaller amounts of substrate and enzyme can be used, (2) partial products are distinguishable from final products, and (3) the endpoint can be determined with greater precision. Satisfactory substrates for *Dpn*I are phage λ DNA, although not all potential sites in λ DNA are methylated,[1] and phage f1 replicative form DNA.[11] Both

[7] A. Cato, Jr. and W. R. Guild, *J. Mol. Biol.* **37,** 157 (1967).
[8] P. A. Sharp, B. Sugden, and J. Sambrook, *Biochemistry* **12,** 3055 (1973).
[9] M. W. McDonell, M. N. Simon, and F. W. Studier, *J. Mol. Biol.* **110,** 119 (1977).
[10] A. R. Dunn and J. Sambrook [55], this volume.
[11] G. F. Vovis and S. Lacks, *J. Mol. Biol.* **115,** 525 (1977).

phages are grown in *E. coli.* A good substrate for *Dpn*II is phage T7 DNA (Lacks and Greenberg[2]); even though this phage is grown in ordinary strains of *E. coli,* its DNA is not methylated at GATC sites. Phage λ and f1 (replicative form) DNAs when grown in *E. coli dam-3* mutants are fully susceptible to *Dpn*II, as are also DNAs from mammalian viruses, such as SV40.[2,11]

To prepare samples for gel electrophoresis, 0.5 to 2.0 μg of DNA in 30 μl buffer are placed into centrifuge tubes. Then 150 μl of assay mix are added. Control samples receive 20 μl of saline buffer; others receive enzyme in saline buffer. Samples are incubated for 2 hr at 37°, chilled, and the DNA is precipitated by addition of 1.2 ml of ethanol. After 20 min at 0°, samples are centrifuged, the supernatant fluid is discarded and the precipitate is dried with a stream of air. It is then dissolved in 10 μl of buffer, 5 μl of electrophoresis sample mix are added, and the entire volume is placed in a well in the gel.

Transformation. This method allows a quantitative determination of restriction enzyme activity. Pneumococcal DNA containing a readily detected genetic marker, such as streptomycin resistance, is treated with enzyme and the DNA is then used to transform cells of a streptomycin-sensitive strain of *S. pneumoniae.* To assay *Dpn*I, methylated DNA from strain 697 can be used; to assay *Dpn*II, DNA from strain 645 can be used. Both of these strains are streptomycin resistant. Procedures for preparing biologically active transforming DNA have been described.[12,13]

To sterile tubes containing 0.3 ml of sterile assay mix are added various amounts, up to 20 μl, of enzyme solution. These tubes then receive 0.1 ml of the appropriate DNA, at 0.5 μg/ml, in 0.15 *M* NaCl. The tubes are incubated for 2 hr at 37°, then chilled. Meanwhile, a culture of cells competent for transformation, at a concentration of ~4 × 10[6] CFU/ml, is prepared by diluting a previously frozen culture of strain 606 into fresh medium. Even though strain 606 contains *Dpn*I, it is transformed well by methylated DNA, because the donor DNA is converted to single strands on entry and so is not susceptible to cleavage by *Dpn*I *in vivo.*[11] The medium used for growing transformable cultures has been previously described.[14] Other details of the transformation procedure have also been presented.[1,2,15] In brief, the culture is incubated for 30 min at 30°. Then, 1.6 ml portions are added to the samples of treated DNA, which have been brought to temperature by holding for 2 min at 30°. The transforming mixtures are incubated at 30° for 20 min, at which time pancreatic deoxyribonuclease is added to terminate DNA uptake, and the cultures

[12] R. D. Hotchkiss, this series, Vol. 3 [102].
[13] J. Marmur, this series, Vol. 6 [100].
[14] S. Lacks, *Genetics* **53,** 207 (1966).
[15] R. D. Hotchkiss, this series, Vol. 3 [105].

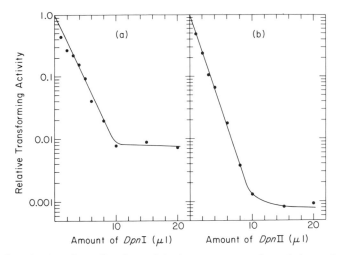

FIG. 2. Inactivation of transforming activity by pneumococcal restriction endonucleases. (a) *Dpn*I. (b) *Dpn*II. Relative transforming activities are indicated for a streptomycin-resistance marker in (a) methylated DNA from strain 697 treated with *Dpn*I and (b) non-methylated DNA from strain 645 treated with *Dpn*II. The enzymes added were BioGel purified fractions diluted tenfold. Without enzyme treatment the DNA in (a) yielded 1.0×10^5 and in (b) 3.8×10^5 transformants per milliliter.

are incubated at 37° for 90 min. At this time cultures are chilled, and the frequency of streptomycin-resistant transformants is determined by plating samples in agar[14] or scoring in liquid medium.[16] The transformed cultures can also be frozen in the presence of 10% (v/v) glycerol[17] and plated on subsequent days.

Inactivation of transforming activity by *Dpn*I and *Dpn*II, respectively, is shown for various amounts of enzyme in Fig. 2a and b. The inactivation is exponential over at least two decades. With increasing amounts of enzyme, the residual transforming activity eventually levels off. Results shown are for a particular streptomycin-resistance marker. Other markers are inactivated at slightly different rates and to different levels, which probably depend on the proximity of the marker to cleavage sites in the DNA.

From the inactivation dependencies shown in Fig. 2 it is possible to determine the relative amount of enzyme present in an unknown sample. When such a sample gives inactivation in the exponential range, the fraction of surviving transforming activity S can be related to the amount of enzyme x by the expression $x = -\ln S/100$, with x in arbitrary units. The unit of activity is based on the amount of enzyme required to inactivate

[16] R. D. Hotchkiss, *Proc. Natl. Acad. Sci. U.S.A.* **40**, 49 (1954).
[17] M. S. Fox and R. D. Hotchkiss, *Nature (London)* **179**, 1322 (1957).

$$\begin{array}{ll}
\text{(a)} & \overset{\text{me}}{5'....GATC....3'} \\
& 3'....CTAG....5' \\
& \qquad\;\;\underset{\text{me}}{}
\end{array} \qquad \begin{array}{l}
5'....GATC....3' \\
\text{(b)}\;\; 3'....CTAG....5'
\end{array}$$

FIG. 3. Recognition sequences for the action of pneumococcal restriction endonucleases. (a) *Dpn*I. (b) *Dpn*II.

the streptomycin-resistance marker to 37% (e^{-1}) survival. The unit is defined as 100 times this amount, so that it is approximately equivalent to the amount of enzyme that can hydrolyze 1 μg of DNA to completion in 1 hr at 37°. This latter amount of activity has been used by others[18] to define a unit of restriction endonuclease. It, thus, corresponds approximately to the unit selected for the transformation assay.

The transformation assay is a useful quantitative tool because it will accurately measure activity in samples that vary in amount over more than a tenfold range. It can be used to measure the restriction enzymes even in crude extracts of pneumococcal strains, provided that the strains lack the major endonuclease.[19] It should be pointed out that the transformation assay can be used to measure the activity of any restriction enzyme.

Properties

The enzymes *Dpn*I and *Dpn*II are distinct molecular entities. From their elution behavior in agarose gel filtration, *Dpn*I appears to have a molecular weight of ~20,000, whereas *Dpn*II has a molecular weight of ~70,000.[19] The enzymes differ, of course, with regard to their substrate specificity, as indicated in Fig. 3. The site of cleavage by *Dpn*I has been shown to be between the methylated adenine and thymine residues on each strand.[20] The precise site of cleavage by *Dpn*II has not been determined. Neither enzyme shows any detectable action on the substrate of the complementary enzyme.

In other properties the enzymes are similar. They both require magnesium ion, at 2 m*M*, for activity. No other cofactors appear to be necessary. Salt concentrations higher than 0.1 *M* are inhibitory. Neither enzyme requires the presence of sulfhydryl compounds. *Dpn*I is much less stable than *Dpn*II, as indicated above, but the reason for this is not known.

[18] R. E. Gelinas, P. A. Myers, and R. J. Roberts, *J. Mol. Biol.* **114**, 169 (1977).
[19] S. Lacks, unpublished (1977).
[20] G. E. Geier and P. Modrich, *J. Biol. Chem.* **254**, 1408 (1979).

The mechanism of action on DNA appears to be similar in the two cases. Neither *Dpn*I nor *Dpn*II will attack heteroduplex DNA in which only one strand is methylated at the recognition site.[11] Not even the appropriately methylated (or unmethylated) strand is broken in the half-methylated substrate. When *Dpn*I or *Dpn*II do act, they appear to cleave both strands simultaneously.[1] No DNA with single-strand breaks appears as an intermediate.

Distribution

The *Dpn*I enzyme is unique in its recognition of a methylated DNA sequence. So far it has been observed only in *S. pneumoniae,* but it has been found in two strains isolated independently from the wild. These strains, incidentally, differed in capsular type.[19] *Dpn*II has been found in a number of independent isolates of *S. pneumoniae.*[21] Enzymes with the specificity of *Dpn*II have been reported also in *Moraxella bovis,*[18] *Moraxella osloensis,*[18] and *Staphylococcus aureus.*[22]

Methylation of adenine in GATC sequences occurs in the DNA of *Dpn*II-producing strains of *S. pneumoniae,* but not in *Dpn*I-producing strains.[1] Strains of other *Streptococcus* species including *sanguis* and *cremoris* were not methylated in this sequence.[19] The sequence is presumably methylated in other bacterial species that produce a *Dpn*II-like enzyme. However, some bacteria, such as *Escherichia coli* and *Haemophilus influenzae* (and *parainfluenzae*), methylate their DNA at this sequence even though they harbor no corresponding nuclease.

Applications

*Dpn*I can be used to test for the occurrence of N^6-methyladenine at GATC sites in any DNA. The extent of fragmentation of the DNA will indicate whether all or only a subset of GATC sequences are methylated. A valuable cross-check to susceptibility to *Dpn*I would be susceptibility to *Dpn*II, because the presence of unusual bases, as in phage T4 DNA, will prevent action of either enzyme whether the DNA is methylated or not.[1] Not only can *Dpn*I determine whether GATC sequences are methylated but, because it cleaves at the methylated sequence, it can be used to localize the sites of methylation in the genome.[11]

*Dpn*II can be used like any restriction endonuclease for sequence-specific cleavage of DNA. Because it cuts phage T7 DNA near the boundary of early and late expressing genes and releases the early fragment intact,[9] it can be useful in studies of this virus.

---

[21] S. Lacks and H. Bernheimer, unpublished (1977).
[22] J. S. Sussenbach, C. H. Monfoort, R. Schiphof, and E. E. Stobberingh, *Nucleic Acids Res.* **3,** 3193 (1976).

[18] Purification and Properties of the *Bam*HI Endonuclease

By GARY A. WILSON and FRANK E. YOUNG

Site-specific endonucleases have been isolated from a diverse variety of microorganisms and used extensively in analyzing and manipulating complex genomes. In fact, the utility of these enzymes has resulted in an emphasis on their application and inadvertently hampered a thorough examination of the properties and biological functions of these endonucleases. For example, the role of site-specific endonucleases *in vivo* has not been determined, although, restriction and modification properties have been confirmed by genetic studies for *Eco*RI,[1] *Eco*RII,[1] *Bam*HI,[2] BsuI,[3] *Bst*1503,[4] *Bgl*I,[5] and *Bgl*II.[5] Although the methods of isolation and optimal reaction conditions have not been systematically examined, most of the enzymes can be readily isolated and purified from contaminating endonucleases and exonucleases.

*Bam*HI proved to be a useful enzyme because it produced a single-stranded end after cleaving between the guanine bases at the site of 5' GGATCC 3',[6] thus providing a useful substrate for ligation. In addition, a number of cloning vectors existed or were developed that contained a single recognition site for *Bam*HI including vectors SV40,[7] pMB9,[8] pBR313,[8] and pBR322.[9] These advantages contributed to the extensive use of this enzyme and to the development of modifications of our original purification procedure[10] that were kindly communicated to us by numerous investigators. This report will summarize the methods of isolation of *Bam*HI and related enzymes from the *Bacillus* genospecies as well as indicate what progress has been made in characterizing these enzymes.

[1] R. Yoshimori, D. Roulland-Dussoix, and H. W. Boyer, *J. Bacteriol.* **112**, 1275 (1972).

[2] G. A. Wilson and F. E. Young, *in* "Microbiology-1976" (D. Schlessinger, ed.), p. 305. Am. Soc. Microbiol., Washington, D.C., 1976.

[3] T. A. Trautner, B. Pawlek, S. Bron, and C. Anagnostopoulos, *Mol. Gen. Genet.* **131**, 181 (1974).

[4] J. F. Catterall and N. E. Welker, *J. Bacteriol.* **129**, 1110 (1977).

[5] C. H. Duncan, G. A. Wilson, and F. E. Young, *J. Bacteriol.* **134**, 338 (1978).

[6] R. J. Roberts, G. A. Wilson, and F. E. Young, *Nature (London)* **265**, 82 (1977).

[7] S. P. Goff and P. Berg, *in* "Recombinant Molecules: Impact on Science and Society" R. F. Beers, Jr. and E. G. Bassett, eds.), p. 285. Raven, New York, 1977.

[8] F. Bolivar, R. L. Rodriguez, M. C. Betlach, and H. W. Boyer, *Gene* **2**, 75 (1977).

[9] F. Bolivar, R. L. Rodriguez, P. J. Greene, M. C. Betlach, H. L. Heyneker, H. W. Boyer, J. H. Crosa, and S. Falkow, *Gene* **2**, 95 (1977).

[10] G. A. Wilson and F. E. Young, *J. Mol. Biol.* **97**, 123 (1975).

Isolation of *Bam*HI

Bacillus amyloliquefaciens RUB 500 is the organism used to produce *Bam*HI. This strain was selected as a spontaneous rifampin-resistant mutant[11] of *B. amyloliquefaciens* H described by Welker and Campbell.[12] It grows well on either minimal salts media,[11,13] or more nutritious media such as Penassay broth, brain heart infusion broth, and Luria broth. However, Bingham and Atkinson[14] report that the strain does not produce endonuclease in Luria broth. Although high yields of the enzyme can be obtained from cells grown in either brain heart infusion broth or Penassay broth, we now routinely use SLBH medium which was communicated to us by H. Boyer. This medium contains per liter: 11 g tryptone, 22.5 g yeast extract, 4 ml glycerol, 51 ml of $1 M$ K_2HPO_4, and 15.7 ml of $1 M$ KH_2PO_4. Salts are autoclaved and added separately. This medium yields 4 to 6 g of cells per liter. Addition of 0.5 ml Polyglycol P2000 (Dow Chemical) per liter reduces foam and permits cells to be grown with rapid aeration at 37°. We have found that it makes little difference on the final amount of enzyme produced if the cells are harvested during logarithmic or early stationary phase of growth. Nevertheless, we routinely harvest cells near the end of logarithmic growth between 200 and 400 Klett units. Following centrifugation, the pellet is suspended in 4 to 6 ml/g wet weight of cells in S buffer (50 mM potassium phosphate buffer, pH 7.0, 1 mM EDTA, 7 mM β-mercaptoethanol, 1 mM NaN_3) containing $1.0 M$ KCl. Legault-Demare and Chambliss[15] demonstrated that this high salt wash removed most of the proteases bound to *Bacillus subtilis*. This wash and all subsequent steps are carried out at 4°.

The cell paste usually consisting of 50 g of wet cells per 100 ml of S buffer can be disrupted by a variety of methods. Although grinding with alumina[10] adequately disrupts the cells, we now generally use sonication (three 30-sec treatments using maximum setting of Bronwill Biosonic III) after the cells have been treated with 100 μg/ml lysozyme for 15 min on ice. Cellular debris is removed by centrifugation for 10 min (Sorvall SS34 rotor, 12,000 rpm) and the turbid supernate clarified by centrifugation for 2 hr at 33,000 rpm in a type 35 ultracentrifuge rotor. The clear, amber supernate is further purified by chromatography on phosphocellulose (Whatman P11). P11 is precycled by suspending the phosphocellulose in 10 volumes of 0.5 N NaOH, washing with H_2O, suspending the phos-

[11] G. A. Wilson and F. E. Young, *J. Bacteriol.* **111,** 705 (1972).
[12] N. E. Welker and L. L. Campbell, *J. Bacteriol.* **94,** 1124 (1967).
[13] F. E. Young and J. Spizizen, *J. Bacteriol.* **81,** 823 (1961).
[14] A. H. A. Bingham and T. Atkinson, *Biochem. Rev.* **6,** 315 (1978).
[15] L. Legault-Demare and G. H. Chambliss, *J. Bacteriol.* **120,** 1300 (1974).

phocellulose in 0.5 N HCl, and washing with S buffer until the phosphocellulose is equlibrated to pH 7.0.

We have often found that a substance in the supernate derived from the cells will form a pellicle on the surface of the P11 if it is applied directly to the column. In order to maximize flow rates, we absorb the supernate to the phosphocellulose by gently mixing the two together in a beaker. The mixture is then poured into a column with a final bed dimension of 2.5 × 40 cm, washed with approximately 250 ml of S buffer and eluted with a 1000 ml linear gradient of 0 to 1.0 M KCl in S buffer at a flow rate of 30 drops/min. *Bam*HI elutes between 0.3 and 0.5 M KCl. Active fractions are pooled, and diluted at least twofold with 1.0 mM EDTA, 1.0 mM NaN$_3$, and 10.0 mM β-mercaptoethanol, and loaded directly on a hydroxylapatite (BioGel HTP catalogue No. 13846) column (1.2 × 8 cm) that has been equilibrated with 25 mM potassium phosphate buffer, pH 7.0, containing 10 mM β-mercaptoethanol, 1 mM EDTA, and 1 mM NaN$_3$. After washing with 100 ml of the same equilibration buffer, enzymatic activity is eluted with a 200 ml linear gradient of 25 to 250 mM potassium phosphate buffer, pH 7.0, in the equilibration buffer. *Bam*HI elutes between 100 and 150 mM potassium phosphate. The active fractions are pooled and dialyzed against 20 mM potassium phosphate buffer, pH 7.0, containing 10 mM β-mercaptoethanol, 1 mM EDTA, 200 mM NaCl, and 50% glycerol. This procedure is essentially the same as one communicated to us by H. Boyer.

Contaminating exonuclease activity can be determined by digesting DNA with samples from column fractions and allowing the reaction to proceed for 24 hr. Because ϕX174 DNA lacks a *Bam*HI site[16] it is particularly useful for detecting nucleases that can introduce single-strand breaks or "nicks." We have found that there are no other nucleases present if one selects only the most active fractions from chromatography on hydroxyapatite.

Smith and Chirikjian[17] have purified *Bam*HI to apparent homogeneity as determined by SDS polyacrylamide gel electrophoresis. This was accomplished by further fractionation after chromatography on phosphocellulose and hydroxylapatite. They estimate that the enzyme was purified 267-fold with a 57% yield after these two steps. A subsequent fractionation on BioRex-70 yielded a further 3-fold purification with only a 36% loss of total activity. Their final fractionation on aminopentyl-Sepharose yielded a preparation that was 1750-fold enriched over the crude extract and retained 19% of the original total activity.

[16] F. Sanger, G. M. Air, B. G. Barrell, N. L. Brown, A. Q. Coulson, J. C. Fiddes, C. A. Hutchinson III, P. M. Slocombe, and M. Smith, *Nature (London)* **265,** 687 (1977).
[17] L. A. Smith and J. G. Chirikjian, *J. Biol. Chem.* **254,** 1003 (1979).

George and Chirikjian[18] have also shown that two novel biospecific matrices, pyran–Sepharose[19] and Blue–CNBr–Sepharose, are useful for isolating BamHI. They found, however, that purified BamHI could not be effectively fractionated on heparin–Sepharose. This appears to be in contrast to the observations of Bickle et al.,[20] who reported that heparin–agarose could be used to purify a number of restriction endonucleases including BamHI, BglI, and BglII.

Smith and Chirikjian[17] also report that some preparations of BamHI have a second endonuclease activity that copurifies with BamHI in trace amounts. They have termed this second activity BamHII. Other investigators have reported to us that they have also noticed a second activity in some preparations but have not been able to isolate the enzyme. Smith and Chirikjian indicate that BamHII cuts ϕX174 at a single site and that this trace activity can be inhibited by 100 mM NaCl.

Characterization of BamHI

In our initial studies of BamHI, we observed that the enzyme was inhibited by NaCl when it was present at more than 300 mM. Cleavage by the enzyme was quite good under a variety of assay conditions, and generally we performed all assays in 6 mM Tris-HCl, pH 7.4 to pH 7.9, 6 mM MgCl$_2$, 6 mM β-mercaptoethanol.[10]

Smith and Chirikjian[17] have examined BamHI in detail using an enzyme preparation that has been purified to homogeneity. They report that the enzyme exists in two active forms as a dimer or tetramer of a 22,000 ($\pm$500) molecular weight subunit. The enzyme has a rather broad pH range in Tris-HCl with optimum activity at pH 8.5. In addition, BamHI is totally inhibited in 250 mM NaCl. The enzyme requires Mg^{2+} at a optimal concentration of 10 mM. Magnesium ion can be replaced with Mn^{2+} with a loss of 80% activity but with no change in the recognition properties of the enzyme. They also reported that Ca^{2+}, Zn^{2+}, and Cu^{2+} could not substitute for Mg^{2+}. The enzyme appears to be stable for over 1 year at 4°C. Aggregation of the enzyme to the tetrameric form occurs in low salt, a property that has also been observed for EcoRI[21,22] and Bst 1503[4] endonucleases. The enzyme is stable to thermal denaturation up to 45°, but stability is increased to 70° in the presence of 100 mM NaCl. We have observed that this increased thermal resistance is also exhibited when the substrate

[18] J. George and J. G. Chirikjian, Nucleic Acids Res. 5, 2223 (1978).
[19] J. G. Chirikjian, L. Rye, and T. S. Papas, Proc. Natl. Acad. Sci. U.S.A 72, 1142 (1975).
[20] T. A. Bickle, V. Pirrotta, and R. Imber, Nucleic Acids Res. 4, 2561 (1977).
[21] P. J. Green, M. C. Betlach, and H. W. Boyer, Methods Mol. Biol. 7, 87 (1974).
[22] P. Modrich and D. Zabel, J. Biol. Chem. 251, 5866 (1976).

DNA is present. Catterall and Welker[4] determined that *Bam*HI was not inactivated after 2 hr exposure at 37° or 40°, however, after 1 hr at 45° the enzyme was completely inactivated.

Lee *et al.*[23] reported that coupling *Bam*HI to Sepharose increased the thermal stability of the enzyme. In addition these insoluble enzymes are reuseable and afford the novel property that they can be removed completely from the reaction mixture merely by centrifugation, thus making it possible to easily recover endonuclease-treated DNA that is free from inhibitors found in agarose and chemicals such as phenol.

Endonucleases of the *Bacillus* Genospecies

A large number of restriction endonucleases have been isolated from related bacilli. In one of the most extensive investigations, Shibata *et al.*[24] examined 62 strains and found that 20 of them contained site-specific endonucleases. All of the strains of *B. amyloliquefaciens* including N, F, K, and H, had a *Bam*HI type enzyme. Another strain that has an iso-schizomer,[25] an enzyme that recognizes the same nucleotide sequence, of *Bam*HI is *Bacillus stearothermophilus*. Catterall and Welker[4] have characterized this enzyme *Bst* 1503 and found it to be a thermostable iso-schizomer of *Bam*HI. It has optimal activity between 60° and 65°, is stable at these temperatures for 10 hr, and is composed of a single subunit with a molecular weight of 46,000. The enzyme exists in two active forms having molecular weights of 180,000 and 96,000.

In collaboration with R. J. Roberts, we determined that *Bam*HI cleaved double-stranded DNA at the sequence 5' GGATCC 3' leaving a 5'-tetranucleotide extension 5' pGATC. The 5' end terminated in a phosphate group and the 3' end in a hydroxyl group providing a substrate for DNA ligase.[6] Following the isolation of *Bam*HI, a number of enzymes were isolated from bacilli that recognize sequences related to the one recognized by *Bam*HI. These include *Bcl*I from the thermophile *Bacillus caldolyticus* that cleaves the nucleotide sequence 5' TGATCA 3'[26] and *Bgl*II[2] from *Bacillus globigii* that recognizes the nucleotide sequence 5' AGATCT 3'.[27] As discussed in more detail elsewhere[14,28] all of these

[23] Y.-H. Lee, R. W. Blakesley, L. A. Smith, and J. G. Chirikjian, *Nucleic Acids Res.* **5**, 679 (1978).

[24] T. Shibata, S. Ikawa, C. Kim, and T. Ando, *J. Bacteriol.* **128**, 473 (1976).

[25] R. J. Roberts, *in* "Handbook of Biochemistry and Molecular Biology" (G. D. Fasman, ed.), p. 532. CRC Press, New York, 1975.

[26] A. H. A. Bingham, T. Atkinson, D. Sciaky, and R. J. Roberts, *Nucleic Acids Res.* **5**, 3457 (1978).

[27] V. Pirrota, *Nucleic Acids Res.* **3**, 1747 (1976).

[28] F. E. Young and G. A. Wilson, *in* "Genetic Engineering" (A. M. Chakrabarty, ed.), p. 145. CRC Press, New York, 1978.

enzymes leave the same single-strand tetranucleotide extension 5' pGATC 3' after cleavage and have been extremely useful tools to select for recombinant molecules. Because the hybrid recognition site is altered such that neither enzyme recognizes it, selection is against the reformation of parental molecules in the presence of the endonuclease. We have taken advantage of this property in constructing hybrid molecules by forming *Bam/Bgl* fusions[29] between plasmids that had been linearized with *Bam*HI and DNA that had been cleaved with *Bgl*II. After ligation, the hybrid Bam/Bgl fusion has the sequence 5' AGATCC 3' which is resistant to both *Bam*HI and *Bgl*II.

Often, bacterial strains contain more than one restriction endonuclease that hamper efforts to purify the enzymes. Duncan *et al.*[5] selected mutants of *B. globigii* that either lacked both enzymes or only produced *Bgl*I or *Bgl*II. These strains have greatly facilitated the isolation of these enzymes and in addition clearly demonstrated that the enzymes did not have common subunits. Even in the strain that lacks both enzymes, DNA is completely modified indicating that the modification enzymes are genetically distinct from the restriction endonucleases. During these studies, a rapid crude assay procedure was developed that could qualitatively show which endonucleases were being produced by *B. globigii*. The assay consists of growing the cells to the middle of the exponential phase of growth in 100 ml of supplemented L broth. Cells are harvested by centrifugation at 10,000 rpm for 10 min at 4° in a Sorvall GSA rotor and suspended in 1.25 ml of a buffer consisting of 50 mM potassium phosphate (pH 7.0), 10 mM β-mercaptoethanol, 1 mM EDTA, 1 mM NaN$_3$, and 10% glycerol. This suspension is disrupted by sonic treatment (three 10-sec treatments using the red probe of Bronwill Biosonic III) and clarified by centrifugation for 30 min at 4°C in a Sorvall SP table top centrifuge. Usually a 4-μl sample of the supernate is used to digest 1 μg of plasmid DNA at 37° for 10 min. The mixture is then analyzed by ethidium bromide–agarose gel electrophoresis.

The restriction endonucleases for the *Bacillus* genospecies have been used extensively in a variety of studies. The relative stability of the enzymes, their ease of isolation and the existence of appropriate vectors with a limited number of recognition sites have contributed to their general applicability in the field of nucleic acid research. In addition, thermal-resistant forms of the enzymes have been found as well as strains which can be selected to produce only one enzyme. Although characterization of the enzymes has not received as much attention, it is clear that there are interesting parameters to study. The existence of a variety of the

---

[29] C. H. Duncan, G. A. Wilson, and F. E. Young, *Proc. Natl. Acad. Sci. U.S.A.* **75**, 3664 (1978).

enzymes that recognize subsets of the basic sequence recognized by
*Bam*HI should provide fruitful ground for investigations on the biochemical properties of the enzymes in the *Bacillus* genospecies.

Acknowledgments

This study was supported in part by funds from Grant VC-27 from the American Cancer
Society. Gary A. Wilson is a Faculty Research Fellow of the American Cancer Society.

[19] Preparation and Properties of the *Hpa*I and *Hpa*II Endonucleases

By Jerome L. Hines, Thomas R. Chauncey, and Kan L. Agarwal

Two restriction endonuclease activities, *Hpa*I and *Hpa*II, have been
isolated from *Haemophilus parainfluenzae*.[1] Endonuclease *Hpa*I has been
isolated in homogeneous form, while *Hpa*II has been purified free of contaminating nuclease and phosphatase activities. *Hpa*I recognizes the
DNA sequence

$$5'\ \ G—T—T\overset{\downarrow}{—}A—A—C$$
$$C—A—A\underset{\uparrow}{—}T—T—G$$

and cleaves the phosphodiester bonds in both strands as indicated. *Hpa*II
recognizes the DNA sequence

$$5'\ \ C\overset{\downarrow}{—}C—G—G$$
$$G—G—C\underset{\uparrow}{—}C$$

and cleaves the phosphodiester bonds in a staggered fashion as indicated.
The reaction products, in both cases, contain 5'-phosphoryl and 3'-
hydroxyl termini.[2]

Assay Methods

Principle. Two assays have been used during the purification of *Hpa*I
and *Hpa*II. The first assay is qualitative and involves the cleavage of λ
DNA by the respective enzyme into fragments of discrete size. These
fragments are separated by agarose gel electrophoresis[1] and visualized by

[1] P. A. Sharp, B. Sugden, and J. Sambrook, *Biochemistry* **12**, 3055 (1973); R. Gromkova and
S. H. Goodgal, *J. Bacteriol.* **109**, 987 (1972).
[2] D. E. Garfin and H. M. Goodman, *Biochem. Biophys. Res. Commun.* **59**, 108 (1974).

staining with ethidium bromide. Characteristic sets of fragments are seen for each of the enzymes. *Hpa*I cleaves λ DNA eleven times while *Hpa*II cleaves λ DNA more than fifty times. The second assay is quantitative and measures the conversion of closed circular RF fd DNA, which contains a single *Hpa*I site, to the linear form which is sensitive to degradation by RecBC DNase.[3]

Reagents for Assay I

Tris-HCl buffer, 100 mM, pH 7.5
MgCl$_2$, 100 mM
2-Mercaptoethanol, 100 mM
Crystalline bovine plasma albumin (Sigma), 1 mg/ml
EDTA, 1 mM
λ DNA, 175 µg/ml
Reaction termination mixture
   EDTA, 500 mM
   Sucrose, 60%
   Bromphenol blue, 0.5%
Agarose gel electrophoresis buffer
   50 mM Tris-HCl, pH 7.9
   20 mM sodium acetate
   2 mM EDTA
   10 mM NaCl

Reagents for Assay II

Tris-HCl buffer, 200 mM, pH 7.5
MgCl$_2$, 100 mM
Dithiothreitol (Calbiochem), 10 mM
EDTA, 1 mM
Crystalline bovine plasma albumin, 1 mg/ml
ATP (Calbiochem), 2 mM
Rf fd [^{3}H]DNA,[4] 1 × 10^5 cpm/µg, 110 µg/ml
Calf thymus DNA,[5] 8.0 mM
Trichloroacetic acid, 50%
*Rec*BC DNase,[6] 1000 units/ml

[3] A. E. Koru, V. MacKay, P. J. Goldmark, and S. Linn, *J. Biol. Chem.* **248**, 4874 (1973); P. Modrich and D. Zabel, *ibid.* **251**, 5866 (1976).
[4] M. Sugiura, T. Okamoto, and M. Takanami, *J. Mol. Biol.* **43**, 299 (1969).
[5] The concentration of calf thymus DNA is given in millimoles of nucleotide based on an absorbance (A_{260}) of 1 equal to 50 µg/ml.
[6] D. C. Eichler and I. R. Lehman, *J. Biol. Chem.* **252**, 499 (1977).

Procedure—Assay I. The assay mixture (40 μl) contains 10 mM Tris-HCl, pH 7.5, 10 mM MgCl$_2$, 0.1 mM EDTA, 100 μg/ml BSA, 10 mM 2-mercaptoethanol, 1 μg of DNA, and enzyme (2–5 μl). The reaction mixture is incubated at 37° for 1 hr. The reaction is stopped by adding 5 μl of the termination mixture, and then layered on an agarose slab gel. The gel is electrophoresed for 4 hr at a current of 100 mA. The gel is then stained with ethidium bromide (1 μg/ml) for 1 hr and destained with H$_2$O for 30 min. Polaroid pictures are taken with Polaroid type 667 film under UV light. One percent agarose gels are used when assaying for *Hpa*I; 2% agarose gels are used when assaying for *Hpa*II.

Procedure—Assay II. The assay mixture (50 μl) contains 50 mM Tris-HCl, pH 7.8, 50 mM NaCl, 5 mM MgCl$_2$, 0.2 mM EDTA, 2 mM dithiothreitol, 200 μM ATP, 0.4 μg of RF fd [^{3}H]DNA, enzyme solution (5 μl, diluted if necessary), and *Rec*BC DNase (5 units). After incubation at 37° for 15 min, the mixture is chilled to 0°, and 3 μl of 8 mM calf thymus DNA and 15 μl of 50% trichloroacetic acid are added. The mixture is allowed to stand for 30 min at 0° and then spun at 12,000 g in a microcentrifuge (Eppendorf) for 20 min at 4°. The supernatant (50 μl) is spotted on a 3MM filter paper (2 × 2 cm), dried under an infrared lamp, and immersed in 5 ml of toluene scintillant. Radioactivity was then determined in a scintillation counter.

Enzyme Assays during Purification. Enzyme assays are not performed until the SP-Sephadex column chromatographic step due to the presence of other nucleases in extracts of *Haemophilus*. The earlier purification steps have been found to be sufficiently reproducible so that they need not be monitored by enzyme assays.

Unit. A unit of enzyme activity is defined as that amount of enzyme which catalyzes the cleavage of 1 μg of λ DNA to completion in 1 hr at 37°.

Purification of the Enzymes HpaI and HpaII. Activities of both endonucleases *Hpa*I and *Hpa*II are present in the first four steps of purification. The bulk of *Hpa*I is adsorbed to SP-Sephadex C-25 (Step 4) whereas none of the *Hpa*II is adsorbed. The small amount of *Hpa*I which is not adsorbed to the cation exchanger is removed from *Hpa*II during the subsequent heparin agarose, BioGel A 0.5 M, and DEAE-Sephadex column chromatographic steps.

Unless otherwise indicated all procedures are carried out at 0°–4° and all centrifugations are at 16,000 g for 50 min. The results of a typical purification for *Hpa*I and *Hpa*II are given in Tables I and II (100 g scale).

Growth of Bacteria. Haemophilus parainfluenzae[7] is grown at 37° in a

---

[7] The strain of *Haemophilus parainfluenzae* was kindly provided by Dr. Rich Roberts. Occasional replating was necessary to maintain viability.

TABLE I
PURIFICATION OF *Hpa*I

Fraction number and step	Volume (ml)	Specific activity (units/mg)	Total protein (mg)
I. Lysate	900	—	—
II. Streptomycin sulfate	1100	—	4000
III. Ammonium sulfate	200	7×10^1	3000
IV. SP-Sephadex I	80	6×10^4	2
V. SP-Sephadex II	20	1.4×10^5	0.8
VI. DEAE-Sephadex	15	9×10^5	0.09
VII. Hydroxylapatite	0.8	1.2×10^6	0.06

Fermacell fermentor (New Brunswick Scientific Co., Inc., New Brunswick, New Jersey) with an automated addition of NaOH to maintain the pH at 7.0. Eighty liters of 3.7% brain heart infusion supplemented with 2 μg/ml of NAD is the normal medium for growth. Cells are harvested in late log phase by quickly chilling the culture by adding approximately 10 liters of ice. The cells are collected with steam-driven Sharples centrifuges. The cell paste made to 50% (w/w) in glycerol is stable for at least 10 months when stored at $-70°$.

Purification of *Hpa*I

Step 1—Preparation of Cell Extracts. Frozen cells (100 gm) are suspended in 400 ml of 50 mM Tris-HCl, pH 8.0, 10 mM 2-mercaptoethanol, and 5% glycerol and allowed to stir overnight at 4°. The suspended cells are disrupted by passing twice through a Manton-Goulen homogenizer at 7,500 psi. Between passes the suspension is chilled in an ice–salt water

TABLE II
PURIFICATION OF *Hpa*II

Fraction number and step	Volume (ml)	Specific activity (units/mg)	Protein (mg)
I. Lysate	900	—	—
II. Streptomycin sulfate	1100	—	4000
III. Ammonium sulfate	200	—	3000
IV. SP-Sephadex I	800	(3.5×10^2)	2800
V. Heparin–agarose	250	8×10^3	90
VI. Hydroxylapatite	12	2.8×10^4	25
VII. Agarose A-0.5 m	60	7.5×10^4	6
VIII. DEAE-Sephadex	4	2.1×10^5	0.5

slurry. For small scale preparations, the cells can be ruptured with sonic bursts delivered by a Branson Sonifier.[1] The extracts are clarified by centrifugation at 40,000 rpm for 60 min. The supernatant (600 ml) is carefully removed from the gummy soft pellet and adjusted to an optical density (A_{260}) of 60 by adding the above buffer (fraction I).

Step 2—Streptomycin Sulfate Fractionation. Streptomycin sulfate (765 units/mg, ICN Pharmaceuticals Inc.) is freshly prepared (5%, w/v) in 50 mM Tris-HCl, pH 8.0, 10 mM 2-mercaptoethanol, and 5% glycerol. The above solution (280 ml) is added dropwise to fraction I (900 ml) over a period of 45 min with gentle stirring. After addition is complete (final streptomycin sulfate concentration, 1.2%) stirring is continued for 30 min. The supernatant is collected by centrifugation (fraction II).

Step 3—Ammonium Sulfate Precipitation. Powdered ammonium sulfate (480 g) is added gradually with stirring to fraction II (1100 ml) to 70% saturation. Addition is over a period of 45 min followed by gentle stirring overnight. The precipitate is collected by centrifugation and resuspended in 500 ml of 55% saturated ammonium sulfate in 50 mM Tris-HCl, pH 8.0, 10 mM 2-mercaptoethanol, and 5% glycerol. After 1 hr at 4°, the precipitate is collected by centrifugation and resuspended in 200 ml of buffer containing 20 mM sodium phosphate, pH 7.8, 0.1 mM EDTA, 10 mM 2-mercaptoethanol, and 10% glycerol (fraction III).

Step 4—SP-Sephadex (C-25) Chromatography. Fraction III (200 ml) is dialyzed against three changes of 10 volumes of 20 mM sodium phosphate, pH 7.8, 0.1 mM EDTA, 10 mM 2-mercaptoethanol, and 10% glycerol for a total of 12 hr. The dialysate is diluted to 500 ml with the above buffer and applied to a SP-Sephadex (C-25) column (5 × 30 cm), previously equilibrated with the above buffer, at 60 ml/hr. After application of the sample, the column is washed with one column volume of the above buffer. All of the *Hpa*II activity is eluted during the application and wash. A 1-liter 0.0 to 0.2 M sodium chloride gradient in the above buffer is applied to the column, and 5 ml fractions are collected. *Hpa*I activity, which elutes at 0.06 to 0.08 M NaCl, is determined using both assay I and II and is pooled as shown in Fig. 1 (fraction IV).

Step 5—Rechromatography of Fraction IV on SP-Sephadex (C-25). A SP-Sephadex column (1 × 18 cm) is prepared and washed with 10 volumes of 20 mM sodium phosphate, pH 7.8, 0.1 mM EDTA, 10 mM 2-mercaptoethanol, and 10% glycerol. Fraction IV (80 ml) is dialyzed against 4 liters of the above buffer for 4 hr and applied to the column at a flow rate of 20 ml/hr. The column is washed with one volume of the equilibrating buffer and the adsorbed enzyme is eluted with a linear gradient (total volume 200 ml) of 0.0 to 0.2 M NaCl in the column buffer. Three milliliter fractions are collected at 15-min intervals. The peak of *Hpa*I activity is at 0.06 to 0.08 M NaCl (fraction V).

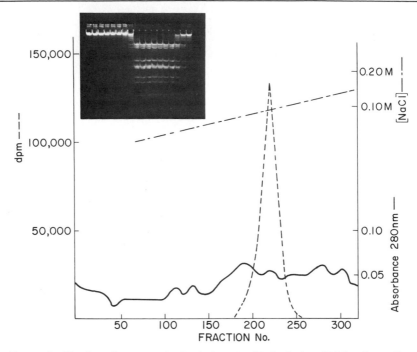

FIG. 1. Purification of *Hpa*I on SP-Sephadex. An SP-Sephadex (C-25) column (5 × 30 cm) was prepared, and fraction III applied as described in the text. A linear gradient 0.0 to 0.2 *M* NaCl in column buffer was used to elute the enzyme activity (—— · —— · ——). Results of assay I (insert) and assay II (· · · · · ·) are shown. Protein was followed by absorbance at 280 nm.

Step 6—DEAE-Sephadex (A-50) Chromatography. Fraction V (50 ml) is dialyzed against 1 liter of 20 m*M* sodium phosphate, pH 7.8, 0.1 m*M* EDTA, 10 m*M* 2-mercaptoethanol, and 10% glycerol for 2 hr. A column (2 × 5.5 cm) of DEAE-Sephadex (A-50) is prepared and washed with 200 ml of the above buffer. Fraction V is applied to the column at a flow rate of 10 ml/hr. The column is washed with 1 volume of the equilibrating buffer and eluted with a linear gradient (total volume 100 ml) of 0.0 to 0.35 *M* NaCl in the above buffer. Three milliliter fractions are collected at 20-min intervals. *Hpa*I activity elutes at 0.18 to 0.21 *M* NaCl (fraction VI).

Step 7—Concentration of Enzyme. A column (1.2 × 1 cm) of hydroxylapatite is prepared and washed with 50 ml of 20 m*M* sodium phosphate, pH 7.8, 0.1 m*M* EDTA, 10 m*M* 2-mercaptoethanol, 10% glycerol, and 0.2 *M* NaCl. Fraction VI (15 ml) was applied to the column at a flow rate of 3 ml/hr. The column was washed with 2 volumes of the equilibrating buffer and then eluted with 20 m*M* sodium phosphate, pH 7.8, 0.1 m*M* EDTA, 10 m*M* 2-mercaptoethanol, 10% glycerol, and 1 *M* NaCl. Most of the activity eluted in a volume of 3 ml. This was dialyzed against 500 ml of 20 mM sodium phosphate, pH 7.8, 0.1 mM EDTA, 10 mM 2-mercaptoethanol, and 10% glycerol for 4 hr, followed by dialysis against 50% glycerol in the same buffer for 24 hr. The enzyme is stored at $-20°$ (fraction VII).

Purification of the Endonuclease HpaII

The purification of *Hpa*II has not been pursued as extensively as the purification of *Hpa*I. The procedure described below, however, yields *Hpa*II activity free of any contaminating nuclease or phosphatase activity. Steps 1–4 are carried out as described above for *Hpa*I.

Step 5—Heparin–Agarose Chromatography. A column (4 × 17.5 cm) of· heparin–agarose[8] is prepared and washed with 2 liters of 20 mM sodium phosphate, pH 7.6, 0.1 mM EDTA, 10 mM 2-mercaptoethanol, and 10% glycerol. The flowthrough and wash from step 4 is applied directly to this column at a flow rate of 40 ml/hr. The column is washed with 1 volume of the equilibrating buffer and eluted with a 2-liter linear gradient of 0.2 to 0.7 M NaCl in the equilibrating buffer. Six milliliter fractions are collected every 12 min. *Hpa*II activity elutes at approximately 0.35 M NaCl. *Hpa*I activity which did not adsorb on SP-Sephadex trails the *Hpa*II activity. *Hpa*I activity, however, is completely removed from *Hpa*II activity in the following steps of chromatography (fraction V).

Step 6—Concentration of the Enzyme Pool. Fraction V (250 ml) is dialyzed against 4 liters of 20 mM sodium phosphate, pH 7.8, 0.1 mM EDTA, 10 mM 2-mercaptoethanol, and 10% glycerol for 4 hr and applied to a hydroxylapatite column (1.2 × 3 cm) which has been preequilibrated with the above buffer containing 0.2 M NaCl. The column is washed with the equilibration buffer and eluted with 20 mM sodium phosphate, pH 7.6, 10 mM 2-mercaptoethanol, 0.1 mM EDTA, 20% glycerol, and 1 M NaCl. *Hpa*II activity elutes in 10 ml (fraction VI).

Step 7—BioGel A-0.5m Chromatography. A BioGel A-0.5m column (2.5 × 92 cm) is prepared and washed with 20 mM sodium phosphate, pH 7.6, 10 mM 2-mercaptoethanol, 0.1 mM EDTA, 10% glycerol, and 1 M NaCl. Fraction VI (10 ml) is applied to the column at a flow rate of 6 ml/hr. *Hpa*II activity, free from contaminating *Hpa*I activity, elutes at 0.67 to 0.73 column volume as shown in Fig. 2 (fraction VII).

Step 8—DEAE–Sephadex A-50 Chromatography. A column (1 × 4 cm) of DEAE–Sephadex is purified and washed with 20 mM sodium phosphate, pH 7.6, 10 mM 2-mercaptoethanol, 0.1 mM EDTA, and 10%

---

[8] T. A. Bickle, V. Pirrotta, and R. Imber, *Nucleic Acids Res.* **4**, 2561 (1977).

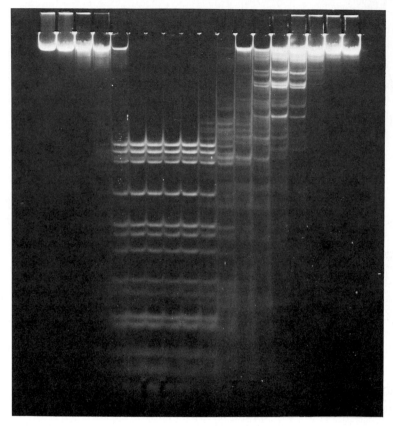

FIG. 2. Purification of *Hpa*II on BioGel A-0.5m. The cleavage of λ DNA by fractions from a BioGel A-0.5m column (2.5 × 92 cm) was followed by agarose (2%) gel electrophoresis. *Hpa*II activity eluted at 0.67 to 0.73 column volume, slightly ahead of small amounts of contaminating *Hpa*I activity. Details of the BioGel filtration step are given in the text.

glycerol. Fraction VII (60 ml) is dialyzed against three changes of 10 volumes of the above buffer and applied to the column at 10 ml/hr. The column is washed with 10 ml of the equilibrating buffer and eluted with a 50 ml linear gradient of 0.0 to 0.3 *M* NaCl in the equilibrating buffer. *Hpa*II

FIG. 3. Sodium dodecyl sulfate electrophoresis of *Hpa*I endonuclease. (A) Ten micrograms of fraction VII was subjected to electrophoresis on a 10% polyacrylamide gel in the presence of sodium dodecyl sulfate (SDS). The incubation buffer included 10 m*M* sodium phosphate, pH 7.0, 1% SDS, and 1% 2-mercaptoethanol. The running buffer included 10 m*M* sodium phosphate, pH 7.0, and 0.1% SDS. The gel was stained with Coomassie Blue and scanned at 550 nm. (B) The molecular weight of denatured and reduced *Hpa*I endonuclease was determined by polyacrylamide gel electrophoresis and compared with protein standards of known molecular weight.

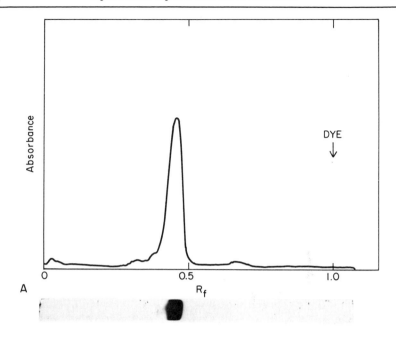

A

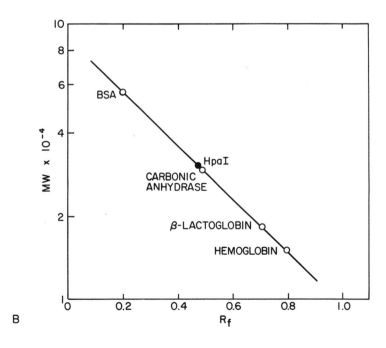

B

elutes at approximately 0.10 to 0.12 M NaCl. Peak fractions are pooled and dialyzed against 250 ml of the equilibrating buffer, followed by dialysis against 50% glycerol in equilibrating buffer. The enzyme is stored at −20° (fraction VIII).

Properties of the Enzymes

Stability. Both *Hpa*I and *Hpa*II are stable for at least 8 months when stored at −20° in 50% glycerol. Enzyme preparations containing less than 0.02 mg/ml protein are labile; the protein concentration, therefore, should be maintained as high as possible.

Absence of Other Enzyme Activities. During the purification of *Hpa*I endonuclease, fractions IV–VII are assayed for contaminating exonuclease, other endonuclease, and phosphatase activities.[9] Fraction IV is determined to be free of contaminating activities and therefore is suitable for most purposes.[10] In the *Hpa*II endonuclease purification, fraction VIII is free of other contamination enzyme activities, while fractions IV–VII are contaminated with exonuclease.

Homogeneity. The hydroxylapatite fraction VII of *Hpa*I is a homogeneous protein as judged by a single protein band obtained on SDS–polyacrylamide gel electrophoresis (10 μg protein band) as shown in the insert of Fig. 3. *Hpa*II endonuclease has not been purified to homogeneity.

Physical Properties. The *Hpa*I endonuclease consists of a single polypeptide chain with a molecular weight of 30,500 as determined by SDS–polyacrylamide gel electrophoresis. However, the higher value of 34,000 is obtained when determined by Sephadex G-100 chromatography. The physical properties of *Hpa*II have not been studied.

DNA Substrate Specificity. Native DNAs containing specific nucleotide sequences recognized by both *Hpa*I and *Hpa*II enzymes are cleaved efficiently. While single-stranded DNAs are not cleaved by *Hpa*I, *Hpa*II has been reported to cleave single-stranded phage DNA.[11] Single-stranded and double-stranded RNA and RNA–DNA hybrids are not cleaved by either enzyme. A chemically synthesized self-complementary octanu-

---

[9] For phosphatase, exonuclease, and endonuclease assays, [^{32}P]oligo(dT$_{15}$) is incubated with the enzyme solution in assay I buffer. Aliquots are taken from the mixture at various times up to 24 hr and spotted on DEAE–cellulose plates. The plates are chromatographed on an RNA homochromatography system and then autoradiographed. The RNA homochromatography mix consists of 3% RNA in 7 M urea.

[10] Fraction IV would be useful for chromosome mapping and for recombinant DNA purposes where the absence of contaminating nuclease and phosphatase activities is essential.

[11] K. Horiuchi and N. D. Zinder, *Proc. Natl. Acad. Sci. U.S.A.* **72**, 2555 (1975).

cleotide, dG—G—T—T—A—A—C—C, is cleaved by *HpaI* endonuclease in a manner identical to the cleavage of native DNA.[12]

pH Optimum. The optimal pH range for the purified *HpaI* enzyme is 7.7 to 8.1 in 10 mM Tris-HCl buffer.[13] The pH dependence of *HpaII* enzyme has not been studied.

Effect of Temperature on Enzymatic Rate. The *HpaI* endonuclease is stable at 37° and cleaves DNA with maximum rate at 45°.[13] At temperatures higher than 45° and lower than 37°, the rate of cleavage of DNA is less than optimal. Temperature-dependent cleavage of DNA with *HpaII* enzyme has not been studied.

Metal Ion Requirement. In the absence of added MgCl$_2$, there is no detectable *HpaI* or *HpaII* endonuclease activity. Under standard assay conditions for DNA cleavage, the optimal Mg^{2+} concentration for *HpaI* activity is 5 mM. Manganese ions can only partially replace Mg^{2+} in both *HpaI* and *HpaII* assays.

Sulfhydryl Requirement. Although 2-mercaptoethanol is included in all buffers during the purification, its presence is not required for maximal enzyme activity. This is true for both *HpaI* and *HpaII*.

Effect of Salt Concentration. The *HpaII* activity is inhibited at salt concentrations above 60 mM, whereas *HpaI* activity is not affected. Salt concentrations as high as 200 mM can be used without affecting the activity of *HpaI*. Thus, selective inhibition of *HpaII* activity can be accomplished by the use of 200 mM NaCl in an assay containing both *HpaI* and *HpaII* endonucleases.

Acknowledgments

This work was supported by National Institutes of Health Grant GM-22199. J.L.H. was supported by a National Institutes of Health Predoctoral Fellowship (GM 07183). K.L.A. received Research Career Development Award GM 224 from the National Institutes of Health.

[12] J. L. Hines and K. L. Agarwal, recent observations (1978).
[13] F. M. DeFilippes, *Biochem. Biophys. Res. Commun.* **58**, 586 (1974).

[20] Purification and Properties of the *HphI* Endonuclease

By DENNIS G. KLEID

```
5'  C G G T G A T A C T G A G C A↓C A T C A G
3'  G C C A C T A T G A C T C G₁T G T A G T C
                              ↑
```

The nucleotide sequence-specific endonuclease from *Haemophilus parahaemolyticus* has properties that differ from the majority of type II restriction endonucleases. This enzyme recognizes an unsymmetric nucleotide sequence and cleaves the DNA at a site approximately one turn of the helix from the center of the recognition sequence.[1] The recognition sequence has been demonstrated to be a five base pair sequence

$$
\begin{array}{l}
5'\ \ \text{G G T G A} \\
3'\ \ \text{C C A C T}
\end{array}
$$

Purification Procedure[1,2]

The endonuclease activity was obtained from *H. parahaemolyticus* cells grown nearly to stationary phase in brain heart infusion supplemented with 10 μg/ml hemin and 2 μg/ml NAD. Ten grams of cells were disrupted by sonication in 10 ml of 10 mM Tris, pH 7.4, 10 mM β-mercaptoethanol. The supernate fluid from a 100,000 g centrifugation (1 hr) was made 1 M in NaCl and layered onto a BioGel A 0.5 M column (200–400 mesh, 2.5 × 50 cm). Fractions were assayed as described in Sharp *et al.*[3] using 10 mM Tris, pH 7.4, 10 mM MgCl$_2$, 10 mM β-mercaptoethanol, 6 mM KCl (*Hph* buffer). Active fractions were made to 50% saturation with ammonium sulfate at 0°, and centrifuged. The precipitated protein was resuspended and dialyzed against 10 mM K$_2$HPO$_4$, pH 7.5, 1 mM dithiothreitol, 100 mM KCl, then loaded on a (1 × 10 cm) phosphocellulose column. The activity was eluted with a 0.1 to 1 M KCl gradient, made in the same buffer. Active fractions, eluting between 0.25 and 0.30 M KCl, were concentrated by dialysis against *Hph* buffer containing 50% glycerol and stored at −20°. Ten grams of cells yielded approximately 5–20 units of enzyme, where 1 unit is the amount of enzyme that cuts 1 mg of λ DNA to a limit product in 1 hr at 37°.

Identification Procedure

When λ DNA is cleaved with a restriction endonuclease and the resulting fragments separated by acrylamide gel electrophoresis, a distinctive pattern of bands is observed. λ DNA (50 μg) was restricted with the *Hph*I endonuclease in *Hph* buffer. After 30 min at 37°, bacterial alkaline phosphatase (Worthington) (0.1 units) was added and the incubation continued for 15 min. The solution was extracted three times with equal volumes of neutralized phenol. On the last extraction the aqueous layer

[1] D. Kleid, Z. Humayun, A. Jeffrey, and M. Ptashne, *Proc. Natl. Acad. Sci. U.S.A.* **73**, 293 (1976).

[2] This procedure gives *Hph*I activity uncontaminated with other nucleases. The final units per milligram of protein varies for each preparation. A similar preparation is available from New England Biolabs, Beverly, Massachusetts.

[3] P. A. Sharp, B. Sugden, and J. Sambrook, *Biochemistry* **12**, 3055 (1973).

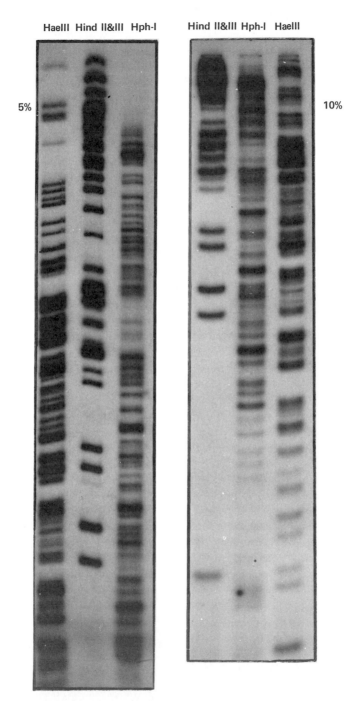

FIG. 1. Polyacrylamide gel electrophoresis separation of restricted λ DNA labeled *in vitro* with [γ-^{32}P]ATP.

was removed and made $0.3 M$ in NaOAc. To this three volumes of ethanol were added, the solution shaken and allowed to stand 1 hr at $-20°$. The DNA was pelleted by centrifugation, washed with ethanol, then dried under vacuum. To this [γ-^{32}P]ATP was added and lyophilized to dryness (20 μl of 1000–3000 Ci/mmole in 20% ethanol at a concentration of 1–2 nmole/ml, New England Nuclear). The DNA was dissolved in 50 μl of 60 mM Tris (pH 8), 15 mM β-mercaptoethanol, 10 mM MgCl$_2$ then 20 units T$_4$-polynucleotide kinase were added. After 1 hr at $37°$ the DNA was ethanol precipitated as above, dried, then dissolved in 20 μl H$_2$O and 20 μl of 50% glycerol containing suitable acrylamide gel electrophoresis dyes. This was electrophoresed through 5 and 10% acrylamide gels in Tris–borate–EDTA buffer as described by Maniatis et al.[4]

The results of a comparison of the restriction of λ DNA with HindII and III, HaeIII and HphI are shown in Fig. 1.

Properties of HphI

Three lines of evidence have been reported[1] suggesting that the HphI endonuclease recognizes the sequence

$$5'\ \ T\ C\ A\ C\ C$$
$$3'\ \ A\ G\ T\ G\ G$$

(1) This sequence is found 8 or 9 base pairs to one side of six completely sequenced HphI cleavage sites. (2) Methylation of purines within the sequence prevent HphI cutting. (3) Mutations within this sequence prevent HphI cutting. Recently, Sanger et al.[5] have shown that this sequence and only this sequence is recognized by HphI on ϕX174 DNA.

Nothing is known about the physical properties of the enzyme, its subunit structure, amino acid composition, or sequence. These properties may prove to be of interest in ascertaining the mechanism of sequence-specific DNA–protein interactions in light of the fact that the nucleotide sequence recognized by this enzyme overlaps with those recognized by the λ repressor,[1] the cro protein of λ,[6] and RNA polymerase.[7]

[4] T. Maniatis, A. Jeffrey, and J. H. van de Sande, Biochemistry 14, 3787 (1975).
[5] F. Sanger, A. R. Coulson, T. Friedmann, G. M. Air, B. G. Barrell, N. L. Brown, J. C. Fiddes, C. A. Hutchison III, P. M. Slocombe, and M. Smith, J. Mol. Biol. 125, 225 (1978).
[6] A. Johnson, B. J. Meyer, and M. Ptashne, Proc. Natl. Acad. Sci. U.S.A. 75, 1783 (1978).
[7] D. Pribnow, Proc. Natl. Acad. Sci. U.S.A. 72, 784 (1975).

[21] Purification and Properties of the Bst1503 Endonuclease

By JAMES F. CATTERALL and N. E. WELKER

Site-specific endonucleases have become invaluable to the study of the structure and function of DNA from both prokaryotic and eukaryotic organisms.[1] Some enzymes of this class, the restriction endonucleases, have been shown to have a specific *in vivo* function acting to protect bacterial cells from infection. An associated enzyme, a modification methylase, must be present in cells producing a restriction enzyme to protect cellular DNA from digestion.[2] It has recently been shown that restriction enzymes may be capable of promoting site-specific recombination *in vitro*.[3]

The obligate thermophile *Bacillus stearothermophilus* exhibited typical host-specific restriction and modification during infection by thermophilic bacteriophages.[4] A restriction endonuclease (endo R·Bst1503[5]) was purified from *B. stearothermophilus* strain 1503-4R.[6] The purified Bst1503 retains its thermostability and is optimally active at temperatures near the optimal growth temperature of the organism. The enzyme is stable to long incubation at 65° in the absence of substrate and forms limit digests after similar incubations in the presence of DNA. Bst1503 is also stable to long term storage at 5°.

Assay of Enzyme Activity

Bst1503 reaction mixtures contained 10 mM Tris-HCl, 0.2 mM MgCl$_2$, and 6.6 mM 2-mercaptoethanol at pH 7.8. DNA was added to 20 μg/ml in a final reaction volume of 0.05 ml. This mixture was heated to 55° prior to the addition of 1–5 μl of the appropriate enzyme preparation. Incubation at 55° was continued for 1 hr. For assay of crude preparations, the enzyme was diluted with TEMP buffer (40 mM Tris-HCl, pH 7.5, 10 mM EDTA, 6.6 mM 2-mercaptoethanol, 0.1 mM phenylmethylsulfonyl fluoride) prior to its addition to the reaction mixture.

Limit products of Bst1503 reactions were separated by electrophoresis

[1] R. J. Roberts, *Crit. Rev. Biochem.* **4**, 123 (1976).
[2] W. Arber, *Prog. Nucleic Acid Res. Mol. Biol.* **14**, 1 (1974).
[3] S. Chang and S. N. Cohen, *Proc. Natl. Acad. Sci. U.S.A.* **74**, 4811 (1977).
[4] J. F. Catterall, N. D. Lees, and N. E. Welker, *in* "Microbiology, 1976" (D. Schlessinger, ed.), p. 358. Am. Soc. Microbiol., Washington, D. C., 1976.
[5] This nomenclature follows the suggestions of H. O. Smith and D. Nathans [*J. Mol. Biol.* **81**, 419 (1973)] and refers to the same enzyme as that called BstI by Roberts.[1]
[6] J. F. Catterall and N. E. Welker, *J. Bacteriol.* **129**, 1110 (1977).

METHODS IN ENZYMOLOGY, VOL. 65

through agarose gels.[7] Reaction mixtures were loaded onto a 1% agarose slab gel (16 × 16 cm; Blaircraftline, Cold Spring Harbor, New York). Electrophoresis was for 6 hr at 50 V in buffer containing 50 mM Tris-HCl, pH 8.4, 20 mM sodium acetate, 18 mM NaCl, 2 mM EDTA, and 0.5 μg/ml ethidium bromide. Fragment bands were visualized immediately with a short wave Mineralight (UV Products, San Gabriel, California). The gels were photographed with a Polaroid MP-3 camera through a Tiffen Series 6 15G filter.

Purification of Endo R · Bst 1503

Cell Growth and Lysis. Cultures of *B. stearothermophilus* strain 1503-4R were grown in TYG medium[6] to the late logarithmic phase of growth (4–5 × 10⁸ cells/ml) and harvested by centrifugation. Cell pellets were washed once with TNC buffer (50 mM Tris-HCl, pH 7.5, 10 mM NaCl, 1 mM CaCl$_2$) and stored at −20°. Forty grams (wet weight) of cells were obtained from a 12-liter culture.

Cell pellets were resuspended in TEMP buffer to an OD$_{525nm}$ of 12–30 and washed twice in the same buffer. The cells were finally suspended in 0.2 volume of TEMP buffer and lysozyme was added to a final concentration of 0.75–1.5 mg/ml. The mixture was incubated at 37° until the solution was clear and viscous. Sodium chloride (1 M) and streptomycin sulfate (10%) were added to final concentrations of 0.045 M and 1.5%, respectively. The solution was placed in ice until a milky white precipitate formed. The precipitate is removed by centrifugation in a table top centrifuge at maximum speed.

Ammonium Sulfate Fractionation. Solid ammonium sulfate (enzyme grade) was added with stirring to 50% saturation. The solution was stirred for 1 hr at 5°. The precipitate was collected by centrifugation at 35,000 g for 20 min and discarded. Solid ammonium sulfate was added to the supernatant fluids to 72% saturation and stirred for 1 hr at 5°. The precipitate was collected by centrifugation at 35,000 g for 20 min.

DEAE–Cellulose Chromatography. The precipitate was resuspended in 10 ml of 20 mM Tris-HCl, 0.5 mM ditiothreitol, 0.1 mM phenylmethylsulfonyl fluoride, 0.5 mM EDTA, 50 mM NaCl, and 10% glycerol, pH 7.4. The solution was layered onto a DEAE–cellulose column (2.5 × 20 cm) and the enzyme eluted with the same buffer.

Phosphocellulose Chromatography. The eluate containing the enzyme was dialyzed twice (2 and 12 hr) against 30 volumes of 20 mM potassium phosphate, pH 6.8, 0.2 mM EDTA, 0.5 mM dithiothreitol, and 10% glycerol and applied to a column (1.2 × 15 cm) of phosphocellulose

---

[7] A. R. Dunn and J. Sambrook, this volume, Article [54].

TABLE I
PURIFICATION OF *Bst* 1503

Step	Units/ml ($\times 10^{-3}$)[a]	Total units ($\times 10^{-3}$)	Protein (mg/ml)	Specific activity (units/mg)	Fold purified	Yield (%)
Crude[b]	16	912	3.4	4,706	1	100
Ammonium sulfate	23.3	848.2	20.6	11,294	2.4	93
DEAE– cellulose	29.7	474.2	11.3	17,412	3.7	52
Phospho- cellulose	33.4	200.6	1.3	25,712	5.5	22

[a] A unit of *Bst* 1503 activity is that amount of enzyme required to cleave 1 μg of phage TP-1C DNA (8.3×10^{-8} μmole) in 30 min at 55°.

[b] If cells are disrupted by sonication[6] protein is 35.8–42.0 mg/ml in the crude and the phosphocellulose step is purified 57.5-fold.

(Whatman P11). The enzyme was eluted with a linear gradient (0 to 1.0 M KCl). A single protein peak having *Bst* 1503 activity was detected in the eluate. *Bst* 1503 eluted between 0.27 and 0.33 M KCl.

The purified enzyme has a specific activity of approximately 2.6×10^4 units/mg of protein (Table I). The enzyme has been stored at 5° for at least 3 years with no appreciable loss in activity.

Properties of Endo R·*Bst* 1503

The *Bst* 1503 endonuclease is optimally active at pH 7.8 in buffer containing 10 mM Tris-HCl, 0.2 mM MgCl$_2$, and 6.6 mM 2-mercaptoethanol. Manganese ions support about 10% of the optimal activity. Other divalent cations were ineffective. The concentration of 2-mercaptoethanol was found not to be critical.

Bst 1503 is active over a relatively narrow temperature range. The main peak of activity falls between 50° and 70°. The optimum temperature is 65°. The stability of *Bst* 1503 to prolonged incubation at high temperatures was investigated by incubating the enzyme in reaction buffer at various temperatures prior to the addition of substrate. At 65° the enzyme retains its activity for at least 14 hr. However, at 70°, 65% of its activity is lost in 2 hr, and after 2 hr at 75° enzyme activity is undetectable.

The protein has a pI of 6.6–6.8, forming a single protein band upon isoelectric focusing over a range of pH 3–10. Enzyme activity comigrates with the protein band on isoelectric focusing gels.

The major active form of the enzyme has a sedimentation coefficient of 8.3 S and consists of four subunits each 46,000 daltons in SDS–

polyacrylamide gels. A minor active form of the enzyme was observed on acrylamide gels and in glycerol gradients. This minor species was 96,000 daltons and resolved in a single band (46,000 dalton) on SDS–acrylamide gels.

The data are consistent with a tetrameric and a dimeric composition for the two active forms, respectively. Interconversion of the two forms does not occur during storage. The conversion may occur in dilute solution as suggested by ultracentrifugation data.[6]

Endo R·*Bst*1503 is functionally associated with a specific modification methylase in the cell. Partially purified preparations of the *Bst*1503 modification methylase completely protect DNA from digestion by *Bst*1503 endonuclease.[8] Although *Bst*1503 is a multisubunit enzyme, it does not appear to share a common subunit with the methylase.

Endo R·*Bst*1503 is identical in its cleavage specificity on various DNA substrates to endo R·*Bam*H1.[9] The site of recognition for the two enzymes is, therefore, the same. It has not been rigorously shown that *Bst*1503 produces staggered ends at the site of cleavage. However, the annealing of *Bst*1503-produced fragments suggests the presence of single-stranded termini (unpublished observation).

[8] W. Levy, personal communication.
[9] G. A. Wilson and F. Young, this volume, Article [19].

[22] Purification and Properties of the *Sst*I Endonuclease

By ALAIN RAMBACH

The *Sst*I endonuclease is a restriction enzyme purified from a *Streptomyces* species which has been named *Streptomyces stanford*[1] and is available from the American Type Culture Collection as ATCC No. 29415. A crude lysate of these cells appears to cleave phage λ DNA into two large fragments[1] and phage λ plac₅ DNA into three large fragments[2]. *Sst*I endonuclease is the major endonuclease which can be found in the lysate. The enzyme produces cohesive ends and seems to cut most DNAs tested at rare sites.

Assay Method

The assay measures the cleavage of phage λ plac₅ DNA (or, alternatively, *Eco*RI cleaved phage λ DNA).

[1] S. Goff and A. Rambach, *Gene* 13, 347 (1978).
[2] C. Pourcel and P. Tiollais, *Gene* 1, 281 (1977).

Reagents

Reaction buffer: 10 m*M* Tris-HCl, pH 7.5, 10 m*M* 2-mercaptoethanol, 10 m*M* MgCl₂, 100 m*M* NaCl

Sample buffer: 1 *M* sucrose, 1% SDS, 50 m*M* EDTA, 1% saturated bromphenol blue

Tris–EDTA–borate buffer: 90 m*M* Tris-OH, 90 m*M* H₂BO₃, 2.5 m*M* EDTA, pH 8.2, 0.5 μg/ml ethidium bromide

Phage λ plac₅ DNA

Agarose

Procedure. Aliquots of 10 μl of phage λ plac₅ DNA (20 μg/ml) in reaction buffer are incubated with 1 μl of extract or column fraction to be assayed at 37° for 1 hr. To each aliquot 5 μl of sample buffer are added and the mixture is applied to a horizontal 0.7% agarose gel in Tris–EDTA–borate buffer. The DNA is electrophoresed at about 3 V/cm for 8 hr and photographed, with an orange filter, under uv light. When there is *Sst*I endonuclease activity, the pattern shows, in addition to the two larger fragments which are not easily separated from uncut DNA, a band corresponding to a 4 kilobases fragment and a less intense band corresponding to a 1 kilobase fragment.

Purification Procedure

Reagents

Buffer I: 10 m*M* Tris-HCl, pH 7.5, 10 m*M* 2-mercaptoethanol, 10 m*M* NaCl

Buffer II: 10 m*M* Tris-HCl, pH 7.5, 10 m*M* 2-mercaptoethanol, 1 *M* NaCl

Buffer III: 10 m*M* KPO₄, pH 7.4, 1 m*M* EDTA, 10 m*M* 2-mercaptoethanol, 10% glycerol

Buffer IV: 10 m*M* Tris-HCl, pH 8.3, 1 m*M* EDTA, 10 m*M* 2-mercaptoethanol, 10% glycerol

Buffer V: 10 m*M* KPO₄, pH 6.8, 5 m*M* 2-mercaptoethanol, 10% glycerol

Storage buffer: 0.15 *M* KCl, 20 m*M* Tris-HCl, pH 8.0, 50% glycerol

BioGel A, 0.5 M, 2.5 × 50 cm column

Phosphocellulose (P-11, Whatman), 0.9 × 15 cm column

DEAE–cellulose (DE-52, Whatman) 1.5 × 30 cm column

Hydroxyapatite (Bio-Rad HTP), 1.5 × 10 cm column

Aquacide

Procedure. Streptomyces stanford is grown in standard Luria broth (5 g/liter yeast extract, 10 m/liter Bacto-tryptone, 5 g/liter NaCl) at 30° with vigorous shaking for 2 days. The cells are pelleted at 15,000 g at 0° for 15 min, or filtered by aspiration on a filter paper. They are washed once in buffer I and frozen at $-20°$ for storage.

A cell pack (10 g) is thawed at 4° in 1.5 volumes of buffer I and then sonicated, with chilling in an ice-salt bath for eight 30-sec intervals, at the maximum setting of a Bronson sonicator. The sample is allowed to cool 30 sec between bursts of sonication. The sonicate is spun at 100,000 g for 45 min in an angle 40 rotor at 0° and the pellet discarded; all subsequent steps are performed at 4° Dry NaCl is added to the clear brown supernatant to a final concentration of 1 M and the sample is applied to a 2.5 × 50 cm column of BioGel A, 0.5 M previously equilibrated with buffer II. Sixty fractions of 4 ml each are collected while the column is washed with buffer II at a flow rate of 25 ml/hr. The fractions are assayed as described above, and the active fractions (25–35) are pooled and dialyzed with one change of buffer against 1 liter of buffer III.

The sample is applied to a 0.9 × 15 cm column of phosphocellulose (P-11, Whatman) previously equilibrated with buffer III at a flow rate of 10 ml/hr. The column is washed with 50 ml of buffer III while 5-ml fractions are collected and assayed. All the activity washes through the column and does not bind. The active fractions (2–11) are pooled, dialyzed twice against 1 liter of buffer IV and applied to a 1.5 × 30 cm column of DEAE–cellulose (DE-52, Whatman) previously equilibrated with buffer IV. The column is washed with 50 ml of buffer IV, then with a 500 ml linear gradient of 0–0.6 M NaCl in buffer IV, and then finally with 50 ml of 1 M NaCl in buffer IV. Sixty fractions of 10 ml are collected. An assay of this column shows a predominant activity eluting at 0.3 M NaCl. In addition to SstI, a minor activity elutes at 0.15 M NaCl.

The active SstI fractions (24–29) are pooled, dialyzed twice against 1 liter of buffer V and applied to a 1.5 × 10 cm column of Bio-Rad HTP hydroxyapatite in buffer V. The column is then eluted with a 400 ml linear gradient of 0–0.4 M KPO₄ pH 6.8 in buffer V while forty 10-ml fractions are collected.

The assay shows a single peak of activity eluting at about 0.15 M PO₄ (fractions 19–23). The active fractions are pooled, concentrated by dialysis against dry Aquacide (Calbiochem) dialyzed into storage buffer and stored at $-20°$. The enzyme retains activity for several months. The final yield is usually 3–4 ml of solution of which 1 μl cleaves 1 μg of DNA in 2 hr.

General Properties

The SstI endonuclease cleavage sites in phage λ DNA are at 0.51 and 0.53 map units on the genome. The phage λ plac₅ DNA carries a third site, at 0.42 map units.[2] *Sst*I does not cleave SV40, ColEI, or pSC101 DNA, but it cleaves adenovirus 2 (Ad2) DNA greater than 10 times.[1] *Sst*I cleaves polyoma DNA at three positions (0.53, 0.83, and 0.98 on the map).[3] The enzyme produces cohesive ends which can be sealed covalently after treatment with DNA ligase at low temperatures, as judged by electron microscopic studies.[4] An interesting bacteriophage vector system has been constructed to clone *Sst*I-cleaved DNA fragments. This is a derivative of the phage λ plac₅, having, by deletion, kept only one *Sst*I site, that one located in the middle of the *lac* gene.[2] When an *Sst*I fragment is inserted in this *lac*⁺ vector, the hybrid phage produced is *lac*⁻ and the corresponding plaque can be recognized by its color on a lactose indicator plate.

Another *Streptomyces* species, *Streptomyces achromogenes,* produces an enzyme termed *Sac*I, with identical specificity to that of *Sst*I (R. Roberts, personal communication).

[3] K. Shishido, personal communication.
[4] P. W. J. Rigby, personal communication.

[23] Preparation and Properties of Immobilized Sequence Specific Endonucleases

By YAN HWA LEE, ROBERT BLAKESLEY, LEONARD A. SMITH, and JACK G. CHIRIKJIAN

The type II restriction endonucleases have become valuable reagents for research in molecular biology. This is primarily due to their unique property of cleaving DNAs at a limited number of specific sites.[1-3] These enzymes have been employed for use in physical DNA mapping, plasmid construction, and gene cloning.[3-5] Most of these enzymes recognize and cleave DNA at specific sequences, which are usually in the form of a palindrome. Since there is evidence implying that these endonucleases are membrane bound,[5] one approach to studying them is to determine what effect insolubilization (i.e., mimicking membrane binding) has on their activity.

[1] K. J. Danna, G. H. Sack, Jr., and D. Nathans, *J. Mol. Biol.* **78**, 363 (1973).
[2] T. J. Kelly, Jr. and H. O. Smith, *J. Mol. Biol.* **51**, 393 (1970).
[3] W. Arber, *Angew Chem. Int. Ed. Engl.* **17**, 73 (1978).
[4] R. L. Rodriguez, F. Bolivar, H. M. Goodman, H. W. Boyer, and M. Betlach, *in* "Molecular Mechanisms in the Control of Gene Expression" (D. P. Nierlich, W. J. Rutter, and C. F. Fox, eds.), p. 471. Academic Press, New York, 1976.
[5] H. O. Smith and M. L. Birnstiel, *Nucleic Acids Res.* **3**, 2387 (1976).

The immobilized enzymes we have prepared retain their activity for at least 6 months. They can be packed in the form of a column or rapidly pelleted from solutions by centrifugation. These procedures allow quick removal of restriction endonucleases from reaction mixtures. Such recovered enzyme are reuseable several times over, permitting cleavage of large amounts of DNA with relatively few units of enzyme. Codigestion by matrix-bound restriction endonucleases avoids the inhibition frequently observed upon subsequent digestion by a second endonuclease. In addition, no phenol extraction is required, resulting in increased recovery of digestion products and eliminating tedious and time-consuming manipulations. The following section describes the catalytic and functional characterization of covalently bound restriction endonucleases.

Materials

Enzymes. EcoRI and BamHI, restriction endonucleases from *Escherichia coli* RY13 and *Bacillus amyloliquefaciens* H, respectively, were obtained from Bethesda Research Laboratories, Inc., or prepared in our laboratory using traditional fractionation procedures[6,7] or biospecific affinity matrices developed in our laboratory.[8] When prepared in our laboratory, by standard methods, cells (10 g) were suspended in 50 mM phosphate (K$^+$), pH 7.5, 0.5 mM EDTA, and 5 mM 2-mercaptoethanol, then broken by sonication. The supernatant from a 2-hr, 100,000 g centrifugation of the sonicate was fractionated on a phosphocellulose column equilibrated with 20 mM phosphate (K$^+$), pH 7.5, 0.1 mM EDTA, 5 mM 2-mercaptoethanol, and 10% glycerol (column buffer). The adsorbed endonuclease activity was eluted upon application of a gradient of 0 to 1 M KCl in column buffer. Peak fractions containing restriction activity were pooled and freed of contaminating nuclease by fractionation on an hydroxylapatite column, using 0 to 1 M phosphate (K$^+$) gradient elution in column buffer. The biospecific matrices used were pyran covalently coupled to hydrazine-activated Sepharose and Cibacron Blue 3FGA coupled to Sepharose by a cyanogen bromide linkage. Selected fractions containing active enzyme from either of the described procedures or commercial preparations were used for coupling restriction endonucleases to activated Sepharose.

[6] P. Modrich and D. Zabel, *J. Biol. Chem.* **251**, 5866 (1976).
[7] L. A. Smith and J. G. Chirikjian, *J. Biol. Chem.* **254**, 1003 (1979).
[8] J. George, J. G. Chirikjian, *Nucl. Acids Res.* **5**, 2223 (1978).

Methods

CNBr-Activated Sepharose. In order to activate the Sepharose 4B (Pharmacia) 100 g were suspended in 100 ml of distilled water. The pH was monitored continuously throughout the reaction. To the suspended Sepharose was added 100 mg of finely divided solid cyanogen bromide. The pH was adjusted to 11 by addition of 8 *N* NaOH with vigorous stirring. After 10 min at 20°, 200 ml of cold 1 *M* phosphate (K$^+$) buffer (pH 7.2) was added. The suspension was washed by filtration on a scintered glass funnel with 20–30 volumes of cold buffer. Activated Sepharose was stored at 4° in an equal volume of buffer and routinely used within 24 hr.

Enzyme Insolubilization and Storage Conditions. The enzymes were insolubilized by combining approximately 2000 units of *Eco*RI or *Bam*HI in 0.5 ml of buffer containing 10 m*M* phosphate (K$^+$), pH 7.2, 0.5 m*M* EDTA, and 50% glycerol with 0.5 ml of packed CNBr-activated Sepharose. The resulting slurry was adjusted to 0.15 *M* phosphate (K$^+$) (pH 7.5), then gently mixed "end over end" at 4° for 16 hr. Residual reactive groups in the Sepharose were blocked by suspension of the gel slurry in 0.1 *M* Tris-HCl (pH 7.5) for 2 hr at 4°. The excess uncoupled endonuclease was removed from the gel by centrifugation, resuspension of the gel in 0.5 *M* NaCl, 25 m*M* Tris-HCl (pH 7.5), 10% glycerol, and 1 m*M* dithiothreitol followed by recentrifugation. This washing procedure was repeated four more times. Finally, the enzyme coupled gel was equilibrated with 50 m*M* Tris-HCl (pH 7.5), 20% glycerol, and 1 m*M* dithiothreitol by washing five times in this buffer. The final gel slurry was stored at 4° and no soluble enzyme was detected upon storage. The procedure described has been applied successfully to four other restriction endonucleases (*Hin*dIII, *Taq*I, *Hpa*I, and *Hha*I).

Matrix-bound enzyme, lyophilized to dryness, retained greater than 90% of its activity. Such lyophilized enzyme samples could be stored at room temperature without substantial loss of activity for at least 6 months.

Coupling Efficiency and Enzyme Activity Recovery. The efficiency of covalent binding of enzyme to Sepharose is sensitive to the ratio of protein in the starting enzyme solution and the extent of cyanogen bromide activation of Sepharose. We define coupling efficiency as the ratio of the total units of enzyme recovered bound to Sepharose to the total units of initial activity. The efficiency varied from 20–90% for *Eco*RI and *Bam*HI coupling experiments. To achieve optimal coupling efficiency, we strongly recommend experimentation on the small scale for a specific enzyme preparation to be coupled to activated Sepharose. Coupling efficiency i*

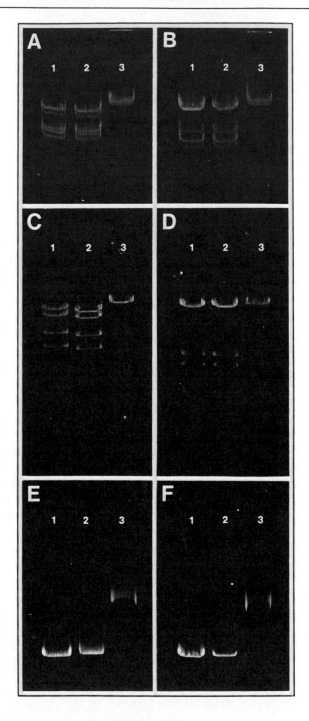

reproducible and can be scaled when the proper combination of enzyme and Sepharose is determined.

Enzyme Assay. Assay for endonuclease activity was carried out in a reaction mixture (40 μl) containing 0.4 to 2 μg of either λ, adenovirus, or SV40 DNA, 50 mM Tris-HCl (pH 7.5), 5 mM MgCl$_2$, 2 mM 2-mercaptoethanol, 50 mM NaCl, and 100 μg/ml autoclaved gelatin. Reaction mixtures were incubated for 1 hr at 37°. Reactions containing matrix-bound enzyme were then centrifuged and the supernatants removed. To these supernatants or to reaction mixtures containing unbound enzyme was added one-fourth volume of a solution containing 50% glycerol and 0.02% bromphenol blue.

Adenovirus and λ DNA fragments were separated on a 1.4% agarose slab gel (10 × 12 cm) at 100 V for 3 hr following procedures previously described.[9] SV40 DNA digests were fractionated on a 1.2% agarose slab gel at 200 V for 2 hr using Tris-borate–EDTA buffer (90 mM Tris-borate, pH 8.3, 2.5 mM Na$_2$EDTA).[10] The gels were stained with ethidium bromide (1 μg/ml) for 10 min and photographed during illumination with a short-wave ultraviolet light.

A unit of enzyme activity is defined as that amount of enzyme required to completely digest 1 μg of λ DNA at 37° in 60 min. Routinely, enzymatic activity is measured by serial dilution of the enzyme into the linear range to determine the minimum amount of enzyme necessary to obtain complete digest.

Comparative Properties of Soluble and Matrix-Bound Bam*HI and* EcoRI—*Analysis of DNA Digestion Patterns.* In order to demonstrate that coupling of restriction endonucleases to Sepharose did not alter their catalytic properties three standard DNAs were tested as substrates. The fragmentation pattern of λ, adenovirus, or SV40 DNA produced by incubation with the insolubilized enzymes was identical to that produced with free enzymes (Fig. 1). Each panel of Fig. 1 is a gel electrophoregram of three reactions: viral DNA alone, viral DNA incubated with soluble enzyme, and viral DNA incubated with insoluble enzyme. Phage λ DNA digestion with Bam HI (Fig. 1A) or Eco RI (Fig. 1B) produced the expected fragment patterns. Adenovirus DNA was cleaved to four fragments by

[9] B. Sugden, B. DeTroy, R. J. Roberts, and J. Sambrook, *Anal. Biochem.* **68**, 36 (1975).
[10] A. C. Peacock and C. W. Dingman, *Biochemistry* **7**, 668 (1968).

FIG. 1. Comparison of viral DNA digests of soluble and insoluble Bam HI or Eco RI. (A), (C), and (E) are reactions with Bam HI and (B), (D), and (F) are reactions with Eco RI, each assayed in the standard reaction mixture as described in the text. The substrates were (A) and (B) phage λ DNA, (C) and (D) adenovirus type II DNA, and (E) and (F) SV40 component I DNA. In all panels, lane 1 was 1 unit of soluble enzyme, lane 2 was 1 unit of insolubilized enzyme, and lane 3 was no enzyme added to the reaction mixture.

TABLE I
STABILITY STUDIES OF SOLUBLE AND INSOLUBLE *Bam*HI AND *Eco* RI

Enzyme	Heat treatment	Activity[a] (%)
*Bam*HI (soluble)	None	100
	45°, 5 min	100
	55°, 5 min	10–20
	60°, 2 min	0
*Bam*HI (insoluble)	None	100
	60°, 2 min	90–100
	65°, 5 min	80–90
*Eco*RI (soluble)	None	100
	40°, 5 min	0
*Eco*RI (insoluble)	None	100
	40°, 5 min	100
	45°, 5 min	90
	50°, 5 min	80

[a] Activity based on the minimum amount of enzyme from a series of dilutions necessary to completely digest 1 μg of λ DNA with appearance of expected fragment pattern.

*Bam*HI and five fragments by *Eco*RI (Fig. 1D). Both enzymes converted SV40 DNA from superhelical component I to linear component III (Fig. 1E and F). In all cases, when insoluble replaced soluble enzyme in the reaction, no difference in the patterns of DNA fragments could be detected.

These results indicate that both linear and superhelical DNAs could be effectively digested by insolubilized enzymes. In addition, when appropriate incubation conditions were imposed on the *Eco*RI-coupled Sepharose, the *Eco*RI* pattern[11] also was obtained. Thus, coupling via the cyanogen bromide linkage apparently does not interfere with critical elements of the active site of these enzymes. This is particularly important since stereochemical hinderances upon coupling of enzymes are well documented in other systems.[12,13]

Thermal Stability Studies. The effect of thermal denaturation on the activities of the soluble and insoluble enzymes demonstrated that coupling enhanced the thermal stability by at least 10° for each enzyme (Table I). Exposure of stock solutions of the insolubilized *Bam*HI to 65° for 5 min resulted in little loss of activity, while the soluble form treated for 5 min at 55° lost nearly all activity. In fact, treatment of soluble *Bam*HI at 60° for only 2 min caused complete inactivation, whereas the activity of bound

[11] B. Polisky, P. Greene, D. E. Garfin, B. J. McCarthy, H. M. Goodman, and H. W. Boyer, *Proc. Natl. Acad. Sci. U.S.A.* **72**, 3310 (1975).
[12] A. Bar-Eli and E. Katchalski, *J. Biol. Chem.* **238**, 1690 (1963).
[13] W. E. Hornby, M. D. Lilly, and E. M. Crook, *Biochem. J.* **98**, 420 (1966).

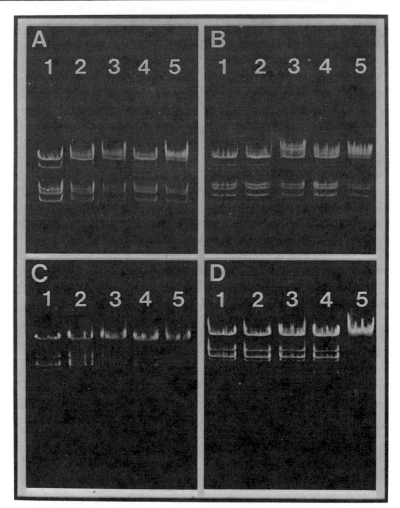

FIG. 2. Comparison of time course assays for soluble and insoluble forms of *Bam* HI and *Eco* RI. (A) and (B) are reactions with matrix bound and free *Bam* HI, respectively. (C) and (D) are reactions with matrix bound and free *Eco* RI, respectively. Each assay mixture containing 1 μg of λ DNA was treated in the standard reaction mixture as described in the text. In each panel the time of incubation was as follows: lane 5, 10 min; lane 4, 20 min; lane 3, 30 min; lane 2, 45 min; lane 1, 60 min.

Bam HI remained stable to incubation at 60° for 2 min. Free *Eco* RI, when incubated at 40°, lost all activity, whereas *Eco* RI bound to Sepharose lost only 20% of its activity when treated at 50°. Although one cannot generate from these data reaction temperature optima for either enzyme, the results clearly demonstrate an improved thermal stability upon insolubiliza-

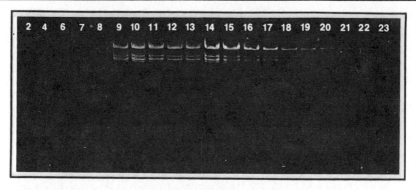

FIG. 3. A flowthrough system for *Eco*RI-Sepharose. Phage λ DNA at a concentration of 300 μg/ml was digested as described in the text. Reaction products were displayed on 1.4% agarose using Tris-acetate buffer, pH 7.8. Electrophoresis was at 13 V/cm at constant voltage for 2 hr. Fractions were 25 μl and total elution volume fraction 8–23 was 0.36 ml. Fractions 1–7 represent buffer displaced by the DNA solution.

tion. Such increased stability also makes this form of these enzymes attractive for study of DNA structures at high temperatures.[14]

Effect of Insolubilization on Reaction Kinetics for EcoRI *and* BamHI. In order to assess the kinetic activity of *Bam*HI and *Eco*RI upon insolubilization, a time course assay for each enzyme was performed. Aliquots of each reaction were removed at various times, then analyzed by agarose gel electrophoresis. As controls, soluble *Bam*HI and *Eco*RI were treated in an identical manner. As shown in Fig. 2, results for the matrix-bound form of each enzyme was similar to that of its respective soluble form. Thus, no apparent change in the time course of DNA degradation was observed upon insolubilization of the enzyme.

Usage of an EcoRI–Sepharose *Column for DNA Digestion.* An advantage of insolubilized enzymes is their usage as chromatographic matrices in continuous flow-through systems. The feasibility of using *Eco*RI–Sepharose in the form of a chromatographic column (0.4 × 1.4 cm) was tested for the digestion of λ phage DNA. In a typical experiment, λ DNA in 0.8 column volume of *Eco*RI assay buffer was applied to the *Eco*RI–Sepharose column previously equilibrated with 50 mM Tris-HCl, pH 7.4, 2 mM β-mercaptoethanol, and 20% glycerol. The column containing λ DNA (50 μg) at a concentration of 300 μg/ml was incubated at 37° for 15 min. The column was then cooled to room temperature and digestion products were recovered by elution with 0.36 ml of the *Eco*RI assay buffer (Fig. 3). Several batches of λ phage DNA have been pro-

[14] R. W. Blakesley, J. B. Dodgson, I. F. Nes, and R. D. Wells, *J. Biol. Chem.* **252**, 7300 (1977).

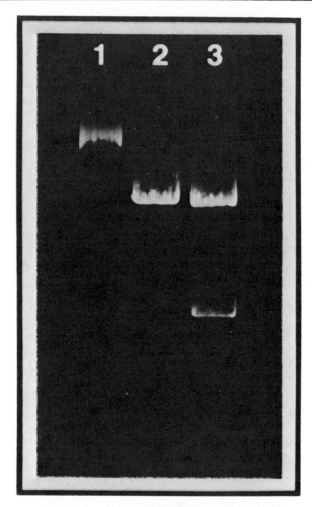

FIG. 4. Codigestion of DNA with insolubilized EcoRI Sepharose and BamHI-Sepharose. SV40 DNA (1 µg) was codigested by EcoRI-Sepharose and BamHI-Sepharose as described in the text. Both enzymes cut SV40 DNA at a single site and yield the two fragments which are displayed on 1.4% agarose using Tris-borate EDTA (pH 8.3) buffer. Lane 1, SV40 DNA, lane 2, BamHI-Sepharose digest of SV40 DNA, lane 3, BamHI-Sepharose and EcoRI-Sepharose codigest of SV40 DNA. Electrophoresis was for 1.5 hr at 13 V/cm at constant voltage.

cessed yielding digestion to the expected fragments. The packed column remains stable when stored at 4° between DNA digestions.

Codigestion of DNA with EcoRI–Sepharose and BamHI–Sepharose. A procedural advantage for employing insolubilized restriction endonucleases has been elimination of the phenol extractions frequently needed in sequential digestions of a specific DNA by more than one enzyme. Losses of digested DNA is usually incurred during the phenol extraction and subsequent dialysis. Alternatively, in the absence of extractions, inhibition is observed upon codigestion of DNAs by more than one enzyme. Using EcoRI–Sepharose and BamHI–Sepharose we have reproducibly achieved codigestion of DNA to the expected fragments (Fig. 4). The reaction was carried out in the EcoRI reaction buffer which contains salt to repress EcoRI* activity. SV40 DNA (1 μg) was codigested with EcoRI-Sepharose and BamHI-Sepharose at 37° for 15 min. The reaction was carried out batchwise in a 0.5-ml plastic minifuge tube. DNA digestion products were recovered by centrifugation of the mixture to remove the insolubilized enzymes and the supernatant was submitted to electrophoresis.

Acknowledgments

The research at Georgetown University was supported by a Grant (CA 16914) for the National Cancer Institute (DHEW). J. G. C. is a scholar of the Leukemia Society and to whom requests should be addressed.

Section IV

Purification and Properties of Enzymes Acting at Sites with Altered Bases

Articles 24 through 31

A. Single-Stranded Specific Endonucleases
Articles 32 through 34

B. Exonucleases
Article 35

C. DNA *N*-Glycosylases
Articles 36 and 37

[24] Purification and Properties of Pyrimidine Dimer Specific Endonucleases from *Micrococcus luteus*[1]

By SHEIKH RIAZUDDIN

Assay Method

Principle. The assay measures conversion of superhelical ϕX174 RFI DNA into relaxed RFII or RFIII forms as a consequence of endonucleolytic incision of closed circular DNA. When the treated DNA is denatured by raising the pH to 12, followed by neutralization to pH 8, incised DNA is converted to single-strand species, whereas covalently closed circular DNA is preserved. During filtration through B84-A (Schleicher and Schuell) membranes, single-stranded DNA is retained and thus separated from unincised DNA.

Reagents

1. Reaction mixture (total volumn 0.3 ml) contains
   Tris·HCl, 10 mM, pH 7.4
   EDTA, 1 mM
   NaCl, 50 mM
   ^{3}H-Labeled ϕX174 RFI DNA, 10,000 cpm, 25 pmole of mononucleotide equivalent
2. Denaturing buffer contains
   Sodium phosphate, 100 mM, pH 12.0
   NaCl, 30 mM
   EDTA, 25 mM
3. Renaturing buffer contains
   Tris-HCl, 2 M, pH 4.0

Procedure. After incubation at 37° for 30 min, the reaction mixture is brought to room temperature and 2 ml of denaturing buffer are added followed by the addition of 0.4 ml of renaturing buffer. After 5 min, the reaction contents are filtered through B 84-A membrane. The reaction tubes are washed twice with 4 ml of cold saline-citrate and the filters are dried and counted in 2 ml of toluene containing scintillation fluid. The radioactivity retaining on the membrane filter is taken as a measure of endonuclease incising activity. The pyrimidine dimer damage specific ac-

---

[1] S. Riazuddin and L. Grossman, *J. Biol. Chem.* **252**, 6280 (1977).

METHODS IN ENZYMOLOGY, VOL. 65

tivity is measured by the difference in activities against ultraviolet irradiated (20 J/m^2) and unirradiated DNA.[2]

One unit of enzyme is that amount which produces, on the average, one break per ϕX174 RFI DNA molecule in a total of 25 pmole of substrate in 30 min at 37°.

Purification Procedure

All operations are performed at 0°–4°. Centrifugations are at 20,000 g for 30 min. Unless otherwise stated, all solvents contain 1 mM EDTA, 1 mM 2-mercaptoethanol, and 10% glycerol.

Growth of Cells. *Micrococcus luteus* cells are grown at 34° under forced aeration in a Microferm laboratory fermenter (New Brunswick Scientific) in 12 liters of Columbia broth (Difco). Cells are harvested at a call density of about 5 × 10^9/ml by centrifugation in a Sharples continuous flow centrifuge and washed with 50 mM phosphate buffer (pH 7.6).

Crude Extract. Cells (75 g) are suspended by homogenization in a Waring Blendor in 50 mM potassium phosphate (pH 7.6) in a final volume of 400 ml. The cell suspension containing 100 mg of lysozyme is incubated at 37° for 10 min and then quickly cooled by the addition of 200 ml of ice. The dark brown viscous mass so obtained is broken by sonic irradiation for 7 min at full power with a Branson Sonifier. Cellular debris are removed by centrifugation to yield 550 ml of a supernatant fluid (fraction A).

DEAE–Cellulose Chromatography. A column of Whatman DE52 (30.5 cm^2 × 12 cm) is washed with 6 liters of 50 mM potassium phosphate at pH 7.0 (buffer X). Fraction A is loaded onto the column at 2 ml/min and washed with 300 ml of the same buffer. The flow-through and wash are combined to a final volume of about 900 ml (fraction B).

Phosphocellulose Chromatography. A column of Whatman P-11 (4.9 cm^2 × 23 cm) is equilibrated by washing with 4 liters of buffer X. Fraction B, after dialysis against 10 volumes of the same buffer, is applied to the column with an adjusted flow rate of 1 ml/min. After washing the column with 200 ml of buffer X, the protein is eluted in 15-ml fractions with a 1600 ml linear gradient of 50–300 mM potassium phosphate at pH 7.4. The pyrimidine dimer damage specific activities eluting between 125 and 175 mM potassium phosphate (Fig. 1) are pooled for further purification (fraction C).

Sephadex G-75 Chromatography. A column of Sephadex G-75 (4.9 cm^2 × 100 cm) is prepared and washed with 100 mM potassium phosphate

[2] S. Riazuddin and L. Grossman, *J. Biol. Chem.* **252,** 6287 (1977).

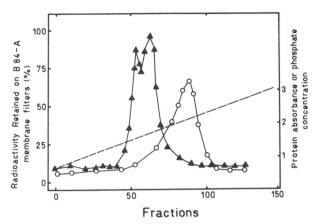

FIG. 1. Elution pattern of Py<>Py correndonucleases from phosphocellulose (P-11): (▲——▲) UV damage-specific endonuclease activities, (O——O) absorbance at 280 nm, (---) concentration of the eluting phosphate buffer.

(pH 7.6), 5 mM 2-mercaptoethanol, 5 mM EDTA, and 10% glycerol (buffer Y). Fraction C after concentration by precipitation with 70% ammonium sulfate is layered onto the column and eluted with buffer Y at a rate of 0.3 ml/min. Eighty 5-ml fractions are collected and assayed. The major activity eluting as a nonsymmetrical peak between fraction 44 and 51 are pooled and concentrated to 5 ml by adsorption to a 2.5 ml column of phosphocellulose and eluting with 10 ml of 300 mM potassium phosphate at pH 7.6 (fraction D).

DNA–Cellulose Chromatography. A column of uv-irradiated DNA–cellulose (0.28 cm² × 8 cm) is washed with 100 ml of 10 mM Tris-HCl at pH 7.6 (buffer Z). Fraction D is dialyzed for 4 hr against 4 liters of buffer Z and applied to the column at a rate of 1 ml in 20 to 30 min. The column is washed with 5 ml of the same buffer and eluted with a 30 ml linear gradient of 10 mM and 2 M NaCl in buffer Z. Twenty 1.5-ml fractions are collected and assayed. The endonuclease activities are eluted as two symmetrical peaks at fractions 3 to 5 and fractions 10 to 14. The two peaks are pooled separately, dialyzed against 3 liters of buffer Z and stored in liquid nitrogen.

Isoelectric Focusing. Fraction D is also resolved into two endonucleases when focused in a vertical isoelectric focusing apparatus (LKB, type 8101) containing a 104 ml linear gradient of 5 to 50% glycerol with 2% Ampholine in the pH range 3 to 10. After 72 hr at 600 V, two activities with isoelectric points 4.7 and 8.7 are separated. The respective fractions are pooled and dialyzed against buffer Z and stored in liquid nitrogen.

Nomenclature

The two enzymes are designated as Py<>Py correndonucleases I and II in that they are endonucleases whose action on pyrimidine dimer-containing DNA eventuates in correctional repair mechanisms. Py<>Py correndonuclease I is eluted first from a column of uv-irradiated DNA–cellulose and has a pI value of 4.7.

Comments on the Purification Procedure. The DEAE–cellulose column is used to remove the DNA and a major part of the exonucleases. Major overall purification of the enzymes is obtained during chromatography on phosphocellulose and Sephadex G-75. The phosphocellulose step separates these enzymes from activities acting on nonpyrimidine dimer photoproducts. The Sephadex step removes a high molecular weight nonspecific endonuclease from the damage-dependent ones which are not resolved on either of the two columns. Complete separation of Py<>Py correndonucleases I and II is achieved either by affinity chromatography on a uv-irradiated DNA–cellulose column or by isoelectric focusing.

A summary of the purification procedure is given in Table I.

The purified enzymes are not contaminated with endonucleolytic activities against native ϕX RFI or single-stranded circular ϕX174 DNA. They further resulted in no loss to transforming activity of *Bacillus subtilis* DNA when incubated with the DNA either singly or together under the standard assay conditions. There is only one peak of activity recovered when either enzyme is individually subjected to isoelectric focusing.

Properties of Purified Enzymes

Stability. The two correndonucleases after separation on a DNA–cellulose column or isoelectric focusing column are very unstable, undergoing complete inactivation in 3 to 4 weeks in 50% glycerol at $-20°$. The losses in activity, however, are considerably minimized when the enzymes are stored in liquid nitrogen. Freeze-thawing does not inactivate either correndonuclease.

When incubated at 45° and 55°, both enzymes exhibit monophasic inactivation. Py<>Py correndonuclease I is more stable than Py<>Py correndonuclease II at either temperature. During a 5-min incubation at 55°, the activity of the latter enzyme is reduced to less than 25%, whereas the former correndonuclease I still retains 80% of its original activity.

General Requirements. Both enzymes exhibit broad pH optima between 7.0 and 7.4 in either Tris-HCl or potassium phosphate. The enzymes show no obligatory cofactor requirement and there is no absolute

TABLE I

PURIFICATION OF Py<>Py CORRENDONUCLEASES FROM 75 G OF *M. Luteus* CELLS

Fraction	Correndo-nucleases	Volume (ml)	Activity (BAP units^a × 10⁻³)		Protein (mg/ml)	Specific activity^c	Yield
			uv-dependent^b	Nonspecific			
Extract	I + II	580	216	193.5	13.8	26.9	86
DEAE–cellulose	I + II	900	250	105.9	5.5	49.8	100
Phosphocellulose	I + II	120	104	4.9	0.6	1370	41
Sephadex G-75	I + II	40	54.3	0.3	80.04	27200	22
DNA–cellulose	I	4	1.7				0.6
	II	7	11.5				4.6

^{a} A BAP unit is defined as the amount of activity which incised 10 pmole of pyrimidine dimers in 30 min at 37°.
^{b} uv-Dependent activity is determined by the difference in total activities against uv-damaged DNA and native DNA.
^{c} Specific activity is defined as bacterial alkaline phosphatase units of ultraviolet-dependent activity per milligram of total proteins.

requirement for cations. Higher ionic strength is required, however, for normal levels of Py<>Py correndonuclease I activity which is optimally active in either 10 mM Tris-HCl supplemented with 50 mM NaCl or in 50 mM potassium phosphate. Py<>Py correndonuclease II requires lower ionic strength and is fully active in 10 mM Tris-HCl. Addition of 100 mM NaCl or KCl or 15 mM MgCl₂ reduces the activity of Py<>Py correndonuclease II to 25%. The enzymes do not require 2-mercaptoethanol in the standard assay mixture, and the activities of two enzymes are not significantly changed by the presence of 25 mM N-ethylmaleimide. This indicates, therefore, there is no sulfhydryl requirement for enzyme activity.

Substrate Specificity. The two enzymes have no appreciable activity against X-irradiated, alkylated, cross-linked, or OsO₄-treated DNA. Both enzymes are highly specific for uv-radiation damage. However, their incision activities as a function of the extent of photochemical damage to φX174 RFI DNA are different. Py<>Py correndonuclease II enzyme produces 63% incision of the molecules at 7 J/m² as opposed to 11 J/m² required by the other correndonuclease to produce a similar level of breaks.

The activities of the two correndonucleases are markedly dependent on the secondary structure of the DNA. While double-stranded DNA is an effective substrate for either enzyme, single-stranded DNA is attacked only by correndonuclease I. Synthetic homopolymers are effective substrates for both enzymes only under specific ionic environment and must be at least ten nucleotides long in chain length.

Products of the Reaction. Both enzymes make single-strand incision in uv-irradiated DNA and catalyze *in vitro* removal of T<>T, T<>C, C<>T, and C<>C cyclobutane dimers by the exonucleolytic activity of DNA polymerase I or *E. coli* exonuclease VII. Although incision in either case leaves a 3'-hydroxyl and 5'-phosphoryl termini with the cyclobutane dimer lying 5' to the break, the nature of the termini generated seems to be different. These 5'-phosphoryl termini resulting from the action of Py<>Py correndonuclease I are easily dephosphorylated by the action of bacterial alkaline phosphatase, whereas enzyme II–DNA product is insensitive to the phosphatase action. Dephosphorylation of sites generated by the latter enzyme is achieved only when the DNA is melted by raising the temperature to above 65°. Furthermore, polynucleotide ligase cannot form a phosphodiester bond between the termini produced by Py<>Py correndonuclease I, and these termini have affinity for *E. coli* binding protein which can protect this region from the hydrolytic action of *E. coli* exonuclease VII. In contrast to it, the incision produced by Py<>Py correndonuclease II is sensitive to polynucleotide ligase and is not protected by the binding protein against exonuclease VII.

Involvement of the Two Correndonucleases in Cellular Repair

Py<>Py correndonucleases I and II are present in the ratio of $1:5$ in wild-type *M. luteus* cells and are completely absent in a uv-sensitive mutant (DB-7) obtained by treatment of the wild-type cells with *N*-methyl-*N'*-nitro-*N*-nitrosoguanidine. The absence of both these enzymes from DB-7 is suggestive of their role *in vivo*. The presence of Py<>Py correndonuclease II in a transformant (DB-200), obtained by treatment of DB-7 cells with wild-type DNA, leads to perceptible increases in ultraviolet resistance. Restoration of normal levels of the second enzyme in a transformant (DB-400), obtained through another transformation step, results in a recovery in uv resistance comparable to the wild type.

All these strains.of *M. luteus* can be made permeable to small molecular weight DNA precursors by treatment with 1% toluene. The uv-induced repair-type DNA synthesis investigated in the permeabilized strains is an expression of the decreased correndonuclease levels which are correlated with the amount of repair synthesis. In the transformant DB-200 and the mutant DB-7 which lacks one or both the enzymes, these values are 45 and 12% of wild type. In DB-400, however, in which both the activities are restored, amount of repair synthesis is comparable to that in the wild-type parent. It seems evident, therefore, that both correndonucleases are involved in cellular repair.

[25] Purification and Properties of a Pyrimidine Dimer-Specific Endonuclease from *E. coli* Infected with Bacteriophage T4[1]

By Errol C. Friedberg, Ann K. Ganesan, and Patricia C. Seawell

Introduction

The pyrimidine dimer-specific endonuclease of bacteriophage T4 is coded by the *v* (*denV*) gene.[2] The enzyme has been called endonuclease V of phage T4[3] or the T4 uv endonuclease. In recent years this enzyme has

[1] This work was supported by grants from the United States Public Health Service (CA 12428 and GM 19010), the American Cancer Society (NP161B and NP 174), and by contracts with the United States Department of Energy [EY-76-5-03-0326 (PA7 and 32)]. We thank Richard Cone for instruction in several of the analytical procedures.
[2] K. Sato and M. Sekiguchi, *J. Mol. Biol.* **102**, 15 (1976).
[3] S. Yasuda and M. Sekiguchi, *Proc. Natl. Acad. Sci. U.S.A.* **67**, 1839 (1970).

been extensively used as a probe for the detection and measurement of pyrimidine dimers in DNA and for studying DNA repair of uv damage in bacterial and mammalian cells.[4-7] Enzyme-catalyzed cleavage of DNA strands containing pyrimidine dimers can be measured by a variety of techniques. For detecting pyrimidine dimers a preparation of enzyme free of activity against unirradiated DNA or DNA containing other types of damage is required. In this article we describe several methods for assaying T4 endonuclease V as well as the partial purification and general properties of the enzyme. Our goal has been to describe in detail a reproducible procedure for obtaining stable preparations of the enzyme which can be used as a specific probe for pyrimidine dimers. Although not homogeneous, these preparations are sufficiently pure for most studies utilizing endonuclease V to measure pyrimidine dimers. In our experience, more extensive purification results in increased lability of this enzyme.

Assay of Endonuclease Activity

For the purpose of purification the enzyme can be assayed by one of three general methods.

Coupled Nuclease Assay

This assay measures the degradation of uv-irradiated radioactive DNA to acid-soluble product by coupling the specific incising activity of T4 endonuclease V with the excising activity of an extract of *E. coli* infected with phage T4v_1. The extract is deficient in endonuclease V activity but contains both host and phage-coded nucleases known to degrade specifically incised uv-irradiated DNA.[8-10]

Reagents

Exonuclease: crude extract prepared by sonication of phage T4v_1 infected *E. coli* in 50 mM Tris-HCl buffer, pH 8.0

DNA: [^{3}H]DNA from *E. coli* or phage T4 (specific activity 10^3–10^5 cpm/μg)[11]

[4] A. K. Ganesan, *Proc. Natl. Acad. Sci. U.S.A.* **70**, 2753 (1973).

[5] M. Paterson, *Adv. Radiat. Biol.* **7**, 1 (1978).

[6] A. A. van Zeeland, *in* "DNA Repair Mechanisms: ICN-UCLA Symposia on Molecular and Cellular Biology" (P. C. Hanawalt, E. C. Friedberg, and C. F. Fox, eds.), p. 307. Academic Press, New York, 1978.

[7] C. A. Smith and P. C. Hanawalt, *Proc. Natl. Acad. Sci. U.S.A.* **75**, 2598 (1978).

[8] S. Yasuda and M. Sekiguchi, *J. Mol. Biol.* **47**, 243 (1970).

[9] E. C. Friedberg and J. J. King, *J. Bacteriol.* **106**, 500 (1971).

[10] E. C. Friedberg, *Photochem. Photobiol.* **21**, 277 (1975).

[11] M. Sekiguchi, S. Yasuda, S. Okubo, H. Nakayama, K. Shimada, and Y. Takagi, *J. Mol. Biol.* **47**, 231 (1970).

E. coli tRNA
50 mM Tris-HCl buffer, pH 8.0
MgCl₂
20% Cold trichloroacetic acid (TCA)
1% Bovine serum albumin

Procedure. Incubation mixtures (1.0 ml) contain 10 μg [³H]DNA irradiated with 1000 J/m² at 254 nm, 100 μg tRNA (to inhibit endonuclease I activity), 1.0 mM MgCl₂, 50 mM Tris-HCl buffer, limiting amounts of T4 endonuclease V activity, and saturating amounts of T4v_1 infected cell extract (usually 1–2 mg of protein). Saturation is defined as the amount of extract that results in maximal release of acid-soluble radioactivity during 15 min of incubation at 37° with the amount of endonuclease V used. Under these conditions the amount of endonuclease V is rate limiting for the reaction with uv-irradiated DNA, and the release of acid-soluble radioactivity bears a linear relationship to endonuclease V activity.[9] After 15 min at 37°, reactions are terminated by placing incubation tubes on ice and adding 0.5 ml cold 1% bovine serum albumin as carrier and 0.5 ml cold 20% TCA. After standing on ice for 15 min the tubes are centrifuged and 1 ml of the supernatant is added to 10 ml of Triton : Omnifluor (1 : 10) for measurement of radioactivity. Samples containing no added endonuclease V are included in each experiment to provide a background value of DNA degradation which is subtracted from the values obtained in the presence of endonuclease. One unit of endonuclease activity is defined as the amount of enzyme which, in the presence of saturating amounts of crude extract of T4v_1 infected *E. coli,* results in the formation of 1 nmole of acid-soluble nucleotide product per hour under these conditions.

Filter Binding Assay

This assay measures the selective binding of T4 endonuclease V to radioactive DNA containing pyrimidine dimers.[12] The resulting DNA–protein complexes are trapped on nitrocellulose filters, and the amount of nonfilterable radioactivity is measured by liquid scintillation spectrometry. This assay can be used to measure T4 endonuclease V in crude lysates as well as in purified fractions providing they do not contain polyethylene glycol (PEG).

Reagents

Filters: Millipore HAWP, 0.45 μm pore size, 25 mm diameter
Filter holders: Tracerlab (ICN) E-8B precipitation apparatus[13]

[12] P. C. Seawell, T. J. Simon, and A. K. Ganesan, *J. Supramol. Struct., Suppl.* **2**, 45 (1978).
[13] In principle any similar filtration unit could be used; however, in our experience, it is difficult to control flow rates in units equipped to filter more than two samples simultaneously.

DNA: ColE1 [^{14}C]DNA (specific activity 57,000 cpm/μg),[14,15] irradiated with 200 J/m^2 at 254 nm

Reaction buffer: 9 mM Tris-HCl (pH 8.0 at 20°), 9 mM EDTA, 90 mM NaCl, 10% (v/v) ethylene glycol (buffer A)

Stop and rinse buffer: 2.1 M NaCl, 0.21 M trisodium citrate (buffer B)

Procedure. The approximate number of filters needed for an experiment is floated upon the surface of distilled water. Any filters that do not wet quickly and uniformly are discarded. The remaining filters are submerged in water until used. The filter holders are set up and connected to a vacuum adjusted to ensure a maximum flow rate of 2–3 ml/min through filters. The filter holders are chilled on ice before use. After each sample they are rinsed with cold distilled water, and the chimneys again placed on ice.

Reaction mixtures (0.3 ml) are maintained in an ice–water bath and contain 0.27 ml of buffer A, 70 ng DNA, and 5–10 μl of the fraction to be tested for endonuclease activity.[16] Each tube should be shaken to ensure complete mixing after the addition of enzyme, but harsh agitation is to be avoided. After 3 min at 0°, 5.0 ml of buffer B (0°) is added, and the mixture is immediately filtered. The filters are rinsed with an additional 5.0 ml of buffer B, carefully removed from the apparatus, and dried. Radioactivity retained by the filters is measured by liquid scintillation spectrometry. One unit of endonuclease activity is defined as the amount of enzyme which, under these conditions, will cause the retention of 1 fmole of ColE1 DNA irradiated with 200 J/m^2 at 254 nm.

DNA Nicking Assay

This assay measures the conversion of form I (supercoiled) DNA to form II (relaxed circular) by electrophoresis of the DNA in agarose gels.[17] It is primarily used to verify results obtained with the filter binding assay and to identify contaminating activities in partially purified endonuclease preparations. It is not suitable for assaying crude extracts which contain significant levels of EDTA-resistant contaminating nucleases.

Reagents

Agarose (Bio-Rad)

Electrophoresis buffer: A stock solution of 400 mM Tris-HCl, 20 mM

---

[14] Any double-stranded radioactive DNA can probably be used for this assay. However, we usually use ColE1 DNA which contains too few superhelical molecules to be suitable for the DNA nicking assay.

[15] D. B. Clewell and D. R. Helinski, *Proc. Natl. Acad. Sci. U.S.A.* **62**, 1159 (1969).

[16] This assay cannot be used for fractions containing PEG because no nonfilterable radioactivity is recovered.

[17] R. B. Helling, H. M. Goodman, and H. W. Boyer, *J. Virol.* **14**, 1235 (1974).

EDTA, 330 mM sodium acetate, 180 mM NaCl (pH 8.17 at 20°) is diluted 1 : 10 just prior to use, giving a pH of 8.05 (buffer A)

DNA: Superhelical ColE1 [^{14}C]DNA, (specific activity 57,000 cpm/μg),[15] irradiated with 20 J/m^2 at 254 nm

Incubation buffer: 10 mM Tris-HCl (pH 8.0 at 20°), 100 mM NaCl, 10 mM EDTA (buffer B)

Sodium dodecyl sulfate (SDS) [10% (w/v) in water]

0.02% Bromphenol blue in 40% sucrose

Ethidium bromide (1 μg/ml in water or buffer A)

Procedure. Agarose (1% in buffer A) is melted in a pressure cooker and cooled to 60°–70° before pipetting into electrophoresis tubes (5 mm i.d. × 13 cm) which are sealed with Teflon tape stretched gently across their bottoms. Upon solidification (usually about 45 min) the gels are partially extruded and the tops evenly sliced to give uniform gels 10 cm long.

Enzyme fractions are added to polystyrene tubes containing approximately 100 ng DNA in buffer B,[18] giving a total volume of 50 μl or less. After 15 min at 37° each reaction is terminated by addition of 5 μl 10% SDS. To this mixture an equal volume of 0.02% bromphenol blue in 40% sucrose is added, and the sample is layered with a micropipette under the electrophoresis buffer atop each gel.

The DNA is electrophoresed at 75 V for 3 hr. Following electrophoresis the gels are stained in ethidium bromide for at least 30 min. Fluorescent bands are visualized by laying the gels on a Blak-Ray Model C-50 viewer or a Chromato-Vue Transilluminator Model C-61 (Ultra-Violet Products, Inc., 5100 Walnut Grove, San Gabriel, California). In this system superhelical DNA migrates most rapidly, followed by unit length linears and relaxed circular molecules, respectively. For measurement of radioactivity, gels are manually sliced into 10 × 1 cm slices or into 45 × 2 mm slices using a Bio-Rad No. 190 gel slicer. The slices are placed into 1 dram shell vials with 1 N HCl (0.1 ml for 1 cm slices, 0.025 ml for 2 mm slices). The vials are autoclaved long enough to melt the agarose (the time it takes for the autoclave to reach an internal temperature of 100°).[19] Instagel (Packard) (3 ml) is added to the cooled vials which are then shaken, capped, and placed in liquid scintillation vials for radioactivity determinations. Samples should be allowed to equilibrate in the dark in the counter until all chemiluminescence has faded (approximately 30 min) before counting. One unit of endonuclease activity is defined as the amount of enzyme which in 15 min at 37° will nick 1 fmole of ColE1 DNA irradiated with 20 J/m^2 254.

---

[18] The presence of 100 μg/ml of bovine serum albumin in the reaction mixture helps to stabilize the enzyme, resulting in greater activity.

[19] A final pH of 4.0 is necessary to prevent resolidification of the agarose upon cooling.

Enzyme Purification

Cell Lysis

A typical purification starts with 20 g of *E. coli* B67 infected with phage T4amN82 (New England Biolabs Inc., Beverly, Massachusetts) stored as a pellet at $-20°$. All procedures are carried out in the cold unless otherwise stated. The pellet is thawed and suspended in 60 ml of cold 5 *M* NaCl in 50 m*M* Tris-HCl buffer at pH 8.0.[20] The cell suspension, in a 150-ml beaker, is given four 30-sec pulses separated by 1-min intervals, with the 1 cm probe of a Branson Sonifer Cell Disruptor (Heat Systems-Ultrasonics Inc., Plainview, New York). The energy intensity is set between 60 and 80 on the output meter. The sonicate is centrifuged in polycarbonate tubes for 90 min at 27,000 rpm in a Beckman 30 rotor at 4° (65,000 g). Following centrifugation the supernatant (62.5 ml) is saved as fraction I (crude lysate).

Phase Partitioning with Polyethylene Glycol/Dextran

Approximately 24 hr prior to use, stock solutions of polyethylene glycol (PEG) [30% (w/w) PEG 6000 (Baker) in 5 *M* NaCl and 50 m*M* Tris, pH 8.0] and dextran [20% (w/w) dextran T500 (Pharmacia) in 5 *M* NaCl and 50 m*M* Tris, pH 8.0] are made and stored at 4°. To 62 ml of fraction I, 20.7 ml each of the PEG and dextran solutions are added to give final concentrations of 6 and 4%, respectively. This mixture is stirred gently for 2 hr. A "wash" solution of PEG and dextran is prepared by adding 10 ml of each of the stock solutions to 30 ml of 5 *M* NaCl in 50 m*M* Tris at pH 8.0. This solution is also stirred in the cold for 2 hr.

Both mixtures are centrifuged at 11,500 rpm for 20 min in the HB swinging bucket rotor in a Sorvall RC2-B centrifuge (21,500 g). The centrifugation separates a top (PEG) phase from a lower (dextran) phase. The top phase is removed from both samples. That from the "wash" is added to the dextran phase from the enzyme-containing sample, and this mixture stirred for 1 hr. The centrifugation is repeated and the top phase combined with the original as fraction IIA (PEG phase) (92 ml). This fraction is dialyzed against two changes (each 6 liters) of 10 m*M* Tris, pH 8.0, containing 10% ethylene glycol (v/v) and 5% PEG 6000 (w/v). Following dialysis a precipitate is usually observed which is sedimented by centrifugation at 21,500 g. The supernatant is saved as fraction IIB (PEG phase-dialyzed) (115 ml).

---

[20] All pH measurements were made with 1 *M* solutions at 20°.

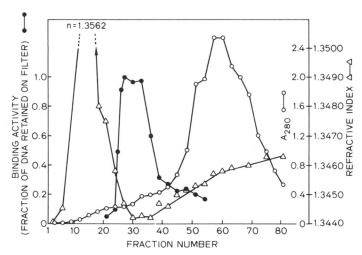

FIG. 1. DEAE-cellulose chromatography of fraction IIB. The column was developed with a linear gradient of 0–0.5 M NaCl beginning at fraction 10, monitored by refractive index (Δ-Δ). (The refractive index of fractions 6–30 reflects the presence of PEG in the flowthrough.) The first nine fractions contained 14.1 ml; the rest contained 9.4 ml. Endonuclease V activity (●——●) was determined by the filter binding assay; protein concentration by A_{280} (O——O).

DEAE–Cellulose Chromatography

Whatman DE-52 is precycled according to manufacturer's instructions with 0.5 N HCl and 0.5 N NaOH (Whatman publication E607A), equilibrated in 100 mM Tris-HCl buffer, pH 8.0, containing 10% (v/v) ethylene glycol, and then rinsed extensively with elution buffer [10 mM Tris-HCl, pH 8.0, containing 10% (v/v) ethylene glycol]. A column (2.5 cm diameter × 25 cm) is packed with DEAE and equilibrated with elution buffer. Fraction IIB (114.5 ml) is loaded onto the column at a flow rate of approximately 30 ml/hr. The column is then rinsed with 30 ml of elution buffer and developed with a linear gradient, 0–0.5 M NaCl, in 1 liter of elution buffer. The flow rate is maintained at approximately 30 ml/hr, and 9.4-ml fractions are collected. Samples (10 µl) are assayed by filter binding. Fractions 25–39 (Fig. 1) are pooled and constitute fraction III (140 ml).

Phosphocellulose Chromatography

Whatman P-11, precycled in 0.5 N NaOH and 0.5 N HCl according to instructions in Whatman publication E607A, is equilibrated in 100 mM sodium phosphate, pH 7.0, containing 10% (v/v) ethylene glycol and then rinsed extensively with elution buffer [10 mM sodium phosphate, pH 7.0,

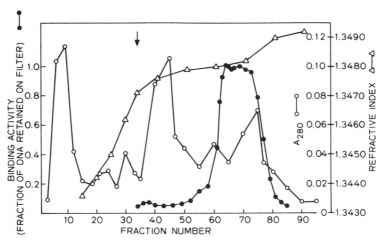

FIG. 2. Phosphocellulose chromatography of fraction III. The column was developed with a linear gradient of 0–1 M NaCl, monitored by refractive index (Δ-Δ). Fractions 1–34 (indicated by an arrow) contained 9.4 ml; the rest contained 2.4 ml. Endonuclease V activity (●——●) was determined by the filter binding assay; protein concentration by A_{280} (○——○).

containing 10% (v/v) ethylene glycol]. A column (1.4 cm diameter × 26 cm) is packed and equilibrated with the same buffer. Sixty-five milliliters of fraction III,[21] previously dialyzed against 3 liters of elution buffer, is loaded onto the column at a flow rate of approximately 20 ml/hr. The column is then rinsed with 40 ml of elution buffer, and developed with a linear gradient, 0–1 M NaCl, in 800 ml of elution buffer. The flow rate is adjusted to 8–10 ml/hr, and 9.4-ml fractions are collected until the refractive index is 1.3469, corresponding to 0.3 M NaCl, after which 2.4-ml fractions are collected. Samples (10 μl) are assayed for endonuclease V activity by the filter binding assay (Fig. 2). Fractions 58–80 (Fig. 2) are pooled, constituting fraction IV (54 ml). This fraction routinely contains 10–20% of the endonuclease V activity originally present in the crude lysate (fraction I), purified 100- to 300-fold. If necessary, the activity can be concentrated five- to eightfold by dialysis against 50 mM Tris-HCl, pH 7.5, containing 10% ethylene glycol and 20% PEG. Attempts to concentrate the enzyme by Aquacide II or Sephadex G50 powder resulted in loss of activity.

Fraction IV can be stored for several months at 4° in 50 mM Tris-HCl, pH 7.5, containing 10% ethylene glycol or 3%PEG. Freezing or storage in 20% glycerol at $-12°$ inactivates the enzyme.

[21] Although in the particular preparation described here only half of fraction III was chromatographed on phosphocellulose, usually the entire fraction is applied to the column. In this case a slightly larger column (1.4 cm diameter × 30 cm) can be used.

When assayed in the presence of 10 mM EDTA, fraction IV selectively nicks uv-irradiated DNA, but not unirradiated DNA. However, in the presence of 5 mM MgCl$_2$ incision or degradation of unirradiated superhelical DNA may be observed, indicating the presence of contaminating nuclease(s). Fraction IV shows no activity against deoxyribouridine-containing PBS-2 DNA in the presence of EDTA, but it does nick AP DNA or DNA alkylated with methyl methane sulfonate.[22]

Further Purification of T4 Endonuclease V

Further purification of the enzyme by the following procedures has been reported.

UV–DNA–Cellulose Chromatography

The use of uv-irradiated DNA–cellulose has been described both by Friedberg and King[9] and by Yasuda and Sekiguchi.[23] In the former procedure T4 DNA–cellulose is prepared by the method of Litman.[24] T4 DNA (75 ml of 0.5–2.0 mg/ml) is mixed with 9.0 g of acid-washed cellulose powder (Munktell No. 410) and irradiated with 10^4 J/m^2 at 254 nm. In the latter procedure DNA–cellulose is prepared by the method of Alberts *et al.*[25] *Escherichia coli* DNA (15 ml of 1.5 mg/ml) is irradiated with a 15-W germicidal lamp for 10 min at a distance of 10 cm and mixed with 5 g of Munktell No. 410 cellulose powder. Chromatography on unirradiated DNA-cellulose has also been described.[22]

Hydroxyapatite Chromatography

This procedure has been described in detail by Yasuda and Sekiguchi.[23]

Gel Filtration through Sephadex or CM-Sephadex

The use of Sephadex G-75 and CM-Sephadex have been described by Minton *et al.*[26] and by Yasuda and Sekiguchi,[23] respectively.

[22] Y. Nishida, S. Yasuda, and M. Sekiguchi, *Biochim. Biophys. Acta* **442**, 208 (1976).

[23] S. Yasuda and M. Sekiguchi, *Biochim. Biophys. Acta* **442**, 197 (1976).

[24] R. Litman, J. Biol. Chem. **243**, 6222 (1968).

[25] B. M. Alberts, F. J. Amadio, M. Jenkins, E. D. Gutman, and F. J. Ferris, *Cold Spring Harbor Symp. Quant. Biol.* **33**, 289 (1968).

[26] K. Minton, M. Durphy, R. Taylor, and E. C. Friedberg, *J. Biol. Chem.* **250**, 2823 (1975).

Sedimentation in Sucrose Density Gradients

Sedimentation of the enzyme in 5–20% sucrose gradients has been described by Friedberg and King.[9]

In our experience all of these procedures result in significant loss of activity when used after phosphocellulose chromatography. This problem can to some extent be averted by collecting enzyme fractions into tubes containing DNA or bovine serum albumin. Because the enzyme is more stable in fractions containing relatively large amounts of protein, for extensive purification we recommend starting with 100 g of T4-infected cells rather than the 20 g described here.

Properties of T4 Endonuclease V

The enzyme has a molecular weight of approximately 18,000 calculated from polyacrylamide gel electrophoresis and filtration through Sephadex G-75.[26]

The enzyme recognizes cyclobutyl dipyrimidines. Within the limits of accurate measurement, saturating amounts of enzyme will produce endonucleolytic incisions equal in number to the pyrimidine dimers present in the substrate DNA. When heteroduplex DNA containing pyrimidine dimers in only one strand is treated with endonuclease V, all detectable nicks occur in the dimer-containing strand.[27] The enzyme incises both native and denatured *E. coli* DNA,[3,23,26] suggesting that it recognizes pyrimidine dimers specifically rather than a conformational distortion in the secondary structure of double-stranded DNA. It attacks glucoslyated and nonglucosylated uv-irradiated DNA.[3,9] Both linear and superhelical DNA are effective substrates if they contain pyrimidine dimers.[28]

The enzyme does not catalyze the excision of pyrimidine dimers from DNA.[8,29]

No substrate other than pyrimidine dimers in DNA has been reported for the enzyme. Specifically, DNA exposed to moderate levels of ionizing radiation, or to methyl methane sulfonate, nitrogen mustard, mitomycin C, or 4-nitroquinoline 1-oxide has been observed not to be attacked.[22,30] DNA containing uracil dimers or hydroxymethyluracil dimers is nicked by the enzyme, but DNA containing 5-thyminyl-5, 6-dihydrothymine (spore photoproduct) is not.[31]

[27] T. J. Simon, C. A. Smith, and E. C. Friedberg, *J. Biol. Chem.* **250**, 8748 (1975).
[28] E. C. Friedberg and D. A. Clayton, *Nature (London)* **237**, 99 (1972).
[29] E. C. Friedberg and J. J. King, *Biochem. Biophys. Res. Commun.* **37**, 646 (1969).
[30] E. C. Friedberg, *Mutat. Res.* **15**, 113 (1972).
[31] F. Makino, H. Tanooka, and M. Sekiguchi, *J. Biochem. (Tokyo)* **82**, 1567 (1977).

The enzyme is not inhibited by 10 mM EDTA and is only slightly stimulated by Mg^{2+} or Mn^{2+}.[3,9] Other divalent cations and monovalent ions have little effect upon its activity. Using the DNA nicking assay we have observed 20–25% inhibition by 10^{-4} M p-chloromercuriphenyl-sulfonic acid, and Yasuda and Sekiguchi[23] have reported 70–80% inhibition by 4×10^{-4} M p-chloromercuribenzoate.

[26] Exonuclease III of *Escherichia coli* K-12, an AP Endonuclease[1]

By STEPHEN G. ROGERS and BERNARD WEISS

Exonuclease III of *E. coli*[2,3] catalyzes the hydrolysis of several types of phosphoester bonds in duplex DNA (Fig. 1). It is a 3′ → 5′ exonuclease, releasing 5′-mononucleotides from the 3′ ends of DNA strands; it is a DNA 3′-phosphatase, hydrolyzing 3′-terminal phosphomonoesters; and it is an AP endonuclease,[4] cleaving phosphodiester bonds at apurinic or apyrimidinic sites to produce new 5′ termini that are base-free deoxyribose 5-phosphate residues.[5] In addition, the enzyme has an RNase H activity; it will preferentially degrade the RNA strand in a DNA–RNA hybrid duplex, presumably exonucleolytically.[5]

Assay Method

Principle. The exonuclease assay is the simplest. It measures the release of acid-soluble radioactive material (mononucleotides) from bacteriophage T7 [³H]DNA. In order to saturate the enzyme and to conserve radiolabeled substrate, the T7 DNA is diluted with unlabeled salmon sperm DNA, and the mixture is sheared by sonication, thereby increasing the concentration of 3′-hydroxyl termini.

[1] Investigations were supported by research grants from the American Cancer Society (NP126) and the National Cancer Institute (CA16519) and by a training grant from the latter (CA90139).

[2] C. C. Richardson, I. R. Lehman, and A. Kornberg, *J. Biol. Chem.* **239**, 251 (1964).
[3] C. C. Richardson and A. Kornberg, *J. Biol. Chem.* **239**, 242 (1964).
[4] This activity has also been called endonuclease VI and is one of at least two different activities that have been referred to as endonuclease II. The preferred designation is "AP endonuclease activity of exonuclease III."
[5] B. Weiss, S. G. Rogers, and A. F. Taylor, *in* "DNA Repair Mechanisms: ICN-UCLA Symposia on Molecular and Cellular Biology" (P. C. Hanawalt, E. C. Friedberg, and C. F. Fox, eds.), p. 191. Academic Press, New York, 1978.

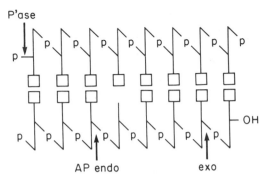

FIG. 1. Actions of exonuclease III on a DNA duplex. Sites of hydrolysis (arrows) are shown for three types of reactions: $3' \rightarrow 5'$ exonuclease (exo), DNA $3'$-phosphatase (P'ase), and endonuclease for apurinic/apyrimidinic sites (AP endo).

Reagents

2× reaction mixture (minus Mg^{2+}): 0.1 mM ^{3}H-labeled sonicated DNA[6] (400–500 cpm/nmole; see below, discussion of preparation of substrate), 100 mM Tris-HCl buffer (pH 8.0), 20 mM 2-mercaptoethanol, and yeast tRNA, 20 μg/ml (to inhibit endonuclease I)[7]

$MgCl_2$, 10 mM

Enzyme diluent: bovine albumin (0.5 mg/ml), 10 mM 2-mercaptoethanol, and 50 mM Tris-HCl buffer (pH 7.6)

Salmon sperm DNA, 2.5 mg/ml in 10 mM Tris-HCl buffer (pH 8.0)[8]

Trichloroacetic acid, 10% (w/v)

Preparation of Substrate. Bacteriophage T7 [3H]DNA is prepared by the methods of Thomas and Abelson.[9] A growth medium containing 3 μCi/ml of [3H]thymidine yields a preparation with a specific radioactivity of 8000–10,000 cpm/nmole. The substrate solution contains 0.05 μmole of [3H]DNA and 0.95 μmole (0.13 ml) of the salmon sperm DNA, which are diluted to a final volume of 1.5 ml in 50 mM NaCl–10 mM Tris-HCl buffer (pH 8.0). The mixture is sonicated for 90 sec in an ice bath with a Branson Sonifier at the highest intensity that can be obtained without producing aeration. To prevent loss during sonication, the solution is sealed in a

---

[6] Concentrations of DNA are expressed throughout as nucleotide equivalents.

[7] The solution is prepared with sterile buffer and water and is stable at 4° for 4–6 weeks if it is kept sterile. Samples may be stored longer if frozen.

[8] The DNA dissolves in about 18 hr at 4° and is stored at 4°.

[9] C. A. Thomas and J. Abelson, *Proced. Nucleic Acid Res.* **1**, 553 (1966).

small tube with Parafilm through which the sonicator microprobe is inserted.

Procedure. The reaction mixture is prepared on the day of use by diluting a portion of the 2× reaction mixture with an equal volume of the MgCl₂ solution. The enzyme solution is diluted with the enzyme diluent immediately before assay. To 0.2 ml of the reaction mixture (containing 10 nmoles of DNA) is added 0.02 ml of the diluent containing up to 4 units of enzyme. After incubation for 30 min at 37°, the tube is placed in an ice bath. Chilled salmon sperm DNA (0.1 ml) and chilled trichloroacetic acid (0.3 ml) solutions are added in succession with mixing, and the tube is centrifuged for 5 min at 10,000–12,000 g.[10] A 500-μl sample of the supernatant is removed with a constriction pipette and added to 5 ml of a liquid scintillation fluid (containing a 2 : 1 mixture of toluene and Triton X-100) for the determination of radioactivity. In the absence of enzyme, less than 2% of the DNA should be acid-soluble. Assays are linear up to about 40% hydrolysis (4 units). The assay may be applied to crude extracts of *E. coli* or of *Haemophilus influenzae*; about 85–90% of the DNase activity measured by this assay will be due to exonuclease III.

A standard unit of exonuclease III DNase activity has been defined[3] as that representing the release of 1 nmole of acid-soluble nucleotides per 30 min at 37°.

Other Assays. DNA 3′-phosphatase activity was originally assayed with 3′-phosphoryl terminated DNA that was uniformly labeled with ^{32}P; the release of Norit-nonadsorbable ^{32}P$_i$ was measured.[2] With crude extracts, this assay is subject to about 15% interference by other enzymes, presumably other exonucleases coupled with 5′-nucleotidases. An alternative assay, employing a substrate uniquely labeled at its 3′ end, is >99% specific for exonuclease III in crude extracts.[11] The most specific assay for the AP endonuclease activity measures the nicking of partially depurinated (acid-treated) supercoiled DNA; about 85–90% of such activity in crude extracts of *E. coli* is due to exonuclease III. Some simpler assays for the endonuclease activity are less specific because they may involve some exonucleolytic degradation as well.[12] Such assays measure the release of acid-soluble fragments from DNA that has been partially depurinated by acid or by methyl methane sulfonate, or they measure the release of large fragments of alkylated DNA from a polyacrylamide gel matrix. The exonucleolytic activity may be minimized in these assays by using bacteriophage T4 DNA as a substrate or by using Ca^{2+} or citrate.

---

[10] A microcentrifuge at room temperature is adequate.
[11] C. Milcarek and B. Weiss, *J. Mol. Biol.* **68,** 303 (1972).
[12] J. E. Clements, S. G. Rogers, and B. Weiss, *J. Biol. Chem.* **253,** 2990 (1978).

Enzyme Overproduction

The purification procedure is based on the use of a strain of *E. coli* that overproduces exonuclease III. It contains the chimeric plasmid pSGR3[13] that contains the following regions: (a) a 3-kilobase fragment of the chromosome of *E. coli* K-12 containing *xth,* the structural gene for exonuclease III, (b) genes for ampicillin and kanamycin resistance, (c) ColEl replication and immunity regions, and (d) a segment of bacteriophage λ extending from the *N* through the *P* gene and specifying a thermolabile repressor (*cI857*). At 32° the plasmid replication is controlled by the ColE1 region, and the cells, by virtue of carrying a multicopy plasmid, have levels of exonuclease III that are 15- to 30-fold higher than wild-type *E. coli*. When the growing cells are heated to 42°, λ transcription and replication are derepressed, and there is an additional increase in plasmid copy number and in exonuclease III production. Final levels of enzyme are between 50 and 120 times that of wild-type *E. coli* and constitute 3 to 6% of the protein in cell extracts.

In the procedure below, the cells are grown to an unusually high density in a bench top fermentor with the aid of a greatly enriched medium and pure oxygen. The use of glycerol rather than glucose as the major carbon source reduces acid production, thereby eliminating the need for pH control and avoiding the harmful effects of low pH on exonuclease III production.[14]

Bacterial Growth. Escherichia coli strain BE257/pSGR3 is grown in SLBH medium[15] in a rotating carboy fermentor (S. M. S. High Density Fermentor, Lab-Line Instruments Co.). SLBH medium contains (per liter) 20 ml glycerol, 37.5 g tryptone (Difco), 22 g yeast extract (Difco), 3 g NaCl, 1 g KCl, 1 beef bouillon cube, and 100 ml of a separately autoclaved 1.0 M potassium phosphate buffer (pH 7.6); the final pH is 7.3–7.4. The temperature of the culture is monitored by withdrawing samples past a hypodermic thermistor probe (Yellow Springs Instrument Co.). The probe is sterilized with ethanol and inserted through a pinhole in a piece of silicone rubber tubing that is interposed between the fermentor's sampling tube and a syringe. Cell growth is followed spectrophotometrically, with samples diluted in 0.15 M NaCl–0.01 M Tris-HCl buffer (pH 8.0) to an A_{525} of 0.1 to 0.5. The culture is "aerated" with pure oxygen at a rate in liters per minute equal to about $1.0 + 0.2 A_{525}$.

An inoculum is prepared by growing BE257/pSGR3 overnight to saturation in a gyratory shaker at 32° in SLBH medium containing 20 μg/ml

[13] R. N. Rao and S. G. Rogers, *Gene* **3**, 247 (1978).
[14] K. Shortman and I. R. Lehman, *J. Biol. Chem.* **239**, 2964 (1964).
[15] J. L. Betz and J. R. Sadler, *J. Mol. Biol.* **105**, 293 (1976).

TABLE I
PURIFICATION OF EXONUCLEASE III FROM *Escherichia coli*

Fraction	Units ($\times 10^{-6}$)	Protein[a] (mg/ml)	Specific activity (units/mg $\times 10^{-3}$)	Yield (%)
I. Sonicate	17.0	19.9	5.5	100
II. Polyethyleneimine	12.9	5.2	16.4	76
III. Ammonium sulfate	11.0	2.3	23.4	65
IV. Phosphocellulose	6.9	0.26	175	40
V. DEAE–cellulose	5.4	2.1	162	31

[a] Protein was determined by the method of M. M. Bradford [*Anal. Biochem.* **72,** 248 (1976)] with bovine albumin as a standard.

of kanamycin and 100 μg/ml of ampicillin.[16] The inoculum (15 ml) is added to 3 liters of SLBH medium (without antibiotics) in the fermentor and grown at 32° to an A_{525} of 2.0 (6 $\times$ 10^8 cells/ml). The culture is then brought up to 42° by replacing 3 liters of the 12-liter water bath with boiling water. When the culture has been at 42° at least 20 min, it is cooled to 37° and maintained at that temperature for about 3–4 hr until an A_{525} of about 35 is reached or until there is a sharp reduction in growth rate and in heat production.[17] The final pH will be 7.0–7.1. The cells are then chilled in an ice bath and harvested by centrifugation. About 100 g of cell paste are obtained and stored at −70°.[18]

Escherichia coli strain W grows better than K-12 at high densities and may also be used as a vehicle for pSGR3. *E. coli* W/pSGR3 is induced at $A_{525} = 3$ and harvested about 6 hr later at $A_{525} = 77$. Additional sterile glycerol (60 ml of a 60% solution) is added at $A_{525} = 25$. The tendency of *E. coli* W cells to aggregate and to sediment interferes with their aeration. The rotating carboy fermentor must therefore be slowed to permit churning of its contents rather than complete centrifugal layering. A sterile 10-fold dilution of Dow Corning Antifoam B is added as needed in 1-ml amounts. The yield from 3 liters of medium is 200 g of cell paste containing 100 times the enzyme activity of wild-type cells.

[16] The antibiotics are made up as 1000-fold concentrated solutions, filter-sterilized, and stored at −20°.

[17] After heat induction, the cells continue to grow and divide, but with an increasingly prolonged generation time. Beyond a 20-fold increase in cell mass, there is no further increase in exonuclease III activity.

[18] Other growth media or equipment may be used provided the following general conditions are met: (a) The cells should be grown at 32° for at least three generations before heat induction. (b) They should then be heated to 42° for 20 min and permitted to grow at 37° until their cell mass (turbidity) is at least 15-fold higher than at the point of induction and equivalent to that of a late log phase culture of wild-type cells.

Purification Procedure

The results of a typical purification are outlined in Table I. From 30 g of cells were obtained 33 mg of exonuclease III (at a current market value of $0.7 million[19]). The cells had 55 times the normal wild-type level of this enzyme, and a 30-fold purification yielded a >98% homogeneous preparation. In the purification procedure, nucleic acids and coprecipitating DNA enzymes are first removed from the cell extract by precipitation with polyethyleneimine. Excess soluble polyethyleneimine is then separated from exonuclease III by precipitation of the latter with ammonium sulfate. Phosphocellulose chromatography yields a >90% homogeneous preparation, and subsequent DEAE–cellulose chromatography achieves concentration of the enzyme and removal of some minor contaminating proteins visible on electrophorograms.

Preparation of the Cell Extract. In this and subsequent steps, purification procedures are performed at 0°–5°, and all centrifugations are at 15,000 g for 10 min. Frozen cells (30 g wet weight) are suspended in 150 ml of a solution containing 50 mM Tris-HCl buffer (pH 8.0), 50 mM KCl, 1 mM Na₃EDTA, and 0.002% phenylmethyl sulfonyl fluoride (PMSF).[20] The suspension is sonicated for 5 min at maximum intensity with a Branson Sonifier in a rosette cooling cell immersed in an ice bath so that the temperature does not rise above 5°. Cell debris is removed by centrifugation and supernatant fluid (fraction I, 155 ml) is saved.

Polyethyleneimine Fractionation. Polyethyleneimine is supplied by BDH Chemicals in an approximately 50% aqueous solution (Polymin-P). Water and concentrated HCl are added to bring the volume to ten times that of the original and the pH to 7.6. To fraction I (155 ml) are added 13.4 ml of this solution, yielding a final concentration of 0.8% Polymin-P (or about 0.4% polyethyleneimine).[21] The mixture is stirred gently for 15 min and centrifuged. The supernatant (fraction II, 150 ml) is saved.

Ammonium Sulfate Precipitation. Ammonium sulfate (47 g) is added slowly with stirring to fraction II (150 ml) over a 20-min period. The mixture is then stirred for 20 min and centrifuged. An additional 23.4 g of ammonium sulfate are added as before to the supernatant (171 ml). The mixture may be left at 0° overnight while the phosphocellulose column is equilibrated. The precipitate is recovered by centrifugation and drained thoroughly. It is then dissolved in 200 ml of 10 mM potassium phosphate

---

[19] Catalog, Miles Biochemicals (1978). This price predates commercial use of the over-producing strain.

[20] PMSF, a protease inhibitor, is added from a 5% (w/v) stock solution in ethanol.

[21] This concentration was determined from trials to be within the wide limits of that needed to give the maximum ratio of A_{280} to A_{260} in the supernatant without precipitating additional enzyme.

buffer (pH 6.5) containing 1 mM 2-mercaptoethanol and 0.002% PMSF. The final volume is 205 ml (fraction III).

Phosphocellulose Chromatography. Phosphocellulose (Whatman P-11) is washed by precycling,[22] and the fines are removed by decantation. It is then suspended in 0.3 M potassium phosphate buffer (pH 6.5), packed in a column (2.5 × 20 cm) and washed with 1 liter of 20 mM potassium phosphate buffer (pH 6.5) containing 1 mM 2-mercaptoethanol. Fraction III is applied to the column which is then washed with 100 ml of the equilibrating buffer. Elution is performed with a solution (1200 ml) containing 1 mM 2-mercaptoethanol and a linear gradient of 0.02 to 0.30 M potassium phosphate buffer (pH 6.5). Sixty 20-ml fractions are collected, and the A_{280} is measured in 1-cm path length cuvettes. Two peaks of absorbance are seen, at about fractions 19 and 45, respectively. The latter contains exonuclease III activity. Fractions containing >85% of the specific activity of the peak exonuclease III fraction are pooled to yield fraction IV. Fraction IV (151 ml) contains 88% of the eluted exonuclease III activity and is >90% homogeneous by gel electrophoresis.

DEAE–Cellulose Chromatography. After removal of the fines from Whatman DE-52, it is suspended in 0.1 M Tris-HCl buffer (pH 8.0), packed into a column (0.9 × 8 cm) and washed with 100 ml of 20 mM Tris-HCl buffer (pH 8.0) containing 1 mM 2-mercaptoethanol. Fraction IV is placed in dialysis tubing that has been boiled for 20 min in 1 mM Na$_3$EDTA, and it is dialyzed twice against 2 liters each of the equilibrating buffer. The enzyme solution is then applied to the column, and the column is washed with 10 ml of 50 mM Tris-HCl buffer (pH 8.0) and 0.1 mM dithiothreitol. The enzyme is eluted with 60 ml of the same buffer containing a linear gradient of from 0 to 0.2 M KCl, and 2-ml fractions are collected. The enzyme peak (about tube No. 20) may be detected by A_{280} in a 1-mm path length cuvette. Tubes containing about 90% of the eluted activity are pooled to yield fraction V (16 ml).

Enzyme Storage and Dilution

Fraction V is diluted with an equal volume of cold glycerol in an ice bath, divided into small samples that are then fast-frozen in a Dry Ice–ethanol bath, and stored at −70°, where it remains stable for at least several years. A working stock may be stored unfrozen at −20° and should retain at least 80% of its activity after a year. Exonuclease III is also relatively stable at 0°–4°,[2] but bacterial contamination becomes a

---

[22] The phosphocellulose is suspended in 15 volumes of 0.5 N NaOH for 30 min, washed with water to an effluent pH of 8, resuspended in 0.5 N HCl for 30 min, and washed with water to a pH of 6.

problem on prolonged storage. The enzyme is diluted before use with the assay diluent (see Assay Method). To minimize contamination by exogenous nucleases, the diluent may be prepared with sterile components and sterile glassware, and the bovine albumin may be heated in water for 1 hr at 80°. The dilutions may be stored for at least several weeks at −20° if they are mixed with an equal volume of glycerol, but the stability is variable and depends on the extent of dilution and handling.

General Properties

Purity. Fraction V is >98% homogeneous as determined by SDS–polyacrylamide gel electrophoresis. It contains no measurable nonspecific endonucleolytic activity on supercoiled DNA, i.e., <0.5% of AP endonuclease activity on ϕX174 RFI DNA containing two apurinic sites per molecule. It has no measurable nucleolytic activity on poly(dA), i.e., <0.005% of its activity on poly(dA) · poly(dT).

Physical Properties.[23] Exonuclease III is a monomeric protein with a molecular weight of 28,000. It has a sedimentation coefficient ($s_{20,w}$) of 2.92 and is globular (f/f_0 = 1.15).

pH Optimum. The optimum pH for the endo- and exonucleolytic reactions is between 7.6 and 8.5 in Tris-HCl buffer. A lower pH optimum (6.8–7.4) has been reported for the phosphatase activity in phosphate buffer.[2]

Requirement for Divalent Cations. For optimal activity, Mg^{2+} or Mn^{2+} are required, and they can reverse the inhibition by EDTA. Calcium ion (5 mM) can substitute for Mg^{2+} in the AP endonuclease reaction but not in the exonuclease reaction; it is, therefore, useful for reducing the exonucleolytic degradation that follows endonucleolytic cleavage.[24] A similar effect has been reported for citrate in the presence of Mg^{2+}.[25] Zinc ion inhibits exonuclease III (>90% at 10^{-4} M).[2,3]

Requirement for Sulfhydryl Compounds.[2,3] Exonuclease III is inhibited by p-chloromercuribenzoate (50–90% at 10^{-4} M). The presence of 2-mercaptoethanol has little effect (<10%) on the results of the standard assay, but this compound stabilizes the enzyme during more prolonged incubations.

Substrate Specifities and Reaction Mechanisms

AP Endonuclease Reaction. The enzyme cleaves at apurinic sites introduced either by acid treatment of the DNA or by the spontaneous elimination of bases that have been alkylated by methyl methanesulfonate.[5,23,26]

[23] B. Weiss, *J. Biol. Chem.* **251**, 1896 (1976).
[24] S. G. Rogers and B. Weiss, unpublished results (1978).
[25] S. Ljungquist, B. Nyberg, and T. Lindahl, *FEBS Lett.* **57**, 169 (1975).
[26] F. Gossard and W. G. Verly, *Eur. J. Biochem.* **82**, 321 (1978).

Apurinic sites are substrates even after reaction with $NaBH_4$,[26] a treatment that reduces the free aldehyde groups on the base-free sugar residues and renders their 3'-phosphoesters stable to alkali. Apyrimidinic sites, which may be specifically generated by the enzymatic removal of uracil residues from uracil-containing DNA, are also cleaved by the enzyme.[5] These properties suggest that the enzyme recognizes the topology of an AP site rather than a specific chemical group.[23]

Exodeoxyribonuclease Reaction. The enzyme is specific for bihelical DNA; it will not work on single-stranded homopolymers. It will, however, degrade single-stranded DNA because of its intrastrand hydrogen bonding, but only at one-fourth to one-third the rate of duplex DNA. The DNA of bacteriophage T4, which is a substrate for the AP endonuclease and phosphatase activities of the enzyme, is largely resistant to exonucleolytic degradation because of its modification by glucosyl residues.[27]

When a DNA duplex is treated with exonuclease III, digestion begins at the 3' ends of each strand so that a partially digested molecule is a duplex with projecting 5' tails. Under standard assay conditions, one molecule of exonuclease III hydrolyzes 150 phosphodiester bonds per minute. As the digestion proceeds, the length of the duplex region becomes progressively shorter until there is an insufficient number of remaining base pairs to maintain a bihelical structure, and the undigested 5' ends of the strands then separate. Thus, after 40–50% digestion of a duplex DNA preparation, the rate of hydrolysis slows abruptly because the remaining DNA is mostly single stranded.

When duplexes are composed of homopolymers such as $(dA)_n \cdot (dT)_n$ or of alternating copolymers such as $d(A-T)_n$, the kinetics are different from those obtained with DNA from natural sources. The strands of such synthetic polymers will slip or creep over each other by the spontaneous breakage and reformation of hydrogen bonds so that the substrate remains maximally base-paired. The exonuclease reaction will, therefore, proceed unabated until almost 100% of the substrate is digested. At 25°, a strand can be degraded up to its 5'-terminal dinucleotide; at higher temperatures, longer 5'-terminal oligonucleotides are obtained as limit products, presumably because of the decreased stability of oligonucleotide duplexes.[28]

At 23–28°, when DNA termini are saturated with an excess of enzyme, the digestion of a homogeneous preparation of linear duplexes proceeds synchronously ($\pm 5\%$) for the first 250 nucleotides.[29] At 5°, in a reaction mixture containing at least 70 mM NaCl, an enzyme molecule will remain bound to the DNA and will remove only six nucleotides from a terminus;

[27] C. C. Richardson, *J. Biol. Chem.* **241**, 2084 (1966).
[28] R. Roychoudhury and R. Wu, *J. Biol. Chem.* **252**, 4786 (1977).
[29] R. Wu, G. Ruben, B. Siegel, E. Jay, P. Spielman, and C. D. Tu, *Biochemistry* **15**, 734 (1976).

the addition of more enzyme leads to the removal of six more nucleotides by each newly bound enzyme molecule.[30]

Under some conditions, at least, exonuclease III is a processive enzyme, i.e., a molecule of exonuclease III will proceed down a DNA chain and catalyze the release of many nucleotides before it dissociates and attacks another chain. From a substrate competition experiment, Wu *et al.*[29] concluded that at 23° exonuclease III removes at least 100 nucleotides processively from a chain terminus. Thomas and Olivera,[31] using substrates with uniquely labeled 3' and 5' termini, concluded that at 37°, exonuclease III worked mainly in a nonprocessive (or distributive) fashion, or at least removed less than 50 nucleotides before dissociating from a chain terminus.

Exonuclease III will attack the 3' end of a DNA strand even if it is terminated by a 3'-phosphoryl group (which it first removes as P_i), one or two paired or mismatched ribonucleotides,[28] or up to three mispaired deoxyribonucleotides.[32] It has been suggested, therefore, that the enzyme recognizes the end of a DNA duplex by the tendency of the latter to unwind and to create an interstrand space similar to that at an AP site.[23] The enzyme will also initiate hydrolysis at a single-strand break, but not at the low temperatures and moderate salt concentrations that would favor base-stacking of residues bounding a nick.[30]

DNA 3'-Phosphatase Reaction.[3] Exonuclease III hydrolyzes 3' terminal phosphomonoesters from duplex DNA at about twice the rate of heat-denatured DNA. It has no measurable activity on 3'-phosphoryl terminated RNA or on dTpTpTp, but it will act on a mixed polymer terminated at its 3' ends by ribonucleoside 3'-phosphate residues. After removal of the terminal phosphates, nucleotides are removed by the 3' → 5' exonuclease activity.

Ribonuclease H Reaction.[5,24] Poly(rA) · poly(dT) is degraded at one-fifth the rate of poly(dA) · poly(dT). The RNA strand of the synthetic hybrid duplex is degraded 100 times faster than the DNA strand, and the enzyme has no measurable activity on either homopolymer strand in the absence of the other. The specificity of exonuclease III can, therefore, be explained by its recognition of a deoxyribose residue on the strand complementary to the one it cleaves.

The RNase activity of the highly purified enzyme cannot be separated from its other activities by gel filtration.[5] The above findings are extensions of previous observations that purified preparations of exonuclease III catalyzed (a) digestion past ribonucleotide residues misincorporated into DNA duplexes,[2] (b) digestion of both strands of a synthetic RNA–

[30] J. E. Donelson and R. Wu, *J. Biol. Chem.* **247**, 4661 (1972).
[31] K. R. Thomas and B. M. Olivera, *J. Biol. Chem.* **253**, 424 (1978).
[32] D. Brutlag and A. Kornberg, *J. Biol. Chem.* **247**, 241 (1972).

DNA hybrid to mononucleotides,[33] and (c) the consecutive cleavage of two ribonucleotides from the 3' end of a DNA duplex.[28] These results suggest a 3' → 5' exonucleolytic mechanism for the RNase H activity of exonuclease III. This activity is clearly distinguishable from other *E. coli* enzymes with RNase H activity: (a) RNase H proper (an endonuclease) and (b) DNA polymerase I (a 5' → 3' exonuclease).

Exonuclease III as an Enzymatic Reagent

Through cleavage by exonuclease III, AP sites in DNA can be identified and distinguished from misincorporated ribonucleotides, which are also alkali labile. The endonuclease product has base-free deoxyribose 5-phosphate groups at its new 5' ends. Such DNA can be used for studies on the specificity of various DNA enzymes (e.g., ligases, kinases, exonucleases).

Limited exonucleolytic digestion of a DNA duplex leaves a 3'-OH primer strand bound to an intact template; it is a repairable molecule in which no sequence information has been lost. Such DNA is widely used as a substrate for the assay of most DNA polymerases. The enzyme has also been used to generate gaps in DNA at specific sites such as restriction enzyme cleavage sites. These gaps may then be treated with single-strand specific mutagens such as bisulfite or repaired with mutagenic base analogues, thereby producing a directed localized mutagenesis.[34] The gaps may also be subjected to sequence analysis after repair by DNA polymerase and radiolabeled triphosphates. Wu and his colleagues have described methods for the control of exonucleolytic degradation, thus enabling the determination of sequences at the 3' and 5' ends of DNA strands.[28−30]

The phosphatase activity of exonuclease III can be used to determine the presence of 3'-phosphoryl end groups in DNA, such as those produced by certain endonucleases. This application is based on the principle that many enzymes that work at 3' ends are inhibited by 3'-phosphoryl end groups. Such enzymes include exonuclease I and the DNA polymerases of *E. coli*. If, for example, a nicked DNA molecule becomes a substrate for DNA polymerase I only after limited hydrolysis by exonuclease III, this fact may be used as partial evidence for the existence of 3'-phospho-monoesters in the substrate.

In conjunction with nonspecific endonucleases and single-strand specific DNases, exonuclease III may also be used to isolate specific protein-binding sites in DNA through the digestion of unprotected regions in protein–DNA complexes.[35]

[33] W. Keller and R. Crouch, *Proc. Natl. Acad. Sci. U.S.A.* **69**, 3360 (1972).
[34] D. Shortle and D. Nathans, *Proc. Natl. Acad. Sci. U.S.A.* **75**, 2170 (1978).
[35] D. Riley and H. Weintraub, *Cell* **13**, 281 (1978).

[27] Endonuclease IV from *Escherichia coli*

By Siv Ljungquist

Endonuclease IV specifically catalyzes the introduction of single-strand breaks at apurinic and apyrimidinic sites in double-stranded DNA.

Assay Method

Principle. The assay of endonuclease IV is based on the release of acid-soluble material from heavily depurinated radioactive DNA.[1]

Reagents

4-(2-Hydroxyethyl)-1-piperazineethanesulfonic acid
(HEPES)–KOH, 1 M, pH 8.2
NaCl, 2 M
EDTA–NaOH, 0.2 M, pH 7.5
Dithiothreitol, 1 M
Bovine serum albumin, 5 mg/ml
DNA, depurinated alkylated *E. coli* [32P]DNA (4 × 10⁵ cpm/μg, 50 μg/ml)

Preparation of Substrate. The substrate is made essentially as described by Paquette *et al.*[2] as follows. *Escherichia coli* [32P]DNA (4 × 10⁵ cpm/μg, 50 μg/ml) in 0.05 M potassium phosphate, pH 7.4, is treated with 0.3 M methyl methanesulfonate at 37° for 45 min. Remaining methyl methanesulfonate is removed by dialysis against two changes of 1000 ml of 0.2 M NaCl, 0.01 M trisodium citrate, and 0.01 M sodium phosphate, pH 7.0, at 2° for 16 hr. The alkylated DNA is incubated at 50° for 6 hr to release methylated purine bases. The partly depurinated DNA is dialyzed against 0.1 M NaCl and 0.01 M sodium citrate, pH 6.0, at 2° for 20 hr, and stored in several small aliquots at −70°.

This substrate will give a plateau level of 15–20% acid-soluble material released after treatment with a saturating amount of endonuclease IV under standard assay conditions (Fig. 1).

Incubation Procedure. The standard reaction mixture (0.1 ml) contains 0.2 M NaCl, 0.05 M HEPES-KOH, pH 8.2, 10^{-3} M EDTA, 10^{-4} M dithiothreitol, 1 μg of depurinated alkylated *E. coli* [32P]DNA, 50 μg bovine serum albumin, and a limiting amount of enzyme (2 to 20 units). Incubation time is 30 min at 37°. The reaction is terminated by chilling the

[1] S. Ljungquist, *J. Biol. Chem.* **252**, 2808 (1977).
[2] Y. Paquette, P. Crine, and W. G. Verly, *Can. J. Biochem.* **50**, 1199 (1972).

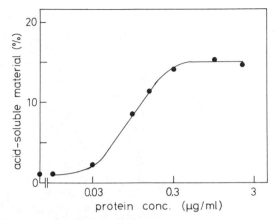

FIG. 1. Release of acid-soluble material from *E. coli* [^{32}P]DNA by endonuclease IV (fraction V) under standard assay conditions.

reaction mixture followed by the addition of 0.1 ml of cold 0.8 *M* perchloric acid. Acid-insoluble material is removed by centrifugation at 15.000 *g* for 10 min, and the acid-soluble fraction is determined by measuring the radioactivity of a 100-μl aliquot of the supernatant solution. A control reaction mixture without enzyme and treated as described has 0.5 to 1% of the added radioactivity in acid-soluble form, depending on the age of the substrate.

Definition of Enzyme Unit. One unit of the enzyme, originally determined by a transformation assay,[3] is defined as the amount catalyzing removal of 20 pmole of ^{32}P in acid-soluble form from the depurinated alkylated [^{32}P]DNA substrate in 30 min at 37°.

Purification Procedure[1]

Growth of Bacteria. Escherichia coli 1100 (*endA*⁻) is grown in a glucose–salts medium supplemented with 0.2% casamino acids and 0.1% yeast extract and harvested in the logarithmic growth phase. The cell paste is frozen at −70° and stored at that temperature. All purification steps are performed at 0° to 4°. All buffers contain 10^{-3} *M* dithiothreitol. Centrifugations are at 20,000 *g* for 15 min. Protein concentrations are determined by the biuret reaction[4] (fractions I–III) or by A_{280} measurements (fractions IV–V). A typical purification from 80 g of cells is outlined in Table I.

[3] S. Ljungquist and T. Lindahl, *J. Biol. Chem.* **249**, 1530 (1974).
[4] A. G. Gornall, C. J. Bardawill, and M. M. David, *J. Biol. Chem.* **177**, 751 (1949).

TABLE I
PURIFICATION OF ENDONUCLEASE IV FROM *Escherichia coli*[a,c]

Fraction	Volume (ml)	Total protein (mg)	Specific activity[b] (units/mg)	Total activity (units × 10⁻⁶)
I. Crude extract	327	7780	300	2.3
II. Ammonium sulfate	61	2690	608	1.64
III. Sephadex G-75	133	370	3,810	1.41
IV. Heat treatment	130	280	4,260	1.17
V. DNA-cellulose	2.2	0.22	1,060,000	0.48

[a] Conditions of purification and assay is described in text.

[b] The enzyme activity of the crude cell extract (fraction I) has been estimated from data obtained by a more specific but tedious transformation assay[3] since other nucleases will interfere with the simple standard assay used here.

[c] Reproduced from ref. 1 with permission from the *Journal of Biological Chemistry*.

Crude Cell Extract. Frozen cells (80 g) are disintegrated in a modified Hughes press at $-25°$. Disrupted cells are thawed and stirred with 300 ml 0.05 M Tris-HCl, pH 8.0. After stirring for 30 min, debris is removed by centrifugation leaving a supernatant solution (327 ml) (fraction I).

Streptomycin Treatment and Ammonium Sulfate Fractionation. To fraction I, 1 volume of 1.6% (w/v) streptomycin sulfate in 0.05 M Tris-HCl, pH 8.0, is added dropwise under gentle stirring. After 1 hr the resulting precipitate is removed by centrifugation. To the supernatant solution (620 ml), 164 g of ammonium sulfate (special enzyme grade) is slowly added and dissolved. After 1 hr the precipitate is removed by centrifugation and discarded. An additional 85 g of ammonium sulfate is added to the supernatant solution (670 ml) and slowly dissolved. The precipitate formed is collected by centrifugation and dissolved in 50 ml of 0.2 M NaCl, 0.05 M potassium phosphate, pH 7.4, and dialyzed for 3 to 4 hr against the same buffer prior to gel filtration (fraction II).

Sephadex G-75 Chromatography. The dialyzed ammonium sulfate fraction is applied to a column (4 × 117 cm) of Sephadex G-75 equilibrated with 0.2 M NaCl and 0.05 M potassium phosphate, pH 7.4. A total volume of 2 liters of the same buffer is immediately applied and 20-ml fractions are collected. Endonuclease IV is eluted after most of the protein and active fractions are pooled (fraction III).

Heat Treatment. The heating step is performed at 65° for 5 min. Fraction III (2–3 mg protein/ml) is divided into 5-ml portions in 20-ml test tubes. A rack of tubes is immersed into a 70° water bath. When the temperature of the enzyme solution reaches 65° controlled with a precalibrated thermometer in one of the test tubes, the rack is transferred into a

65° water bath and the incubation period of 5 min is started. The heat treatment is stopped by rapid chilling in an ice bath. After 30 min at 0°, precipitated protein is removed by centrifugation and the supernatant (130 ml) is dialyzed against 3 l of 0.02 M Tris-HCl, pH 8.0, 5% glycerol for 10 hr (fraction IV).

DNA–Cellulose Chromatography. The dialyzed heat-treated fraction IV is applied to a column (2.2 × 7.5 cm) of native DNA–cellulose[5] previously equilibrated with 0.02 M Tris-HCl, pH 8.0, and 5% glycerol. By washing the column with 100 ml of the same buffer most of the protein applied to the column is eluted. Endonuclease IV, adsorbed to the column, is then eluted with 0.3 M NaCl, 0.02 M Tris-HCl, pH 8.0, and 5% glycerol in 3-ml fractions, and the active fractions are pooled (18 ml). Traces of nucleic acid are removed by applying the active pool onto a DEAE–cellulose column (2 × 2 cm) equilibrated with 0.3 M NaCl, 0.02 M Tris-HCl, pH 8.0, and 5% glycerol. Endonuclease IV does not adsorb under these conditions and is recovered in the effluent after washing with 0.3 M NaCl, 0.02 M Tris-HCl, pH 8.0, and 5% glycerol. Two milliliter fractions are collected and the endonuclease IV active fractions are pooled and concentrated tenfold with a Diaflo ultrafiltration apparatus (Amicon, filter PM 10). The concentrated endonuclease IV fraction is then dialyzed against 0.3 M NaCl, 0.02 M Tris-HCl, pH 8.0, and 5% glycerol for 20 hr (fraction V). The dialyzed fraction is mixed with 1 volume of 87% glycerol and is frozen and stored in 0.5 ml aliquots at −70°. By using an alternative concentration method, trace quantities of uracil–DNA glycosylase present in the enzyme preparation can be removed. The DNA–cellulose eluate is supplemented with 0.02 volume of 0.5 M potassium phosphate, pH 7.5, and applied to a column (1.1 × 2 cm) of hydroxyapatite[6] previously equilibrated with 0.3 M NaCl, 0.02 M Tris-HCl, and 0.01 M potassium phosphate, pH 7.5. The column is washed with 15 column volumes of this buffer to remove the uracil-DNA glycosylase activity. The endonuclease IV adsorbed to the column is eluted with 0.3 M NaCl and 0.3 M potassium phosphate, pH 7.5, in fractions of 0.5 ml. The endonuclease IV active fractions were pooled (2–3 ml), 1 volume of 87% glycerol was added, and the fractions divided into 0.5-ml aliquots and frozen at −70°. This procedure gives a lower recovery (30 to 40% yield) than the ultrafiltration method.

Properties

Stability. In preparations stored for 6 months in 45% glycerol at −70° no loss of activity was observed. On heating in 0.2 M NaCl, 0.05 M

---

[5] B. Alberts and G. Herrick, this series, Vol. 21, p. 198.
[6] G. Bernardi, this series, Vol. 22, p. 325.

potassium phosphate, and 10^{-3} M dithiothreitol, pH 7.4, the enzyme retains full activity after 5 min at 60°, while 80–90% is retained at 65°, indicating that endonuclease IV is a fairly heat resistant protein.

Purity. By analysis by sodium dodecyl sulfate–polyacrylamide slab gel electrophoresis experiments five to six bands are observed. The preparation is devoid of detectable contaminating DNA exonuclease and phosphatase activities. After hydroxylapatite chromatography the preparation is also free from detectable amounts of 3-methyladenine–DNA glycosylase and uracil–DNA glycosylase.

The purified enzyme has a molecular weight of 30,000–33,000 as estimated by gel filtration and sucrose gradient centrifugation.

Substrate Specificity. Endonuclease IV catalyzes the formation of single-strand breaks in double-stranded DNA containing apurinic sites, introduced, e.g., by pH 5 treatment, or apyrimidinic sites introduced by treatment of DNA containing dUMP residues with uracil-DNA glycosylase. There is no detectable activity at lesions different from apurinic and apyrimidinic sites introduced by ultraviolet irradiation, X irradiation, or alkylation of DNA.

Reaction Requirements. Endonuclease IV has a broad pH optimum at pH 8 to pH 8.5. It shows no requirement for divalent metal ions and is fully active in the presence of 10^{-3} M EDTA. Further, endonuclease IV activity appears unusually resistant to the presence of high concentrations of NaCl in reaction mixtures. In the assay procedure used here, the enzyme is optimally active at 0.2 to 0.3 M NaCl and retains 50% of its maximal activity at 0.56 M, a salt concentration that strongly suppresses the activity of most enzymes acting on DNA.

[28] Purification and Properties of the Human Placental Apurinic/Apyrimidinic Endonuclease

By NANCY L. SHAPER and LAWRENCE GROSSMAN

General Introduction

Apurinic/apyrimidinic endonuclease (AP endonuclease) participates in an excision repair mechanism by catalyzing the hydrolysis of phosphodiester bonds in DNA at apurinic or apyrimidinic sites. This enzyme isolated from prokaryotic cells has been purified to apparent homogeneity and chemically characterized[1,2]; however, it has only been partially purified from eukaryotic cells.[3,4] The following section describes the purification of AP endonuclease to apparent homogeneity from human placenta.

Assay Method

Principle

The assay, which measures the cleavage of phosphodiester bonds at apurinic sites in depurinated replicative form I ϕX174 DNA (ϕX174 RFI DNA), is a modification of the assay of Center and Richardson.[5] If ϕX174 RFI DNA is denatured by alkali, two interlocked single-stranded circular molecules will be produced. Upon subsequent neutralization, the two molecules will resume their double-stranded structure. When ϕX174 DNA containing apurinic sites is incubated with AP endonuclease, incision at the apurinic site occurs; after denaturation a single-stranded circle and a single-stranded linear molecule will be produced, and upon neutralization, both molecules will remain single stranded. Under appropriate salt conditions, nitrocellulose filters bind single-stranded DNA and the amount of AP endonuclease present can be determined by measuring the amount of DNA which is bound to the filter.

Preparation of ϕX174 RFI [³H]DNA. The following protocol was developed by Dr. G. Nigel Godson, Radiobiology Laboratory, Yale University School of Medicine. Approximately 15 ml of an overnight culture of *Escherichia coli* HF4704, thy⁻, in TPG[6] medium containing 1 mM CaCl₂, 1% casein hydrolysate, and 2 μg/ml thymidine is inoculated into 500 ml of the same medium, and growth is continued at 37° until a density of 2×10^8 cells/ml is reached. The cells are then infected with ϕX174 *am*3 (fresh phage stocks, i.e., less than 6 weeks old, give higher yields of DNA) at a multiplicity of infection of 3–5 and the speed setting on the shaker water bath (New Brunswick) is reduced to 75 rpm. Exactly 5 min after infection 30 μg/ml of chloramphenicol (Sigma) is added, and 5 min later 1 mCi of [methyl-³H]thymidine (5 Ci/mmole, Amersham/Searle) is added. At 75 min after infection the cells are harvested and resuspended in 8 ml of cold

[1] B. Weiss, *J. Biol. Chem.* **251**, 1896 (1976).
[2] T. Inoue and T. Kada, *J. Biol. Chem.* **253**, 8559 (1978).
[3] S. Ljungquist and T. Lindahl, *J. Biol. Chem.* **249**, 1530 (1974).
[4] W. S. Linsley, E. E. Penhoet, and S. Linn, *J. Biol. Chem.* **252**, 1235 (1977).
[5] M. S. Center and C. C. Richardson, *J. Biol. Chem.* **245**, 6285 (1970).
[6] R. L. Sinsheimer, B. Starman, C. Nagler, and S. Guthrie, *J. Mol. Biol.* **4**, 142 (1962).

10% sucrose containing 50 mM Tris-HCl, pH 8.0. The cell suspension is transferred to a cellulose nitrate tube (Beckman Type 40 rotor), 0.4 ml of 0.25 M Na$_2$EDTA, pH 8.0, and 0.8 ml of lysozyme (10 mg/ml in 0.25 M Tris-HCl, pH 8.0) are added and the mixture is kept at 0°. Ten minutes later 0.4 ml of 10% sodium N-laurylsarcosinate (ICN/K&K Labs) is added by placing a 1-ml plugged pipette containing the detergent at the bottom of the tube and blowing continuously through the pipette for about 45 sec to effect thorough mixing of the detergent with the entire cell population. (A small amount of material will spill over the top of the tube due to the bubbling.) The cell suspension will clear in about 5 min; the tube is then filled with parafin oil and immediately centrifuged at 35,000 rpm for 45 min at 4°. The supernatant fluid is poured into a sterile tube and the gelatinous pellet is discarded. The sample is extracted twice with phenol, and the phenol–aqueous layer is washed several times with ether to remove the phenol and bubbled free of ether with nitrogen. It is then incubated with 0.1 ml of ribonuclease A (Worthington, 4.5 mg/ml in 0.3 M sodium acetate heated to 70° for 15 min) and 250 units of ribonuclease T$_1$ (Calbiochem) for 60 to 90 min at 37°. The sample is layered on two 35-ml 5–20% sucrose density gradients (10 mM Tris-HCl, pH 8.0, 1 mM Na$_2$EDTA, 1 M NaCl) and centrifuged in a Beckman SW27 rotor at 27,000 rpm for 22 hr at 5°. Fractions (1 ml) are collected, and 3-μl aliquots are counted in 3 ml Aquasol (New England Nuclear) in a liquid scintillation counter. The A_{260} values are determined over the peak area to verify that this peak coincides with the labeled peak. If the RNase digestion is not complete the upper side of the A_{260} peak will continue to increase. The appropriate fractions are precipitated with 2.5 vol of 95% ethanol and dissolved in 2–3 ml of 1 mM Tris-HCl, pH 7.5, 1 mM Na$_2$EDTA.

Approximately 150–200 μg of ϕX174 RFI DNA can be obtained from one preparation; 2% of this DNA is present as form II as determined by the nitrocellulose filter assay.

Reagents. The assay mixture (0.05 ml) contains 100 mM HEPES–NaOH, pH 8.25, 3 mM MgCl$_2$, 0.025 mM ϕX174 RFI [^{3}H]DNA nucleotide, and 1 μl of an appropriate dilution of enzyme.

The enzyme is diluted immediately before use in 10 mM Tris-HCl buffer, pH 7.5, 2 mM Na$_2$EDTA, 1 mM 2-mercaptoethanol, 0.5 mg/ml bovine serum albumin (GIBCO) and 0.2 mg/ml cytochrome c (Sigma).

Apurinic DNA is prepared immediately before use as follows: ϕX174 RFI [^{3}H]DNA (2–4 × 10^7 cpm/μmole), stored at 0.25 μmole/ml in 1 mM Tris-HCl, pH 7.5, 1 mM Na$_2$EDTA, is mixed with an equal volume of depurination buffer (20 mM NaCl, 20 mM sodium citrate, 20 mM NaH$_2$PO$_4$ which has been adjusted to pH 5.2 with 6 N HCl). The mixture is heated at 70° for 15 min and then cooled to 0°. This treatment introduces, on the average, one apurinic site per ϕX174 DNA molecule (see below, discussion of quantitation of apurinic sites).

Procedure. The assay mixture is incubated at 37° for 10 min. The reaction is terminated by the addition of 2 ml of denaturation buffer (100 mM Na$_2$HPO$_4$, 300 mM NaCl, 25 mM Na$_2$EDTA adjusted to pH 12.1 with solid NaOH). After 1 min at room temperature, 0.2 ml of 2 M Tris-HCl is added, the mixture is vortexed and filtered through a premoistened nitrocellulose filter (Schleicher and Schuell, type BA85). The reaction tubes are rinsed twice, and the filter well is rinsed once with 0.6 M NaCl–0.06 M sodium citrate. The filters are dried under an infrared lamp and counted in a liquid scintillation counter after the addition of 3 ml Instafluor (Packard).

Although depurinated DNA is itself susceptible to alkaline hydrolysis at the apurinic site, this is a slow reaction which requires 4-hr incubation at room temperature to reach completion.

(*Note:* The pH of the DNA denaturation buffer is critical. Upon standing the buffer adsorbs CO_2 and the small drop in pH is enough to cause incomplete denaturation. Also if the pH is too high, irreversible denaturation occurs causing all the ϕX174 DNA molecules to bind to the nitrocellulose filter.)

Quantitation of Apurinic Sites. The average number of apurinic sites per DNA molecule was determined by incubating the apurinic DNA in the denaturation buffer for 4 hr at room temperature. After renaturation and filtration the number of nicks introduced was determined by Poisson analysis.[7]

Unit. One unit of enzyme catalyzes the hydrolysis of 1 nmole of phosphodiester bonds per min.

Other Methods. Protein concentrations were determined using the Bio-Rad Protein Assay Kit. Due to interference by the ampholines, protein concentrations in the isoelectric focusing fractions were determined spectrophotometrically at 280 nm. Protein concentrations in fractions VI and VII were determined by the *o*-phthalaldehyde procedure.[8] Bovine γ-globulin was used as the protein standard.

Purification Procedure

All manipulations were carried out at 0°–4° unless otherwise stated. The results of a typical purification are outlined in Table I.

Preparation of the Crude Placental Extract. One human placenta was obtained as soon after birth as possible (usually 5–20 min) and immediately placed in 500 ml of chilled buffer A containing 30 mM Tris-maleate

[7] U. Kuhnlein, E. E. Penhoet, and S. Linn, *Proc. Natl. Acad. Sci. U.S.A.* **73,** 1169 (1976).
[8] S. A. Robrish, C. Kemp, and W. H. Bowen, *Anal. Biochem.* **84,** 196 (1978).

TABLE I
Purification of the Apurinic/Apyrimidinic Endonuclease from Human Placenta

Procedure	Fraction number	Volume (ml)	Total units	Total protein (mg)	Specific activity (units/mg protein)	Purification (-fold)	Recovery (%)
Crude extract	I	180	223	24,000	0.0093	0	100
OSA	II	240	67	2,500	0.027	3	(30)
Concentration and dialysis	III	5	156	2,400	0.065	7	70
Isoelectric focusing	IV	50	83	290	0.29	31	37
Sephadex G-75	V	12	58	12	4.8	516	26
Sephadex G-75	VI	10	33	2.6	12.7	1,366	15
DNA–agarose	VII	8	25	0.7	35.7	3,839	11

buffer, pH 6.8, 1 mM Na$_2$EDTA, 20% glycerol, 0.3 M sucrose, 5 × 10^{-4} M phenylmethyl sulfonyl fluoride, and 200 units/ml of Trasylol (Mobay Chem. Corp., FBA Pharmaceuticals, New York, New York). The latter two compounds, both protease inhibitors, were added to the buffer just before use.[9] The placenta was brought back to the laboratory cold room (within 5 min), removed from the buffer, and placed membrane side down on a sheet of aluminum foil. Approximately 200 g of tissue, cut from the placenta with scissors, were added to 200 ml of buffer A containing 0.15 M NaCl and homogenized in a Waring Blender at high speed for 1.5 min. The debris was removed by centrifugation at 16,300 g for 30 min. The supernatant fluid was immediately dialyzed against buffer B (30 mM Tris-maleate buffer, pH 6.8, 1 mM Na$_2$EDTA, 20% glycerol) for 2 hr and then recentrifuged at 16,300 g for 30 min. The upper layer of this pellet is easily dislodged and interferes with the subsequent chromatography step so care must be taken in decanting the supernatant fluid. This material (fraction I) is immediately chromatographed on the following albumin affinity absorbent.

OSA Affinity Chromatography. Octyl succinic anhydride coupled to Sepharose 4B by a diaminohexane spacer (OSA) efficiently binds serum albumin, the major contaminating protein in the crude placental extract. The procedure of Aslam *et al.*[10] was followed for preparing the absorbent except that it was made in 100-ml batches. About 300 ml of absorbent is prepared and a column (4.2 × 20 cm) is poured and equilibrated with buffer B, pH 6.8. Fraction I (180 ml) is applied to the column at a flow rate of 40 ml/hr. The column is washed overnight with buffer B, and the eluant is collected as a single fraction. Approximately 20% of the protein comes through in the wash, whereas the AP endonuclease remains bound to the column. A linear gradient (total volume 900 ml) of 0 to 1 M NaCl in buffer B, pH 7.8, is established, and 7.5-ml fractions are collected. The AP endonuclease elutes at 0.43 M NaCl (just after a large nonalbumin protein peak). The peak of activity is pooled to yield fraction II. Approximately 10% of the total protein applied to the column is recovered in this peak. Inhibition of the AP endonuclease occurs at this step, since it appears that only 30% of the total activity applied to the column is recovered. However, in the subsequent step, which involves concentration and dialysis, the activity recovered from the OSA column increases to 70%.

Preparation of the Sample for Isoelectric Focusing. Fraction II is prepared for isoelectric focusing (IEF) by concentration in a 200 ml Amicon stirred cell apparatus under N$_2$ pressure using a PM 10 membrane. When the volume is reduced to approximately 20 ml the sample is transferred to

[9] G. T. James, *Anal. Biochem.* **86,** 574 (1978).
[10] S. Aslam, D. P. Jones, and T. R. Brown, *Anal. Biochem.* **75,** 329 (1976).

a 50 ml Amicon apparatus and concentrated to a volume of 5 ml. The nonionic detergent Ammonyx-LO[11] is added to a final concentration of 0.5% by dilution from a 5% stock solution in water. (Ammonyx-LO is obtained from the Onyx Chem. Co., Jersey City, New Jersey and is lyophilized to dryness before use.) The concentrated sample is then dialyzed 2 hr against 500 ml 1% glycine, 15% glycerol, 0.5% Ammonyx-LO, pH 6.7. A small precipitate usually appears after dialysis, consequently the sample is centrifuged at 12,000 g for 10 min and then applied immediately to the IEF bed (fraction III).

Isoelectric Focusing. An LKB Multiphor apparatus is used for the preparative isoelectric focusing (IEF) run. The bed is prepared according to the instructions given in LKB application note 198. A solution (150 ml) containing 1% glycine, 15% glycerol, 2% ampholines (LKB) pH range 6–9, is degassed and Ammonyx-LO is added to a final concentration of 0.5%. Then 6 g Ultrodex (LKB) is added, the bed is poured, allowed to dry, and placed on the Multiphor plate previously cooled to 4° with a circulating water bath. The sample (5–6 ml) is applied and electrofocused at 4° for a period of 14 hr at 8 W constant power. The pH gradient is determined, and the samples are eluted in LKB columns with 5–6 ml of buffer B. The fractions containing AP endonuclease at pI of 7.4 are pooled (fraction IV) and concentrated, first in a 50 ml Amicon and then in an 8 ml Amicon, using PM 10 membranes, to a final volume of 1.5 ml.

Observance of Multiple Peaks Upon IEF. In the absence of protease inhibitors a total of five peaks of AP endonuclease activity were observed after IEF (pI's of 6.9, 7.1, 7.4, 7.6, and 7.8). In an experiment in which the activity with pI of 7.4 was recovered from the IEF bed, incubated at 4° for 24 hr and then refocused, multiple peaks were again obtained. In the presence of protease inhibitors, only two peaks of AP endonuclease activity were observed (pI of 6.9 and 7.4). The activity of pI 7.4 is the predominant peak (60–100% of the total activity). These results suggest that the other activities arise as a consequence of proteolysis of the predominant protein species with pI of 7.4.

Sephadex G-75 Chromatography. Fraction IV was applied to a G-75 superfine (Pharmacia) column (1.2 × 110 cm) which contained a 1-cm plug of G-10 at the bottom of the column. The column was equilibrated with 50 mM KHPO$_4$ buffer, pH 7.0, 0.3% Ammonyx-LO, and 20% glycerol (buffer C). The flow rate was 2 ml/hr, and 1-ml fractions were collected. The void volume of this column is 55 ml, and the AP endonuclease elutes at 80 ml, just after the major protein peak. The entire peak is pooled (fraction V), concentrated to 1 ml in the Amicon using a PM 10 membrane, and

[11] M. L. Applebury, D. M. Zuckerman, A. A. Lamola, and T. M. Jovin, *Biochemistry* **13**, 3448 (1974).

rechromatographed on the same column with buffer C. The AP endonuclease now elutes free of the contaminating protein peak. The activity is pooled (fraction VI) and stored at 0°.

DNA–Agarose Chromatography. Single-stranded DNA–agarose is prepared according to the method of Arndt-Jovin et al.[12] A column (7 × 1 cm) is poured and equilibrated with buffer C at pH 6.7. Fraction VI was applied to this column, and the column was washed with 30 ml of buffer C, pH 6.7, at a flow rate of 4 ml/hr. The AP endonuclease (fraction VI) was eluted with 1 M KCl in buffer C. The activity was stored at 0°.

Properties

Homogeneity and Molecular Weight. Due to the limited quantities of protein in fraction VII, charge and size heterogeneity was assessed after radiolabeling with $Na^{125}I$. An aliquot of fraction VI was radiolabelled with $Na^{125}I$ (13–17 $\mu Ci/\mu g$, Amersham/Searle) according to the method of Fraker and Speck[13] using the iodogen 1,3,4,6-tetrachloro-3a,6a-diphenylglycoluril (Pierce Chem. Co.). In order to remove unbound $Na^{125}I$ the labeled sample was chromatographed on a small DNA–agarose column equilibrated with buffer C. Unbound $Na^{125}I$ is eluted in the column wash; the ^{125}I-labeled enzyme is subsequently eluted with 1 M KCl in buffer C (fraction VII). The homogeneity of ^{125}I-labeled fraction VII was assessed by neutral polyacrylamide gel electrophoresis. Electrophoresis was performed on a 7.5% neutral polyacrylamide slab gel (pH 8.7) essentially as described by Davis.[14] After electrophoresis, the slab gel was dried for autoradiography or, alternatively, sectioned and assayed for enzymatic activity. Autoradiography revealed a single, sharp band with an R_f of 0.25 (relative to bromphenol blue). Enzymatic activity also migrated as a single, sharp peak with an R_f of 0.25.

Molecular weight heterogeneity of ^{125}I-labeled fraction VII was assessed on a 15% SDS–polyacrylamide slab gel in the presence of reducing agent.[15] Autoradiography revealed a single sharp ^{125}I-labeled band corresponding to an apparent molecular weight of 16,000.

Requirements. The enzyme has a pH optimum of 8.25. Enzymatic activity is inhibited 50% in the presence of 50 mM NaCl. Calcium ions do not activate the enzyme. Magnesium ions were required for maximum activity with an optimal concentration at 3 mM. In the absence of Mg^{2+} some activity was seen (0.4% of that obtained with 3 mM Mg^{2+}) and this

[12] D. J. Arndt-Jovin, T. M. Jovin, W. Bahr, A. Frischauf, and M. Marquardt, *Eur. J. Biochem.* **54**, 411 (1975).

[13] P. J. Fraker and J. C. Speck Jr., *Biochem. Biophys. Res. Commun.* **80**, 849 (1978).

[14] B. J. Davis, *Ann. N.Y. Acad. Sci.* **121**, 404 (1964).

[15] U. K. Laemmli, *Nature (London)* **227**, 680 (1970).

residual activity was inhibited by 2 mM EDTA. Manganese ions could partially replace Mg^{2+}; its optimal concentration was 1 mM, and at this concentration the activity was 50% of that observed with 3 mM Mg^{2+}.

Stability. Fraction VI has been stored at 0° for a period of 3 months with no loss in activity. The stability of fraction VII has not been assessed.

Additional Substrates. Fraction VI contains no nonspecific endonucleolytic activity when tested in 4000-fold excess on ϕX174 RFI DNA. Exonuclease activity (as determined by release of acid-soluble material) was not observed when either nicked ϕX174 DNA or single-stranded *E. coli* DNA were used as substrates. Phosphomonoesterase activity was undetectable.

[29] Purification and Properties of *Escherichia coli* Endodeoxyribonuclease V

By BRUCE DEMPLE, FREDERICK T. GATES III, and STUART LINN

Assay I

Principle. This assay measures the conversion of phage fd single stranded, circular DNA to a linear form which is hydrolyzed to acid-soluble products by *Escherichia coli* exonuclease I.

Reagents

1. The reaction mixture contains in 150 μl
   67 mM glycine-NaOH, pH 9.5
   25 mM KCl
   10 mM $MgCl_2$
   27 μM ^{3}H-labeled fd DNA[1] (4 nmole total DNA-nucleotide; 3 to 6 $\times$ 10[3] cpm/nmole)
   0.85 to 2 units exonuclease I[2]

---

[1] Prepared as described by P. J. Goldmark and S. Linn, *Proc. Natl. Acad. Sci. U.S.A.* **67,** 434 (1970). Unless otherwise noted, all DNA concentrations are expressed as DNA nucleotide residues.

[2] I. R. Lehman and A. L. Nussbaum, *J. Biol. Chem.* **239,** 2628 (1964). One unit of exonuclease I hydrolyzes 10 nmoles of DNA nucleotide to acid-soluble material in 30 min at 37°. The amount of exonuclease I added to the endonuclease V reaction mixture should be sufficient to render at least 4 nmoles of DNA nucleotide acid soluble under the conditions of the assay, while introducing a minimal number of endonucleolytic cleavages into the fd DNA. For our preparations of exonuclease, this is 0.85 to 2 units.

2. Bovine serum albumin, 10 mg/ml
3. Trichloroacetic acid, 7% (w/v)
4. Aqueous scintillation fluor: 9.1 g of 2,5-diphenyloxazole (PPO) and 0.61 g of 1,4-bis[2-(5-phenyloxazoyl)]benzene (POPOP) are mixed for 1 hr with 2140 ml toluene. 1250 ml Triton X-100 is then added and mixing is continued for 5 hr

Procedure. An appropriate amount of endonuclease is added, then the reaction mixture is incubated for 30 min at 37°C. It is then chilled on ice, and 0.05 ml of bovine serum albumin (10 mg/ml) and 0.25 ml of 7% TCA are added with mixing following each addition. After 5 min at 0°, the precipitate is removed by centrifugation at 10,000 g for 5 min; a 300-μl aliquot of the resultant supernatant is mixed with 3.3 ml aqueous fluor in a "minivial," and the radioactivity determined in a scintillation counter. (If a standard 20-ml vial is used, the aliquot should be brought to 0.9 ml and 10 ml of the fluor should be utilized.)

Calculations. One unit of endonuclease activity is defined as the amount which produces 1 nmole of acid-soluble product during the 30 min incubation. The assay is linear over the range of 0.15 to 1.5 units. A blank value of 0.3 nmole or less of acid-soluble material is observed in the absence of endonuclease, due to linear fd DNA and/or contamination of the exonuclease I by endonuclease.

The number of endonucleolytic cleavages can be estimated by determining the number of fd DNA molecules opened during the incubation. An fd DNA molecule contains 6389 nucleotides, so the acid-solubilization of 1 nmole of DNA is taken to be due to the conversion of 0.157 pmole of DNA molecules from the circular to the linear form. Since only one endonucleolytic cleavage is required for each such conversion, a minimum number of such cleavages is estimated unless one utilizes a Poisson correction (see assay II).

Assay II

Principle. This assay measures the conversion of closed circular duplex DNA from phage PM2 to the nicked form which can be selectively denatured and bound to nitrocellulose filters.

Reagents

1. The reaction mixture contains in 50 μl
   67 mM Tris-HCl, pH 8.2
   25 mM KCl
   10 mM MgCl$_2$

20–100 μM ^{3}H-labeled PM2 DNA[3] (1–5 nmoles total DNA nucleotide; 3–10 × 10[3] cpm/nmole
2. 2.5 mM EDTA, pH 8.0–0.01% sodium dodecyl sulfate
3. CHCl$_3$: n-octanol (9 : 1)
4. 0.3 M potassium phosphate adjusted to pH 12.3 as measured at 0.3 M with 5 N KOH
5. 1.0 M potassium phosphate adjusted to pH 4.0 as measured at 1.0 M with 5 N H$_3$PO$_4$
6. 5 M NaCl
7. 50 mM Tris-HCl, pH 8.2–1 M NaCl
8. 0.3 M NaCl–0.03 M sodium citrate
9. Nitrocellulose filters (Schleicher and Schuell type BA85, 0.45 μm pore size), equilibrated with the Tris–NaCl solution at least 12 hr before use
10. Nonaqueous scintillation fluor: 24 g of 2,5-diphenyloxazole (PPO) and 0.6 g of 1,4-bis[2-(5-phenyloxazoyl)]benzene (POPOP) are mixed for 1 hr with 6 liters of toluene

Procedure. An appropriate amount of endonuclease V is added, then the reaction mixture is incubated for 30 min at 37°. The reaction mixture is chilled on ice, then 0.5 ml EDTA–SDS solution is added. For fractions I–V, the mixture is then extracted at room temperature with an equal volume of chloroform–octanol solution. (For more purified fractions the extraction step is unnecessary.) A 0.2-ml aliquot of the supernatant is mixed with 0.1 ml water and counted for total radioactivity in 3.3 ml aqueous fluor (see Assay I). Another 0.2-ml aliquot is added to 0.2 ml of 0.3 M potassium phosphate, pH 12.3, at room temperature. After 2 min, 0.15 ml of 1.0 M potassium phosphate, pH 4.0, is added, followed by 0.14 ml of 5 M NaCl and 5.0 ml of Tris–NaCl solution. The sample is then passed through a nitrocellulose filter. The filter is washed with 5 ml of the NaCl–sodium citrate solution once before and once after the sample has been passed over it. The filter is then dried under an infrared lamp, and radioactivity determined by counting in 5 ml of nonaqueous scintillation fluor.

Calculations. The number of nicks formed during the incubation with enzyme can be calculated by assuming that the target sites are distributed among the DNA molecules according to a Poisson distribution and that all sites have an equal probability of becoming nicked. The equation $U = e^{-\mu}$ can then be used to determine the average number of nicks per molecule, μ, where U is the fraction of molecules not nicked, i.e., the fraction *not*

[3] Prepared by the method of R. T. Espejo and E. S. Canelo [*Virology* **34**, 738 (1968)] as modified by U. Kühnlein, E. E. Penhoet, and S. Linn [*Proc. Natl. Acad. Sci. U.S.A.* **73**, 1169 (1976)].

TABLE I
PURIFICATION *E. coli* ENDONUCLEASE V[a]

Fraction[b]	Volume (ml)	Protein (mg)	Enzyme activity (units $\times$ 10^{-3})	Specific activity (units/mg)
I. Crude extract	453	5890	398	69
II. Streptomycin supernatant	482	5300	304	57
III. Ammonium sulfate precipitate	222	4880	357	73
IV. Potassium phosphate precipitate	80	1400	302	216
V. DEAE–cellulose	455	114	69.6	611
VI. Phosphocellulose	80	2.24	38.0	17,000
VII. CM-Sephadex concentrate	4	0.12	6.92	57,000
VIII. Glycerol gradient	28	<0.03	6.92	>230,000

[a] Activity was measured by assay I.

[b] Protein in Fractions I through VI was determined by the method of O. H. Lowry, N. J. Rosebrough, A. L. Farr, and R. J. Randall, *J. Biol. Chem.* **193**, 265 (1951). Protein in Fractions VII and VIII was determined by the microassay system of M. M. Bradford, *Anal. Biochem.* **72**, 248 (1976). Bovine serum albumin was used as the standard in all protein determinations.

bound to the nitrocellulose filter. (The total DNA is obtained from the first aliquot removed from the reaction mixture.) It should be noted that U is always ≤ 1 and that μ increases exponentially as U approaches 0. Therefore the most accurate determinations of enzyme activity are obtained when less than 1 nick per molecule is formed, i.e., $U \geq 37\%$. Of the DNA molecules in the preparation, 5–10% are nicked after incubation without enzyme; this value is measured and used to correct the calculations.

Purification Procedure

Unless otherwise noted, all operations are carried out at 0°–4°.

Crude extract. *Escherichia coli* strain JC4583 (an endonuclease I-deficient K12 strain) is grown to mid- or late-log phase in L broth[4] at 37° with forced aeration, then harvested by centrifugation at $4000\,g$ for 5 min. The cell paste can be stored for up to 3 months at −20° after quick-freezing in liquid nitrogen. In one such preparation (Table I), 80 g of cell paste were suspended in 50 m*M* glycylglycine-NaOH, pH 7.0—1 m*M* potassium phosphate, pH 6.8—0.1 m*M* dithiothreitol (buffer A) to a total

[4] G. Bertani and J. J. Weigle, *J. Bacteriol.* **65**, 113 (1953).

volume of 200 ml. The cells were then broken by five, 45-sec pulses from a Branson Sonifier at 110 W, the temperature being maintained between 2° and 10° by an ice–salt bath. After centrifugation at 25,000 g for 20 min, the supernatant was diluted with buffer A to an absorbance of 200 at 260 nm (fraction I).

Streptomycin Precipitation. To fraction I was added slowly with stirring 0.085 volume of 5% streptomycin sulfate (Schwarz-Mann) in buffer A. After 30 min, the precipitate was removed by centrifugation at 25,000 g for 25 min and discarded (fraction II).

Ammonium Sulfate Precipitation. To each 100 ml of fraction II were added with stirring 37.9 g enzyme grade ammonium sulfate (Schwarz-Mann). The pH was maintained at 7.0 by the addition of 7 N NH$_4$OH. After 30 min, the precipitate was collected by centrifugation at 25,000 g for 15 min, and resuspended in buffer A to an absorbance of 350 at 260 nm (fraction III).

Dibasic Potassium Phosphate Precipitation. To fraction III was added slowly with stirring 0.155 volume of 4 M K$_2$HPO$_4$. After 10 min on ice, the precipitate was removed by centrifugation for 10 min at 27,000 g. A second aliquot of 4 M K$_2$HPO$_4$ (0.41 volume based on fraction III) was added slowly with stirring. After 30 min the precipitate was collected by centrifugation for 25 min at 27,000 g and resuspended in a minimal volume of buffer A. This suspension was then dialyzed for 12 hr against 40 volumes of buffer A (fraction IV).

DEAE–Cellulose Chromatography. A column containing 450 ml of packed resin[5] was equilibrated with buffer A. Fraction IV was applied to the column at 1.5 ml/min. The column was washed with 300 ml of buffer A, then a linear gradient from 0 to 0.4 M KCl of buffer A (4 liters total) was applied at 1.5 ml/min. Twenty milliliter fractions were collected into tubes containing 0.8 ml of 0.5 M (NH$_4$)$_2$SO$_4$–0.5 M potassium phosphate (pH 7.0). The active fractions (0.11 to 0.17 M KCl) were pooled and dialyzed for 12 hr against 10 volumes of buffer B (30 mM potassium phosphate, pH 6.8—0.5 mM dithiothreitol) (fraction V).

Phosphocellulose Chromatography. A column containing 150 ml of packed resin[5] was equilibrated with buffer B. Fraction V was then applied at a rate of 1.5 ml/min. The column was washed with 90 ml of buffer B, then a linear gradient from 0 to 0.6 M KCl in buffer B (1.2 liters total) was applied at 1 ml/min. Ten milliliter fractions were collected into polypropylene tubes. The active fractions (0.33 to 0.36 M KCl) were pooled and dialyzed for 10 hr against 50 volumes of buffer B (fraction VI).

CM25-Sephadex Chromatography. A column containing 30 ml of packed CM-Sephadex[5] was equilibrated with buffer B. Fraction VI was

[5] DEAE–cellulose, type 40, is purchased from the Brown Co., Berlin, New Hampshire. Phosphocellulose P-11 is a Whatman product. CM25-Sephadex is from Pharmacia.

TABLE II

RELATIVE RATES OF DIGESTION OF VARIOUS DNAS BY ENDONUCLEASE V[a]

DNA	pH 8.2	pH 9.5
fd		30.3
Untreated PM2	≡1.0	
Relaxed PM2	1.0	
UV-irradiated PM2	4.0	
Bisulfite-treated PM2	0.7	
PM2 exposed to pH 5.2	3.7	
OsO₄-treated PM2	7.5	
PBS-2		~120

[a] PM2 DNA was "relaxed" to untwisted form I with *E. coli* omega protein. PM2 DNA (0.7 to 1.0 mM in 10 mM Tris-HCl, pH 7.5) was irradiated with ultraviolet light in a watch glass on ice for 5 min with 20 erg/mm²/sec from a Westinghouse G15T8 Sterilamp and diluted immediately into the reaction mixture. PM2 DNA was treated with sodium bisulfite so as to produce 1–2 uracil residues per DNA molecule as described by Lindahl *et al.*[7] DNA was depurinated by treatment at pH 5.2 at 70° for 10 min in 0.1 M NaCl–10 mM sodium citrate–10 mM Tris-HCl. This treatment forms roughly 1.5 apurinic sites per PM2 DNA molecule. For OsO₄ treatment, 1.2 mM DNA in 25 μl of 10 mM Tris-HCl, pH 7.5, was partially denatured in 100 μl of 0.3 M potassium phosphate, pH 12.3, at room temperature, then exposed to 15 μl of 1% OsO₄ for 30 min. This treatment produces pyrimidine bases saturated at the 5,6-double bond. The DNA was renatured by the addition of 75 μl of 1 M potassium phosphate, pH 4.0, and 50 μl of 5 N NaCl and then dialyzed against 10 mM Tris-HCl, pH 7.5. Reactions containing PBS-2 DNA[8] were as for fd DNA, except that 1 nmole of DNA nucleotide is added and the exonuclease I is omitted.

applied to the column at 0.6 ml/min. The column was washed with 24 ml of buffer B, then a linear gradient from 0 to 0.5 M KCl in buffer B (400 ml total) was applied at 0.6 ml/min. Six milliliter fractions were collected into polypropylene tubes. The active fractions (0.20 to 0.23 M KCl) were pooled and dialyzed for 10 hr against 100 volumes of 30 mM potassium phosphate, pH 6.8—0.05 mM dithiothreitol (buffer C). The dialyzed material was concentrated by application to a 2-ml phosphocellulose column in a 5-ml polypropylene syringe at 4 ml/hr. After washing with 5 ml of buffer C, the activity was eluted into polypropylene tubes with buffer C containing 0.5 M KCl (fraction VII).

Glycerol Gradient Sedimentation. A 0.10-ml portion of fraction VII was layered onto a 5.2-ml linear glycerol gradient (5 to 20%) in 50 mM potassium phosphate, pH 6.8–10 mM (NH₄)₂SO₄–0.05 mM dithiothreitol in a polyallomer tube. After centrifugation for 24 hr at 50,000 rpm in a Spinco SW-50.1 rotor, fractions were collected into polypropylene tubes by dripping from the bottom of the gradient tube. A single, symmetrical peak of activity was observed approximately 40% down the tube (fraction VIII).

The purified fractions are unstable to freeze-thawing and long-term storage, although they are somewhat stable for several months in 50% glycerol at $-20°$ in polypropylene tubes.[6] Endonuclease V activity is also sensitive to dialysis; losses can be minimized by using Spectrapore 1 dialysis membrane (6,000 to 8,000 dalton cutoff).[6]

Properties

Specificity. The enzyme makes endonucleolytic cleavages in single-stranded DNA, acting at many sites in the fd DNA molecule.[6] Duplex DNA is a substrate of tenfold lower susceptibility (Table II).[7,8] Relaxation of the PM2 DNA to a nonsupercoiled form, or exposure to uv light, sodium bisulfite (to form a few uracil residues), pH 5.2 (to form apurinic sites), or osmium tetroxide (to form pyrimidines saturated at the 5,6-double bond) renders the duplex DNA as much as sevenfold more active as a substrate for endonuclease V[6] (Table II). The enzyme has similar relative activities on linear duplex T7 DNA exposed to the various insults.[6]

The duplex DNA of the *Bacillus subtilis* bacteriophage, PBS-2,[8] in which thymine is totally substituted by uracil, is the best substrate. Indeed only this DNA is rendered acid soluble by the purified enzyme. The enzyme does not appear to have an exceptionally high affinity for uracil in DNA, however, since the PM2 DNA containing a small number of uracil residues introduced by $NaHSO_3$ is cleaved at about the same rate as untreated PM2 DNA (Table II).

Effect of Metals. Endonuclease V exhibits a strong requirement for Mg^{2+} and is completely inactive on fd DNA in its absence. Manganese ions can restore some of the activity on PBS-2 DNA (15% at 2 mM Mn^{2+}), whereas Ca^{2+}, Zn^{2+}, Co^{2+}, or Fe^{2+} cannot activate the enzyme.[6] A unique property of endonuclease V is the strong inhibition by Fe^{3+} with either fd or PBS-2 DNA.[6,9]

Effect of pH. With fd DNA substrate, endonuclease V exhibits a sharp optimum at pH 9.5. The enzyme is also 50-fold more active on duplex substrates at this pH than at pH 7.0.

Molecular Size. When sedimented through a glycerol gradient by the

[6] F. T. Gates and S. Linn, *J. Biol. Chem.* **252,** 1647 (1977).
[7] T. Lindahl, S. Ljungquist, W. Siegert, B. Nyberg, and B. Sperens, *J. Biol. Chem.* **252,** 3286 (1977).
[8] PBS-2 DNA is prepared as described by E. C. Friedberg, A. K. Ganesan, and K. Minton [*J. Virol.* **16,** 315 (1975)], except that 5 mCi of 6-³H-uridine (Schwarz-Mann) are added immediately after infection, and the culture was treated with chloroform and harvested immediately after lysis.
[9] R. V. Blackmore and S. Linn, *Nucleic Acids Res.* **1,** 1 (1974).

method of Martin and Ames,[10] endonuclease V exhibits an $s_{20,w}$ of 2.3. Assuming that the protein is globular, this corresponds to a molecular weight of approximately 20,000.

[10] R. G. Martin and B. N. Ames, *J. Biol. Chem.* **236**, 1372 (1961).

[30] Purification and Properties of an Endonuclease Specific for Nonpyrimidine Dimer Damage Induced by Ultraviolet Radiations

By Sheikh Riazuddin

Assay Method

Principle. The assay measures formation of the enzyme–substrate complex prior to incision of the substrate by the endonuclease.

$$E + S \rightleftharpoons ES \rightarrow E + P$$

Reagents

1. Reaction mixture (total volume 0.1 ml) contains Tris-HCl, 10 mM, pH 7.4
   2-Mercaptoethanol, 1 mM
   NaCl, 100 mM
   Escherichia coli [³H]DNA, 5000 cpm, 250 pmole of mononucleotide equivalent
   Enzyme, 0.2–1.0 units
2. Saline citrate (300 mM NaCl + 30 mM sodium citrate)

Procedure. After incubation at 0° for 10 min. the reaction mixture is diluted with 2 ml of cold saline–citrate and filtered immediately through HAWP Millipore filters. The reaction tubes are washed once with 4 ml of cold saline–citrate, and the filters are dried and counted in 2 ml of toluene-containing scintillation fluid. The radioactivity retained on the Millipore filters is taken as a measure of endonuclease binding activity and the nonpyrimidine dimer uv damage specific activity is determined by the difference in binding activities against heavily uv-irradiated (1000 J/m²) and lightly uv-irradiated (50 J/m²) DNA substrates.

This binding assay is used during the purification procedure to test various fractions for the presence of this enzyme. However, for quantitative purposes, another assay[1] is used. This assay measures the release of

[1] S. R. Kushner and L. Grossman, this series, Vol. 21, Part D [15].

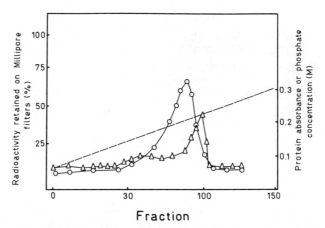

FIG. 1. Elution pattern of DNA-binding activity from phosphocellulose (P-11). The activity is determined using heavily ultraviolet-irradiated (1000 J/m²) DNA (△—△) and has been corrected for binding activities against native or lightly UV-irradiated (50 J/m²) DNA. Absorbance is measured at 280 nm (○—○) and concentration of the eluting phosphate buffer is obtained from a measurement of the conductivity (- - -).

phosphomonoesters produced by endonucleolytic incision of ^{32}P-labeled uv-irradiated (20,000 J/m²) DNA. One unit of activity is defined as that amount of enzyme required to release 10 pmole of ^{32}Pi from heavily uv-irradiated DNA in 30 min at 37°.

Purification Procedure

All operations are performed at 0–4°. Centrifugations are at 20,000 g for 30 min. Unless otherwise stated, all solvents contain 1 mM EDTA, 1 mM 2-mercaptoethanol, and 10% glycerol.

Growth of *Micrococcus luteus* cells, crude extract formation, first DEAE–cellulose chromatography, and phosphocellulose chromatography steps are exactly the same as those for the isolation of pyrimidine dimer damage specific endonucleases. In the phosphocellulose chromatography step, when the adsorbed proteins are eluted from the column with a 50–300 mM potassium phosphate at pH 7.4, this enzyme is eluted immediately after the main protein peak (Fig. 1), whereas the pyrimidine dimer damage specific endonucleases are eluted prior to the main protein peak. The most active fractions are pooled for further purification (fraction C).

Sephadex G-75 Chromatography. When fraction C after concentration by precipitation with 70% ammonium sulfate, is subjected to Sephadex G-75 chromatography, the endonuclease is eluted as a major peak in the vicinity of cytochrome c marker (MW = 13,000). The most active fractions constituting the peak are pooled (fraction D).

TABLE I
PURIFICATION OF AN ENDONUCLEASE SPECIFIC FOR NONPYRIMIDINE DIMER DAMAGE
INDUCED BY UV RADIATION

Fraction	Volume	Activity units ($\times 10^{-3}$)		Protein (mg/ml)	Specific activity[b]	Yield
		UV-dependent[a]	Non-specific			
Extract	580	c	3860	13.8	—	—
DEAE–cellulose	980	c	2100	5.5	—	—
Phospho-cellulose	60	15	1.0	1.0	233	100
Sephadex G-75	30	9.1	—	0.1	1517	75
DEAE–cellulose	2	5.8	—	0.4	7250	41

[a] UV-dependent activity is determined by the difference in total activities against uv-damaged DNA (10,000 J/M) and native DNA as determined by bacterial alkaline phosphatase assay.[1]
[b] Specific activity is defined as bacterial alkaline phosphatase units of ultraviolet-dependent activity per milligram of total protein.
[c] UV-dependent activity could not be determined because of contaminating Py < > Py correndonucleases.

DEAE–Cellulose Chromatography. A column of DEAE–cellulose (2.9 cm² × 10 cm) is equilibrated by washing with 4 liters of 50 m*M* potassium phosphate (pH 7.6) containing 1 m*M* EDTA, 1 m*M* MSH and 10% glycerol. Fraction D after dialysis against 10 volumes of the same buffer, is applied to the column and after washing the column with 50 ml of the same buffer, the protein is eluted in 5 ml fractions with a 300 ml linear gradient of 50 to 200 m*M* potassium phosphate at pH 7.6. The non-pyrimidine dimer uv-damage specific enzyme eluting at 100 to 125 m*M* potassium phosphate is pooled and dialyzed against the column buffer. After concentrating it to 2 ml in an Amicon ultrafiltration cell equipped with a Diaflo PM 10 membrane, the enzyme is stored in liquid nitrogen (fraction E).

Comments on the Purification Procedure. The first DEAE–cellulose column is used to remove the DNA and a major part of the exonucleases. About 30 to 50% of the total proteins are removed with a minimal loss of the endonuclease. This step is necessary in separating this enzyme from the major protein peak on the next column. The phosphocellulose column separates this enzyme from the pyrimidine dimer damage-dependent ones. The Sephadex G-75 chromatography step removes a high molecular weight nonspecific protein. The second DEAE–cellulose step is used under conditions in which this enzyme is adsorbed on the column, separating it completely from Py < > Py correndonucleases which are not retained by DEAE–cellulose. A summary of the purification procedure is given in Table I.

Fraction E contains the uv damage-dependent enzyme, 31-fold purified in 41% yield. The enzyme is isolated as a protein which binds to heavily uv-irradiated DNA in preference to unirradiated or lightly uv-irradiated DNA. Nevertheless, it is an endonuclease which produces chain scission in damaged DNA. The purified enzyme is not a homogeneous protein as revealed by polyacrylamide gel electrophoresis. However, it is not contaminated with endonucleolytic activities against native ϕx174 FRI DNA.

Properties of the Purified Enzyme

Stability. The endonuclease after purification on a DEAE–cellulose column (Fraction E) is very unstable, undergoing complete inactivation in 3–5 weeks in liquid nitrogen. The enzyme is, however, not sensitive to freeze-thawing.

General Requirements. The endonuclease has a broad pH optima at 7.0–7.6 in Tris-HCl. The enzyme shows no obligatory cofactor requirement but addition of α-ATP(0.1–1.0 mM) to reaction mixtures causes a slight stimulation (about 30%). High ionic strength is required for normal levels of enzymatic activity and in 10 mM Tris. HCl the enzyme is completely inactive. Addition of 10–15 mM MgCl$_2$ restores full activity of the enzyme. Requirement for magnesium can be replaced by the presence of 75–100 mM NaCl or KCl or 20 mM MnCl$_2$; CaCl$_2$ or ZnCl$_2$ has no effect on the enzyme activity.

Substrate Specificity. The enzyme is active against heavily uv-irradiated DNA, x-irradiated DNA, or OsO$_4$-treated DNA. The enzyme also acts on alkylated DNA or depurinated DNA possibly due to the presence of contaminating apurinic endonuclease.[2] When uv-irradiated DNA is used as a substrate, this enzyme requires 390 J/m^2 to produce 63% incision of ϕX174 RFI molecules as opposed to 7 to 11 J/m^2 required by Py < > Py correndonucleases.[3] Since OsO$_4$ treated or X-irradiated DNA is also a substrate for this enzyme, its specificity seems to be due to the presence of thymine glycol in the DNA.[4] The purified enzyme makes endonucleolytic incision in uv-irradiated DNA and generates 3′-phosphoryl and 5′-hydroxyl termini. When incised DNA is used as a substrate for DNA polymerase I, it is hydrolyzed at a rate comparable to

[2] J. Laval, *Nature (London)* **269**, 829 (1977).
[3] S. Riazuddin and L. Grossman. *J. Biol. Chem.* **252**, 6280 (1977).
[4] P. V. Hariharan and P. A. Crutti, *Proc. Natl. Acad. Sci. U.S.A.* **71**, 3532 (1974).

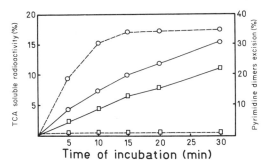

FIG. 2. Excision of pyrimidine dimers by the 5' → 3'-oxonucleolytic activity of DNA polymerase I from uv-irradiated DNA which has been incised by either Py < > Py correndonucleases (O—O) or the endonuclease specific for nonpyrimidine dimer UV-induced damage (□—□). Solid lines represent 5% TCA-insoluble radioactivity and broken lines represent fraction of total dimers excised.

that of DNA treated with Py < > Py correndonuclease. However, there is no release of pyrimidine dimers, whereas 40% of total dimers from DNA treated with Py < > Py correndonucleases are removed by the exonucleolytic functions of DNA polymerase I (Fig. 2).

Occurrence. This enzyme is present at normal levels in the mutant DB7 and transformants DB200 and DB400, which vary in radiosensitivity as compared to the wild-type parent.

[31] The Use of DNA Fragments of Defined Sequence for the Study of DNA Damage and Repair

By WILLIAM A. HASELTINE, CHRISTINA P. LINDAN,
ALAN D. D'ANDREA, and LORRAINE JOHNSRUD

Introduction

A wide variety of physical and chemical agents damage DNA. The biological effects of such agents include cell death, neoplastic transformation, and mutations. Cells are normally capable of repairing some types of DNA damage by enzymatic modification of the DNA. In most cases, the details of the repair pathways will determine the ultimate biological effect of the initial lesions. Therefore, it is desirable to develop new approaches for the characterization of lesions in DNA and of the effects of repair activities on damaged DNA.

We have recently applied some of the remarkable progress that has been made in nucleic acid technology to such problems. The basic approach involves the use of a DNA fragment of defined sequence as a probe to study the damage and repair of DNA. The advantage of using such a sequence is that the precise site of damage and the subsequent fate of the modified DNA can be determined rapidly.

The DNA fragment used for these studies is a fragment of the lactose operon that contains the promotor and operator region (*lac* p-o). This DNA sequence was selected for several reasons:

1. The sequence of the DNA is known.[1,2]

2. The sequence has been cloned in pMB9 by Johnsrud[3] in a form that is ideal for use. Two copies of the sequence are on each plasmid and are bounded by sites for the restriction endonuclease *Eco*RI. The recombinant plasmid, named pLJ3, carries tetracycline resistance and can be easily maintained in bacteria. In addition, the resulting multiple copies of the *lac* operator serve to titrate the repressor so that colonies carrying the sequence are blue on X-gal indicator plates.[4–6] As a result, the plasmid can be easily maintained, amplified and isolated in large quantities.

3. The orientation of the two copies of the *lac* p-o region within pLJ3 permits the rapid isolation of large amounts of the known sequence and the terminal labeling of both strands of the known sequence. A schematic diagram for the isolation of the $5'$-^{32}P end-labeled strands of opposite polarity is shown in Fig. 1.

4. The sequence contains binding sites for two proteins—the *lac* repressor and the *E. coli* RNA polymerase.[7,8] This permits the study of the effect of DNA modifications in protein–DNA interactions, and conversely, the effect of bound proteins on the sites of damage inflicted on DNA by physical and chemical agents.

5. The plasmid pLJ3 carries the uv5 mutation.[1] This mutation is in the promotor region and permits efficient *in vitro* transcription of the DNA by *E. coli* RNA polymerase even in the absence of the cap protein.

What follows are detailed procedures for the purification and use of this sequence. For the most part, the procedures described are adapted

[1] J. Grall, personal communication

[2] R. D. Dickson, J. Abelson, W. M. Barnes, and W. S. Reznikoff, *Science* **187**, 27 (1975).

[3] L. Johnsrud, *Proc. Natl. Acad. Sci. U.S.A.* **75**, 5314–5318 (1978).

[4] K. J. Marians, R. Wu, J. Stawinski, T. Hoaumi, and S. A. Narang, *Nature (London)* **263**, 744 (1976).

[5] H. L. Heyneker, J. Shini, J. M. Goodman, H. W. Boyer, J. Rosenber, R. E. Dickerson, S. A. Narang, K. Itakway, S. Lin, and A. D. Riggs, *Nature (London)* **263**, 748 (1976).

[6] J. R. Sakler, M. Tecklenburg, J. L. Betg, B. V. Boeddel, D. G. Unasura, and M. H. Caruthers, *Gene* **1**, 305 (1977).

[7] W. Gilbert, A. Maxam, and A. Mirzabekov, *Control Ribosome Synth. Proc. Alfred Benzon Symp., 9th, 1976* p.   (1976).

[8] A. Humayn, D. Kleid, and M. Ptashne, *Nucleic Acids Res.* **4**, 1595 (1977).

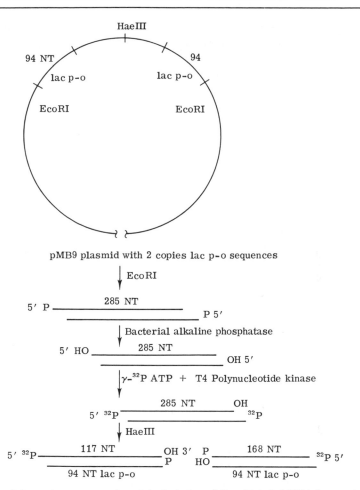

FIG. 1. Schematic representation of the isolation of 5' end-labeled DNA fragments of the *lac* p-o region.

from methods that have been developed by others and are included here for the convenience of those wishing to use this plasmid. Several examples of the use of the plasmid for analysis of problems of DNA damage and repair are described.

Materials and Methods

A diagram of the insertion of the DNA sequence within the pMB9 plasmid, its isolation by *Eco*RI restriction and subsequent 5'-^{32}P labeling with polynucleotide kinase and [γ-^{32}P]ATP is given in Fig. 1. Following terminal labeling, a fragment that is approximately 285 base pairs long is cleaved asymmetrically by *Hae*III to produce a 117 base pair fragment and a fragment that is approximately 168 base pairs long. These two sequences can be separated by electrophoresis. Only one 5' terminus of each of the resulting fragments is labeled. Since the polarity of the *lac* p-o insertions within the plasmid is the same, opposite strands of the same sequence are labeled. The nucleotide sequence of the *lac* p-o region is provided in Fig. 2.

Plasmid Isolation. The pMB9 plasmid pLJ3 is grown in *E. coli* strain MM 294 (obtained from M. Meselson), endo I⁻, B1⁻, r_k−m_k+ and kept in a stab at room temperature. Bacteria are grown and the plasmid harvested using techniques similar to Tanaka and Weisblum[9] as modified by Rubin.[10] X-gal plates with 10 μg/ml tetracycline and made according to Miller[11] are streaked with an inoculum from the stab. After 2 to 3 days at 37°, several blue colonies are added to 200 ml of YT medium[11] (8 g Bactotryptone, 5 g Bacto yeast, and 5 g NaCl per liter) and incubated overnight at 37° without shaking. Fifteen milliliters of this saturated culture is added to 760 ml of YT medium in each of five 2-liter flasks and grown shaking at 37° approximately 3.5 hr to an O.D.$_{590}$ of 1.2. One milliliter of chloramphenicol (Sigma) (600 mg stored in 5 ml of ethanol) is added to each culture flask and left shaking for 12–15 hr. Cells are harvested in 1-liter bottles at 10,000 rpm for 45 min at 0°. The cell pellets are rinsed with 0.5 liter of 10 m*M* Tris-HCl, pH 7.4, and 1 m*M* EDTA and kept on ice throughout the remaining procedures. Each cell pellet is resuspended in 10 ml of 25% sucrose, 50 m*M* Tris-HCl, and 40 m*M* EDTA, pH 8.0, and treated with 3 ml of 10 mg/ml lysozyme in 50 m*M* Tris-HCl pH 8.0 and 40 m*M* EDTA and left to stand on ice for 5 min with occasional gentle swirling. Three milliliters of 0.5 *M* EDTA pH 8.0 is added to each flask, and the mixture is again left on ice for 5 min. To each flask, 27 ml of Triton-X mix (2 ml of 10% Triton-X, 25 ml of 0.5 *M* EDTA, pH 8.0, 10 ml of 1.0 *M* Tris-HCl, pH 8.0, H₂O to 200 ml) is added. The mixture becomes viscous during this period but not before. After the suspension has been left on ice for 10 min, it is centrifuged in polyallomer tubes at 25,000 rpm for 30 min at 0° in an SW27 Beckman rotor. The supernatant is decanted into a graduated vessel (*note*, do not remove the viscous portion just above the pellet), and 0.95 g of CsCl and 0.1 ml of 10 mg/ml ethidium bromide per milliliter of supernatant are added to bring the refractive index to 1.390–1.395. The mixture is centrifuged in polyallomer tubes in a Beckman vertical Ti50 rotor at 42,000 rpm for 24 hr at 10°–15° or in a fixed angle 40, 50, 50Ti, or

[9] T. Tanaka and B. Weisblum, *J. Bacteriol.* **121**, 354 (1974).
[10] G. Rubin, personal communication.
[11] J. Miller, "Experiments in Molecular Genetics," p. 47. Cold Spring Harbor Lab., Cold Spring Harbor, New York, 1972.

FIG. 2. Nucleotide sequence of the *lac* p–o region. The unprimed numbers refer to the distance in nucleotides from the 5' terminus of the 117 fragment, and the primed numbers refer to the distance in nucleotides from the 5' terminus of the 168 fragment.

65 rotor at 38,000 rpm for 3 days. If the centrifugation is performed in a vertical rotor, then a brake should not be applied at the termination of the spin.

Plasmid Purification. DNA corresponding to less dense, nicked, or chromosomal DNA and more dense, covalently closed, circular plasmid are clearly visible as distinct bands under long wavelength uv light. The lower band is removed using needle side puncture. The ethidium bromide is removed from the DNA by passage over a small Dowex 100 column. The Dowex 100 is first thoroughly washed with 0.5 M NaOH, then with 0.5 M HCl, and finally with distilled H$_2$O and stored in 0.5 M NaCl, 0.1 M Tris-HCl pH 7.4. The columns, 1-ml total volume in a Pasteur pipette, are equilibrated with elution buffer 10 mM Tris-HCl, pH 7.4, 0.1 mM EDTA, and 0.2 M NaCl. One to 2 ml of plasmid band are layered onto each column, and the column is washed with two volumes of elution buffer. Nine volumes of ethanol are added to the collected fractions which are then vortexed and left to precipitate overnight at −20°. The mixture is centrifuged in a Sorval RC5 rotor at 10,000 rpm for 30 min at 4°. The supernatant is carefully decanted, and the pellets are gently rinsed with cold 95% ethanol. The plasmid is lyophylized to dryness, and each pellet is resuspended in a small volume (200–300 μl per tube) of 10 mM Tris, pH 7.4, and 0.1 mM EDTA. One to 2 μl is taken from each aliquot and used for spectrophotometric quantitation of the amount of DNA isolated. An O.D. of 2 at 260 nM corresponds to a concentration of 0.1 mg/ml. For 4 liters of harvested cells, 1.5 to 2.0 mg of plasmid are usually obtained. In order to obtain pure plasmid DNA a further step is needed to separate plasmid DNA from contaminating nucleic acid species. This can be done by either sucrose gradient centrifugation or Sepharose 4B or BioGel A15M column chromatography.

*Eco*RI *Restriction of the Plasmid.* To prepare the substrate for ^{32}P end labeling, the plasmid is first cleaved with the restriction endonuclease *Eco*RI. The 285 base pair fragment that results may then be stored in 10 mM Tris-HCl, pH 7.4, and 0.1 mM EDTA at −20° until further treatment. For this procedure all precipitations of nucleic acid are performed by raising the salt concentration of the suspension to 0.3 to 0.4 M with 4 M NaOAc, pH 5.0, and adding three times the volume of 95% ethanol. The reaction vessels are vortexed and chilled for 5 min in a Dry Ice–ethanol bath, and centrifuged in an Eppendorf centrifuge in the cold for 5 min. The supernatant is decanted, and the pellet rinsed with cold 95% ethanol. The DNA is then lyophylized and resuspended in the desired buffer. All reactions are performed in siliconized 1.5-ml Eppendorf capped tubes. All glass pipettes and test tubes are also siliconized.

For *Eco*RI restriction, reaction volumes are typically 200 μl containing 200 g DNA plasmid and 250 units of New England Bio Laboratories or

Boehringer-Manneheim *Eco*RI, in the buffer specified by the manufacturer of the enzyme. Incubation proceeds for 60 min at 37° after which the mixture is phenol-extracted twice, and ether-extraced four to five times. The ether is removed by gentle blowing, and the mixture is precipitated and rinsed with ethanol. The pellet is lyophylized to dryness and resuspended in 20 μl of electrophoresis buffer (TEB, 50 mM Tris-HCl, pH 8.3, 50 mM boric acid, and 1 mM Na$_2$EDTA) to which is added 10 μl of 15% glycerol, 0.06% each of xylene–cyanol and bromphenol blue suspended in 1× electrophoresis buffer. Electrophoresis is performed using a 7.5% polyacrylamide slab gel with slots 8 × 8 × 0.2 mm. The 7.5% polyacrylamide gel is prepared from a stock solution of 38% acrylamide and 2% *N,N'*-methylenebisacrylamide (Eastman) which is prewashed for 2 hr in 0.5% charcoal and 1.5% analytical grade mixed bed resin (Bio-Rad AG 501-X8, 20–40 mesh). The acrylamide mix is filtered through Whatman No. 1 filter paper and through a Millipore 0.45 μm filter before use. The 7.5% gel is made in TEB buffer with 0.07% ammonium persulfate. *N,N,N',N'*-Tetramethylethylenediamine (Eastman) is used to catalyze polymerization. Approximately 200 μg of DNA can be loaded per slot. Electrophoresis is performed at 150 V for approximately 4 hr or until the slow-moving dye has run halfway down the gel. The plates are removed from the gel, and the DNA bands visualized by uv shadowing.[12] The gel, which must be aged 24 hr before use, is placed on a thin layer chromatography plate which contains a fluorescent indicator. DNA is visualized by illumination with a hand-held short-wave uv light. The bands are excised and the acrylamide pieces are crushed with a siliconized glass rod in an Eppendorf tube and eluted with 600 μl of double-distilled water. After 24 hr at 37°, the acrylamide is spun down in an Eppendorf microcentrifuge, the supernatant carefully removed, and the pellet washed with 200 μl of double-distilled water. The supernatants are pooled and filtered through a 0.45 μm Millipore "Swinex" filter. The filtrate is then lyophylized to dryness and resuspended in a small volume of Tris-EDTA buffer for storage or 30 mM Tris-HCl, pH 8.0, for treatment with bacterial alkaline phosphatase.

Incorporation of 5'-[32]*P End Label.* The DNA eluted from one track of the *Eco*RI preparative gel, approximately 5 μg, is resuspended in 30 μl of 30 mM Tris-HCl, pH 8.0, to which is added 0.1 unit of bacterial alkaline phosphatase (Worthington). Incubation proceeds for 60 min at 37°. Several phenol extractions are performed followed by five to six ether extractions, after which the ether is gently blown off and the DNA precipitated. The following treatment of the 285-base pair fragment with polynu-

---

[12] S. M. Hassur and H. S. Whitlock *Anal. Biochem.* **59**, 162 (1974).

cleotide kinase and [γ-^{32}P]ATP is performed according to the protocol for DNA sequencing by Maxam and Gilbert[13] (also see this volume, Article 57). Total reaction volumes are 100 μl with 20 μl of denaturation buffer (10 mM Tris-HCl, pH 9.5, 1 mM spermidine, and 0.1 mM EDTA), to which is added 10 μl of 10× kinase buffer (500 mM Tris-HCl, pH 9.5, 100 mM MgCl$_2$, and 50 mM dithiothreitol), 60 μl of γ-labeled ATP (made according to the protocol given in this article), and 10 μl of polynucleotide kinase (Boehringer-Manneheim). Incubation should proceed at 37° for 60 min.

*Hae*III *Restriction and Isolation of 117 and 168 Base Pair Fragments.* One-half to 5 μg of ^{32}P end-labeled 285 bp fragment is incubated with 40 units of *Hae*III (Bio Labs or Boehringer-Manneheim) for 60 min at 37° in 100 μl of buffer specified by the enzyme manufacturer. Cleavage produced two fragments, approximately 117 and 168 base pairs in length. The DNA is precipated, washed with ethanol, lyophylized to dryness, and resuspended in 20 μl of 1× TEB and 10 μl of glycerol–dye solution. Samples are loaded on 7.5% polyacrylamide slab gels as described and run for approximately 6 hr, or until the faster moving dye has just run off the bottom of the gel. The gel is then removed carefully from the gel plates, wrapped in cellophane wrap, taped onto a jacketed X-ray film, and pressed between glass plates. A 5-min exposure is usually sufficient to visualize the bands. The gel is then aligned properly with the developed film, and the radioactive bands with 117- and 168-base pair fragments are excised. The acrylamide bands are crushed with a glass rod and eluted with 600 to 800 μl of 10 mM Tris-HCl, pH 7.4, 0.1 mM EDTA, 0.2 M NaOAC, and 100 μg RNA carrier for 12 to 36 hr at 37°. The pellets are centrifuged, washed with 200 μl of 0.2 M NaOAC, and the pooled supernatants for each DNA fragment are filtered through a Millipore "Swinex" apparatus with a 0.45 μm filter. The fractions are then brought to 0.4 M NaOAC and precipitated with ethanol. The DNA pellets are washed with ethanol and resuspended in Tris–EDTA and stored at −20°.

Alternative Procedure for Plasmid Isolation. Rapid methods for isolation of plasmid DNA without centrifugation have recently been developed. These include phase partitioning of a deproteinized cleared lysate[14] and the passing of such a lysate over a high salt BioRad A-50 column.[14] Both procedures, however, require modifications to match the capacity of the centrifugation method.

DNA Sequencing Analysis; Synthesis of [γ-^{32}P]ATP, Electrophoretic Gels. DNA sequencing is performed according to the Maxam and Gilbert protocol (see [57] this volume). In addition, neocarzinostatin, an an-

[13] A. Maxam and W. Gilbert, *Proc. Natl. Acad. Sci. U.S.A.* **74,** 560 (1977).
[14] R. Ohlsson, C. C. Hentschel, and J. G. Williams, *Nucleic Acids Res.* **5,** 583 (1978).
[15] F. Bolivar, R. Rodriguez, M. Betlack, and H. Boyer, *Gene* **2,** 75 (1977).

titumor drug which cleaves DNA at positions of thymines and at adenines[16-18] is also used here as a rapid sequencing technique. The 5' end-labeled DNA to be cleaved with neocarzinostatin is added to 200 μl of freshly made buffer (50 mM Tris-HCl pH 7.4, 10 mM mercaptoethanol, 20 μg of carrier DNA, and 20 μl of neocarzinostatin, 1 unit/μl). Incubation proceeds for 30 min at 37° followed by two phenol extractions, four to five ether extractions, precipitation of the DNA, an ethanol rinse, lyophyliza-tion, and resuspension in loading buffer.

Synthesis of [γ-^{32}P]ATP is performed as described by Maxam and Gilbert[13] modified as described by Haseltine and Pedersen (see [64] this volume). Both preparative 7.5% and 20% sequencing electrophoretic gels are made according to the Maxam and Gilbert protocol.[13]

Examples

Three examples of the use of this DNA sequence for studies of DNA damage and repair are given. The first illustrates the use of the probe to define the site of damage by agents known to break DNA-neocarzinosta-tin and bleomycin. The second example is chosen to demonstrate the use of the DNA for obtaining precise information about the mechanism of action of a DNA repair enzyme, the uv correndonuclease of *M. luteus*. The third example shows how microsomal enzymes can be used in con-junction with the sequence to study the site of damage of chemicals that require metabolic activation, in this case, aflatoxin B$_1$.

Site-Specific Cleavage of DNA by Neocarzinostatin and Bleomycin. It was previously reported that the antitumor antibiotic neocarzinostatin and bleomycin break DNA *in vivo* and *in vitro*.[19,20] To investigate whether these agents exhibited base or sequence specificity, 5'-^{32}P end-labeled 117-nucleotide long *lac* p-o DNA fragments were incubated with these drugs under conditions known to break DNA.[19] The results of such an experiment are shown in Fig. 3. Inspection of this figure shows that both neocarzinostatin and bleomycin cleave DNA at specific sites. The lengths of the degradation products are different for the different drugs. Compari-son of the lengths of the cleavage products with these products of the standard chemical DNA sequencing procedure reveals the nucleotide or sequence specificity of the scission events. Neocarzinostatin breaks the DNA at positions of thymines and to a lesser extent at adenines. The

[16] C. W. Haidle, K. K. Weiss, and M. T. Kuo, *Mol. Pharmacol.* **8**, 531 (1972).
[17] R. Poon, T. A. Beerman, and T. H. Goldberg, *Biochemistry* **16**, 486 (1977).
[18] A. D'Andrea and W. Haseltine, *Proc. Natl. Acad. Sci. U.S.A.* **75**, 3608 (1978).
[19] T. A. Beerman and T. H. Goldberg, *Biochim. Biophys. Acta* **475**, 281 (1977).
[20] I. Ohtsuke and N. Ishida, *J. Antibiot.* **28**, 143 (1975).

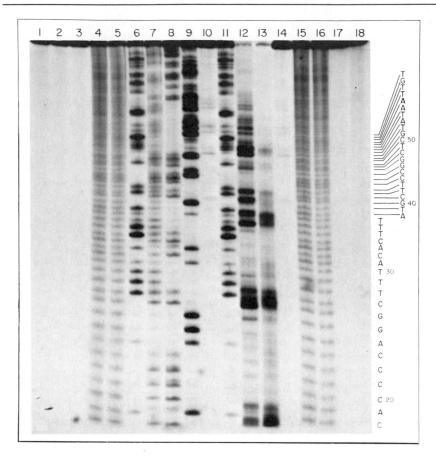

FIG. 3. Effect of Fe^{2+}, neocarzinostatin, and bleomycin on DNA. The 117 nucleotide 5′ end-labeled DNA sequence of the *lac* p-o region (60,000 cpm, 1 ng DNA per reaction) was incubated with neocarzinostatin (NCS) and bleomycin in the presence and absence of Fe^{2+}. The reactions included untreated DNA (lane 1); neocarzinostatin alone (lane 2); plus 2-mercaptoethanol (2-ME) alone (lane 3); 1 mM Fe^{2+}, and 250 μg/ml purified NCS, with (lane 4) and without (lane 5); 2-Me, and 250 mg/ml purified NCS (lanes 6, 11); 2-Me, and 250 μg/ml bleomycin (lane 10); 250 μg/ml bleomycin, 1 mM Fe^{2+} with (lane 15) and without (lane 16) 2-Me; and 1 mM Cu^{2+} with (lane 17) and without (lane 18) 250 μg/ml bleomycin. All reactions were stopped by two phenol extractions and ethanol precipitation of the DNA. Lanes 7, 8, and 9 were the C + T, C, and G > A reactions of the Maxam and Gilbert protocol.[13] Electrophoresis was for 20 hr at 1000 V.

reaction is dependent upon added β-mercaptoethanol. Analysis of the experiment pictured here and of other experiments using the 168 nucleotide fragment demonstrates that bleomycin breaks DNA at the sequence GC and GT and to a lesser extent at AT. This reaction takes place in the absence of β-mercaptoethanol, but is stimulated by β-mercaptoethanol. In

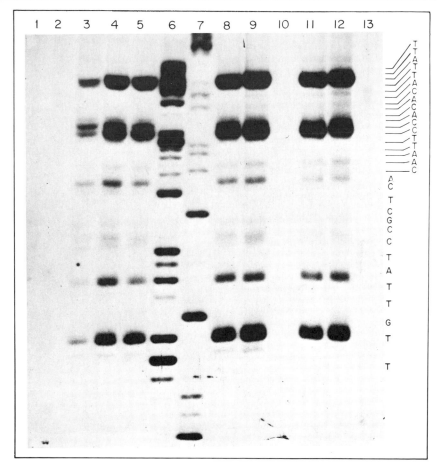

FIG. 4. Site-specific cleavage by purified *Micrococcus luteus* correndonucleases of DNA with uv-induced pyrimidine dimers. The 168 base pair ^{32}P 5' end-labeled DNA sequence was irradiated with 254 nm uv light at varying doses, incubated with saturating amounts of purified *M. luteus* correndonuclease peaks I and II G-75 fraction (21), and loaded onto a denaturing 20% polyacrylamide sequencing gel alongside sequencing reactions of the Maxam and Gilbert protocol. The reaction contained: unirradiated DNA treated with 5 μl (lane 1) and 15 μl (lane 2) of purified enzyme. DNA irradiated at 120 J/m^2 (lane 3), 480 J/m^2 (lane 4), 960 J/m^2 (lane 5), 1440 J/m^2 (lane 8), 4320 J/m^2 (lane 9), 7200 J/m^2 (lane 11) and incubated with 5 μl of correndonuclease (lane 12) to ensure enzyme saturation cleavage sites; irradiated DNA at 1400 J/m^2 (lane 10) and 10,080 J/m^2 (lane 13) underwent the same manipulations as all other reactions but with no enzyme present. Lanes 6 and 7 contained C + T and G > A reactions of the Maxam and Gilbert sequencing protocol. Electrophoresis was performed for 11 hr at 1000 V.

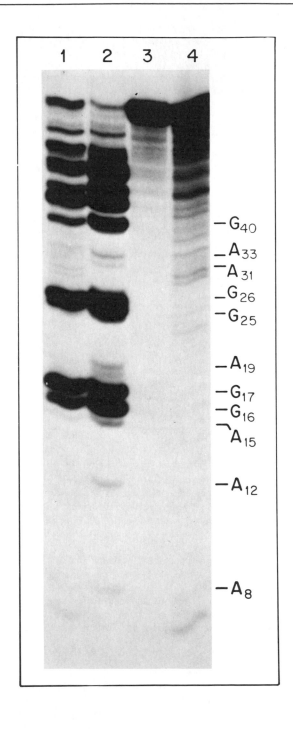

the presence of Fe^{2+}, bleomycin cleaves the DNA at TT, TA, and AT sequences as well as at GC and GT sequences. Figure 3 also shows that incubation of DNA with 1 mM ferrous sulfate results in breaking of the DNA at every nucleotide.

Site-Specific Cleavage of DNA Containing Pyrimidine Dimers by Purified M. luteus *Correndonuclease.* Two correndonucleases that introduce strand scissions into DNA that contain pyrimidines dimers induced by irradiation with uv light, have been purified from *M. luteus.*[21] To investigate the site of scission of the DNA relative to the position of possible pyrimidine dimers, the 5′-^{32}P labeled 117-nucleotide long *lac* p-o sequence was subjected to increasing doses of uv irradiation followed by incubation with a highly purified mixture of the two correndonuclease enzymes. Figure 4 shows that the enzymes cleave the irradiated, but not the unirradiated DNA, at specific sites. The location of these breaks within the sequence can be deduced by comparison of the mobility of the 5′ end-labeled degradation products with products of chemical degradation of DNA seen on parallel slots of the same gel. The correndonucleases cleave irradiated DNA at all possible sites of pyrimidine dimers, TT, CT, TC, and CC. The precise site of enzymatic cleavage relative to the thymine dimers is under investigation using this system.

The Use of Microsomal Enzymes to Activate Procarcinogens. Many carcinogens and other agents that damage DNA require metabolic activation. We chose to use aflatoxin B_1 to determine whether the *lac* p-o fragment of defined sequence could be used to investigate the site of damage by such agents. Aflatoxin B_1 was selected because it requires metabolic activation and because the principal DNA adduct is formed at N^7 of guanine.[22,23]

[21] S. Riazzudin and L. Grossman, *J. Biol. Chem.* **252**, 6286, (1977).

[22] J. J. Essigman, R. G. Croy, A. M. Nadzan, W. F. Busby, V. N. Reginhold, G. Buck, and G. N. Wogan, *Proc. Natl. Acad. Sci. U.S.A.* **74**, 1870 (1977).

[23] J. Lin, J. A. Miller, and E. C. Miller, *Cancer Res.* **37**, 4430 (1977).

FIG. 5. Extensive degradation of DNA by purified-microsome activated AFB$_1$. A 5′ end-labeled DNA fragment of the *lac* p-o region 117 nucleotides long (60,000 cpm) 1 ng DNA per reaction was incubated with 1 mM AFB$_1$ in the presence of a preparation of 10 mg/ml microsomal protein and purified by the method of Kinoshita *et al.*[24] prepared from the liver of phenobarbital-treated rats. After the reactions, the DNA was deproteinized and incubated in 1 M piperidine 30 min at 90° before it was layered onto a 20% polyacrylamide–urea gel. The cleavage products of the reaction (lane 2) were compared to those produced by incubation of the unmodified DNA fragment with 1 μl of dimethyl sulfate (DMS) in 200 μl of 50 mM sodium cacodylate, pH 8.0, 10 mM MgCl$_2$, 0.1 mM EDTA. After incubation with DMS, the DNA was also treated in 1 M piperidine for 30 min at 90°. Also shown are controls in which DNA was incubated with the microsomes in the absence of AFB$_1$ (lane 3) and with 1 mM AFB$_1$ in the absence of the microsomes (lane 4). The DNA of both reactions was treated with 1 M piperidine as described above.

Such adducts generally result in labilization of the N-glycosylic bond. Accordingly, the 5'-^{32}P end-labeled 117-nucleotide fragment was incubated with aflatoxin B_1 in the presence or absence of a rat liver microsome preparation. After the reaction, the DNA was treated with 1 M piperidine at high temperatures, a procedure that is expected to break DNA at sites of modified guanines.[22] DNA that was incubated with both aflatoxin B_1 and the microsomal preparation was broken by such treatment, whereas DNA incubated with aflatoxin B_1 alone or the microsomal preparation was not (Fig. 5).[24] Comparison of the mobilities of these DNA degradation products with those produced by similar treatment of DNA that had been modified by dimethyl sulfate shows that strand scission occurred at position of guanine and to a much lesser extent at position of adenine. This experiment demonstrates that a DNA fragment of defined sequence may be used together with microsomal enzymes to detect the sites of damage to DNA caused by chemicals that require metabolic activation.

[24] N. Kinoshita, B. Shears, and J. V. Gelboin, *Cancer Res.* **33**, 1937 (1973).

[32] Purification and Properties of S_1 Nuclease from *Aspergillus*

By VOLKER M. VOGT

The most convenient way to distinguish double helical from single-stranded nucleic acids is by digestion of the single-stranded components with specific nucleases. Several ribonucleases, notably pancreatic RNase and RNase T_1, have been routinely used to hydrolyze single-stranded RNA. Nucleases capable of specifically degrading single-stranded DNA have been characterized only more recently. Among this class of nucleases is the single-strand specific nuclease from *Aspergillus oryzae*, which has come to be known as S_1. This enzyme was first characterized by Ando,[1] who showed that it is stimulated by Zn^{2+}, that it yields 5'-mononucleotides, and that it degrades RNA as well as DNA. The utility of S_1 in measuring reannealing of nucleic acids was first demonstrated by Sutton.[2] Complete purification and further enzymological properties of the enzyme were reported by Vogt.[3] The advantages of S_1 are that it can be prepared easily and cheaply in large quantities, that it is stable to low concentrations of denaturants often used in annealing,[3] and that its

[1] T. Ando, *Biochim. Biophys. Acta* **114**, 158 (1966).
[2] W. D. Sutton, *Biochim. Biophys. Acta* **240**, 522 (1971).
[3] V. M. Vogt, *Eur. J. Biochem.* **33**, 192 (1973).

specificity for single-stranded nucleic acids is extremely high under the right conditions. A disadvantage for some types of studies is that it is active only at acid pH.

The first part of this article presents a highly reproducible partial purification of S_1 yielding enzyme that is perfectly adequate for all uses involving precipitating and quantitating nucleic acids from a mixture of double and single-stranded species treated with S_1. The second part of the article discusses further purification of the enzyme to rid it of low levels of contaminating nucleases capable of introducing nicks into double-stranded DNA. S_1 nuclease of this higher purity is needed, for instance, in studies that involve cleavage at mismatched bases[4] or preexisting single-stranded breaks,[5-7] or that involve accurate measurement of the double-stranded sizes of DNA fragments remaining after digestion with S_1.[8] Since S_1 itself can introduce breaks into double helical DNA at high enzyme and low salt concentration, such studies require much more carefully controlled conditions than the routine use of S_1 for acid-solubilizing single-stranded DNA or RNA.

Materials and Assay

Materials

Buffer A: 0.03 M sodium acetate, pH 4.6, 1 mM ZnSO$_4$, 0.05 M NaCl

Buffer B: 0.03 M sodium acetate, pH 4.6, 0.1 mM ZnSO$_4$, 5% glycerol

Buffer C: 0.02 M Tris-HCl, pH 7.5, 0.1 mM ZnSO$_4$, 5% glycerol

Buffer D: 0.05 M acetic acid titrated to pH 3.80 with NaOH, 0.01 M NaCl, 0.1 mM ZnSO$_4$, 5% glycerol.

α-Amylase powder (Sigma Biochemicals, catalogue No. 6630), glass fiber filters (such as GF/C from Whatman), DEAE-cellulose (DE-52, Whatman), SP-Sephadex C-50 (Pharmacia), calf thymus DNA (highly polymerized), radioactive phage λ DNA (or other intact high molecular weight linear DNA of defined length)

Assay. The standard reaction consists of 30 μg of heat-denatured DNA in 0.2 ml buffer A, plus an appropriate amount of enzyme. Incubations are carried out at 37° and terminated by chilling and addition of 0.8 ml 10%

[4] T. E. Shenk, C. Rhodes, P. W. J. Rigby, and P. Berg, *Proc. Natl. Acad. Sci. U.S.A.* **72**, 989 (1975).

[5] P. Beard, J. F. Morrow, and P. Berg, *J. Virol.* **12**, 1303 (1973).

[6] J. E. Germond, V. M. Vogt, and B. Hirt, *Eur. J. Biochem.* **43**, 591 (1974).

[7] R. C. Wiegand, G. N. Godson, and C. M. Radding, *J. Biol. Chem.* **250**, 8848 (1975).

[8] A. J. Berk and P. A. Sharp, *Cell* **12**, 721 (1977).

perchloric acid. After at least 10 min at 0°, the samples are passed through glass fiber filters and the absorbance of the filtrate at 260 nm measured. Up to five to ten samples can be passed through the same filter without appreciable decrease in flow rate. The filtrate can be collected conveniently in a test tube placed inside a small suction flask. If RNA is used as a substrate for the enzyme, it is preferable to use trichloroacetic acid instead of perchloric acid to avoid hydrolysis of the RNA while the samples are waiting to be processed. The adsorbance then should be read at 280 nm (since TCA absorbs strongly at 260 nm). The solubilization of absorbing material is plotted as a function of time and the initial rate calculated. One unit is defined as the amount of enzyme required to hydrolyze 1 μg of DNA per minute at 37°.

Purification

Heat Step. Four grams of crude amylase powder is dissolved by stirring for 1 hr at 0° in 200 ml buffer B plus 0.05 M NaCl. Insoluble material is removed by centrifugation for 10 min at 10,000 g. The dark brown supernatant is titrated to pH 5.0. It is then heated to 70° in a water bath at 75° by vigorous swirling in a 500-ml flask. After immediate chilling in ice water, the precipitated protein is removed by two centrifugations as above. All subsequent operations are carried out at 0°–4°.

Ammonium Sulfate Precipitation. The supernatant after heating is adjusted exactly to a volume of 200 ml and 90 g of ammonium sulfate is added. After 1 hr of stirring, precipitated proteins are removed by centrifugation and discarded. An additional 50 g ammonium sulfate is added to the supernatant and stirring continued for 3 hr. After decanting of the resulting saturated ammonium sulfate solution from undissolved crystals, the finely divided protein precipitate is collected from the liquid by centrifugation and redissolved in 50 ml of buffer C. In order to reduce the concentration of residual ammonium sulfate, this solution is diluted further with buffer C (approximately 3 volumes) until the conductivity is equal to that of buffer C plus 0.05 M NaCl.

DEAE–Cellulose Chromatography. The enzyme solution is adsorbed onto a DEAE–cellulose column (1.8 × 17 cm) equilibrated with buffer C plus 0.05 M NaCl and then rinsed with a further 200 ml. A 400 ml gradient linear in NaCl to 0.35 M and containing buffer C serves to elute absorbed proteins. All DNase as well as RNase activity should appear in a single peak between 0.18 and 0.22 M NaCl.

Concentration of the Enzyme. The pooled activity peak is dialyzed overnight against buffer B plus 0.02 M NaCl, and then passed through a 1.5-ml DEAE–cellulose column equilibrated with the same buffer. The

TABLE I
SUMMARY OF PURIFICATION[a]

Step	Total protein (mg)	Specific activity (units/μg)	DNase[b] ――― RNase	SS-DNase[c] ――― DS-DNase	DS-Nicking[d]
Crude	1200	0.2	0.4	>400	40
Heated to 70°	700	0.3	6	>400	35
Ammonium sulfate	50	3.5	6	>400	20
DEAE–cellulose	8	20	6	>400	1

[a] The data in columns 2–4 are expected values as judged from a number of independent preparations. The data in the last two columns represent the outcome of experiments on one particular preparation. One unit of S₁ hydrolyzes 1 μg of DNA per minute at 37°.

[b] The DNase activity was measured on calf thymus DNA and the RNase activity on total RNA from *E. coli.*

[c] SS, single stranded; DS, double stranded. The DS activity was estimated by substituting for the denatured DNA usually used in the assay native calf thymus DNA that had been pretreated with S₁ purified through the SP-Sephadex step described below.

[d] The units are the approximate number of nicks introduced into radioactive, native phage λ DNA (50 kilobase pairs), as measured by alkaline sucrose sedimentation. Samples containing 1 μg of DNA and 1 unit of S₁ were incubated in 0.1 ml assay buffer plus 0.2 M NaCl for 60 min at 37°.

adsorbed protein, which is brown in color, is eluted in concentrated form by buffer B plus 0.5 M NaCl. An equal volume of glycerol is added and the enzyme is stored at −20°, where it is stable for more than 6 months.

Comments on the Purification and Use of S₁. Table I summarizes some expected values of parameters measured during purification of the enzyme. More detailed data on a particular purification can be found in Vogt.[3] The purification scheme described here differs slightly with respect to that reported earlier[3] in that the ammonium sulfate concentrations have been adjusted to give more reproducible yields. Also, the pH of the crude α-amylase dissolved in buffer B has been found to differ from batch to batch. Titration to pH 5.0 before the heat step may be required in either direction. Typically, the cumulative yield of S₁ activity through the DEAE step is from 50 to 100%.

As can be seen in Table I, the additional RNase activity in crude extracts that is not inherent in S₁, which probably corresponds to T₂ ribonuclease, is inactivated by the heat step. The removal of traces of T₂ ribonuclease from S₁ is discussed in Rushizkey *et al.*[9] The table also shows that even as a crude extract, the S₁ preparation is highly specific for single

[9] G. W. Rushizkey, V. A. Shaternikov, J. H. Mozejko, and H. A. Sober, *Biochemistry* 14, 4221 (1975).

strands, as measured by acid solubilization of native calf thymus DNA that has been pretreated with purified S_1 to remove any single-stranded regions. By a more sensitive assay, as described below, the crude extract can be shown to contain an activity that nicks intact double-stranded DNA. The bulk of this contaminating nuclease activity is not removed until the DEAE–cellulose column, to which it does not bind. Hence in this step it is important, after adsorption of the protein to the resin, to rinse the column as described before starting the salt gradient.

S_1 is employed typically in measurements of reannealing of radioactive DNA with unlabeled DNA or RNA. Annealed mixtures to be assayed, which are usually in buffer with a high concentration of salt, are diluted with buffer A to reduce the salt concentration to 0.05 M and to adjust the pH so that it is near 4.6. This standard pH is above that at which S_1 shows maximal activity, 4.3, in order to reduce chances of depurination[3,7] in reactions that require double-stranded DNA to remain intact (see below). At pH values of 4.6, 5.2, and 5.8 the activity of the enzyme is about 70, 25, and 10%, respectively, of the maximal activity. For reproducible digestions, the dilution of annealing mixtures into buffer A should be sufficient to reduce any concentrations of dodecyl sulfate or formamide to 0.05 and 5%, respectively, conditions under which the activity of S_1 is unaltered.[3] If only small quantities of nucleic acid are present in the annealing mix, it is advisable to add 20 μg/ml denatured calf thymus DNA in buffer A as a carrier. Phosphate buffers should be avoided or appropriately diluted, since phosphate inhibits the enzyme at concentrations as low as 10 mM.[3,7] If a chelating agent is present, the dilution should be sufficient to bring its concentration well below the 1 mM $ZnSO_4$ in buffer A. For analyses requiring maximum digestion of single-stranded nucleic acid, the amount of enzyme used typically is about 20 times that calculated to be necessary to degrade the carrier (or other) single-stranded DNA in the reaction. Since S_1 is stable under the assay conditions, any convenient time of incubation can be used. After digestion, samples are usually precipitated with 10% trichloroacetic acid and the precipitate collected on glass fiber filters and counted for radioactivity.

The "background," or percentage of radioactive denatured DNA that upon digestion with S_1 remains precipitable by trichloroacetic acid, ranges near 1%. This value is a function of the size and type of DNA, of salt concentration, and of the digestion temperature. Increasing the salt concentration of buffer A to 0.3 M can raise the background to near 5% for intact single strands of λ DNA, although this change in salt has little effect on the activity of S_1. Presumably the salt acts to stabilize short regions of DNA with secondary structure. Hence for routine quantitation of annealing, a low salt concentration is recommended (usually 0.05 M NaCl), although a few single-stranded breaks may be introduced into the DNA under these conditions. Increasing the temperature has the same effect on the background as decreasing the salt concentration. S_1 remains fully active at least to 50°.

The background after S_1 digestion can be significantly reduced by assaying double-stranded DNA remaining after digestion by adsorption to DEAE filters rather than by TCA precipitation.[10] Maxwell *et al.*[10] have shown that this procedure, which is very convenient, yields the same value for the extent of reannealing as does TCA precipitation.

Further Purification of S_1

The pooled concentrated S_1 fraction from the DEAE column, which is about 5% pure, contains a barely measurable contaminating DNase that can introduce single-strand breaks into native DNA. When 1 μg of λ phage DNA in 0.1 ml is incubated for 60 min at 0.2 M NaCl with 10 units of S_1 from this step in the purification, an average of two to six nicks per double-stranded DNA of 50 kilobase pairs is found. Since the majority of the nicking activity detected in the ammonium sulfate fraction (Table I) appears in the DEAE column wash fraction, it is plausible that the residual nicking in DEAE-purified S_1 results from trailing of this DNase from the wash fractions. The residual extraneous DNase can be removed from S_1 by either of two procedures.

Method A: DEAE–Cellulose Chromatography. The pooled fractions from the first DEAE column are dialyzed against buffer B plus 0.05 M NaCl and then adsorbed to a DEAE column identical to the first. Elution and concentration is carried out as described.

Method B: SP-Sephadex Chromatography. The pooled fractions from the first DEAE column are dialyzed overnight against two changes of buffer D. The resulting solution is adsorbed onto a small (0.8 × 9 cm) column of SP-Sephadex washed previously with 0.05 M EDTA and equilibrated with buffer D. After adsorption of the protein, the column is rinsed with 25 ml of buffer, and the protein is eluted with a 50 ml gradient of NaCl from 0.01 to 0.20 M in buffer D. The S_1 activity appears at 0.04 to 0.06 M NaCl. After assay of the column fractions, the S_1 activity is immediately pooled and adjusted to pH 4.6 with NaOH, dialyzed against buffer B plus 0.02 M NaCl, and concentrated on a small column as described.

Comments on the Purification and Uses of Highly Purified S_1

Both methods above remove detectable nicking activity from S_1. Ten units of enzyme from either purification introduce less than one nick for

[10] I. H. Maxwell, J. VanNess, and W. E. Hahn, *Nucleic Acids Res.* **5**, 2033 (1978).

TABLE II
EFFECT OF SALT ON NICKING BY PURIFIED S_1[a]

Ionic strength	Nicking
0.52	<0.5
0.32	<0.5
0.12	2
0.07	7
0.02	14
0.02, no enzyme	<0.5

[a] Radioactive phage λ DNA was incubated with 2 units of S_1 nuclease for 60 min at 37° in 0.02 M Na acetate, pH 4.6, 1 mM $ZnSO_4$, plus varying amounts of NaCl. The number of nicks per double-stranded DNA (size 50 kilobase pairs) was estimated by alkaline sucrose gradient sedimentation. The enzyme had been purified over two consecutive DEAE columns plus an SP-Sephadex column.

every two λ DNA molecules in 60 min at 37° in 0.2 M NaCl. For comparison, three commercial S_1 preparations were purchased and examined by the same assay. Two of them were as free of extraneous DNase as enzyme prepared by methods A or B, while one contained slightly more nicking activity than S_1 from the first DEAE column (i.e., 1 nick per 1 unit S_1). There was no correlation of nicking with price or purity claimed by the manufacturer. Of the two methods, A consistently gives better yields, with no significant loss of enzymatic activity apparent. However, the specific activity of S_1 is not increased by this step. Hahn and VanNess[11] have also observed that repeated chromatography on DEAE eliminates nicking activity from S_1. Method B yields S_1 that is at least 70% pure, but often leads to loss of a large fraction of enzymatic activity, presumably because the pH required for absorption to the sulfo-Sephadex is close to that where the enzyme is inactivated.[3,9] It is essential in this step to use a small column to minimize the decrease in pH that results as the salt gradient removes protons bound to the resin.

At high enzyme and low salt concentration, extensively purified S_1 still is able to introduce nicks into native DNA, a reaction that probably is catalyzed by the enzyme itself rather than by a contaminating activity. The dependence of nicking on ionic strength is shown in Table II, and has also been observed previously.[3-5,7] Several uses of S_1 require such high enzyme concentrations, and for these it is important to adjust the ionic strength to a value high enough (usually to 0.2) to suppress nonspecific nicking of the particular DNA studied. However, since the experiment in the table was performed on a large phage DNA, the nicking observed, even at the low salt concentration, may not be significant for much smaller DNA species.

[11] W. E. Hahn and J. VanNess, *Nucleic Acids Res.* **3**, 1419 (1976).

Among these other and more exacting uses of S_1 are double-stranded cleavage of superhelical DNA,[5-7] specific removal of deletion loops in DNA heteroduplexes,[4,7] cleavage of poly(dA/poly(dT) duplex regions near their melting temperature,[12] cleavage at a single base mismatch,[4] and cleavage opposite to preexisting nicks.[5-7] All the reactions involving scission of duplex DNA, including these specific cleavages as well as the nonspecific nicking at low salt, probably depend on transient melting of a short stretch of base pairs.

In summary, S_1 nuclease for use in quantitating double-stranded forms of either RNA or DNA is simple and inexpensive to prepare. One standard preparation yields enough enzyme to degrade grams of DNA. A small amount of nicking of double-stranded DNA may occur under these conditions. For studies that require double-stranded DNA to remain intact, the enzyme should be purified further as described, and the exact ionic conditions of the assay should be monitored carefully.

[12] H. Hofstetter, A. Schamböck, J. VandenBerg, and C. Weissmann, *Biochim. Biophys. Acta* **454**, 587 (1976).

[33] Purification and Properties of *Neurospora crassa* Endo-exonuclease,[1] an Enzyme which Can Be Converted to a Single-Strand Specific Endonuclease

By M. J. FRASER

Assay Method

Principle. The assay with denatured or native DNA as substrate[7] measures the formation of acid-soluble material that absorbs at 260 nm.

[1] The tentative name "endo-exonuclease" has been assigned to this enzyme[2] simply to indicate that it has high levels of both endonuclease activity and exonuclease activity[2,3] and to distinguish it from two other nucleases of *Neurospora* which also have specificity for single-stranded nucleic acids, a single-strand specific endonuclease[4,5] and a single-strand specific exonuclease.[6] The latter enzyme was also found to have a low level of endonuclease activity. The nomenclature for these nucleases will have to be revised after the relationships between the enzymes have been fully clarified.

[2] S. Kwong and M. J. Fraser, *Can. J. Biochem.* **56**, 370 (1978).

[3] M. J. Fraser, R. Tjeerde, and K. Matsumoto, *Can. J. Biochem.* **54**, 971 (1976).

[4] S. Linn and I. R. Lehman, *J. Biol. Chem.* **240**, 1287 and 1294 (1965).

[5] S. Linn, Vol. 12, Part A, p. 247.

[6] E. Z. Rabin, B. Preiss, and M. J. Fraser, *Prep. Biochem.* **1**, 283 (1971); E. Z. Rabin, H. Tenenhouse, and M. J. Fraser, *Biochim. Biophys. Acta* **259**, 50 (1972); H. Tenenhouse and M. J. Fraser, *Can. J. Biochem.* **51**, 569 (1973).

[7] C. Mills and M. J. Fraser, *Can. J. Biochem.* **51**, 888 (1973).

METHODS IN ENZYMOLOGY, VOL. 65

Reagents

Heat-denatured and native calf thymus or fish roe DNA, each 2 mg/ml in 5 mM Tris-HCl, pH 8.0
Tris-HCl buffer pH 8.0, 1.0 M
MgCl$_2$, 0.1 M
"Carrier" DNA, native calf thymus or fish roe DNA, 2 mg/ml
Perchloric acid, 1.0 N

Procedure. Since several assays are usually run at any one time, it is most convenient to prepare stocks of "assay mixes" consisting of 10 ml denatured or native DNA, 4 ml H$_2$O, 3 ml Tris-HCl buffer, and 3 ml MgCl$_2$. Each reaction mixture contains 400 μl "assay mix" plus enzyme and H$_2$O to a total volume of 600 μl. Thus, the final concentrations of DNA, Tris-HCl and MgCl$_2$ are, respectively, 670 μg/ml, 0.1 and 0.01 M. The mixture is pre-warmed for 3 min at 37° and enzyme added to start the reaction. Aliquots of 100 μl are removed from the reaction mixture at appropriate times and transferred to Microfuge tubes containing 100 μl "carrier" DNA. The contents of each tube are briefly mixed with a Vortex mixer and 150 μl perchloric acid added and mixed thoroughly. The Microfuge tubes are placed in an ice bath for at least 15 min and then the precipitated DNA is centrifuged in a Microfuge for 1.5 min at maximum speed. An aliquot (200 μl) of the supernatant is diluted into 800 μl H$_2$O and the absorption at 260 nm measured with a spectrophotometer in cuvettes with a 1 cm path length. The linear portion of the rate curve (A_{260} versus time) is used to calculate the activity. Blanks without enzyme can be run as a check on nuclease contamination in the reaction components.

Definition of Unit and Specific Activity. One unit of enzyme activity is defined as the amount of enzyme that releases 1.0 A_{260} unit of acid-soluble material from denatured DNA in 30 min at 37° under the conditions described above. The units of activity per milliliter of enzyme solution added to the reaction mixture are calculated from the A_{260} increase measured in 30 min × 10,500/μl enzyme added. The factor 10,500 takes into account the dilution of the enzyme and acid-soluble material in the assay. The unit of activity used here is approximately equivalent to the degradation of 1.25 μmoles DNA nucleotides per minute and is 24 times the unit used by Linn[5] for the *Neurospora* single-strand-specific endonuclease. Specific activity is expressed as units per milligram of protein. Protein is determined by the method of Lowry *et al.*[8] In fractions up to the DEAE–cellulose step this procedure is used only after the protein is precipitated with trichloroacetic acid.[4]

[8] O. H. Lowry, N. J. Rosebrough, A. L. Farr, and R. J. Randall, *J. Biol. Chem.* **193,** 265 (1951).

Purification Procedure

Source of Enzyme. The enzyme is prepared from mycelia of the Oak Ridge wild-type strain of *Neurospora crassa,* 74-OR-23-1V A, obtainable from the Fungal Genetics Stock Center (Humboldt State University Foundation, Arcata, California). Conidia are grown for 7 days at room temperature on enriched agar slants[9] and two to three loops of the dry conidia innoculated into each flask of autoclaved medium. Mycelia are grown in 1-liter amounts of Difco N. 0817-01 *Neurospora* minimal medium in 2.8-liter Fernbach flasks for 5 days at room temperature (21°) on a reciprocating shaker under constant laboratory fluorescent lighting. Under these conditions, the cultures are at the end of "log-phase" growth.[3] The mycelia are collected on a Buchner funnel, washed with cold water, pressed dry between paper towels,[10] weighed, cut into 2.5 cm strips with scissors, wrapped in aluminum foil and stored frozen at $-90°$ until extracted. No loss of activity occurs within 1 month of storage under these conditions. A yield of about 60 g wet weight of light yellow mycelia per flask is obtained.

The first two steps of the following purification are minor modifications of the methods which Linn[5] used for the preparation of *Neurospora* single-strand-specific endonuclease.

Extraction of Mycelia. Frozen mycelia (120 g) are ground to a fine powder with a porcelain mortar and pestle cooled with solid CO_2. The mycelial powder is suspended in $0.05\,M$ glycylglycine buffer, pH 7.0 (10% suspension, wet weight per volume) and sonicated for 5 min in 60-ml batches in 100-ml beakers partially immersed in an ice bath. Sonication is carried out with a Blackstone sonicator using the 1.1-cm diameter probe at full power. The temperature of the sonicate is not allowed to exceed 20°. The pooled sonicates are centrifuged for 5 min at $2100\,g$ in a refrigerated centrifuge. The supernatants are pooled (crude extract) and made $0.04\,M$ in $MgCl_2$ by the addition of 1.35 ml 3 M $MgCl_2$ per 100 ml extract. The extract is allowed to remain 18 hr at 0° and then centrifuged for 20 min at $10,000\,g$. The inactive sediment is discarded. All subsequent operations are carried out at 0° and all centrifugations for 20 min at $10,000\,g$.

Ammonium Sulfate Precipitation. The clear supernatant is brought to pH 7.5 by the addition of 0.1 volume of 1 M Tris-HCl, pH 7.5, and then 568 g/liter solid ammonium sulfate is added with stirring over a 20-min

---

[9] *Neurospora* culture agar from the BBL Division of BioQuest. The agar slants and liquid culture medium can be prepared in the laboratory according to the methods described by R. H. Davis and F. J. de Serres (Vol. 17, Part A, p. 79).

[10] Since this purification scheme was developed, it has been found that washing the mycelial mats with cold acetone not only dries them more thoroughly, but also increases the efficiency of extraction by approximately sevenfold.

TABLE I
PURIFICATION OF ENDO-EXONUCLEASE FROM 120 g LOG PHASE MYCELIA OF *Neurospora crassa*

Fraction	Protein (mg)	Activity[a] (units)	Specific activity (units/mg)	Ratio ss/ds[b]
Extract[c]	3340	15,800	4.7	3.1
MgCl$_2$ supernatant	2320	11,000	4.7	1.9
(NH$_4$)$_2$SO$_4$ fraction	1210	15,300	12.6	2.2
Phosphocellulose filtrate	428	185,000[d]	432	4.8
DEAE–cellulose gradient fraction	151	57,000	377	13.1
Hydroxyapatite gradient fraction	46	29,000	648[e]	20.4

[a] Activity with heat-denatured DNA at pH 8.0 in the presence of 0.01 M Mg^{2+}.

[b] Ratio of activity with heat-denatured DNA to activity with native DNA measured under the conditions of assay. ss, single stranded; ds, double stranded.

[c] Fresh extracts of end of log phase mycelia have approximately 20 times as much activity with denatured DNA per gram of mycelia as extracts of stationary phase starved mycelia (cf. Linn[5]).

[d] Proteolytic activation of endo-exonuclease precursor occurs at this step. The losses which occur in the two subsequent steps are believed to be due mainly to proteolytic degradation of the enzyme.[3] This is accompanied by an increasing single-stranded (ss) to double-stranded (ds) ratio.

[e] This specific activity is 70% higher than that obtained previously by Linn[2] for the enzyme purified from stationary phase starved mycelia and the yield per gram mycelia is approximately 500 times higher. It is estimated that this enzyme in its precursor and active forms constitutes approximately 2% of the protein of log phase mycelia.[16]

period. The suspension is then stirred for 1 hr, allowed to stand for 3 hr, and finally centrifuged. The precipitate is suspended in 0.1 volume of 0.02 M potassium phosphate buffer, pH 6.5, and dialyzed overnight against two changes of 2 volumes of the same buffer to remove ammonium sulfate. Insoluble material is removed by centrifugation.

Chromatography on Phosphocellulose. The centrifuged dialysate is passed through a phosphocellulose column equilibrated with 0.02 M potassium phosphate buffer, pH 6.5 at a flow rate of 2 ml/min. The phosphocellulose[11] (1 ml packed adsorbent per milliliter of ammonium sulfate fraction) is used in 2.5-cm diameter columns, usually 15 to 18 cm in height

[11] Phosphocellulose (cation exchange powder, P11) and DEAE–cellulose (anion exchange powder, DE-23) are Whatman products. They are washed as described by E. A. Peterson and H. A. Sober [*J. Am. Chem. Soc.* **78**, 751 (1951)]. The phosphocellulose is prepared fresh for each purification and is not reused. The DEAE–cellulose, on the other hand, may be regenerated and reused.

depending on the volume of fraction to be filtered. After applying the sample, the column is washed with buffer until the A_{280} falls to at least one-tenth of the maximum reached as the unadsorbed material passes through the column. The activity with denatured DNA is recovered in a very high yield in the pass-through fraction[12] (12 to 16 times that observed in the original extract) and a substantial amount of protein remains adsorbed on the column (see Table I).

Chromatography on DEAE–Cellulose. The combined pass-through and wash fractions from the phosphocellulose step are loaded on a DEAE–cellulose column equilibrated with 0.02 M potassium phosphate buffer, pH 7.5. The maximum amount of protein loaded on a 1 × 30 cm DEAE–cellulose column is 500 mg as estimated from the absorption at 280 nm (A_{280}). Fractions (4 ml) are collected at 8-min intervals. The column is washed with the equilibration buffer until the A_{280} reaches that of the buffer and then a linear 0.02–0.13 M potassium phosphate buffer (pH 7.5) gradient of 360 ml is applied. About 80–90% of the activity with denatured DNA activity is eluted in the range 0.06–0.10 M buffer. The active fractions (approximately 120 ml) are pooled, concentrated eight- to tenfold by dialysis against 1 liter of 30% (w/v) polyethylene glycol[13] and then dialyzed overnight against two changes of 1 liter each of 0.05 M potassium phosphate buffer, pH 6.5.

Chromatography on Hydroxyapatite. A 1 × 15 cm column of hydroxyapatite[14] is equilibrated with 1 liter of 0.05 M potassium phosphate buffer, pH 6.5, under a gravity flow of at least 0.4 ml/min, with the buffer reservoir no more than 40 cm above the top of the column.[15] The dialyzed, concentrated DEAE–cellulose fraction (about 15 ml containing 150 mg protein) is applied to the column in the same manner and the column then washed with 80 ml starting buffer. A linear 0.05–0.45 M potassium phosphate (pH 6.5) gradient of 320 ml is used to elute the enzyme. The elution occurs in the range 0.1–0.2 M buffer. Active fractions are pooled (approximately 80 ml), concentrated as described above, and dialyzed against two

---

[12] It should be noted that the single-strand-specific endonuclease activity extracted previously from stationary phase starved mycelia adsorbed strongly on phosphocellulose and was recovered in much lower yield at this step.[5] The high yield seen here with enzyme from end of log phase mycelia results from the activation of the inactive precursor of the endo-exonuclease by the action of endogenous proteinases.[2] The precursor is not present in stationary phase starved mycelia and present only at low levels in early log phase mycelia (M. J. Fraser and E. Käfer, unpublished results). The lack of adsorption of the enzyme from log phase mycelia on phosphocellulose appears to be due to less extensive proteolysis of the enzyme after activation.[3]

[13] Carbowax 6000, Union-Carbide Chemical Company of Canada, Ltd., Montreal, Canada.

[14] Hypatite C, Clarkson Chemical Company, Williamsport, Pa.

[15] Greater pressure can cause mechanical breakage of the hydroxapatite crystals which leads to a reduction in flow-rate.

changes of 1 liter each of 0.02 M potassium phosphate buffer, pH 6.5. The concentrated enzyme is passed through 0.45 μm Millipore filters into sterile vials and stored at 0°. This preparation contains approximately 30–50% pure endo-exonuclease which acts on denatured DNA in an endonucleolytic manner and on native DNA in an exonucleolytic manner (see Properties). Only three to four major polypeptides are seen in fresh preparations by SDS-gel electrophoresis.[3] The ratios of the activity with denatured DNA to that with native DNA in fresh preparations vary from 5:1 to 20:1, increasing with the time utilized for the purification. Although the endonuclease activity of the preparation is stable for at least 2 years, the exonuclease activity is gradually lost on aging.

Conversion of Endo-exonuclease to Single-Strand Specific Endonuclease. At 0° the ratio of activities with denatured and native DNA increases to 40–50:1 in 1 month. The loss of exonuclease activity with native DNA can be greatly hastened by preincubation at 37° for 1 hr. In this case the ratio of the two activities increases to 260:1. Evidence has been obtained[3] which indicates that the end-product of this "aging" process is very similar to the single-strand specific endonuclease activity described by Linn[5] and that the aging results from proteolysis due to the presence of small amounts of proteinases.

The most rapid and effective way to eliminate the double-strand DNase activity is to incubate the enzyme in 0.02 M potassium phosphate buffer, pH 6.5, overnight at 0° with 10 mM EDTA. Such treatment results in a preparation with no detectable activity with double-stranded DNA provided that the assays are performed as described here but in the absence of Mg^{2+} and with the additions of 0.1 M NaCl to stabilize the duplex structure of native DNA and 1 mM EDTA. Under these conditions single-stranded DNA is degraded without the addition of divalent metal ions.[3]

Preparation of Pure Endo-exonuclease. Endo-exonuclease preparations yielding only a single polypeptide when subjected to SDS-polyacrylamide gel electrophoresis can be prepared by polyacrylamide gel electrophoresis at pH 8.0 in the presence of 6 M urea.[2] The front-running (very acidic) protein component is eluted from the gels and dialyzed in turn to remove urea and to restore divalent metal ions, Co^{2+} and Mg^{2+}. Recoveries of activity of up to 40% have been obtained, and the ratio of activities with denatured and native DNA was the same as that observed for the enzyme derived from hydroxyapatite. The single polypeptide has endonucleolytic activity with RNA as well. The specific activity of the purified enzyme, measured with single-strand DNA as substrate, is 700–800 units/mg. Since these procedures have not yet been carried out on a preparative scale, they are not described in detail here. Purified enzyme is not required to obtain preparations with only a single-strand specific endonuclease (see above).

Properties

Specificity. Endo-exonuclease acts on single-stranded DNA and RNA in an endonucleolytic manner and on linear single- and double-stranded DNA in an exonucleolytic manner.[3] It has no action on double-stranded RNA. Circular single-stranded DNA (bacteriophage ϕX174 viral DNA) is degraded by the enzyme, while superhelical covalently closed circular double-strand DNA (ϕX174 RFI DNA) is "nicked," but not further degraded at low enzyme concentration.[3] However, at high enzyme concentrations circular double-stranded DNA is converted to linear double-stranded DNA and subsequently degraded to acid-soluble material.[16] Similar results were found previously[17] for the *Neurospora* endonuclease which is believed to be derived from this enzyme through proteolysis, except that the linear double-stranded DNA (RFIII form) was not further degraded. The *Neurospora* endonuclease can also sense distortions in uv light irradiated DNA, making single-strand breaks at low enzyme concentrations and double-strand breaks at high enzyme concentrations.[18] It seems highly probable that the endonuclease activity associated with the endo-exonuclease can also sense distortions in uv light irradiated DNA, since the enzyme converts irradiated RFI DNA to acid-soluble material faster than unirradiated RFI DNA.

Effect of pH and Salt. The enzyme degrades DNA and RNA at maximum rates in 0.1 M Tris-HCl at pH 7.5–8.5. Salt concentrations of 0.1–0.2 M have only slight inhibitory effects on the activity with denatured DNA (mainly endonucleolytic), but strongly inhibit the activity with native DNA (see Table II).

Effect of Divalent Cations. The activity with denatured DNA is stimulated somewhat by 10 mM Mg^{2+}, but is not dependent on Mg^{2+} (see Table II). The activity with RNA (endonucleolytic only), on the other hand, is inhibited by 10 mM Mg^{2+}, presumably because the secondary structure of RNA is stabilized under these conditions. The activity with native DNA is dependent on Mg^{2+} (see Table II). Calcium substitutes poorly for Mg^{2+} in supporting the two activities with DNA. Zinc preferentially inhibits the exonuclease activity of the enzyme (Table II). It is possible that the enzyme contains strongly bound Co^{2+} associated with the endonuclease ac-

---

[16] D. M. Gáler, M. Sc. Thesis, McGill University, Montreal (1978).

[17] A. C. Kato, K. Bartok, M. J. Fraser, and D. T. Denhardt, *Biochim. Biophys. Acta* **308**, 68 (1973).

[18] A. C. Kato and M. J. Fraser, *Biochim. Biophys. Acta* **312**, 645 (1973).

TABLE II
DIFFERENTIAL INHIBITION OF THE EXONUCLEASE ACTIVITY OF *Neurospora*
ENDO-EXONUCLEASE

Concentration of component added to reaction mixture[a]	Activity (% of control)		Ratio ss/ds[b]
	Activity with denatured DNA	Activity with native DNA	
No addition	100	100	5.8
0.1 M NaCl	98	30	19
0.2 M NaCl	93	16	31
No addition	100	100	4.0
0.05 mM ZnCl$_2$	81	41	8.0
0.10 mM ZnCl$_2$	54	9	25
After overnight preincubation with 10 mM EDTA[c]			
No EDTA, 10 mM MgCl$_2$ (control)	100	100	4.6
1 mM EDTA, 0 mM MgCl$_2$	81	0	4000
1 mM EDTA, 10 mM MgCl$_2$	141	251	2.6
1 mM EDTA, 10 mM CaCl$_2$	21	16	6.3

[a] The reaction mixtures contained 667 μg/ml denatured or native DNA 0.1 M Tris-HCl (pH 8.0), 0.01 M MgCl$_2$, and enzyme.
[b] ss, single stranded; ds, double stranded.
[c] The reaction mixtures contained DNA and buffer as described above and enzyme which had been preincubated either without (control) or with 10 mM EDTA and diluted tenfold into the reaction mixture. Additions of MgCl$_2$ and CaCl$_2$ were made as indicated.

tivity of the enzyme as suggested by Linn.[5] Cobalt and magnesium are required for the recovery of activities after gel electrophoresis in the presence of 6 M urea.

Inhibitors. Physiological concentrations of ATP (0.1–0.5 mM) inhibit both nuclease activities of endo-exonuclease.[3] No other specific inhibitors are known. This inhibition was found previously for the *Neurospora* endonuclease[19] as well. Phosphate inhibits the exonuclease activity of the endo-exonuclease to a greater extent than the endonuclease activity (50% inhibitions, respectively, by 20 and 80 mM potassium phosphate buffer, pH 8.0).

Molecular Size. Endo-exonuclease activity prepared by urea–gel electrophoresis has been found in association with single polypeptides ranging in molecular weight from 33,000 to 53,000 as determined by SDS–gel electrophoresis.[2] Presumably the different sizes arise from different ex-

[19] E. Z. Rabin, M. Mustard, and M. J. Fraser, *Can. J. Biochem.* **46**, 1285 (1968).

tents of proteolysis during purification. The largest of these polypeptides has the specific activity listed above and has approximately the same molecular weight (53,000) as that determined previously (55,000) for the *Neurospora* endonuclease by sucrose density gradient centrifugation. The smaller fragments have lower specific activities.[2] The inactive endo-exonuclease precursor is also a single polypeptide with a native molecular weight of 88,000. It is converted by treatment with trypsin *in vitro* into an active enzyme with a native molecular weight of 61,000.[16]

Commercial Preparation of Endo-exonuclease. A preparation of intracellular nuclease sold[20] under the name of "endonuclease (from *Neurospora crassa*)" is endo-exonuclease by all tests that we have applied. One preparation that we examined in detail had a specific activity of 720 units/mg and a ratio of activities with denatured to native DNA in the range 5.0–8.2:1. Both activities were inhibited by 0.1–0.5 mM ATP. The exonuclease activity with native DNA was more sensitive to salt than the endonuclease activity with denatured DNA and was abolished by dialysis overnight against 10 mM EDTA as described in Table II. The preparation, supplied as an ammonium sulfate suspension, was found to be remarkably stable when stored at 0°–4°. No activity with denatured DNA was lost during a 5-year period. However, after dialysis against 0.02 M potassium phosphate buffer, pH 6.5, to remove the ammonium sulfate, the activity was less stable. About 80% of the activity was lost during the EDTA treatment versus 10% for the preparation described here (cf. Table II).

[20] By the Boehringer-Mannheim Corporation, New York, New York.

[34] Purification and Properties of the Mung Bean Nuclease[1]

By M. LASKOWSKI, SR.

Until recently we have called this enzyme mung bean nuclease I because mung bean sprouts contain additional nucleases.[2] Since no other laboratory uses the qualifying number, we have decided to drop it.

Mung bean nuclease cleaves single-stranded DNA endonucleolytically to form 5'-P-terminated mono- and oligonucleotides[3] ending predominantly in 3'-OH A(60%) and 3'-OH T(30%). With an excess of enzyme a mixture of mononucleotides is obtained.[4] The enzyme has no specificity

[1] Supported by the United States Department of Energy (EY76S023225), the National Science Foundation (BMS 73-06750) and the National Institutes of Health (GM 17788 and HL 15892). This article is dedicated to the memory of Dr. Andrzej J. Mikulski.

[2] P. H. Johnson and M. Laskowski, Sr., *J. Biol. Chem.* **243**, 3421 (1968).
[3] S.-C. Sung and M. Laskowski, Sr., *J. Biol. Chem.* **237**, 506 (1962).
[4] A. J. Mikulski and M. Laskowski, Sr., *J. Biol. Chem.* **245**, 5026 (1970).

with respect to the sugar moiety and hydrolyzes derivatives of deoxyribose, ribose,[2,4] and arabinose.[5] Preference for single-stranded versus double-stranded DNA is approximately 30,000 : 1 with T4 DNA but drops to 2 : 1 with synthetic AT polymer.[6] It is strongly dependent on Mg^{2+} concentration,[6] pH, and ionic strength.[7] The same enzyme molecule[4] catalyzes hydrolysis of the 3'-monophosphoryl group from mono- and oligonucleotides (ω-monophosphatase activity).

With native viral DNA as substrate mung bean nuclease inflicts a limited number of endonucleolytic cleavages at specific AT-rich sites. For example, Chan et al.[8] demonstrated that the lac repressor binding capacity of λ plac DNA was maximally reduced by 2–5 nicks inflicted by mung bean nuclease, whereas 300 nonspecific cleavages were required. During the first phase of the reaction with linear duplex DNA (T7, PM2, gh-1) transient, but fairly long-lasting products are formed. They range from 0.2 × 10^6 to 1.4 × 10^6 daltons with a mean value of about 0.6 × 10^6 daltons.[7,9] During the second phase of the reaction these products are further hydrolyzed by the exophilic action of the enzyme to acid-soluble material.[7]

Chromatin (chicken erythrocyte nuclei) is hydrolyzed to a stage of large soluble fragments ranging from 30 to 100 S with a mean value of about 60 S. These fragments when digested with micrococcal nuclease generate smaller products among which dinucleosomes and mononucleosomes were identified.[10]

Assays

Liberation of Acid-Soluble Material. An assay particularly suitable for following the purification[11,12] of mung bean nuclease depends on the liberation of acid-soluble material from denatured DNA.[3] A solution of calf thymus or salmon sperm DNA (2 mg/ml) is dialyzed against a large volume of distilled water. This serves a dual purpose: to denature DNA and to remove an excess of salt often present in commercial preparations of DNA. The dialyzed solution is heated for 10 min in a boiling water bath and rapidly cooled on ice. The reaction mixture consists of 0.5 ml of the DNA solution, 0.5 ml of 0.1 M ammonium acetate, pH 5.0, enzyme solu-

---

[5] W. J. Wechter, A. J. Mikulski, and M. Laskowski, Sr., *Biochem. Biophys. Res. Commun.* **30**, 318 (1968).

[6] P. H. Johnson and M. Laskowski, Sr., *J. Biol. Chem.* **245**, 891 (1970).

[7] W. D. Kroeker, D. Kowalski, and M. Laskowski, Sr., *Biochemistry* **15**, 4463 (1976).

[8] H. W. Chan, J. B. Dodgson, and R. D. Wells, *Biochemistry* **16**, 2356 (1977).

[9] W. D. Kroeker and D. Kowalski, *Biochemistry* **17**, 3236 (1978).

[10] M. Fujimoto, A. Kalinski, A. E. Pritchard, D. Kowalski, and M. Laskowski, Sr., *J. Biol. Chem.* **254**, 7405 (1979).

[11] W. Ardelt and M. Laskowski, Sr., *Biochem. Biophys. Res. Commun.* **44**, 1205 (1971).

[12] D. Kowalski, W. D. Kroeker, and M. Laskowski, Sr., *Biochemistry* **15**, 4457 (1976).

tion (preferably containing from 0.003 to 0.006 units), and water to make 2 ml (final concentration of buffer 25 mM). The reaction mixture is incubated for 30 min at 37°, then the reaction is terminated by the addition of 2.0 ml of acid lanthanum reagent composed of 0.02 M La(NO$_3$)$_3$ in 0.2 M HCl. The precipitate is removed by centrifugation, and the absorbancy of the supernatant solution read at 260 nm against a blank incubated without enzyme. The amount of enzyme required to cause an increase in absorbancy of 1.0/min is defined as 1 unit of activity. The enzyme protein is not measured in terms of weight, but in terms of absorbancy at 280 nm. One A_{280} unit of protein is that amount which if dissolved in 1 ml would have an absorbance of unity at 280 nm with a 1 cm light path. Potency is analogous to specific activity and is defined as activity (ΔA_{260}/min) per A_{280} unit. This assay method can be adapted to a much lower scale by using ^{33}P-labeled denatured T7 DNA[7] and measuring the amount of radioactive acid-soluble material.

Spectrophotomatic Determination of Nuclease Activity. Nuclease activity is conveniently measured by the Kunitz[13] spectrophotometric method. This method, however, cannot be used with crude extracts which show high blanks and low activity. With purified enzyme this method has many advantages. It is fast; the reaction is continuously recorded; and several samples can be assayed simultaneously. The method has often been used in our laboratory to measure kinetic parameters.[2,4,6,7] The assay mixture consists of 2 ml of substrate solution (80 to 500 μg of DNA per milliliter) in 25 mM sodium acetate buffer, pH 5.0, containing 0.01 mM zinc acetate, 1 mM cysteine, and 0.001% Triton X-100. Enzyme solution (0.01 to 0.05 unit) in 2 ml of the same buffer kept at the same temperature (usually 23°) is added and the recording started. Alternatively, a small volume of enzyme (not greater than 10 μl) can be added to 3.0 ml of reaction mixture with a Hamilton syringe. If RNA is used instead of DNA, the same conditions can be followed.

3'-(ω)-Monophosphatase Activity. This catalytic activity is an intrinsic property of mung bean nuclease molecule. The name ω-monophosphatase has been suggested[4] to emphasize the difference between this activity and 5'-nucleotidase,[14] which is specific for mononucleotides, whereas ω-monophosphatase attacks mononucleotides, varied length oligonucleotides,[4] and probably high molecular weight polynucleotides. The liberated orthophosphate may be determined by any of the commonly used methods described by Fisk and SubbaRow,[15] Lowry and Lopez,[16]

[13] M. Kunitz, *J. Gen. Physiol.* **33**, 349 (1950).
[14] E. Sulkowski, W. Björk, and M. Laskowski, Sr., *J. Biol. Chem.* **238**, 2477 (1963).
[15] C. H. Fiske and Y. SubbaRow, *J. Biol. Chem.* **66**, 375 (1925).
[16] O. H. Lowry and J. A. Lopez, *J. Biol. Chem.* **162**, 421 (1946).

King,[17] and Chen et al.[18] A more sensitive and simpler method involves coupling the monophosphatase reaction with adenosine deaminase and measuring the formation of inosine. It was proposed by Ipata[19] for the determination of 5'-nucleotidase, by Kroeker et al.[20] for the determination of 5'-AMP produced by the digestion of polyadenylic acid, and was used by Kowalski et al.[12] as a parallel method to follow the purification of mung bean nuclease. The reaction mixture consists of 0.9 ml of 25 mM sodium succinate, pH 6.0, 50 μl of 1 mM 3'-AMP and 30 μl of adenosine deaminase (type I, Sigma Co., diluted 1 : 500 in 25 mM ammonium acetate adjusted to pH 8.0). The mixture is placed in a 1-ml cell with a 1-cm light path in a thermostated (25°) compartment of a Gilford model 2400 spectrophotometer. Mung bean nuclease ($\leq$50 μl) is added and the decrease in absorbency at 265 nm is recorded against a blank containing water instead of 3'-AMP. Under these conditions ΔA_{265}/min = -0.114 is equal to 1.0 μmole of 3'-AMP hydrolyzed per minute. One unit is defined as that amount of enzyme which converts 1.0 μmole of 3'-AMP to adenosine per minute per milliliter of reaction mixture at 25°.

Supporting Methods. Initially the interest in mung bean nuclease was centered on its ability to preferentially hydrolyze single-stranded DNA and leave double-stranded DNA intact. For example, the enzyme has been used for the determination of base composition and sequence[21] of the cohesive ends in λ phage DNA, and the composition of easily denaturable regions of a number of different DNAs.[22] During this period the supporting methods dealt with identification and determination of mono- and oligonucleotides.[23]

Subsequently, the interest in mung bean nuclease has shifted to its use as a reagent for probing the structure of native viral DNA.[6-9] Chromatin[24] has also been used as a substrate for the study of the effect of structure.[10] Methods which allow the determination of size,[23] such as viscosimetry,[9,25-27] have thus regained their significance. The original method using an Ostwald type viscosimeter[26,27] has only an historical

[17] E. J. King, *Biochem. J.* **26,** 292 (1932).
[18] P. S. Chen, T. Y. Torribara, and H. Warren, *Anal. Chem.* **28,** 1756 (1956).
[19] P. L. Ipata, *Anal. Biochem.* **20,** 30 (1967).
[20] W. D. Kroeker, D. M. Hanson, and J. L. Fairley, *J. Biol. Chem.* **250,** 3767 (1975).
[21] Ghangas and R. Wu, *J. Biol. Chem.* **249,** 7550 (1974).
[22] W. Kedzierski, M. Laskowski, Sr., and M. Mandel, *J. Biol. Chem.* **248,** 1227 (1973).
[23] This series, Vol. 12B, several articles.
[24] J. Bonner, G. R. Chalkley, M. Dahmus, D. Fambrough, F. Fujimura, R.-C. C. Huang, J. Huberman, R. Jensen, K. Marushige, H. Ohlenbusch, B. Olivera, and J. Widholm, Vol. 12B, p. 3.
[25] J. Eigner, this series, Vol. 12B, p. 386.
[26] M. Laskowski, Sr. and M. K. Seidel, *Arch. Biochem.* **7,** 465 (1945).
[27] E. J. Williams, S.-C. Sung, and M. Laskowski, Sr., *J. Biol. Chem.* **236,** 1130 (1961).

value; it is too slow. However, the method using an ultra low shear rheometer, Contraves LS-100, was useful in determining the size of the large products generated early in the reaction with native viral DNA.[9,28] Other methods for evaluating the size of molecules also became widely used. Sucrose gradient centrifugation,[28,29] gel filtration on high porosity gels, e.g., Sephadex A15m[29] and particularly electrophoresis on agarose[29-32] became standard methods for evaluating the size distribution of reaction products of single-strand-specific and double-strand-specific nucleases.

One of the additional supporting methods was introduced by Richardson and his colleagues.[33-35] The method allows separate determination of single-strand nicks and double-strand scissions (total cleavages minus single-strand cleavages). The principle is to mark the newly formed 5' termini with ^{32}P. The originally present phosphoryl group is removed with alkaline phosphatase[36] at 25° from the cleaved ends and at 65° from the nicks. The ^{32}P is introduced by polynucleotide kinase in the presence of a great excess of [γ-^{32}P]ATP. Occasional resistance to polynucleotide kinase has been encountered.[34,37] Recently, the method has been modified[38] by simplifying the preparation of polynucleotide kinase and by introducing four additional precautions in performing the reaction.

1. Performing the kinase-catalyzed reaction at pH 8.6 in order to prevent the exchange of the ^{32}P label with 5'-phosphoryl termini which can occur at lower pH values[39]

2. Irreversibly inactivating alkaline phosphatase by pH 12 treatment[40]

3. Ensuring completeness of the end-labeling reaction by using high salt and polynucleotide concentration[41]

4. Removing the unreacted γ-^{32}P]ATP efficiently by diffusion from agarose pellets. This procedure is described below because it may have a rather general application

[28] W. D. Kroeker and J. L. Fairley, *J. Biol. Chem.* **250**, 3773 (1975).

[29] This series, Vol. 30, several articles.

[30] R. B. Helling, H. M. Goodman, and H. W. Boyer, *J. Virol.* **14**, 1235 (1974).

[31] P. A. Sharp, B. Sugden, and S. Sambrook, *Biochemistry* **12**, 3055 (1973).

[32] P. H. Johnson and L. I. Grossman, *Biochemistry* **16**, 4217 (1977).

[33] C. C. Richardson, *Proced. Nucleic Acid Res.* **2**, 815 (1971).

[34] B. Weiss, T. R. Live, and C. C. Richardson, *J. Biol. Chem.* **243**, 4530 (1968).

[35] B. Weiss, Vol. 21, p. 319.

[36] A. Toriani, Vol. 12B, p. 212.

[37] P. Modrich and D. Zabel, *J. Biol. Chem.* **251**, 5866 (1976).

[38] W. D. Kroeker and M. Laskowski, Sr., *Anal. Biochem.* **79**, 63 (1977).

[39] J. H. van de Sande, K. Kleppe, and H. G. Khorana, *Biochemistry* **12**, 5050 (1973).

[40] N. W. Y. Ho and P. T. Gilham, *Biochim. Biophys. Acta* **308**, 53 (1973).

[41] J. R. Lillehang and K. Kleppe, *Biochemistry* **14**, 1225 (1975).

A solution of 2% agarose in 0.2 M potassium phosphate, 0.3 M NaCl, and 0.01 M EDTA (pH 7.8) was kept at 100° in a boiling water bath. Using a pasteur pipette, two drops of the melted agarose are placed on a sheet of Parafilm. Within 20 sec the reaction mixture is blown into the agarose bubble with a disposable 50 μl pipette, drawn back into the pipette, and blown back into the bubble to rinse the pipette and ensure mixing with the agarose. After 60 sec the bubble hardened into a pellet and is placed in the compartment of a wire basket (made from window screen) and submerged in 2 liters of agarose buffer. After 3 hr the buffer is changed, and, after an additional 2 hr, each pellet is removed, rinsed briefly in water, and placed in a scintillation vial; Cerenkov radiation is determined in a Packard Tri-Carb liquid scintillation spectrometer. The ^{32}P counting efficiency is approximately the same as that obtained by counting the same amount of radioactivity in 5 ml of water. Figure 1 illustrates the efficiency of this procedure and is self-explanatory.

Method of Preparation. Methods of preparing partially purified mung bean nuclease have been described[2-5] as well as a procedure resulting in a homogeneous enzyme preparation.[11] A modification[12] which eliminates the chromatography step on hydroxylapatite is now being used since it is simpler and faster. The preparations obtained by either method appear to be identical by several criteria. The simplified modification is described below.[12]

Step 1: Extraction. In the past few years mung bean sprouts have become a popular food and, judging by the standards of food stores in Buffalo, freshly sprouted mung bean (3–5 days' sprouting) are easily available. If the sprouting of large amounts of beans has to be done in the laboratory the procedure described below may be followed.

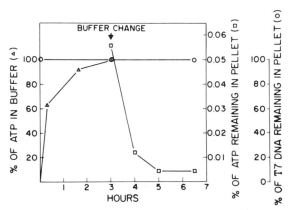

FIG. 1. Diffusion of [γ-^{32}P]ATP from the pellet. Reprinted from Kroeker and Laskowski.[38]

Ten pounds of seeds were washed thoroughly with tap water and placed into two galvanized wash tubs. The seeds were overlayed with 1 inch of warm water (30°) and allowed to soak for 3 hr. The seeds were drained, and the soaking procedure was repeated twice with gentle stirring every hour. After the last soak, the seeds were drained thoroughly. Sprouting was done for 3 days (or until roots are about 3 cm long) watering three to four times per day and draining thoroughly after each watering.

At the time we were devising the method of preparation we used frozen sprouts donated by R.J. Reynolds Foods, Inc., Jackson, Ohio. The frozen sprouts were treated with 2 liters of warm (55°) water per 2 kg, were ground in the cold room (4°) in a large Waring blender (1 gallon size), and squeezed through cheesecloth. Cold water is used for extraction of fresh sprouts. The extract was adjusted with solid ammonium sulfate to 20% saturation (109 g/liter). The mixture was filtered through a large Buchner funnel with the aid of Celite 545. The precipitate was discarded. The filtrate was adjusted to 50% saturation with ammonium sulfate (179 g/liter of solution). The mixture was filtered and again the precipitate was discarded. Solid ammonium sulfate was added (198 g/liter) to attain 80% saturation, and the precipitate collected on hard filter paper covered with a thin layer of Ceilte 545. Approaching 50% saturation in two steps decreases the amount of enzyme lost due to coprecipitation.

Step 2: Ethanol Fractionation after Preliminary Dialysis. The 80% saturation precipitate was dissolved in 200 ml of 30 mM ammonium acetate buffer, pH 5.9, and dialyzed against two changes of the same buffer at 4° for 24 hr. The precipitate that formed was removed by centrifugation at 4°.

The supernatant solution was adjusted to 0.1 M with respect to NaCl and was cooled to 0°. Ethanol (0°) was added to reach 35% (v/v) concentration (0.583 ml of 95% ethanol per milliliter of enzyme solution). The mixture was left at 4° for 20 min and centrifuged at 4°. The precipitate was discarded. The supernatant solution was cooled to −17° and 95% ethanol (−17°, 0.5 ml per milliliter of solution) was added to attain 55% concentration (v/v). The mixture was stirred 15 min at −17° and centrifuged at the same temperature for 90 min at 9500 rpm in the GSA rotor of a Sorvall centrifuge. The precipitate was suspended in 40 ml of cold water and centrifuged for 30 min (SS-34 rotor at 0°) at 15,000 rpm. The supernatant was dialyzed 24 hr against two changes of 0.02 M ammonium acetate, pH 6.0, and lyophilized (can be kept for years).[11]

Step 3: Heat Treatment. Two grams of material from the ethanol step was suspended in 200 ml of 0.02 M ammonium acetate, pH 6.0. The mixture was adjusted to pH 6.0 with ammonium hydroxide and stirred at room temperature until dissolution was complete. Next, 37 ml of 0.5 M sodium acetate, pH 4.5, containing 5 mM zinc acetate and 10 mM cys-

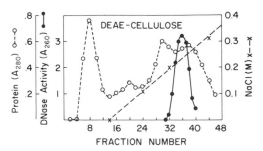

FIG. 2. DEAE-cellulose chromatography of the dialyzed material from the $(NH_4)_2SO_4$ step. The column (3.2 × 15 cm) was eluted by means of a peristaltic pump at 2 ml/min, and 15-ml fractions were collected. The DNase activity (●——●) was measured by the production of acid-soluble material absorbing at 260 nm. The protein (○---○) was estimated by measuring the absorbance at 280 nm. The NaCl gradient is represented by the dashed line. D. Kowalski et al.[12] reprinted with permission from *Biochemistry* **15**, 4457 (1976). Copyright by the American Chemical Society.

teine was added. The mixture was placed in a 500-ml flask in a 70° water bath for 15 min, cooled in ice, centrifuged, and filtered through Whatman No. 1 paper. Precipitates were discarded.

Step 4: Ammonium Sulfate Fractionation. The solution from the previous step was brought to 50% saturation by a slow addition (20 min) with stirring of solid ammonium sulfate (298 mg/ml). After an additional 20 min of stirring the precipitate was removed by centrifugation and discarded. The supernatant was brought to 80% saturation with solid ammonium sulfate (198 mg/ml) and was stirred 20 min. The mixture was centrifuged and the supernatant discarded. The precipitate was dissolved in 40 ml of 0.05 *M* sodium acetate, pH 6.0, containing 1 m*M* cysteine and 0.1 m*M* zinc acetate (buffer A). The solution was dialyzed overnight against 4 liters of the same buffer.

Step 5: DEAE–Cellulose Chromatography. A column (3.2 × 15 cm) was packed with DE-52 cellulose (Whatman) and equilibrated with buffer A. The dialyzed enzyme solution from the previous step was applied to the column and washed with 60 ml of buffer A. Elution was performed with a 500 ml linear gradient from 0 to 0.35 *M* NaCl in buffer A resulting in the protein-activity profile shown in Fig. 2. As a consequence of the NaCl gradient the pH of the effluent increased from 6 to 6.8. This superimposed pH gradient enhanced the resolution. The central fractions representing greater than 80% of the eluted activity were pooled and dialyzed overnight against 6 liters of buffer B (0.001 *M* sodium acetate, pH 5.0, containing 1.0 m*M* cysteine and 0.1 m*M* zinc acetate).

Step 6: Chromatography on CM-Cellulose. A column (2.1 × 15 cm) was packed with CM-52 cellulose (Whatman) and equilibrated with buffer B. The dialyzed solution from the previous step was charged and the column was washed with 50 ml of buffer. Elution was performed with a 400 ml

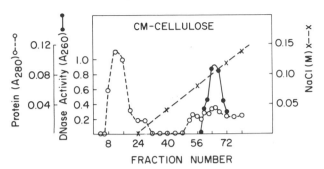

FIG. 3. Carboxymethyl cellulose chromatography of the pooled and dialyzed material from the DEAE-cellulose step. The column (2.1 × 15 cm) was eluted by means of a peristaltic pump at 2 ml/min, and 7-ml fractions were collected. Prior to collection, 0.35 ml of 1 M sodium acetate, pH 5.0, was placed into each tube. DNase activity (●——●) was determined as in Fig. 2. (O---O) absorbance at 280 nm. The NaCl gradient is represented by the broken line. Reprinted with permission from Kowalski et al., Biochemistry 15, 4457 (1976). Copyright by the American Chemical Society.

linear gradient from 0 to 0.15 M NaCl in buffer B (Fig. 3). A 0.35-ml aliquot of 1 M sodium acetate pH 5.0 was placed in all tubes prior to collection in order to maintain a pH of 5 in the collected fractions. The pH of the column effluent dropped from 5.0 to 3.9 as a function of the increased salt gradient. The formation of a pH gradient again markedly improved separation. The central fractions were pooled and accounted for more than 75% of eluted activity. The pooled fractions were concentrated to about 5 ml by dialyzing against solid Ficoll 400.

Step 7: Gel Filtration on Sephadex G-100. A column (1.6 × 96 cm) was packed with Sephadex G-100 (Pharmacia) and equilibrated with at least 1 liter of buffer C (0.05 M sodium acetate, pH 6.0, 0.05 M NaCl, 0.1 mM zinc acetate). The extensive washing was necessary to remove small amounts of carbohydrate from the swollen Sephadex beads. The pattern (Fig. 4) shows that the enzyme is eluted between 1 and 2 column void volumes and that the symmetrical distribution of the eluted enzyme has an approximately constant potency across the peak. The pooled fractions 21–25 were concentrated to about 1 ml by dialysis against Ficoll 400 and dialyzed against 10 mM sodium acetate, pH 5.0, 0.1 mM zinc acetate, and 1 mM cysteine. The purified enzyme is stored in 50% glycerol (v/v) at −20°, and shows no loss of activity after 6 months.

Table I summarizes the purification procedure. The enzyme is stable at 5° at each step of purification.

Properties. The enzyme is a glycoprotein containing about 29% carbohydrate[12] as estimated by the method of Dubois et al.[42] Without prior

---

[42] M. Dubois, K. A. Gilles, J. K. Hamilton, P. A. Rebers, and F. Smith, *Anal. Chem.* **28**, 350 (1956).

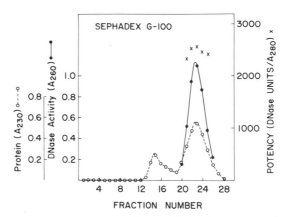

Fig. 4. Sephadex G-100 gel filtration of the pooled and concentrated material from the carboxymethyl cellulose step. Flow rate was maintained at 12 ml/hr by means of a Mariotte flask, and 5-ml fractions were collected. DNase activity (●——●) was determined as in Fig. 2; (○---○) absorbance at 230 nm; (×) potency (i.e., units/A_{280} unit; $A_{280} = A_{230}/5.0$) of the fractions eluting in the active peak. The void volume of the column is 60 ml (fraction 12). Reprinted with permission from Kowalski *et al.*, *Biochemistry* **15**, 4457 (1976). Copyright by the American Chemical Society.

reduction of disulfide bonds, mung bean nuclease shows a single band in sodium dodecyl sulfate, corresponding to a molecular weight of 39,000. After reduction, it shows three bands corresponding to molecular weights 39,000, 25,000, and 15,000. Only 30% of the molecules are intact, whereas 70% are cleaved into 25,000 and 15,000 moieties.[12] Since the intact species and the cleaved species migrate as a single band prior to reduction, it suggests that the cleaved species is held together by a disulfide bond(s). When stained with the periodic acid–Schiff (PAS) reagent all three bands show the presence of carbohydrate. The removal of carbohydrate by N-acetyl-β-glucosoaminidase H[43] reduces the apparent molecular weight to 31,000 but does not significantly change the nuclease to ω-monophosphatase ratio. The amino acid composition is shown in Table II[12,44–46] and is based on the molecular weight 38,600. The enzyme contains one buried sulfhydryl residue per mole that reacts slowly with 5,5′-dithiobis-(2-nitrobenzoic acid), and three S—S bridges. $E_{1\ cm}^{0.1\%} = 2.6$ at 280 nm, not corrected for carbohydrate. The enzyme requires Zn^{2+} (0.1 mM), cysteine (1 mM), and 0.005% Triton X-100 for stability at pH 5.0 (optimum).[12] L-Serine can replace cysteine in stabilizing the enzyme and circumvents the problem of nicking DNA which can occur in the presence of

[43] R. B. Timble and F. Maley, *Biochem. Biophys. Res. Commun.* **78**, 935 (1977).
[44] C. H. W. Hirs, W. H. Stein, and S. Moore, *J. Biol. Chem.* **211**, 941 (1954).
[45] G. Reeck, *in* "Handbook of Biochemistry" (H. A. Sober, ed.). 2nd ed. C-281 Chemical Rubber Publ. Co., Cleveland, Ohio, 1970.
[46] H. H. Edelhoch, *Biochemistry* **6**, 1948 (1967).

TABLE I

SUMMARY OF THE MUNG BEAN NUCLEASE PURIFICATION PROCEDURE[a]

Step	Total A_{280} units	Total activity (units)	Potency (activity units/ A_{280} unit)	Yield (%)	A_{280}/A_{260}	Activity remaining after 6-week storage in solution at 5° (%)
Crude extract	537,000	32,200	0.06	100		96
Ammonium sulfate and ethanol	2,820	22,600	8	70		99
70° treatment	520	21,000	40	65		97
Ammonium sulfate	186	17,300	93	54	1.36	100
DE-52 cellulose	27.9	10,500	378	33	1.43	82
CM-52 cellulose	3.26	6,100	1870	19	1.55	91
G-100 Sephadex	1.66	4,760	2870	15	1.69	

[a] Values listed are based on averages of five purifications and were found to vary only slightly between preparations. Units of activity on denatured DNA are defined on p. 265. Reprinted with permission from Kowalski et al., Biochemistry 15, 4457 (1976). Copyright by the American Chemical Society.

TABLE II
AMINO ACID COMPOSITION OF MUNG BEAN NUCLEASE[a]

Amino acid	No. of residues[b]
Cysteic[c]	7.3
Aspartic	45
Methionine sulfone[d]	10
Threonine[e]	21
Serine[e]	27
Glutamic	21
Proline	8.2
Glycine	16
Alanine	30
Valine[f]	28
Methionine	0.9
Isoleucine[f]	18
Leucine	24
Tyrosine[e]	9.6
Phenylalanine	18
Histidine	9.1
Lystine	14
Arginine	13
Tryptophan[g]	14
Total	334

[a] Reprinted with permission from Kowalski et al., Biochemistry 15, 4457 (1976). Copyright by the American Chemical Society.
[b] Based on a molecular weight of 38,600 for the protein portion of the glycoprotein.
[c] Determined after performic acid oxidation.
[d] Found in hydrolysates of the native protein.
[e] Corrected for destruction during hydrolysis using the factors of Hirs et al.[44]
[f] Corrected for incomplete hydrolysis using the factors of Reeck.[45]
[g] Determined spectrophotometrically by the method of Edelhoch.[46]

sulhydryl compounds during long incubation at pH5.[9] At pH range 7 to 8 the enzyme is stable without additions.

The kinetic properties of mung bean nuclease have a strong dependence on the nature of the substrate. With denatured DNA the enzyme is a typical endonuclease showing preference for A↓pN and T↓pN linkages.[3] During the course of hydrolysis intermediates of all sizes are observed.[2-6] With the usual amounts of enzyme final products are small oligo- and mononucleotides. The oligonucleotides can all be converted to mononucleotides with a 1000-fold excess of enzyme.[4] The trinucleotides are digested beginning from the 3' end.[5]

With native DNA mung bean nuclease exhibits a dual action.[7,9] First, nicks and double-strand scissions are introduced at specific sites. These are followed by an exophilic liberation of acid-soluble material from the newly created termini. It is possible to stimulate the exophilic activity by

TABLE III
TIME COURSE FOR THE EARLY PHASE OF NATIVE T7 DNA HYDROLYSIS BY MUNG BEAN
NUCLEASE[a]

	Original reaction conditions					Modified reaction conditions		
Time (hr)	0	6	12	22	25	28	28.33	28.83	29.33
Acid soluble (%)	<0.2	<0.2	0.9	1.8	2.1	2.9	5.1	7.9	9.9
MW ($\times 10^6$)	26	5.8	4.2	1.8	1.5	1.4	1.0	0.9	0.8

[a] The original reaction conditions were 0.11 ml of T7 [^{33}P]DNA (480 μg/ml, 800 cpm/μg in 0.01 M KCl, 0.01 M Tris-HCl, pH 7.1) and 0.11 ml of 0.1 M sodium acetate, pH 5.0, 0.08 M NaCl, 0.02 mM zinc acetate, 2 mM 2-mercaptoethanol, 0.002% Triton X-100. The reaction was initiated with 0.9 units of mung bean nuclease (in 1.8 μl containing 50% glycerol) and run at 22°. At 28 hr, the reaction conditions were modified by diluting threefold with water and raising the temperature to 30°. An aliquot (0.01 ml) was removed at the times indicated and the percentage of acid-soluble DNA was determined. Also, at the times indicated, 0.01 ml was removed and diluted to 0.1 ml to attain a final Tris-HCl concentration of 0.02 M, pH 8.1. Neutral sucrose gradient sedimentation was performed with these samples and molecular weight were calculated as previously described[28] from the equation $S = 0.0882 \times MW^{0.346}$. Reprinted with permission from Kroeker et al., Biochemistry 15, 4463 (1976). Copyright by the American Chemical Society.

lowering the ionic strength (Table III), but as yet no condition has been found that eliminates the exophilic action. The dual action prohibits the determination of the base sequence at cleavage sites. Wells et al.[47] in a recent review discussed several possible models for site recognition by single-strand-specific nucleases. The indirect evidence suggests that early cleavages by mung bean nuclease occur at AT rich, easily denaturable regions (thermolabile model). The effects of Mg^{2+}, ionic strength, and temperature are consistent with this model. When T7 DNA is used as substrate, the reaction starts with the production of nicks.[9] Nicks do not accumulate, however, but reach a maximum of only three per molecule. At this point the number of double-strand scissions increases at a constant rate and reaches a limit value of 37 per molecule. These results suggest that cleavage across from a nick is favored. Preferential cleavage opposite the naturally occurring nicks in T5 st(0) DNA is not observed. Only after the nicks which are in G + C-rich regions, are enlarged to gaps (an average of five residues) with the aid of exonuclease III is cleavage opposite the original nick highly favored.[9] In contrast, S_1 nuclease is reported to cleave T5 DNA preferentially across the nicks.[48-50]

[47] R. D. Wells, R. W. Blakesley, S. C. Hardies, G. R. Horn, J. E. Larson, E. Selsing, J. F. Burd, H. W. Chan, J. B. Dodgson, K. F. Jensen, I. F. Nes, and R. M. Wartell, Crit. Rev. Biochem. 4, 305 (1977).
[48] K. Shishido and T. Ando, Biochim. Biophys. Acta 390, 125 (1975).
[49] R. C. Weigand, G. N. Godson, and C. M. Radding, J. Biol. Chem. 250, 8848 (1975).
[50] P. Beard, J. F. Morrow, and P. Berg, J. Virol. 12, 1303 (1973).

Acknowledgments

I am indebted to Drs. W. Ardelt, M. Fujimoto, P. H. Johnson, D. Kowalski, W. D. Kroeker, A. J. Mikulski, A. E. Pritchard, and E. Sulkowski for critical reading of the manuscript.

[35] Purification and Properties of Venom Phosphodiesterase[1]

By M. LASKOWSKI, SR.

The previous reviews[2-4] on this subject by the same author became hopelessly outdated during the past 3 years, when it was found that venom phosphodiesterase exhibits some properties common with single-strand specific endonucleases. Until that time phosphodiesterase had been a model of $3' \rightarrow 5'$ exonuclease which consecutively released a 5'-P-mononucleotides.[4]

Only after a homogeneous preparation of phosphodiesterase that was free of interfering enzymes had been obtained[5-7] was it possible to inquire whether the enzyme is capable of opening native circular DNA. It was found that supercoiled PM2 DNA (form I) is opened 10^4 times faster[7] than the relaxed circular DNA (form I').[8] With single-stranded linear T7 DNA transient large molecular weight fragments are formed by endonucleolytic cleavages. The relative rates of hydrolysis of various substrates are shown in Table I.

With PM2 DNA as substrate and under favorable conditions (Mg^{2+} concentration and ionic strength) phosphodiesterase produces not only the full-length linear double-stranded DNA (form III) but also a series of fragments of form III that result from two almost simultaneous cleavages, possibly occurring at the base of the supercoiled branch. Figure 1 shows schematic representation of the action of phosphodiesterase.[7] After the

[1] Supported by the United States Department of Energy (EY76S023225), the National Science Foundation, (BMS 73-06750), and the National Institutes of Health (GM 17788 and HL 15892).
[2] M. Laskowski, Sr., *Proced. Nucleic Acid Res.* **1,** 154 (1966).
[3] M. Laskowski, Sr., *Adv. Enzymol.* **29,** 165 (1967).
[4] M. Laskowski, Sr., *in* "The Enzymes" (P. D. Boyer, ed.), 3rd ed., Vol. 4, p. 313. Academic Press, New York, 1971.
[5] L. B. Dolapchiev, E. Sulkowski, and M. Laskowski, Sr., *Biochem. Biophys. Res. Commun.* **61,** 273 (1974).
[6] L. E. Perry, A. E. Pritchard, and M. Laskowski, Sr., *Abstr., 5th Int. Symp. Anim., Plant Microb. Toxins,* p. 29 (1976).
[7] A. E. Pritchard, D. Kowalski, and M. Laskowski, Sr., *J. Biol. Chem.* **252,** 8652 (1977).
[8] The selective phosphodiesterase digestion of form I in the presence of form I' is the basis of a sensitive assay for the DNA nicking-closing enzyme [D. Kowalski, *Anal. Biochem.* **93,** 346 (1979)].

TABLE I
RELATIVE REACTION RATES ON VARIOUS SUBSTRATES

DNA substrate[b]	Product	Relative rate[c]	Method
SS T7	Acid-soluble DNA	1×10^6	A_{260} acid-soluble
DS T7	Acid-soluble DNA	2×10^5	A_{260} acid-soluble
Form I PM2	Form II	2×10^4	Agarose gels
Form I PM2	Form II	1×10^4	Fluorescence assay
SS T7	Large fragments	10^3-10^4	Agarose gels
Form II PM2	Form III	2×10^3	Agarose gels, initial substrate form I
Form II PM2	Form III	2×10^3	Agarose gels, initial substrate form II
Form I' PM2	—[d]	1	Fluorescence assay

[a] Reprinted from Pritchard et al.[7]
[b] SS and DS are single-stranded and double-stranded DNA, respectively.
[c] Initial reaction rates are expressed relative to a rate of 6×10^{10} cleavages per minute per unit of enzyme on form I' DNA, covalently closed circular duplex DNA.
[d] The products are presumably monomucleotides but the fluorescence assay measures only the rate of disappearance of form I' PM2.DNA. No form II or form III DNA intermediates were detected by this assay or gel electrophoresis.

linear double-stranded fragments are formed they are shortened from both termini without leaving single-stranded ends. Lately, it was established[9] that of the five sites in PM2 DNA susceptible to phosphodiesterase four are located in the previously mapped[10,11] regions of easy denaturation, whereas the fifth is not (Fig. 2), suggesting that phosphodiesterase may be a more sensitive probe for tertiary structure of DNA than some physical methods.

With ϕX174 (form I) only a full-length form III is produced by the action of phosphodiesterase.[12] The cleavages occur in at least five different locations with different frequency at each of the locations. The locations can be identified with an accuracy of about ±40 base pairs. The most frequent break occurs at about position 2258[13] located within gene G. The next most frequent break occurs at about 2300[13] located between genes F and G, where the major hairpin loop was postulated.[14] Increasing the superhelical density from −0.074 to −0.110 had no effect on the pattern.[12]

[9] A. E. Pritchard and M. Laskowski, Sr., J. Biol. Chem. 253, 6606 (1978).
[10] C. Brack, T. A. Bickle, and R. Yuan, J. Mol. Biol. 86, 693 (1975).
[11] C. Brack, H. Eberle, T. A. Bickle, and R. Yuan, J. Mol. Biol. 110, 119 (1976).
[12] A. E. Pritchard and M. Laskowski, Sr., J. Biol. Chem. 253, 7989 (1978).
[13] F. Sanger, G. M. Air, B. G. Barrell, N. L. Brown, A. R. Coulson, J. C. Fiddes, C. A. Hutchison III, P. M. Slocombe, and M. Smith, Nature (London) 265, 687 (1977).
[14] J. C. Fiddes, J. Mol. Biol. 107, 1 (1976).

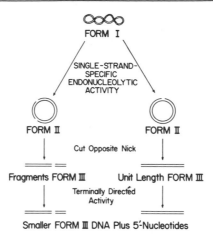

FIG. 1. Mechanism of action of phosphodiesterase on form I PM2 DNA. Reaction pathways are shown for both the production of unit length and fragmented form III DNA intermediates from Pritchard *et al.*[7]

Single-stranded T_7 DNA is rendered acid soluble about 20-fold faster than native linear DNA (Table I). With single-stranded oligonucleotides as substrates all previous findings[2-4] are still valid. A new addition deserves mentioning.[15] It was long known that the free 5'-monophosphoryl group enhances hydrolysis[16] of deoxyribo derivatives and that the effect decreased with increasing chain of oligonucleotide.[17] With a very long poly (rA) chain no acceleration of hydrolysis by 5'-phosphate was observed.[15] The purified enzyme also hydrolyzes nucleotide triphosphates with the liberation of pyrophosphate.[5]

$$NTP \rightarrow NMP + PP$$

The ratio between the nuclease and pyrophosphatase activities remains constant throughout the whole process of purification.[5] From structural considerations it seems probable that the same active center is involved in both manifestations of activity.[18]

The purified phosphodiesterase also hydrolyzes chromatin in a manner similar to other single-strand specific nucleases, e.g., mung bean nuclease and nuclease P_1.[19] It produces large fragments in the range from 30 to 100 S, with an average of 60 S.

[15] A. Stevens, *Biochem. Biophys. Res. Commun.* **81**, 656 (1978).
[16] M. Privat de Garilhe and M. Laskowski, Sr., *J. Biol. Chem.* **223**, 661 (1956).
[17] W. E. Razzell and H. G. Khorana, *J. Biol. Chem.* **234**, 2105 (1959).
[18] H. G. Khorana, *in* "The Enzymes" (P. D. Boyer, H. Lardy, and K. Myrbäck, eds.), 2nd ed., Vol. 5, p. 79. Academic Press, New York, 1961.
[19] M. Fujimoto, A. Kalinski, A. E. Pritchard, D. Kowalski, and M. Laskowski, Sr., *J. Biol. Chem.* **254**, 7405 (1979).

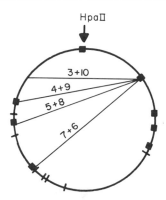

FIG. 2. A map of PM2 DNA showing the positions of the *Hind*III and *Hpa*II cleavage sites, and the early denaturation regions (squares) determined by Brack *et al.*[10,11] The *Hind*-III sites are located at 33, 54, 59, 61, 72, and 76% of the genome clockwise from *Hpa*II site (0%). The early denaturation regions are located at 0, 16, 24, 28, 32, 64, 75, and 79%. The positions of the coordinated phosphodiesterase cleavage sites are indicated by the lines with the numbers of the corresponding paired fragments above each line. The larger fragment of each pair has the lower assigned number. The actual positions of double cleavages are 15,62; 15,72; 15,78; 15,85. From Pritchard and Laskowski.[9]

Assay

A modest advantage of bis-*p*-nitrophenyl phosphate is its low cost and general use by commercial suppliers of phosphodiesterase. The use of *p*-nitrophenyl-pT is advantageous because of specificity toward 5′-monoester forming phosphodiesterases and exclusion of 3′-monoester formers. In addition it has 1000-fold higher V_{max}.[18] Several other substrates have been proposed.[2–4] For histochemical work α-naphthyl-pT is recommended.[20]

For routine purification and standardization we use one of two procedures.

Nonrecording Procedure.[5] The assay mixture contains in 1 ml of 5 μmoles bis-*p*-nitrophenylphosphate, 10 μmoles $MgCl_2$, 100 μmoles Tris-HCl, pH 9.0. The mixture is preincubated 5 min at 37°, enzyme is added with a Hamilton syringe (10 μl), and after 5-min incubation the reaction is stopped by the addition of 2 ml 0.1 M NaOH containing 10 mM EDTA,[8] the absorbancy is read at 400 nm, using 17,600 as molar extinction coefficient.[21] One unit is defined as the liberation of 1 μmole of nitrophenyl per minute per 1 ml volume. If the additions of enzyme and subsequent addi-

[20] H. Sierakowska and D. Shugar, *Biochem. Biophys. Res. Commun.* **11**, 70 (1963).
[21] E. Sulkowski and M. Laskowski, Sr., *Biochim. Biophys. Acta* **240**, 443 (1971).

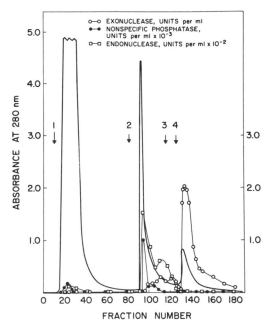

FIG. 3. Chromatography on Con-A Sepharose 4B. The column (35 × 1.5 cm) was equilibrated with 0.2 M sodium acetate buffer, pH 6.0, at room temperature (~22°). The dialyzed
enzyme preparation from step 1 (~25 to 30 ml) was charged on the column at a rate of 20
ml/hr and was followed (the first arrow) by the original buffer until the absorbance of the
effluent at 280 nm fell to 0.05 (the second arrow). At that time 0.05 M α-methyl-
D-mannopyranoside in equilibrating buffer was started. This eluted endonuclease and
the nonspecific phosphatase with about 10–12% of exonuclease. When absorbance at 280 nm
fell below 0.1, the buffer was changed to 0.05 M sodium phosphate, pH 7.2, containing 1 M
NaCl and 0.3 M α-methyl-D-mannopyranoside. After 30 ml of this solution were passed
through (1½ hr), arrow 3, the flow was stopped for 6 hr and was resumed again, arrow 4, with
a rate reduced to 10 ml/hr. Volume of each fraction was 3 ml. The right ordinate refers to
activities of the three enzymes. Coefficients are given on the figure. From Dolapchiev et al.[5]

tion of base are scheduled at 30-sec intervals, a series of up to eight
samples can be run conveniently. The blank contains water instead of
enzyme solution.

   Recording Procedure. The reaction is recorded in either a Gilford
model 2400 or Cary model 16 instrument thermostated at either 37° or 25°.
A substrate is either bis-*p*-nitrophenyl phosphate or *p*-nitrophenyl-pT.
With Ca^{2+}, solubility does not allow a substrate concentration higher than
1 m*M*; with Mg^{2+}, it may be 5 m*M*. Total volume should be not less than 3
ml for convenient reading; it is adjusted with buffer pH 9, preferably 100
m*M* Tris-HCl. Spectrophotometer is set at 400 nm. If only two samples

STEP 4 NADP-SEPHAROSE

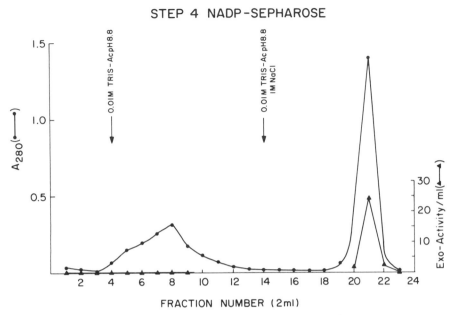

FIG. 4. Chromatography on NADP-Sepharose.[6]

are measured, and with relatively potent enzyme (not with crude venom), the recording procedure is recommended.

Method of Preparation

Step 1.[22] *Acetone Fractionation.* Lyophilized venom of *Crotalus adamanteus* (5 g) is dissolved in 300 ml of cold water with stirring (30 min). The small amount of insoluble material is removed by centrifugation. The clear solution is placed in an ice bath and treated with 200 ml of cold 0.5 *M* acetate buffer, pH 4.0, and then with 362.5 ml of acetone (−20°) to attain a concentration of 42%. The mixture was stirred for 30 min and centrifuged at 0° for 15 min (Sorvall, 13,000 *g*). The heavy yellow precipitate I is dissolved in water adjusted to pH 9 with dilute ammonia and lyophilized for future use. The clear supernatant solution is transferred to the bath at −17°, stirred 2 hr, and centrifuged at −17°. The precipitate II is also saved. The supernatant solution is transferred to the −17° bath and 138 ml of −20° acetone are added with stirring to attain 50% concentration. The mixture is stirred for 1 hr and centrifuged at −17°. The precipitate is dissolved in cold 0.05 *M* sodium acetate buffer, pH 6, and dialyzed against the same buffer overnight to remove acetone.[5] The solu-

[22] E. J. Williams, S. -C. Sung, and M. Laskowski, Sr., *J. Biol. Chem.* **238**, 1130 (1961).

TABLE II
SUMMARY OF THE PURIFICATION PROCEDURE

Step	Total activity units	Total A_{280} units	Potency activity per A_{280}	Activity recovered (%)
Crude	540	6550	0.08	100
Acetone, ppt. III	255	1061	0.24	47
pH 3.6 exposure	277	694	0.40	51
Con-A	127	32	4.0	24
Bio-Gel	125	8	17	23
NADP	116	5	19	21

tion is lyophilized. The purification in this step is about 2.5- to 3-fold, the yield ~50%, and contamination with 5'-nucleotidase is decreased by a factor of about 100.

Step 2.[21] The lyophilized precipitate III is dissolved in 10 volumes of water, pH is adjusted to 3.6 with 1 M acetic acid and the mixture is incubated at 37° for 3 hr, after which pH is readjusted to 6 with 2 M NH$_4$OH. The solution is dialyzed overnight against cold (4°), 0.2 M sodium acetate buffer, pH 6.0, and clarified by centrifugation. This reduces contaminating 5'-nucleotidase by another factor of about 100.

Step 3.[5] The dialyzed solution is absorbed on a concanavalin A–Sepharose column (Pharmacia) and developed as described in the legend to Fig. 3.

Tubes 130–180 (Fig. 3) were pooled. They contained almost 50% of phosphodiesterase (exonuclease) activity charged on the column. The 5'-nucleotidase is held stronger than phosphodiesterase, and, therefore, this step contributes an additional safety measure in removing 5'-nucleotidase.

Step 4.[5,6] The peak containing phosphodiesterase (exonuclease) is concentrated to a volume of 10 ml on Amicon ultrafilter 402 using PM10 membrane. The solution is dialyzed against 0.2 M sodium acetate, pH 6 and charged on Bio-Gel P-150 column (2.5 × 80 cm) previously equilibrated with 0.2 M sodium acetate, pH 6, at room temperature. The first (larger) peak contains some inactive material followed by phosphodiesterase. The material containing activity is pooled and concentrated on Amicon.

Step 5.[6] The phosphodiesterase solution from the previous step is dialyzed against 0.01 M Tris-acetate, pH 8.8, and placed on the NADP–Agarose column (0.7 × 15 cm). NADP–sorbant was prepared according

[23] A. M. Janski and A. E. Oleson, *Anal. Biochem.* **71**, 471 (1976).

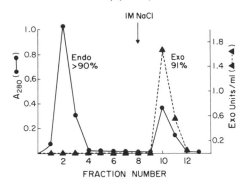

Ppt. Ⅲ, inactivated at pH3.6
NADP-column, pH8.8, 0.0I-Tris-Ac.

FIG. 5. Separation of phosphodiesterase (exonuclease) from endonuclease.[6]

to Janski and Oleson[23] and is now available commercially. The results are shown in Fig. 4. The product is homogeneous in electrophoresis at pH 4.3 and with SDS at pH 7.8. Phosphodiesterase is stored in 5 mM Tris-acetate, pH 8.8, containing 0.005% Triton X-100, adjusted to 50% with glycerol, and kept at $-20°$. Under these conditions it was stable for at least 1 year. Phosphodiesterase is a glycoprotein with a molecular weight of about 120,000. This figure varies depending on the method used and is not corrected for carbohydrate. The composition of neither the amino acids nor the carbohydrates moieties is known. Phosphodiesterase stains heavily with the periodic acid–Schiff reagent on polyacrylamide gel. The results of the purification procedure are summarized in Table II.

The first four steps were devised primarily to eliminate monophosphatases, whereas step 5 (Fig. 4) is aimed primarily at the elimination of endonuclease. The effectiveness of separation of endonuclease[24,25] from exonuclease (phosphodiesterase) using NADP–Agarose is shown in Fig. 5. The material used in this experiment was precipitate III, obtained from acetone fractionation (*step 1*) which was further subjected to inactivation of 5′-nucleotidase by exposure to pH 3.6 for 3 hr at 37° (*step 2*). Precipitate III as obtained from the crude venom still contains all three enzymes in detectable amounts. The destruction of 5′-nucleotidase was necessary since NADP–Agarose does not efficiently separate 5′-nucleotidase and phosphodiesterase, as both compete for NADP. However, endonuclease and exonuclease are separated from each other almost quantitatively.

[24] J. G. Georgatsos and M. Laskowski, Sr., *Biochemistry* 1, 288 (1962).
[25] D. J. Tutas, Ph.D. Thesis, SUNY at Buffalo (1973).

Several methods leading to purified venom phosphodiesterase have been worked out in different laboratories.[26-33] However, to the best of our knowledge only the preparation described above has been rigorously tested for both physical homogeneity and absence of all three interfering enzymes: 5'-nucleotidase,[32] the nonspecific phosphatase,[32,33] and venom endonuclease.[24,25]

Note Added in Proof. The paper of Oka *et al.*[34] describes one-step purification of phosphodiesterase. The method utilizes affinity column of Blue Sepharose CL-6B or Blue Dextran Sepharose 4B. The results show that phosphatase and 5'-nucleotidase appear with the bulk of contaminating proteins while phosphodiesterase is retained and then eluted with buffer of an increased phosphate concentration. We repeated this experiment[35] with Blue Sepharose CL-6B and obtained a heterogeneous preparation. However, inserting this step between steps 4 and 5 of our scheme improved the potency of the final preparation by about 4% and sharpened the band in acrylamide electrophoresis.[35]

Acknowledgments

I am indebted to Dr. Pritchard, Dr. Perry, Dr. Kowalski, Dr. Fujimoto, Dr. Kalinski, Dr. Sulkowski, and Mr. David A. Hartman for critical reading of the manuscript.

[26] E. B. Keller, *Biochem. Biophys. Res. Commun.* **17**, 412 (1964).
[27] S. D. Ehrlich, G. Torti, and G. Bernardi, *Biochemistry* **10**, 2000 (1971).
[28] G. R. Philipps, *Hoppe-Seyler's Z. Physiol. Chem.* **356**, 1085 (1975).
[29] G. R. Philipps, *Biochim. Biophys. Acta* **432**, 237 (1976).
[30] A. M. Fischauf and F. Eckstein, *Eur. J. Biochem.* **32**, 479 (1973).
[31] T. Tatsuki, S. Iwanaga, and T. Suzuki, *J. Biochem. (Tokyo)* **77**, 831 (1975).
[32] E. Sulkowski, W. Björk, and M. Laskowski, Sr., *J. Biol. Chem.* **238**, 2477 (1963).
[33] G. M. Richards, D. J. Tutas, W. J. Wechter, and M. Laskowski, Sr., *Biochemistry* **6**, 2908 (1967).
[34] J. Oka, K. Uedo, and O. Hayaishi, *Biochem. Biophys. Res. Commun.* **80**, 841 (1978).
[35] D. A. Hartman and M. Laskowski, Sr., unpublished.

[36] Uracil-DNA Glycosylase from *Escherichia coli*

By Tomas Lindahl

General Introduction

DNA glycosylases catalyze the hydrolysis of base-sugar bonds (glycosyl bonds, previously called N-glycosidic bonds) in DNA. Several different DNA glycosylases have been detected, each of which is specific for a certain type of unusual or damaged residue in DNA. A repair process, termed "base excision-repair," is initiated *in vivo* by the enzymatic cleavage of unconventional nucleotide moieties to a free base and a deoxyribose phosphate residue remaining in DNA as an apurinic or apyrimidinic site. The DNA glycosylase studied in most detail is uracil-DNA glycosylase, which catalyzes the release of free uracil from DNA containing dUMP residues.

Assay Method

Principle. The assay measures the release of free uracil in the form of acid-soluble material from DNA containing ^{3}H- or ^{14}C-labeled dUMP residues. Degradation of the DNA substrate by nucleases in crude enzyme fractions and phosphorolysis of released mononucleotides is minimized by the inclusion of EDTA in the reaction mixture and the absence of added divalent metal ions and phosphate. While this simple assay is adequate for measurements of enzyme levels in crude cell extracts from *E. coli,* the identity of the released material with free uracil should be chromatographically verified in initial work on other systems.

Reagents. The reaction mixture (total volume 50 μl prior to addition of enzyme) contains

HEPES–KOH, 70 mM, pH 8.0
EDTA (neutralized with KOH), 1 mM
PBS1 [^{3}H]DNA, heat-denatured, 0.4 μg (10,000–20,000 cpm)
Enzyme, 2×10^{-7} to 5×10^{-6} units in 2 μl

Uracil-DNA glycosylase-containing protein fractions are diluted in the following buffer immediately before use

NaCl, 300 mM
HEPES–KOH, 50 mM, pH 7.4
EDTA (neutralized), 1 mM
Dithiothreitol, 1 mM
Bovine serum albumin, 0.1 mg/ml

The other assay reagents are

Carrier DNA (heat-denatured calf thymus DNA), 1 mg/ml
HClO$_4$, 0.8 M

Procedure. The assay mixture is incubated at 37° for 30 min. The reaction is terminated by chilling to 0°, followed by addition of 10 μl of carrier DNA solution and 60 μl of cold 0.8 M HClO$_4$. After 10 min at 0°, each sample is centrifuged at 15,000 g for 15 min, and the radioactivity of 60 µl of the supernatant is determined.

Unit. One unit of uracil-DNA glycosylase is defined as the amount of enzyme that catalyzes the release of 1 µmole of free uracil per minute under the reaction conditions described. The reaction mixture contains 0.4 µg PBS1 DNA, in which 36% of the total nucleotides are dUMP residues, so the substrate has close to 460 pmoles dUMP residues. In 30 min, 10^{-6} enzyme unit releases 30 pmole uracil from the DNA.

Preparation of Substrate. Uracil-DNA glycosylase acts effectively on both double-stranded and single-stranded DNA, so any uracil-containing DNA or polydeoxynucleotide may be employed as substrate, e.g., DNA from the *Bacillus subtilis* bacteriophage PBS1 or its derivative PBS2, which occurs naturally with uracil replacing thymine, bacterial DNA with [^{14}C]cytosine residues deaminated by bisulfite treatment,[1] or homopolymers or copolymers containing dUMP residues synthesized with terminal deoxynucleotidyltransferase. However, preparations of the latter enzyme suitable for synthesis of other polydeoxynucleotides may be contaminated with traces of mammalian uracil-DNA glycosylase, and the most convenient substrate in our experience is PBS1 DNA, first employed in this fashion by Friedberg *et al.*[2]

Radioactive labeling of PBS1 phage and purification of the virus and its DNA follow standard procedures developed for *E. coli* T4 and T7 phage, and a detailed protocol has been published.[3] One complication in work with PBS1 is that lysates of reasonably high titer (10^{10} PFU/ml) are most readily obtained with very mobile host cells that "swarm" over conventional agar plates. Such host cells will overgrow the phage plaques in plaque assays, making accurate titrations difficult. It has been noted[4] that arsenate-sensitive mutants of *B. subtilis* (*asa*) are useful as host cells in plaque assays of PBS1, because in this case "swarming" can be prevented by inclusion of 10^{-3} M arsenate in the media.

Purification Procedure

Uracil-DNA glycosylase has been isolated from several different *E. coli* K12 strains in our laboratory. Frozen cell pastes of commercially grown bacteria, harvested in the logarithmic growth phase, have been satisfactory as starting material.

[1] R. Shapiro, B. Braverman, J. B. Louis, and R. E. Servis, *J. Biol. Chem.* **248,** 4060 (1973).
[2] E. C. Friedberg, A. K. Ganesan, and K. Minton, *J. Virol.* **16,** 315 (1975).
[3] T. Lindahl, S. Ljungquist, W. Siegert, B. Nyberg, and B. Sperens, *J. Biol. Chem.* **252,** 3286 (1977).
[4] A. Adams-Lindahl, Ph.D. Thesis, New York University School of Medicine (1971).

Crude Extract. Eighty grams of frozen bacteria are mechanically disintegrated. We have routinely employed an X press[5] that permits disintegration of the bacteria in the frozen state, but other standard methods of disruption or lysis of *E. coli* can probably be used equally well. All subsequent operations are performed at $0°-4°$. The thawed, disintegrated cells are added to 400 ml of 50 mM Tris-HCl, 1 mM EDTA, and 0.1 mM dithiothreitol, pH 8.0, and the mixture is stirred gently for 30 min. The debris is removed by centrifugation at 20,000 g for 30 min (fraction I).

Streptomycin Treatment and Ammonium Sulfate Fractionation. One volume (about 400 ml) of 1.6% (w/v) streptomycin sulfate in 50 mM Tris-HCl, 1 mM EDTA, and 0.1 mM dithiothreitol, pH 8.0, is added slowly under gentle stirring. After 1 hr, the precipitate is removed by centrifugation at 20,000 g for 30 min. Solid ammonium sulfate (Serva p.A. enzyme grade, or equivalent grade) is slowly added to 40% saturation (1.56 M according to the table published by Wood[6]), and the pH is kept at 7.0 to 7.4 by dropwise addition of concentrated NH$_4$OH. After 30 min, the precipitate is removed by centrifugation at 15,000 g for 30 min. Additional ammonium sulfate is added to the supernatant to a final concentration of 60% saturation (2.35 M), and the precipitate is collected by centrifugation as above. The supernatant solution usually contains about 20% of the total uracil-DNA glycosylase activity of fraction I and is discarded. While slightly higher yields of enzyme activity may be obtained by collecting the protein fraction precipitating between 35 and 65% saturation with ammonium sulfate, this procedure yields a less highly purified final product. The 40–60% ammonium sulfate fraction is suspended in 30 ml of 1 M NaCl, 10 mM HEPES–KOH, 10 mM 2-mercaptoethanol, 1 mM EDTA, and 5% glycerol, pH 7.4, and dialyzed for 4 hr against the same buffer. During this dialysis, the volume almost doubles in size and the suspended precipitate goes into solution. If the solution remains slightly turbid after dialysis, it is clarified by centrifugation (fraction II).

Gel Chromatography. Fraction II is applied to a column (4 × 117 cm) of Sephadex G-75 equilibrated with 1 M NaCl, 10 mM HEPES–KOH, 10 mM 2-mercaptoethanol, 1 mM EDTA, 5% glycerol, pH 7.4, and eluted with the same buffer. Because of its relatively low molecular weight, uracil-DNA glycosylase is eluted as a single symmetrical peak of activity after most of the protein. The fractions containing 75–80% of the enzyme activity are pooled and concentrated about fourfold in an Amicon ultrafiltration cell equipped with a Diaflo PM10 membrane (fraction III).

Hydroxyapatite Chromatography. The purification scheme relies on the unusual property of uracil-DNA glycosylase to bind only very weakly to

[5] L. Edebo and C. G. Hedén, *J. Biochem. Microbiol. Technol. Eng.* **2,** 113 (1960).
[6] W. I. Wood, *Anal. Biochem.* **73,** 250 (1976).

hydroxyapatite while it binds effectively during affinity chromatography on columns containing immobilized DNA. A small-scale pilot experiment should, therefore, be performed to establish the lowest phosphate concentration that allows recovery of the enzyme in the fraction of nonadsorbed protein passing through the column. Several different preparations of hydroxyapatite made according to Bernardi[7] very reproducibly have given 12–14 mM phosphate as a suitable concentration under the conditions described. Commercially obtained hydroxyapatite appears much more variable in quality.

Fraction III is dialyzed for a total of 15 to 20 hr against two changes of 200 mM KCl, 12 mM potassium phosphate, and 1 mM dithiothreitol, pH 7.4, and applied to a column (2 × 10 cm) of hydroxyapatite equilibrated with the same buffer. The uracil-DNA glycosylase is recovered in the effluent, usually as a peak of activity appearing slightly later but overlapping with the main peak of eluted protein. Active fractions are pooled (fraction IV). Most of the protein (60–80%) applied to the column is adsorbed to the hydroxyapatite under these conditions. This step also removes many contaminating enzyme activities, including 98–100% of the endonuclease activities for apurinic and apyrimidinic sites present in fraction III.

DNA–Agarose Chromatography. Heat-denatured calf thymus DNA (Worthington) is covalently bound to cyanogen bromide-activated Sepharose (Pharmacia, Inc.) according to Arndt-Jovin et al.[8] Fraction IV is dialyzed for a total of 15 to 20 hr against two changes of 30 mM Tris-HCl, 1 mM EDTA, 1 mM dithiothreitol, and 5% glycerol, pH 7.4, and slowly applied to a column of denatured DNA–Sepharose (1 × 10 cm) equilibrated with the same buffer. The column is washed with 30 ml of the buffer, and the enzyme is eluted with the same buffer supplemented with 60 mM NaCl. The active fractions are pooled, concentrated about fivefold in a small ultrafiltration cell equipped with a Diaflo PM10 membrane, and stored in several small aliquots at −70°. The enzyme retains unchanged activity for at least 1 year under these conditions, and only small decreases in activity occur after a few freezing-thawing cycles. The purification schedule is summarized in Table I.

Properties of the Purified Enzyme

Molecular Weight. The active enzyme has a molecular weight of 24,000–25,000, as estimated by its hydrodynamic properties, and the same

---

[7] G. Bernardi, this series, Vol. 22, p. 325.
[8] D. J. Arndt-Jovin, T. M. Jovin, W. Bähr, A. M. Frischauf, and M. Marquardt, *Eur. J. Biochem.* **54**, 411 (1975).

TABLE I
PURIFICATION OF URACIL-DNA GLYCOSYLASE FROM 80 g OF *E. coli* CELLS

Fraction	Volume (ml)	Total protein (mg)	Specific activity (units/mg)	Total activity (units)
I. Crude extract	395	8290	0.0035	29
II. Ammonium sulfate	55	3210	0.0061	20
III. Scphadex G-75	32	240	0.061	15
IV. Hydroxyapatite	34	31	0.30	9.3
V. DNA–agarose	1.0	0.16	38	6.1

molecular weight is obtained by polyacrylamide gel electrophoresis in the presence of sodium dodecyl sulfate. Thus, the catalytically active enzyme has a monomeric structure. The amino acid composition and ultraviolet absorption of the purified enzyme show no surprising features.

Purity. In typical preparations of uracil-DNA glycosylase, the enzyme in fraction V is 95–99% pure, as estimated by slab gel electrophoresis under denaturing conditions. Such preparations are free from significant amounts of nuclease activities and other DNA glycosylases. When a particular preparation is used as a reagent, it may be useful to verify the inability of the enzyme to introduce chain breaks in covalently closed circular DNA containing apurinic or apyrimidinic sites.[3]

Requirements. Uracil-DNA glycosylase has no requirement for divalent metal ions or other cofactors. It has a broad pH optimum around 8.0, and is moderately inhibited (45%) when 100 mM NaCl is added to the standard reaction mixture. Enzyme activity is resistant to the inclusion of 5 mM N-ethylmaleimide in the assay mixture.

Specificity. The enzyme releases uracil from single-stranded DNA, poly(dU), or double-stranded DNA containing either A · U or G · U base pairs. The K_m value for dUMP residues in DNA is 4×10^{-8} M. Uracil is not released from RNA, dUMP, or deoxyuridine. The minimal chain length of an oligodeoxynucleotide serving as substrate seems to be four nucleotide residues. Further, the enzyme cannot release thymine, adenine, guanine, 5-methylcytosine, 5-bromouracil, pyrimidine dimers, hypoxanthine, xanthine, or 3-methyladenine from DNA.

Mechanism of Action. Uracil-DNA glycosylase cleaves dUMP residues in DNA by hydrolysis in an essentially irreversible reaction. Chain cleavage at the resulting apyrimidinic sites in DNA may be obtained subsequently by exposure to an endonuclease that selectively attacks DNA at such sites or by alkali treatment.

The enzyme is product inhibited. Thus, uracil acts as a noncompetitive inhibitor ($K_i = 1.2 \times 10^{-4} M$) of the enzymatic cleavage of dUMP residues in double-stranded PBS1 DNA.

The structural gene for uracil-DNA glycosylase has been located to about 55.6 min at the *E. coli* K12 genetic map, and *E. coli* mutants (*ung*) with strongly reduced amounts of the enzyme have been isolated.[9]

[9] B. K. Duncan, P. A. Rockstroh, and H. R. Warner, *J. Bacteriol.* **134,** 1039 (1978).

[37] Purification and Properties of 3-Methyladenine-DNA Glycosylase from *Escherichia coli*[1]

By SHEIKH RIAZUDDIN

Assay Method

Principle. The assay measures the liberation of ethanol-soluble material from calf thymus DNA modified by treatment with ³H-labeled dimethyl sulfate.

Reagents

1. Reaction mixture (total volume 0.1 ml) contains:
   HEPES–KoH, 70 mM, pH 7.6
   EDTA, 1 mM
   2-Mercaptoethanol, 1 mM
   Glycerol, 5%
   ³H-Dimethyl sulfate-treated DNA, 4000 cpm, 60 nmole of mononucleotide equivalent
   Enzyme, 0.02–0.10 μunits
2. Ethanol, 99%
   NaCl, 5 M
   Carrier DNA (thermally denatured calf thymus DNA), 0.2%

Procedure. The reaction mixture is incubated at 37° for 15 min and then chilled to 0°. To each sample, 10 μl of carrier DNA, 10 μl of (5 M) sodium chloride and 240 μl of 99% ethanol are added. After 10 min at 0°, the samples are centrifuged at 23,000 g for 30 min, and 300 μl of each supernatant are recovered for determination of radioactivity. The ethanol-soluble material can be further analyzed by paper chromatography either by direct analysis of a 100-μl aliquot or after concentration by adsorption to and elution from acid-washed charcoal.

[1] S. Riazuddin and T. Lindahl, *Biochemistry* **17,** 2110 (1978).

Unit. One unit of enzyme catalyzes the release of 1 μmole of free 3-methyladenine per minute under the standard reaction conditions.

Preparation of Substrate. Calf thymus DNA (1.5 mg/ml) is treated with ^{3}H-dimethylsulfate (1 mCi/ml) in 0.25 M potassium cacodylate, pH 7.4 and 10^{-3} M EDTA at 37° for 1 hr in the dark. The reaction is terminated by chilling and addition of 2 volumes of 99% cold ethanol. After spooling on a glass rod, the DNA is washed in 70, 80, and 90% ethanol, dissolved in 1 M NaCl, 10^{-2} M Tris-HCl, and 10^{-3} M EDTA, pH 7.4, at 0° and dialyzed against the same buffer for 16 hr, followed by an additional 3-hr dialysis against buffer without salt. The treated DNA has a specific radioactivity of 60–70 cpm/nmole and is stored at $-70°$.

Purification Procedure

All operations are performed at 0°–4° and centrifugations are carried out for 30 min at 20,000 g. Unless otherwise stated, all buffers contain 1 mM 2-mercaptoethanol, 1 mM EDTA, and 5% glycerol.

Crude Cell Extract. *Escherichia coli* 1100 cells (80 g) are mechanically disintegrated, e.g., at $-25°$ in an X-press, and gently stirred with 320 ml of 50 mM HEPES–KoH (pH 7.8) for 30 min. After complete thawing, the debris are removed by centrifugation to give 315 ml of a clear supernatant (fraction I).

Streptomycin Treatment and Ammonium Sulfate Fractionation. To fraction I, 315 ml of 5% streptomycin sulfate solution is slowly added under gentle stirring. After 1 hr, the precipitate formed is removed by centrifugation, and solid ammonium sulfate is slowly added to a final concentration of 46% saturation (1.80 M). During this treatment, care is taken to maintain pH of the solution between 7.0 and 7.4 by dropwise addition of NH$_4$OH. After 30 min, the precipitate is removed by centrifugation and additional solid ammonium sulfate is added to the supernatant to a final concentration of 67% saturation (2.60 M). The precipitate thus obtained is removed by centrifugation, suspended in 40 ml of buffer A (1 M NaCl and 10 mM HEPES–KoH, pH 7.4) and finally dialyzed for 4 hr against the same buffer to give 62 ml of a clear solution (fraction II).

Gel Filtration. Fraction II is divided in two equal parts and chromatographed on two columns (3.8 × 105 cm) of Sephadex G-75 equilibrated with buffer A. 3-Methyladenine-DNA glycosylase is eluted just after the main protein peak. Active fractions from both columns are pooled and concentrated to 35 ml in an Amicon ultrafiltration cell equipped with a Diaflo PM 10 membrane (fraction III).

Phosphocellulose Chromatography. Fraction III is dialyzed for 12 hr against two changes of buffer B (10 mM Tris-HCl, pH 6.7, at 4°), then the dialysis bag is opened to adjust the pH of the protein solution to 6.7 (at 4°), and dialysis continued for another 2 hr. The protein solution is then applied at an adjusted flow rate of 0.5 ml/min to a phosphocellulose column (1.2 × 10 cm) previously equilibrated by washing with 8 liters of buffer B. After washing the column with 75 ml of the same buffer, the adsorbed proteins are eluted by a linear gradient of 0–250 mM NaCl in buffer B. The DNA glycosylase activity is eluted as a single symmetrical peak at 100 mM NaCl. The most active fractions are pooled and concentrated by ultrafiltration to a final volume of about 10 ml (fraction IV).

Native DNA–Cellulose Chromatography. Fraction IV is dialyzed for 4 hr against buffer B at pH 7.6 (at 4°) and applied to a column of DNA–cellulose (0.7 × 5 cm) equilibrated with the same buffer. The column is washed with 5 ml of the buffer, and the enzyme is eluted with a linear gradient of 0–1 M NaCl in the column buffer. The DNA glycosylase activity is eluted as a single peak of activity at 0.5 M NaCl, and the most active fractions are pooled, dialyzed against 70 mM HEPES–KoH (pH 7.8) and stored in several small aliquots at −70° (fraction V).

Comments on the Purification Procedure. At the high concentration of streptomycin sulfate used in the purification scheme, a large fraction of exonuclease III is removed in the streptomycin precipitate while the 3-methyladenine-DNA glycosylase remains in the supernatant. This step, therefore, removes an important contaminant, although it does not give a very large purification. Major overall purification of the DNA glycosylase is obtained during chromatography on phosphocellulose and native DNA cellulose. The phosphocellulose step removes a considerable part of the total protein as well as important contaminating enzymes, such as most of the uracil-DNA glycosylase, exonuclease III, endonuclease IV, and an endonuclease specific for X-ray-induced damage. However, recovery of the purified enzyme is dependent on several factors. The column and the protein solution (fraction III) need to be thoroughly equilibrated against buffer B, and the protein concentration should not exceed 10 mg/ml. Under the conditions of these experiments, 60–90% of the enzyme is adsorbed to the phosphocellulose. The native DNA cellulose chromatography step, however, is quite easy and reproducible.

When 3-methyladenine-DNA glycosylase is partly purified up to fraction III, with buffers containing 10^{-3} M phenylmethylsulfonylfluoride, the chromatographic properties of the enzyme (on Sephadex G-75), and the yield are not detectably affected. It seems unlikely, therefore, that the isolated enzyme is a proteolytic fragment of a larger protein present *in vivo.*

A summary of the purification procedure is given in Table I.

Fraction V contains 3-methyladenine-DNA glycosylase purified 2800-

TABLE I
PURIFICATION OF 3-METHYLADENINE-DNA GLYCOSYLASE FROM 80 gm OF *E. coli* CELLS

Fraction	Protein (mg)	Specific activity (μU/mg)	Total activity (mU)
I. Crude extract	8100	3.3	27
II. Ammonium sulfate	2900	6.1	18
III. Sephadex G-75	310	34	11
IV. Phosphocellulose	6	670	4.0
V. DNA–cellulose	0.2	9300	1.9

fold in 7% yield. It is not a homogeneous enzyme, as polyacrylamide gel electrophoresis under nondenaturing conditions reveals the presence of five protein bands, and only one of these contains enzyme activity after elution. The enzyme activity migrates at 0.65 times the rate of bovine serum albumin indicating that it is not a basic protein.

Absence of Interfering Enzymes. Fraction V appears to be free from endonuclease activities against alkylated, depurinated, X-irradiated, uv-irradiated or 7-bromomethyl-12-methyl benz(*a*)anthracene-treated DNA. There seems to be no exonuclease activity against single-stranded DNA. Most preparations, however, still contain traces of uracil-DNA glycosylase which may be removed by either of two methods. Fraction V may be chromatographed on a small hydroxyapatite column as described by Lindahl,[2] and 3-methyladenine-DNA glycosylase free from uracil-DNA glycosylase is obtained in about 30% yield. Alternatively, an *E. coli* *ung*⁻ mutant is used as the source of enzyme. Since uracil-DNA glycosylase does not release 3-methyladenine from DNA, the fraction V enzyme may be used without any interference.

Properties of 3-Methyladenine-DNA Glycosylase

Physical Parameters. The sedimentation coefficient of 3-methyladenine-DNA glycosylase is 2.3 S as determined by cosedimentation with lysozyme, carbonic anhydrase, and bacterial alkaline phosphatase in a sucrose gradient. The Stokes radius is similarly determined by cochromatography with the three reference enzymes on a Sephadex G-75 column to be 20 Å. Assuming a value of $\bar{V} = 0.725$ gm/cm³ for the partial specific volume of the protein, these data yield a molecular weight of $19,000 \pm 2000$ for 3-methyladenine-DNA glycosylase by the Svedberg equation.

[2] T. Lindahl, *Nature (London)* **259**, 64 (1976).

General Requirements. The 3-methyladenine-DNA glycosylase has a broad pH optimum at pH 7.2–7.8 and displays similar activity in either HEPES–KoH or Tris-HCl buffers. The enzyme shows no obligatory cofactor requirement, but addition of $MgCl_2$ (5–15 mM) to reaction mixtures causes a slight stimulation (about 30%) over the activity observed in the presence of 1 mM EDTA. No stimulatory effect is found with $CaCl_2$. Addition of 1 mM P_i or ATP to reaction mixture has no detectable effect. The enzyme contains essential sulfhydryl groups, because addition of 5 mM N-ethylmaleimide to the standard reaction mixture causes 32% inhibition, while 1 mM p-mercuribenzoate gives 93% inhibition (in the absence of mercaptoethanol). The activity is not very sensitive to addition of neutral salts, and 50% inhibition of the activity is obtained with 300 mM KCl in the reaction mixture.

Substrate Specificity. Fraction V enzyme is highly specific for alkylated DNA and removes 3-methyladenine residues from native DNA modified by treatment with methyl methane sulfonate, dimethyl sulfate, N-methyl-N'-nitro-N-nitrosoguanidine, or N-methyl-N-nitrosourea. The enzyme can also remove 3-ethyladenine residues from DNA after treatment with corresponding ethylating agents. However, there is no release of other modifications in the DNA, such as 7-methylguanine, 3-methylguanine, 7-methyladenine or O^6-methylguanine. The enzyme removes 3-methyladenine as a free base leaving an apurinic site susceptible to endonuclease IV of *E. coli* and, therefore, resembles a previously purified DNA glycosylase specific for dUMP residues in DNA.[3] There is no associated endonuclease activity with the result that no simultaneous chain breaks are introduced in the alkylated DNA. There is no activity of the enzyme on arylalkylated DNA,[4] irradiated (uv- or γ-), depurinated, or deaminated DNA.

The release of 3-methyladenine from alkylated DNA is markedly dependent on the secondary structure of the DNA. While double-stranded DNA is an effective substrate, single-stranded DNA is only attacked at a very slow rate. In this respect, this DNA glycosylase is different from uracil-DNA glycosylase which removes uracil residues from denatured DNA at two- to threefold higher rate than from native DNA.

K_m Value and Mechanism of Action. The K_m value for 3-methyl-dAMP residues in DNA is quite low, 6×10^{-9} M, as determined by varying the concentration of alkylated DNA containing a known amount of modified adenine residues in the standard reaction mixture. The enzyme removes modified adenine by a hydrolytic, not a phosphorolytic, mechanism by

[3] T. Lindahl, S. Ljungquist, W. Siegert, B. Nyberg, and B. Sperens, *J. Biol. Chem.* **252**, 3286 (1977).
[4] M. P. Rayman and A. Dipple, *Biochemistry* **12**, 1202 (1973).

cleaving the glycosyl bond of the modified nucleotide. Increasing amounts of 3-methyladenine in the reaction mixture inhibit the enzyme, and 50% inhibition is observed with $8 \times 10^{-4} M$ 3-methyladenine. On the other hand, $2 \times 10^{-3} M$ 3-methylguanine or $5 \times 10^{-3} M$ of either O^6-methylguanine or 7-methyladenine cause less than 10% inhibition, in agreement with the notion that such residues in DNA are not effective substrates for the enzyme. Free adenine, 1-methyladenine, 7-methylguanine, caffeine, or uracil at a concentration of $5 \times 10^{-3} M$, also fail to inhibit the enzyme.

Occurrence. Extracts from several DNA repair-defective *E. coli* strains contain a normal amount (80–120% of the wild-type level) of this enzyme activity. These strains include four exonuclease III deficient mutants, *E. coli* BW9101 (*xthA*), BW2001 (*xthA*), AB3027 (*xthA, polA*), and NH5016 (*xthA*); the uv-endonuclease-deficient mutant AB1886 (*uvrA*); and the uracil-DNA glycosylase deficient mutant BD10 (*ung*).

Section V

Endonuclease Cleavage Mapping Techniques

[38] Fractionation of Low Molecular Weight DNA or RNA in Polyacrylamide Gels Containing 98% Formamide or 7 M Urea

By TOM MANIATIS and ARGIRIS EFSTRATIADIS

The chain length of small, single-stranded DNA or RNA molecules can be accurately determined by electrophoresis in polyacrylamide gels containing 7 M urea[1] or 98% formamide.[1,2] Under the appropriate conditions, the electrophoretic mobility of these molecules is independent of base composition and secondary structure.[1,2] Figure 1 shows the relative electrophoretic mobility of a number of RNA and DNA molecules with chain lengths of 10–150 nucleotides (nt) in a 12% polyacrylamide gel containing 7 M urea.[1] RNA and DNA molecules with identical chain lengths comigrate in this gel. Molecules with chain lengths greater than 150–200 nt do not completely denature at room temperature in 7 M urea, so it is necessary to use polyacrylamide gels containing 98% formamide to size larger molecules. Formamide is known to completely denature nucleic acids at room temperature.[2] Figure 2 shows the relative electrophoretic mobility of the *Hin*dII fragments of bacteriophage ϕX174 DNA in a 5% polyacrylamide gel containing 98% formamide. The exact sizes of the *Hin*dII fragments are known from the complete nucleotide sequence of ϕX174 DNA.[3] DNA markers *cannot* be used to size RNA in formamide-polyacrylamide gels because RNA molecules migrate more rapidly than DNA of the same size. The chain length of even larger DNA molecules can be estimated by electrophoresis in low percentage agarose gels containing alkali[4] or methyl mercury hydroxide.[5] Methyl mercury agarose gels can also be used to size single-stranded RNA. Alternatively, DNA and RNA can be sized by denaturing in 1 M glyoxal and 50% dimethyl sulfoxide at 50° followed by electrophoresis in ordinary polyacrylamide gels.[6]

Denaturing polyacrylamide gels containing 7 M urea or 98% formamide are also used frequently to purify nucleic acids because of the high resolution afforded by the technique and the ease and efficiency by which nucleic acids can be recovered from the gels. We will describe

[1] T. Maniatis, A. Jeffrey, and H. Van de Sande, *Biochemistry* **14**, 387 (1975).

[2] J. C. Pinder, D. Z. Staynov, and W. B. Gratzer, *Biochemistry* **13**, 5367 (1974).

[3] F. Sanger, G. M. Air, B. G. Barrell, N. L. Brown, A. R. Coulson, J. C. Fiddes, C. A. Hutchinson III, P. M. Slocombe, and M. Smith, *Nature (London)* **265**, 687 (1977).

[4] M. W. McDonnel, M. N. Simon, and W. F. Studier, *J. Mol. Biol.* **110**, 119 (1977).

[5] T. Maniatis, G. K. Sim, A. Efstratiadis, and F. C. Kafatos, *Cell* **8**, 163 (1976).

[6] G. K. McMaster and G. G. Carmichael, *Proc. Natl. Acad. Sci. U.S.A.* **74**, 4835 (1977).

METHODS IN ENZYMOLOGY, VOL. 65

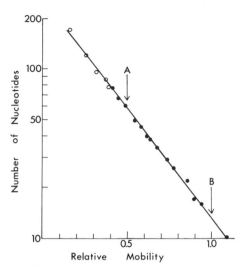

FIG. 1. Chain length calibration of a 12% polyacrylamide-TBE gel containing 7 *M* urea. The closed circles represent synthetic DNA markers.[1] The open circles represent RNA molecules of known chain length. RNA markers are (from smallest to largest) glycine tRNA, mixed tRNA, and serine tRNA from *Staphylococcus epidermidis* provided by R. Roberts; 5 S and 7 S RNA purified from 60 SrRNA subunits from *A. polyphemus* by R. Gelinas. The chain lengths of these molecules are 75, 85, 89, 120, and 168 nt, respectively. The mobility is plotted relative to that of bromphenol blue.

procedures for preparing polyacrylamide gels containing 7 *M* urea or 98% formamide and a procedure for extracting nucleic acids from these gels.

Slab Gels Apparatus

The apparatus described by DeWachter and Fiers[7] or any commercially available slab gel apparatus can be used.

Polyacrylamide Gels Containing 7 M Urea

Solutions and Reagents

1. Gel buffer [½ X Tris-borate-EDTA; (TBE)[8]] 10 X TBE:
   Tris-base (Trizma), 180 g
   Boric acid, 55 g
   Na_2-EDTA, 9.3 g

[7] R. DeWachter and W. Fiers, this series, Vol. 21, p. 167.
[8] A. C. Peacock and C. W. Dingman, *Biochemistry* **8**, 608 (1969).

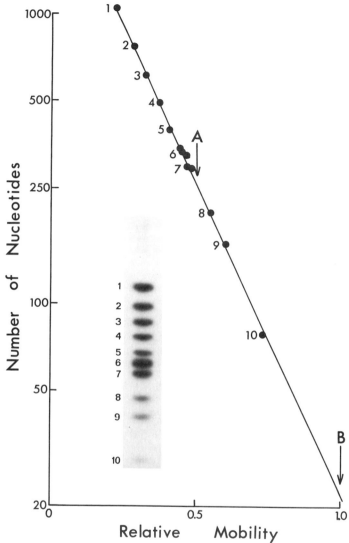

FIG. 2. Relative mobilities of denatured DNA fragments on a 5% polyacrylamide gel containing 98% formamide. The mobility of ^{32}P-labeled ϕX174 DNA digested with HindII is plotted relative to the bromphenol blue dye marker. (A) Xylene cyanol. (B) Bromphenol blue. The inset shows the autoradiogram of the gel used to construct the graph. The sizes of the fragments in nucleotides are 1049, 770, 609, 495, 393, 335–340–345 (triplet), 297–291 (doublet), 163, 79 for fragments 1–10, respectively.

Bring to 1 liter, filter and store at room temperature. If a precipitate forms with storage, the buffer should be discarded.

 2. Acrylamide/bisacrylamide
     Acrylamide (Bio-Rad), 29 g
     Bisacrylamide (Bio-Rad), 1 g
Dissolve in 100 ml of distilled water and filter through fluted paper.

 3. Ammonium persulfate, 1 g in 10 ml H_2O. Make up fresh daily.
 4. Urea, Schwarz/Mann ultrapure.
 5. Temed, $N,N,N'N'$-tetramethylethylenediamine (Bio-Rad)
 6. Marker dyes: Xylene cyanol FF (Edward Gurr Ltd., London) and bromphenol blue (Matheson, Coleman Bell)
     For a 20 × 20 × 0.15 cm slab gel mix:
       30 ml acrylamide/bisacrylamide
       31.5 g urea
       3.75 ml 10 X TBE buffer
       0.5 ml 10% ammonium persulfate

Bring to a volume of 75 ml with distilled water. After degassing, the gel is polymerized by adding 25 μl of Temed. Once the gel has polymerized, the sample wells are rinsed with gel buffer and pieces of acrylamide are removed by using a Pasteur pipette connected to a vacuum aspirator. The gel can be used immediately.

Preparing Samples for Electrophoresis

Samples are ethanol precipitated prior to electrophoresis to remove salt and buffer. This is accomplished by placing the sample in a 1.5 ml polypropylene tube (Walter Sarstedt, Inc., Princeton, New Jersey; order No. 39/10A), adding 0.1 volume 20% sodium acetate, 2 volumes of 95% ethanol, mixing thoroughly and placing the tube in an ethanol–Dry Ice bath for 15 min. The precipitate is recovered by centrifuging for 5 min in an Eppendorf model 3200 centrifuge, rinsed two times with 70% ethanol, dried thoroughly, and resuspended in 25 μl of deionized formamide (see below). Two to 5 μl each of 0.5% xylene cyanol FF and bromphenol blue (marker dyes) are added and the sample placed in a boiling water bath for 2 min. The sample is layered onto the gel using a drawn out capillary, and the gel is run at room temperature at constant voltage (10 V/cm). The bromphenol blue and xylene cyanol dye markers comigrate with oligonucleotides approximately 13 and 58 long, respectively.

Polyacrylamide Gels Containing 98% Formamide

Additional Solutions and Reagents

1. Formamide (FX420 Matheson, Coleman Bell), 99%
2. Mixed bed ion exchange resin (Bio-Rad AG 501-X8) 20–50 mesh
3. $Na_2HPO_4 \cdot 7H_2O$
4. $NaH_2PO_4 \cdot H_2O$

Procedure

Deionized Formamide. Five grams of the mixed bed resin is mixed with 100 ml of formamide, stirred for 0.5–1 hr and filtered through a paper filter. The formamide must be deionized a second time if it is not used within 48 hr.

Weigh out acrylamide and bisacrylamide (see tabulation below).

Gel (%)	Acrylamide (g)	Bis (g)
4.0	2.55	0.45
5.0	3.29	0.56
6.0	3.82	0.68
7.0	4.48	0.79
10.0	6.38	1.13
15.0	9.55	1.70

Weigh out into a small tube:
$Na_2HPO_4 \cdot 7H_2O$, 0.32 g
NaH_2PO_4, 0.04 g
Ammonium persulfate, 0.10 g

The acrylamide plus bis are dissolved in a final volume of 74 ml deionized formamide. One milliliter of water is added to the sodium phosphate, ammonium persulfate mixture and the buffer solution is mixed with the acrylamide–formamide solution and filtered through fluted paper. (Degassing is not necessary.) The gel is polymerized with 150 μl Temed. The gel should stand for at least 10 hr before using. Pieces of acrylamide are cleaned from the sample wells by using a Pasteur pipette connected to a vacuum aspirator.

Running Buffer

Na_2HPO_4, 6.8 g, 0.016 M
$NaH_2PO_4H_2O$, 1.6 g, 0.004 M
Water to 3 liters

Recovery of DNA or RNA from Polyacrylamide Gels

Principle. The same procedure is used to extract DNA or RNA from gels except that sterile solutions and equipment are used for RNA. In general, nucleic acids can be eluted almost quantitatively from a crushed gel piece suspended in a high ionic strength solution. A concentrated solution of ammonium acetate[9] is preferable to other salts because of its high solubility in ethanol. Sodium dodecyl sulfate (SDS) (0.1–0.2%) and/or EDTA (no more than 10 mM) are usually added to prevent nucleolytic degradation. If the nucleic acid is present in very small amounts (1 μg or less) and carrier nucleic acid cannot be added during the final ethanol precipitation step, the addition of Mg^{2+} facilitates effective precipitation.[9]

During the extraction, gel material (presumably water-soluble, non-cross-linked polyacrylamide) is invariably eluted with the nucleic acid. The amount of this material (gel impurities) increases as the percentage of acrylamide used in the gel decreases. Gel impurities are ethanol precipitable and cannot be removed by dialysis, gel filtration, or centrifugation. It is possible (but cumbersome) to remove impurities by hydroxyapatite or DEAE–cellulose chromatography. Fortunately most enzymes (with the exception of reverse transcriptase) are not inhibited by the gel impurities. If mRNA extracted from a gel is to be used for reverse transcription, the gel impurities can be removed by binding the mRNA to an oligo(dT) cellulose column, washing the column thoroughly and eluting the RNA.

Reagents and Equipment

> Gel extraction buffer (GEB): 0.5 M ammonium-acetate, 10 mM magnesium acetate, 0.1% SDs (pH adjustment is not necessary).
> Cold ethanol, 95 and 70%
> 2-Butanol (secondary butyl alcohol, Eastman)
> Water bath
> Small disposable petri dishes (Falcon, Oxnard, California)
> Teflon-coated spatula
> Disposable plastic syringes
> Forceps
> Cellulose acetate filters (Schleicher and Schuell OE67)
> Swinnex filter holders (Millipore)

The filter holders are recycled by soaking 1 hr or overnight in a solution of detergent containing 3% bleach. The pieces are rinsed ten times with tap water and ten times with distilled water and air-dried. (Do not oven dry.)

[9] W. Gilbert and A. Maxam, *Proc. Natl. Acad. Sci. U.S.A.* **70**, 3581 (1973).

Siliconized Scintillation Vials and Corex Tubes. Siliconization is performed in a hood as follows: Clean, dry vials and tubes are filled with a reusable solution of 5% (v/v) dichlorodimethylsilane (Fisher) in chloroform. After 1 min, the tubes are drained, left in the hood to dry (approximately ½ hr), and then washed thoroughly with distilled water and dried. Dichlorodimethylsilane is toxic; inhalation and contact with the skin should be avoided.

Procedure. Nucleic acid bands are detected by autoradiography or ethidium bromide staining, excised and each gel piece agitated in a small petri dish containing distilled water for 15–30 min. This removes the formamide or urea and gel buffer from the gel fragment without eluting the nucleic acid. This step is especially important for formamide gels which contain phosphate buffer. If this presoak is not performed, the phosphate buffer precipitates with the gel impurities, forming an insoluble pellet which traps all of the nucleic acid. The gel piece is transferred to a siliconized scintillation vial and crushed to small pieces with a Teflon spatula. If the gel piece is large (a strip from a 6 mm × 20 cm gel), the gel is crushed in a siliconized 30 ml Corex tube. Depending on the size of the gel fragment, 4–10 ml of GEB are added and the elution mix incubated overnight in a 60° water bath. Alternatively, the vials are placed on a rotary shaker in the warm room (37°) overnight. The solution is filtered through a cellulose acetate filter which is wetted with GEB and placed in a Swinnex holder connected to a disposable plastic syringe. If large gel pieces are extracted, the gel suspension is placed in a 30-ml Corex tube, spun at top speed for 2 min in a clinical centrifuge, and the supernatant filtered as above. More GEB is added, the crushed gel pellet is resuspended by vortexing, and the spin-filtration step repeated. It is sometimes useful to reduce the volume before ethanol precipitation by consecutive extractions with 2-butanol.[10] This is accomplished by adding an equal volume of 2-butanol to the gel filtrate, vortexing, and spinning in a clinical centrifuge. The upper phase is discarded and the extraction repeated as many times as necessary to reduce the volume to the desired amount.

[10] D. Stafford and D. Bieber, *Biochim. Biophys. Acta* **378**, 18 (1975).

[39] Separation and Isolation of DNA Fragments using Linear Polyacrylamide Gradient Gel Electrophoresis

By Peter G. N. Jeppesen

Restriction endonuclease cleavages of virus and plasmid DNAs lead in general to mixtures of products of very widely varying sizes, from several

METHODS IN ENZYMOLOGY, VOL. 65

thousands of base pairs to small fragments of 20 base pairs or less. A current problem in restriction enzyme technology is the development of methods for fractionating these fragments of DNA at high resolution. Particularly difficult are the separation of very large fragments and the separation of fragments differing very little in molecular weight. Gel electrophoresis has become the method of choice because of its high resolving power, but the technique is not without its failings.

Polyacrylamide gels of different concentrations and degrees of cross-linking have been extensively employed in fractionating single- and double-stranded nucleic acids and are capable of very high resolution. The relative electrophoretic mobilities of single-stranded RNA and DNA species in a uniform gel are approximately linearly related to the logarithms of their molecular weights, until, with increasing size, molecules are progressively excluded owing to the effective "pore" size of the gel. The upper molecular weight range of a particular gel may be extended by decreasing the concentration of cross-linker and/or decreasing the total acrylamide concentration, at the expense of resolving power at lower molecular weights. There are, of course, minimum concentrations of acrylamide and cross-linker below which gels will not form, and these define the upper molecular weight range of single-stranded nucleic acids for which polyacrylamide gels can be usefully employed. Agarose has been used either alone or in conjunction with concentrations of polyacrylamide which would not otherwise form gels in order to extend further the high molecular weight range which may be resolved by gel electrophoresis, but with an even greater loss of resolving power for smaller fragments.

In the case of double-stranded nucleic acids (e.g., DNA) the relatively stiff rodlike structure behaves somewhat differently on gels from the globular structure of single-stranded nucleic acid. A similar logarithmic relationship between molecular weight and mobility holds approximately at low molecular weights, but, with increasing size, resolving power decreases without the DNA being totally excluded from the gel; since the DNA can migrate as a rod with its axis parallel to the direction of migration, the pore size would have to be less than the cross section of the rod to cause complete exclusion. Above a limiting molecular weight, DNA molecules migrate together as a nonresolvable band. This limit is inversely related to the concentration of the gel and is approximately 10^7 daltons in the weakest polyacrylamide gels. There are other factors which cause small deviations from the mobilities expected according to size and are probably due to base composition effects. (An example of this will be described later in this article.)

Thus, both in the cases of single- and double-stranded nucleic acids, the conflicting requirements of high resolution (high gel concentration) and high molecular weight limit (low gel concentration) necessitate a compromise for any particular set of fragments, and in certain cases no one concentration of acrylamide will give complete separation of all the fragments in a set. It was in an attempt to overcome this problem encountered in restriction endonuclease digests of DNA[1] that the author tried electrophoresis through gels containing a linear concentration gradient of acrylamide; different-sized fragments should be optimally resolved at different positions in the gradient. Thus, smaller fragments would be increasingly retarded by decreasing pore size as they migrate through the gel, whereas larger fragments would not see such small pore size and their resolution should not suffer as a consequence. Linear concentration gradient polyacrylamide gels had been previously described for the analysis of proteins.[2] In addition to extending the fragment size range which can be fractionated on a single gel, there is another advantage of gradient gels. Because at any one time the front of a particular migrating species is moving fractionally more slowly than its tail, owing to the progressively increasing gel concentration, each band undergoes a continuous sharpening effect during electrophoresis. The sharpness of bands obtained on gradient gels is quite noticeable when compared with uniform concentration gels, and this also results in improved resolving power. Once the apparatus has been set up, the preparation of gradient gels takes little longer than the preparation of uniform concentration gels. The technique has been used in restriction site mapping using DNA polymerase specifically to label nearest neighbors,[3] and also for the mapping of deletion mutations[4,5] where the ability to resolve large numbers of restriction fragments has been particularly useful.

Apparatus

Many types of slab gel electrophoresis apparatus are now in widespread use and are either commercially available or produced in laboratory workshops. Most of these are suitable for forming and running concentration gradient gels, as long as the mold can be adapted to fill from the bottom for forming the gradient. The apparatus originally described by the author[1] has been modified slightly for ease of pouring and running the gels and is similar to that available through Raven Scientific Ltd., Haverhill,

[1] P. G. N. Jeppesen, Anal. Biochem. **58**, 195 (1974).
[2] J. Margolis and K. G. Kenrick, Anal. Biochem. **25**, 347 (1968).
[3] P. G. N. Jeppesen, L. Sanders, and P. M. Slocombe, Nucleic Acids Res. **3**, 1323 (1976).
[4] B. Allet, P. G. N. Jeppesen, K. J. Katagiri, and H. Delius, Nature (London) **241**, 120 (1973).
[5] B. Allet and R. Solem, J. Mol. Biol. **85**, 475 (1974).

England. Briefly, the gel mold consists of two glass plates of overall size 17 × 20 cm ("short" gel) or 17 × 40 cm ("long" gel), with one having a 14 × 2 cm rectangle cut out at the top to allow the gel to make contact with the upper buffer reservoir. The plates are separated at their edges and bottom by strips of Perspex (Plexiglas) 1.5 cm wide and 0.15 cm thick, and the whole is held together by spring clips. The bottom Perspex strip extends beyond the glass plates and along its upper edge at one side is introduced the filling tube of 0.15 cm (o.d.) polyethylene tubing, so that the outlet just enters the gel space. The filling tube is held in position between the side and bottom Perspex strips; at the other side of the mold the Perspex strips abut one another. The sides and bottom of the mold are then sealed by allowing a warm 12% (w/v) solution of gelatin to seep in and fill the air spaces between Perspex strips, tubing, and glass plates, and allowing the gelatin solution to cool and gel. Gelatin forms an excellent seal and is easily removed for cleaning purposes by washing the plates in warm water. After the acrylamide gels have been formed (see discussion on gel preparation below) the bottom Perspex strip and filling tube are removed and the glass plates containing the gel are inserted into the electrophoresis apparatus, with the bottom of the gel submerged in the lower (anode) buffer reservoir and the top of the gel making contact with the upper (cathode) buffer reservoir via the cutout described above.

Gel Preparation

The gel is formed in the mold in two stages. First, the acrylamide concentration gradient is introduced and allowed to polymerize, after which a low concentration (2.5%) starting gel is formed on top, in which the sample wells are cast. If the sample well template were inserted directly into the top of the acylamide gradient, mixing would occur, leading to difficulties in obtaining reproducibility between gels with respect to the exact starting concentration of the gradient. In order that the samples should migrate as fast as possible through the starting gel, the buffer concentration is arranged to be one-fifth that of the main gel. Thus, the voltage gradient in the starting gel is five times higher, and the bromphenol blue tracer dye completely traverses the starting gel in 10–15 min after beginning electrophoresis. Although larger DNA fragments are increasingly retarded, depending on their size, in an electrophoretic run of several hours, the time spent by a fragment in the starting gel may effectively be ignored.

An added advantage of the decrease in voltage gradient at the junction between the starting and gradient gels is that samples concentrate as they decelerate across the junction, leading to band sharpening upon entry into the gradient gel. By loading the samples in even lower ionic strength

TABLE I

COMPONENTS FOR 3.5–7.5%, 5–10%, AND 10–20% POLYACRYLAMIDE CONCENTRATION GRADIENT GELS[a]

| | Separating gel (polyacrylamide concentration gradient) | | | | | | 2.5% starting gel (All gradients) |
| | 3.5–7.5% | | 5–10% | | 10–20% | | |
	B	A	B	A	B	A	C
Stock solutions							
40% (w/v) acrylamide (ml)[b]	1.31	2.81	1.88	3.75	3.75	7.5	0.94
2% (w/v) bisacrylamide (ml)[b]	0.87	0.78	0.83	0.72	0.72	0.50	0.89
50% (w/v) sucrose (ml)	3	6	3	6	3	5.5	—
10 X E buffer (ml)[c]	1.5	1.5	1.5	1.5	1.5	1.5	0.3
Water to make up to (ml)	15	15[d]	15	15[d]	15	15[d]	15
Catalysts							
10% (w/v) ammonium persulfate (ml)[e]	0.040	0.040	0.040	0.040	0.040	0.040	0.040
10% (v/v) TEMED (ml)[f]	0.085	0.025	0.060	0.020	0.040	0.015	0.150

[a] Quantities in the Table are for the preparation of "short gels" in the apparatus described in the text. For different volume gel molds, quantities should be scaled appropriately.
[b] Deionized. See text.
[c] 1 X E buffer is 0.04 M Tris-acetate, pH 8.3, 0.02 M sodium acetate, 0.002 M EDTA. This is also the buffer used in the electrophoresis reservoirs.
[d] Use 14.5 ml. See text.
[e] Freshly prepared.
[f] N,N,N',N'-Tetramethylethylenediamine. The solution should be freshly prepared.

buffer, a similar concentrating effect is obtained upon entering the starting gel.

Reagents. Acrylamide and "bisacrylamide" (N,N'-methylene-bisacrylamide) should preferably be of electrophoretic grade. Before use the stock solutions (see Table I) are deionized by stirring with a small quantity of Amberlite MB-1 (Rohm & Haas Co., Philadelphia, Pennsylvania) for 30 min, filtered, and stored at 4°. All other reagents should be of the purest grade available. Solutions of polymerization catalysts (Table I) are freshly prepared each time before use.

Separating Gel (Polyacrylamide Concentration Gradient). The author has used gels containing a variety of concentration gradients depending on the particular set of DNA fragments to be separated. Table I gives the conditions for making concentration gradients of 3.5–7.5%, 5–10%, and 10–20% acrylamide as models. The 3.5–7.5% gels have been found to give good separations of the products of many restriction endonuclease cleavages, over a wide range of molecular weights and are used as the first trial for uncharacterized digestions. Where greater resolution of small fragments is required, 5–10% or even 10–20% gels have often proved useful. Further adjustments to the concentration range may be made by trial to optimize the fractionation of a particular fragment set if necessary. There are three general principles which apply to the production of any desired concentration gradient.

1. The simultaneous establishment of a 10–20% (w/v) sucrose gradient greatly increases the stability of the acrylamide gradient while filling the mold, by reducing vertical mixing and diffusion.

2. To avoid mixing of the gradient by convection currents during the exothermic polymerization reaction it is important to ensure that gelation occurs first at the top of the gradient and spreads downward. The concentrations of polymerization catalysts are determined by experiment to fulfil this condition and also to allow sufficient time to complete pouring the gradient before polymerization begins.

3. The proportion of cross-linker required for the production of physically strong, clear gels of good resolving power varies with the total acrylamide concentration.[6] To keep the proportion of bisacrylamide constant for any given concentration of acrylamide encountered in different concentration gradients, the following linear relationship linking bisacrylamide and acrylamide concentrations was used in formulating gradient gel components

$$\text{Bisacrylamide conc. (\%)} = 0.126 - \text{acrylamide conc. (\%)} \times 0.003$$

[6] D. P. Blattler, F. Garner, K. van Slyke, and A. Bradley, *J. Chromatogr.* **64**, 147 (1972).

This gives suitable proportions of bisacrylamide for production of gels between 2 and 40% acrylamide concentration.

The acrylamide concentration gradient is formed as follows. An ordinary two-compartment sucrose gradient mixer of total volume matching as nearly as possible the acrylamide gradient to be poured is arranged above a magnetic stirrer in such a way that a small stirring bar can be operated simultaneously in each compartment. The objectives of this are twofold: (1) to counteract the slight nonlinearity of gradient which accompanies the use of a single stirring bar of finite volume and (2) to ensure that any initial backflow due to unequal hydrostatic pressures in the compartments at the start (see the following) is rapidly mixed into the bulk. The outlet from the gradient former is fed into the gel mold via the inserted filling tube, using a variable rate peristaltic pump to control the flow rate, adjusted to ensure pouring the complete gradient in 10–20 min, depending on the size.

Components A and B before addition of catalysts, and 10 ml of E buffer (see Table I) are degassed in Buchner flasks attached to an electric vacuum pump until formation of gas bubbles ceases. The correct amounts of TEMED and ammonium persulfate solutions (Table I) are then added to components A and B and mixed briefly by swirling. The less dense component, B, is then added to the outlet side of the gradient mixer, and the valve connecting both compartments is opened momentarily to fill the interconnecting channel. This is to avoid a trapped air pocket which can prevent proper mixing at the start of pouring the gradient. Component A is then added to the other compartment. (*Note:* a slightly smaller volume of A than B is used to allow for the greater density of the former. If equal volumes were used there would be a flow of A into B on opening the interconnecting channel, to equalize the hydrostatic pressures, thus giving an initial concentration of acrylamide greater than that desired. A small initial backflow, on the other hand, slightly affects the final concentration of acrylamide, which is not as important.) The stirring bars are then set in motion at a suitable speed to ensure efficient mixing without turbulence, and the interconnecting valve is opened, allowing the chambers to reach equilibrium. Two to 3 ml of degassed E buffer are then pipetted into the bottom of the mold, the peristaltic pump started, and the gradient pumped into the mold below the surface of the buffer. Displacement filling from the bottom and buffer layering in this manner were found to give the minimum of mixing during pouring and lead to a straight upper polymerization line. After delivery of the gradient the filling tube is clamped off close to the gel mold, and the gradient mixer and peristaltic pump tubing are disconnected and thoroughly rinsed with water to prevent gel formation by acrylamide soluting remaining in the apparatus. A polymerization line at the top of the gradient should be evident within 15–30 min of pouring and a second, lower, line may also be seen, indicating the extent of polymerization. This line slowly descends as polymerization proceeds, a process usually taking a further 30–60 min for completion.

Starting Gel. In the apparatus originally described for the preparation of gradient gels,[1] the sample wells were formed entirely within the thickness of the gel so that species migrated down the center of the gel without coming into contact with the glass plates. Although this was found to prevent artifacts resulting from irregularities in polymerization at the glass surface due to dirt on the glass, etc., the thick slab gels utilized for this purpose proved less suitable for autoradiographic location of ^{32}P-labeled samples than thinner gels. Consequently, the thickness of the gel mold was reduced to 0.15 cm, and the increased likelihood of surface effects tolerated. The template used for forming the wells was cut from 0.15-cm Perspex sheet, to give 1-cm deep wells. The widths of the individual wells depend on the volumes and quantities of the DNA samples to be applied (see the following).

Fifteen milliliters of component C (Table I), without catalysts added, is degassed as described earlier, and the surface of the lower gel is rinsed two or three times with small volumes of this solution from a Pasteur pipette. The catalysts are then added, the solution mixed by swirling and then poured into the remaining space in the gel mold. The sample well template is wetted and carefully inserted to avoid trapping air bubbles, and anchored if necessary with a small piece of adhesive tape. The exposed surface of acrylamide solution is then sprayed with a fine aerosol mist of 0.1% sodium dodecyl sulfate (SDS), before leaving the gel to polymerize. The SDS reduces absorption of oxygen and reduces the tendency to give poor polymerization close to the surface. This is especially liable to occur with the low concentration of acrylamide used in the starting gel.

Polymerization of the upper gel should be apparent within 10–15 min: it is important to ensure that good polymerization occurs right at the junction of the two gel layers to prevent a nongelled "dead space," and hence loss of resolution when migrating species cross the junction. Therefore the upper gel is left about 1 hr to completely polymerize before removing the template. After its removal, the sample wells are rinsed two or three times with small volumes of one-fifth concentration E buffer from a Pasteur pipette.

Electrophoresis

After setting the gel in the electrophoresis apparatus and filling the buffer reservoirs, the gel is allowed to reach its running temperature (see below). Prerunning has not been found to improve subsequent separations using this system.

Conditions for performing restriction endonuclease digestions are detailed elsewhere in this volume and will not be repeated here. After incubation of the DNA sample with the appropriate restriction endonuclease, the fragments are prepared for electrophoresis as follows. The incubation mixture is first extracted twice with equal volumes of redistilled phenol saturated with 10 mM Tris-HCl, pH 8.0, 1 mM EDTA, followed by three ether extractions to remove excess phenol. Sodium chloride is then added to give a final concentration of 0.5 M, and the DNA is precipitated by the addition of 2.5 volumes of 95% ethanol and leaving at $-20°$ overnight. After collection of the precipitate by centrifugation, the pellet is washed with ethanol and dried *in vacuo*. The digested DNA is then taken up in a small volume of loading buffer in a length of polyethylene capillary tubing connected to a microsyringe. Loading buffer is 1/10 X E buffer containing 10% sucrose and 0.1% bromphenol blue as a tracer dye. DNA is usually loaded at a concentration of up to 1–2 μg/μl in the loading buffer, using about 10 μl for a 1-cm wide well. The polyethylene tubing is inserted into the well below the surface of the buffer, and the DNA sample is injected into the well by use of the syringe. Electrophoresis with constant current is then carried out at either 4° or 22° with a voltage gradient of approximately 15 V/cm across the gel. Using a gel of the described cross-sectional area, currents are 20–25 mA at 4° and 40–50 mA at 22°. In trial runs the bromphenol blue marker dye is run to within 1 cm of the bottom of the gel; for short gels this takes 6–8 hr at 22° or 12–16 hr at 4°. The shorter time required for running gels at room temperature (22°) is an advantage, although preparative gels are usually run at 4° to minimize the chance of "nicking" the DNA which might occur during a warm run. Slight alterations in relative positions of bands when runs are made at 4° or 22° (see the following) may also influence the choice of running temperature.

Location and Extraction of DNA

After electrophoresis, ^{32}P-labeled DNA bands may be located by direct autoradiography, after removing one of the glass plates and covering the gel with cellophane. For nonradioactively labeled DNA, fragment bands may be seen by uv fluorescence using ethidium bromide or, alternatively, after staining with methylene blue. The use of ethidium bromide is rapid and sensitive, but methylene blue staining is sometimes more convenient, especially in preparative gels where bands are to be cut out and the DNA fragments eluted. The gel is carefully removed from the mold, and the upper gel is usually cut off and discarded at this stage. (The gel is more easily handled under water, when it has less tendency to fracture.)

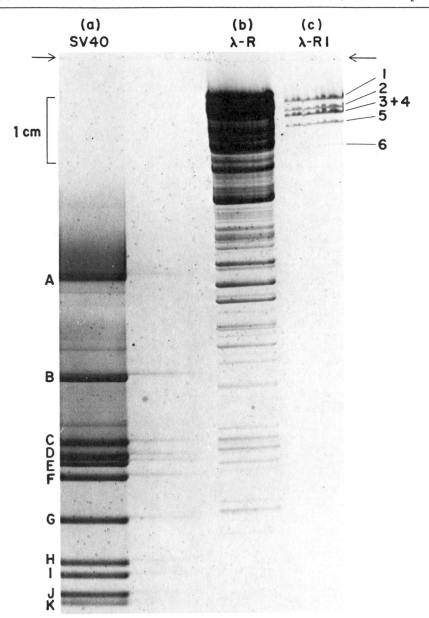

FIG. 1. A 3.5–7.5% "short" polyacrylamide concentration gradient gel, run at 4° for 16 hr and using a current of 25 mA. The DNA fragments are stained with methylene blue. (a) Restriction endonucleases HindII+III digest of SV40 DNA. The molecular weights of fragments A–K are as follows: A, 6.5×10^5; B, 4.2×10^5; C and D, 3.2×10^5; E and F, 2.3×10^5; G, 2.1×10^5; H, 1.2×10^5; I, 1.0×10^5; J, 8.7×10^4; K, 7.4×10^4. [K. Danna and

For ethidium bromide fluorescence the gradient portion of the gel is covered with a 0.001% solution of ethidium bromide, and the fluorochrome is allowed to penetrate the gel for 15–30 min. Fluorescent bands are then visible when the gel is illuminated with near-uv light. Alternatively, the gel is stained in 0.02% methylene blue containing 0.01 M Tris-acetate, pH 8.3, for 1–2 hr at 4°. All subsequent operations on stained DNA are performed as much as possible in the cold and out of direct sunlight, in order to minimize photodegradation of the dye complexed DNA. Excess stain is washed out by several changes of distilled water over a period of 5–8 hr, after which band patterns formed by the DNA fragments are clearly visible. Bands containing as little as 0.25 μg DNA per centimeter can be seen with methylene blue.

For extraction of the DNA from bands in preparative gels, the electrophoretic method described in Allet et al.[4] has been found the most satisfactory, giving efficient extraction of intact DNA as assayed by electron microscopy. The region of gel containing the required material is excised and transferred to an eluting tube filled with E buffer. The eluting tube consists of a short Perspex tube whose lower end is plugged with glass wool and sealed with a bag formed from dialysis membrane. Electrophoresis is then carried out with the dialysis bag immersed in an anode buffer reservoir, and the top of the Perspex tube inserted through a rubber seal into a cathode reservoir, both reservoirs being filled with E buffer. After about 12 hr at 50–100 V, the eluted DNA which has collected in the dialysis bag is removed and concentrated by ethanol precipitation.

Gel Position and Molecular Weight

Figure 1 shows the separation on a 3.5–7.5% polyacrylamide gradient gel of restriction endonucleases HindII+III cleavage products from simian virus 40 (SV40) DNA and bacteriophage λ DNA and the restriction endonuclease EcoR1 products of λ DNA. In this particular experiment the HindII+III digestions have not quite gone to completion, as evidenced by the presence of faint bands in the SV40 DNA digest corresponding to partial digestion products and the rather streaky appearance of the higher molecular weight regions of both the SV40 and the λ DNA digests. The

D. Nathans, Proc. Natl. Acad. Sci. U.S.A. **68**, 2913 (1971).] (b) Restriction endonucleases HindII+III digest of λ DNA. (c) Restriction endonuclease EcoR1 digest of λ DNA. The molecular weights of fragments 1–6 are as follows: 1, 13.69 × 10⁶; 2, 4.49 × 10⁶; 3, 3.54 × 10⁶; 4, 3.04 × 10⁶; 5, 3.54 × 10⁶; 6, 2.31 × 10⁶. [B. Allet, P. G. N. Jeppesen, K. J. Katagiri, and H. Delius, Nature (London) **241**, 120 (1973); B. Allet, K. J. Katagiri, and R. F. Gesteland, J. Mol. Biol. **78**, 589 (1973).] Arrows indicate beginning of gradient. [From P. G. N. Jeppesen, Anal. Biochem. **58**, 195 (1974).]

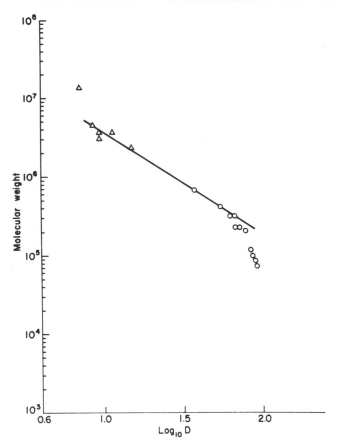

FIG. 2. A double-logarithm plot of molecular weights against distances migrated D in the gradient gel illustrated in Fig. 1. D is measured in millimeters. Open circles represent fragments A-K obtained by digesting SV40 DNA with restriction endonucleases HindII+III, and open triangles represent fragments 1-6 obtained by digesting λ DNA with restriction endonuclease EcoR1. [From P. G. N. Jeppesen, *Anal. Biochem.* **58,** 195 (1974).]

material closest to the origin in the HindII+III-treated λ DNA is probably undigested λ DNA, molecular weight 30.8×10^6. The molecular weights for the HindII+III fragments of SV40 DNA and the EcoR1 fragments of λ DNA are given in the legend to Fig. 1.

Although the relationship between molecular weight and distance migrated on the gradient gel system is complex, it has been found empirically that if the logarithms of the molecular weights of the fragments are plotted against the logarithms of the distances they have moved from the beginning of the gradient, the points approximate to a sigmoid-shaped curve

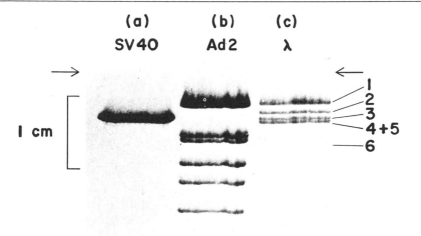

FIG. 3. A 3.5–7.5% polyacrylamide concentration gradient gel, run at 22° for 8 hr and using a current of 45 mA. The DNA fragments are stained with methylene blue. (a) Linear SV40 DNA (molecular weight 3.0×10^6) [L. V. Crawford and P. H. Black, *Virology* **24**, 388 (1964)] obtained by cleavage of circular SV40 DNA with restriction endonuclease *Eco*R1 [R. N. Yoshimuri, Ph.D. Thesis, University of California (1971)]. (b) Restriction endonuclease *Eco*R1 digest of adenovirus type 2 DNA. The molecular weights of the six fragments in descending order down the gel are 13.6×10^6, 2.7×10^6, 2.3×10^6, 1.7×10^6, 1.4×10^6, 1.1×10^6 [U. Pettersson, C. Mulder, H. Delius, and P. A. Sharp. *Proc. Natl. Acad. Sci. U.S.A.* **70**, 200 (1973)]. (c) Restriction endonuclease *Eco*R1 digest of λ DNA. The molecular weights of fragments 1–6 are given in the legend to Fig. 1. Arrows indicate beginning of gradient. [From P. G. N. Jeppesen, *Anal. Biochem.* **58**, 195 (1974).]

having a central portion which is effectively linear over a large range of molecular weights. Figure 2 shows such a plot for the fragments of Fig. 1 whose molecular weights are known. The linearity may be extended to the lower molecular weight region of Fig. 2 by raising the higher concentration limit of the gel (not shown here), although this results in a lower overall separation of the larger fragments, unless a longer gel is used. In general, for the molecular weight region below approximately 5×10^6, it has been found that the more nearly the terminal velocities of the migrating species approach zero because of increasing gel concentration (i.e., the closer the bands come to equilibrium in the gel), the more closely the double logarithm plot of their positions resembles a straight line. Thus in Fig. 2, the larger fragments which lie closest to the straight line do not migrate

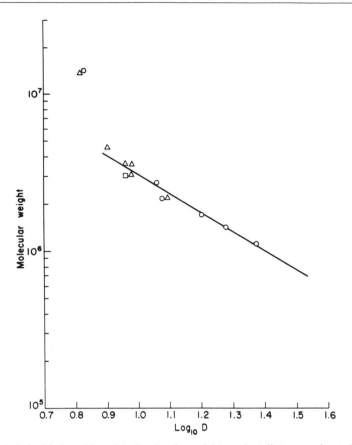

FIG. 4. A double-logarithm plot of molecular weights against distances migrated, D (mm), for the restriction endonuclease EcoR1 fragments separated in the gel of Fig. 3. The open square represents linear SV40 DNA, open circles represent adenovirus type 2 DNA fragments, and open triangles represent λ DNA fragments. [From P. G. N. Jeppesen, *Anal. Biochem.* **58,** 195 (1974).]

much farther if electrophoresis is continued for longer periods of time, whereas the four smaller fragments continue to migrate and are lost from the bottom of the gel.

The linear portion of the curve may be extended slightly toward the higher molecular weight region by reducing the lower concentration limit of the gradient. However, this cannot be reduced below 2.5% as concentrations of acrylamide less than this value give gels which are too weak for formation in slabs of the type described here.

It can be seen that the data for two of the fragments produced by restriction endonuclease EcoR1 digestion of λ DNA deviate rather strikingly from the expected, as shown in Fig. 2. Fragments 4 and 5 have molecular weights of 3.04 and 3.54 $\times$ 10^6, respectively (see legend to Fig. 1), but fragment 4 runs slow (hardly distinguishable from fragment 3, molecular weight 3.54 $\times$ 10^6) whereas fragment 5 runs fast, ahead of fragments 3 and 4. The unpredicted behavior of these fragments is further complicated by the change in their relative positions if the gel run is made at 22° rather than at 4°. An *Eco*R1 digest of λ DNA fractionated at 22° is shown in Fig. 3. Fragment 4 now runs relatively faster as compared to the 4° separation and is difficult to resolve from fragment 5. An explanation of the anomalous behavior of these fragments with respect to molecular weight, and the changes in their relative positions as a function of temperature, can only be surmised, although it is clear that factors other than molecular weight alone are involved, such as rigidity. Such properties would depend on base sequences in the DNA,[7] and in this respect it is interesting to note that λ DNA has been shown to contain regions of highly atypical G-C content.[8]

Also run on the 3.5–7.5% gel illustrated in Fig. 3 are restriction endonuclease *Eco*R1 digests of SV40 DNA (i.e., linearized SV40 DNA) and adenovirus type 2 DNA. A double logarithm plot for the fragments on this gel is given in Fig. 4. With the exception of the two highest molecular weight fragments, the points all lie in the central portion of the sigmoid curve relationship and closely resemble a straight line. Thus, although DNA base sequence effects as discussed above preclude the accurate estimation of molecular weights using this system of gradient gel electrophoresis, Figs. 2 and 4 show that approximate molecular weights for unknown DNA species may be deduced if they are run in parallel with an appropriate set of standards, e.g., characterized restriction endonuclease digests of viral DNAs such as those described here.

[7] R. S. Zeigler, R. Salomon, C. W. Dingman, and A. C. Peacock, *Nature (London)*, *New Biol.* **238**, 65 (1972).
[8] A. D. Hershey, ed., "The Bacteriophage Lambda," Cold Spring Harbor Lab, Cold Spring Harbor, New York, 1971.

[40] Use of Preparative Gel Electrophoresis for DNA Fragment Isolation

By MARSHALL H. EDGELL and FRED I. POLSKY

Large amounts of DNA may be fractionated with high resolution by means of preparative electrophoresis utilizing discontinuous collection.

METHODS IN ENZYMOLOGY, VOL. 65

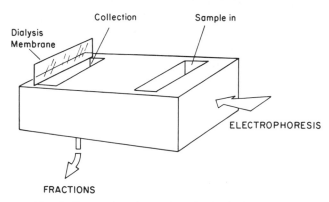

FIG. 1. Preparative electrophoresis: the basic concept.

A very sharp band of DNA when electrophoresed into a collection chamber, which is well mixed and through which there is a continuous flow of collection buffer, will emerge from the chamber with a sharp leading edge but with an exponential tail due to continuous dilution in the collection chamber. In order to avoid this problem we limit the electrophoresis to a suitable period of time and then drain the contents of the collection chamber into a fraction collector, refill the chamber and resume electrophoresis.[1] This cycle can be repeated under control of a programmable timer which switches the power to the electrophoresis power supply and to the fill and drain pumps. The high resolution attained with this collection system does not appear to be reduced significantly by going to very large gel electrophoresis slabs.

The Preparative Apparatus

The electrophoresis apparatus is designed to generate a large rectangular slab with a well at one end in which to apply the sample and a well at the other end in which to collect the fractionated DNA (Fig. 1).

Nucleic acids migrate very rapidly across the collection chamber, since it contains only buffer so the apparatus must support a dialysis membrane across the rear of the chamber to retain the fractionated DNA. This dialysis membrane (D) is clamped between two gasketed pieces of Lucite (A, B) which form the collection chamber (Fig. 2). The chamber is filled and drained through a stainless steel pipe (E). The assembled collection chamber (F) is placed in the electrophoresis apparatus and the drain pipe (E) is sealed against leaks with an O ring compression fitting (G) (Fig. 3). Agarose gels do not adhere well to the plastic walls of the apparatus, so

[1] F. Polsky, M. H. Edgell, J. G. Seidman, and P. Leder, *Anal. Biochem.* **87**, 397 (1978).

prior to pouring the gel we glue strips of cellulose paper (Whatman 3MM) to the floor and walls with silicon glue to provide a matrix into which the gel can bind. It is important to let the agarose gel set well (about 4 hr minimum), otherwise considerable deformation can occur during electrophoresis.

Electrophoresis is carried out for a suitable period of time and then stopped. The collection chamber is drained into a fraction collector and then the chamber is refilled and electrophoresis is resumed. These operations are under control of an electronic timer (Chronometrics, 2614 Silverdale Dr., Silver Spring, Maryland; model C-502). The collection chamber should not over- or underfill. Therefore, two electrodes, one in the collection chamber (I) and one in contact with the gel surface (H), detect conductivity between the chamber and the gel and signal the timer to cease filling the chamber. We find inexpensive clock motor peristaltic pumps suitable for filling and draining (Markson Science Inc. Box 767, Del Mar, California; model 13002 or 13003).

Buffer is recirculated between the two buffer chambers by means of a small pump. We prevent the establishment of a hydraulic head by connecting the two chambers with a 2-cm plastic tube (M). A chloride-free buffer is preferable to minimize chlorine formation during long runs. We use 40 mM Tris-acetate (pH 7.8), 5 mM sodium acetate, and 1 mM EDTA.

All of our experience has been with agarose gels. It is possible, how-

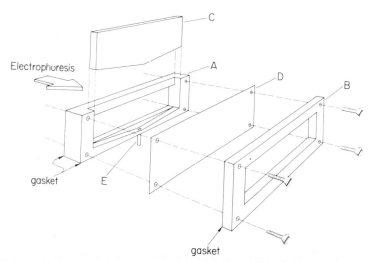

FIG. 2. Collection chamber detail. The collection chamber well former (C) is assembled in place between the gasketed end pieces (A and B). It is removed after the gel solidifies to form the collection chamber. The analysis membrane (D) is wet when the chamber is assembled. Sample collection and chamber refilling is through the stainless steel tube (E).

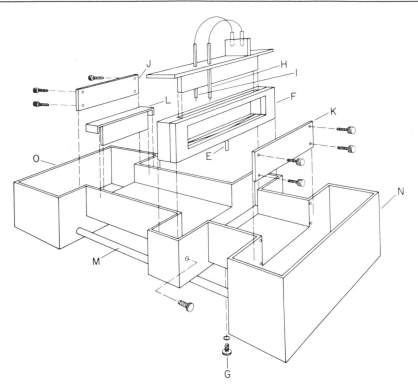

FIG. 3. Preparative electrophoresis apparatus. This is a sketch of the apparatus we are currently using. Detailed drawings are available from Dr. Edgell. The assembled collection chamber (F) is installed and the drain pipe (E) sealed with the O ring compression fitting (G). Temporary end pieces (J, K) are installed and gel is cast in the shallow (6 × 25 × 15 cm) bed so created. A sample well is generated by means of a well former (L). After the gel solidifies pieces J, K, L, and C are removed and the level sensing assembly (H and I) is installed. Lids cover the buffer chambers (N and O) and the gel during electrophoresis. Tube M reduces the hydraulic head between the buffer chambers generated by buffer recirculation.

ever, to polymerize acrylamide in this apparatus without difficulty. One simply carrys out the polymerization in a nitrogen or CO_2 atmosphere. This may be achieved with lids which allow one to purge to oxygen from the apparatus or by covering the apparatus with a cardboard box with a few holes near the bottom and blowing nitrogen in to displace the oxygen.

Operating Conditions

The voltage applied across the collection chamber "pumps" the liquid from the chamber at a rate proportionate to that voltage (presumably due to electroendoosmosis). As a consequence of this effect, we have been operating the preparative apparatus at about 40 V, 150 mA (~1 V/cm).

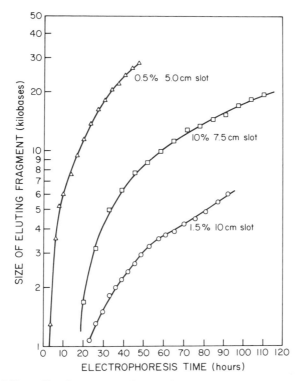

FIG. 4. Mobility calibrations. *Eco*R1 digests of mouse DNA were electrophoresed under different conditions. The plot is of the size of the DNA fragments eluting after various times of electrophoresis. The sizes were determined by analytical gel electrophoresis with λ *Eco*R1, pRSF2124 *Eco*R1, and pSC101 *Eco*R1 fragments as standards. The time for sample collection and refilling the chamber has been subtracted from the data. All the gels were 2 cm thick. The buffer was 40 mM Tris acetate (pH 7.8), 5 mM sodium acetate, and 1 mM EDTA. The gels were run at ~1 V/cm (electrode to electrode distance). The slot distance refers to the distance between sample and collection wells.

During long runs condensing moisture can short the level sensing electrode to the dialysis membrane. We solve this problem by leaving the lid slightly ajar near the collection chamber to reduce condensation. We also place a sealed glass capillary between the electrode and the dialysis membrane to prevent physical contact between the electrode and the membrane.

The resolution as a function of DNA fragment size that one sees with this system differs from that obtained with analytical systems, since each fragment is run to the bottom of the gel. We vary the resolution by altering both the agarose gel concentration and the distance of the sample well from the collection chamber (Fig. 4). For example, the time between

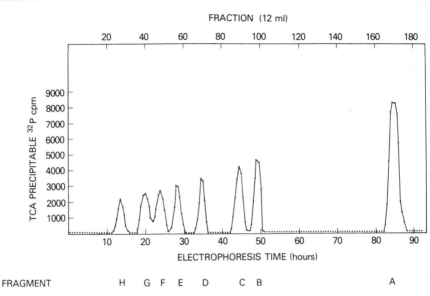

FIG. 5. Lamdoid *Eco*R1 fragments eluted from preparative gel. 4.5×10^5 cpm of TCA-precipitable DNA and 5 mg of high molecular weight calf thymus DNA were digested to completion with *Hin*dIII and *Eco*R1, respectively. The reactions were stopped by heating to 70° for 10 min and extracted with phenol buffered with 20 mM Tris and 1 mM EDTA, after making the restriction reaction mixtures 1 M NaCl. The digests were mixed and coprecipitated with ethanol, resuspended in 10 ml preparative gel electrophoresis buffer, which was 5% sucrose and 0.015% bromphenol blue. The sample was applied to the gel in a sample well formed 8 cm from the elution chamber and allowed to enter the gel at 3 V/cm which drew 3 mA/cm². When the tracking dye had entered the gel the voltage was increased to 10 V/cm which drew 6 mA/cm² for the duration of the analysis. The 12-ml elution chamber was drained and refilled at intervals of 30-min electrophoresis time. Sample aliquots were counted after TCA precipitation onto Millipore filters.

elution of a 2 and a 3 kb fragment can be increased from 2 hr (0.5% gel, 5.0 cm) to 20 hr (1.5% gel, 10 cm).

This preparative apparatus has a high capacity for DNA. We have run as much as 20 mg of *Eco*R1-cut mouse DNA on a 2-cm thick gel (6.5 μg/mm²) without any detected loss in resolution. The carrying capacity of the gel depends of course on the sequence complexity of the DNA which one is running. The carrying capacity of the gel (μg/mm²) should be determined by running a dilution series of the DNA mixture in analytic gels and using the staining pattern to indicate the concentration at which unacceptable overloading occurs. The high resolution recovery of a lamdoid phage DNA digest shown in Fig. 5 was carried out in the presence of 5 mg of *Eco*R1-digested calf thymus DNA. The recovery of DNA has been high, 80% or better,[1] but we do not often get this data quantitatively from our runs. We have not found it necessary to apply any reverse voltage just prior to each collection. This may be due to the low voltage gradient forced on us by the effects described earlier. Preparative runs of complex DNA mixtures extend from 3 to 7 days.

Processing the Fractions

We usually find the fractions contaminated with bacteria ($\sim 10^4$ CFU/ ml). Our attempts to eliminate this problem have so far been unsuccessful. Consequently we add a few drops of chloroform to the samples shortly after collection. The DNA appears to remain intact. Once the fractions of interest have been identified those fractions are pooled, phenol-extracted, precipitated by ethanol with tRNA carrier,[1,2] and stored frozen. Fractions stored in the cold as long as 2 months prior to identification have been shown to possess mostly intact sticky ends as assayed by ligation and high levels of transformation activity (*B. subtilis* DNA).

We have been using the preparative electrophoresis apparatus to fractionate pools of mouse *Eco*R1 fragments prefractionated by other means.[2] We identify the fractions of interest by coprecipitating aliquots from the preparative electrophoresis fractions with tRNA, running the redissolved DNA on analytical gels, transferring the DNA to nitrocellulose,[3] and probing the filters with nick translated sequences. An *Eco*R1 fragment carrying an adult β-globin gene (C57B1/6/Hbbs) has been fractionated in this fashion (Fig. 6).[4] This preparative electrophoresis system has also been used to fractionate mRNA mixtures with little degradation, high resolution and excellent recovery (G. Wertz and M. H. Edgell, unpublished data).

We have used the apparatus to purify DNA fragments cloned in phage or plasmids. In this case, where one is interested in only one or a few fragments, we can obtain exceptionally clean fragments overnight by running a short distance at 2–4 V/cm. These fragment preparations contain little if any cleavage or ligation inhibitors and very little cross contamination with other sequences.

An Alternate Apparatus

Since this article was written, we have designed and tested a considerably simpler apparatus. Plans are available from one of the authors

[2] S. M. Tilgman, D. C. Tiemeier, F. Polsky, M. H. Edgell, J. G. Seidman, A. Leder, L. W. Enquist, B. Norman, and P. Leder, *Proc. Natl. Acad. Sci. U.S.A.* **74**, 4406–4410 (1977).
[3] E. M. Southern, *J. Mol. Biol.* **98**, 503–517 (1975).
[4] F. Rougeon and B. Mach, *Gene* **1**, 229–239 (1977).

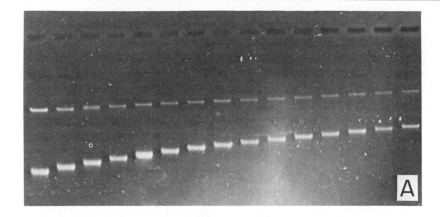

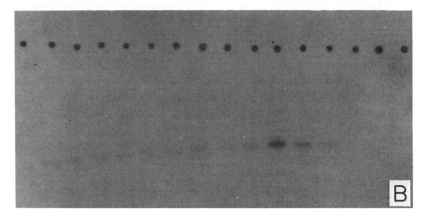

FIG. 6. Assay for an adult β-globin gene. An EcoR1 digest of C57B1/6 mouse DNA (40 mg) was fractionated by RPC5 and fractions which had been shown to contain β-globin sequences were pooled. This pool was electrophoresed on a 2-cm preparative 0.7% agarose gel. The sample well was 5 cm from the collection chamber. 500 λ aliquots were taken from every other 10-ml fraction, coprecipitated with 40 μg of tRNA and electrophoresed in a 1% analytical agarose gel. The even fractions from 30 to 60 were applied and electrophoresed for 1 hr. Then the even fractions from 62 to 90 were applied to the same wells and electrophoresis was continued. The DNA was visualized by ethidium bromide staining (A). Fractions containing β-globin sequences were detected (B) by transferring the DNA to a nitrocellulose membrane and probing with a cDNA clone containing β-globin sequences (pCR1.M$_\beta$Gg;[4]). The hybridization was probe driven and the probe had a specific activity of 50 counts/pg. The nitrocellulose membranes with bound DNA were pretreated with Denhardt's solution (0.02% Ficoll, 0.02% pyrrolidine, and 0.02% bovine serum albumin) for 1 hr. The hybridization reaction contained 0.1% SDS, 6X SSC, 3X Denhardt's solution, 50 μg/ml E. coli DNA, 50 μg/ml salmon sperm DNA , 20 μg/ml poly(A), 20 μg/ml poly(C), and 50 μg/ml tRNA. After hybridization the membranes were washed three times for 30 min at 53° in 0.1% SDS, 0.1 X SSC, and then twice for 30 min at 53° in 0.1 X SSC. Autoradiography was carried out at −70° with a Cronex intensifying screen.

(M. H. Edgell). The new apparatus retains the high capacity and resolution of the apparatus described here but requires less than half the shop time to construct and appears even more trouble-free to operate. It is even possible to use existing laboratory equipment to put together an acceptable apparatus using the alternate design concepts. This new design is available commercially from Doran Instruments, 21 Conant St., West Concord, Massachusetts 01742.

Acknowledgments

The formative work giving rise to this preparative device was carried out by the authors in the laboratory of Dr. P. Leder, NICHD, National Institutes of Health, with the aid of J. G. Seidman.

[41] RPC-5 Column Chromatography for the Isolation of DNA Fragments

By R. D. WELLS, S. C. HARDIES, G. T. HORN, B. KLEIN, J. E. LARSON, S. K. NEUENDORF, N. PANAYOTATOS, R. K. PATIENT, and E. SELSING

Introduction

RPC-5 column chromatography is a useful tool for the high resolution fractionation of single-stranded DNA or RNA oligonucleotides, double-stranded DNA restriction fragments, and the complementary strands of DNA restriction fragments. The technique can be used for fractionation of milligram quantities of nucleic acids. Good recovery is observed and the nucleic acids are recovered in a high state of purity. The order of fractionation is not always the same as produced by gel electrophoresis. Therefore, the sequential use of these two methods is more powerful than either method alone. In some cases RPC-5 column chromatography can completely resolve duplex restriction fragments of the same size.

This article describes the technical details of the fractionation procedures. Emphasis is on practical laboratory methodology. The details of experimental results and the theory behind some separations are presented elsewhere (noted in text).

Fractionation of Single-Stranded DNA and RNA Oligomers

This laboratory first became interested in the use of RPC-5 column chromatography several years ago when it became necessary to devise a high resolution preparative technique for fractionating oligonucleotides in

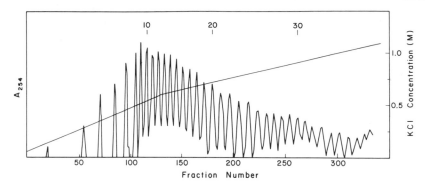

FIG. 1. RPC-5 column chromatographic separation of rA_m oligonucleotides. rA_n (Sigma; molecular weight $\sim 100,000$) was digested to oligonucleotides by incubation in $0.3\,M$ KOH at $37°$ for 30 min. The reaction mixture was dialyzed versus 10 mM Tris-HCl and 0.1 mM EDTA (pH 8.0). This removes smaller oligonucleotides (less than ~ 8 units in length). The dialyzed oligonucleotide mixture was loaded onto a 2.5×100 cm column of RPC-5 which was run essentially as described previously (see references in text). The oligomers were eluted with a gradient of KCl (concentration indicated on figure). This is not a high pressure column; a Buchler peristaltic pump was used. Fractions (15 ml) were collected at 8-min intervals. Oligonucleotide sizes (as indicated in the figure) were determined by analytical 20% polyacrylamide gel electrophoresis.

a homologous series.[1,2] Single-stranded homopolymers, such as dT_n, were degraded with pancreatic deoxyribonuclease to give a series of molecules with varying chain length. We wanted to isolate molecules of defined chain length in the region of 20–30 nucleotides in length. Prior studies with DEAE–cellulose chromatography show that resolution was lost at approximately 10–15 nucleotides in length.[3]

A typical elution profile produced by RPC-5 chromatography of single-stranded oligomers in a homologous series is shown in Fig. 1. In this case, rA_n was digested with alkali to give a random mixture of oligomers. The oligomers were fractionated according to chain length. Twenty percent polyacrylamide gel electrophoresis[4] was used for the determination of the chain length of these oligonucleotides. That the oligonucleotides in fact differed in chain length by 1 monomer unit was previously proved by chemical methods.[1] In a number of cases, fractionation of oligonucleotides up to 60 in length has been observed.[5,6]

[1] J. F. Burd, J. E. Larson, and R. D. Wells, *J. Biol. Chem.* **250**, 6002–6007 (1975).
[2] J. B. Dodgson and R. D. Wells, *Biochemistry* **16**, 2367–2374 (1977).
[3] H. G. Khorana and W. J. Conners, *Biochem. Prep.* **11**, 113 (1966).
[4] J. F. Burd and R. D. Wells, *J. Biol. Chem.* **249**, 7094–7101 (1974).
[5] E. Selsing, J. F. Burd, and R. D. Wells, unpublished work.
[6] G. C. Walker, O. C. Uhlenbeck, E. Bedows, and R. I. Gumport, *Proc. Natl. Acad. Sci. U.S.A.* **72**, 122–126 (1975).

Single-stranded homo-oligonucleotides of the following types have been satisfactorily resolved[4-6] in large amounts: dA_m, dT_m, dC_m, rA_m, rU_m, and rC_m. However, until recently it has not been possible to separate dG_m or rG_m oligonucleotides. Our prior attempts to fractionate dG oligomers according to chain length on RPC-5 columns, even in the presence of 8 M urea or other denaturing agents, were futile. Recent studies have shown that dG_m oligomers can be satisfactorily resolved on RPC-5 columns if they are run in 12 mM NaOH (pH approximately 12).[5] The impetus to attempt this fractionation came from the success of other workers[7] in the separation of complementary strands of DNA restriction fragments by RPC-5 column chromatography at high pH.

dG_n (approximately 10 S in size) was degraded by partial depurination with 50% acetic acid at 37° for 48 hr. The phosphodiester backbone was then cleaved with NaOH at the depurinated sites.[8] Approximately 200 OD_{260} units of this partially degraded dG_n was loaded onto a 1.4 × 20 cm RPC-5 column which was run essentially as described.[1] When a linear gradient (1 liter) of 0.1 M KCl to 1.0 M KCl containing 12 mM NaOH (pH ~12) was applied to the column, an elution profile similar to other oligomer separations on RPC-5 was obtained (see Fig. 1). Approximately 35 peaks were discernible in the profile; thus, we are optimistic that it will be possible to identify and characterize dG oligonucleotides which vary in size by one nucleotide unit.

Fractionation of Duplex DNA Restriction Fragments

The previously described uses of RPC-5 chromatography as well as its successful history for tRNA fractionations[9,10] led to its application as a fractionation procedure for duplex DNA restriction fragments.[11-14] In order to determine how to improve our large-scale fractionations, we devised a small-scale analytical system which was used to study column behavior under various conditions.[14] This analytical setup, which is de-

[7] H. Eshaghpour and D. M. Crothers, *Nucleic Acids Res.* 5, 13–21 (1978).

[8] S. J. Harwood and R. D. Wells, *J. Biol. Chem.* 245, 5625–5634 (1970).

[9] R. L. Pearson, J. F. Weiss, and A. D. Kelmers, *Biochim. Biophys. Acta* 228, 770–774 (1971).

[10] A. D. Kelmers, H. O. Weeren, J. F. Weiss, R. L. Pearson, M. P. Stulberg, and G. D. Novelli, this series, Vol. 20, Part C, pp. 9–34.

[11] S. C. Hardies and R. D. Wells, *Proc. Natl. Acad. Sci. U.S.A.* 73, 3317–3121 (1976).

[12] A. Landy, C. Foeller, R. Reszelbach, and B. Dudock, *Nucleic Acids Res.* 3, 2575–2592 (1976).

[13] R. D. Wells, R. W. Blakesley, J. F. Burd, H. W. Chan, J. B. Dodgson, S. C. Hardies, G. T. Horn, K. F. Jensen, J. E. Larson, I. F. Nes, E. Selsing, and R. M. Wartell, *Crit. Rev. Biochem.* 4, 305–340 (1977).

[14] J. E. Larson, S. C. Hardies, R. K. Patient, and R. D. Wells, *J. Biol. Chem.* 254, 5535 (1979).

scribed below, was ideal for fractionating between 10–100 μg of DNA. The information gained in these studies was then applied to our large scale system which is described later.

Technical Description of Analytical System

A schematic outline of a small-scale system used for the fractionation of duplex DNA restriction fragments on RPC-5 is shown in Fig. 1. This analytical-scale system was designed for short runs using small amounts of DNA to facilitate the study of the influence of various parameters on resolution of DNA fragments. The results of these studies are described elsewhere[14] and are reviewed in "Factors Affecting Fractionation."

The components of the analytical system (Fig. 2) are relatively inexpensive, totaling less than $2000 (except for the spectrophotometer), and are generally available to most research workers.

A 0–1000 psi pressure gauge and a Milton-Roy minipump (purchased from Anspec) are attached by the three-way connector (D in Fig. 2) to the top of the column. The tubing (including the column), fittings, and flanging iron were manufactured by Altex and were purchased from the Anspec Co. Inc., 4126 Packard Road, Ann Arbor, Michigan. Do not force these fittings very tightly since they may strip; finger tight is sufficient. Most

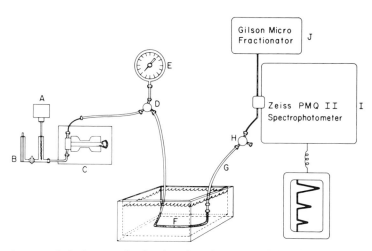

FIG. 2. An analytical system for fractionation of DNA restriction fragments. The system is comprised of the following components: (A) overhead stirring motor, (B) gradient maker (20 ml capacity), (C) Milton Roy high pressure pump, (D) high pressure three-way connector, (E) pressure gauge, (F) column in high-pressure Teflon tubing, (G) short connecting tubing, (H) high-pressure three-way connector (to allow the purging of air bubbles), (I) Zeiss PMQ II spectrophotometer equipped with 8-μl flow cell, (J) Gilson fraction collector. The source of the components is described in the text.

leakage occurs from improper flanging. We have found that plastic tubing does not last if connected directly to the pump. Thus, we have a 4-inch length of stainless steel tubing linking the pump to the Teflon tubing via a Swagelok adaptor. The gradient is formed by a two-compartment gradient maker (manufactured by Norman Erway, Oregon, Wisconsin); the low salt side of the gradient maker is stirred with a spiral glass rod powered by an overhead stirring motor (purchased from Polyscience Corp., Niles, Illinois).

Absorbance of the effluent is constantly monitored with a Zeiss PMQ II spectrophotometer which is equipped with an 8-μl flow cell with a 1-cm path length (purchased from Altex) and an Esterline-Angus recorder. Other monitoring devices may be employed, which are substantially less expensive than the Zeiss, however this unit was used because of its availability.

The three-way connector (H in Fig. 2) which is on the low-pressure side of the column allows air bubbles to be flushed from the flow cell without detaching the column. The blind end plug in H (Fig. 2) is replaced with a Luer syringe adaptor for this purpose.

Temperature is controlled by immersing the column in an ordinary thermostated water bath. For temperatures below ambient, the column is run through a water-filled, jacketed condensor which is attached to a circulating cooling fluid.

Fractions are collected in a Gilson Microfractionator (purchased from Gilson Medical Electronics, Middleton, Wisconsin) (J in Fig. 2). This fraction collector was found to be most appropriate because of its capability of handling fractions with small volumes and because of its compactness.

Preparation of Resin

RPC-5 was purchased from Miles Laboratory, Elkhart, Indiana (lot 7) or was a generous gift from Dr. G. D. Novelli (Oak Ridge, Tennessee). We have been advised that RPC-5 is no longer commercially available. However, Plascon 2300 (from which RPC-5 can be easily prepared[9]) may be obtained in limited quantity from Dr. Novelli. Quantities of good RPC-5 are also available from many investigators around the world who were involved in tRNA fractionations in recent years. Thus, there does not appear to be a problem in obtaining RPC-5 at the present time for investigators who are interested in pursuing this technique. In addition, alternative resins have been studied[14] and further work on this problem is continuing.

Before packing, the RPC-5 resin is suspended in approximately 5 volumes of the lower concentration eluting mixture (see below) which also contains 0.06% sodium azide. Since the Adogen 464 can be stripped from the Plascon 2300 in low salt concentrations, it is advisable to maintain relatively high ionic strengths. The resin is customarily stored in the above solution. After shaking or stirring the suspension for 3–18 hr, just prior to packing, it is degassed with a water aspirator with simultaneous magnetic stirring.

Packing of the Analytical Column

The column is constructed from high-pressure Teflon tubing (1.5 mm i.d. by approximately three times the desired length of the final resin bed) which was flanged and fitted with threaded high pressure Altex fittings on both ends. These fittings screw into either a three-way connector (D in Fig. 2) or a coupling (at the bottom of the column).

After allowing the degassed suspension to settle, the thick slurry of resin is drawn up into a small syringe (1 or 3 ml). The slurry is then syringed into the column through a Luer adaptor and coupling while the other end of the column is open to the atmosphere. It is important that no air bubbles get into the column; if this occurs, the column must be refilled. It is difficult to add more resin to the column after it is once packed, but the column can be shortened by pumping out some resin. A settled slurry usually produces a packed resin bed which is approximately one-third the length of the tubing. If a thicker slurry is used, the resin bed will be somewhat longer.

The bottom of the filled column is screwed approximately halfway into a coupling. A circle of Whatman No. 50 filter paper is inserted into the coupling followed by a short flanged 0.3 mm i.d. length of tubing (G in Fig. 2) with fittings on both ends. The filter paper when wedged between two flanges in this fashion acts as the column bottom and will restrain the resin even with a pressure drop of 1000 psi over the bed. The top of the column is then attached to the pump and pressure gauge through a three-way connector, taking care that no air remains in the closed system.

The column should be packed and equilibrated with about 10 ml of the lower concentration eluting solution. For a resin bed which is 10–15 cm in length, an initial flow rate of approximately 15 ml/hr is reasonable. If the pressure is kept as low as an acceptable flow rate will permit, the life of the column will be prolonged. If a column is run at higher pressures, it packs more tightly and subsequent runs will also require higher pressures.

Preparation of the Eluting Solutions

A number of salts can be used as eluants for RPC-5. For DNA restriction fragments we have tested KCl, NaCl, sodium acetate, and potassium acetate.[14] The results of these determinations are described below. All salt solutions contained 10 mM Tris buffer at the appropriate pH (see later), 2 mM sodium thiosulfate, and 10 mM EDTA. A beneficial effect of the inclusion of EDTA is to reduce the back pressure of the column and thereby increase its lifetime; also as described in "Preparation of DNA Sample," the presence of EDTA protects the DNA sample from possible nucleolytic degradation.

The eluting solutions are passed through Millipore filters into sterile vessels to protect the pump fittings and the column from dust particles and to protect the DNA from bacterial degradation. All eluting solutions are degassed with a water aspirator to prevent air bubbles from forming in the column and the flow cell.

Chloride ion is corrosive toward the inner metal components of the high pressure pump; thus, the pump should be flushed with water immediately after a run in which this ion is employed.

The source of the eluting salt can affect the performance of the RPC-5 fractionation. Satisfactory results[14] were obtained with sodium acetate, KCl, and NaCl purchased from Mallinckrodt, Inc. and with potassium acetate purchased from Baker Co. When sodium acetate purchased from Drake Bros., Inc. was used a loss in resolution, sharpness, and reproducibility of the profile was observed. Interestingly, it did not appear that the Drake Bros. sodium acetate was more impure than that obtained from two other manufacturers when analyzed for trace metal ion contamination. The reason for this variation is unknown.[14]

Preparation of DNA Sample

The progress of the restriction enzyme digest should be checked by analytical polyacrylamide gel electrophoresis to ensure completion of the reaction. The remainder of the DNA sample should be extracted with phenol to remove protein and then with ether to help remove phenol. The DNA solution should be dialyzed into 10 mM Tris-HCl (pH 6.8), 2 mM sodium thiosulfate, and 0.1 mM EDTA. The use of sterile conditions and the incorporation of EDTA is to protect the DNA from nuclease digestion. These steps were found to be necessary in order to obtain reproducible and sharp RPC-5 profiles. With these precautions, a DNA restriction enzyme digest can be stored for many months and still give a sharp column profile.

Bacterial contamination is avoided by treating the gradient maker, the syringes used for filling the column, and the resin with 0.06% sodium azide. All glassware is autoclaved prior to use, and only DNA restriction endonucleases known to be free of other nuclease activities are used for the preparation of fragments. Most of the DNA restriction enzymes used for fractionations were purified and characterized in our laboratory.[15]

[15] S. C. Hardies, R. K. Patient, R. D. Klein, F. Ho, W. S. Reznikoff, and R. D. Wells, *J. Biol. Chem.* **254**, 5527 (1979).

When impure DNA restriction enzymes (containing exonucleases) were used for the preparation of fragments, the high resolution fractionation was not observed. Instead, a great deal of overlap between peaks was found.

Loading the DNA Sample

The size of the load of DNA which is acceptable for the analytical column varies with the number and the size of fragments in the digest. In our analytical system (Fig. 2), 0.5 μg is a lower limit of detection for a given fragment. For a digest of a 4600 bp plasmid DNA containing 17 fragments ranging in size from 43–850 bp (see below), 50 μg is an appropriate load. The fragments are well resolved and are easily detectable in the absorbance profile and by polyacrylamide gel analysis. Twice this amount of DNA can be loaded successfully; an upper limit has not been established. When only a few components are to be separated, the sample should contain at least 1 μg of the smallest fragment.

The DNA solution is loaded directly onto the column with a 1-ml syringe with drawn-out plastic tubing (10–15 cm in length) attached to the needle. The tubing is inserted into the column, and approximately 200 μl of the elution solution is removed. The DNA sample, in 50–200 μl, is drawn into the syringe taking care not to entrap any air bubbles in the tubing. With the tip of the drawn-out tubing just below the liquid level in the column, the syringe is carefully withdrawn as it is emptied. Filling the column is completed with a low salt elution solution.

Undoubtedly the biggest problems in the operation of the analytical column are the presence of air bubbles and leaks which are due to poor flanges at the junctions. If air bubbles enter the column, they are very difficult to remove without the loss of sample. Air bubbles in the flow cell are equally difficult to remove and can obscure the absorbance profile. Fortunately, these problems can be avoided by careful technique.

Running the Analytical Column

The choice of gradient may depend on circumstances as discussed in "Factors Affecting Fractionation in RPC-5." For most cases we recommend a 40-ml gradient from 0.55 to 0.75 M in KCl run at 43° at a pH of 6.8. Under these conditions a peak width of about 1 ml can be expected; therefore we recommend 250-μl fractions. Since a column of 10–15 cm can be run at 15 ml/hr, the entire analysis can be completed in about 3 hr. If sodium acetate is to be used as the eluting salt, a 40-ml gradient between 1.3–1.7 M is recommended. The salt range within which the DNA elutes can vary with the batch of resin, the load, and the column dimensions.

Columns can be rerun. After eluting the restriction fragments, a column can be purged with approximately 10 ml of the eluting solution with either the chloride concentration at 1.0 M or the acetate concentration at 2.1 M. Some absorbing material may elute in this wash. The column is then reequilibrated in the lower salt elution solution (~10 ml). A uv-absorbing peak is commonly observed shortly after the low salt equilibration is begun. This is probably due to the accumulation on the column of one or more of the absorbing components of the eluting solutions. It is not DNA as judged by uv spectra and polyacrylamide gel electrophoresis.

Technical Description of Preparative System

The preparative column system described in this section is essentially a scale-up of the analytical system described above. Also, columns set up for tRNA fractionations may be satisfactory. Alternatively, the setup described in the legend to Fig. 1 may be applicable and has the advantage of developing less back pressure. A system can be constructed from commonly available and relatively inexpensive components with the constraint that the column must withstand the attendant pressure. The use of metal columns has been described.[12]

The main component of our setup is a jacketed 70 × 0.9 cm Altex column (#252-34) fitted with an adjustable plunger; the manufacturer states it is capable of withstanding 500 psi. The requirement of temperature control has been reported.[14] We have used a thermostated, Haake type F heater bath or an LKB cooling system to circulate water at the required temperature through the jacket. The same pump, pressure gauge, lines, and fittings were used as described above for the analytical system. The gauge is essential for setting appropriate flow rates with safe pressures.

Special attention must be paid to the gradient maker. Because of the slow flow rates employed (often <0.5 ml/min) and the large volume of the gradient, gradient makers with short direct connections frequently exhibit diffusion between the chambers which results in equilibration of the two sides. We recommend connecting the chambers by means of a siphon to eliminate this problem. The column profile is continuously recorded with a 0.1-ml, 3-mm flow cell in a LKB Uvicord II type 8303A detector with the lamp and filter set at 254 nm. The TYP 6520-5 recorder is operated by the LKB Uvicord II 8300 control unit. Other instruments suitable for monitoring absorbance or transmittance at 254 or 260 nm could be substituted. Absorbance profiles are easier to quantitate. Fractions are collected in the cold (4°) in an LKB Ultrorac 7000 fraction collector mounted in a Colora cold box. This may help guard against degradation of the fragments over the long run times (8–48 hr) of the experiments.

The amount of resin used depends on the amount of DNA to be fractionated. The upper ratio limit may be around 0.5 mg DNA/ml resin bed, but excess resin does not detract from resolution and may even improve it. We have fractionated between 0.1–5.0 mg of DNA on a 61.5 × 0.9 cm, 39-ml resin bed with good results. We have also fractionated up to 30 mg of DNA on a 120 × 0.9 cm column. With less than approximately 1 mg of DNA per 60 ml of resin the sample losses increase, so we would generally recommend staying in the range of 1–10 mg of DNA per 30 ml of resin.

For packing a column, the dry resin is first stirred (usually overnight) with the starting salt solution and then degassed, allowed to settle into the thickest possible slurry, poured into the column, and packed under flow. The column can be packed by layers, or a column extension can be used if one is available which will contain the back pressure. How tightly the resin is packed will determine the pressure at which the column will have to be run (column diameter, resin bed length, and type of end fitting used will also contribute). We have not found that resolution is influenced by pressure. The resin should be packed under pressures similar to those to be used during elution. We have generally used 200–400 psi. During and after packing, the resin should be equilibrated with several column volumes of starting salt solution.

We reuse a column many time with occasional replacement of the top 0.5 cm of resin if discoloring occurs. The column is washed with a concentrated salt solution (1.0 M KCl or 2.0 M sodium acetate) and stored between runs in high salt containing 0.02% sodium azide.

The end of the useful life of a resin bed is indicated when back pressure approaches the limits of containment. The resin can be regenerated without stripping the Adogen 464 by washing with 0.2 M NaOH plus 0.2 M salt[11] or stripped and recoated.[9] After many uses, the resin may show decreased capacity and peaks may elute at lower salt concentrations indicating the accumulated loss of exchanger (Adogen 464), thus necessitating recoating. Whenever possible, avoid exposing the resin to salt concentrations of less than 0.4 M to prevent accidental stripping.

The choice and preparation of eluting solutions is the same as described for the analytical system. We have used gradients with total volumes of between 1 to 2 liters for a salt increment of 0.2–0.4 M. A convex gradient may be useful to balance a reduction in resolving power at higher fragment sizes. Conductivity or refractive index measurements of the fractions have been used to monitor the true shape of our gradients. This is particularly helpful in diagnosing problems with a new setup.

After restriction enzyme digestion, the sample is usually extracted with phenol, dialyzed against the starting solution, and loaded through the pump. We have found that inclusion of 0.1 mM EDTA in the sample at all times is necessary for reproducible column performance. The gradient is applied immediately at a flow rate of 40–100 ml/hr cm² (~5 psi/cm bed). Fraction volume depends on the resolution required, but we have found 4 ml appropriate for the column described above.

It is advisable to analyze many of the fractions by polyacrylamide gel electrophoresis. This, in conjunction with the elution profile, will identify the fractions to be pooled for optimum purity and recovery of the required fragment. Gel analysis is important since column artifacts are known which cause peaks to appear in the elution profile which are not due to fragment fractionation.[14,16]

We generally obtain 60–90% recovery of the loaded DNA. Resolution is comparable to or better than that obtained on the analytical systems described above. When the recovered fragments are dialyzed into an appropriate salt solution, they can be digested easily with other restriction enzymes.[11,16] We have not observed any difficulties due to contamination or degradation of fragments during runs under these conditions.

Examples of Restriction Fragment Fractionations

Figure 3 shows a typical fractionation of the analytical system. The sample was a *Hae*III digest of a miniCol E1 derivative, pRZ2[15] which gave 17 fragments ranging in size from 43–850 bp, including three 425 bp fragments. Figure 3 shows that virtually all 17 fragments can be separated from each other when using a KCl gradient at pH 6.8 at 43°. Even the 98 and 102 bp fragments, which differ in size by as little as 4%, were separated.

Large-scale RPC-5 columns have been used for separating quantities of DNA restriction fragments from several sources, including λ*plac*5 DNA,[11] φX174 replicative form DNA,[13] recombinant plasmid DNAs,[16–19] λ DNA,[12] mouse DNA,[20] and Col E1 DNA.[7]

An example of the type of fractionation which can be achieved on a preparative column is shown in Fig. 4. The large-scale fractionation of a restriction digest of pRW307 DNA on a preparative column is illustrated. pRW307 is a chimeric plasmid composed of mCol E1, a tetracycline resis-

[16] S. C. Hardies and R. D. Wells, *Gene* **7**, 1 (1979).
[17] R. K. Patient, S. C. Hardies, J. E. Larson, R. B. Inman, L. E. Maquat, and R. D. Wells, *J. Biol. Chem.* **254**, 5548 (1979).
[18] R. K. Patient, S. C. Hardies, and R. D. Wells, *J. Biol. Chem.* **254**, 5542 (1979).
[19] N. Panayotatos, *Fed. Proc., Fed. Am. Soc. Exp. Biol.* **37**, 772 (1978); N. Panayotatos and R. D. Wells, *J. Biol. Chem.* **254**, 5555 (1979); *J. Mol. Biol.*, in press (1979).
[20] S. M. Tilghman, D. C. Tiemeier, F. Polsky, M. H. Edgell, J. G. Seidman, A. Leder, L. W. Enquist, B. Norman, and P. Leder, *Proc. Natl. Acad. Sci. U.S.A.* **74**, 4406–4410 (1977).

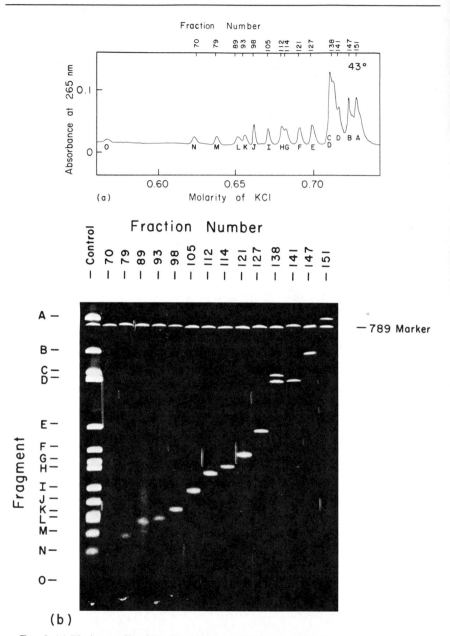

FIG. 3. (a) Elution profile of HaeIII digest of pRZ2 DNA on RPC-5. The fractionation was performed as described in the text at 43° and pH 6.8. The composition of each of the peaks was determined by gel electrophoresis (b) of the fractions indicated across the top of (a). The fragments which were found in each fraction are designated by letters; the sizes of the fragments are the following: A, 850; B, 575; C, 465; D, 425 (three fragments); E, 255; F, 203;

tance element, and T7 DNA.[19] Restriction of this plasmid with HaeIII and EcoRI generates 16 blunt-ended fragments and 4 fragments with one blunt end (HaeIII) and one 4-nucleotide sticky end (EcoRI).[19] In the fractionation shown in Fig. 4, nine out of the twenty fragments were well resolved; the rest were somewhat contaminated by overlapping fragments.

Fragments with EcoRI sticky ends bind to the column more tightly than blunt-ended fragments of the same size, as discussed above. This phenomenon is clearly observed in Fig. 4 for the four fragments (370, 255, 165, and 75 bp in length) that bear single sticky ends. The 370-bp fragment eluted later than a much larger blunt-ended fragment 510 bp in length. Similarly, the 165- and 75-bp fragments also eluted later than expected on the basis of their sizes; the 75-bp fragment eluted 11 fractions later than an 82-bp fragment. Of particular interest is the behavior of the 255-bp fragment. Three fragments comigrated in a single 255-bp band on the control gel; two of them are blunt-ended, but the third carries an EcoRI sticky end. As expected, one of these three fragments eluted later (by approximately 5 fractions) than the other two.

Thus, the presence of sticky ends can be used to an advantage for resolving fragments of identical lengths that could not be resolved by gel electrophoresis.

Figure 5 is a composite of data from three preparative columns used to fractionate restriction fragments which all contain blunt ends. ϕX174 replicative form DNA was digested with HaeIII or with HindII, and pRZ2 DNA was digested with HaeIII. The resolution on two of these columns was reported previously[13,17] in a somewhat different form.

In general, fractionation of the fragments was according to size, however, as has been repeatedly discussed, several properties besides fragment size affect the elution of DNA on RPC-5. For example, note the separation of the three 425-bp fragments in the HaeIII digest of pRZ2 DNA (triangles). With linear gradients as used in Fig. 5 the larger fragments eluted closer together than the smaller ones. This observation suggests that resolution may be improved by employing a gradient with a gradually decreasing slope.

G, 180; H, 169; I, 135; J, 117; K, 102; L, 98; M, 85; N, 69; O, 43. The designation D represents three fragments, each of which is 425 bp in length, that comigrate on polyacylamide gel electrophoresis.

(b) 5% Polyacrylamide gel electrophoretic analysis of fractions from the RPC-5 column (a). Samples (0.050 ml) of the peak fractions shown above were analyzed. The "control" gel shows the unfractionated digest which was applied to the RPC-5 column. A 789 bp marker was coelectrophoresed with the samples to facilitate aligning of the gels. Fragment 0 (43 bp) was too small to be detected by gel analysis under our routine conditions but was consistently observed in the absorbance profiles.

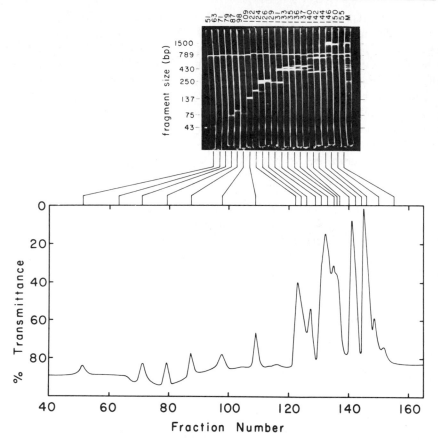

Fig. 4. Fractionation of a HaeIII-EcoRI digest of pRW307 DNA. DNA (2.5 mg) was restricted with HaeIII and EcoRI into 20 fragments ranging between 43 and 1500 bp in size. The products were extracted with phenol and ether and then loaded (in 7 ml) onto a 0.9 × 61 cm RPC-5 column equipped with high-pressure fittings and Teflon lines. The column was eluted at a rate of 1 ml/min with a 1-liter, degassed, linear gradient of 0.5–0.8 M KCl containing 20 mM Tris-HCl, 0.1 mM EDTA (pH 7.2) under approximately 200 psi at 43°. Fractions of 4 ml were collected. The absorbance at 254 nm was monitored in the Uvicord flow cell of an LKB fraction collector. Aliquots (20 μl) of the indicated fractions were analyzed on 5% polyacrylamide tube gels. The gel marked M shows an unfractionated digest. A 789 bp marker fragment was also loaded on every gel for calibration purposes.

Factors Affecting Fractionation on RPC-5

Column Parameters. A recent study[14] has explored a variety of parameters which affect column operation. Among those tested were temperature, salt and pH.

Potassium chloride, sodium chloride, sodium acetate, and potassium

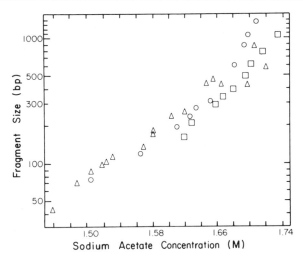

FIG. 5. Preparative fractionation of DNA restriction fragments on RPC-5. The size of DNA restriction fragments is plotted versus the salt concentration at which the peak elution occurred. Three separate experiments are summarized to show the typical behavior of RPC-5 in fractionating double-stranded, blunt-ended restriction fragments. Although details varied between experiments, the usual conditions were as follows: the digested DNA (0.8 to 1.5 mg) was dialyzed against the equilibration buffer, loaded onto a 0.9 × 60 cm column of RPC-5 equilibrated with a solution containing 10 mM Tris-acetate (pH 7.5), 2 mM sodium thiosulfate, and 1.3 M sodium acetate. The DNA was then eluted with 1300 ml of the above buffer solution containing a gradient of 1.30 to 1.85 M sodium acetate. The flow rate averaged 65 ml/hr at a maximum pressure of 300 psi, and 4-ml fractions were collected. Salt concentration was measured by conductivity of the fractions, and fragment elution was visualized by absorbance at 254 nm and by analytical polyacrylamide gel electrophoresis. ○, ϕX174 replicative form DNA digested with *Hae*III; □, ϕX174 replicative form DNA digested with *Hin*dII; △, pRZ2 DNA digested with *Hae*III.

acetate were tested as eluting salts. Columns using KCl elution buffer were run at temperatures from 13° to 53° (in 10° increments) and at pH values from 6.0 to 9.4. The conditions for best resolution with this salt were 43° at pH 6.8. Sodium acetate was investigated at temperatures from 8° to 33° and at pH values from 7.0 to 9.1. The best temperature here was 23°. Changing the pH in this range had little effect; we commonly used pH 8.2. Resolution was much better with KCl than with sodium acetate. There are, however, some instances when sodium acetate may be more effective. The order of elution in sodium acetate is not always strictly according to size.[11,14,17,18] In Fig. 3, the 575-bp fragment elutes from the column later than the 850. On preparative columns,[17] the three 425-bp fragments are separated, two of them eluting later than predicted for their size. There are several other examples of delayed elution. Most of these cases can be related to fragments containing regions of high AT content.[17,18] This characteristic of sodium acetate columns sometimes enables the separation of fragments that coelute in KCl solution.

Sodium chloride behaved in a similar fashion to KCl, and potassium acetate behaved similarly to sodium acetate showing that the anion had the dominant influence.

Size. The size of a DNA restriction fragment is the dominating factor in the fractionation (see Figs. 4 and 5). We presume that this is due to an ion exchange phenomenon between the restriction fragment and the resin.

DNAs as large as intact λ DNA (31×10^6) have been successfully chromatographed. However, for large fragments (above several million daltons) the influence of size becomes less important.[11] This may reflect an increase in relative importance of a hydrophobic component to the fractionation.

Base Composition. It has become apparent[17,18] that a number of fragments bind more tightly to the resin than expected on the basis of their size. Examples of this are shown in Fig. 5. AT-rich fragments elute at higher salt concentrations than fragments of equivalent size which are not AT rich.[17] This was demonstrated by analyzing a number of purified fragments for their nucleotide composition by direct determination of their constituent mononucleotides and by analytical cesium chloride and cesium sulfate density gradient centrifugation. In addition, denaturation mapping studies by electron microscopy indicated that an AT-rich run within an otherwise GC-rich fragment can give rise to delayed elution.

Interestingly, all of the *Hae*III fragments of pRZ2 DNA which elute later than predicted from their size[17] either contain known genetic regulatory sites or specifically bind *E. coli* RNA polymerase.

Sticky Ends. The effect of a single *Eco*RI end on a fragment is shown in Fig. 4 and was discussed above. The role of sticky ends in binding to RPC-5 has been carefully evaluated on the smaller analytical columns (described above) and was reported previously.[14] RPC-5 column chromatography was performed on the products of restriction enzymes producing 2, 3, and 4 base "sticky ends." Among the cases examined, there were examples of both 3' and 5' protruding ends and of various base compositions in the exposed bases. Compared to the blunt-ended *Hae*III fragments as a control, some of the sticky ended fragments (*Eco*RI and *Hin*dIII, both 4 base ends) eluted from the column later. "Sticky ends" produced by some other enzyme had little or no effect. They were: *Hha*I (GCG↓C), *Hae*II (PuGCGC↓Py), *Hin*fI (G↓ANTC), and *Taq*I (T↓CGA). Behavior on the column cannot as yet be correlated with the nature of the ends. However, knowledge of the relative behavior of the different kinds of fragments can be very useful in separating some DNA sequences. Otherwise difficult to separate regions of a molecule may be resolved by using appropriate double digests.

Other Factors. There must be at least one other parameter which influences the fractionation of DNA restriction fragments, since nucleotide analyses on two 425-bp fragments, which separate on RPC-5, are identical.[17] Mass spectroscopic analysis demonstrates a difference in minor nucleotide content between the two fragments (J. L. Wiebers, personal communication). This may be the basis of their separation. Alternatively, the distribution of nucleotides on the two fragments could give rise to different DNA conformations which influence the RPC-5 behavior.

Separation of Complementary Strands of DNA Restriction Fragments

It was recently reported[7] that the complementary strands of the smaller DNA fragments generated by a *Hae* III digest of Col E1 DNA could be separated by RPC-5 chromatography at pH 12.2. The fragments studied were 70, 172, 250, and 440 bp in length. Fractionation of each of these purified fragments under the alkaline denaturing conditions yielded two elution peaks. These workers[7] provided evidence that the two peaks were in fact the complementary strands.

This observation is quite important, since the separation of complementary strands in large quantities has been limited to whole viral genomes or to large DNA restriction fragments. In alkaline cesium chloride gradients small fragments form somewhat diffuse bands. Renaturation of alkali-denatured fragments on heavily loaded polyacrylamide gels has prevented the bulk separation of strands from smaller fragments. Hence, this technique should be quite useful for a variety of applications where large quantities are required.

Figure 6 shows the separation of the complementary strands of a 210-bp fragment generated by *Hin*dII digestion of ϕX174 replicative form DNA. Purification of the duplex 210-bp fragment was described above (Fig. 5). The strand separation technique[7] described in the legend to Fig. 6 is readily reproducible under the conditions tested (i.e., using 7.5 μg of duplex DNA fragment). However, we (N. Panayotatos and R. D. Wells, unpublished) and others (D. Bastia, personal communication) have experienced difficulty in the fractionation of the complementary strands of trace quantities of radioactively labeled DNA restriction fragments. It can be anticipated that further experimental developments will resolve these problems.

Reproducibility

There can be little doubt concerning the reproducibility of the techniques described herein. The high resolution fractionation of single-stranded oligonucleotides according to chain length has been reproduced by a number of workers in this laboratory[1,2,5] as well as by others.[6] In

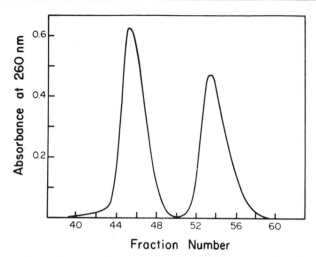

Fraction Number

Fɪɢ. 6. Separation of the complementary strands of a 210 bp restriction fragment. The 210 bp *Hin*dII fragment from ϕX174 replicative form DNA was isolated by RPC-5 chromatography (Fig. 5) and dialyzed versus 0.25 M KCl–5 mM Tris (pH 7.4)–0.5 mM EDTA. The pH of the solution was raised to 12.5 by the addition of 1 M NaOH, and 7.5 μg of the fragment (in 0.2 ml) was loaded onto a 1.5 × 200 mm RPC-5 column equilibrated with 0.25 M KCl–12 mM NaOH (pH 12.5). All solutions were degassed immediately before use and kept under a nitrogen atmosphere to maintain the high pH. The column was washed with 2 ml of the equilibration buffer and was eluted with a 20-ml linear gradient of 0.90–0.95 M KCl in 12 mM NaOH (pH 12.5) at 0.23 ml/min and 150 psi. Fractions of 0.2 ml were collected. The elution profile was continuously monitored at 265 nm. The separated strands eluted between 0.92 and 0.93 M KCl. Appropriate fractions were pooled and dialyzed, and aliquots were analyzed on 5% polyacrylamide gels.

addition, the fractionation of DNA restriction fragments on RPC-5 column chromatography has been reproduced by a number of workers in this laboratory[11,13,14,17,19] as well as by others.[7,12,20]

Prospects for the Future of RPC-5 Column Chromatography

It is likely that this technique will be used much more widely in the future for a number of the following applications.

1. In cases where it is necessary to isolate milligram quantities of defined segments of chromosomes, RPC-5 column chromatography is the only available high resolution technique at the present time. Specific applications are the following.

a. The isolation of defined regions of complex chromosomes (such as eucaryotic chromosomal DNAs) in order to partially fractionate the genome. The fractionation may be followed by hybridization with specific probes such as RNA transcripts or viral nucleic acids.

b. The isolation of defined segments of genomes, such as operators, promoters and origins of DNA replication etc., for biochemical and physical studies.

c. Isolation of regions of certain tumor virus genomes, such as promoters and origins of replication, which cannot be cloned under the present NIH Guidelines.

2. The separation of some fragments from other fragments of similar chain length but with different base compositions. It is not possible to accomplish these separations, even on small scale, by polyacrylamide gel electrophoresis.

3. The separation of some fragments which have approximately the same size (and, therefore, cannot be separated on gels) but one of which contains an *Eco*RI or *Hin*dIII end and the other has ends which are or behave like blunt ends.

4. Separation of the complementary strands of DNA restriction fragments.

5. The high resolution fractionation of single-stranded oligonucleotides. To the best of our knowledge, this is the only high resolution technique for separating quantities of oligonucleotides. Separations of up to approximately the 60-mer have been observed in some cases (DEAE–cellulose chromatography loses resolution at approximately the 10–15-mer[3]). These fractionations are very important for synthetic studies on oligonucleotides as well as for the isolation of single-stranded oligomers such as intermediates in DNA replication (i.e., RNA primers or short Okazaki fragments).

6. As a probe for AT-rich fragments or for AT-rich regions within GC-rich fragments. In addition, present studies[18] indicate that fragments which bind more tightly to the column contain known regulatory regions. If further work bears out this correlation, these chromatographic properties could be used as an indication of biological function.

Potential future uses of RPC-5 are the following.

1. The separation of DNAs which have small differences in their content of nonpaired nucleotides, such as the following.

a. DNA containing a short nick or gap.

b. Intermediates in DNA replication, such as growing fork regions which may contain several nonpaired nucleotides.

c. Transcription complexes consisting of the DNA template, which is predominantly in the duplex form, but with a nascent RNA still attached.

d. Intermediates in DNA recombination.

e. The separation of covalently closed circular DNAs containing different numbers of supercoiled turns.

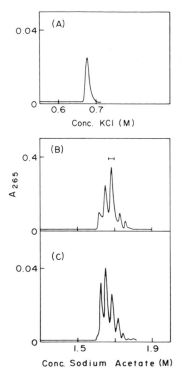

FIG. 7. Analysis of 789 bp fragment by HPLC on RPC-5. Analytical scale RPC-5 columns were run essentially as described in the text except that column loads varied between 10–50 μg total DNA. (A) 20 ml of 0.55–0.75 M KCl gradient at 43°C. In a repeat of this experiment the gradient was extended to 0.95 M with no other material eluting. (B) 40 ml of 1.3–2.1 M sodium acetate gradient at 13°C. The peak under the bar was pooled and rerun (C) under the same conditions.

      f. Replicative form I from form III DNA.

      g. DNAs with frayed ends (i.e., possibly produced purposely by ExoIII or the λ exonuclease) versus identical fragments which are fully double stranded.

      2. The separation of certain protein–nucleic acid complexes (such as nucleosomes or transcription complexes) from naked DNA.

Appendix

      When sodium acetate was used as an eluting salt, it was frequently found that more than one peak appeared in the absorbance profile for each fragment. This was particularly true when low temperature[14] or heavy loads[16] were used. The following experiment was done to determine whether or not this peak splitting reflected heterogeneity in the DNA molecules.

A 789-bp fragment was purified from a plasmid by use of preparative RPC-5 column chromatography.[16] Numerous peaks were associated with this fragment in the preparative column profile. We chromatographed this DNA on our previously described analytical system at 43° using a KCl gradient. Figure 7A shows that a single sharp peak was observed.

If the same fragment was chromatographed at 13° using a sodium acetate gradient, multiple peaks were observed (Fig. 7B). In similar experiments, as many as 19 peaks were observed under these chromatographic conditions. Also, these same conditions produced multiple peaks for other pure fragments (data not shown) which, like the 789-bp DNA, were otherwise found to be virtually homogeneous by a variety of criteria.[16]

When a single peak from the 789 fractionation in sodium acetate solution was pooled (see bar in Fig. 7B for pooled fractions) and then rechromatographed under the same conditions (Fig. 7C), the peak did not run true but instead the entire population of peaks reappeared. Moreover, the extent of heterogeneity suggested by RPC-5 chromatography with the sodium acetate gradients cannot be due to nicks or damaged ends because our previous results showed that at least half of the fragments were completely undamaged.[16]

Thus, we believe this peak splitting is not due to heterogeneity in the DNA fragments but is due to a column artifact.

[42] Fractionation of DNA Fragments by Polyethylene Glycol Induced Precipitation

By JOHN T. LIS

Precipitation by polyethylene glycol (PEG) has proved to be a general method of concentrating a variety of biological macrostructures. Bacterial cells, bacteriophage, plant and animal viruses, ribosomes, proteins, and DNA are all precipitable by similar procedures; however, the minimally required concentration of PEG in general is less for larger and/or more anisometric structures. Bacterial cells and rod-shaped viruses (the large tobacco mosaic virus and the filamentous bacteriophage fd) require only 1% PEG for quantitative precipitation[1,2]; structures of intermediate size

[1] D. L. Eshenbaugh, D. Sens, and E. James, *Anal. Biochem.* **58**, 390 (1974).
[2] K. R. Yamamoto, B. M. Alberts, R. Benzinger, L. Lawhorne, and G. Treiber, *Virology* **40**, 734 (1970).

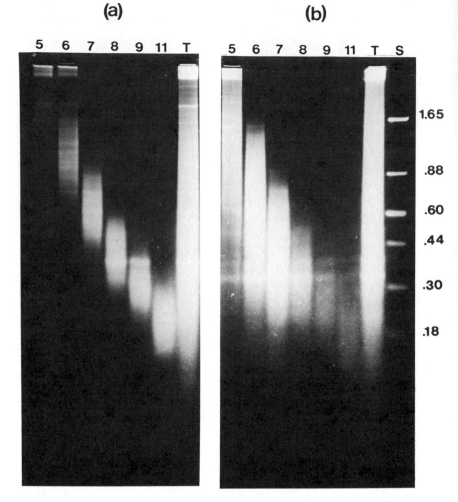

FIG. 1. Size analysis of PEG precipitated DNA by agarose gel electrophoresis. The numbers above each track correspond to the percentage PEG used to obtain the respective precipitates. (a) and (b) show the size distribution of precipitated *Hae* III and *Mbo* I cleaved total *D. melanogaster* DNA, respectively. The total mixture of DNA fragments before fractionation is shown in tracks labeled T. Track S contains size standards, *Hha* I cleaved ColE1, and the size of each band is written in units of kilobase (1000 base pairs). The values 0.60, 0.30, and 0.18 kb each represent the average size of an unresolved pair of fragments. DNA was precipitated as described in under Method from a 500-μl solution containing 0.5 M NaCl, 10 mM Tris-HCl, pH 7.4, 10 mM EDTA, 4.4 mM MgCl$_2$, DNA at 100 μg/ml, and PEG at 5%. The incubation was at 0° for 24 hr and the precipitate was collected by centrifugation for 2 min in a microfuge (15,000 g). Additional PEG was added to the supernatant to give a final concentration of 6%. The process was repeated for each of the indicated PEG concentrations. The precipitated DNA pellets from (a) and (b) were resuspended in 10 mM Tris-HCl, pH 7.4, 1 mM EDTA, and 6% of the total precipitate was fractionated by electrophoresis on a 2.5% agarose gel.

and anisometry (ribosomes and bacteriophage φX174) require 10%[2]; while some proteins require PEG concentrations exceeding 20%.[3] The relationship of macromolecular size and anisometry to the threshold PEG concentration is only approximate for this chemically diverse collection of macrostructures (for exceptions, see Table 1 of Yamamoto et al.[2]). However, a chemically homogeneous set of macromolecules, native DNA fragments, do show a clear inverse relationship of size to the threshold PEG concentration.[4] A statistical thermodynamic basis for these recent empirical observations resides in a treatise put forth by Onsager in 1949[5] on the effects of shape on the interaction of colloidal particles.

The striking dependence of the PEG concentration required for precipitation on the length of DNA allows a mixture of different size DNAs to be fractionated into classes by selective precipitation.[4,6,7] The ability of the method to fractionate DNA is demonstrated by the results shown in Fig. 1a. In this experiment the starting material is the HaeIII generated restriction fragments of *Drosophila melanogaster* which provide the size spectrum of DNA fragments shown in track T. These fragments are fractionated by size using successive precipitation steps. After each step the DNA precipitate is collected and additional PEG is added to the supernatant. The addition of PEG at a concentration of 5% causes precipitation of DNA fragments larger than 1650 base pairs (track 5). The smaller fragments, as well as the PEG,[2] remain in the supernatant. Another size class of fragments is precipitated by increasing the PEG concentration to 6%, track 6. Repetition of this process in steps of PEG concentration yield the DNA classes shown in Fig. 1.

The method possesses several virtues: high capacity, simplicity of both the fractionation and recovery of DNA fragments, and applicability to crude lysates. The high capacity of the method makes it an attractive first step in large-scale preparations of specific fragments, since precipitations can be performed at DNA concentrations at least as high as 1 mg/ml.[4]

The method requires no sophisticated equipment, expensive chemicals, or complex procedures as can be seen from the standard protocol presented in the discussion of Method below. The separated fractions that result are in the form of DNA precipitates which can be resuspended in a buffer suitable for subsequent use of the DNA. The precipitates contain

[3] W. Hönig and M. Kula, *Anal. Biochem.* **72,** 502 (1976).
[4] J. T. Lis and R. Schleif, *Nucleic Acids Res.* **2,** 383 (1975). (Correction, The photograph of Fig. 1 should be interchanged with the photograph of Fig. 2.) J. T. Lis and R. Schleif, *Nucleic Acids Res.* **2,** 757 (1975).
[5] L. Onsager, *Ann. N.Y. Acad. Sci.* **5,** 627 (1949).
[6] J. T. Lis and R. Schleif, *J. Mol. Biol.* **95,** 409 (1975).
[7] R. Ogata and W. Gilbert, *Proc. Natl. Acad. Sci. U.S.A.* **74,** 4973 (1977).

small amounts of PEG resulting from solvent trapped in the pellet. If this needs to be removed, the pellet can be resuspended in $0.2 M$ NaCl and the DNA reprecipitated with 2 volumes of ethanol. Polyethylene glycol is efficiently removed from DNA by several other procedures as well: chromatography on DEAE–cellulose,[6] gel electrophoresis,[8] or CsCl density gradient centrifugation.[2]

Size fractionation of DNA fragments by PEG precipitation should prove applicable even in the presence of crude cell lysates. The ability of the method to separate DNA fragments by size under such conditions has not been reported; however, $E.$ $coli$ plasmid DNA is precipitated from crude lysates at the expected threshold concentration (unpublished observations). Although a multitude of structures in a crude lysate are precipitated by PEG, the minimal concentration of PEG required varies considerably for different structures. It, therefore, should be possible to separate DNA fragments not only from other size fragments but also from other types of cellular components. Indeed, the PEG precipitation protocol has been included as a step in a rapid $E.$ $coli$ plasmid purification procedure.[9] After this step the plasmid DNA is usually suitable as a substrate for restriction endonucleases and for transformation.[9]

The size fractionation of DNA by PEG precipitation does possess limitations: the resolution is only fair in comparison with gel electrophoresis and RPC-5 chromatography[10,11]; fractionation of fragments above 2000 bp is at present poor; and at least a moderate DNA concentration is required for efficient precipitation. In the stepwise precipitation shown in Fig. 1a the resolution of two species is essentially complete when their sizes differ by more than a factor of 2.5. In this experiment increments of 1 or 2% PEG were used; however, decreasing the size of the increment should increase resolution. Reprecipitation should also improve resolution as well as remove low levels of contaminating fragments (not detectable in this figure) arising from solvent trapped in the pellet.

A special problem occurs when fragments possess cohesive ends such as those generated by restriction endonucleases EcoRI or MboI. This problem arises because the PEG precipitation is usually performed by incubating high concentrations of the DNA fragments and salt at 0° for several hours. These conditions permit the joining of fragments via their cohesive ends to form an assortment of multimers. Multimers could be-

[8] P. A. Albertsson, "Partition of Cell Particles and Macromolecules." Wiley, New York, 1960.
[9] A. Rambach and D. Hogness, $Proc.$ $Natl.$ $Acad.$ $Sci.$ $U.S.A.$ **74**, 5041 (1977).
[10] A. Landy, C. Foeller, R. Reszelbach, and B. Dudock, $Nucleic$ $Acids$ $Res.$ **3**, 2575 (1976).
[11] S. C. Hardies and R. D. Wells, $Proc.$ $Natl.$ $Acad.$ $Sci.$ $U.S.A.$ **73**, 3117 (1976).

have like single fragments of equivalent size in response to precipitation by PEG. To test whether this occurs, the PEG induced precipitation of fragments with cohesive ends was compared to the precipitation of fragments possessing flush ends. *Mbo*I was used for this test since it makes single-strand breaks adjacent to a specific 4-bp sequence thereby generating cohesive ends. The spectrum of fragment sizes resulting from the cleavage of total *D. melanogaster* DNA with *Mbo*I is shown in track T of Fig. 1b and is similar to the spectrum generated by *Hae*III (Fig. 1a, track T). However, the *Mbo*I fragments precipitated by 5% PEG (Fig. 1b track 5) cover a broad range of sizes from the largest down to 300-bp fragments. This precipitation of the small fragments is in striking contrast to the size dependent precipitation of *Hae*III generated, flush-ended fragments (Fig. 1a, track 5). It should be noted that the electrophoretic separation of DNA fragments is performed above the melting temperature of the cohesive ends. The results of Fig. 1b demonstrate that small fragments, which possess cohesive ends, do indeed contaminate fractions that contain larger fragments; however, the larger fragments do not contaminate fractions that are expected to contain only smaller fragments (Fig. 1b tracks 6–11). Thus, the method can be used to remove large fragments from preparations of smaller fragments, albeit with reduced yield.

A solution to this problem associated with cohesive ends is to perform the incubation step for the PEG precipitation at temperatures above the melting temperature of annealed cohesive ends. The melting temperature of cohesive ends of fragments produced by known restriction endonucleases is well below room temperature and depends on the length of the cohesive ends, base composition, and salt concentration. Although I have routinely incubated the PEG–DNA mixture at 0° to minimize degradation of DNA by contaminating nucleases, Lerman has precipitated DNA with PEG at room temperature.[12] Furthermore, the partitioning coefficient of calf thymus DNA between PEG and dextran phases (the partitioning system from which simplified PEG precipitation method evolved) is the same at both 4° and 20°.[8]

Method

The following specific directions should be regarded as a recipe that can be modified in accordance with information in the Appendix.

1. The DNA solution is prepared for precipitation by adding NaCl to 0.5 M and Tris, pH 7.4, and EDTA to 10 mM each. The DNA concentration should be greater than 10 μg/ml, preferably 100 μg/ml. Add an amount of PEG 6000 (Union Carbide) required to fractionate the DNAs of

[12] L. S. Lerman, *Proc. Natl. Acad. Sci. U.S.A.* **68**, 1886 (1971).

interest using the results of Fig. 1a and the results in Lis and Schleif[4] as guides. Polyethylene glycol is conveniently added as a 50% (w/w) solution (p = 1.05) on a volume-to-volume basis.

2. Incubate the mixture at 0°. Incubations lasting 12 hr or more are best for DNA in the size range of 500 base pairs and larger. However, 1-hr incubations are sufficient for fractionation of lower molecular weight species which are precipitated at PEG concentrations greater than 7.5%.

3. Collect the precipitated DNA by low speed centrifugation; 8000 g for 5 min is more than sufficient to sediment the precipitated DNA into a firm pellet.

4. The supernatant is carefully removed with a hand-controlled pipetting devise or by decanting.

If another size class of DNA is to be precipitated, additional PEG is added to the supernatant, and operations 2–4 are repeated.

Appendix

A summary of data on the sensitivity of the PEG precipitation method to various parameters is as follows.

1. PEG Concentration. The minimum PEG required to precipitate DNA fragments (the threshold concentration) shows an inverse dependence on fragment size as is demonstrated in Fig. 1a and in a previous publication.[4]

2. Salt Concentration. Efficient precipitation requires high salt concentration, 0.5 M or higher. If the NaCl concentration is lowered to 0.2 M, no DNA of any size is detected as a rapidly sedimenting form following attempted precipitations by PEG concentrations as high as 12%.[4] At 0.35 M NaCl no DNA is detected following precipitation by 6.5% PEG; however, 12% PEG causes precipitation of DNA from 49,000 to 240 bp. Very high salt concentration, 1.1 M, allows precipitation of DNA down to 375 bp by as little as 6.5% PEG.[4]

3. DNA Concentration. The size of DNA fragments precipitated by PEG is relatively independent of DNA concentration in the range of 10–1000 μg/ml. However, below 50 μg/ml the recovery of DNA is less than quantitative and is 50% at 10 μg/ml.[4]

4. Incubation Time. Incubations as short as 5 min allow fractionation of fragments into size classes; however, the resolution is less than with longer incubations, especially in the high molecular weight range. This stems from the fact that the efficiency of precipitation of high molecular weight DNA near the threshold PEG concentration increases with time. λ DNA is precipitated with 10, 70, and 100% efficiency after incubation for 2, 9, and 25 hr, respectively, at a PEG concentration which was determined to be threshold in an 18-hr incubation.[4]

5. *Centrifugal Force.* Both the size of the DNA fragments and the efficiency with which they are precipitated is independent of the centrifugal force used to collect precipitated DNA at least in the range of 1900 to 27,000 g for a 10-min centrifugation.[4] Centrifugation at 480 g can result in a 50% loss of fragments of all sizes.

6. *Divalent Ions.* Preparations of specific DNA fragments usually require digestion with either of two classes of enzymes, restriction endonucleases, or single-strand-specific nucleases. The presence of divalent ions at concentrations normally required by both classes of enzyme, ≤ 10 mM, do not interfere with the size fractionation. However, divalent ions at 10 mM extend the size range of the precipitated DNA to include fragments that are approximately one-half the size of those precipitated in the absence of divalent ions.[4]

7. *pH.* The method is insensitive to pH at least in the range of pH 5.0–8.3.[4]

[43] A Photographic Method to Quantitate DNA in Gel Electrophoresis

By ARIEL PRUNELL

The purpose is to determine the relative amount of DNA in bands obtained upon electrophoresis in polyacrylamide or agarose gels.

The procedure[1] is, in outline, as follows: The gel is stained with ethidium bromide and photographed under ultraviolet illumination. The photograph is traced with a microdensitometer, and pen deflections converted into fluorescence intensities (which are proportional to the amounts of DNA; see below) with the use of the *characteristic curve* of the film. This curve relates optical densities of the film to light intensities or exposures. Practically, a wide range of exposures is obtained with a *step tablet,* and the characteristic curve consists of a plot of densities in a photograph of the step tablet (this photograph is termed *sensitogram*) against the logarithm of exposures.

Procedure

Staining. Gels are stained at room temperature for 2 hr in electrophoresis buffer[2] supplemented with 2 μg/ml of ethidium bromide

---

[1] A. Prunell, F. Strauss, and B. Leblanc, *Anal. Biochem.* **78**, 57 (1977).

[2] U. E. Loening, *Biochem. J.* **102**, 251 (1967).

METHODS IN ENZYMOLOGY, VOL. 65

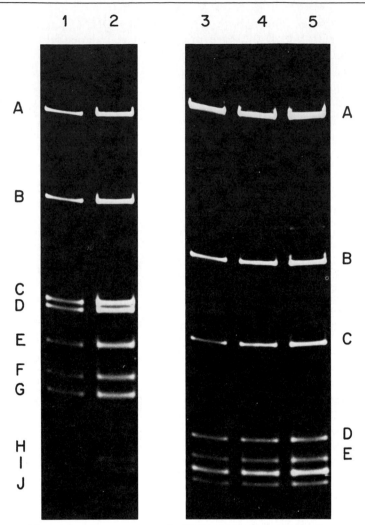

FIG. 1. Electrophoretic patterns of SV40 DNA restriction digests on 2% polyacrylamide–0.5% agarose slab gels. Conditions are as described in Prunell et al.[3] In lanes 1 and 2 were loaded 10 and 20 μl of a HindII + III digest containing 0.2 μg of DNA/μl. In lanes 3, 4, and 5, were loaded 10, 15, and 20 μl of a HaeIII digest also containing 0.2 μg of DNA/μl. HindII + III and HaeIII fragments are lettered according to Danna et al.[4] and Lebowitz et al.[5]

(Sigma, St. Louis, Missouri). In staining and photography, the slab gels (0.3 × 16 × 40 cm) are left on one of the two glass plates used to cast them. Alternatively, electrophoresis buffer may be replaced by water. This will result in a slight expansion of the gel.

Ultraviolet Illumination and Photography. Gels are illuminated from above by two short-wavelength ultraviolet lamps (CS 215; UV Products, San Gabriel, California; alternatively, two CS 15 lamps may be used) positioned on either side parallel to the lanes of the gel and oriented at 45° to it. Photographs are taken on Kodak Ektapan 4 × 5 inch sheet films through a Kodak Wratten 23A red filter with an exposure time of 3 min. Films are developed at 21° for 8 min in Kodak D76 developer and fixed for 4 min, at the same temperature, in Kodak Rapid Fixer. Figure 1[3-5] shows such a photograph in which the amount of DNA is to be quantitated. A sensitogram (see below) is routinely processed together with the gel photographs.

Sensitograms. These are contact prints of a No. 2 photographic step tablet (Eastman Kodak) containing 21 steps of density between 0.05 and 3.05. The step tablet is held in contact with the emulsion side of the film and illuminated from above through a Wratten 23A filter. The distance from the light box to the step tablet should be sufficient to ensure a uniform illumination. Neutral filters added to the light box allow a suitable exposure of the film in 3 min. Exposure times for both sensitograms and gel photographs must be the same in order to avoid artifacts known as reciprocity failures.[6] Many sensitograms are prepared at one time, and stored, undeveloped, at −20° until use. Before being developed, they are kept at room temperature for about 2 hr.

Densitometry. Photographs are traced with a Joyce-Loebl MKIII C microdensitometer, as follows: Suitable gray wedges provide optimum pen deflections for each band of the pattern. Gel photographs and sensitograms are traced with the same wedges, which avoids having to take their slope into account. In addition, for each wedge used, a step of appropriate density in the sensitogram is chosen as a fixed reference of pen deflection. This ensures the same deflection for regions having the same density in both sensitogram and gel photograph and suppresses any disturbance which may arise from the nonlinearity of the wedges.

Densities of the steps of the tablet [D in Eq. (1), see below] are measured with a digital densitometer (TD 504; Macbeth Products Division, Newburg, New York). They can also be measured, although less conveniently, with any kind of spectrophotometer, including the Joyce-Loebl microdensitometer, if the wedge slope is accurately known. In this latter case, only differences of density between the steps are obtained.

Characteristic Curve. If I_i is the intensity of the incident light on the

[3] A. Prunell, H. Kopecka, F. Strauss, and G. Bernardi, *J. Mol. Biol.* **110**, 17 (1977).
[4] K. J. Danna, G. H. Sack, and D. Nathans, *J. Mol. Biol.* **78**, 363 (1973).
[5] P. Lebowitz, W. Siegel, and J. Sklar, *J. Mol. Biol.* **88**, 105 (1974).
[6] P. Glafkides, "Chimie et Physique Photographique." Publications Paul Montel, Paris, 1967.

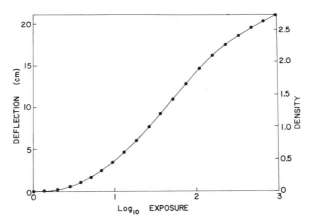

FIG. 2. Microdensitometer pen deflections, obtained in the tracing of a sensitogram, are plotted as a function of the logarithm of the exposure, $\log_{10}I$. The value of I_i [see text and Eq. (1)] has been chosen equal to 1000. Points refer to the steps of the tablet. Densities are calculated from the slope of the gray wedge used (0.13 OD units/cm).

tablet, and I the intensity transmitted to the film through the step of optical density D, then

$$\log_{10}I = \log_{10}I_i - D \qquad (1)$$

I is the exposure. I_i is not known, but can practically be given any arbitrary value. This results from the fact that only relative amounts of DNA between bands, and not absolute ones, are to be measured. Figure 2 shows the characteristic curve obtained by plotting the densities of the steps of a sensitogram, in terms of pen deflections, against $\log I$, as specified by Eq. (1). Such a curve has to be constructed for each wedge used. The curve of Fig. 2 is approximately linear in the intermediate range of intensity. Curvatures in the under and overexposure regions are related to the sensitivity and saturation levels of the film.

Fluorescence Quantitation. Each peak of the tracing is fitted with a baseline and subdivided by an odd number (usually from 9 to 15) of vertical lines at regular intervals. Deflection values, at their intercept with the peak and the baseline, are converted into intensities by means of the corresponding characteristic curve. The differences between peak and baseline intensities are integrated over the entire peak, giving the fluorescence intensity of the DNA contained in the band.

All calculations may be computerized: Characteristic curves are represented by series of third degree polynomials and numerical integrations done by standard procedures. These calculations can be conveniently performed with programs available in the library of any desk computer. Figure 3A shows the fluorescence of bands A to G in lanes 1 and 2 of Fig. 1, measured at two apertures of the camera objective, as a function of the

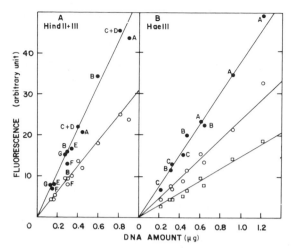

FIG. 3. Plots of fluorescence against the amount of DNA in the bands of Fig. 1. Objective apertures were $f/5.6$ (●), $f/8$(○), and $f/11$ (□). The amount of DNA was calculated from the size of the fragments, expressed as a fraction of SV40 genome size (see Danna et al.,[4] Lebowitz et al.,[5] and Fiers et al.[7]) and multiplied by the amount of DNA (in micrograms) loaded in the lanes. (A) Data from bands A to G in lanes 1 and 2 of Fig. 1. (B) Data from bands A to C in lanes 3–5 of Fig. 1.

amount of DNA. Figure 3B shows the same plot with bands A to C in lanes 3 to 5 of Fig. 1, at three apertures. An examination of Fig. 3 leads to the conclusion that fluorescence intensities are directly proportional to the amount of DNA—most of the points fall on straight lines passing through the origin. Some points, however, lie off the lines, but to a similar extent at all apertures. Such deviations are not related to the photographic procedure, but may give some idea of the accuracy achieved in the fluorescence measurement. In the particular case of Fig. 3A, the deviations are similar in both lanes, and may be due to an error in estimations of the amount of DNA (see legend of Fig. 3)[4,5,7] or to differential binding of ethidium bromide to the fragments.

Comments

1. The filters on uv lamps become opaque or solarized with use, significantly faster, in our experience, than indicated by the manufacturer. Solarized filters may be replaced at considerable expense, or they may be simply removed, in which case the Wratten 23A filter on the camera (see above) must be replaced by a 24.

[7] W. Fiers, R. Contreras, G. Haegeman, R. Rogiers. A. Van de Voorde, H. Van Heuverswyn, J. Van Herreweghe, G. Volckaert, and M. Ysebaert, Nature (London) 273, 113 (1978).

2. Photodecomposition of ethidium bromide can reduce the fluorescent intensity by as much as 40% in about 10 min, when CS 215 lamps are used. However, the reduction of intensity is the same for all bands, provided the uv illumination is uniform. This can be checked *in situ* with a shortwave uv counter (J225; UV Products). In our experiments the uv illumination was found to be uniform in the full width but not in the full length (40 cm) of the gel, the illumination decreasing 10 cm beyond the middle of the lamps.

3. The objective should have a focal distance sufficient to minimize photographic artifacts known as vignetting, which result in a lower exposure of the edges as compared to the center of the film. An objective of $f = 150$ mm provides a uniform exposure in a circle in the center of the film of diameter equal to about 70% of its length.

4. Since fluorescence is not integrated over the total amount of DNA in a band, but only over the portion of it that is densitometered, only gels in which DNA migration is uniform and band widths identical within a lane should be quantitated.

5. A similar procedure has been reported[8] in which only the linear part of the characteristic curve of Fig. 2 is used. This simplifies the computations. In addition, a special device was used to superpose the gel pattern and the sensitogram on the same film. This may not be of great advantage since, in our experience, development and fixing are very reproducible from one film to the other so that the development of a sensitogram in every experiment is no longer required when films from the same lot are used.

6. Finally, it is important to note that measurements of the area under the peaks in the densitometer tracing, or of their height,[8] which do not take into account the logarithmic nature of the photographic response, cannot be substituted for the procedure described here.

Acknowledgments

I am indebted to F. Strauss and B. Leblanc who contributed to the method and who coauthored the original report quoted in Prunell *et al.*[1]

[8] D. E. Pulleyblank, M. Shure, and J. Vinograd, *Nucleic Acids Res.* **4**, 1409 (1977).

[44] Two-Dimensional Agarose Gel Electrophoresis "SeaPlaque" Agarose Dimension

By RICHARD C. PARKER and BRIAN SEED

There are many techniques currently available for mapping the restriction endonuclease sites of various DNAs. Most of these involve gel electrophoretic fractionation of the products of enzymatic digestion. Although the simplicity and resolution of gel electrophoresis are unrivalled, it is often difficult to manipulate or isolate DNA electrophoretically embedded in gel media. When large numbers of restriction fragments are to be analyzed, enzymatic redigestion of DNA electrophoresed through gels becomes an unprofitable approach to restriction mapping. Methods for the *in situ* digestion of fragments trapped in gels have been presented elsewhere. In general, these techniques have not become popular because they require large amounts of enzyme for complete digestion, and either have not been shown to be applicable or have been shown to be inapplicable to a broad range of enzymes with different specificities.

In the following we present a simple method for redigestion of restriction fragments electrophoresed through a low melting temperature agarose. The technique is both easy to use and economical of the second-dimension enzyme. It is applicable to all the enzymes we have tested and does not require extensive modification of conventional electrophoretic apparatuses. The resolving characteristics of the low melting-temperature (LMT) agarose used are essentially identical to those of ordinary agaroses; however, the LMT agarose is both more expensive and more fragile than commonly used agaroses.

In the following procedure, restriction endonuclease products are fractionated in a gel made from hydroxyethyl agarose (SeaPlaque Agarose, Marine Colloids, Inc.). SeaPlaque Agarose melts at 65° and remains in solution at 37°, allowing the gel to be dissolved without denaturing DNA. It is possible to find conditions which permit complete digestion of DNA with restriction endonucleases in the presence of SeaPlaque Agarose at 37°. Enzyme efficiencies are not greatly reduced under these conditions. Following fractionation, the bands of interest can be visualized either by ethidium bromide fluorescence or autoradiography. Discrete regions of the gel are excised and placed in tubes, the agarose melted, buffers and restriction endonucleases added, and the DNA cleaved by the added enzymes.

Pouring the Gel

Gels for the first dimension of two-dimensional DNA electrophoresis are made by dissolving hydroxyethyl agarose in E buffer (40 mM Tris, 5 mM sodium acetate, 1 mM EDTA, pH adjusted to 7.4 with glacial acetic acid) by heating to at least 65°. Before the gel is poured, the agarose should be cooled to 37° to minimize shrinkage while the gel solidifies. If this is not done, the gel may crack or detach from the apparatus.

Pouring an LMT agarose gel in a horizontal apparatus presents no special problems. When pouring vertical gels in forms with insertable

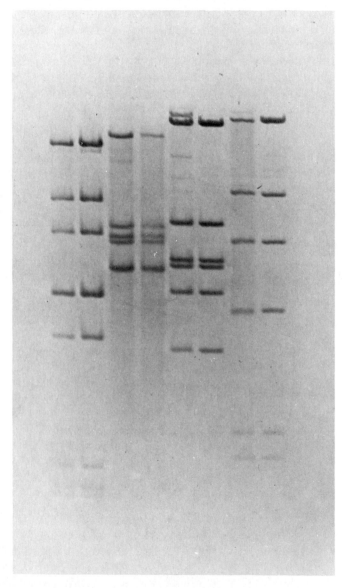

combs, however, care must be taken to remove the comb without tearing the gel in a refrigerator before removing the comb. After the gel has solidified, buffer can be pipetted around the exposed teeth and the comb the gel in a refrigerator before removing the comb. After the gel has solidified, buffer can be pipetted around the exposed teeth and the comb removed without damage to the gel.

Low melting temperature agarose gels are more slippery than ordinary gels and tend to slide from vertical holders more easily. Depending on the design of the apparatus, a mechanical support at the bottom of the gel, or a conventional agarose plug, may be necessary.

Running the Gel

The gel should be run for at least 10 min before samples are applied. During this time the voltage should be gradually increased. Low melting temperature agarose gels frequently crack when high voltage gradients (5 V/cm) are applied to the gel without this gradual increase. If the voltage is slowly increased, the gels are stable to at least 6.6 V/cm. In our apparatus, this gradient spans a 15-cm long, 4-mm thick gel and requires approximately 60 mA of current. Horizontal apparatuses with thin (e.g., paper) wicks may exhibit sufficient ohmic heating at the ends of the gel to melt LMT agarose. Running at lower voltages, or in the cold, can avoid melting.

Samples are layered on LMT agarose gels in the same manner as on other agarose or acrylamide gels. Mobilities of both circular and linear DNA molecules in this agarose are similar to their mobilities in gels made from more commonly used agaroses.

Band Detection and Isolation

Depending upon the quantity of DNA available, it may be more convenient to determine the location of the DNA in the gel by autoradiography of ^{32}P-labeled DNA or by ethidium bromide (EtdBr) fluorescence. If bands are to be detected by autoradiography, the gel should be soaked for 10 min in dH$_2$O or in buffer. The bands can be cut out of the gel by measuring their mobility on the autoradiogram and removing that region of the gel. If bands are to be detected by EtdBr fluorescence, the gel

---

FIG. 1. A 0.7% agarose gel was used to electrophoretically resolve the products of restriction endonuclease digests of λ CI$_{857}$ DNA in the presence (slots 1, 3, 5, 7) and absence (slots 2, 4, 6, 8) of SeaPlaque agarose. The enzymes used were *Ava*I (slots 1, 2), *Bam*HI (slots 3, 4), *Eco*RI (slots 5, 6), and *Hin*dIII (slots 7, 8).

should be stained for 10 min in 0.2 μg/ml of EtdBr and then destained for 10 min without EtdBr. The solutions for staining and destaining can be made in distilled water or in a buffer appropriate for the second restriction endonuclease.

The bands can then be visualized by either long wave or short wave uv light depending upon the quantity of DNA in the bands. It is preferable to use long wave light in order to minimize nicking of the DNA. The bands should be cut from the gel in slices that are as thin as possible so that the entire volume of the slice can be loaded on the next gel.

Restriction Endonuclease Digestion in SeaPlaque

After slices of the gel have been placed into individual test tubes, their volume should be estimated and $\frac{1}{9}$ volume of a stock of ten times concentrated enzyme buffer added. This step is unnecessary if the gel was already soaked in buffer.

The samples are then placed at 65° until the agarose melts. When liquid, the sample is placed at 37° and allowed to thermally equilibrate. After it has equilibrated, a sterile solution of nuclease-free bovine serum albumin (Pentex BSA, Miles Laboratories) is added to a final concentration of 0.1%. The addition of BSA is essential for complete digestion.

The presence of agarose does not seem to greatly inhibit restriction endonucleases under these conditions. To ensure complete digestion, two to three times more enzyme should be added than would be used to digest the DNA under standard conditions. If the number of restriction sites in the DNA is known, we generally normalize the amount of enzyme per microgram of DNA so that the ratio of enzyme to concentration of sites is equivalent to that found in the standard assay with λ DNA. Where the number of sites in lambda is known, we calculate the amount of enzyme required by the following formula

$$\frac{\lambda_{mw}}{X_{mw}} \frac{X_{sites}}{\lambda_{sites}} \frac{\text{units of enzyme}}{1 \ \mu\text{g of } \lambda \text{ DNA}} = \frac{\text{units of enzyme}}{1 \ \mu\text{g of X DNA}} \tag{1}$$

The Second Dimension

If the second gel is a vertical gel, it is important that the LMT agarose remain as a liquid until the DNA has migrated into the gel. If the agarose is allowed to set up, the bands in the second dimension will be very broad. If the second gel is a horizontal gel, the agarose may be allowed to set up. In the event that a vertical gel is used, there are two ways to ensure that the agarose stays in solution. The first, and easiest, is to heat both the samples and the running ("E") buffer for the upper reservoir to 65°. Alternatively the samples may be loaded in 33% formamide. The latter has the disadvantage of increasing the sample volume.

Results

Four of the enzymes that work well under our standard second dimension conditions were used to obtain the patterns shown in Fig. 1. This gel contains the electrophoretically resolved products of digestions of λCI_{857} DNA in the presence or absence of SeaPlaque agarose. The restriction endonucleases used were: AvaI, BamHI, EcoRI, and HindIII. In all cases, the complete digest products obtained in the presence of LMT agarose were identical to those in the parallel digestion.

At the time of writing, we have tested the following seventeen enzymes: AluI, AvaI, BamHI, EcoRI, HaeII, HaeIII, HhaI, HincII, HindIII, HpaI, HpaII, KpnI, PstI, PvuII, SalI, SmaI, and XhoI. All were capable of producing complete digests in the presence of LMT agarose in less than two hours.

The potential uses of LMT agarose are clearly broader than those presented here. We anticipate that as the need arises, various other applications will be devised. In our experience, the single most important step for obtaining good enzymatic activity, in the presence of LMT agarose, has been the addition of bovine serum albumin to 1 mg/ml. This knowledge may be helpful in the design of future applications.

Acknowledgments

We would like to thank Norman Davidson and Tom Maniatis for their suggestions and encouragement.

[45] The Use of Intensifying Screens or Organic Scintillators for Visualizing Radioactive Molecules Resolved by Gel Electrophoresis

By RONALD A. LASKEY

General Principles

This article describes methods for obtaining enhanced autoradiographic images of radioactive molecules in agarose or acrylamide gels. The methods described make use of solid state scintillation to overcome two problems. (a) Low-energy β particles, e.g., those from ^3H, ^{14}C, or ^{35}S, are quenched within gels. Conversion of the emitted energy to visible light by an organic scintillator (PPO) infused into the gel increases penetration range and therefore increases the proportion of the emitted energy which

TABLE I
SENSITIVITIES OF METHODS DESCRIBED IN THE TEXT FOR ISOTOPE DETECTION IN AGAROSE
OR ACRYLAMIDE GELS[a]

Isotope	Type of method	dpm/cm² required for detectable image ($A_{540}=0.02$) in 24 hr	Enhancement over direct autoradiography
^{125}I	Screen	100	16
^{32}P	Screen	50	10.5
^{14}C	PPO	400	15
^{35}S	PPO	400	15
^{3}H	PPO	8000	>1000

[a] Data are from Laskey and Mills[1,2] for exposure at $-70°$ using film preexposed to reach an absorbance increment of 0.15 above the background absorbance of unexposed film. Direct autoradiography for comparison was performed on Kodirex film.

is absorbed by the adjacent X-ray film. (b) Conversely, high-energy β particles (e.g., from ^{32}P) or γ rays (e.g., from ^{125}I) pass through and beyond the film. Their excess energy can be captured and returned to the film in the form of visible light by placing a high density fluorescent "intensifying screen" beyond the film.

Table I[1,2] summarizes the enhancement obtained by use of the methods described below, compared to conventional direct autoradiography. For maximum efficiency of all these methods and for quantitation of the images obtained it is necessary to bypass the reversible first stage of latent image formation in the film. This is most easily achieved by preexposing film to an *instantaneous* flash of light ≤1 msec). A procedure for performing this is described below. Its rationale has been discussed previously.[1] (See Appendix for a discussion of a conflicting claim in Swanstrom and Shank.[3])

For optimum sensitivity and quantitative accuracy film should be exposed to the sample at $-70°$.[1,4,5] Provided that film has been preexposed correctly the loss of sensitivity at $+22°$ compared to $-70°$ is only twofold. However the advantages of the methods described here are not obtained with unfogged film at $+22°$.

Neither preexposure nor exposure at low temperature increases the efficiency of conventional direct autoradiography. They are required only

[1] R. A. Laskey and A. D. Mills, *Eur. J. Biochem.* **56**, 335 (1975).
[2] R. A. Laskey and A. D. Mills, *FEBS Lett.* **82**, 314 (1977).
[3] R. Swanstrom and P. R. Shank, *Anal. Biochem.* **86**, 184 (1978).
[4] U. Lüthi and P. G. Waser, *Nature (London)* **205**, 1190 (1965).
[5] K. Randerath, *Anal. Biochem.* **34**, 188 (1970).

when the isotopic emission energy is divided into many smaller quanta of light.[1]

This article is concerned with isotope detection only after gel electrophoresis. Related procedures for visualizing weak β emitters on paper or thin layer chromatograms have been published[4,6] but are outside the scope of this paper. However the intensifying screen procedure described here for visualizing strong β emitters or γ emitters in gels is equally applicable to paper or thin layer chromatograms.

Choice of Film. All of the methods described below require "screen-type" X-ray film. "Direct" films such as Kodirex or No Screen are inefficient at recording the visible light produced by organic scintillators or intensifying screens. The sensitivities of various commercially available screen-type films for the methods described here have been compared previously.[2] Kodak X-Omat R and Fuji RX were found to be the most suitable of those tested.

Procedure for Preexposing Film to a Hypersensitizing Light Flash

Film is hypersensitized immediately before use by exposure to a single instantaneous flash of light from an electronic photographic flash unit or stroboscope. The intensity of the flash is adjusted to increase the absorbance of the film to 0.15 (A_{540}) above the absorbance of unexposed film.[1] It is essential that the duration of the flash is short (in the order of 1 msec). Flashes of longer duration only increase the fog level of film without hypersensitizing it. The wavelength of light used is unimportant, but adjusting the wavelength provides a convenient means of decreasing the effective intensity of the emitted light, since the sensitivity of screen-type X-ray films is greatest in the blue region of the spectrum. Therefore orange filters (e.g., Kodak Wratten number 21 or 22) decrease the light output from most simple photographic flash units to approximately the correct intensity for preexposure.

The exact light intensity required depends on the type of film and flash unit and must, therefore, be determined by experimental exposures. Fine adjustments are most easily made by (a) varying the distance between the film and the light source (this distance should be greater than 50 cm to obtain even illumination), (b) adding additional neutral density filters, or (c) varying the diameter of an aperture in an opaque mask. If the fog level obtained on the preexposed film is unevenly distributed, it may be necessary to include a translucent diffuser with the filters. Whatman No. 1 filter paper is suitable for this purpose.

[6] J. Shine, L. Dalgarno, and J. A. Hunt, *Anal. Biochem.* **59**, 360 (1974).

Either side of the double-coated film may be exposed to the sample (or screen), but we have adopted the convention of applying the side nearest the light source to the intensifying screen or to samples impregnated with scintillator. Storage of films after preexposure is not recommended because the shelf life is decreased and the fog level rises more rapidly.

An alternative method of hypersensitizing film has been suggested in Kodak publicity information. This is a prolonged exposure to hydrogen gas which is likely to offer the advantage of a slightly more uniform fog level but which may be less convenient to perform in the average photographic darkroom.

Procedure for Using Intensifying Screens to Visualize γ-Emitting or Strong
    β-Emitting Isotopes[2]

The procedure to be described is applicable to wet or dry gels (consisting of agarose or acrylamide), to nitrocellulose filters, or to chromatograms on paper or thin layer plates.[2]

A screen-type X-ray film is preexposed to increase its absorbance (A_{540}) to 0.15 above that of unexposed film. It is placed between the sample and a calcium tungstate medical X-ray intensifying screen and clamped either in a radiographic cassette or between glass plates. Exposure is performed at $-70°$ for dehydrated samples, or above $0°$ for hydrated gels (see comments).

Comments on Choice of Intensifying Screen. The relative merits of alternative screens have been discussed previously.[2] Two types of calcium tungstate screen were found to be more sensitive than others for this purpose. They were Du Pont Cronex Lightning Plus and Fuji Mach 2. Screens consisting of either europium-activated barium fluorochloride or terbium-activated mixtures of lanthanum oxysulfide and gadolinium oxysulfide were more efficient than calcium tungstate for ^{32}P detection, but were found to be unsuitable for long exposures to hypersensitized film because their endogenous isotope content blackened hypersensitized film spontaneously, causing an increased absorbance of ≥ 0.2 per week at $-70°$ compared to <0.1 per week for calcium tungstate.

Since refinements are continuously made to the range of commercially available intensifying screens, it is probable that screens with more suitable properties will be developed in the future.

Comments on the Intensifying Screen Procedure. The use of intensifying screens offers slightly less resolution than direct autoradiography. For thin dehydrated samples the loss of resolution is not serious. However when hydrated gels are frozen and exposed at $-70°$ the loss of resolution becomes more serious. The loss of resolution in this circumstance is probably not caused by the screen itself, but by physical distortion of the gel when ice crystals form, since the loss is also observed for direct autoradiographs of frozen, hydrated gels exposed without intensifying screens. Therefore, to obtain maximum resolution when using intensifying screens, it is advisable to dry gels before exposure. For acrylamide gels this is facilitated by preparing gels according to formula I of Blattler *et al.*[7] Where gel drying is inconvenient exposure at room temperature (but necessarily using preexposed film) should be considered as an alternative.

Efficiency of ^{32}P detection is increased further by enclosing the film between two screens (sample : screen A : film : screen B), but this procedure is not recommended because it causes severe loss of resolution through scattering by the screen which lies between the film and the sample.[2] The detection efficiency for ^{125}I is not increased by using two screens.

Procedure for Using PPO to Visualize Weak β-Emitting Isotopes in
Polyacrylamide Gels[8]

This procedure exploits the solvent properties of dimethyl sulfoxide to carry the organic scintillator PPO (2,5-diphenyloxazole) into the lattices of polyacrylamide gels.[8] The procedure is suitable for hydrated gels. Dried gels must first be reswollen in water.

The hydrated gel is equilibrated with dimethyl sulfoxide by soaking it in approximately 20 volumes of dimethyl sulfoxide for 30 min (or longer for gels >3 mm thick) followed by a second 30 min immersion in fresh dimethyl sulfoxide. The separate tanks of used dimethyl sulfoxide are retained for reuse in the same sequence. It is essential to remove all water from the gel at this stage or PPO will not enter the gel.

After equilibration the gel is immersed in 4 volumes of a 22% (w/v) solution of PPO in dimethyl sulfoxide for 3 hr (longer times may result in diffusion of bands). (*CAUTION:* Dimethyl sulfoxide penetrates skin, therefore contact of this solution with skin should be avoided by the use of good rubber gloves.) The impregnated gel is soaked in excess water for 1 hr (longer for gels >3 mm thick) to remove dimethyl sulfoxide. It is then dried under vacuum.

The dried gel is exposed to a hypersensitized screen-type X-ray film (see above) at −70°.

Comments on the Procedure for Applying PPO to Polyacrylamide Gels. A major advantage of PPO over alternative scintillators is its high solubility. Dimethyl sulfoxide appears to be the only solvent which has been used

[7] D. P. Blattler, F. Garner, K. Van Slyke, and A. Bradley, *J. Chromatogr.* **64,** 147 (1972).
[8] W. M. Bonner and R. A. Laskey, *Eur. J. Biochem.* **46,** 83 (1974).

successfully for this purpose. It is completely miscible with water, yet it dissolves high concentrations of PPO (which is nonpolar) and it enters gels freely without causing serious physical distortion. Gels shrink during immersion in dimethyl sulfoxide but swell again during the final immersion in water. If gels crack during drying under continuous vacuum refer to Blattler *et al.*[7]; gels prepared according to formula I[7] dry without cracking. The impregnation procedure can be interrupted by storing gels containing dimethyl sulfoxide ($\pm$PPO) at or below $-20°$. Excess PPO can be recovered rapidly from used solutions by adding 1 volume of PPO in dimethyl sulfoxide to 3 volumes of 10% ethanol followed by filtering after 10 min.[9]

The optimum concentration of PPO is sharp for gels containing $\geqslant 10\%$ acrylamide but much broader (and probably lower) for gels containing less acrylamide. A 10% (w/v) solution of PPO in dimethyl sulfoxide is more suitable for gels containing less than 5% acrylamide. The choice of procedure for gels consisting of both agarose and acrylamide is discussed in the next section.

Procedure for Using Organic Scintillators to Visualize Weak β-Emitting Isotopes in Agarose Gels[10]

The procedure described here[10] is based on an earlier method[9] which was subsequently found to be unsuitable for high concentrations of agarose and which gave variable results depending on the extent of the vacuum used to dry gels. In the revised procedure ethanol is used to carry PPO into the lattices of the gel.

Note that agarose gels dissolve in dimethyl sulfoxide, but that some gels which consist of mixtures of agarose and acrylamide can be impregnated by the procedure described in this section. When using agarose–acrylamide mixed gels the choice between ethanol and dimethyl sulfoxide as solvent should be determined by soaking a fragment of gel in each solvent. Only if it dissolves in dimethyl sulfoxide should ethanol be used. Gels which resist immersion in dimethyl sulfoxide should be impregnated by the method described in the previous section [using 10% (w/v) PPO in dimethyl sulfoxide for gels of less than 5% acrylamide].

Before impregnation with PPO the agarose gel is equilibrated with ethanol by soaking it in approximately 20 volumes of absolute ethanol for $\geqslant 30$ min followed by a second (and if necessary third) 30-min immersion in fresh ethanol to remove water. Then the gel is soaked in a 3% (w/v) solution of PPO in ethanol for $\geqslant 3$ hr. Nucleic acids will not diffuse in

---

[9] R. A. Laskey, A. D. Mills, and J. S. Knowland, Appendix in R. A. Laskey and A. D. Mills, *Eur. J. Biochem.* **56**, 335 (1975).

[10] R. A. Laskey, A. D. Mills, and N. R. Morris, *Cell* **10**, 237 (1977).

ethanol. Therefore, any of these steps can be elongated indefinitely. Finally the gel is immersed in water for ≤ 1 hr to remove ethanol and to precipitate PPO *in situ* and then it is dried under vacuum. The gel can be *gently* heated during drying. If it is exposed directly to steam before most of the water has been removed the agarose will melt!

The dried gel is exposed to a hypersensitized screen-type X-ray film (see above) at −70°.

Comments on the Procedure for Applying PPO to Agarose Gels. The optimum concentration of PPO for this procedure is only 3% (w/v). Higher concentrations quench and scatter the emissions resulting in lower sensitivity and poorer image quality. The earlier (now obsolete) procedure[9] referred to above used 10% PPO in methanol (which dissolves more PPO than ethanol) without the final aqueous precipitation step. However, the final PPO concentration was being drastically decreased inadvertently during gel drying, since the solution of PPO in methanol, rather than methanol alone, was sucked from the gel under vacuum. Methanol can be used instead of ethanol in the revised procedure if preferred.

Solutions of PPO in ethanol can be reused several times, since the optimum PPO concentration is reasonably broad. Excess PPO can be recovered from used solutions by adding water and filtering.

Procedure for Using Organic Scintillators to Visualize Weak β-Emitting Isotopes after Transfer to Nitrocellulose Filters[11]

Nitrocellulose filters can be impregnated with PPO by the method of Southern.[11] First the filter must be dried thoroughly in air to remove all traces of water. It is then dipped through a 20% (w/v) solution of PPO in toluene and dried in air.

The dried filter is exposed to hypersensitized screen-type X-ray film at −70°.

Comments on the Procedure for Applying PPO to Nitrocellulose Filters. It is essential that all water is removed from the filter before exposure to the PPO solution, or PPO will not enter. Therefore, the efficiency of PPO impregnation should be monitored, before exposure by looking for uniform fluorescence of the filter under uv illumination.

Quantitative Interpretation of Film Images

The film images obtained by any of the procedures described here can be quantitated by microdensitometry. Provided that the film has been hypersensitized by preexposure before use (to achieve absorbance incre-

---

[11] E. M. Southern, *J. Mol. Biol.* **98**, 503 (1975).

ments between 0.1 and 0.2), the absorbance of the film image will be proportional to the amount of radioactivity in the corresponding region of the gel and to the time of exposure[1] (for all exposure times between 1 hr and 3 months at $-70°C$). For maximum accuracy exposure at $-70°C$ is preferable.[1]

Film images obtained using intensifying screens or organic scintillators cannot be interpreted quantitatively when unfogged film is used because smaller amounts of radioactivity produce disproportionately faint images. Conversely they cannot be interpreted quantitatively if preexposure is too intense so that the increase in absorbance above the background absorbance of untreated film is greater than 0.2. This causes smaller amounts of radioactivity to produce disproportionately dense images.[1] Consequently it also decreases resolution substantially. Quantitative interpretation is also unreliable for film images which exceed an absorbance of approximately 1.5 (A_{540}) because the number of silver halide crystals which are available to record an emission decreases as more become saturated during exposure.

To confirm the accuracy of quantitation by microdensitometry, the amounts of radioactivity in selected slices of gel can be measured in a liquid scintillation spectrometer. Slices need not be hydrolyzed. Slices containing ^{32}P or ^{125}I are counted efficiently by surrounding the dried slice by conventional scintillants, whereas weak β emitters can be measured in slices of gel which have been impregnated with PPO and then dried without use of any additional scintillant in the vial.

Appendix

A paper describing the use of intensifying screens at low temperature has been published by Swanstrom and Shank.[3] The data presented in their careful study are generally in good agreement with Laskey and Mills[2] after correction for differences between the types of film used as standards for direct autoradiography. However, Swanstrom and Shank reported that preexposing the film had little if any effect on sensitivity. R. Swanstrom and P. R. Shank (personal communication) have now generously acknowledged that their observations were made using only large amounts of radioactivity in relatively short exposures, i.e., conditions where the effect of preexposure is minimal,[1] and that they have now confirmed that preexposure is required to achieve full enhancement when smaller amounts of radioactivity or longer exposure times are used.

Swanstrom and Shank also reported[3] that under some circumstances a second screen can substantially enhance detection efficiency when the screens are arranged as follows: screen A, sample, film, screen B. Although the distance between screen A and the film causes slight loss of resolution when compared to the resolution from a single screen, the image obtained is superior to that obtained when the screens are arranged: sample, screen A, film, screen B. Therefore the arrangement of screens suggested by Swanstrom and Shank[3] should be considered seriously when maximum sensitivity rather than maximum resolution is required.

[46] Recovery of DNA from Gels

By HAMILTON O. SMITH

Introduction

Agarose or polyacrylamide gel electrophoresis is widely used as a high resolution technique for fractionation of DNA molecules by size. It is valuable both as an analytical procedure and as a preparative procedure for isolating individual species of DNA molecules from a mixture. A major problem in preparative use is to recover DNA from individual gel bands in good yield and free of gel matrix contaminants. This article provides an overview of the methodology for such recovery.

General techniques for DNA separation on gels are detailed elsewhere in this volume. A first consideration in recovery of a particular DNA species from a gel is to locate the band of interest. This can be done by fluorescent staining with ethidium bromide. The band may then be excised from the gel with a razor blade under direct visualization by long wave uv light. (A short wave uv lamp introduces significant photochemical damage and should usually be avoided.) If the DNA is ^{32}P-labeled, an autoradiogram may be obtained, and the DNA bands can be located by superimposition of the gel over the film on a light box. A radioactivity survey meter is useful for monitoring excision accuracy. A gel slice containing the band is cut out and held over the meter probe for assay. Thin neighboring slices on either side of the excised region may then be cut from the gel and their radioactivity compared as a check on the excision accuracy.

There are four basic methods for recovery of DNA from gel slices. These are (1) electroelution, (2) elution by diffusion, (3) gel dissolution, and (4) extrusion of DNA by gel compression. No single method is best in all cases. Choice of a method depends on a number of factors including (1) the type of gel matrix, (2) gel concentration, (3) DNA size, (4) scale of the procedure, and (5) the level of gel contamination that can be tolerated in the recovered DNA solution.

METHODS IN ENZYMOLOGY, VOL. 65

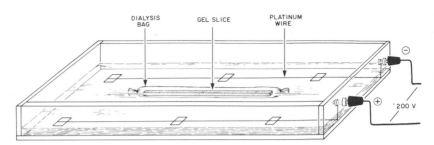

FIG. 1. Apparatus for elution of DNA from gel slices by electrophoresis.

With agarose gels, contamination of recovered DNA with varying amounts of agarose is almost a certainty, and in many cases this will interfere with subsequent enzymatic treatments or analysis. Methods of agarose removal include (1) equilibrium density gradient centrifugation, (2) extraction, and (3) chromatographic separation. Polyacrylamide contamination has not been noted as a significant problem in the literature.

In the treatment to follow, methods of DNA recovery will be described individually, followed by a section on methods of purification and concentration of recovered DNA, and a section on choice of method.

Methods of DNA Recovery from Gels

Electroelution. A simple and convenient procedure for electrophoretic elution of DNA from gel slices is that of McDonell *et al.*[1] A gel slice is placed inside a dialysis bag together with a small amount of low conductivity buffer, such as 5 mM Tris and 2.5 mM acetic acid (pH 8). (A long gel slice from a preparative slab gel is easily slipped into a long dialysis bag that is tied at one end and filled with buffer.) A shallow, rectangular plastic box with platinum electrodes running along opposite sides serves as the electroelution chamber (Fig. 1). The bag is tied at both ends, placed in the box between and parallel to the electrodes, and enough buffer is added to establish electrical contact with the bag, but not to cover it, as this simply adds to the conductance and electrical heating. Usually 15–30 min at a gradient of 10–15 V/cm is sufficient to transfer the DNA from the gel to the buffer in the bag. The buffer is then drained or pipetted from the bag and treated further if necessary to purify or concentrate the DNA solution.

[1] M. W. McDonell, M. N. Simon, and F. W. Studier, *J. Mol. Biol.* **110**, 119 (1977).

Excessive electrophoresis results in adherance of DNA to the bag. This has formed the basis of one method for purification of DNA from agarose contaminants which is worth brief mention. In this method,[2] DNA is purposely electrophoresed onto a dialysis membrane. DNA is then recovered by eluting the membrane with a small amount of the electrophoresis buffer for 1–2 hr at 4° with shaking.

Elution by Diffusion. This procedure, developed by Maxam and Gilbert,[3] is most commonly used for recovery of relatively small single- or double-stranded [^{32}P]DNA restriction fragments from polyacrylamide gels in preparation for DNA sequencing. However, it is also useful for labeled or unlabeled DNA recovery from agarose and is suitable even for relatively large-scale use. The procedure consists of crushing the gel slice into fine particles in a small amount of buffer and eluting by incubation for several hours at 25°–42°. DNA and buffer are then freed from gel particles by filtration through glass wool.

For small gel slices, Maxam and Gilbert recommended carrying out the elution in 1-ml plastic Eppendorf pipette tips. A small wad of glass wool is stuffed tightly into the point of each tip using a 3-mm glass rod. The glass wool should be siliconized to prevent DNA loss by adsorption. Tips are placed in the rims of 10 × 75-mm glass test tubes, 1 ml of 1% dichlorodimethylsilane (Eastman No. 9650) in toluene is added to each, and then gently centrifuged through the glass wool. (A Whisperfuge Model 1385, Damon/IEC Division or similar light weight, low speed instrument is recommended.) The recovered siliconizing solution is returned to the stock bottle. The tips are then rinsed twice with distilled water by centrifugation, placed in a beaker, and dried for 2 hr or more at 60°. Each pipette tip point is then carefully sealed by melting with a small flame.

To elute DNA from a gel slice, it is transferred to one of the pipette tips and ground to a paste with a siliconized 5-mm glass rod. Alternatively, the slice may be placed in a tuberculin syringe barrel and expressed through the syringe tip (without a needle) into a second syringe barrel. This is repeated several times between the two syringes before finally transferring the finely divided gel into the pipette tip. Excessive crushing or syringing of low percentage polyacrylamide gel slices (6% or less) is not advisable since the gel particles may plug the glass wool during centrifugation and some linear polyacrylamide may become solubilized. The eluting solution (0.6 ml), containing 0.5 M ammonium acetate, 0.01 M magnesium

[2] A. Ya. Stronglin, Yu. I. Kozlov, V. G. Debabov, R. A. Arsatians, and M. L. Zlochevsky, *Anal. Biochem.* **79**, 1 (1977).

[3] A. M. Maxam and W. Gilbert, *Proc. Natl. Acad. Sci. U.S.A.* **74**, 560 (1977).

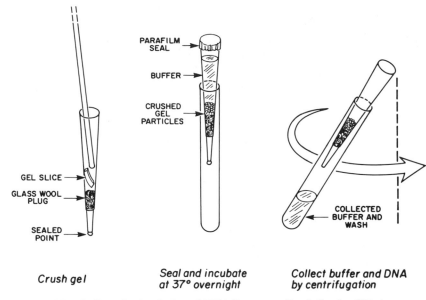

Fig. 2. Steps in the elution of DNA from a small gel slice by diffusion.

acetate, 0.1% sodium dodecyl sulfate, and 0.1 mM EDTA, is added and mixed evenly with the crushed gel. The pipette tip is then covered with Parafilm and incubated at 37° for several hours (usually overnight). Buffer and DNA are recovered by cutting off the sealed point, placing the pipette tip in a siliconized 10 × 75 mm test tube, centrifuging for a few minutes, and rinsing once with 0.2 ml of eluting solution (Fig. 2).

Recoveries are generally >80% despite the frequently unfavorable ratio of gel to elution buffer volume. Apparently, the configurations available to the DNA in the gel are sharply limited compared to free solution so that partitioning into the buffer is highly favored; >95% recoveries are not uncommon.

The DNA is most conveniently purified from the SDS-containing buffer by addition of 2.5 volumes of ethanol, chilling at −70° in a Dry Ice–alcohol bath for 10 min, and centrifuging at 8000 g for 10 min. The DNA pellet is redissolved in 0.4 ml of 0.3 M sodium acetate, reprecipitated with the addition of 1 ml of ethanol, the precipitate rinsed with ethanol without disturbing the pellet, and dried by vacuum before redissolving in a suitable buffer.

For large gel slices, the elution may be carried out in 5 or 10 ml B-D Plastipak syringes. A disc of appropriate diameter is cut from a 1.6 mm thick porous polyethylene sheet (F 13638, Bel-Art Products, Pequannock, New Jersey) with a cork borer and inserted snugly in the bottom of the syringe to serve as a bed support. The syringe tip may be sealed by melting in a flame. After overnight incubation at 37°, DNA and buffer are recovered by cutting off the sealed tip, suspending the syringe in the rim of a tube so that the tip is above the bottom, and centrifuging at low speed (200 rpm) for a few minutes. Suitable tubes for use in centrifugation are a Corex No. 8441, 15-ml tube with the 5-ml syringe and a conical No. 8140 50-ml Pyrex tube with the 10-ml syringe.

DNA Recovery by Gel Compression. Agarose gels are compressible, particularly after disruption of the gel matrix by freezing. This makes it possible to recover a good portion of the interstitial buffer and DNA by squeezing the gel mechanically or in a centrifugal field.

The freeze-squeeze procedure of Thuring *et al.*[4] is simple, rapid and very useful for recovering high molecular weight DNAs from low percentage agarose gels. A gel slice is enclosed in a small envelope of Parafilm and cooled to $-20°$ (several minutes in a deep freeze or 15–20 sec on Dry Ice). The frozen gel slice is then firmly squeezed by thumb pressure against a table top for about 10 sec. For large gel slices, a Parafilm-coated garlic press has been suggested. The expressed buffer forms a droplet on the hydrophobic Parafilm surface and is readily separated and recovered from the flattened gel slice with a Pasteur pipette. Small particles of agarose are removed by centrifugation in 1.5-ml tubes of an Eppendorf centrifuge model 5412. Recovery of fluid may be monitored by weighing. Typically, fluid (DNA) recoveries are 60–70% with 0.5% gels and 40–45% with 2% gels.

A slight modification of this simple procedure is equally good. The gel slice is placed directly into an 1.5-ml Eppendorf tube, rapidly frozen and thawed twice in a Dry-Ice–alcohol bath, and then centrifuged 5 min in the Eppendorf centrifuge (15,000 rpm). Supernatant fluid recovery is 85–90% with a 0.5% gel and 40–50% with a 2% gel.

Ultracentrifugation (100,000 g for 30 min) of unfrozen gel slices suspended in buffer has been reported to give 30–50% recoveries.[5] Various combinations of diffusion and compression techniques have also been employed to improve yields.[6,7]

DNA Recovery by Gel Dissolution. In an agarose gel, the strands of

[4] R. W. J. Thuring, J. P. M. Sanders, and P. Borst, *Anal. Biochem.* **66,** 213 (1975).
[5] C. Brack, H. Eberle, T. A. Bickle, and R. Yuan, *J. Mol. Biol.* **108,** (1976).
[6] F. C. Wheeler, R. A. Fishel, and R. C. Warner, *Anal. Biochem.* **78,** 260 (1977).
[7] S. G. Clarkson, V. Kurer, and H. O. Smith, *Cell* **14,** 713 (1978).

agarose form a three-dimensional network held together by double helical junctions containing interchain hydrogen bonds.[8] Chaotropic agents, such as potassium iodide[9] or sodium perchlorate,[10] readily dissolve the gel matrix when present in high concentration. Thus, it is possible to solubilize a gel slice completely under relatively mild conditions that do not damage or denature the DNA. DNA and dissolved agar can then be separated by chromatography or isopycnic centrifugation.

Potassium iodide is currently the agent of choice.[9] A stock solution in equilibrium with solid KI is prepared and stored in a dark bottle at room temperature. To prevent oxidation of I^- to free I_2, 1 mM $Na_2S_2O_3$ should be included; the presence of free I_2 is indicated by a yellowish color. The gel slice is placed in a tube, crushed with a glass rod, and mixed with several volumes of the saturated KI solution to give an agarose concentration of 0.2% or less. The mixture is incubated at 37°–45° for about 1 hr until solubilization is complete. Low percentage gels (<0.8%) are quite readily soluble, while higher percentage gels require longer incubation, and in some cases, addition of more saturated KI solution or adjustment of the KI concentration with solid KI to compensate for the gel buffer content. The mixture can be aspirated with a Pasteur pipette from time to time to test for completion of solubilization. Agarose at a concentration of 1% (w/v) will remain soluble in 2 M KI at room temperature.

Two methods are commonly used to separate the dissolved agarose and DNA: isopycnic centrifugation in a KI gradient[9] and hydroxylapatite chromatography.[10,11] For the former, the DNA and agarose containing KI solution is adjusted to a final density of 1.5 g/ml and an agarose concentration of <0.2%. The adjustment may be made using the relationship, $n_{20}^D = 0.1731\rho + 1.1617$, which is valid in the range $1.3 < \rho < 1.7$, where n_{20}^D is the refractive index and ρ is the density of KI in g/ml. The solution is placed in a 5-ml Spinco tube and centrifuged for 40 hr at 40,000 rpm in an SW50.1 rotor at 20°. DNA bands at $\rho \sim 1.46$ g/ml while agarose bands at $\rho \sim 1.6$ g/ml. At least 50 μg of DNA may be run in a 5 ml gradient and the DNA band may be readily visualized with a long wave uv light if ethidium bromide, 50 μg/ml, is included in the gradient (Fig. 3). Alternatively, 10 μl of each fraction may be spotted onto glass fiber filters,[9] dried, and stained for 5 min in a 0.5 μg/ml ethidium bromide solution. Spots containing >0.02 μg of DNA are readily visible by short wave uv light.

[8] D. A. Rees, *J. Polym. Sci., Part C* **28**, 261 (1969).
[9] N. Blin, A. V. Gabain, and H. Bujard, *FEBS Lett.* **53**, 84 (1975).
[10] N. M. Wilkie and R. Cortini, *J. Virol.* **20**, 211 (1976).
[11] R. Wu, E. Jay, and R. Roychoudhury, *Methods Cancer Res.* **12**, 87 (1976).

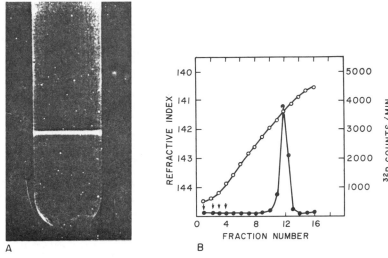

A   B

FIG. 3. (A) A band of DNA in a potassium iodide density gradient visualized by ethidium bromide fluorescence. The gradient (mean density of 1.45 g/ml) contains 25 μg of DNA and 50 μg/ml of ethidium bromide and was centrifuged at 20° for 40 hr at 40,000 rpm in a Spinco rotor SW56. (B) Banding pattern of T5[³²P]DNA and agarose in a potassium iodide density gradient. Arrows indicate the position of the agarose as detected by precipitation in 50% ethanol. (Courtesy of N. Blin, A. V. Gabain, and H. Bujard and the North-Holland Publishing Company, Amsterdam.)[9]

After collection, ethidium may be removed by extraction with *n*-butanol or isoamyl alcohol, and KI may be removed by dialysis.

Hydroxylapatite chromatography is an alternate procedure to centrifugation. BioGel HTP is suspended in 0.01 M sodium phosphate buffer, pH 6.7, and a 0.5-ml bed volume is packed into a small column, e.g., a Pasteur pipette plugged with glass wool. BioGel HTP has a capacity of 100–200 μg native DNA per milliliter of bed volume. The DNA and agarose-containing KI solution is adjusted to 0.01 M phosphate buffer and 1 mM EDTA and loaded onto the column at room temperature. The column is rinsed free of agarose by elution with 10 column bed volumes of saturated KI solution in phosphate–EDTA buffer. Potassium iodide may then be rinsed away with 5–10 volumes of phosphate–EDTA buffer prior to elution of DNA with 0.4 M phosphate buffer (pH 6.7) and 1 mM EDTA. Phosphate may be removed by dialysis against a suitable buffer if desired. Wu[11] has suggested the use of Tris phosphate buffer for elution instead of sodium phosphate because of its greater solubility in ethanol. It is then possible to concentrate eluted DNA fractions conveniently by ethanol precipitation. A 1 M stock solution of Tris phosphate (pH 7.2) contains 6.9 ml of 85% (by weight) phosphoric acid and 25 g of Tris base in a final volume of 100 ml. One volume of 0.2 M Tris phosphate is soluble in 2.5 volumes of ethanol. The same elution schedule may be used as with sodium phosphate.

Removal of Gel Contaminants from DNA

DNA solutions recovered from polyacrylamide or agarose gels generally contain variable amounts of gel matrix material. Vigorous grinding or syringing of polyacrylamide releases soluble linear polyacrylamide molecules which may interfere with electron microscopy and DNA sequencing protocols, but which are not particularly inhibitory to enzymes. Contaminating polyacrylamide may be evident as a white precipitate on addition of ethanol. With agarose, the contaminating polysaccharide molecules may be visible as a flocculent precipitate on addition of ethanol. Furthermore, the contaminated DNA solutions are frequently refractory to treatment by a variety of enzymes. The inhibitory component seems to be polysaccharide sulfate (agaropectin) which is present in small amounts in purified agarose,[12] although some interference from the neutral agarose has not been ruled out for individual enzymes. Wu[13] has shown that polynucleotide kinase is strongly inhibited by a variety of polysaccharide sulfates, including dextran sulfate, heparin, amylopectin sulfate, crude agar, and to a lesser degree, purified agarose. Other workers have noted that many of the restriction endonucleases will not cleave DNA taken from agarose gels unless efforts are made to remove the agarose contaminants.[2,9,14] The mechanism of inhibition appears to be enzyme binding; heparin, for example, is known to bind many DNA enzymes.

Several methods for DNA purification are in use. These include (1) extraction with phenol,[1] (2) chromatography on DEAE cellulose[7,15] or on hydroxylapatite,[6,10,11] and (3) centrifugation in KI equilibrium density gradients.[9] (In KI, polysaccharide is considerably more dense than DNA; with CsCl, separation is apparently poor.) Procedures for hydroxylapatite chromatography and KI density gradients are described above under dissolution of gels.

Phenol extraction is a simple and efficient procedure for removing agarose.[1] The DNA solution in 10 mM Tris-Cl (pH 8), 1 mM EDTA, or other suitable buffer is extracted with an equal volume of water-saturated

[12] B. Russel, T. H. Mead, and A. Polson, *Biochim. Biophys. Acta* **86**, 169 (1964).
[13] R. Wu, *Biochem. Biophys. Res. Commun.* **43**, 927 (1971).
[14] P. P. Stepien, U. Bernard, H. J. Cooke, and H. Küntzel, *Nucleic Acids Res.* **5**, 317 (1977).
[15] H. O. Smith and M. L. Birnstiel, *Nucleic Acids Res.* **3**, 2387 (1976).

phenol. Phases are separated by centrifugation at 10,000 g for 5 min. Agarose is found at the interface or in the phenol layer. The supernatant is carefully pipetted off and dialyzed, or extracted twice with ether and ethanol-precipitated to remove phenol. Blin *et al.*[9] report continued presence of enzyme inhibitory material after phenol extraction. It seems likely that the ionic species in agarose are less effectively removed than is neutral agarose.

DEAE–cellulose chromatography gives high purified DNA.[15] DNA in 0.15 M NaCl, 50 mM Tris-HCl, (pH 8), and 1 mM EDTA is loaded onto a small DE52 cellulose (Whatman) column equilibrated with the same buffer. Capacity of DE52 is greater than 200 μg DNA/ml bed volume. Inclusion of ethidium bromide as a DNA label, at a 5–10% dye to DNA (w/w) ratio in the DNA sample and at 1 μg/ml in the eluting buffers, aids in visual monitoring of DNA binding and elution. The loaded column is washed with at least 10 volumes of the 0.15 M NaCl-containing buffer to remove neutral and charged species of agarose before eluting the DNA with buffer containing 1 M NaCl. Eluted DNA may be precipitated with 2.5 volumes of ethanol, collected by centrifugation, dried, and redissolved in a small volume of a suitable buffer. Recovery is 60–80% for small DNAs (up to 5 kb), but may be less than 50% for large DNAs (10–50 kb).

Occasionally DNA is recovered in volumes too large for convenient or efficient concentration of DNA by ethanol precipitation. A useful procedure in such cases is extraction with 2-butanol. The 2-butanol forms a two-phase system such that a large fraction of the water is partitioned into the upper 2-butanol layer, while the lower aqueous phase retains DNA, salts, and a small amount of the 2-butanol.[16] To apply the procedure, the DNA solution is initially shaken with an equal volume of 2-butanol. The phases are separated by brief centrifugation and the upper 2-butanol phase is removed by a pipette and discarded. The first extraction saturates the aqueous phase with 2-butanol, and little concentration occurs. Subsequent extractions result in rapid reduction of the aqueous volume. Concentration to more than 100-fold is readily achieved. The ease of concentration depends on buffer and salt concentration as well as the relative volume of 2-butanol used at each step. Since salts are concentrated in the aqueous phase along with DNA, the procedure is best carried out with DNA solutions in a low molarity buffer, e.g., 10 mM Tris-HCl (pH 8), 1 mM EDTA. High salt also decreases the solubility of the water in the 2-butanol layer, e.g., 5 M CsCl solutions are refractory to concentration. After concentration, 2-butanol is easily removed from the DNA solution by ether extraction or dialysis.

[16] D. W. Stafford and D. Bieber, *Biochim. Biophys. Acta* **378**, 18 (1975).

Choice of Method

Polyacrylamide gel slices are not soluble by any mild procedure nor are they easily compressible; thus one is limited to diffusion or electroelution methods for DNA elution. The Maxam-Gilbert procedure is designed for small-scale recovery from polyacrylamide gels, gives excellent yields for submicrogram quantities of low molecular weight (less than 1000 base pairs), native, or denatured DNA, and is convenient for multiple samples. The recovered DNA is of satisfactory purity for enzymatic treatments or DNA sequencing. Electroelution should be considered for higher molecular weight DNA, large gel slices, or low percentage gels (less than 6%) which may compact on centrifugation making recovery of buffer difficult. The procedure as described by McDonell et al.[1] is simple, efficient, and fast and deserves wider application than it has received.

For agarose gels, any of the four basic methods may be used. Solution in KI followed by banding in KI density gradients gives the highest recoveries and very pure DNA. The major drawbacks are expense and time. The freeze-squeeze procedure and its variations are useful for recovery of large DNA from small slices of low percentage gels, particularly when one is not concerned with moderate losses, when convenience and speed are desired, and when mild agarose contamination can be tolerated. The diffusion methods and electroelution are adaptable to large slices, and when combined with hydroxylapatite or DEAE–cellulose chromatography for removal of contaminants, have been used for large scale preparations (a few hundred micrograms).[7] Final recoveries of about 50% should be expected.

[47] The Analysis of Nucleic Acids in Gels Using Glyoxal and Acridine Orange

By GORDON G. CARMICHAEL and GARY K. MCMASTER

Introduction

For accurate determination of the molecular weights of nucleic acids using gel electrophoresis techniques, the molecules to be compared must have equivalent conformations. This can be achieved by removing native secondary and tertiary structure with a variety of chemical denaturants, thus reducing the electrophoretic mobility to a simple function of polynucleotide molecular weight, gel composition, and voltage gradient applied.

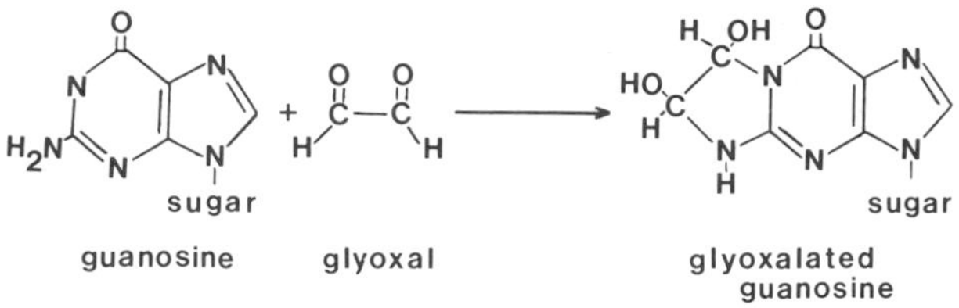

FIG. 1. The glyoxalation of guanosine. The structure of the adduct was determined by Shapiro and Hachmann.[6]

Most currently available methods of analyzing denatured nucleic acids in gels rely on the continual presence of denaturants in the gel to maintain this disruption of ordered conformation. For some applications, however, it would be advantageous to irreversibly denature the nucleic acid molecules before running, so that the gel could be constructed in the absence of such denaturants. This would allow for greater flexibility in the type of gel which could be used, as well as in the choice of samples to be analyzed at the same time. One reagent which accomplishes this is glyoxal (ethanedial), which denatures RNA and DNA virtually irreversibly at neutral or slightly acidic pH.[1-3] Glyoxal may react reversibly with all bases of both DNA and RNA,[4,5] but forms a stable adduct with guanosine at or below pH 7 (Fig. 1).[4-6] It can be seen that glyoxalation sterically hinders the formation of G-C base pairs, thus preventing the renaturation of treated RNA or DNA molecules.

In this article we describe in detail a previously published method for denaturing RNA or DNA molecules with glyoxal, followed by electrophoresis through either polyacrylamide- or agarose-containing slab gels in a low ionic strength buffer (which may help to maintain the fully denatured, extended conformation[2]).[7] Using this method reliable molecular weight estimates for both RNA and DNA molecules of widely varying sizes and G + C contents were obtained; glyoxalated DNA and RNA molecules were shown to lie on the same log molecular weight versus mobility curve, or on very similar curves.[7] This means that DNAs of known sequence, and therefore of precisely known molecular weights,

[1] B. R. Brooks and O. L. Klammerth, *Eur. J. Biochem.* **5**, 178 (1968).

[2] M. J. Hsu, H. J. Hung, and N. Davidson, *Cold Spring Harbor Symp. Quant. Biol.* **38**, 943 (1973).

[3] J. R. Hutton and J. G. Wetmur, *Biochemistry* **12**, 558 (1973).

[4] K. Nakaya, O. Takenaka, H. Horinishi, and K. Shibata, *Biochim. Biophys. Acta* **161**, 23 (1968).

[5] N. E. Broude and E. I. Budowsky, *Biochim. Biophys. Acta* **254**, 380 (1971).

[6] R. Shapiro and J. Hachmann, *Biochemistry* **5**, 2799 (1966).

[7] G. K. McMaster and G. G. Carmichael, *Proc. Natl. Acad. Sci. U.S.A.* **74**, 4835 (1977).

can be used as size standards for the estimation of RNA molecular weights. Furthermore, both glyoxal-denatured and untreated DNA and RNA can be analyzed on the same slab gel; glyoxal in one sample does not interfere with migration of a polynucleotide in an adjacent lane. Thus, for example, two RNA molecules held together by secondary or tertiary structure may move as one band in one lane and as two bands when glyoxal-denatured and electrophoresed in a neighboring lane.

Also in this article we describe the use of the metachromatic stain, acridine orange, for the visualization of nucleic acids in gels. Acridine orange can interact in two different ways with polynucleotides. When intercalated between the stacked bases of double-helical nucleic acids, acridine orange manifests a green fluorescence at 530 nm.[8,9] When the polynucleotide is predominantly single stranded, however, acridine orange can bind electrostatically to, and stack along, the phosphate backbone, fluorescing red at 640 nm.[10,11] Since the commonly used stain, ethidium bromide, stains glyoxalated nucleic acids poorly,[7] acridine orange has proved ideal for detecting native and glyoxalated nucleic acids in gels.[7] Its metachromasy not only visually confirms glyoxal denaturation but also makes it a powerful reagent for rapid determination of gross nucleic acid structure after gel electrophoresis.

Materials

Reagents

Glyoxal, a 30% aqueous solution (technical grade) was obtained from Fluka AG Chemische Fabrik, Buchs, Switzerland. After opening it is stored in a brown bottle at 4° for up to 1 year.

Acridine orange, standard for microscopy, was from Fluka. A stock solution of 10 mg/ml in H_2O is stored in the dark at 4°.

Dimethyl sulfoxide, analytical grade, was from Fluka.

Mixed-bed ion exchange resin, AG-501-X8(D), containing a blue pH indicator dye, was from Bio-Rad.

Standard buffer: 0.01 M NaH_2PO_4/Na_2HPO_4, pH 7.0. This buffer is prepared by dilution (with twice-distilled, deionized H_2O) from a stock of 0.5 M buffer, which has been filtered through Millipore nitrocellulose membrane filters.

[8] L. S. Lerman, J. Mol. Biol. 3, 18 (1961).
[9] L. S. Lerman, Proc. Natl. Acad. Sci. U.S.A. 49, 94 (1963).
[10] D. R. Bradley and M. K. Wolf, Proc. Natl. Acad. Sci. U.S.A. 45, 949 (1959).
[11] A. Blake and A. R. Peacocke, Biopolymers 6, 1225 (1968).

Glyoxal Deionization. Commercial 30% or 40% aqueous solutions of glyoxal are technical grade and consist of various hydrated forms of glyoxal as well as small amounts of oxidation products and chemically related substances, such as glycolic acid, glyoxylic acid, and formic acid.[12] Unless deionized, such glyoxal solutions cause extensive degradation of RNA.[7,13] Deionization is achieved by passing a solution of glyoxal through a column of mixed bed ion exchange resin, AG-501-X8(D), containing a pH indicator dye. Pour 1–2 cm^3 of dry mixed-bed resin beads into a small-tipped 10-ml glass pipette which has been truncated at the top. Load onto this column 1–2 ml of concentrated glyoxal and collect the flow through. Reload the flow through onto the column a second and a third time. At this point the pH of the glyoxal solution should have risen from about 1 to about 6, and the ion exchange resin should still be blue. If the pH indicator dye in the beads has turned yellow, the glyoxal is not yet fully deionized and must be repassed through fresh ion exchange resin until the pH indicator remains blue. Complete deionization is especially important before glyoxalation of very large RNAs. It is recommended that glyoxal be deionized just before each use (to minimize RNA degradation by easily formed air oxidation products[14]), although such purified material has been stored at 4° and used for 48 hr with no noticeable ill effects. Deionized glyoxal has also been stored safely indefinitely at −20°, in completely filled, capped plastic tubes (Eppendorf, 1.5 ml).

Glyoxal Methods

Sample Preparation

Deproteinization. Protein, as well as nucleic acids, reacts with glyoxal, and its presence may result in anomalous migration or sticking of DNA or RNA to the top of the gel if it is not removed prior to glyoxalation. Therefore, all samples should be deproteinized by phenol extraction and ethanol precipitation.

Denaturation and Glyoxalation. DNA and RNA samples (about 0.5–1.0 μg per band for nonisotopically labeled material) are denatured by incubation for 1 hr at 50° in small, tightly capped Eppendorf plastic tubes (250 μl size), in standard buffer containing 50% (v/v) dimethyl sulfoxide and 1.0

[12] A. Rose and E. Rose, eds., "The Condensed Chemical Dictionary," 6th ed., p. 545. Van Nostrand-Reinhold, Princeton, New Jersey, 1961.
[13] G. K. McMaster, Ph.D. Thesis, University of Lausanne (1978).
[14] J. B. Hendrickson, D. J. Cram, and G. S. Hammond, "Organic Chemistry," 3rd ed., p. 757. McGraw-Hill, New York, 1970.

M glyoxal, freshly diluted from the concentrated, deionized stock solution.[15] After incubation the samples are immediately loaded onto the gel; the presence of dimethyl sulfoxide makes them sufficiently dense for easy loading.

Gel Electrophoresis

The gel composition and running conditions may be chosen from among the plethora of available techniques so as to best suit the experimental needs. The gel matrix may consist of polyacrylamide, agarose, or a composite of the two, such as described by Peacock and Dingman.[16] Gels may be cast in either a vertical or a horizontal[17] apparatus. Several considerations, however, should be kept in mind.

1. For best results with RNA, all solutions should be sterilized by Millipore filtration, and acrylamide and N,N'-methylenebisacrylamide should be recrystallized by the method of Loening.[18]

2. For accurate molecular weight determinations very low ionic strength (0.01 M) buffer at neutral pH (such as our standard buffer) should be used as the gel buffer and the electrophoresis running buffer so as to maintain the denatured molecules in an extended conformation. Since the low ionic strength phosphate buffer is slowly consumed during electrophoresis, it should be recirculated, if possible, and for longer electrophoresis times it should be changed every 2.5 hr. Current should not exceed 45 mA. Specific electrophoresis conditions which have proved useful for RNA molecular weight determinations will be discussed below.

3. The reversibility of the glyoxal–guanosine adduct is an important phenomenon above pH 8 (see below), so high pH buffers should be avoided.

RNA Molecular Weight Determinations

Standards. Molecular weight standards should be carefully chosen; they should be of precisely known size, and there should be enough of

---

[15] Maximal denaturation results in about a twofold reduction in the electrophoretic mobility of RNAs. Increasing the glyoxal concentration above 1 M does not affect the subsequent electrophoretic mobility of polynucleotides, but lower concentrations may result in incomplete denaturation of RNAs containing extensive secondary structure and consequently less polynucleotide extension and anomalously fast mobilities in gels. Incubation at temperatures above 50° is not recommended, as this may result in glyoxal decomposition and nucleic acid degradation. Glyoxalation and denaturation are generally not complete in 1 hr at only 37°.

[16] A. C. Peacock and C. W. Dingman, *Biochemistry* **7**, 668 (1968).

[17] M. W. McDonell, M. W. Simon, and F. W. Studier, *J. Mol. Biol.* **110**, 119 (1977).

[18] U. E. Loening, *Biochem. J.* **102**, 251 (1967).

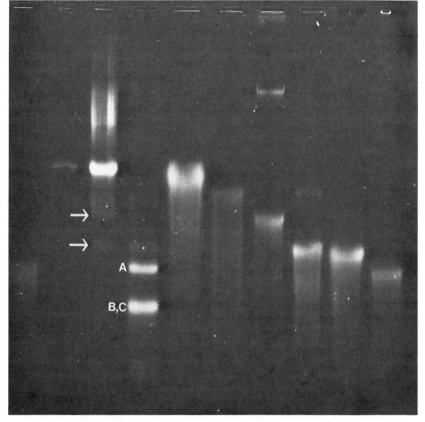

FIG. 2. Glyoxal-denatured nucleic acids electrophoresed in a horizontal 1% agarose gel. Samples were denatured with dimethyl sulfoxide and glyoxal as described in the text and then electrophoresed for 2 hr at 100 V in standard buffer. All bands appeared red after acridine orange staining. (a) One microgram of 16 S *E. coli* rRNA. (b) Four-tenths micrograms of SV40 full-length linear DNA. (c) Two micrograms of SV40 DNA partially digested with restriction endonucleases *Hpa*II and *Bam*HI to yield fragments of 57 and 43% unit length. [From M. Fried and B. Griffin, *Adv. Cancer Res.* **24,** 67 (1977).] (d) Two micrograms of SV40 *Hin*dIII DNA fragments A, B, and C (34, 22.5, and 20.5% genome length, respectively. [From K. Danna, G. H. Sack, Jr., and D. Nathans, *J. Mol. Biol.* **78,** 363 (1973).] B,C are not resolved, due to overloading. (e) Two micrograms of 28 S mouse rRNA. (f) One microgram of 26 S *Physarum polycephalum* rRNA. (g) One microgram of 23 S *E. coli* rRNA. Upper band is high molecular weight DNA. (h) One microgram of 19 S *Physarum polycephalum* rRNA, slightly contaminated with 26 S rRNA. (i) One microgram of 18 S mouse rRNA. (j) One microgram of 16 S *E. coli* rRNA. [Reprinted from G. K. McMaster and G. G. Carmichael, *Proc. Natl. Acad. Sci. U.S.A.* **74,** 4835 (1977).]

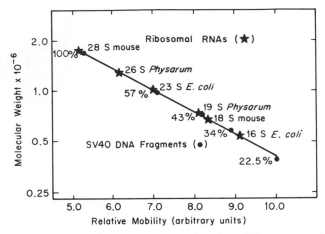

FIG. 3. Semilogarithmic plot of relative electrophoretic mobility versus molecular weight for the samples shown in Fig. 2. DNA bands, represented by dots, were used as standards to define a straight line. The numbers beside the dots refer to the relative sizes of these fragments to full-length SV40 DNA.[19,20] The stars represent the mobilities of the RNAs analyzed which have been placed on this line in order to interpolate molecular weights. [Reprinted from G. K. McMaster and G. G. Carmichael, *Proc. Natl. Acad. Sci. U.S.A.* **74**, 4835 (1977).]

them to accurately determine the log molecular weight versus mobility relationship. Since it has been shown that glyoxal-denatured DNAs and RNAs of the same size migrate identically, or nearly so, convenient standards may be derived by cleavage of viral DNAs of known sequence with various bacterial restriction endonucleases. Simian virus 40 (SV40) DNA[19,20] and coliphage ϕX174 DNA[21] are two excellent and commercially available sources of such standards.

Electrophoresis and Interpolation of Molecular Weights. Both DNA standards and RNA samples are denatured with glyoxal and dimethyl sulfoxide as previously described, electrophoresed through a gel matrix in standard buffer, and the bands detected by acridine orange staining (see below) or by autoradiography if they are radioactively labeled. An example of a horizontal 1% agarose gel containing denatured restriction fragments of SV40 DNA as markers and a variety of denatured rRNAs is shown in Fig. 2. The distance from the sample well to the top of each band

[19] V. B. Reddy, B. Thimmappaya, R. Dhar, K. N. Subramanian, B. S. Zain, J. Pan, P. K. Ghosh, M. L. Celma, and S. M. Weissman, *Science* **200**, 494 (1978).
[20] W. Fiers, R. Contreras, G. Haegeman, R. Rogiers, A. Van de Voorde, H. Van Heuverswyn, J. Van Herreweghe, G. Volckaert, and M. Ysebaert, *Nature (London)* **273**, 113 (1978).
[21] F. Sanger, G. M. Air, B. G. Barrell, N. L. Brown, A. R. Coulson, J. C. Fiddes, C. V. Hutchinson, III, P. M. Slocombe, and M. Smith, *Nature (London)* **265**, 687 (1977).

TABLE I

COMPARISON OF PUBLISHED ESTIMATES FOR RNA MOLECULAR WEIGHTS WITH THOSE
DETERMINED BY THE GLYOXAL METHOD[a,b]

| | MW × 10⁻⁶ | |
| | Glyoxal gel | Published |
RNA		
16 S *E. coli*	0.55 ± 0.03	0.54 ± 0.02[c,d]
23 S *E. coli*	1.03 ± 0.03	1.07[d]
18 S mouse	0.68 ± 0.02	0.68 ± 0.06[d,e]
28 S mouse	1.75 ± 0.05	1.74 ± 0.1[e]
19 S *Physarum*	0.70 ± 0.02	0.76 ± 0.01[f]
26 S *Physarum*	1.29 ± 0.03	1.37 ± 0.05[f]

[a] Adapted from G. K. McMaster and G. G. Carmichael, *Proc. Natl. Acad. Sci. U.S.A.*
74, 4835 (1977).

[b] RNA molecular weights were determined after gel electrophoresis from standard
curves derived using denatured SV40 DNA restriction endonuclease fragments of
known size (Fig. 2). Molecular weights are expressed as the sodium salt of the polynucleotide. Each value shown reflects the average of at least five separate determinations
using different gel systems.

[c] P. Fellner, *in* "Ribosomes" (M. Nomura, A. Tissieres, and P. Lengyel, eds.), p. 169.
Cold Spring Harbor Lab, Cold Spring Harbor, New York, 1974.

[d] W. M. Stanley and R. M. Bock, *Biochemistry* **4**, 1302 (1965).

[e] P. K. Wellauer, I. B. Dawid, D. E. Kelley, and R. P. Perry, *J. Mol. Biol.* **89**, 397 (1974).

[f] Determined under nondenaturing conditions. A. Zellweger and R. Braun, *Exp. Cell
Res.* **65**, 413 (1971).

is measured. RNA molecular weights are determined by plotting log molecular weight versus relative mobility for the DNA standards, drawing a
line through these points, and then placing the RNA mobilities on this line
and interpolating their sizes. In Figure 3 the DNAs shown in Fig. 2 were
used to define such a standard curve, then the rRNA mobilities were
placed on this line. Table I shows the values obtained for the RNA species
analyzed in Fig. 2 and compares them with published values. The variability shown for the glyoxal-determined molecular weights represents the
highest and lowest values obtained after electrophoresing these RNA
species in gels of different composition, at different voltages, and for
different lengths of time.

It is crucial for accurate RNA molecular weight estimation that the
interpolation be made from a standard curve where relative mobilities are
directly proportional to the logarithms of their molecular weights, as is the
case in Fig. 3. Table II shows this useful "linear range" for various gels
electrophoresed for the given times at the voltages shown.

A major problem for very large nucleic acid molecules is a nonlinear
migration versus log molecular weight relationship. This probably results

TABLE II
GELS USED FOR MOLECULAR WEIGHT ESTIMATIONS[a]

| Gel[b] | Running conditions | | Approximate "linear range"[c] $(\times 10^{-6})$ |
	Time (min)	Voltage (V)	
6.0% Acrylamide	195	100	0.025–0.20 (4 S–9 S)
2.5% Composite	105	130	0.025–0.20 (4 S–9 S)
2.5% Composite	180	130	0.20 –0.70 (9 S–18 S)
2.5% Composite	300	130	0.50 –1.4 (16 S–26 S)
2.0% Composite	300	130	0.50 –1.75 (16 S–28 S)
1.75% Composite	180	140	0.50 –1.75
1.5% Composite	115	130	0.20 –1.0
1.5% Agarose	120	100	0.20 –1.0
1.0% Agarose horizontal	120	100	0.40 –1.75

[a] Adapted from G. K. McMaster and G. G. Carmichael, *Proc. Natl. Acad. Sci. U.S.A.* **74,** 4835 (1977).

[b] Composite gels always contained 0.5% agarose and the indicated percentage of acrylamide. Gels were 10 cm long and 3 mm thick.

[c] Values are expressed as molecular weight, with sedimentation values in parentheses. Within this "linear range," the relative mobilities of glyoxal-denatured nucleic acids are directly proportional to the logarithms of their molecular weights.

when they are deformed from their equilibrium conformations by migration through the gel matrix at a voltage gradient which is too steep.[17,22]

Although we have not yet determined the optimum conditions for molecular weight determinations of glyoxal-denatured RNAs larger than about 2×10^6, it is likely that the combination of very low percentage horizontal agarose gels and very low voltage gradients[17,23] will greatly extend the range of the glyoxal method.

Remarks on Glyoxal Reversibility

As mentioned earlier in this article, the glyoxal–guanosine adduct is known to be stable at neutral or acidic pH. This adduct is completely stable for days at pH <6,[6] and at pH 7 the adduct is stable for at least 20 hr at 20°.[5] However, at alkaline pH the adduct may be easily reversed. Broude and Budowsky[5] measured the rate constant of decomposition of

[22] H. Lehrach, D. Diamond, J. M. Wozney, and H. Boedtker, *Biochemistry* **16,** 4743 (1977).
[23] W. L. Fangman, *Nucleic Acids Res.* **5,** 653 (1978).

the glyoxal–guanosine adduct to the starting products as a function of pH in phosphate or bicarbonate buffers at 20°. This rate constant is about 0.001 min⁻¹ at pH 8, about 0.02 min⁻¹ at pH 10, and about 0.06 min⁻¹ at pH 11. Complete reversion of the glyoxal–guanine adduct to guanine was observed after 21 hr at pH 7.8, 25°, in a sodium phosphate buffer,[6] while glyoxal-denatured DNA could be reannealed after treatment for 76 hr at pH 8, 45°.[24] In addition, it has been shown that glyoxal is rapidly converted to glycolic acid by the Cannizzaro reaction at elevated pH.[25] Thus, as it is removed from RNA or DNA at high pH, glyoxal is destroyed. Borate ions markedly stabilize the adduct; no decomposition was observed in 48 hr at pH 10.0 in the presence of sodium borate.[5]

Another 1,3-dicarbonyl reagent, Kethoxal,[26] interacts with guanosine analogously to glyoxal[6,27,28] and could perhaps be substituted for glyoxal in the gel techniques described here. After 95% of the Kethoxal was removed from treated tRNA by incubation at 37° for 20 hr at pH 7.6, about 50% of the original tRNA accepting activity was recovered.[27] Kethoxal was also removed from RNA in 2 hr at 37° in 0.013 M Tris base.[28]

These data on the reversibility of the adduct suggest the potential suitability of the glyoxal gel method for experiments where the recovery of nucleic acids in a biologically active form, or in a form capable of molecular hybridization, is desired. This greatly expands the usefulness of the technique. Glyoxal denatured RNAs have been successfully transferred to diazobenzyloxymethyl paper after a brief treatment at pH 11 to remove the glyoxal[29] and have been subsequently detected by hybridization with specific DNA probes according to the method of Alwine et al.[30]

Acridine Orange Staining

Staining Procedure

Acridine orange is added to standard buffer to a final concentration of 30 μg/ml. Gels containing polyacrylamide are stained by immersion in this solution for 15 min in the dark at 22°. Agarose gels are stained by immer-

[24] J. R. Hutton and J. G. Wetmur, *Biochemistry* **12**, 558 (1973).
[25] P. Salomaa, *Acta Chem. Scand.* **10**, 311 (1956).
[26] Kethoxal (β-ethoxy-α-ketobutyrate) is the registered trademark of the Upjohn Company, Kalamazoo, Michigan.
[27] M. Litt and V. Hancock, *Biochemistry* **6**, 1848 (1968).
[28] H. F. Noller, *Biochemistry* **13**, 4694 (1974).
[29] L. Villareal, personal communication.
[30] J. C. Alwine, D. J. Kemp, and G. R. Stark, *Proc. Natl. Acad. Sci. U.S.A.* **74**, 5350 (1977).

sion for 30 min in the dark. After staining, the gels are transferred to a flat, enameled pan and allowed to destain, again in the dark. Removal of the very high background fluorescence of the gel proceeds most satisfactorily when carried out overnight at 4°, although 2 hr at 22° is generally sufficient for acrylamide-containing gels, and 1 hr at 22° is often sufficient for agarose gels. As excess dye diffuses out of the gel it is adsorbed by enamel, thus reducing the destaining time required. After the destaining process, the adsorbed acridine is removed from the enameled pan by rinsing it with 95% ethanol, or with hot running tap water (5–10 min). After destaining, RNA and DNA bands are often visible without ultraviolet illumination, appearing orange–red.

Bands are best visualized by illumination with short wavelength ultraviolet light. When placed on a uv transilluminator ($\lambda_{max} = 254$ nm; model C-61, Ultraviolet Products, San Gabriel, California), double-stranded polynucleotides are seen as bright green bands. Single-stranded material (most RNAs, denatured DNAs) appears as bright reddish-orange bands. If destaining is incomplete, the gel background will appear very green, and the red bands will appear dull, or even black. Faint bands may not be visible at all at this stage. It should be pointed out that the red fluorescence is somewhat less intense than is the green fluorescence for a given amount of polynucleotide; therefore, more single-stranded material than double-stranded material is required for visualization. If the gel is well destained, 0.1 μg of single-stranded, and 0.05 μg of double-stranded polynucleotide should be easily visible in a photographed gel band. Destained gels may be stored in the dark at 4° for 1 week with no noticeable deterioration in color or band sharpness.

Photography

One should not judge the success of the staining/destaining merely by appearance to the naked eye. Using black and white photography with Polaroid type 107C film, or type 105 positive/negative film, a red filter greatly enhances the ability to detect red bands in a green background. If color photographs are desired, Polaroid type 108 color film is used, with a yellow filter to enhance color resolution. Several exposure settings should be tried to obtain the brightest possible reds.

Remarks on Acridine Orange Metachromasy

The interaction of acridine orange with nucleic acids is complex, and the fluorescence emission spectrum is affected not only by the gross secondary structure of the polynucleotides with the dye but also by the concentration of acridine orange, pH, presence of organic solvents, temperature, and ionic strength.[11] At very high polynucleotide–dye ratios, most molecules can be made to fluoresce green.[11] Thus, the centers of heavily overloaded bands may appear green, while the fringes, containing lower polynucleotide concentrations, fluoresce red. One should, therefore, exert caution when inferring nucleic acid structure when staining is carried out under very different conditions from those described here. For example, acridine orange staining of gels containing 6 M urea at pH 3.5[31] reveals all bands, regardless of structure, as green.[32]

Acknowledgments

This research was supported by Grants 3.538.75 (to Dr. S. Modak) and 3.738.76 (to Dr. B. Hirt) from the Swiss National Foundation. G. G. C. was the recipient of a postdoctoral fellowship from The Jane Coffin Childs Memorial Fund for Medical Research.

[31] J. M. Rosen, S. L. C. Woo, J. W. Holder, A. R. Means, and B. W. O'Malley, *Biochemistry* **14**, 69 (1974).
[32] L. Schmidt and J. Ruderman, personal communication.

[48] A General Method for Defining Restriction Enzyme Cleavage and Recognition Sites

By Nigel L. Brown and Michael Smith

Introduction

Class II restriction endonucleases cleave double-stranded DNA into specific fragments.[1,2] These enzymes are powerful tools for the dissection of DNA, and they are essential for the construction of recombinant DNA molecules and for DNA sequence determination.[3] The enzymes characterized to date recognize specific sequences of nucleotides, four to six nucleotide pairs long. Over forty enzymes recognizing different sequences are now known. While all the known class II restriction enzymes cleave both strands of DNA, the particular phosphodiester bonds hydrolyzed vary from enzyme to enzyme. Thus, in some cases DNA fragments are produced with flush ends (i.e., the cleaved internucleotide bonds are between the same nucleotide pairs in duplex DNA), some produce termini

[1] H. O. Smith and K. W. Wilcox, *J. Mol. Biol.* **51**, 379–391 (1970).
[2] T. J. Kelly, Jr. and H. O. Smith, *J. Mol. Biol.* **51**, 393–409 (1970).
[3] R. J. Roberts, *Crit. Rev. Biochem.* **4**, 123–164 (1976).

with 5' extensions and others produce termini with 3' extensions. Extensions with one, two, three, four, or five protruding nucleotides have been described.[3,4] The cleavages can be within or immediately adjacent to the recognition sequence,[3] but in some cases the cleavages are displaced from the recognition sequence by a specific number of nucleotides.[4-6] All restriction enzymes characterized to date (sometimes with difficulty[7]) produce fragments with 5'-phosphorylated (and 3'-hydroxyl) termini.

The wide variety in types of fragment termini have resulted in a number of strategies for characterizing the cleavage sites. In general these are based on the principle of ^{32}P-labeling the ends of a mixture of fragments produced by a given enzyme followed by determination of the nucleotide sequence at the labeled ends. Because of the wide variety of types of fragment ends, there is no universal strategy for this approach.[3] These methods are of little use when applied to an enzyme which cleaves away from its recognition sequence, since there are no nucleotides common to the ends of all fragments.

Clearly, a completely general strategy is required and such a method is the subject of this article. The procedure, for a particular cleavage site, aligns the position of the cut in each strand of DNA alongside the "ladder" sequence of the DNA in the region of the cleavage; the ladder being generated by one of the rapid gel methods for DNA sequence analysis.[8-10] The method was developed for the characterization of the restriction endonuclease *Pst*I which recognizes the sequence

5'-CTGCAG-3'
3'-GACGTC-5'

cleaving between the A and G residues to produce a 3'-tetranucleotide extention.[11] In its present form the method has been used to characterize *Hga*I[4] together with a number of known and new restriction enzymes, which together represent all the catagories described above. The method was developed and has been used most extensively in conjunction with the "plus and minus" enzymatic DNA sequencing method.[8] However, it

[4] N. L. Brown and M. Smith, *Proc. Natl. Acad. Sci. U.S.A.* **74**, 3213–3216 (1977).
[5] D. Kleid, Z. Humayen, A. Jeffrey, and M. Ptashne, *Proc. Natl. Acad. Sci. U.S.A.* **73**, 293–297 (1976).
[6] N. L. Brown, C. A. Hutchison III, and M. Smith, unpublished results.
[7] V. Pirrota, *Nucleic Acids Res.* **3**, 1747–1760 (1976).
[8] F. Sanger and A. R. Coulson, *J. Mol. Biol.* **94**, 441–448 (1976).
[9] F. Sanger, S. Nicklen, and A. R. Coulson, *Proc. Natl. Acad. Sci. U.S.A.* **74**, 5463–5467 (1977).
[10] A. M. Maxam and W. Gilbert, *Proc. Natl. Acad. Sci. U.S.A.* **74**, 560–564 (1977).
[11] N. L. Brown and M. Smith, *FEBS Lett.* **65**, 284–287 (1976).

has been used successfully in conjunction with the "terminator" enzymatic method[9] and with the chemical DNA sequencing method.[10]

Principle

Aligning the Cut in Each Strand Alongside an Enzymatically Produced DNA Ladder Sequence. The method is illustrated in Fig. 1. A fragment of

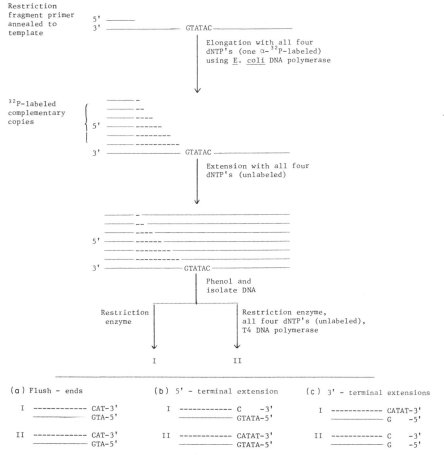

Fig. 1. Principle of the method for determining the position of the cuts in each DNA strand at a restriction endonuclease cleavage site in conjunction with an enzymatic method for DNA sequence analysis. The examples chosen show the hypothetical symmetrical recognition site.

$$5'\text{-CATATG-}3'$$
$$3'\text{-GTATAC-}5'$$

cleaved symmetrically to give (a) flush ends, (b) tetranucleotide 5'-terminal extensions, and (c) tetranucleotide 3'-terminal extensions.

DNA (primer) is annealed with a single-stranded, complementary template DNA chosen to contain the restriction enzyme cleavage site within 20 to 200 nucleotides of the priming site. The restriction fragment maps required to select the correct primer–template combination are obtained by conventional methods[12] using any suitable DNA which is cleaved by the restriction enzyme under investigation. DNA polymerase is used with [α-^{32}P]deoxynucleoside 5'-triphosphate(s) followed by unlabeled deoxynucleotide 5'-triphosphates to pulse label the region immediately adjacent to the priming site (Fig. 1). The DNA is isolated and divided into two fractions. One of these (sample I) is cleaved by the restriction enzyme under investigation. When subjected to electrophoresis alongside either the "plus and minus"[13] or the "terminator"[14] ladder sequence pattern of the strand which is ^{32}P-labeled, it defines the position of cleavage in that DNA strand. The cleavage in the unlabeled complementary strand is defined by treating the second sample (II) of pulse-labeled DNA with the restriction enzyme in the presence of T4 DNA polymerase and deoxynucleoside 5'-triphosphates. If the cleavage produces a 5' protruding terminus, the polymerase extends the 3' end of the pulse-labeled strand until it is the same length as the complementary strand. If the restriction enzyme produces a 3' protruding end, then the protruding end, which is the pulse-labeled strand, will be trimmed to the same length as the complementary strand by the 3'-exonuclease activity of T4 DNA polymerase. There is no change in the length of the pulse-labeled strand if there is a flush-ended cleavage. Thus if the T4 DNA polymerase-treated fragment is subjected to gel electrophoresis, as described above, it indirectly but precisely defines the position of cleavage in the complementary strand. These experiments are illustrated in Fig. 1 and 2.

It is important that the DNA used as substrate in the site location experiment (samples I and II) contains the minimum amount of uncopied template DNA, because the double-stranded cleavage activity of some restriction enzymes (e.g., *Alu*I, *Hph*I) is inhibited by single-stranded DNA. The "chase" step with unlabeled triphosphates reduces the amount of single-stranded DNA present.

Aligning the Cut in Each Strand Alongside a Chemically Produced Ladder Sequence. In this case the same strategies described above are applied to a duplex fragment of DNA which has been ^{32}P-labeled in one of the 5'-phosphate groups. Because the restriction enzyme cleavage and the chemically generated ladder sequence[15] do not have the same 3' termini,

[12] K. J. Danna, this volume, Article [53].
[13] N. L. Brown and M. Smith, *J. Mol. Biol.* **116**, 1–28 (1977).
[14] A. J. H. Smith, this volume, Article [58].
[15] A. M. Maxam and W. Gilbert, this volume, Article [57].

(a)

(b)

(c)

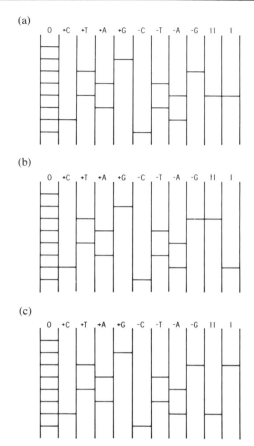

FIG. 2. Diagram of the autoradiography pattern expected for enzymes which cleave the symmetrical duplex sequence

5'-CATATC-3'
3'-GTATAC-5'

to give fragments with (a) flush ends, (b) tetranucleotide 5'-terminal extensions, and (c) tetranucleotide 3'-terminal extensions (channels I and II as defined in Fig. 1). The reference ladder pattern is produced by the "plus and minus" method (only the pattern generated by the hexanucleotide is shown). The 0 channel is the graticule containing all ladder bands. When the "terminator" method is used to generate the sequence pattern it would be equivalent to the plus pattern. When the chemical method is used it is similar to the minus pattern, but may be displaced slightly.[15]

the fragments may not align as precisely as in the enzymatic ladder sequencing. This nonalignment is most pronounced if the restriction enzyme cleavage site is within 40 nucleotides of the labeled 5' end; however, the results are easily interpretable.

Mapping of Cleavage Sites for the Unknown Restriction Enzyme

As mentioned above the mapping is carried out using a small, well characterized DNA which is a substrate for the enzyme. Suitable mapping and sizing techniques are described in Article [53] of this volume. Ideally the DNA used should be one whose sequence is completely known (e.g., bacteriophage φX174 or G4 DNA, SV40 DNA, or the plasmid pBR322), since any postulated recognition sequence can be tested, most conveniently by computer analysis,[16–18] for its predicted fragmentation pattern which can be compared with the experimentally determined pattern.[19] Usually only a limited number of sites cleaved by the unknown enzyme have to be analyzed to determine the recognition sequence. Because of this only partial mapping information is required in most cases. This can be obtained by double-digestion, preferably using as the second enzyme one which cleaves the substrate only a small number of times. This minimizes the complexity of the double-digestion fragment pattern and may provide unambiguous information on the location of several sites cleaved by the unknown enzyme. Frequently a site for the unknown enzyme can be very rapidly mapped to one of two locations, and it may be quicker and easier to perform the DNA sequencing experiment at both locations (only one of which will show the cleavage site) rather than first mapping the site unambiguously.

Determination of the Cleavage Sites

The procedures for use with either the "plus and minus"[8,13] or the terminator[9,14] sequencing methods are essentially identical. The only differences being in the preparation of the labeled DNA substrate prior to cleavage with the unknown restriction enzyme. These labeling methods will be described separately. The procedure for use with the chemical method[10,15] is also described.

The experimental details given below may have to be modified to fit specific experimental circumstances. For example, some restriction enzymes may require a different buffer composition to that used in the experiments described here. If the buffer is one in which T4 DNA polymerase is not active, the restriction enzyme cleavage can be performed, and then the DNA precipitated and resuspended in a buffer suitable for T4 DNA polymerase activity. Optimal conditions for restriction

[16] D. McCallum and M. Smith, *J. Mol. Biol.* **116**, 29–30 (1977).
[17] R. Staden, *Nucleic Acids Res.* **4**, 4037–4051 (1977).
[18] C. L. Queen and L. J. Korn, this volume, Article [60].
[19] P. G. N. Jeppesen, this volume, Article [39]; K. J. Danna, this volume, Article [53].

enzyme cleavage are not necessary; suboptimal conditions often generate sufficient cleavage product to give satisfactory data.

The Plus and Minus Method. In the plus and minus DNA sequencing method,[13] a DNA primer is annealed to a DNA template and extended asynchronously such that a staggered series of ^{32}P-labeled extension products are obtained, in which, say, between 0 and 300 nucleotides have been added to the primer. The labeled extension products (still annealed to the template) are purified from the DNA polymerase and unincorporated deoxyribonucleoside triphosphates by phenol extraction and column chromatography. After purification, the labeled DNA is used as substrate for the base-specific plus and minus reactions which define the DNA sequence. For the determination of the cleavage site of an unknown restriction enzyme, samples of the purified, labeled extension product can be taken and extended with unlabeled triphosphates to convert the DNA to duplex structure. This duplex DNA is then used as substrate for the cleavage site location experiment. This method will be described in this section. Alternatively a separate sample of the annealed primer–template DNA can be pulse labeled with [α-^{32}P]deoxyribonucleoside triphosphate and used as a substrate for the base-specific reactions (as described below for use in conjunction with the "terminator" method).

EXPERIMENTAL DETAILS: (a) Annealing: Template DNA (0.4 pmole of a suitable single-stranded DNA, e.g., bacteriophage ϕX174 DNA) and primer DNA (1.0–1.5 pmole of an appropriate duplex restriction fragment, prepared by electrophoresis in polyacrylamide as described in Article [40] of this volume) in buffer A (100 mM NaCl, 10 mM MgCl$_2$, 1 mM dithiothreitol, 20 mM Tris-HCl, pH 7.5), total volume 15 μl, are mixed and heated in a sealed capillary tube at 100° for 3 min. The mixture is then incubated at 67° for 15–30 min. This procedure first denatures the duplex DNA fragment and then allows annealing with the single-stranded template. Bacteriophage ϕX174 DNA and its double-stranded replicative form (RF) DNA are available commercially.[20]

(b) Extension: The primer–template mixture is made up to 25 μl total volume containing 50 μM dCTP, 50 μM dTTP, and 50 μM dGTP together with 10–20 μCi [α-^{32}P]dATP (specific activity approximately 350 Ci/mmole) in buffer A. The [α-^{32}P]dATP is commercially available[21,22] or can be synthesized.[23] The mixture is cooled to 5° in an ice water bath and 1 unit of subtilisin-treated *E. coli* DNA polymerase I (Klenow enzyme[24,25]) is

[20] New England Biolabs, 283 Cabot St., Beverly, MA 01915.
[21] New England Nuclear, 549 Albany St, Boston MA 02118.
[22] The Radiochemical Centre, Amersham, Bucks U.K.
[23] R. H. Symons, this series, Vol. 29 [11].
[24] Boehringer Mannheim Biochemicals, 7941 Castleway Drive, P.O.Box 50816, Indianapolis, IN 46250.
[25] P. Setlow, this series, Vol. 29 [1].

added. After 1 min half the reaction mixture is removed into 25 μl of 0.1 M EDTA. After 3 min the remainder of the reaction mixture is pooled with the first sample.

(c) Removal of polymerase and triphosphates: Phenol (25 μl, water-saturated) is added to the extension mixture, and the mixture is vigorously shaken (Vortex mixer) for 30 sec. The phenol is removed by extraction into ether (twice with 250 μl), and residual ether is removed from the aqueous phase in a stream of air. The aqueous layer is applied to a column of Sephadex G-100 (3 $\times$ 200 mm), and the column is eluted with 0.05 mM EDTA, 2 mM Tris-HCl (pH 7.5) at a flow rate of approximately 50 μl/min. Elution is monitored using a Geiger counter and the faster moving polynucleotide fraction (100-150 μl) is collected in a small polypropylene tube and freeze-dried.

(d) Sequence analysis of the cleavage site: The labeled polynucleotide is then dissolved in R buffer (50 mM NaCl, 7 mM MgCl$_2$, 1 mM dithiothreitol, 7 mM Tris-HCl, pH 7.5 in 20–30 μl). At this stage the primer DNA may be removed from the newly synthesized labeled DNA by incubation with a suitable restriction enzyme (the "datum" enzyme, often that enzyme originally used to generate the primer). Otherwise the datum enzyme can be added at a later stage, or a short primer may not be removed. Samples (usually 2 μl) are removed for the plus and minus reactions[8,13] a further sample (normally 1 μl) is taken for the cleavage site location experiments.

This sample (1 μl) is incubated with dCTP, dTTP, dATP, and dGTP (all at 200 μM final concentration) and T4 DNA polymerase (0.1 unit)[26–28] in R buffer (10 μl total volume) at 37° for 30 min. The sample is then divided into two.

The first sample (I) is vigorously mixed (30 sec) with phenol (5 μl, water-saturated) and extracted with ether (four times with 250 μl, water-saturated). The residual ether is removed in a stream of air and 10X R buffer (approx. 0.2 μl) is added (fresh R buffer must be added after phenol and ether extractions for some restriction enzymes to be fully active). The restriction enzyme to be characterized (approx. 0.2 unit) is then added (together with the datum restriction enzyme if the primer is to be removed at this stage) and the mixture is incubated at 37° for 30 min. The reaction is stopped in the same way as the plus and minus incubations[8,13] [e.g., by adding a mixture (20 μl) containing bromphenol blue (0.03%) xylene cyanol (0.03%), and EDTA (0.025 M) in 90% formamide].

The second sample (II) is incubated with restriction enzyme(s) as de-

[26] I. R. Lehman, this series, Vol. 29 [6].
[27] Bethesda Research Laboratories, 605 S. Stonestreet Ave., Rockville, MD 20850.
[28] Miles Laboratories Inc., P.O. Box 2000, Elkhart, IN 46515.

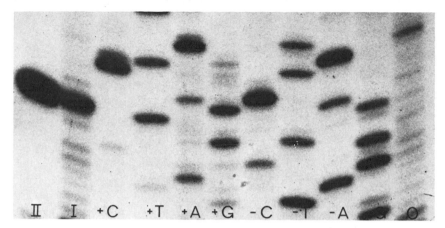

FIG. 3. Determination of a *Hin*dII cleavage site using the plus and minus method. This enzyme recognizes and cleaves the sequence

$$\overset{\downarrow}{\text{5'-GTYRAC-3'}}$$
$$\underset{\uparrow}{\text{3'-CARYTG-5'}}$$

producing flush ends.[3] The sequence cleaved in the experiment shown here is GTTGAC.

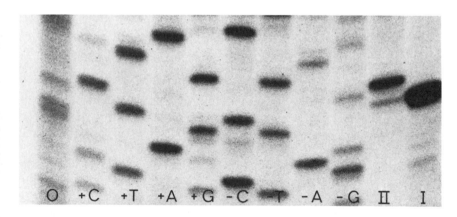

FIG. 4. Determination of a *Hpa*II cleavage site using the plus and minus method. The enzyme recognizes and cleaves the sequence

$$\overset{\downarrow}{\text{5'-CCGG-3'}}$$
$$\underset{\uparrow}{\text{3'-GGCC-5'}}$$

i.e., it produces a dinucleotide 5'-terminal extension.[3]

scribed for sample I, but with no prior phenol inactivation of the T4 DNA polymerase. After incubation at 37° for 30 min, the reaction is stopped as described above.

Both samples (or aliquots thereof) are denatured by heating at 100° for 3 min and analyzed by electrophoresis on an acrylamide-urea gel alongside the products of the plus and minus reactions using the same primer–template–datum enzyme combination. The electrophoresis and autoradiography are carried out as described elsewhere.[14] The results of representative experiments are shown in Fig. 3 and 4.

The "Terminator" Method. In the terminator method[14] a DNA primer is annealed to a DNA template and divided into four samples, each of which is extended in the presence of all four deoxyribonucleoside triphosphates (one of which is α-^{32}P-labeled) and one dideoxyribonucleoside (or arabinoside) triphosphate. These four samples when analyzed on an acrylamide gel allow the DNA sequence to be read. For the determination of the cleavage site of an unknown restriction enzyme, a fifth sample of the annealed primer–template combination must be extended and "pulse-labeled" adjacent to the primer. This pulse-labeled DNA is then used as a substrate for the site-location experiments.

EXPERIMENTAL DETAILS. Template DNA (0.2 pmole of a suitable single-stranded DNA, e.g., bacteriophage ϕX174 DNA) and primer DNA (0.5 pmole of an appropriate duplex restriction fragment, prepared by electrophoresis as described in Section III of this volume) in R buffer (50 mM NaCl, 7 mM MgCl$_2$, 1 mM dithiothreitol, 7 mM Tris-HCl, pH 7.5), total volume 5–10 μl, are mixed and heated in a sealed capillary tube at 100° for 3 min. The mixture is then incubated at 67° for 15–30 min, and divided into five samples. Four samples are used for the chain-termination incubations.[14]

The remaining sample is incubated at 20°–25° in R buffer (5 μl total volume) containing dCTP, dTTP, and dGTP (all at 25 μM final concentration) together with 1 μCi [α-^{32}P]dATP (specific activity approximately 350 Ci/mmole) and 0.2 unit subtilisin-treated *E. coli* DNA polymerase (0.2 unit; Klenow enzyme).[24,25] After 1 min unlabeled dATP (1 μl, 0.5 mM) is added and the incubation continued for 10 min. Phenol (5 μl, water-saturated) is added and the sample vigorously mixed for 30 sec. The phenol is removed with ether (four times with 250 μl; water-saturated), then the residual ether is removed in a stream of air. The volume of the mixture is made up to 10 μl with R buffer, and the mixture divided into two samples.

The first sample (I) is incubated with the restriction enzyme being characterized (approximately 1 unit) at 37° for 15 min. If the primer is to be removed from the extended product, the datum restriction enzyme

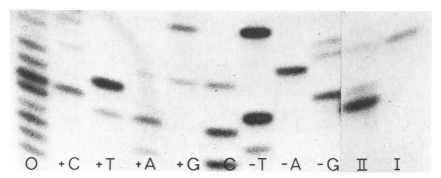

FIG. 5. Determination of a *Hgi*AI cleavage site using the "terminator" method. The enzyme recognizes and cleaves the sequence

$$5'\text{-G}\begin{pmatrix}T\\A\end{pmatrix}\text{GC}\begin{pmatrix}T\\A\end{pmatrix}^{\downarrow}\text{C-3}'$$

i.e., it produces a tetranucleotide 3'-terminal extension.[29]

$$3'\text{-C}_{\uparrow}\begin{pmatrix}A\\T\end{pmatrix}\text{CG}\begin{pmatrix}A\\T\end{pmatrix}\text{G-5}'$$

The sequence cleaved in the experiment shown here is GTGCTC. Dideoxyribonucleotides were used as the chain terminators.

(approximately 1 unit is added at the same time). The reaction is stopped in the same way as the four "chain termination" reactions used to define the sequence [e.g., by adding a mixture (20 μl) containing bromphenol blue (0.03%), xylene cyanol (0.03%), and EDTA (0.025 M) in 90% formamide].

The second sample (II) is incubated with the restriction enzyme(s) as described for sample I, together with T4 DNA polymerase (0.1 unit).[26–28] After incubation at 37° for 15 min, the reaction is stopped as described for sample I.

Both samples (or aliquots thereof) are denatured by heating at 100° for 3 min, and analyzed by electrophoresis alongside the corresponding chain termination reactions used to define the sequence, as described elsewhere.[14] A typical experiment is shown in Fig. 5.[29]

The Chemical Method. The procedure for the determination of the sites in each DNA strand cleaved by an unknown restriction enzyme for use in conjunction with the chemical method can only be applied to duplex DNA labeled at the 5' end of one of the strands. This 5' end-labeled DNA is also the substrate for the chemical reactions used in determining the DNA

[29] N. L. Brown, M. McClelland, and P. R. Whitehead, unpublished results.

sequence. Both single-stranded DNA (5' or 3' end-labeled) and 3' end-labeled duplex DNA can be sequenced by the chemical method, but single-stranded DNA will not be a substrate for many restriction enzymes, and 3' end-labeled duplex DNA will only allow information to be obtained on the position of the cleavage site in the labeled strand.

EXPERIMENTAL DETAILS. A restriction fragment (approximately 2.5 pmole, chosen such that the cleavage site of the unknown restriction enzyme lies within, say, 200 nucleotides of one of the ends of the fragment) is labeled in its 5'-phosphate groups by using [γ-^{32}P]ATP and polynucleotide kinase.[15] The fragment is then cleaved asymmetrically with another restriction enzyme, and the resulting two label fragments are separated by electrophoresis and eluted as described elsewhere in this volume.[15] The 5' end-labeled fragment containing the cleavage site of the restriction enzyme to be characterized is taken up in water, and two small samples (each approximately one-twentieth to one-tenth of the amount of DNA used per chemical reaction, i.e., per channel, in the sequence analysis) are removed for the site location experiments. Sonicated carrier DNA can then be added to the remainder of the sample, and this divided into four to be used in the DNA sequence determination, as described elsewhere.[15]

One of the small samples (I) is made up to 5 μl in R buffer (50 mM NaCl, 7 mM MgCl$_2$, 1 mM DTT, 7 mM Tris-HCl, pH 7.5) and the restriction enzyme to be characterized (1 μl; approximately 1 unit) is added. The mixture is incubated at 37° for 15 min. The reaction is stopped by heating at 70° for 10 min, and the contents of the capillary are mixed with 0.3 M sodium acetate (25 μl) in a siliconized Eppendorf tube. Carrier tRNA (20 μg) and cold ethanol (90 μl; 95%) are added. The mixture is cooled at $-60°$ for 5 min and centrifuged to precipitate the nucleic acids (10,000 g, 5 min). The nucleic acids are resuspended and precipitated from 0.3 M sodium acetate (25 μl) a further twice. The pellet is washed briefly with 95% ethanol and then dried in vacuo. This procedure effectively removes all salts from the sample, ensuring that the sample is of the same composition as the samples from the four chemical reactions used to determine the DNA sequence.

The other small sample (II) is made up to 5 μl in R buffer containing dCTP, dTTP, dATP, and dGTP (all at 50 μM). The restriction enzyme to be characterized (approximately 1 unit) and T4 DNA polymerase (0.1 unit) are added, and the mixture is incubated at 37° for 15 min in a sealed capillary tube. The reaction is stopped by heating at 70° for 10 min, and the sample is then ethanol precipitated and dried as described above.

Both samples I and II are dissolved up as described for the products of the four chemical reactions used to define the DNA sequence.[15] The sam-

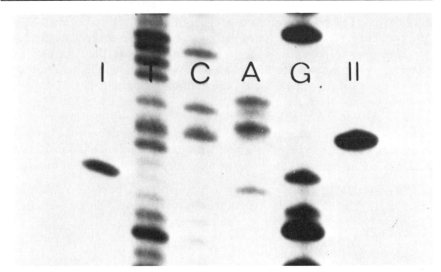

FIG. 6. Determination of a *Hin*fI cleavage site using the chemical method. The enzyme recognizes and cleaves the sequence

$$\downarrow$$
$$5'\text{-GANTC-3'}$$
$$3'\text{-CTNAG-5'}$$
$$\uparrow$$

i.e. it produces a trinucleotide 5'-terminal extension.[3] The sequence cleaved in the experiment shown here is GAGTC. The precise alignment of the band in channel II was determined from a shorter autoradiographic exposure. The channels labeled C and G are from reactions specific for those nucleotides; channel T shows bands due to both T and C residues; channel A shows bands due to A residues (strong) and C residues (weak). The chemical reactions used are described elsewhere.[15]

ples, or aliquots thereof, are then analyzed by electrophoresis alongside the products of the chemical reactions on a denaturing acrylamide gel. An experiment is shown in Fig. 6.

Interpretation of the Autoradiographs

The experiments described here to locate the cleavage sites of a restriction enzyme generate the same products whichever DNA sequencing method is used. Samples I and II contain a labeled DNA fragment with a 3'-hydroxyl group, and the length of the fragment is a measure of the distance from the 5' end to the phosphodiester bond cleaved by the restriction enzyme (for sample I) or to the phosphodiester bond opposite that cleaved in the unlabeled DNA strand (sample II). The length of the products of the DNA sequencing reactions bear a relationship to the nucleotide specified by the reaction which is different in the different methods, i.e., the "plus" and "terminator" methods have the specified nucleotide as the 3'-nucleotide in the product, whereas the "minus" and chemical methods have products which terminate one nucleotide to the 5' side of the specified nucleotide. An additional complication is that the products of the chemical method contain a 3'-phosphate group.

The following simple rules may assist in the interpretation of autoradiographs, and these may be followed by reference to Figs. 3–6.

If the band in channel I is aligned with band x in the channels defining the DNA sequence, and the band in channel II is aligned with band y

(i) For the plus reactions, the cleavage sites are between nucleotides x and $(x + 1)$ in the labeled strand and y and $(y + 1)$ in the unlabeled strand.

(ii) For the minus reactions the cleavage sites are between nucleotides $(x - 1)$ and x and $(y - 1)$ and y.

(iii) For the terminator reactions the cleavage sites are between nucleotides x and $(x + 1)$ and y and $(y + 1)$.

(iv) For the chemical method, the cleavage sites are between nucleotides $(x - 1)$ and x and $(y - 1)$ and y. However, bands in channels I and II may not align exactly with bands in the channels defining the DNA sequence, especially if the cleavage site is within 40 nucleotides of the labeled 5' end of the restriction fragment. This is due to the 3'-phosphate in the products of the chemical reactions.

Conclusion

The basic principle of the method described above can be easily adapted to fit specific experimental circumstances.[4] Often the characterization of one cleavage site is sufficient to define a restriction enzyme recognition site, especially if the prediction is confirmed by the fragments produced using a DNA substrate whose sequence is completely known. In other cases, particularly where the cleavage site is offset from the recognition site, a small number of sites must be compared.[4]

[49] Rapid DNA Isolations for Enzymatic and Hybridization Analysis

By Ronald W. Davis, Marjorie Thomas, John Cameron, Thomas P. St. John, Stewart Scherer, and Richard A. Padgett

The ability to rapidly isolate DNA from a large number of individual organisms has greatly facilitated the characterization of the genome of these organisms. In particular, ability to rapidly isolate DNA from organisms containing cloned DNA segments using recombinant DNA techniques has greatly facilitated the identification of particular clones. The methods described here are for the isolation of DNA from bacteria, yeast, and tissue culture cells. However, the methods are very general and probably can be adapted for the isolation of DNA from any organism. The DNA preparations are not pure DNA but are sufficiently pure for most enzymatic and hybridization analyses. These methods have been most frequently used for identifying the size of a particular cloned DNA fragment or for surveying the genome organization in a large number of individual organisms or strains. The general approach is to remove first enzymatically or physically any rigid cell wall and then to lyse the cells with a detergent, usually SDS. Nucleases are inactivated with diethyl oxydiformate and most of the cell debris, proteins, and SDS are precipitated by addition of potassium. The DNA is then recovered by ethanol precipitation. RNA which is present in these preparations generally interferes with most restriction endonucleases, and it is eliminated by ribonuclease treatment. The digested RNA generally does not interfere with analysis and is not removed. To be described are the details of the procedure for the isolation of DNA from particular organisms.

Abbreviations and Reagents

Diethyl oxydiformate (Calbiochem)
SDS: Sodium dodecyl sulfate (Accurate Chemical and Scientific Corp., Hicksville, New York)
EDTA: Disodium ethylenediaminetetraacetic acid (Mallinckrodt)
Tris: Tris(hydroxymethyl)aminomethane (BRL)
KAc: Potassium acetate (Baker)
EtOH: Ethanol
λdil: 10 mM Tris, pH 7.5, 10 mM MgSO$_4$
TE: 10 mM Tris, 1 mM Na$_3$EDTA
14 mM β-mercaptoethanol: 0.1% v/v (Baker)
Lyticase: Prepared from *Oerskovia xanthinolyticia*. Zymolyase (Kirin brewery) also can be used.

Rapid λ DNA Isolation[1]

1. Make a normal λ plate stock using media prepared with agarose (unpurified agar greatly inhibits restriction enzymes).

[1] J. R. Cameron, R. Philippsen, and R. W. Davis, *Nucleic Acids Res.* **4**, 1429 (1977).

2. Cool the plates and then overlay them with 5 ml cold λ dil. Let plates remain overnight at 4°.

3. Pipette 0.4 ml of the clear lysate into a 1.5 ml microfuge tube. The remainder of the lysate can be saved for culture purposes.

4. Add 1 μl diethyl oxydiformate at room temperature.

5. Add 10 μl 10% SDS. Invert tube to mix.

6. Add 50 μl 2 M Tris, 0.2 M EDTA, pH 8.5, and mix.

7. Incubate at 70° for 5 min in the hood.

8. Add 50 μl 5 M KAc and cool tubes.

9. Wait at least 30 min with tubes in ice.

10. Sediment precipitate in microfuge for 15 min.

11. Decant supernatant into new tube.

12. Fill tube with EtOH at room temperature.

13. Sediment precipitate in a microfuge for 5 min.

14. Discard supernatant and invert tube on paper towel to drain. *All* liquid should either drain off or evaporate (use vacuum desiccator if necessary) but do not overdry.

15. Dissolve precipitated DNA in 50 μl 10 mM Tris, pH 7.5, 1 mM Na_2 EDTA, 1 μg/ml RNase A. Five microliters is usually sufficient for one restriction digest and gel track.

Rapid Plasmid and/or Bacterial DNA Isolation from Colonies or Broth

1. Pick colonies on agar (or agarose) plates with flat toothpicks into 1.5-ml centrifuge tubes containing the buffer below, or sediment 1 ml saturated culture and resuspend pellet in 0.5 ml, 50 mM Tris (pH 8.5), 50 mM Na_2 EDTA, 15% sucrose, and 1 mg/ml lysozyme (added just prior to use).

2. Keep 10 min at room temperature.

3. Add 1 μl diethyl oxydiformate at room temperature.

4. Add 10 μl 10% SDS. Invert tube to mix.

5. Heat to 70° for 5 min in a hood to obtain bacterial DNA. Do not heat to obtain just plasmid DNA.

6. Add 50 μl 5 M KAc and cool tubes.

7. Wait at least 30 min with tubes on ice.

8. Sediment precipitate in microfuge for 15 min.

9. Decant supernatant into new tube.

10. Fill tube with EtOH at room temperature.

11. Sediment precipitate in microfuge for 5 min.

12. Discard supernatant and invert tube on paper towel to drain. *All* liquid should either drain off or evaporate (use vacuum desiccator if necessary) but do not overdry.

13. Dissolve precipitated DNA in 50 μl of 10 mM Tris, pH 7.5, 1 mM Na_2 EDTA, and 1 μg/ml RNase A.

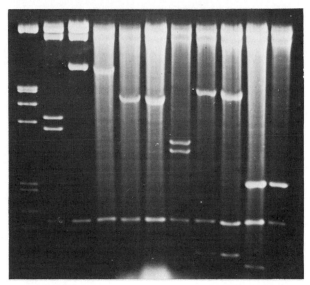

FIG. 1. Rapid λ DNA preparations. Characterization of the number and size of DNA inserted into a λ vector. The first lane on the left contains size standards. The next two lanes contain EcoRI-cleaved DNA from purified phage. The remaining lanes contain rapid λ DNA preparations. The background is from cleaved E. coli DNA. All of these λ strains contain yeast DNA that can complement an E. coli trpC mutant. As can be seen they also all contain one common EcoRI fragment.

The amount of DNA obtained is generally sufficient for two gel tracks. The supercoiled plasmid DNA from step 13 can be used directly for gel electrophoresis. The RNase is used because RNA generally inhibits restriction endonucleases. All of the steps in this, protocol can be conducted at room temperature except the potassium—SDS precipitation (step 7). This step requires long incubation at 0° and long centrifugation to sediment all cell debris and potassium–SDS-protein precipitate.

Rapid Plasmid DNA Isolation for 10 ml Culture

1. Grow cells to saturation in 10 ml of L broth plus 0.4% glucose and selective drug if applicable.

2. Sediment cells in culture tube or 10-ml centrifuge tube and resuspend pellet in 1.4 ml TE, pH 8.5.

3. Put into 1.5 ml microfuge tube and centrifuge for 0.5 min.

4. Resuspend in 0.4 ml of 15% sucrose, 50 mM Tris-HCl, 50 mM Na$_2$EDTA, pH 8.5. Vortex thoroughly.

5. Cool on ice and add 0.1 ml of a freshly prepared 5 mg/ml lysozyme solution in above buffer at 0°.

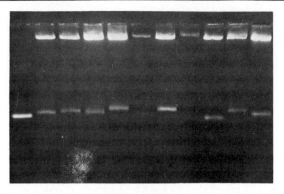

FIG. 2. Rapid uncleaved plasmid DNA preparation from colonies. The size of the covalently closed supercoiled DNA extracted from colonies is determined. The initial vector pMB9 DNA is in the gel track on the left. The other gel tracks contain rapid DNA preparations from colonies of transformants containing pMB9 DNA linked to small cDNA copies of mRNA. The approximate size of the inserted sequence can be determined by comparison of the mobility to that of pure pMB9 DNA. The lower bands are from covalently closed circular plasmid DNAs. The upper bands are from uncleaved cellular DNA.

6. Invert gently and occasionally for 10 min on ice.

7. Add 0.3 ml of 0.1% Triton X-100, 50 mM Tris-HCl 50 mM Na$_2$EDTA, pH 8.5, at 0°.

8. Invert gently and occasionally for 10 min on ice.

9. Centrifuge 2 min in microfuge located in cold room.

10. Decant supernatant into new microfuge tube.

11. Add 2 μl diethyl oxydiformate and shake briefly.

12. Heat to 70° for 15 min, cool 15 min on ice and centrifuge 2 min in microfuge.

13. Decant supernatant into new microfuge tube.

14. Fill tube with EtOH at room temperature.

15. Sediment precipitate in a microfuge for 5 min.

16. Discard supernatant and invert tube on paper towel to drain. All liquid should either drain off or evaporate (use vacuum desiccator if necessary) but do not overdry.

17. Dissolve precipitated DNA in 10 mM Tris, pH 7.5, 1 mM Na$_2$ EDTA, 1 μg/ml RNase A. Five microliters is sufficient for one agarose gel track. The DNA may be cleaved with most restriction enzymes.

Rapid Yeast DNA Isolation[2]

1. Grow 5 ml yeast cells into stationary phase.

2. Wash cells by sedimenting, resuspending in 1 ml 1 M sorbitol, transferring to 1.5 ml microfuge tube, and resedimenting.

[2] K. Struhl, D. T. Stinchcomb, S. Scherer, and R. W. Davis, *Proc. Natl. Acad. Sci. U.S.A.* **76**, 1035 (1979).

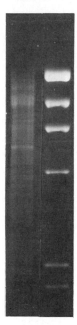

FIG. 3. Rapid cellular DNA preparations from colony cleaved with *Eco*RI endonuclease. DNA was prepared from a single colony of *Salmonella typhimurium* LT2 and cleaved with *Eco*RI endonuclease (track on left). The track on the right contains λDNA cleaved with *Hin*dIII endonuclease and is used as a size standard.

3. Resuspend cells in 0.5 ml of 1 M sorbitol, 50 mM potassium phosphate, pH 7.5, 14 mM β-mercaptoethanol, and 25 units lyticase.

4. Incubate 30 min at 30°.

5. Sediment and resuspend spheroplasts in 0.5 ml of 50 mM Na$_3$ EDTA, pH 8.5, 0.2% SDS.

6. Add 1 μl diethyl oxydiformate at room temperature.

7. Heat to 70° for 15 min in hood.

8. Add 50 μl 5 M KAc and cool tube.

9. Wait at least 30 min with tubes on ice.

10. Sediment precipitate in microfuge for 15 min.

11. Decant clear supernatant into new tube.

12. Fill with EtOH at room temperature.

13. Sediment precipitate in microfuge for 15 sec.

14. Discard supernatant and invert tube on paper towel to drain. *All* liquid should either drain off or evaporate (use vacuum desiccator if necessary) but do not overdry.

15. Dissolve precipitated DNA in 50 μl of 10 mM Tris, pH 7.5, 1 mM Na$_2$EDTA, 10 μg/ml RNase A.

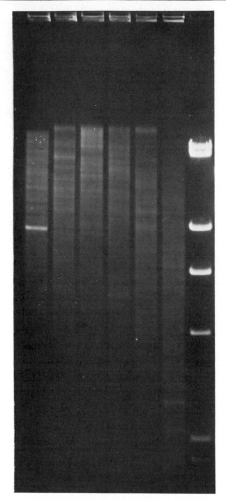

FIG. 4. Rapid DNA preparation of yeast DNA. DNA was prepared from a small culture of *Saccharomyces cerevisiae*. The track on the right contains λ DNA cleaved with *Hin*dIII endonuclease and is used as a size standard. The other tracks contain the yeast DNA cleaved with various combinations of restriction endonucleases.

Cells grown in minimal media or rich media with galactose lyse more readily than cells grown in rich media with glucose. Ethanol precipitate should form immediately and be quite large, small precipitation indicates inefficient spheroplasting and lysis.

Rapid Animal Cell DNA Isolation

1. Two 100-mm plates of confluent cells yields about 100 μg DNA.
2. Add to each of two plates washed with isotonic saline 2.5 ml of

0.1 M Na$_2$EDTA, 0.2 M Tris, pH 8.5, 1% SDS, and 100 μg/ml Proteinase K (Merck).

3. Gently scrape the viscous solution off the two plates with a rubber policeman and pour into a 15-ml uncapped Corex centrifuge tube (for JA20 rotor).

4. Incubate at 60° for 2 hr.

5. Add 10 μl diethyl oxydiformate and continue incubation for 30 min in a hood.

6. Cool tube and add 1 ml 5 M KAc.

7. Wait for at least 30 min with tube in ice.

8. Sediment precipitate in JA20 rotor for 15 min at 15,000 rpm and 2°.

9. Decant supernatant into new tube and add or fill with cold EtOH (5–10 ml) at 0°.

10. Sediment precipitate at 15,000 rpm for 5 min at 2°C.

11. Discard supernatant and invert tube on paper towel to drain. All liquid should either drain off or evaporate (use vacuum desiccation if necessary) but do not overdry.

12. Dissolve precipitated DNA in 0.5 ml of 10 mM Tris, pH 7.5, 1 mM EDTA, 1 μg/ml RNase.

The volumes in this procedure can be reduced by a factor of 10 and the reactions conducted in 1.5-ml microfuge tubes and sedimentations conducted in a microfuge. Approximately 5×10^6 cells lysed in 0.5 ml will give about 10 μg of DNA.

The procedures just described are now routinely used. Certain individuals have had difficulty in obtaining DNA that can be cleaved with restriction endonucleases. The source of these difficulties is probably from the manner in which the operations are conducted. If the ethanol precipitation of the DNA is allowed to occur for a long time or at a low temperature some cellular components also appear to precipitate which interfere with subsequent analysis. Also the KAc and ethanol precipitation may leave traces of SDS which can inactivate the restriction endonucleases. A second ethanol precipitation may eliminate this problem. It should be noted that treatment with diethyl oxydiformate at high temperature results in DNA molecules that are inviable in transfection assays. The diethyl oxydiformate step alone or the high temperature step alone can be used to inactivate nucleases and give viable DNA molecules. The fact that the DNA molecules prepared as described are inviable has not resulted in any ambiguity in the endonuclease cleavage or hybridization analysis. The combination of diethyl oxydiformate treatment and high temperature have been observed to give more reproducible results and higher molecular weight DNA.

Section VI

Determination of DNA Fragment Sizes

[50] Conversion of Circular DNA to Linear Strands for Mapping

By RICHARD C. PARKER

Circular DNA molecules are a suitable substrate for restriction endonucleases. When properly treated, covalently closed circular DNAs (form I DNA) can be converted by a multihitting restriction endonuclease to full-length linear molecules (form III DNA). After a subsequent reaction, the resulting molecules contain the information necessary to form a restriction endonuclease site map of the DNA and to ascertain the relative molecular weights of DNA fragments resolved by gel electrophoresis without the introduction of external standards.[1]

A restriction endonuclease digestion of DNA can be inhibited by ethidium bromide (EtdBr). Altering the concentration of EtdBr in a reaction permits covalently closed circular DNA to be singly nicked or singly cleaved by an enzyme that in the absence of EtdBr would cut the DNA many times.

The singly cleaved DNA molecules form a complete set of full-length permuted linear molecules. This set of molecules, each of which contains the entire genome, contains members with ends formed by cleavage at any site recognized by the restriction endonuclease. If there are six sites in the circular DNA there will be six types of full-length linears, each with a different 5' end, formed by cleavage in the presence of EtdBr. This is the case with SV40 DNA and HindIII.

Digestion of the permuted linears with a restriction endonuclease that has only one recognition site in the DNA (a single-hitting enzyme) will yield twice as many fragments as there are types of linear molecules in the population. Therefore, digestion of a circular DNA in the presence of EtdBr with an enzyme that can cleave the DNA n times yields n types of permuted linears. Cleavage of the n types of permuted linears with a single-hitting enzyme will give $2n$ fragments.

After electrophoretic separation of the $2n$ fragments, the inherent size relationships of the fragments makes it possible to determine a restriction endonuclease map of the DNA. It is also possible to independently determine the relative size of the fragments to within $\pm 2.5\%$.

Conversion of Form I DNA to Form III DNA

The optimal concentration of ethidium bromide required to limit the cleavage of covalently closed circular DNA by a multihitting restriction

---

[1] R. C. Parker, R. M. Watson, and J. Vinograd, *Proc. Natl. Acad. Sci. U.S.A.* **74**, 851 (1977).

endonuclease can only be determined by titration. In order to obtain optimal yields it is necessary to determine the temperature at which the restriction endonuclease is most active. A simple examination showed that different enzymes have enhanced activities at widely varying temperatures: *Pst*I (23°), *Eco*RI (37°), *Hin*dIII (55°).

After determining the optimal temperature the yield of form III can be maximized by titrating the EtdBr while maintaining all other variables (temperature, buffer, enzyme, and DNA concentrations) constant. Increasing the level of EtdBr minimizes the number of molecules receiving more than one cut. The same effect can be achieved by decreasing the amount of enzyme or DNA in the reaction. If too much EtdBr is present very little double-stranded cleavage will occur and most of the molecules formed in the reaction will be singly nicked noncovalently closed circles (form II DNA). It is possible to obtain over 90% of the population as form II DNA. These molecules can be shown to be singly nicked by the presence of approximately equal amounts of single-stranded linear molecules and single-stranded circles in alkaline CsCl velocity experiments.

If the temperature is changed to optimize enzymatic activity, the level of EtdBr will also have to be changed. At temperatures with greater enzymatic activity, more EtdBr is required to produce similar effects to those achieved at other temperatures. In no case does preincubation of the enzyme with EtdBr effect the results.

Sizing and Mapping of Cleaved Permuted Linears

Cleavage of the permuted linears with a single-hitting enzyme, followed by electrophoretic separation of the resulting fragments, makes it possible to determine the relative molecular weights of the fragments in the gel as a function of electrophoretic mobility. These sizes are determined independent of other techniques, assuming that the mobility of a linear DNA in a constant concentration gel is a smooth function of its molecular weight.

The resolved bands can be analyzed in pairs. The smallest DNA fragment migrates faster than the others. Its complement, the largest DNA fragment, migrates slower than the others. When these two fragments are paired the result is a full-length molecule. The next to the fastest can be paired with the next to the slowest to form a full-length molecule, etc. Therefore, the following n equations can be formulated:

$$MW_i + MW_{(2n+1-i)} = MV_{form\ III} \qquad (i = 1, 2, \ldots, n) \qquad (1)$$

where i equals 1 for the slowest band, MW_i is the molecular weight of band i, and n is the number of sites recognized by the restriction endonuclease in the EtdBr limited digest.

After exploring many functions that related molecular weight to mobility we adopted the general form

$$MW_i = \exp(a_0 + a_1 x_i + a_2 x_i^2 + a_3 x_i^3) = f(x_i) \quad (i = 1, 2, \ldots, n) \quad (2)$$

where x_i is the distance migrated by band i. The mobility of full-length linear molecules (form III DNA) can be applied to Eq. (2) yielding

$$MW_{\text{form III}} = \exp(a_0 + a_1 x_{\text{III}} + a_2 x_{\text{III}}^2 + a_3 x_{\text{III}}^3) = f(x_{\text{III}}) \quad (3)$$

The preceding equations and the information from the electrophoretic separation of the n cleaved permuted linears lead to the following n equations:

$$f(x_i) + f(x_{(2n+1-i)}) = f(x_{\text{III}}) \quad (i = 1, 2, \ldots, n) \quad (4)$$

These n equations can be used to determine the four coefficients a_0, a_1, a_2, and a_3. In different experiments these coefficients will change as agarose, buffer, run length, voltage, etc., vary.

To obtain the best curve relating molecular weight to mobility it is necessary to introduce one additional equation. This equation is derived from fractionating the products of a complete digest of the DNA with the multihitting enzyme that is used in the EtdBr limited reaction (sample A) and the products of a complete digest from a reaction containing both the multihitting enzyme and the single-hitting enzyme that was used to cleave the permuted linears (sample B). The two samples are run in two slots of the gel that was used for fractionating and determining the mobilities (x_i above) of the cleaved permuted linears.

One of the bands (p) present in sample A will not be present in sample B. In sample B, band p will have been cleaved by the single-hitting enzyme giving rise to two bands not present in sample A, p' and p''. The smaller of these bands, p', will be present in the cleaved permuted linears as band $2n$. These relationships lead to the Eq. (5):

$$f(x_{\text{p}}) = f(x_{\text{p}'}) + f(x_{\text{p}''}) \quad (5)$$

Equations (3)–(5) provide ($n + 2$) relationships that were used (with the normalization that $MW_{\text{III}} = 1$) to determine the coefficients in Eq. (2), a_0, a_1, a_2, and a_3. This is an overdetermined nonlinear set of equations. It was solved by a least-squares technique with the aid of a computer program (see Appendix).

In order to solve for the coefficients a_0, a_1, a_2, and a_3, the program first solves the linear fit:

$$f(x) = \exp(a_0 + a_1 x_i) \quad (6)$$

After determining the best coefficients for this equation the program adds another term and using the predetermined a_0 and a_1, solves the equation:

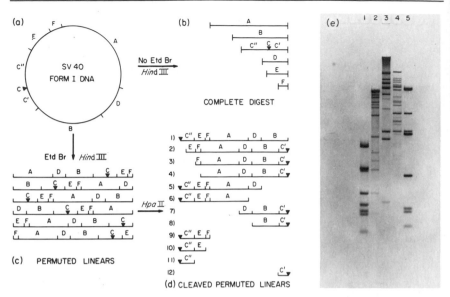

FIG. 1. (a) *Hind*III (slashes) and *Hpa*II (triangle) restriction enzyme sites on SV40 DNA. (b) Complete digest products from a *Hind*III digestion of SV40 DNA. (c) Permuted linears from an ethidium bromide (EtdBr) limited *Hind*III digestion of SV40 DNA. (d) The permuted linears after digestion with *Hpa*II. The 12 fragments are ordered, as they would be resolved in a gel, from largest to smallest. Note that fragment 12 differs from fragment 8 by complete digest product B; similar reasoning yields the map that appears in (a). (e) Gel photograph. Slot 1: SV40/*Hind*III and SV40/*Hind*III/*Hpa*II complete digests. Slot 2: SV40/(EtdBr) *Hind*III after cleavage of permuted linears with *Hpa*II, the slowest migrating band is SV40 form III. Slot 3: λ/*Eco*RI and λ/*Hind*III complete digests. Slot 4: PM2/(EtdBr) *Hind*III after cleavage of permuted linears with *Hpa*II, the slowest migrating band is PM2 form III. Slot 5: PM2/*Hind*III and PM2/*Hind*III/*Hpa*II complete digests.

$$f(x) = \exp\left(a_0 + a_1 x_i + a_2 x_i^2\right) \tag{7}$$

In the process of solving this quadratic exponential the values of a_0 and a_1 are not maintained as they were when solving Eq. (6). However, their values for Eq. (6) are the starting point, along with $a_2 = 0$, for solving Eq. (7).

Finally, the values of a_0, a_1, and a_2 that best fit Eq. (7) are used as the starting point, along with $a_3 = 0$, for solving the cubic exponential that is Eq. (2). The quality of the fit improves greatly between Eqs. (6) and (7) but changes very little from the quadratic Eq. (7) to the cubic Eq. (2).

The computer output consists of the coefficients that solve Eqs. (2), (6), and (7), the relative sizes of each of the cleaved permuted linears, form III DNA, and band p of Eq. (6), an error term which equals the difference between 1 and the sum of the relative sizes of the pair of bands that form a full-length molecule, and the sum of the error squared for the entire population. A sample of the output is presented with the program in the Appendix.

Mobilities of the relevant DNA fragments were determined from negative photographs of EtdBR-stained gels (Fig. 1). The negatives were optically scanned and mobilities were measured as the distance from the top of the gel to the peak of the band on the trace (Fig. 2).

Each set of coefficients was tested by using them to solve the equation $MW_i = (a_0 + a_1x_i + a_2x_i^2 + a_3x_i^3)$ for each of the cleaved permuted linears. The molecular weights for the appropriate pairs were then summed. The deviation of this sum from 1 (the desired total) was then squared and the squares of the deviations were summed. The program then changed the values for $a_0, \ldots, a_3$ until the sum of the deviations squared was minimized.

Using this approach it is possible to obtain molecular weights that sum to within $\pm 1.5\%$ of the expected value. If it is assumed that the logarithm of the molecular weight is a linear function of the electrophoretic mobility [as opposed to the cubic function in Eq. (2)] it is only possible to sum within 6% in the systems tested.

HindIII cleaves SV40 DNA six times, while HpaII cleaves this DNA only once. Digestion of 1 μg of SV40 form I DNA with 5.5 units of HindIII at 55° in the presence of 18.4 μg of EtdBr in a 50-μl reaction maximizes the yield of permuted linears. These molecules when cut with HpaII produce 12 cleaved permuted linears that can be electrophoretically resolved on a 1% agarose gel (Fig. 1). The sizes of the fragments formed, along with similar data from experiments with other systems, are presented in Table I.

The fragment sizes, determined by the third-order exponential function, can be used to construct a restriction endonuclease map. Each of the $2n$ molecules formed in the second digest has one end in common, the site of the single-hitting enzyme. The other end of the molecule is a restriction endonuclease site for the first, multihitting enzyme (Fig. 1). Each of the possible permutations is represented. Additionally, any given molecule differs in size from one of the other molecules by the size of a complete digest product of the DNA with the first enzyme. In Fig. 1d cleaved permuted linear No. 12 differs in size from cleaved permuted linear No. 8 by the size of complete digest product "B." Similarly, fragments No. 10 and No. 11 differ only by complete digest product "E."

By calculating the differences in the relative molecular weights of the cleaved permuted linears and by knowing approximate molecular weights of the complete digest products [these can be determined from Eq. (2)], it is possible to construct a restriction endonuclease map of the DNA. There may be some ambiguity in the final map if complete digest products are

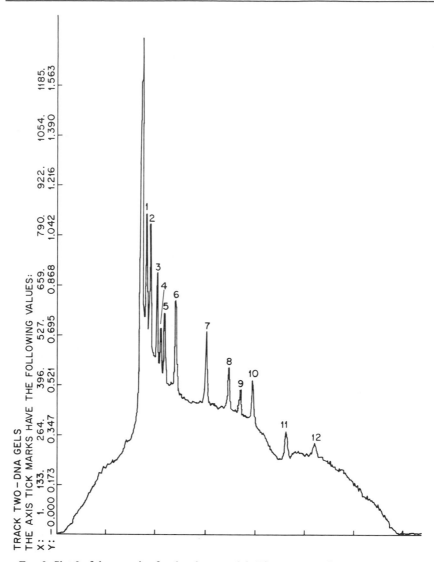

FIG. 2. Slot 2 of the negative for the photograph in Fig. 1e was optically scanned yielding this trace of the SV40/HindIII permuted linears after cleavage with HpaII. Note that the peaks corresponding to fragments 4 and 9 are disproportionately small. This indicates that the rate of the EtdBr limited cleavage of SV40 DNA with HindIII is not the same at each of the enzyme's recognition sites.

similar in size. If such ambiguities arise they can be resolved by other methods, such as the isolation of a specific partial digest product.

Equation (2) can also be used to size additional DNA fragments run on the same gel as the cleaved permuted linears. The molecular weights of

TABLE I
MOLECULAR WEIGHTS OF BACTERIOPHAGE λ RESTRICTION ENDONUCLEASE PRODUCTS[a]

	Thomas[b]	Wellauer[c]	Our values
*Hin*dIII B	—	5.84	5.97
*Eco*RI B	4.74	—	4.79
*Hin*dIII C	—	4.05	4.22
*Eco*RI C	3.73	—	3.73
*Eco*RI D	3.48	—	3.59
*Eco*RI E	3.02	—	3.07
*Hin*dIII D	—	2.67	2.73
*Eci*RI F	2.13	—	2.18
*Hin*dIII E	—	1.40	1.47

[a] The measurements by Thomas and Davis were done by electron microscopy using ϕX174 DNA as a standard. Wellauer *et al.* also used electron microscopy to determine molecular weights with SV40 DNA as a standard. They used 3.28×10^6 as the molecular weight of SV40; we use 3.27×10^6. Our values are averages from five PM2 calibration curves.
[b] M. Thomas and R. W. Davis, *J. Mol. Biol.* **91**, 315 (1975).
[c] P. K. Wellauer, R. H. Reeder, D. Carroll, D. D. Brown, A. Deutch, T. Higashinakagawa, and I. B. Dawid, *Proc. Natl. Acad. Sci. U.S.A.* **71**, 2823 (1974).

the *Eco*RI and the *Hin*dIII complete digest products of λ DNA were determined using the data from the analysis of cleaved permuted linears of PM2 DNA. These fragments were resolved on a 1% agarose gel (Fig. 1), and the fragment molecular weights were determined according to Eqs. (1)–(5) (Tables I and II). The form of Eq. (2) as determined by these data is shown in Fig. 3.

The data presented in Table II allow for the determination of three restriction endonuclease maps. The *Hin*dIII map of SV40 DNA presented in Fig. 1 can be ascertained from these data. The size of smallest fragment, 0.24×10^6 daltons (which equals 0.36 kilobases), implies that there is a *Hin*dIII site that distance from the *Hpa*II site. This fragment is also found in the double enzyme complete digest and is labeled C'. Fragment No. 11 of the cleaved permuted linears is fragment C'' of the double enzyme digest and implies that there is a *Hin*dIII site approximately 0.54 kb from the *Hpa*II site. These two *Hin*dIII sites define the "C" fragment.

Cleaved permuted linear No. 10 has a molecular weight of 0.62×10^6 (0.93 kb). That is the distance between a *Hin*dIII site and the sole *Hpa*II site in SV40 DNA. Between those two sites, however, there is another *Hin*dIII site which is either 0.54 kb away from the *Hpa*II site (fragment No. 11) or 0.36 kb away (fragment No. 12). Therefore, placement of a *Hin*dIII site 0.93 kb away from the *Hpa*II site will form a *Hin*dIII to *Hin*dIII distance of either 0.39 kb (0.93–0.54) or 0.57 kb (0.93–0.36). Table II indicates that there is a *Hin*dIII complete digest product of 0.25×10^6

TABLE II

MOLECULAR WEIGHTS OF CLEAVED PERMUTED LINEARS AND COMPLETE DIGEST PRODUCTS[a]

PM2/(EtdBr) HindIII/HpaII		PM2/ HindIII		SV40/(EtdBr) HindIII/HpaII		SV40/ HindIII		SV40/(EtdBr) HpaI/EcoRI		SV40/ HpaI		SV Hp
III	6.27	A	3.53	III	3.26	A	1.08	III	3.22	A	1.33	1.3
1	4.95	A″	1.98	1	3.07	B	0.70	1	2.77	B	1.24	1.2
2	4.65	A′	1.48	2	2.89	C	0.66	2	2.53	A″	0.75	0.7
3	4.35	B	1.42	3	2.63	C″	0.37	3	2.08	C	0.63	0.6
4	3.99	C	0.61	4	2.50	D	0.29	4	1.20	A′	0.53	0.5
5	3.80	D	0.265	5	2.35	E	0.25	5	0.75			
6	3.74	E	0.245	6	2.01	C′	0.24	6	0.53			
7	3.45	F	0.18	7	1.28	F	—					
8	2.86	G	—	8	0.91							
9	2.59			9	0.76							
10	2.52			10	0.62							
11	2.34			11	0.37							
12	1.98			12	0.24							
13	1.72											
14	1.48											

[a] All data have been converted to daltons $\times 10^{-6}$. We determined from our curves that SV40 = 51.6% of PM2. The form III values in the table are less than 100% because of the mathematical function used. All fragments smaller than 1.42×10^6 were measured from SV40 calibration curves. The SV40 curves slightly overestimated molecular weights of fragments greater than 50% of SV40 and slightly underestimated smaller fragments. The PM2 data is accurate to within ±2.5%. It is not possible to determine the molecular weights of the smallest HindIII complete digest products because they are smaller than any of the cleaved permuted linears. Columns 1, 3, and 5 are cleaved permuted linears; 2, 4, 6, and 7 are complete digests.

[b] K. N. Subramanian, J. Pan, S. Zain, and S. M. Weissman, Nucleic Acids Res. 1, 727 (1974).

(approximately 0.39 kb) and there is no HindIII complete digest product of 0.38×10^6. Therefore, the next HindIII site has been mapped; fragment E (0.25×10^6) is beside fragment C″ (0.37×10^6) and their molecular weights sum to 0.62×10^6 the size of cleaved permuted linear No. 10. The next largest cleaved permuted linear has a molecular weight of 0.76. When complete digest fragment F is placed adjacent to cleaved permuted linear 10, the result is a fragment that is approximately 0.76×10^6. The size of fragment F is not shown in Table II. It is smaller than 0.24×10^6 and is outside of the range of accurate fragment sizing. It is apparent, however, that it should be placed adjacent to cleaved permuted linear number 10 in order to form cleaved permuted linear No. 9 because the alternative would require forming fragment No. 9 by placing a complete digest fragment beside fragment No. 12. This would be impossible for the complete

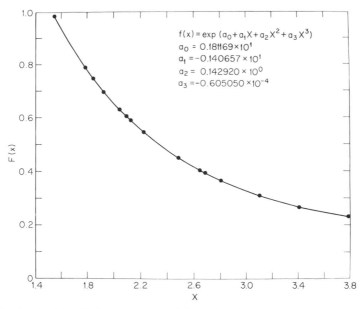

FIG. 3. Graph of the third-order exponential function describing the electrophoretic mobility molecular weight relationship. The data used to derive this curve come from the PM2 fragments in slots 4 and 5 of Fig. 1e.

digest fragment would have to be 0.52×10^6 (0.76×10^6–0.24×10^6). This assignment can be confirmed by other approaches or by more accurate sizing of complete digest F. The next piece to be added must form a cleaved permuted linear of 0.91×10^6. This can be accomplished by placing fragment B (0.70×10^6) beside fragment C' (0.24×10^6) which is cleaved permuted linear No. 12. At this point, the following order has been established:

$$-F–E–C''–C'–B–$$

where C' and C'' are formed by cleavage of HindIII fragment C with HpaII.

By continuing this analysis of the data in Table II and extending it to the other information provided, it is possible to determine the restriction maps for SV40 DNA with HindIII, HpaI, and HpaII; additionally, the restriction map of PM2 DNA with HindIII can be determined (Fig. 4). The data for the latter map leaves one assignment ambiguous. That is the positioning of fragments D and E which are very similar in size. As published before, the final location of these fragments places D beside A and E beside B.

PM2/*Hind* III Maps

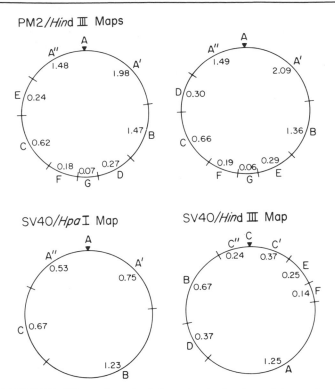

FIG. 4. (a) PM2/*Hind*III map from the information in the sizes of the smaller half of the cleaved permuted linears. (b) PM2/*Hind*III map from the information in the sizes of the larger half of the cleaved permuted linears. Note that fragments D and E are reversed compared to Fig. 2a. (c) SV40/*Hind*III map from the information in the sizes of the smaller half of the cleaved permuted linears; the information in the larger half gives the same map.

Discussion

The third-order exponential function that is used in this paper to study the relationship between electrophoretic mobility and molecular weight was arbitrarily chosen. Other, nonexponential functions could have been employed; a cubic polynomial met with only slightly less success.

Regardless of which mathematical relationship is used, the methodology contains the inherent assumption that mobility is a smooth function of molecular weight. If this assumption is valid (to date there is no sufficient evidence to assess the quality of the assumption) the method presented here offers a simple method for gel calibration and restriction endonuclease site mapping of close circular DNAs.

A limitation of the technique is that it only allows for gel calibration over a defined range of molecular weight. Fragments that migrate faster than the smallest cleaved permuted linear, or slower than form III DNA, cannot be accurately sized using Eq. (2). It is possible, however, to size DNAs of widely differing origins when they are electrophoresed simultaneously in different slots of a gel.

If DNAs are to be sized in this manner, it is necessary to minimize slot-to-slot variation in fragment mobility. Many factors involved in the mobility of a DNA fragment in a gel must be controlled before sample mobilities can be used to infer molecular weights.

Uneven heating in gels frequently leads to samples in center slots migrating faster than identical samples in side slots. Heating can be reduced by using thin glass plates, by having water or air circulation, and by running at low voltages. High sample concentrations can lead to overloaded bands which migrate anomalously rapidly. To prevent rapid migration due to overloading, we tried to load less than 50 ng of DNA with a sample run halfway (running distance ~7.5 cm) into our 4 mm thick, 1% agarose gels.

It is also possible to alter DNA mobility by placing too great a sample volume on the gel. Band shape after the run is a function of the sample volume applied. With sample wells having a cross-sectional area of 10 mm², it is desirable to load no more than 25 μl. Finally, band shape is also altered by salt concentration. When sample mobility is going to be compared, the different DNAs should be layered on in approximately the same salt and it should be less than 100 mM.

The technique presented allows for gel calibration and restriction endonuclease site mapping of circular DNAs without the introduction of external standards. This can most easily be accomplished by the following.

1. An EtdBr titration designed to find a reasonable level of conversion to Form III DNA. "Reasonable" is partially determined by the availability of the DNA. Greater than 50% conversion to form III was obtained in the systems described in this article.

2. Fractionation of the permuted linears formed by EtdBr limitation of the restriction endonuclease digest on a low melting temperature agarose gel.

3. Cleavage of the permuted linears in the presence of the low melting temperature agarose (see Parker and Seed [44], this volume).

4. Fractionation of the cleaved permuted linears and complete digest products by gel electrophoresis.

5. Analysis of mobilities and determination of fragment sizes by a nonlinear, least-squares analysis with the aid of a computer.

6. Determination of restriction endonuclease sites from the relative fragment sizes.

Acknowledgments

I would like to thank Robert Watson and Jerome Vinograd with whom I originally collaborated in doing much of this work (see Parker et al.[1]). I would also like to thank Brian Seed for his invaluable help and suggestions.

Appendix

```
00020   C THIS PROGRAM FINDS THE BEST FIT OF AN EQUATION OF THE FORM
00030   C Y=EXP(A0+A1*X+A2*X*X+...) BETWEEN THE MOLECULAR WEIGHTS, Y, OF THE
00040   C PERMUTED CLEAVAGE PRODUCTS AND THE DISTANCES MIGRATED, X.  TO USE
00050   C PROGRAM IT IS NECESSARY TO ENTER THE FOLLOWING VARIABLES:
00060   C    NX=THE NO. OF PERMUTED CLEAVAGE PRODUCTS + 1
00070   C    NN=THE HIGHEST POWER OF X DESIRED + 1
00080   C    NC1=THE ORDERING NO. OF ONE OF THE TWO PERMUTED CLEAVAGE PRODUC
00090   C       WHICH ARE THE SAME LENGTH AS TWO OF THE COMPLETE PRODUCTS
00100   C       FOUND IN A DOUBLE DIGEST
00110   C    NC2=THE ORDERING NO. OF THE OTHER DOUBLE DIGEST PRODUCT
00120   C    XC=THE MOBILITY OF THE SINGLE DIGEST PRODUCT CORRESPONDING TO T
00130   C       SUM OF PIECES NC1+NC2
00140   C    X(I)=AN ARRAY OF THE MOBILITIES OF THE PERMUTED CLEAVAGE PRODUC
00150   C       IN ORDER OF DECREASING MOLECULAR WEIGHT.  X(1) IS THE MOBIL
00160   C       OF THE FULL LENGTH MOLECULE (FORM 3).  THERE MUST BE NX
00170   C       ENTRIES IN THIS ARRAY.
00180   C
00190         DIMENSION X(100),A(20),Y(100),XP(101),YP(101),P(20,24),DA(22),
00200        ,DOC(3),ERR(100)
00210         EXTERNAL F
00220         COMMON NA,NH,NX,N1,NC1,NC2,XC,YC,ERC,X,ERR
00230   C
00240   C    DOC IS AN ARRAY OF PLOT CONTROL CHARACTERS
00250   C
00260         DATA DOC,EPS/3*0.0,1.0E-6/
00270      10 CONTINUE
00280         WRITE(5,3)
00290       3 FORMAT(' ENTER NX,NN,NC1,NC2,XC, FORMAT(4I5,F10.0)')
00300         READ (5,1,END=100)NX,NN,NC1,NC2,XC
00310       1 FORMAT(4I5,F10.0)
00320         IF(NN .LT. 2 .OR. NN .GT. 20) GO TO 100
00330         IF(NX .LE. 0 .OR. NX .GT. 99) GO TO 100
00340         N1=NX+1
00350         NH=(NX-1)/2
00360         WRITE(5,4)
00370       4 FORMAT(' ENTER X(I), FORMAT(F10.0)')
00380         READ (5,2)(X(I),I=1,NX)
00390       2 FORMAT(F10.0)
00400         DO 95 I=1,NX
00410         IF(X(I) .LT. 10.0)GO TO 95
00420         WRITE(5,96)
00430      96 FORMAT('0 X IS LARGER THAN 10.0, POSSIBLE DATA ERROR')
00440         STOP
00450      95 CONTINUE
00460         DO 205 NA=2 ,NN
00470         NC=NA+2
00480         IF(NA .GT. 2) GO TO 210
00490         DO 25 I=1,NA
00500         P(I,1)=0.5**I
00510      25 CONTINUE
00520         GO TO 220
00530     210 DO 215 I=1,NA
00540     215 P(I,1)=A(I)
00550         P(NA,1)=0.0
00560     220 DO 225 I=1,NA
00570     225 DA(I)=0.2
00580         CALL AMOEBA(P,DA,NA,EPS,F)
00590         DO 45 I=1,NA
```

```
00600          A(I)=P(I,NC)
00610       45 CONTINUE
00620          CALL F(A,RES)
00630          DO 5 I=1,NX
00640          Y(I)=FEXP(X(I),NA,A)
00650        5 CONTINUE
00660          WRITE(5,6)(A(I),I=1,NA)
00670        6 FORMAT('1   THE COEFFICIENTS A0, A1, A2, ... ARE'/(5X,5E15.6))
00680          WRITE(5,8) RES
00690        8 FORMAT('0   SUM OF ERROR SQUARE = ',E14.6)
00700          WRITE(5,7)(X(I),Y(I),ERR(I),I=1,NX),XC,YC,ERC
00710        7 FORMAT(1H0,12X,'X',12X,'Y',12X,'ERROR'/(3X,3F15.6))
00720          DX=(X(NX)-X(1))/100
00730          DO 15 I=1,101
00740          XP(I)=X(1)+(I-1)*DX
00750          YP(I)=FEXP(XP(I),NA,A)
00760       15 CONTINUE
00770      205 CONTINUE
00780          GO TO 100
00790   C
00800   C      SYSTEM PLOTTING SUBROUTINES
00810   C
00820          CALL SCALE(X(NX),X(1),XMX,XMN,15,IE)
00830          CALL LABEL(0.,0.,XMN,XMX,15.,15,'X',1,0)
00840          CALL LABEL(0.,0.,0.,1.,10.,10,'F(X)',4,1)
00850          CALL XYPLT(NX,X,Y,XMN,XMX, 0.0,1.0,DOC,0,4)
00860          CALL XYPLOT(101,XP,YP,XMN,XMX, 0.0,1.0,DOC,1)
00870      100 STOP
00880          END
00890          FUNCTION FEXP(X,NA,A)
00900          DIMENSION A(1)
00910          ARG=A(1)
00920          DO 5 I=2,NA
00930          ARG=ARG+A(I)*X**(I-1)
00940        5 CONTINUE
00950          FEXP=EXP(ARG)
00960          RETURN
00970          END
00980          SUBROUTINE F(A,RES)
00990          DIMENSION A(1),Y(100),X(100),ERR(100)
01000          COMMON NA,NH,NX,N1,NC1,NC2,XC,YC,ERC,X,ERR
01010          DO 5 I=1,NX
01020          Y(I)=FEXP(X(I),NA,A)
01030        5 CONTINUE
01040          ERR(1)=Y(1)-1.0
01050          RES=ERR(1)**2
01060          DO 15 I=1,NH
01070          ERR(I+1)=Y(I+1)+Y(N1-I)-1.0
01080          ERR(N1-I)=ERR(I+1)
01090          RES=RES+ERR(I+1)**2
01100       15 CONTINUE
01110          IF(NC1*NC2 .LE. 0) RETURN
01120          YC=FEXP(XC,NA,A)
01130          ERC=YC-Y(NC1)-Y(NC2)
01140          RES=RES+ERC**2
01150          RETURN
01160          END
01170          SUBROUTINE AMOEBA(P,Y,N,E,F)
01180   C
01190   CCCCCCCCCCCCCCCCCCCCCCCCCCCCCCCCCCCCCCCCCCCCCCCCCCCCCCCCCCCCCCCCCCCCC
01200   C
01210   C   UPON ENTRY THE FOLLOWING PARAMETERS MUST BE PASSED:
01220   C
01230   C      N -- THE NUMBER OF VARIABLES FOR THE MINIMIZATION.
01240   C      E -- ABSOLUTE DELTA FUNCTION VALUE FOR DETERMINING CONVERGENCE.
01250   C      F -- THE NAME OF A SUBROUTINE WHICH WHEN CALLED BY CALL F(V,X)
01260   C           WHERE V IS AN ARRAY OF N VALUES WILL RETURN WITH THE COR-
01270   C           RESPONDING FUNCTION VALUE IN X.
01280   C      P -- AN ARRAY OF DIMENSION (N,N+4) WITH INITIAL VALUES FOR THE
01290   C           VARIABLES X(I) IN P(I,1) FOR I=1(1)N.
01300   C      Y -- A VECTOR OF DIMENSION (N+2) CONTAINING N DISPLACEMENTS
01310   C           DX(I) IN Y(I) FOR I=1(1)N.  THESE VALUES DX(I) WILL BE USED
01320   C           TO CONSTRUCT THE INITIAL SIMPLEX IN THE ARRAY P:
01330   C               P(I,J)=X(I)+DELTA(J-1,I)*DX(I) FOR I=1(1)N,J=1(1)N+1,
01340   C           WHERE DELTA(I,J) IS THE KRONECKER DELTA FUNCTION.
01350   C
01360   CCCCCCCCCCCCCCCCCCCCCCCCCCCCCCCCCCCCCCCCCCCCCCCCCCCCCCCCCCCCCCCCCCCCC
```

```
01370      C
01380            INTEGER N,I,J,NS,NH,NL,NC,NR,NT,NEW
01390            REAL P,Y,E,F,YH,YL,YC,YR,YT,X,FLN,ERR
01400            REAL RFACT1,RFACT2,CFACT1,CFACT2,EFACT1,EFACT2
01410            DIMENSION P(20,1),Y(1)
01420            DATA RFACT1,RFACT2/-1.0,2.0/
01430            DATA CFACT1,CFACT2/0.5,0.5/
01440            DATA EFACT1,EFACT2/2.0,-1.0/
01450      C
01460      CCC  INITIALIZE SUBROUTINE PARAMETERS.
01470      C
01480            FLN=FLOAT(N)
01490            NS=N+1
01500            NC=N+2
01510            NR=N+3
01520            NT=N+4
01530      C
01540      CCC  CONSTRUCT INITIAL SIMPLEX.
01550      C
01560        100 DO 101 J=1,N
01570            DO 102 I=1,N
01580            P(I,J+1)=P(I,1)
01590            IF(I.EQ.J) P(I,J+1)=P(I,J+1)+Y(I)
01600        102 CONTINUE
01610        101 CALL F(P(1,J),Y(J))
01620            CALL F(P(1,NS),Y(NS))
01630      C
01640      CCC  FIND CURRENT MAX AND MIN.
01650      C
01660        200 NH=NS
01670            NL=NS
01680            YH=Y(NS)
01690            YL=Y(NS)
01700            DO 201 I=1,N
01710            IF(Y(I).LE.YH) GO TO 202
01720            YH=Y(I)
01730            NH=I
01740            GO TO 201
01750        202 IF(Y(I).GE.YL) GO TO 201
01760            YL=Y(I)
01770            NL=I
01780        201 CONTINUE
01790      C
01800      CCC  COMPUTE CENTROID.
01810      C
01820        300 DO 301 I=1,N
01830            X=0.0
01840            DO 302 J=1,NS
01850            IF(J.EQ.NH) GO TO 302
01860            X=X+P(I,J)
01870        302 CONTINUE
01880        301 P(I,NC)=X/FLN
01890            CALL F(P(1,NC),YC)
01900      C
01910      CCC  REFLECT CURRENT MAX THROUGH CENTROID.
01920      C
01930        400 DO 401 I=1,N
01940        401 P(I,NR)=RFACT1*P(I,NH)+RFACT2*P(I,NC)
01950            CALL F(P(1,NR),YR)
01960            IF(YR.LT.YL) GO TO 500
01970            X=YL
01980            DO 402 I=1,NS
01990            IF(I.EQ.NH) GO TO 402
02000            IF(Y(I).GT.X) X=Y(I)
02010        402 CONTINUE
02020            IF(YR.GT.X) GO TO 600
02030            NEW=NR
02040            GO TO 800
02050      C
02060      CCC  EXPAND.
02070      C
02080        500 DO 501 I=1,N
02090        501 P(I,NT)=EFACT1*P(I,NR)+EFACT2*P(I,NC)
02100            CALL F(P(1,NT),YT)
02110            NEW=NT
```

```
02120          IF(YT.GE.YL) NEW=NR
02130          GO TO 800
02140    C
02150    CCC   CONTRACT.
02160    C
02170      600 NEW=NH
02180          IF(YR.GT.YH) GO TO 601
02190          NEW=NR
02200          YH=YR
02210      601 DO 602 I=1,N
02220      602 P(I,NT)=CFACT1*P(I,NEW)+CFACT2*P(I,NC)
02230          CALL F(P(1,NT),YT)
02240          IF(YT.GT.YH) GO TO 700
02250          NEW=NT
02260          GO TO 800
02270    C
02280    CCC   SHRINK SIMPLEX TOWARD CURRENT MIN.
02290    C
02300      700 DO 701 J=1,NC
02310          IF(J.EQ.NL) GO TO 701
02320          DO 702 I=1,N
02330      702 P(I,J)=0.5*(P(I,J)+P(I,NL))
02340          CALL F(P(1,J),Y(J))
02350      701 CONTINUE
02360          YC=Y(NC)
02370          GO TO 802
02380    C
02390    CCC   TEST FOR CONVERGENCE.
02400    C
02410      800 DO 801 I=1,N
02420      801 P(I,NH)=P(I,NEW)
02430          YH=YR
02440          IF(NEW.EQ.NT) YH=YT
02450          Y(NH)=YH
02460      802 ERR=0.0
02470          DO 803 I=1,NS
02480          ERRI=(Y(I)-YC)**2
02490          IF(ERRI .GT. ERR)ERR=ERRI
02500      803 CONTINUE
02510          ERR=SQRT(ERR)
02520          IF(ERR.LT.E) GO TO 1000
02530    C
02540    CCC   UPDATE CYCLE DATA.
02550    C
02560      900 GO TO 200
02570    C
02580    CCC   END OF MINIMIZATION.
02590    C
02600     1000 RETURN
02610          END
02620    //
02630

ENTER NX,NN,NC1,NC2,XC, FORMAT(4I5,F10.0)
   15    4   13   15   3.10

ENTER X(I), FORMAT(F10.0)
2.44
2.72
2.79
2.86
2.96
3.04
3.06
3.14
3.39
3.53
3.56
3.66
3.92
4.15
4.46
```

```
THE COEFFICIENTS A0, A1, A2, ... ARE
    0.183334E+01  -0.768442E+00

SUM OF ERROR SQUARE =    0.785824E-02

        X              Y              ERROR
     2.440000       0.959198       -0.040802
     2.720000       0.773506       -0.023364
     2.790000       0.732998       -0.009234
     2.860000       0.694611        0.002212
     2.960000       0.643233        0.018861
     3.040000       0.604881        0.010511
     3.060000       0.595656        0.010746
     3.140000       0.560141        0.022377
     3.390000       0.462237        0.022377
     3.530000       0.415090        0.010746
     3.560000       0.405630        0.010511
     3.660000       0.375627        0.018861
     3.920000       0.307601        0.002212
     4.150000       0.257768       -0.009234
     4.460000       0.203130       -0.023364
     3.100000       0.577625        0.066895


THE COEFFICIENTS A0, A1, A2, ... ARE
    0.292751E+01  -0.147393E+01   0.111100E+00

SUM OF ERROR SQUARE =    0.384683E-03

        X              Y              ERROR
     2.440000       0.992605       -0.007395
     2.720000       0.771356        0.009176
     2.790000       0.726201        0.005391
     2.860000       0.684434        0.003272
     2.960000       0.630088        0.005847
     3.040000       0.590679       -0.007451
     3.060000       0.581348       -0.008424
     3.140000       0.545957       -0.001252
     3.390000       0.452791       -0.001252
     3.530000       0.410228       -0.008424
     3.560000       0.401869       -0.007451
     3.660000       0.375759        0.005847
     3.920000       0.318838        0.003272
     4.150000       0.279190        0.005391
     4.460000       0.237820        0.009176
     3.100000       0.563275        0.006616


THE COEFFICIENTS A0, A1, A2, ... ARE
    0.292544E+01  -0.147298E+01   0.111210E+00  -0.585645E-04

SUM OF ERROR SQUARE =    0.384422E-03

        X              Y              ERROR
     2.440000       0.992637       -0.007363
     2.720000       0.771454        0.009065
     2.790000       0.726304        0.005368
     2.860000       0.684539        0.003308
     2.960000       0.630191        0.005942
     3.040000       0.590776       -0.007341
     3.060000       0.581443       -0.008309
     3.140000       0.546044       -0.001118
     3.390000       0.452838       -0.001118
     3.530000       0.410248       -0.008309
     3.560000       0.401883       -0.007341
     3.660000       0.375751        0.005942
     3.920000       0.318769        0.003308
     4.150000       0.279063        0.005368
     4.460000       0.237611        0.009065
     3.100000       0.563366        0.006986
ENTER NX,NN,NC1,NC2,XC, FORMAT(4I5,F10.0)


STOP
```

Section VII

Determination of Fragment Ordering

[51] Denaturation Mapping

By Santanu Dasgupta and Ross B. Inman

Introduction

In recent times, physical isolation of genes or their regulatory components has emerged as a routine tool for genetic analysis. This approach has been particularly useful in eukaryotic systems where collection and mapping of mutants is laborious, if not impossible, in some cases. Most of the eukaryotic genomes seem to have highly ordered patterns of mutually interspersed unique and repetitive nucleotide sequences.[1-3] A clustering of genes have also been found to occur in prokaryotic systems[4,5] and might have a regulatory role in coordinate expression of the genes. In view of these results, it is important to develop reliable tools for isolation and characterization of specific DNA fragments so that their position in the sequence organization of the genome can be identified and related to their phenotypic expression.

Denaturation mapping of DNA molecules, or DNA fragments, provides a reliable frame of reference which can be used to identify, align, or map different regions on a genome or fragment of a genome.[6,7] Used in conjunction with site-specific restriction endonucleases, it can provide a direct method for ordering the fragments of a large DNA molecule. This article will discuss the general principles and techniques involved and a few examples where they have been employed with reasonable success.

General Strategy

Denaturation mapping of DNA is based on the one-to-one correlation between helical stability and base composition at the intramolecular level and thus reflects the localized compositional heterogeneity along a large DNA molecule. Measurements on purified bacteriophage DNA indicate that long pieces of DNA can possess a unique pattern of localized dena-

[1] E. H. Davidson, B. R. Hough, W. H. Klein, and R. J. Britten, *Cell* **4**, 217 (1975).
[2] J. E. Manning, C. W. Schmid, and N. Davidson, *Cell* **4**, 141 (1975).
[3] M. M. Wilkes, W. R. Pearson, J.-R. Wu, and J. Bonner, *Biochemistry* **17**, 60 (1978).
[4] S. Spadari and F. Ritossa, *J. Mol. Biol.* **53**, 357 (1970).
[5] P. S. Sypherd and S. Osawa, *in* "Ribosomes" (M. Nomura, A. Tissières, and P. Lengyel, eds.), p. 669. Cold Spring Harbor Lab., Cold Spring Harbor, New York, 1974.
[6] R. B. Inman and M. Schnös, *J. Mol. Biol.* **49**, 93 (1970).
[7] D. K. Chattoraj and R. B. Inman, *Methods Mol. Biol.* **7**, 33 (1974).

turation under partial melting conditions. The general approach for ordering the fragments from a larger DNA molecule would therefore essentially involve matching the denaturation pattern of the parent molecule with those of the individual components. This should be done at a degree of denaturation that produces the most characteristic denaturation pattern. In difficult cases it may be necessary to make the comparison at two or more degrees of denaturation. This would become necessary, for instance, when several fragments of similar length occur in regions that do not have characteristic denaturation. In principle, the fragments from any genome could be ordered in this way although the maximum size of DNA that can be analyzed may be limited by the size and number of DNA species present during spreading.

Preparation of DNA Samples for Electron Microscopy

Denaturation. Among the possible denaturation constraints (high temperature, high pH, high formamide concentration, or organic solvents), the high pH condition seems to yield the most reproducible result.[8] The degree of denaturation that produces the most characteristic denaturation pattern will have to be determined by trials for each type of DNA and its fragments. This may be done by either incubating the DNA in denaturing buffer at various pH values for a constant period of time or by using denaturing buffer at optimum pH for various periods of time. A typical denaturation buffer contains 67.8 mM Na$_2$CO$_3$, 10.7 mM EDTA, and 31.5% formaldehyde (Matheson, Coleman, and Bell). A total of 4.7 ml of this buffer is adjusted to the desired pH with 0.2–0.3 ml of 5 M NaOH. Three microliters of this buffer is then mixed with 7 μl of DNA solution (A_{260} = 0.01–0.03) in 20 mM NaCl and 5 mM EDTA (the final solution contains 9% formaldehyde and 0.13 M Na$^+$). The mixture is allowed to stand at 23° for 10 min or more, depending on degree of denaturation required, and then chilled in an ice bath before spreading for electron microscopy. The total amount of DNA required to produce a moderate to high density distribution of molecules on the specimen grid is 0.1–0.3 ng.

Spreading for Electron Microscopy. A clear resolution between single-strand and double-strand regions is crucial for the success of this mapping technique. The formamide spreading technique[9] with minor modifications consistently yields reasonable degrees of resolution in this laboratory. The method consists of forming a protein monolayer (containing attached DNA) over an aqueous hypophase. Samples of the monolayer are then picked up on a carbon film evaporated onto mica discs (details have been

[8] R. B. Inman, this series, Vol. 29, p. 451.
[9] R. W. Davis, M. Simon, and N. Davidson, this series, Vol. 21, Part D, p. 413.

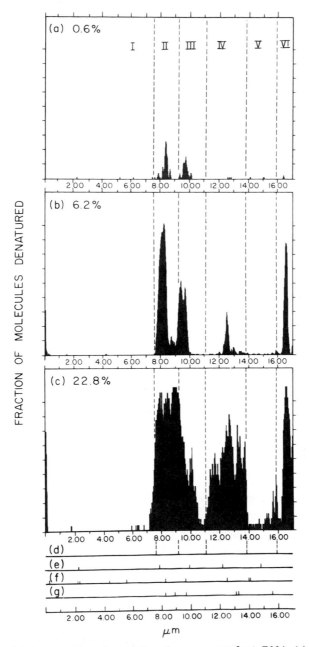

Fig. 1. Partial denaturation and restriction cleavage maps for λ DNA. (a)–(c) Denaturation maps obtained at different degrees of denaturation (0.6, 6.2, and 22.8%, respectively). (d)–(g) Restriction cleavage sites mapped for *Eco*RI, *Bam Hpa*I and *Hin*dIII, respectively (see footnote 13, p. 436). The dotted vertical lines indicate *Eco*RI cleavage sites and all maps are normalized to a λ length of 17.06 μm.

discussed previously[8]). The DNA solutions, partially denatured as described above, are diluted with equal volumes of formamide (Matheson, Coleman and Bell) and adjusted to 0.01% cytochrome c (twice crystallized and lyophilized, Calbiochem). Five microliters of DNA–protein solution is then gently drained down a wet and thoroughly cleansed stainless steel rod (0.5 cm diameter with pointed tip) and the surface monolayer is picked up on the carbon film by gently touching the surface. The film is subsequently washed with ethanol, dried in a stream of warm and dry nitrogen gas, and rotary shadowed with 2.8 cm of 0.02 cm diameter platinum wire at a distance of 7 cm and at an angle of about 7°. The film is floated off the mica at an air–water interface and picked up on 200 mesh copper grids for observation in the electron microscope.

Measurements and Analysis of Data. Clean untangled molecules are photographed in large numbers, and the denaturation site positions are measured and aligned using a projector-digitizer-calculator system as previously described.[8]

Examples

1. PARTIAL DENATURATION AND RESTRICTION CLEAVAGE MAPS. In Fig. 1 we show how denaturation mapping can, in principle, provide information about the location of restriction cleavage products. Three commonly used degrees of partial denaturation for λ DNA are shown in Fig. 1a–c where there is a progression from denaturation located at only the center (Fig. 1a) to the more complex pattern consisting of three broad zones of denaturation spread over the right half of the molecule (Fig. 1c). It can be seen that, in the case of the six *Eco*RI cleavage products of linear λ DNA (Fig. 1d and dotted line in Fig. 1a–c), they can easily be identified on the basis of their characteristic partial denaturation patterns. For instance at 6.2% denaturation (Fig. 1b), fragment I would not have any denatured sites except for an infrequent small site located at the extreme left end. Fragments II, III, IV, and VI would be recognized by characteristic denatured sites and finally section V would have no denaturation.

A similar ordering could be made for *Bam* cleavage products (Fig. 1e), but the method would begin to become much less useful for the *Hpa*I and *Hin*dIII products (Fig. 1f–g) because of the small size of some of the segments.

2. CHARACTERIZATION OF COMPOSITE AND CHIMERIC PLASMID DNA. The following example shows how denaturation mapping can be used to order the components within a complex chimeric plasmid DNA.[10] The plasmid analyzed below was constructed by first forming a composite of *Col*E1 and *RSF*1010 DNA (each was linearized with *Eco*RI before liga-

[10] T. Tanaka, B. Weisblum, M. Schnös, and R. B. Inman, *Biochemistry* **14**, 2064 (1975).

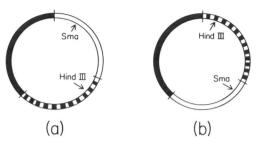

(a)   (b)

FIG. 2. Two alternative forms of the chimeric plasmid. Closed section, *RSF*1010 sequence; open section, *Col*E1 sequence; dotted section, *Drosophila melanogaster* sequence. The lines drawn across the molecule represent *Eco*RI sites and arrows show the position of *Sma* and *Hin*dIII cleavage positions.

tion). The resulting composite was then incompletely digested with *Eco*RI so that, on the average, only one of the two *Eco*RI sites was cleaved. This material was then ligated with *Drosophila melanogaster* DNA fragments obtained by *Eco*RI digestion. The following analysis was then carried out on a clone obtained from the above ligated mixture. Due to the method used to make the chimeric plasmid, there are two alternate structures possible because of the two possible *Eco*RI cleavage positions in the *Col*E1-*RSF*1010 composite DNA and these are shown in Fig. 2. The actual structure selected after the cloning procedure corresponds to model (a) in Fig. 2, and the proof for this rested on two independent methods, one of which involved a denaturation map analysis of the positions of the three DNA segments within the chimeric unit. The denaturation map data are presented in Fig. 3. First the average positions of denatured sites at several degrees of denaturation for the two original plasmids (cleaved with *Sma*) are shown in Fig. 3a and b. The histogram average for the *Sma*-cleaved composite is given in Fig. 3c and the dotted lines running between Fig. 3a–c show the deduced positions of the components in the composite structure. Figure 3d shows the histogram average of the chimera, and again the dotted lines indicate how the denaturation pattern can be matched up with the composite. It is to be noted that the degree of denaturation in Fig. 3d is somewhat lower than in Fig. 3c, and thus the second denatured zone from the left in *Col*E1 [observed in Fig. 3b (III) and in Fig. 3c] is missing in Fig. 3d. Finally in Fig. 3e is the histogram average of the chimera, this time linearized with *Hin*dIII. For ease of comparison, the histogram average in Fig. 3e has been permuted so that it has a starting point corresponding to the average shown in Fig. 3d. The actual starting point of *Hin*dIII cleaved molecules is shown by the arrow situated at 2.16 μm. Again the degree of denaturation shown in Fig. 3e is somewhat higher than that of Fig. 3d.

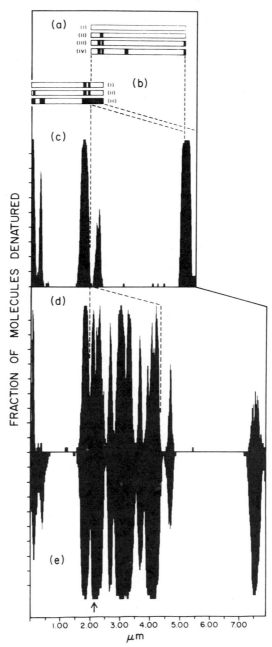

FIG. 3. Partial denaturation analysis of a chimeric plasmid DNA. (a) Average position of denatured sites in *RSF*1010 DNA linearized by *Sma* cleavage. The maps shown in (I)–(IV) were obtained at increasing degrees of denaturation and are normalized to 3.15 μm. (b) Average position of denatured sites in *Col*El DNA linearized by *Sma* cleavage. The maps

A similar approach has been employed to determine the sequence organization of cloned histone DNA from sea urchin *Psaminechinus miliaris* which consists of five GC-rich histone coding regions separated by AT-rich spacer DNA sequences. The arrangement of histone genes cloned into a λ phage DNA was determined by matching the denaturation map of the intact recombinant λ-histone DNA with that of the 5.6 kb histone DNA reclaimed from recombinant DNA by digestion with *Hin*dIII and fractionation by gel electrophoresis.[11]

3. QUANTITATIVE ANALYSIS OF A REPEATED DENATURATION PATTERN. Denaturation mapping can also be used in a statistical analysis of the regularity of repeat units and for determination of the exact repeat length, especially under situations where large numbers of repeating units are involved. The rDNA of *Xenopus laevis* comprises several hundred almost identical genes clustered at a single locus and has been subjected to such an analysis.[12] The method essentially consists of cross-correlating the denaturation pattern of rDNA molecules with a series of sine and cosine waves of varying wavelengths. The correlation function increases to a maximum value when the wavelength of the series matches the length of the repeat unit and becomes zero for a random denaturation pattern. Evidently the accuracy of this correlation method would increase with a higher number of repeat units per molecule. Analysis of rDNA molecules showed that the mean repeat length was 5.4 ± 0.4 μm ($8.7 \pm 0.6 \times 10^6$ daltons). Further estimates of regularity of the repeat units and homogeneity of the population of rDNA molecules were made by visual alignments of the repeating units as described in examples (1) and (2) above. The repeat length from the correlation method corresponds closely with the value estimated by visual alignment and direct length measurement. The denaturation pattern of rDNA is consistent with the known base composition of regions within the repeat (see Wensink and Brown[12] for detailed discussion).

[11] R. Portmann, W. Schaffner, and M. Birnstiel, *Nature (London)* **264**, 31 (1976).
[12] P. C. Wensink and D. D. Brown, *J. Mol. Biol.* **60**, 235 (1971).

(I)–(III) correspond to the same denaturation conditions as (I)–(III) in (a) and have been normalized to 2.40 μm. (c) Denaturation map average of a composite molecule constructed from *Col*E1 and *RSF*1010 DNA. The composite was linearized by *Sma* cleavage and lengths normalized to 5.55 μm. Dotted lines show the position of the *Col*E1 and *RSF*1010 segments within the composite. (d) Denaturation map average of a chimeric molecule constructed from the composite and a *Drosophila melanogaster* DNA fragment. The chimeric molecule was linearized with *Sma* and the dotted lines show the position of composite and *Drosophila melanogaster* DNA segments. Lengths normalized to 7.90 μm. (e) Denaturation map average of the chimeric molecule linearized by *Hin*dIII cleavage. For ease of comparison with (d) above, this map has been artificially permuted so that the actual ends occur at 2.16 μm (shown by the arrow).

Summary

It is apparent from the above examples that the denaturation mapping technique provides a ready and direct method for analyzing the overall sequence organization of DNA. It gives a reasonably accurate measure of repeat lengths, the regularity with which a specific sequence is repeated, the mutual orientation among the repeats, and an approximate measure of A + T content within a repeating sequence. The process requires very little DNA (in nanogram quantities) and is particularly useful under situations where extensive purification of large quantities of DNA is difficult.

[13] S. Gottesman and S. L. Adhya, in "DNA Insertion Elements, Plasmids and Episomes" (A. Bukhari, J. A. Shapiro, and S. L. Adhya, eds.), p. 713. Cold Spring Harbor Lab., Cold Spring Harbor, New York, 1977.

[52] Genetic Rearrangements and DNA Cleavage Maps

By A. I. Bukhari and D. Kamp

A large number of restriction enzymes (endodeoxyribonucleases) are now available. Since the DNA fragments generated by these enzymes can be physically separated it has become routine to manipulate at will different well-defined segments of a DNA molecule. The first step in this manipulation is to locate the sites at which a restriction enzyme cleaves the DNA molecule. A cleavage map is a representation of a linear array of these sites. We describe here how genetic rearrangements can be used to derive a cleavage map. The genetic rearrangements can also help in the isolation and identification of restriction fragments with specific genes or sites without extensive mapping. Conversely, comparison of cleavage products of two species of molecules can lead to the identification of genetic rearrangements.

The common genetic rearrangements are diagrammed in Fig. 1. Deletions (and substitutions) can be very useful in determining the order of fragments generated by restriction enzymes. Simple insertions, duplications, and inversions may also be helpful in the identification of specific fragments and in understanding the spatial arrangement of genes. The changes that can be expected in the cleavage pattern of DNA molecules, because of the genetic rearrangements, are listed in Table I. In this article, we will focus mainly on the use of deletions and substitutions for physical mapping; we will also describe the general rules for using insertions, duplications, and inversions.

A primary example of the use of deletions and substitutions is the

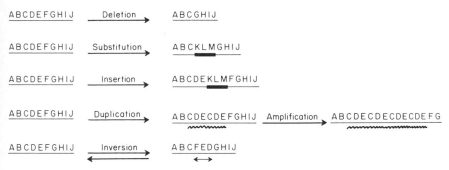

FIG. 1. Common genetic rearrangements. Letters A–J denote hypothetical DNA sequences. Sequences D, E, and F are missing in the deletion and are replaced by new sequences K, L and M in substitution. In insertion, new sequences K, L, and M are added to the existing sequences. The wavy lines show the duplicated and amplified sequences. In inversion, the sequences D, E, and F, shown by a double-headed arrow, are reversed in orientation.

mapping of restriction sites in the bacteriophage λ genome.[1–10] The use of deletions and substitutions for physical mapping is normally limited by their availability. However, the DNA molecules generally amenable to physical mapping are the virus and plasmid genomes. Various methods are available for isolating deletions in viruses and plasmids. Frequently, DNA segments of a complex genome to be mapped are cloned into a virus or a plasmid vector. Common vectors for cloning are the bacteriophage λ, or small plasmids.[11,12] Since deletions or substitutions in these vectors can be isolated easily they can be used to physically dissect the cloned fragments.

Isolation of Deletions and Substitutions

Deletions in bacteriophage λ can be obtained simply by isolating particles that have reduced amounts of DNA. It has been found that λ particles

[1] B. Allet, P. G. N. Jeppesen, K. J. Katagiri, and H. Delius, *Nature (London)* **241,** 120–123 (1973).
[2] B. Allet, *Biochemistry* **12,** 3972 (1973).
[3] M. Perricaudet and P. Tiollais, *FEBS Lett.* **56,** 7 (1975).
[4] M. Thomas and R. W. Davis, *J. Mol. Biol.* **97,** 123 (1975).
[5] D. M. Haggerty and R. F. Schleif, *J. Virol.* **18,** 659 (1976).
[6] D. I. Smith, F. R. Blattner, J. Davies, *Nucleic Acids Res.* **3,** 343 (1976).
[7] V. Pirrotta, *Nucleic Acids Res.* **3,** 1747 (1976).
[8] D. Kamp, R. Kahmann, D. Zipser and R. J. Roberts, *Mol. Gen. Genet.* **154,** 231 (1977).
[9] B. Allet and A. I. Bukhari, *J. Mol. Biol.* **92,** 529 (1975).
[10] L. H. Robinson and A. Landy, *Gene* **2,** 1 and 33 (1977).
[11] K. N. Timmis, S. N. Cohen and F. C. Cabello, *Prog. Mol. Subcellular Biol.* **1,** (F. E. Hahn ed.). Springer-Verlag, Berlin (1978).
[12] M. Zabeau and R. J. Roberts, *in* "Molecular Genetics" (J. H. Taylor, ed.). Academic Press, New York (in press).

TABLE I
CHANGES IN CLEAVAGE PATTERNS AS A CONSEQUENCE OF GENETIC REARRANGEMENTS

	No. of cleavage sites[a]	Missing fragments	New fragments[b]
Deletion	0	1	1
	$\geqslant 1$	$\geqslant 2$	1
Substitution	0/0	1	1
(deletion/insertion)	$\geqslant 1/0$	$\geqslant 2$	1
	$0/\geqslant 1$	1	$\geqslant 2$
	$\geqslant 1/\geqslant 1$	$\geqslant 2$	$\geqslant 2$
Duplication	0	1	1
	$\geqslant 1$	0	1
Insertion	0	1	1
	$\geqslant 1$	1	$\geqslant 2$
Inversion	0	0	0
	$\geqslant 1$	2	2

[a] Cleavage sites in the deleted, inserted, duplicated or inverted DNA. In substitutions, the cleavage sites are given as follows: number of cleavage sites in deleted DNA/ number of cleavage sites in new DNA.
[b] In rare instances a cleavage site can be generated by deletion at the site of rearrangements because of fusion of nucleotide sequences. This would result in the generation of additional fragments.

that carry less than normal amounts of DNA are more resistant than wild type to heat or chelating agents, such as pyrophosphate or EDTA.[13] Thus, repeated treatment of appropriate λ strains with these agents can result in high enrichment of deletions. This method, in principle, can be applied to those viruses in which DNA is cut at two specific sites during maturation and the DNA in between the two sites is packaged into the virions, with certain constraints on the size of DNA. Thus the DNA between the two *cos* sites in λ can be packaged within a range of $+5$ to -20% of the normal λ length.[14,15] This method is not applicable to those viruses that have a headful mode of packaging of DNA and always package a fixed amount of DNA.[15]

Substitutions in bacteriophage λ can be generated by abnormal excision of the prophage.[16] Genes located adjacent to the prophage integration site can be incorporated into the phage genome with a concomitant loss of some phage DNA. In such cases, one end point of the substitutions (and of the deletions in the λ genome) is always at the λ *att* site, which is used by the phage for integrative recombination.[16]

[13] J. S. Parkinson and R. Huskey, *J. Mol. Biol.* **56**, 369 (1971).
[14] M. Feiss, R. A. Fisher, M. A. Crayton, and C. Egner, *Virology,* **77**, 281 (1978).
[15] H. Murialdo and A. Becker, *Microbiol. Rev.* **42**, 529 (1978).
[16] A. Campbell, *in* "Bacteriophage λ" (A. Hershey, ed.), p. 13. Cold Spring Harbor Laboratory, Cold Spring Harbor, New York, 1971.

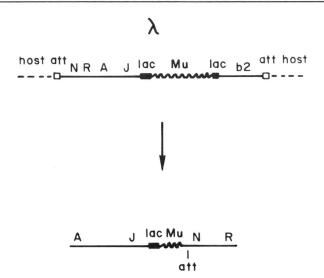

FIG. 2. Generation of λ substitution mutants carrying segments of bacteriophage Mu DNA. Prophage Mu is inserted into the *lac* genes carried on λp*lac*5. The λ*lac* :: Mu structure is too large to be packaged into λ heads. Consequently, the λ particles can only arise if parts of the λ-Mu structure are deleted. Plaque-forming λ-Mu hybrids arise when the deletions span a part of Mu DNA and the λ b2 region but leave the essential λ*g* genes intact.[17]

Another method for isolating deletions or substitutions is to first introduce an insertion element into the viral genome, or the plasmid (or into the DNA cloned in a phage or a plasmid). Complete excision of the insertion element then can lead to deletion formation, whereas partial excision will generate a substitution. One use of such a system is illustrated by the isolation of λ-Mu hybrids.[17] When bacteriophage Mu is inserted in λ DNA (Fig. 2) the λ-Mu structure is too large to be accommodated in the λ phage heads. The λ particles can arise only if a deletion reduces the size of λ-Mu structure to appropriate proportions. Almost any desired deletion in λ can be isolated from such a system by isolating different kinds of λ particles (i.e., plaque-forming or non-plaque-forming phage particles). This method can also be applied to those viruses that have a headful mode of DNA packaging. Thus, deletions in phage P22 genome have been isolated by first inserting transposons for drug resistance into its genome.[18] The deletions in plasmids have been isolated by insertion and excision of a transposon.[19]

[17] A. I. Bukhari and B. Allet, *Virology.* **63**, 30 (1975).
[18] G. M. Weinstock and D. Botstein, *Cold Spring Harbor Symp. Quant. Biol.* **43**, 1209 (1979).
[19] N. Federoff, *Cell* **16**, 697 (1979).

Deletions (and in fact all other rearrangements) in plasmids and phages can also be obtained *in vitro* by digesting the DNA with appropriate restriction enzymes and rejoining the fragments. In particular, an insertion element containing appropriate restriction sites can be inserted into the plasmid and parts of the plasmid and the insertion element can be subsequently removed.[20] Deletions may also be isolated by simply selecting for elimination of certain restriction sites. For example, transformation with a plasmid, such as pBR322, that contains only one *Eco*RI or *Hin*dIII site, after treatment with *Eco*RI or *Hin*dIII can result in the isolation of the plasmid derivatives which have lost large segments spanning the restriction sites.

It should be noted that since deletions in λ can be readily isolated, it is relatively simple to delete parts of the substitutions in λ. If a DNA segment is cloned into λ, then deletions in the cloned segment can be isolated. An illustration of this principle is provided by O'Day *et al.*,[21] who inserted the 18 kilobase *Eco*RI fragment of bacteriophage Mu into λ and then isolated deletions in λ generating an extensive deletion map of the 18 kilobase fragment of Mu. Thus, eukaryotic DNA cloned into λ DNA can be analyzed in a similar manner.

Deletion Mapping

Deletion mapping represents the fastest and easiest way to map restriction nuclease fragments. Digests of a phage or plasmid DNA molecules and deletion variants are subjected to electrophoresis side-by-side and the following principles are used for the evaluation of the gels. Fragments that are missing in the digest of the deletion mutant originate from the part of the DNA molecule that is covered by the deletion. Fragments that are present in both molecules map completely outside the deletion. In general, there will be one fragment that is only found with the deletion mutant. This fragment overlaps the deletion endpoints. As a rare event, a cleavage site can be generated by the deletion when the appropriate sequences have been fused together. In this case two additional fragments will appear in the deletion strain. The more deletion strains are available for this kind of analysis, the more accurate and unambiguous the interpretation will be. A set of deletions, analyzed on the same gel, will be sufficient in many instances to establish the cleavage map for a particular restriction enzyme.

[20] R. Meyer, D. Figurski, and D. R. Helinski, *in* "DNA Insertion Elements, Plasmids and Episomes" (A. I. Bukhari, J. Shapiro, and S. Adhya, eds.), p. 559. Cold Spring Harbor Laboratory, Cold Spring Harbor, New York, 1977.

[21] K. O'Day, D. Schultz, W. Ericsen, L. Rawluk, and M. Howe, *Virology,* **93,** 320 (1979).

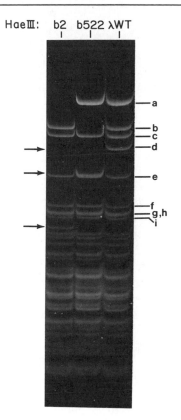

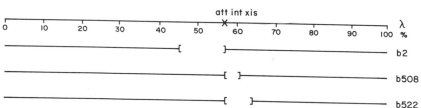

FIG. 3. Cleavage products of λ, λ b2, and λ b522 DNA with *Hae*III. Fragments were separated on a 2% agarose gel and stained with ethidium bromide. The arrows point to the attachment site fragments of λ, λ b2, and λ b522. Drawing at the bottom gives a map of the deletions. Attachment site is indicated by X on the λ genome calibrated in 0–100%.

The kind of information that can be easily extracted by deletion mapping is illustrated in the following example, where fragments in the attachment site region of λ have been identified. Figure 3 shows the fragments that are generated by *Hae*III from λ DNA and two deletion mutants of λ, λ b522 and λ b2. These deletions end at the attachment site and do not overlap each other. Fragment a is missing in λ b2, but is present in λ and λ b522. It must therefore map to the left of the attachment site in the region that is covered by the b2 deletion. Fragment b is present in λ and λ b2, but is missing in λ b522. This fragment must originate from the region to the right of the attachment site of λ. Fragment d of λ, which is neither in λ b2 nor in λ b522, must contain the attachment site. The order of fragments is therefore a − d(*att*) − b. The attachment site fragments of λ b2 and b522 can also be identified. The sizes of the attachment site fragments give some clues to the approximate location of the *Hae*III sites nearest to the λ attachment site. The distance of the *Hae*III site to the right of the attachment site cannot be more than 750 base pairs, because the b2 attachment site fragment is about 750 base pairs long. The b522 attachment site fragment is about 1350 base pairs long, and therefore the *Hae*III site to the left of the attachment site must be located within a 1350 base pair distance to the attachment site. Because the wild-type λ attachment site fragment is about 1700 base pairs long, the *Hae*III sites must be located 950 to 1350 base pairs to the left of the attachment site and 350 to 750 to the right. The same calculation can be made for the *Hae*III sites that flank the b2 and b522 attachment sites. If only one of the four *Hae*III sites can be mapped precisely, for instance by using another cleavage site as a reference point, precise locations for all four *Hae*III sites can be obtained. Fragments a, b, and d constitute only 80% of the DNA that is covered by the deletions b2 and b522, and it is obvious that in this particular example only the large fragments have been identified that can be resolved well by gel electrophoresis.

In general, deletion mapping will be most useful and normally sufficient for the establishment of a cleavage map when a limited number of cleavage sites is present and the fragments can be resolved completely. In cases where more than 50 fragments are generated, only fragments that are larger or smaller than average can be usually analyzed under the appropriate conditions. In these cases it is helpful to isolate restriction fragments that contain deletions and then apply the principle of deletion mapping to these fragments rather than to the whole phage or plasmid genome.[8] For instance, the *Eco*RI fragments that overlap the deletions b2, b508, b522 can be isolated and purified easily. Figure 4 shows an analysis of the *Mbo*II digests of these *Eco*RI fragments. All but one of the *Mbo*II fragments of the b522 *Eco*RI *att* fragment should fall into two groups.

1. Fragments mapping between the right endpoint of the b522 deletion and the right *Eco*RI site are also present in *Mbo*II digests of the b508 and b2 *Eco*RI *att* fragments.

2. Fragments mapping between the left *Eco*RI site and the attachment sites are also present in *Mbo*II digests of b508 *Eco*RI *att* but missing in *Mbo*II digests of b2 *Eco*RI *att*.

b522 *Eco*RI *att-Mbo*II fragments b and g belong to group (1), while fragments a, c, e, and f have to be placed in group (2). Fragment d is the

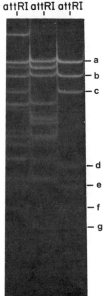

MboⅡ
fragments of: b508 b2 b522
attRI attRI attRI

— a
— b
— c

— d
— e
— f
— g

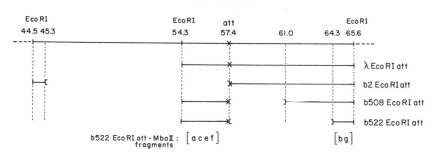

| EcoRI | | | EcoRI | att | | EcoRI | |
| 44.5 | 45.3 | | 54.3 | 57.4 | 61.0 | 64.3 | 65.6 |

λ EcoRI att

b2 EcoRI att

b508 EcoRI att

b522 EcoRI att

b522 EcoRI att - MboⅡ :   [a c e f]   [b g]
            fragments

FIG. 4. Mapping of MboII fragments in the att region of λ. Purified EcoRI att fragments of λ deletions b2, b522, and b508 were cleaved with MboII, and the cleavage products were separated on a 4% agarose gel. A map is given at the bottom. The top horizontal line shows the λ genome between 44.5–65.6% coordinates. Deletion endpoints are indicated by brackets. It can be inferred that fragment d, shown in the gel, contains the b522 attachment site.

only unique fragment in the MboII digest of b522 EcoRI att. It therefore must be the fragment that overlaps the deletion endpoint and contains the b522 attachment site.

This kind of analysis can be carried on for MboII fragments of the b508 and b2 EcoRI att fragments yielding two more groups:

3. The fragments mapping in the region defined by the attachment site and the right endpoint of the b508 deletion (fragments present in the MboII digest of b2 EcoRI att but missing in MboII digests of b508 and b522 EcoRI att), and

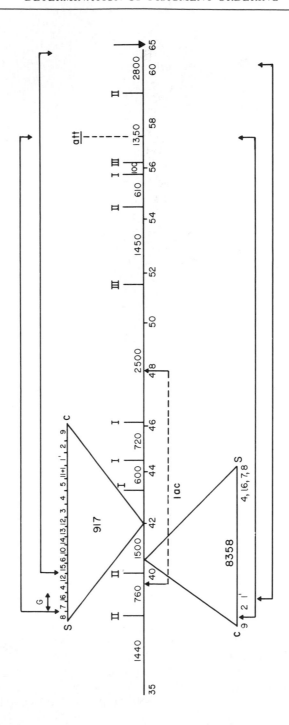

4. Those from the region between the right endpoint of b508 and the right endpoint of b522 (fragments present in the *Mbo*II digest of b2 and b508 *Eco*RI *att* but missing in the *Mbo*II digest of b522 *Eco*RI *att*).

Substitutions

The rules of deletion mapping are also applicable to substitutions, only in this case additional fragments might be generated. If the substitutions are related, i.e., the new DNA in different substitution mutants is derived from a specific DNA molecule, then a cleavage map of the substituting DNA itself can be obtained. An example of this method is provided by the mapping of λ-Mu hybrids, in which a part of λ has been substituted by Mu DNA. By examining the restriction enzyme digests of the DNAs of these hybrids, it was possible to confirm the order of λ fragments obtained by other methods. Furthermore, by comparing the digests to that of Mu DNA it was possible to deduce the order of many of the fragments generated by *Hin*dII and *Hin*dIII cleavage of bacteriophage Mu DNA.[9] Figure 5 depicts the generation of λ-Mu hybrids from two different Mu insertions in the *lacZ* gene in λ p*lac*5. Most of the plaque-forming hybrids arose from deletions beginning within Mu and ending at some point in λ. Thus, as shown in the figure, continuous blocks of λ fragments were missing in the λ-Mu hybrids and the Mu fragments present in the hybrids could be ordered with respect to each other.

FIG. 5. Relationship of the structure of λ-Mu hybrids to the DNA cleavage maps. Figure shows two Mu insertions in the *lacZ* gene carried on λp*lac*5 from which several λ-Mu hybrids were isolated.[17] The central horizontal line represents the 35–65% segment of the λ genome.[9] The *lac* region indicated by broken lines is located between 39 and 48% λ. The numbers above the horizontal line represent the approximate sizes of the fragments, in base pairs, generated by the combined *Hin*d restriction activities (*Hin*dII, *Hin*dIII, *Hpa*I). Both Mu insertions, 917 and 8358, are located in the 1500 base pair fragment originating from the *lac* region, but the insertions are present in opposite orientations. The letters c and S refer to the c end, the left end, and the S end, the right end, of Mu DNA. The expansion of 917 insertion shows the order of the *Hin*d fragments, identified by numbers, which would be expected from bacteriophage Mu DNA.[9,32] The fragments are not drawn to scale. The invertible region G is shown by a double headed arrow. In the insertion 8358 only a few fragments from each end of Mu are listed. The horizontal lines with arrows pointing toward the Mu DNA indicate deletions, originating within Mu DNA and extending to the right into the λ DNA. Thus, the top deletion line in 917 insertion shows that all Mu fragments except fragment 8 are missing in this hybrid (the Mu-*lac* junction fragment at this end is not numbered). The deletion in λ removes the 600, 720, 2500, 1450, 610, 100, and 1350 base pair λp*lac*5 fragments. The second 917 deletion shows that Mu fragments 12, 4, 16, 7 and 8 are present in this hybrid and all of the λ fragments to the right up to 65% λ are missing. In the hybrids originating from 8358 insertion, the Mu fragments are present in the reverse order. Such analyses of a large number of hybrids helped to establish the *Hin*dII cleavage map of Mu.[9,32]

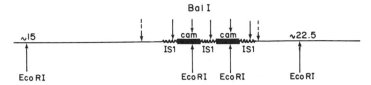

FIG. 6. Proposed structure of Tn9 in Pl*cam*. At least two Tn9 copies are stably present in "tandem" (sharing an IS*1* in the middle), in the Pl*cam* strain in use in our laboratory. The figure shows the Pl segment (defined by the *Eco*RI cleavage sites at coordinates 15 and 22.5) containing Tn9. The arrows pointing upward indicate *Eco*RI cleavage sites; the arrows pointing downward indicate *Bal*I sites. The *Bal*I cuts, outside of Tn9, are indicated by dotted arrows.

Insertions

A large number of spontaneously isolated mutations in *Escherichia coli* are caused by the insertion of IS elements, naturally occurring insertion sequences of various sizes. Insertions can be deliberately generated with transposons, which are complex transposable units carrying selectable markers. To isolate such insertions, the donor molecules carrying the transposons must be present in the cell along with the recipients, such as a plasmid or a bacteriophage genome. The recipient molecules containing transposons can then be obtained by retransferring these molecules to new hosts and selecting for the markers present on the transposons.[22]

Insertions can be used in characterizing DNA molecules in at least three different ways.

1. As described earlier, insertions can be used to produce deletions. The rules of deletion mapping can then be applied for physical mapping.

2. Insertions can provide new restriction enzyme sites for manipulation of DNA. Thus Tn9, a transposon for chloramphenicol resistance, has one *Eco*RI cut and insertion of Tn9 at any given site introduces a new *Eco*RI site.[23,24] This in turn can be used for dissecting a large fragment into subfragments for further mapping.

3. They can help understand the spatial arrangement of various genes on the restriction fragments which is the general goal of most mapping studies. An example is provided by Barth and Grinter,[25] who isolated a large number of insertions at different sites on the plasmid RP4. Based on the mutations produced by these insertions, various genes could be immediately assigned to different fragments by comparing the digests of the insertion mutants with the wild type plasmid. In bacteriophage Mu, the IS*1* and

[22] A. I. Bukhari, J. A. Shapiro, and S. L. Adhya, eds., "DNA Insertion Elements, Plasmids and Episomes." Cold Spring Harbor Laboratory, Cold Spring Harbor, New York, 1977.
[23] A. I. Bukhari and S. Froshauer, *Gene* **3**, 303 (1978).
[24] F. DeBruijn and A. I. Bukhari, *Gene* **3**, 315 (1978).
[25] P. T. Barth and N. J. Grinter, *J. Mol. Biol.* **113**, 455 (1977).

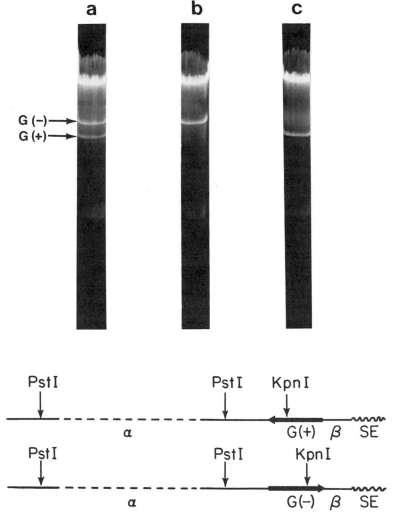

FIG. 7. Cleavage of Mu DNAs with *Kpn*I and *Pst*I. The arrows point to the fragments that are diagnostic for the orientation of the G segment as indicated in the map below. DNAs are obtained from (a) heat induced Mu *cts*62 where both orientations of the G segment are present, (b and c) Mu *cts*62 *gin*⁻.[30] The *gin*⁻ mutation prevents the inversion of the G segment and either orientation of the G segment can be obtained. The orientation of G is indicated as + (also called the *flip* orientation) and − (or *flop*). α, β, SE refer to segments of mature Mu DNA.[26]

Tn9 insertions in the X mutants, eliminating the B gene function, were shown to be located in the 5000 base pair EcoRI fragment of Mu DNA.[23,24]

Duplications

A diagnostic feature of duplicated or amplified sequences is the presence of more than stoichiometric amounts of fragments produced by a restriction enzyme which makes more than one cut in the duplicated sequences. If the enzyme makes only one cut in the duplicated sequences, then a new fragment equal to the size of duplication is produced. Let us consider the example of Tn9 insertion in phage P1cam. Tn9 is a transposon of approximately 2800 base pairs and contains at each end the 800 base pair IS1 element. Thus, of 2800 base pairs, about 1600 base pairs constitute IS1 sequences. It was inferred that Tn9 is present in duplicated copies in P1cam, as shown in Fig. 6, based on the following reasons.[24] Two fragments of approximately 850 base pairs produced by the enzyme BalI, that cuts once within and at either end of Tn9, are present in at least produces three fragments that hybridize with IS1. If only one copy of Tn9 were present, EcoRI would be expected to generate only two such fragments. One of the three fragments is about 2000 base pairs in length and evidently represents Tn9 minus one IS1 element.

Inversions

Inversions can be of three types:

1. Stable inversions, in which inversion of a segment is generated by illegitimate recombination.

2. Genetically unstable inversions, in which general recombination between large inverted repetitious sequences leads to inversion. The frequency of such inversion may not be high enough to be readily detected in restriction enzyme digests of DNA molecules.

3. Flip-flops, in which inversion is promoted at a high frequency by site specific recombination mechanisms and thus a segment of DNA is present in two different orientations in a given population.

The well-known examples of flip-flop include invertible segments of bacteriophage Mu and bacteriophage P1,[26] a segment apparently controlling

[26] L. T. Chow and A. I. Bukhari, *in* "DNA Insertion Elements, Plasmids and Episomes" (A. I. Bukhari, J. Shapiro, and S. Adhya, eds.), p. 295. Cold Spring Harbor Laboratory, Cold Spring Harbor, New York, 1977.

phase variation in *Salmonella*,[27] the herpes simplex genome[28] and the 2 micron circle in *Saccharomyces cerevisiae*.[29] A diagnostic feature of flip-flop is the presence of less than stoichiometric amounts of fragments produced by a restriction site located asymmetrically in the invertible segment and a site located outside of the invertible segment. This principle is illustrated by the fragments generated from bacteriophage Mu DNA by the enzymes which cut within and outside of the invertible G segment.[9,30,31] Figure 7 shows the *Kpn*I-*Pst*I digests of Mu DNAs. *Kpn*I site is located asymmetrically within G whereas the *Pst*I site is outside of G. Two different fragments can be seen when the DNA molecules have the G segment in both orientations. When G is only in one orientation, only one fragment specific to a particular orientation can be seen.

[27] J. Zieg, M. Silverman, M. Hilmen, and M. Simon, *in* "The Operon" (J. H. Miller and W. S. Reznikoff, eds.), p. 411. Cold Spring Harbor Laboratory, Cold Spring Harbor, New York, 1978.

[28] B. Roizman, *Cell* **16**, 481 (1979).

[29] J. R. Broach, J. F. Atkins, C. McGill and L. Chow, *Cell* **16**, 827 (1979).

[30] D. Kamp, R. Kahmann, D. Zipser, T. R. Broach, and L. T. Chow, *Nature (London)* **271**, 577 (1978).

[31] A. Toussaint, N. Lefebvre, J. R. Scott, J. A. Cowan, F. DeBruijn, and A. I. Bukhari, *Virology* **89**, 146 (1978).

[32] M. Magazin, M. Howe, and B. Allet, *Virology* **77**, 677 (1977).

[53] Determination of Fragment Order through Partial Digests and Multiple Enzyme Digests

By KATHLEEN J. DANNA

Many applications of restriction endonuclease cleavage of DNA are possible only if the resulting fragments have been ordered to produce a physical map. This article describes the basic principles of two techniques for fragment ordering: analysis of partial digestion products and multiple enzyme digestion. These were the first methods used to determine a physical map of a DNA genome, the simian virus 40 (SV40) genome,[1] and remain perhaps the most straightforward and easy to interpret procedures. Moreover, aside from apparatus for DNA fragment analysis, these approaches require only the DNA and restriction endonuclease(s) of interest.

Ordering DNA fragments by partial endonuclease digestion is analogous to a sequencing technique for RNA described by Adams *et al.*,[2] who

[1] K. J. Danna, G. H. Sack, Jr., and D. Nathans, *J. Mol. Biol.* **78**, 363 (1973).

[2] J. M. Adams, P. G. N. Jeppesen, F. Sanger, and B. G. Barrell, *Nature (London)* **223**, 1009 (1969).

used partial T1 ribonuclease digestion products of R17 RNA to order T1 oligonucleotides of the limit digest. A partial digest of DNA is obtained by limiting the reaction time so that the endonuclease does not cleave at all possible recognition sites in the DNA. Thus, partial digestion yields some fragments comprised of two or more contiguous complete digestion products. By purifying a partial digestion product, incubating it with excess enzyme to complete the digestion, and identifying the resultant fragments, one can determine which final products are contained in a given partial digestion product. Analysis of several partial digestion products in this way enables one to deduce the order of all fragments of the limit digest.

Multiple enzyme digestion for ordering DNA fragments employs an approach routinely used for sequencing proteins and RNA, namely, sequential digestion with enzymes of different specificity. For DNA, the cleavage products of one endonuclease are characterized with respect to size and are then digested with a second endonuclease. Analysis of the resultant double-digestion products establishes the relationship between the cleavage sites of the two enzymes.

The partial digest and multiple enzyme digest approaches to fragment ordering are best illustrated by example. This chapter presents a model study which develops a physical map of the SV40 genome. Section I describes procedures for digestion of DNA with endonuclease and for analysis of cleavage products, with emphasis on techniques, such as polyacrylamide gel electrophoresis and autoradiography, that are used in the model study. Ordering fragments through analysis of partial digestion products is illustrated in Section II for the two sets of SV40 DNA fragments produced by cleavage with *Hin*cII and with *Hin*dIII. In Section III, *Taq*I and *Bam*H1 are used in multiple enzyme digestions with *Hin*cII and *Hin*dIII to generate a complex physical map that includes the cleavage sites for all four enzymes.

I. Basic Procedures

A. *Digestion of DNA with Restriction Endonucleases*

The first step in ordering DNA fragments is complete digestion of DNA with the endonuclease(s) of choice. Optimal reaction conditions (e.g., pH, salt concentrations, and temperature) for specific restriction endonucleases are described in the catalogues published by suppliers[3] and

---

[3] Bethesda Research Laboratories, Rockville, Maryland; Boehringer Mannheim, Indianapolis, Indiana; Miles Laboratories, Inc., Elkhart, Indiana; New England Biolabs, Beverly, Massachusetts.

elsewhere in this volume.[4] However, the amount of enzyme[5] needed to yield a limit digest must be determined empirically because the number of cleavage sites for a particular endonuclease in a given species of DNA cannot be predicted. A series of pilot reactions, in which both the ratio of enzyme to DNA and the incubation time are varied, is useful for determining the amount of enzyme needed to attain complete digestion. In the model study, each pilot reaction contained 0.2 μg of SV40 DNA in a 20-μl volume with either 0.1 unit, 0.5 unit, or 1 unit of enzyme. A 5-μl aliquot was withdrawn from each reaction mixture at 30 min, 1 hr, 2 hr, and 3 hr, and reaction in each was stopped by the addition of sodium dodecyl sulfate (SDS) to a final concentration of 1% (w/v). Samples were then analyzed electrophoretically, as described in Section I,B, to determine which conditions resulted in complete digestion.

An important characteristic of a limit digest is that all cleavage products are equimolar. Therefore, when uniformly labeled [^{32}P]DNA is cleaved, the amount of radioactivity in each limit product is directly proportional to its size. Complete digestion can be verified by the addition of more endonuclease to a reaction mixture and incubation for a longer time. If digestion is complete, neither the amounts nor the sizes of the products will change.

The same methods can be used to establish conditions for partial digestion. Short reaction times result in large fragments that contain several contiguous complete digestion products, and longer times result in smaller fragments. A preparation of partial digestion products including fragments of all sizes can be obtained by combining several reaction mixtures incubated for different lengths of time.

Preparative reaction mixtures should be exactly scaled to pilot reactions that yield a high proportion of the desired products. A preparative digest containing 1×10^6 to 2×10^6 dpm of [^{32}P]DNA proved sufficient to map SV40 DNA, which is about 5000 nucleotide pairs in length.

B. Analysis of Cleavage Products

DNA fragments produced by restriction endonucleases have been separated by reverse phase chromatography,[6] hydroxylapatite chromatography,[7] agarose gel electrophoresis,[8,9] and polyacrylamide gel

---

[4] See this volume, Section III.
[5] For example, New England Biolabs defines 1 unit of enzyme as the amount that completely digests 1 μg of phage λ DNA in 15 min at the optimal temperature of incubation.
[6] R. D. Wells *et al.*, this volume, Article [41].
[7] D. Davis *et al.*, this volume, Article [49].
[8] P. A. Sharp, B. Sugden, and J. Sambrook, *Biochemistry* **12**, 3055 (1973).
[9] R. C. Parker and B. Seed, this volume, Article [44].

electrophoresis.[10,11] For most of the analyses in the model study, vertical slab gels of polyacrylamide were used because of their high resolving power and high capacity. Visualization of DNA fragments in gels has been achieved by the use of both fluorescent[12] and nonfluorescent stains,[13] by the use of tungstate screens,[14] and by autoradiography.[15] In the model study, fragments of [^{32}P]DNA (specific activity of 5×10^5 dpm/μg of DNA) were visualized by autoradiography, a sensitive method that allows as little as 10^{-3} μg of DNA to be observed in 16 hr. Detailed descriptions of both slab gel electrophoresis and autoradiography have been presented in this series.[9-11,14,16] The remainder of this section reviews only the specific techniques used to prepare the slab gels, samples, and wet- and dried-gel autoradiograms for the model study.

Slab gels (14-cm wide, 13-cm long, 1-mm thick) are routinely prepared by the method of Loening[17] from these stock solutions:

1. Acrylamide (recrystallized from ethyl acetate), 15% (w/v)-N,N'-methylenebisacrylamide (recrystallized from acetone), 0.75% (w/v)

2. 10× electrophoresis buffer: 0.4 M Trizma base, 0.2 M sodium acetate, 0.02 M sodium EDTA, adjusted to pH 7.8 with glacial acetic acid

3. Ammonium persulfate, 5% (w/v), freshly made

4. N,N,N',N'-tetramethylethylenediamine (TEMED), neat

For a 4% polyacrylamide gel (total volume 40 ml), 10.7 ml of stock acrylamide solution are mixed with 4 ml of 10× electrophoresis buffer and 24.9 ml of deionized water. Polymerization is catalyzed by the addition of 0.42 ml of 5% ammonium persulfate and 0.042 ml of TEMED. The solution is poured between two glass plates, as described by DeWachter and Fiers,[16] to form the slab gel.

Prior to electrophoresis, DNA samples containing 1% SDS (w/v) are incubated at 37° for 10 min to disrupt protein–DNA aggregates. Samples are then made 10% (w/v) in sucrose and 0.02% (w/v) in bromphenol blue, are layered into wells in the gel, and are electrophoresed at constant voltage in a buffer of 0.04 M Trizma base, 0.02 M sodium acetate, 0.002 M sodium EDTA, adjusted to pH 7.8 with glacial acetic acid.

The time of electrophoresis and voltage required depend on the range of fragment sizes that must be resolved. A mixture of DNA fragments

[10] T. Maniatis and A. Efstratiadis, this volume, Article [38].
[11] P. G. N. Jeppesen, this volume, Article [39].
[12] G. G. Carmichael and G. K. McMaster, this volume, Article [47].
[13] G. S. Hayward, *Virology* **49**, 342 (1972).
[14] R. A. Laskey, this volume, Article [45].
[15] K. J. Danna and D. Nathans, *Proc. Natl. Acad. Sci. U.S.A.* **68**, 2913 (1971).
[16] R. DeWachter and W. Fiers, Vol. 21, Part D, p. 167.
[17] U. Loening, *Biochem. J.* **102**, 251 (1967).

ranging in size from 200 to 2000 nucleotide pairs can be resolved on a 13-cm long slab gel of 4% polyacrylamide by electrophoresis at 120 V for 3 hr. For adequate resolution of larger fragments, such as partial digestion products, the voltage or the time of electrophoresis or both should be increased. As an alternative, a gel with a larger pore size (i.e., lower percentage of acrylamide or agarose) can be employed.

Autoradiographic analysis of [^{32}P]DNA fragments can be achieved by wet-gel exposure of X-ray film, as described by DeWachter and Fiers,[16] or by dried-gel exposure. Wet-gel autoradiography is essential for purification of [^{32}P]DNA fragments from gels (see Section I,C). For wet-gel autoradiography, one of the glass plates enclosing the gel is removed; the gel, supported by the remaining glass plate, is covered with Saran wrap; a piece of medical X-ray film (e.g., Kodak Blue Brand or Kodak RP Royal X-omat) is laid atop the Saran wrap; and a clean glass plate is clamped on top of the film to ensure uniform contact. After an appropriate exposure time, the film is processed in Kodak D-19 developer (5 min) and Kodak Rapid-Fixer (5 min). As little as 2000 dpm of [^{32}P]DNA in an area of 1 mm^2 produces an easily visible spot on Kodak Blue Brand film after a 30-min exposure.

The alternate procedure of dried-gel autoradiography results in sharper bands because of a reduced scattering angle between the radioactive sample and the film. The gel can be dried on an automatic gel dryer[18] or by Maizel's modification[19] of a method described by Fairbanks et al.[20] The gel is first transferred to a sheet of Whatman 3MM paper, is placed gel-side up on a porous support (either a metal grid or a porous polyethylene sheet), and is covered with Saran wrap. With an automatic gel dryer, the assembly is placed gel-side up onto a prewarmed heating plate (about 80°), which has an integral vacuum manifold. The assembly is covered with a sheet of silicone rubber, which forms a seal about the gel when the vacuum system is activated. The combination of heat and vacuum dries a 14-cm × 13-cm × 1-mm gel in about 35 min. The dried gel is placed tightly against a piece of X-ray film for autoradiography.

C. Purification of DNA Fragments

Individual DNA fragments are conveniently purified by preparative gel electrophoresis, excision of gel segments containing DNA bands,[16] and recovery of the DNA from each segment. For purification of the [^{32}P]DNA

[18] For example, a Gel Slab Dryer, Model 224, manufactured by Bio-Rad Laboratories, Richmond, California.
[19] J. V. Maizel, Jr., Methods Virol. 5, 180.
[20] G. Fairbanks, Jr., C. Levinthal, and R. H. Reeder, Biochem. Biophys. Res. Commun. 20, 393 (1965).

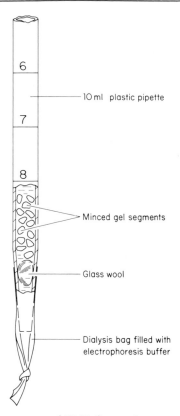

6

7

8

10 ml plastic pipette

Minced gel segments

Glass wool

Dialysis bag filled with
electrophoresis buffer

FIG. 1. Apparatus for recovery of DNA from gel segments by electrophoresis into a dialysis bag.

fragments in the model study, a wet-gel exposure of a preparative gel was made as described in Section I,B, except that labels written with ^{32}P-ink were placed at the corners of the slab before exposure.[16] The developed X-ray film was aligned on top of the gel by means of the radioactive labels, and the outline of the gel was traced onto the film. With the guidance of the tracing, the film was accurately aligned under the glass plate supporting the gel so that gel segments corresponding to DNA bands could be excised with a scalpel or razor blade.

For a description of general methods to recover DNA from gels, see Smith.[21] In the model study, recovery was accomplished by electrophoresis of the sample into a dialysis bag, a reliable method, which results in 80–90% recovery of DNA. A simple apparatus, illustrated in Fig. 1, consists of a short segment of a 10-ml plastic pipette with a glass

[21] H. O. Smith, this volume, Article [46].

wool plug in the tip. Attached to the pipette is a 3–4-cm long dialysis bag filled with electrophoresis buffer. The dialysis tubing should fit tightly over the tapered end of the pipette. After the pipette has been filled with electrophoresis buffer, minced gel segments containing a DNA fragment are transferred into the pipette and allowed to settle. The assembly is placed into a cylindrical gel apparatus with the dialysis bag toward the positive electrode so that the DNA will migrate into the bag during subsequent electrophoresis. The DNA recovered from the bag can be used directly or can be concentrated by precipitation in 0.03 M sodium acetate, pH 6.0, and 70% ethanol at $-20°C$. DNA purified from polyacrylamide gels in this way is suitable for further endonuclease digestion. In the model study, 90% of a partial digestion product about 1000 nucleotide pairs in length was recovered from a 1-ml volume of 4% polyacrylamide gel segments by electrophoresis at 150 V for 3 hr.

II. Ordering of Fragments by Partial Digestion

Analysis of partial digestion products to order fragments produced by cleavage of DNA with a restriction endonuclease employs the techniques described in Section I in the following steps.

1. The electrophoretic profile for products of complete digestion is established.
2. Individual partial digestion products from a large-scale digest are purified.
3. Each partial digestion product is redigested with an excess of enzyme and electrophoresed in parallel with a marker of completely digested DNA.
4. The resulting data are analyzed to construct a physical map.

The method is exemplified by the mapping of cleavage sites for *Hinc*II and *Hind*III on SV40 DNA, a circular molecule with a length of 5224[22] or 5226[23] nucleotide pairs. Figure 2 shows the major products of complete digestion of SV40 DNA with *Hinc*II (lane a) and with *Hind*III (lane b). Because they migrated off the gel, the two smallest fragments in the *Hinc*II digest, F and G, are not shown. The fragments are labeled alphabetically in order of decreasing size, and the length of each, derived from the nucleotide sequence of the DNA, is listed in Table I. Figure 2 (lane c) shows an example of an incomplete digest of SV40 DNA with

[22] W. Fiers, R. Contreras, G. Haegeman, R. Rogiers, A. Van de Voorde, H. Van Heuverswyn, J. Van Herreweghe, G. Volckaert, and M. Ysebaert, *Nature (London)* **273**, 113 (1978).
[23] V. B. Reddy, B. Thimmappaya, R. Dhar, K. N. Subramanian, B. S. Zain, J. Pan, P. K. Ghosh, M. L. Celma, and S. M. Weissman, *Science* **200**, 494 (1978).

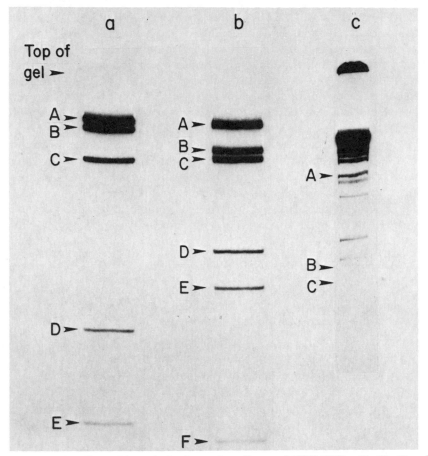

FIG. 2. Autoradiographic analysis of complete digests of SV40 DNA with HincII and HindIII and a partial digest with HindIII. (a) SV40 [^{32}P]DNA (0.1 μg) was digested with 0.25 unit of HincII in 10 μl of 10 mM Tris-HCl, pH 7.9, 7 mM MgCl$_2$, 60 mM NaCl, 7 mM 2-mercaptoethanol, 0.5 mg/ml gelatin for 1 hr at 37°. A 5-μl aliquot was electrophoresed on a 13-cm long 4% polyacrylamide gel at 120 V for 2.5 hr and a dried-gel autoradiogram was prepared. (b) SV40 [^{32}P]DNA (0.1 μg) was digested with 0.25 unit of HindIII in 10 μl of 7 mM Tris-HCl, pH 7.4, 7 mM MgCl$_2$, 50 mM NaCl, 0.5 mg/ml gelatin for 1 hr at 37°. A 5-μl aliquot was analyzed as described for sample a. (c) SV40 [^{32}P]DNA (4 μg) was digested with 7.5 units of HindIII in a volume of 360 μl. At 10, 20, and 35 minutes 120 μl of the sample was removed and the reaction stopped by addition of SDS to a final concentration of 1% (w/v). A mixture of 1 μl from each sample was electrophoresed on a 4% polyacrylamide gel at 75 V for 20 hr adjacent to a HindIII complete digest marker and a dried-gel autoradiogram was prepared. Fragments are labeled alphabetically in order of decreasing size.

HindIII, the positions of the final products A, B, and C indicated by arrows. In theory, an incomplete digest of circular SV40 DNA with HindIII might contain up to thirty partial digestion products, including the

TABLE I
Sizes of SV40 DNA Fragments Produced by Cleavage
with HindIII and with HincII

HindIII product	Nucleotide pairs[a–c]	HincII product	Nucleotide pairs[a–c]
A	1768	A	1961,[b] 1963[c]
B	1169	B	1538
C	1099,[b] 1101[c]	C	1067
D	526	D	369
E	447	E	240
F	215	F	29
		G	20

[a] To account for the staggered breaks produced by HindIII, the number of nucleotide pairs in each fragment was taken to be one-half of the total nucleotides.
[b] Fiers et al.[22]
[c] Reddy et al.[23]

six unit-length linear species. Of these, seven are clearly resolved in the example, and one short fragment migrated off the gel.

Individual products of partial digestion with HincII and HindIII, purified by the method described in Section I,C, were redigested with the appropriate enzyme, and the products derived from each were identified by electrophoresis of the digest in parallel with a complete digest marker. In each case, the intact partial digestion product was also electrophoresed adjacent to the marker so that the distance of migration could be measured. Examples of electrophoretic analysis of partial products of HindIII digestion are shown in Fig. 3. When a partial product gives rise to a set of equimolar fragments, as judged from the intensities of the bands in the autoradiogram, analysis is usually straightforward. For example, the HindIII partial product in lane a of Fig. 3 yields fragments C and E when redigested (lane c), and the partial product in lane d yields E and F (lane f). One can conclude that C is contiguous to E and that E is contiguous to F in the SV40 genome. On the other hand, the fragment in lane g gives rise to A, B, C, and D, but clearly A and D are present in greater amount than B and C. Such a result is expected when the partial digestion product is actually a mixture of two different fragments that happened to comigrate in the preparative gel. Thus, in this example, one of the partial products is comprised of A and D, and the other of B and C. In other cases, redigestion may result in no apparent change in mobility of a putative partial digestion product because the fragment is actually a final product. This conclusion is confirmed if the putative partial product comigrates with a fragment in the complete digest marker.

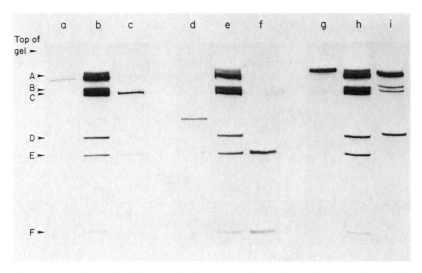

Fig. 3. Analysis of *Hin*dIII partial digestion products. Lanes b, e, and h are *Hin*dIII complete digest markers. Lane c is the result of redigestion of the partial digestion product in lane a; lane f is the digest of the partial product in lane d; and lane i is the digest of the partial product in lane g. Each partial digestion product was digested with *Hin*dIII by incubating 0.01 μg of DNA with 0.1 unit of enzyme in a volume of 30 μl for 1 hr at 37°. Samples were electrophoresed at 120 V for 2.5 hr on a 4% polyacrylamide gel and analyzed by dried-gel autoradiography.

Qualitative results based on comigration should be verified by comparing the size of each partial digestion product with the sum of sizes of the fragments derived from it. This is particularly important for identifying instances in which two partial digestion products that comigrate are also equimolar. In contrast to the example shown in Fig. 3 (lanes d and f), the two sets of final products derived from an equimolar mixture of partial products cannot be distinguished on the basis of intensities of the bands in the autoradiogram. However, the combined sizes of all the final products derived from such a mixture will be twice the estimated length of the putatively homogeneous partial product. Although a limited amount of information can be derived from analysis of an equimolar mixture of partial products, one can usually obtain sufficient data from less ambiguous cases to construct a physical map.

The length of a partial digestion product can be estimated on the basis of electrophoretic mobility,[1,24] using a plot of relative mobility versus log of fragment length. Figure 4 illustrates such a curve for unit-length linear SV40 DNA and the fragments in a *Hin*dIII digest of SV40 DNA, the

[24] A. J. Shatkin, J. D. Sipe, and P. Loh, *J. Virol.* **2**, 989 (1968).

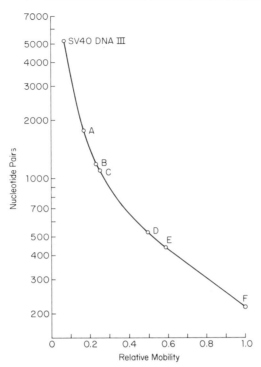

FIG. 4. Standard plot of relative mobility versus log of fragment length in a 4% polyacrylamide gel. Unit-length linear SV40 DNA and the fragments in a complete digest of SV40 DNA with HindIII were used as markers.

fragment lengths taken from the nucleotide sequence of the SV40 genome.[22,23] If DNA fragments of known length are not available as markers, the relative sizes of the fragments in a complete digest of [^{32}P]DNA can be determined from the relative radioactivities of the fragments assuming that radioactivity is directly proportional to length. The radioactivity in individual fragments can be measured by excising the bands from a preparative gel, as described in Section I,C. Each segment is dissolved in 0.2 ml of 30% hydrogen peroxide by incubation at 65° in a tightly capped scintillation vial and counted in a liquid scintillation spectrometer using a fluor for aqueous samples.[25]

Table II lists the results from analyses of several partial digestion products of both HindIII and HincII. Included are the estimated length of each partial digestion product, the final products derived from each, and the sum of lengths of the final products. For these data, the estimated length of each partial product agrees reasonably well with the sum of sizes

[25] For example, Aquasol, manufactured by New England Nuclear, Boston, Massachusetts.

TABLE II

ANALYSIS OF HindIII AND HincII PARTIAL DIGESTION PRODUCTS

HindIII product (relative mobility)[a]	Estimated size[b] (nucleotide pairs)	Final products	Sum of sizes[c] (nucleotide pairs)
0.39	670	E, F	662
0.18	1600	C, E	1546
0.14	2300	B, C	2268
		A, D	2294
0.13	2500	A, D, F	2509
0.12	2900	B, C, D	2794

HincII product (relative mobility)[a]	Estimated size[b] (nucleotide pairs)	Final products	Sum of sizes[c] (nucleotide pairs)
2.2	60	F, G	49
0.19	1500	C, D	1436
0.18	1600	B, G	1558
0.15	2000	A, F	1990
0.14	2300	B, D, E	2147
0.11	3200	A, C, F	3057

[a] Relative mobility was measured as distance migrated by fragment divided by distance migrated by HindIII F on the same gel.
[b] Estimated from electrophoretic mobility, using the plot in Fig. 4.
[c] The size of each final product was derived from the data of Fiers et al.[22]

of the final products. Each set of final products listed in Table II represents a group of fragments contiguous in the original DNA molecule. For example, based on the HindIII partial product of relative mobility 0.39, the final product E is contiguous to F in SV40 DNA. Analysis of the partial product of relative mobility 0.18 indicates that C and E are contiguous. Because these two groups of fragments share fragment E, they can be linked in the order C–E–F. The results in Table II can be arranged such that members common to each group of contiguous fragments are placed in overlapping positions, as shown in Fig. 5, to determine the order of all the fragments. The data thus lead to the construction of the two cleavage maps in Fig. 6, one for HindIII and the other for HincII.

III. Ordering of Fragments through Multiple Enzyme Digestion

One application of multiple enzyme digestion for ordering fragments[26] parallels the use of partial digests discussed in Section II. That is, if the products of enzyme α are to be ordered, purified products of several

[26] R. C. Yang, A. Van de Voorde, and W. Fiers, Eur. J. Biochem. 61, 119 (1976).

```
        HindⅢ Products                        HincⅡ Products

        E  C                                  C  A  F

           C  B                                  A  F

           C  B  D                                 F  G

                 D  A                                G  B

                 D  A  F                                B  E  D
        _____                  _____
                       F  E                                   D  C

        -  E  C  B  D  A  F  E  -            -  C  A  F  G  B  E  D  C  -
```

FIG. 5. Results of analysis of *Hin*dIII and *Hin*cII partial digestion products. Each group of contiguous fragments has been arranged so that members common to each group overlap to determine the complete order.

accessory enzymes (β, γ, δ, . . .) may be used in place of partial digestion products. For example, a fragment produced by enzyme β that includes three cleavage sites for enzyme α will yield, upon digestion with α, two α fragments and two fragments originating from the ends of the β fragment. If the end fragments can be distinguished from all α products, then one can conclude that the two observed α fragments are contiguous. Since the method requires that the β fragment include three or more α cleavage sites, accessory enzymes that produce large fragments are most useful. Moreover, several accessory enzymes are required to establish a sufficient number of sets of contiguous fragments to generate a map. The techniques required are similar to those utilized in Section II.

1. Electrophoretic profiles of fragments produced by several endonucleases (α, β, γ, δ, . . .) are determined.

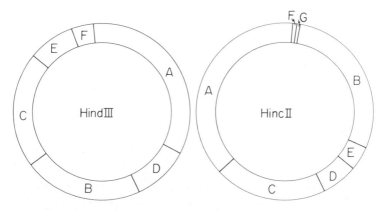

FIG. 6. Cleavage maps for the *Hin*dIII and *Hin*cII products of SV40 DNA.

2. Fragments produced by complete digestion with accessory enzymes β, γ, δ, . . . , are purified by the procedure described in Section I,C.

3. Each purified fragment is digested with enzyme α, and the limit products of α derived from each are identified by electrophoresis in parallel with a marker of α products.

4. Sets of contiguous fragments are arranged such that common members are placed in overlapping positions to construct a map.

This basic approach can also be used to correlate two existing cleavage maps. For example, the *Hinc*II and *Hin*dIII maps in Fig. 6 might be aligned by digesting purified *Hinc*II products with *Hin*dIII. The relative positions of the cleavage sites for the two enzymes can be deduced from the resulting data.

A second application of multiple enzyme digestion allows the correlation of independently constructed cleavage maps of two different enzymes. In contrast to the first approach, this method relies on electrophoretic analysis of double digests and usually requires no purification of fragments. A particularly simple analysis results if each accessory enzyme used in this method recognizes only a single cleavage site in the DNA.

The first step of the procedure involves characterizing the cleavage products of each accessory enzyme. In the model study, the endonucleases *Taq*I and *Bam*HI are used to relate the *Hin*dIII and *Hinc*II cleavage maps that were established in Section II (Fig. 6). *Taq*I and *Bam*H1 each cleaves SV40 DNA at a single site, as shown in Fig. 7. The single bands in lanes a (*Taq*I digest) and b (*Bam*H1 digest) correspond to unit-length linear SV40 DNA. The distance between the cleavage sites for the two enzymes can be estimated as described in Section II by sequential digestion of ^{32}P-labeled SV40 DNA and quantitation of the radioactivity in the products, A and B (Fig. 7, lane c). Fragment A accounts for 58% of the total radioactivity and fragment B for 42%. Since SV40 DNA contains about 5200 nucleotide pairs, fragment A is approximately 3000 nucleotide pairs long and fragment B is about 2200 nucleotide pairs long.

The next step involves double digestion with each accessory enzyme and the enzymes *Hinc*II and *Hin*dIII. If optimal reaction conditions for two enzymes are similar, the enzymes can be used simultaneously. For example, for double digestion of SV40 DNA with *Bam*H1 and *Hin*dIII, 0.02 μg of SV40[32P]DNA was incubated with 0.1 unit of *Hin*dIII and 0.1 unit of *Bam*H1 at 37°C for 1 hr in 20 μl of 7 mM Tris-HCl, pH 7.9, 7 mM MgCl$_2$, 50 mM NaCl, 7 mM 2-mercaptoethanol, and 0.5 mg/ml gelatin. Likewise, *Hinc*II and *Bam*H1 can be used together. On the other hand, since conditions optimal for *Taq*I require incubation at 50° with no NaCl, sequential digestion is necessary.

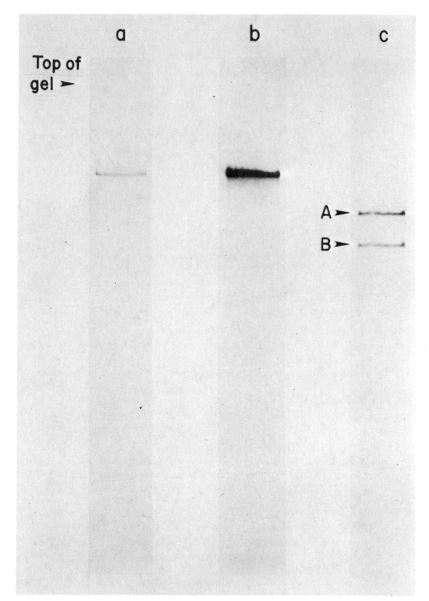

FIG. 7. *Taq*I, *Bam*H1, and *Taq*I/*Bam*H1 digests of SV40 DNA. (a) SV40 [^{32}P]DNA (0.02 μg) was digested with 0.1 unit of *Taq*I in 20 μl of 10 mM Tris-HCl, pH 8.0, 6 mM MgCl$_2$, 6 mM 2-mercaptoethanol, 0.5 mg/ml gelatin at 50° for 1 hr. The sample was electrophoresed on a 1.3% agarose gel for 3 hr at 40 V and analyzed by dried-gel autoradiography. (b) SV40 [^{32}P]DNA (0.02 μg) was digested with 0.1 unit of *Bam*H1 in 20 μl of 6 mM Tris-HCl, pH 7.9, 6 mM MgCl$_2$, 50 mM NaCl, 7 mM 2-mercaptoethanol, 0.5 mg/ml gelatin at 37° for 1 hr. Analysis was the same as for sample a. (c) SV40 [^{32}P]DNA (0.02 μg) was digested with *Taq*I as described for sample a. Then 1 μl of 1 M NaCl and 0.1 unit of *Bam*H1 were added and the sample was incubated at 37° for 1 hr. Analysis was the same as for sample a.

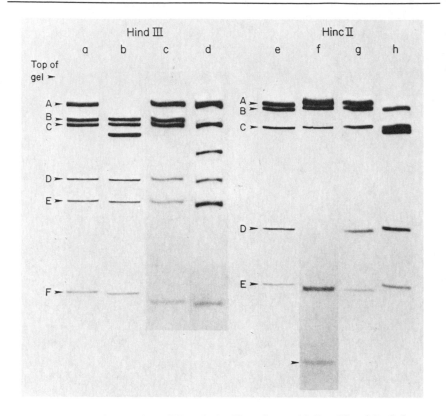

FIG. 8. Double digests of *Hin*dIII and *Hin*cII products with *Bam*HI and *Taq*I. Lanes a and c are complete *Hin*dIII digest markers; lanes e and g are complete *Hin*cII digest markers. Lane b, 0.02 μg of SV40 [^{32}P]DNA was incubated with 0.1 unit of *Hin*dIII and 0.1 unit of *Bam*HI at 37° for 1 hr in 20 μl of 7 mM Tris-HCl, pH 7.9, 7 mM MgCl$_2$, 50 mM NaCl, 7 mM 2-mercaptoethanol, 0.5 mg/ml gelatin. Lane d, 0.02 μg of SV40 [^{32}P]DNA was digested with *Taq*I as for sample a in Fig. 7. One microliter of 1 M NaCl and 0.1 unit of *Hin*dIII were added and incubation was continued for 1 hr at 37°. Lane f, 0.02 μg of SV40 [^{32}P]DNA was digested with 0.1 unit of *Bam*HI and 0.1 unit of *Hin*cII in 20 μl of 6 mM Tris-HCl, pH 7.9, 7 mM MgCl$_2$, 50 mM NaCl, 7 mM 2-mercaptoethanol, 0.5 mg/ml gelatin at 37° for 1 hr. Lane h, 0.02 μg of SV40 [^{32}P]DNA was digested with *Taq*I as described in the legend to Fig. 7. One microliter of 1 M NaCl and 0.1 unit of *Hin*cII were added and incubation was continued at 37° for 1 hr. All samples were electrophoresed on 4% polyacrylamide gels at 120 V for 2.5 hr and analyzed by dried-gel autoradiography.

Fragments resulting from double digestion are analyzed electrophoretically to localize the cleavage sites for *Taq*I and *Bam*H1 within specific *Hin*dIII and *Hin*cII fragments. In Fig. 8, a comparison between a complete *Hin*dIII digest (lane a) and a *Bam*H1/*Hin*dIII double digest (lane b) indicates that in the double digest *Hin*dIII A is missing, and a new band below

TABLE III
PRODUCTS OF MULTIPLE ENZYME DIGESTION

Fragment cleaved by *Bam*H1	Size[a] (nucleotide pairs)	Estimated sizes[b] of products (nucleotide pairs)	
*Hin*dIIIA	1768	900	900
*Hin*cIID	369	240	130
*Taq*I linear DNA	5224	3000	2200

Fragment cleaved by *Taq*I	Size[a] (nucleotide pairs)	Estimated sizes[b] of products (nucleotide pairs)	
*Hin*dIIIB	1169	740	430
*Hin*cIIA	1961	1000	1000

[a] Derived from the data of Fiers *et al.*[22]
[b] Estimated from electrophoretic mobility, using the plot in Fig. 4.

*Hin*dIII C is present. Similarly, for the *Taq*I/*Hin*dIII digest (lane d), *Hin*dIII B is missing and a new band below *Hin*dIII C appears. Therefore, the cleavage site for *Bam*H1 is within the *Hin*dIII A fragment and that for *Taq*I is within the *Hin*dIII B fragment. The other two double digests (lanes f and h) indicate that *Bam*H1 cleaves within the *Hin*cII D fragment and that *Taq*I cleaves within the *Hin*cII A fragment. Since both the *Hin*dIII A and the *Hin*cII D fragments contain the *Bam*H1 recognition site, they must overlap; likewise, the *Hin*dIII B and the *Hin*cII A fragments overlap. These results roughly determine the relative orientations of the two maps.

In order to relate the two maps precisely, the exact locations of the *Bam*H1 and *Taq*I sites on the *Hin*dIII and *Hin*cII maps must be determined. Each double digest contains two new products, the sizes of which can be estimated on the basis of electrophoretic mobility, using the plot in Fig. 4. The simplest result of double digestion, not exemplified in this model study, is that the two new products migrate as distinct bands. A second possibility is shown in Fig. 8, lane b. The new product between the *Hin*dIII C and D fragments has an estimated length of 900 nucleotide pairs, about half the length of the parent fragment, *Hin*dIII A (1768 nucleotide pairs). The fact that the new band is more dense than the band corresponding to the longer *Hin*dIII C fragment indicates that the band is actually a doublet. *Bam*H1 therefore cleaves the *Hin*dIII A fragment near its center to produce two comigrating fragments. The same conclusion can be reached for the *Taq*I/*Hin*cII digest (lane h), in which *Taq*I cleaves the *Hin*cII A fragment (1961 nucleotide pairs) to yield two fragments, each

TABLE IV
CLEAVAGE OF FRAGMENTS A AND B FROM A *Taq*I/*Bam*H1DIGEST
WITH *Hin*cII AND *Hin*dIII

	Products of digestion with *Hin*cII		Products of digestion with *Hin*dIII	
*Taq*I/*Bam*H1 fragment	Identifiable *Hin*cII products	Estimated sizes[a] of additional products	Identifiable *Hin*dIII products	Estimated sizes[a] of additional products
A	B, E, F, G	240 1000	C, E, F	430 900
B	C	130 1000	D	740 900

[a] Estimated from electrophoretic mobility, using the plot in Fig. 4.

about 1000 nucleotide pairs in length. In contrast to these cases, the *Taq*I/*Hin*dIII and *Bam*H1/*Hin*cII digests exemplify another possible result, namely, that only one new band appears but, adjudged from the intensity of the band in the autoradiogram, it cannot be a doublet. Shown in lane d of Fig. 8, the *Hin*dIII B fragment (1169 nucleotide pairs) is cleaved by *Taq*I to produce a new fragment about 740 nucleotide pairs long that migrates between C and D. Since the predicted length of the companion new fragment is about 430 nucleotide pairs, it should migrate near *Hin*dIII E (450 nucleotide pairs long). Consistent with this prediction is the observation that the *Hin*dIII E band is broader and denser than the D fragment in the double digest, indicating that it is indeed a doublet. Likewise, *Bam*Hl cleaves the *Hin*cII D fragment (lane f) to yield the new fast-migrating fragment (about 140 nucleotide pairs), indicated by an arrow and a second fragment that comigrates with the *Hin*cII E fragment (about 240 nucleotide pairs). These results are summarized in Table III.

The data are used to construct a cleavage map by comparing the known distance between the *Taq*I and *Bam*H1 cleavage sites with the possible distances calculated from the lengths of the double digestion products. For example, since *Taq*I produces two fragments from *Hin*dIII B, 740 and 430 nucleotide pairs in length, the cleavage site might be nearer the B–D junction or nearer the B–C junction (see Fig. 6). As summarized in Table III, *Bam*H1 cleaves the *Hin*dIII A fragment near its center. The shorter distance between the *Bam*H1 and *Taq*I cleavage sites, about 2200 nucleotide pairs, should equal the sum of half of *Hin*dIII A (900 nucleotide pairs), *Hin*dIII D (526 nucleotide pairs) and either the 740- or 430-nucleotide pair fragment derived from *Hin*dIII B. The former possibility yields a total of 2166 nucleotide pairs whereas the latter yields only 1856. The *Taq*I site is, therefore, near the B–C junction. These arguments locate

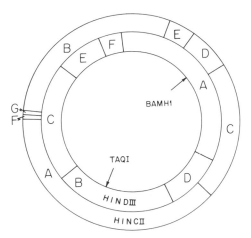

Fig. 9. Composite physical map of SV40 DNA, including the cleavage sites for *Taq*I, *Bam*H1, *Hin*dIII, and *Hin*cII.

the *Bam*H1 and *Taq*I cleavage sites on the *Hin*dIII map. With regard to the *Hin*cII map, *Bam*H1 cleaves the D fragment asymmetrically to yield fragments 130 and 240 nucleotide pairs long and *Taq*I cleaves the A fragment near its center to yield two fragments about 1000 nucleotide pairs in length. The two possible arrangements of the *Taq*I and *Bam*H1 cleavage sites in the *Hin*cII map yield a distance of either 2307 (if *Bam*H1 cleaves near the D–E junction) or 2197 (if *Bam*H1 cleaves near the D–C junction), so that *Bam*H1 probably cleaves *Hin*cII D near the D–C junction. These results can be verified by purifying fragments A and B from a *Taq*I/*Bam*H1 double digest and cleaving each with *Hin*cII and with *Hin*dIII. As shown in Table IV, the digestion of fragment B with *Hin*cII yields *Hin*cII C, a new product of 1000 nucleotide pairs, and a new product of 130 nucleotide pairs. Digestion of fragment B with *Hin*dIII yields *Hin*dIII D, a new product of 900 nucleotide pairs, and a third product 740 nucleotide pairs long. Thus, the conclusions drawn from the original double digests are sound.

The double digest data generate the composite physical map of the SV40 genome shown in Fig. 9. Not only are the *Taq*I and *Bam*H1 sites located within specific *Hin*dIII and *Hin*cII fragments, but also the relationship between the *Hin*dIII and *Hin*cII cleavage sites is established.

[54] Mapping Viral mRNAs by Sandwich Hybridization

By ASHLEY R. DUNN and JOSEPH SAMBROOK

Messenger RNAs are most commonly assigned to specific genomic locations by demonstrating that they are complementary to a restriction fragment of DNA whose position within the genome is known. Hybridization to two adjacent fragments is usually taken to mean that a particular mRNA contains sequences which are contiguous within the genome and which span the restriction endonuclease cleavage site. However, this conclusion is fragile because the possibility exists that the mRNA preparation contains two or more species which happen to be complementary to adjacent genomic fragments. In eukaryotic systems, the interpretation is clouded further because mature mRNAs generally are spliced and consist of sequences derived from noncontiguous genomic regions. The technique of sandwich hybridization[1] eliminates some of these problems, and it provides a biochemical method to determine whether sequences from different regions of a genome are covalently joined to one another in mRNA.

In theory, the technique is of general application and can be used to analyze the transcription products of any segment of DNA whose restriction maps are known. However, we have found the technique particularly valuable for mapping a unique class of viral mRNAs which are generated during the course of lytic infection with a number of Ad2–SV40 hybrid viruses and consist of covalently linked adenoviral and SV40 sequences. The genomic structure and transcription map of one such adeno–SV40 hybrid virus, Ad2+ND1, is shown in Fig. 1.[2,3] When such RNAs are hybridized to defined fragments of SV40 or adenoviral DNA immobilized on nitrocellulose filters, the 3' or 5' end of the RNA protrudes as a single-stranded tail. The sequences contained within the tail can be determined by a second round of hybridization using ^{32}P-labeled fragments of viral DNA as probes. This is illustrated schematically in Fig. 2.

Experimental Procedures

Isolation of RNA.[4] Cytoplasmic extracts are prepared from monolayers of infected cells or from infected suspension cultures. Washed cell pellets

---

[1] A. R. Dunn and J. A. Hassell, *Cell* **12**, 23–36 (1977).
[2] A. M. Lewis, M. J. Levin, W. H. Weise, C. S. Crumpacker, and P. H. Henry, *Proc. Natl. Acad. Sci. U.S.A.* **65**, 1128–1135 (1969).
[3] E. Southern, *J. Mol. Biol.* **98**, 503–518 (1975).
[4] E. A. Craig and H. J. Raskas, *J. Virol.* **14**, 26–32 (1978).

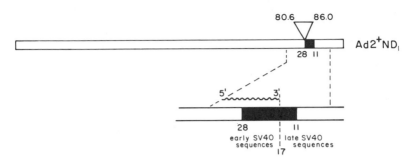

FIG. 1. The genome structure of Ad2⁺ND1. Ad2⁺ND1, originally isolated by Lewis *et al.*,[2] contains a 0.94 kilobase insertion of SV40 DNA (SV40 map coordinates 28–11 shown in black) which replaces 1.9 kilobases of the Ad2 genome located between position 80.6 and 86 on the conventional physical map of adenovirus type 2. In the expanded part of this figure the hybrid transcript is shown with its 5' end in adenovirus 2 and its 3' end in SV40 sequences.

are resuspended in 5 volumes of isotonic buffer (0.15 M NaCl, 10 mM Tris-HCl, pH 7.5, 0.001 M EDTA) and allowed to swell for 10 min at 0°. After the addition of Nonidet P40 (NP40) to a final concentration of 0.5% and mixing by vigorous pipetting, nuclei are pelleted by centrifugation at 1000 rpm for 10 min at 0°. The cytoplasmic extract is cleared by centrifugation at 10,000 rpm for 20 min at 0°, adjusted to 0.2% SDS and 0.001 M EDTA, and extracted twice with phenol (saturated with 0.5 M Tris-HCl, pH 7.5, 0.001 M EDTA, 0.15 M NaCl) and once with chloroform. RNA is stored at $-20°$ as an ethanol precipitate.

Agarose Gel Electrophoresis. Fragments of viral DNA generated by cleavage with restriction endonucleases are separated by electrophoresis through slab gels[5] (17 × 15 × 0.4 cm) cast with 1–1.2% agarose. After electrophoresis, the contents of the gel are denatured *in situ* (see the following) and transferred to nitrocellulose filters using the Southern blotting technique.[3]

Separation of Viral DNA Strands. For some purposes, hybridization to separated strands of DNA is required. In this case, fragments of viral DNA generated by cleavage with restriction endonucleases are separated by preparative electrophoresis through 1% agarose gels.[6,7] Strips of agarose containing specific fragments are cut from the gel and the DNA denatured *in situ* by immersion in 0.3 M NaOH for 30 min at room temperature. After rinsing in several changes of distilled water, the agarose strips are loaded on preformed gels cast with 1.4% agarose such that the length of one side of the agarose strips is in direct contact with the surface of the gel. A little

[5] F. W. Studier, *J. Mol. Biol.* **79**, 237–248 (1973).
[6] G. S. Hayward, *Virology* **49**, 342–344 (1972).
[7] P. A. Sharp, P. H. Gallimore, and S. J. Flint, *Cold Spring Harbor Symp. Quant. Biol.* **39** 457–474 (1974).

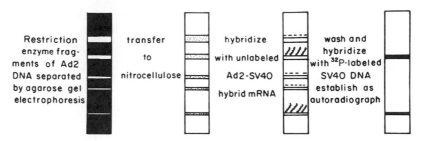

FIG. 2. Schematic representation of sandwich hybridization. Restriction enzyme fragments of adenoviral DNA are separated according to size by agarose gel electrophoresis and transferred to nitrocellulose filters by the Southern blotting technique.[3] Unlabeled RNA from cells lytically infected with an Ad2–SV40 hybrid virus is hybridized to filters containing immobilized fragments of adenoviral DNA. Viral mRNAs hybridize to complementary sequences leaving, in the case of the hybrid mRNAs, their SV40 sequences as single-stranded tails. After washing and a second round of hybridization with ^{32}P-labeled SV40 DNA, the filter is established as an autoradiograph. Bands which appear in the final autoradiograph represent fragments containing adenoviral DNA sequences which, in part, serve as template for the synthesis of the hybrid mRNA.

molten agarose is used to effect a seal between the agarose strip containing denatured DNA and the strand-separating gel. Electrophoresis is carried out using phosphate buffer (36 mM Tris-HCl, 30 mM NaH$_2$PO$_4$, 1 mM EDTA, pH 7.7)[6] at 1.5 V/cm for the appropriate time after which the separated strands of DNA are visualized by staining with ethidium bromide. After denaturation (see below) and transfer to nitrocellulose, filters containing separated viral DNA strands are baked at 80° for 2 hr. Each preparative gel yields a nitrocellulose filter which is dissected to yield 12–15 individual strips for hybridization.

Blotting. Gels containing DNA fragments or separated viral DNA strands are submersed in 0.2 M NaOH, 0.6 M NaCl (gel soak I) for 45 min at room temperature. After rinsing in distilled water, gels are transferred to a bath containing 1 M Tris-HCl (pH 7.4) and 0.6 M NaCl (gel soak II). DNA is transferred onto a sheet of nitrocellulose (B6, Schleicher and Schüell), using essentially the method of Southern.[3] DNA is immobilized by baking nitrocellulose filters at 80° for 2 hr in a vacuum oven.

Isolation of Restriction Enzyme Fragments of Viral DNA. Restriction enzyme fragments of viral DNA separated by agarose gel electrophoresis are visualized by staining with ethidium bromide and examination using ultraviolet illumination. Individual fragments are cut from the gel using a clean scalpel blade. DNA is eluted by the addition of 2 volumes of elution buffer (0.15 M NaCl, 10 mM Tris-HCl, pH 7.6, 0.001 M EDTA). After three strokes in a glass dounce homogenizer the DNA–agarose suspension is extracted for 2 hr at 4° in phenol saturated with 0.5 M Tris-HCl, pH 7.6, 0.15 M NaCl, 0.001 M EDTA. After centrifugation, the supernatant is extracted once more with phenol, once with chloroform, and finally precipitated by the addition of 2 volumes of ethanol. To ensure purity we routinely pass semipurified fragments through a second agarose gel and isolate fragments as described above. Finally the DNA is precipitated by the addition of 2 volumes of ethanol at −20°.

Nick Translation. Fragments of DNA or intact viral DNA are labeled *in vitro* with [α-^{32}P]deoxyribonucleoside triphosphates by the nick-translation reaction of *E. coli* polymerase I[8] using the conditions established by Rigby *et al.*[9] The radiolabeled DNAs are purified by G-50 Sephadex chromatography. Denaturation is achieved by incubation of labeled DNA in 0.2 *M* NaOH for 10 min at room temperature followed by rapid cooling in ice and neutralization with HCl.

Sandwich Hybridization Reagents

Filter presoak rinse (Denhardt[10])
   Polyvinylpyrrolidone, 0.02% (w/v)
   Ficoll, 0.02% (w/v)
   Bovine serum albumin, 0.02% (w/v)
   6 × SSC
First stage (RNA) hybridization mix
   Polyvinylpyrrolidone, 0.02%
   Ficoll, 0.02%
   Bovine serum albumin, 0.02%
   SDS, 0.5%
   EDTA, 0.001 *M*
   6 × SSC
   Unlabeled RNA
Filter washing solution
   2 × SSC
   SDS, 0.5%
Second stage (^{32}P-labeled DNA) hybridization mix
   Polyvinylpyrrolidone, 0.02%
   Ficoll, 0.02%
   Bovine serum albumin, 0.02%
   SDS, 0.5%
   EDTA, 0.001 *M*
   ^{32}P-Labeled denatured viral DNA 1 μg

Sandwich Hybridization. To avoid problems with RNase contamination all glassware is baked in a hot air sterilizer at 160° for 2 hr and all solutions autoclaved at 15 lb for 20 min.

[8] R. B. Kelly, N. R. Cozzarelli, M. P. Deutscher, I. R. Lehman, and A. Kornberg, *J. Biol. Chem.* **245**, 39–45 (1970).
[9] P. W. J. Rigby, M. Dieckmann, C. Rhoades, and P. Berg, *J. Mol. Biol.* **113**, 237–251 (1977).
[10] D. Denhardt, *Biochem. Biophys. Res. Commun.* **23**, 641–646 (1966).

FIG. 3. A convenient hybridization water bath. The hybridization unit, made of Plexiglas (or equivalent), is composed of two compartments. The water-tight outer shell is about 30 cm long, 40 cm wide, and 25 cm high. The inner unit, which can be removed for easy access, houses a wheel with spring clamps to support eight hybridization tubes set in a horizontal position. The wheel is driven at 4 rpm by a suitable gear motor via a connecting chloroprene (neoprene) (O) ring belt. An 800 W laboratory heater-stirrer fitted with a propeller is mounted on the outer shell of the bath and allows accurate ($\pm 1°$) control of temperature.

Baked nitrocellulose filters containing immobilized DNA are wetted with $6 \times$ SSC, rolled into cylinders and inserted into 150×25 mm test tubes. Before hybridization filters are presoaked in "filter presoak rinse" for between 3 and 9 hr at 65°. Pellets of ethanol-precipitated RNA are dried and dissolved in a small volume of 10 mM Tris-HCl, pH 7.5, 0.001 M EDTA which is added to "first stage hybridization mix" to give a final volume of 3–4 ml. The tubes are sealed with Teflon stoppers and tape and set in a horizontal position on a rotating wheel submersed in a water bath at 65°. The apparatus which was constructed in our laboratory for this purpose is shown in Fig. 3. After hybridization for 12–16 hr, tubes are removed and the hybridization fluid discarded and replaced with "filter wash solution." Filters are washed exhaustively for 4–6 hr at 65° in sev-

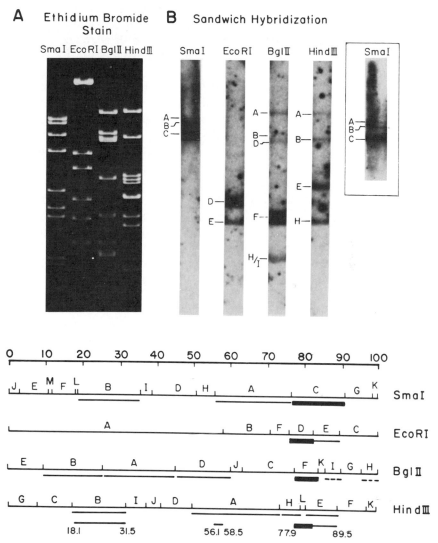

Fig. 4. Analysis of Ad2-SV40 hybrid mRNAs by sandwich hybridization. Late cytoplasmic (20 hr) RNA isolated from CV1 cells infected with Ad2⁺ND1 (see Fig. 1) is hybridized to restriction enzyme fragments of Ad2 DNA immobilized on a nitrocellulose filter. Sandwich hybridization is carried out using ³²P-labeled SV40 DNA as described in the text. (A) represents the original ethidium bromide-stained gel. (B) Autoradiograph of sandwich hybridization. Bands which appear in the final autoradiograph (exposed for 8 days using Kodak No Screen XR-1 film) are identified by reference to the original ethidium bromide-stained gel. Those regions of the viral genome which are complementary to sequences contained within the hybrid mRNA (the main body of the mRNA and the 5′ leader sequences) are indicated on the restriction enzyme maps included at the bottom of this figure. From Dunn and Hassell.[1] Copyright © MIT. Published by The MIT Press.

eral changes of filter wash solution. At the end of this period the washing solution is replaced with 3 ml of "second stage hybridization mix" containing up to 1 μg denatured ^{32}P-labeled viral DNA or specific viral DNA fragments labeled *in vitro* by nick translation (specific activity $5 \times 10^7 - 1 \times 10^8$ cpm/μg).

Hybridization is carried out at 65° for 12–16 hr, after which filters are exhaustively washed in filter wash solution for 4–6 hr at 65° . After finally rinsing in 2 × SSC, filters are air dried, mounted on Whatman 3M filter paper, and established as autoradiographs.

A typical analysis of sandwich hybridization using cytoplasmic RNA prepared from cells infected with Ad2$^+$ND1 is shown in Fig. 4.

Sandwich Hybridization to Separated Strands of Viral DNA Fragments. The capacity to separate the strands of certain viral DNA fragments by electrophoresis in agarose gels[6,7] and subsequently to transfer these sequences to a nitrocellulose support allows hybridization to be localized within the "fast" or "slow" migrating viral DNA strands. In the case of well-studied viruses such as adenovirus it is known whether the fast and slow migrating strands of many specific DNA fragments direct the synthesis of either rightward or leftward transcripts on the complete viral genome.

Although the single-stranded DNA within the strand-separating gels is itself available for hybridization after transfer to nitrocellulose, there is always a certain amount of reannealing of denatured strands which routinely run as a slow migrating duplex. Ordinarily these reannealed sequences are unavailable for further hybridization; however, we have found that denaturation of these sequences and subsequent hybridization to them provides a useful reference point on the final autoradiograph for assignment of analytical hybridization to the fast or slow migrating strands of viral DNA (see Fig. 5). To this end we routinely denature the contents of the strand separating gel in gel soak I (see Experimental Procedures, section on blotting) for 45 min at room temperature and then gel soak II for 45 min at room temperature before transferral to a nitrocellulose filter in precisely the manner described for gels containing denatured fragments of viral DNA.

General Considerations

Several factors contribute to the overall sensitivity of the sandwich hybridization technique. Of primary importance is the integrity of the unpaired RNA tails during and after the first stage hybridization. In this respect, maintenance of the nitrocellulose filters in a RNase-free environment throughout the entire period of hybridization is crucial. Clearly the stability of RNA–DNA hybrids formed under our sandwich hybridiza-

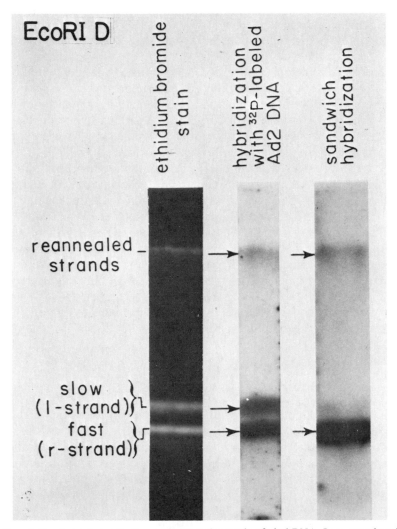

FIG. 5. Sandwich hybridization to separated strands of viral DNA. Late cytoplasmic (20 hr) RNA isolated from CV1 cells infected with Ad2+ND1 (see Fig. 1) is hybridized to the separated strands of the *Eco*RI fragment D of Ad2 DNA. Sandwich hybridization is carried out as described in the text using ^{32}P-labeled SV40 DNA and the autoradiograph exposed for 3 days using No Screen film. Hybridization of the hybrid mRNA occurs only to the fast migrating strand of *Eco*RI D previously shown to share sequences with the "r" strand of intact viral DNA.[7] From Dunn and Hassell.[1]

tion conditions is important, and, although we have not investigated this factor, it seems likely that these molecules obey the same set of rules which apply to RNA–DNA hybrids formed in solution or immobilized on nitrocellulose filters. The minimal length of the RNA tail, or the sequences contained within the tail which are complementary to sequences of the probe, which can form stable RNA–DNA hybrids under the hybridization conditions described here, must be at least 10–15 base pairs (assuming 50% G + C residues).

Because large amounts of viral DNA sequences are contained on the nitrocellulose filters, correspondingly large amounts of RNA can be used during the first stage of hybridization. It is usually feasible, therefore, to achieve the high Cot values for forming RNA–DNA hybrids even when the concentration of specific RNA sequences in the total RNA preparation is low. We have routinely used milligram quantities (1–5 mg/ml) of unfractionated cytoplasmic RNA in our hybridizations which also offers the secondary benefit of protecting the minor concentration of sequences complementary to the probe from degradation with trace amounts of ribonuclease. Where only small quantities of RNA are available, RNase-free carrier RNA (*E. coli* rRNA) can be used as carrier. Ultimately, however, the sensitivity of the technique depends on the concentration of sequences represented in the RNA tails and the specific activity of the ^{32}P-labeled DNA probe used for their detection. In part, high sensitivity can be achieved through the use of DNA fragments labeled to high specific activity with [α-^{32}P]deoxyribonucleoside triphosphates by the nick translation reaction of *E. coli* polymerase.

Future Applications for Sandwich Hybridization

An attractive application of the sandwich hybridization technique takes advantage of the fact that the polyriboadenylic acid residues contained at the 3' end of mRNAs are not encoded within the structural gene. When mRNAs are hybridized to complementary sequences immobilized on nitrocellulose filters, the poly(A) tract remains unpaired and is available for a second round of hybridization with a probe containing long stretches of ^{32}P-labeled poly(dTTP).

This approach seems particularly valuable for detecting and mapping viral transcripts in virally transformed cells. By using "Southern blots" containing restriction endonuclease-generated fragments of viral DNA, it should be possible within a single sandwich hybridization to determine those areas of the viral genome which are transcriptionally active in any line of transformed cells. Several choices of radioactively labeled probe can be considered useful for the detection of polyriboadenylic acid tracts by sandwich hybridization.

Radioactively Labeled Poly(dT). Homopolymers of radioactively labeled poly(dT) can be synthesized using terminal deoxynucleotidyltransferase and α-^{32}P-labeled thymidine triphosphate. For efficient extension terminal transferase normally requires exposed 3' termini which can be provided by a single-stranded homopolymer such as oligo(dT). Alternatively, using the conditions established by Roychoudury et al.[11] it is possible to use restriction enzyme fragments of DNA as a primer for terminal transferase without the necessity of exposing the 3' termini using 5'-exonuclease.

Radioactively Labeled DNA Circles Containing Poly(dT) Tails. When circular DNA molecules, such as SV40, are incubated under certain conditions with terminal deoxynucleotidyltransferase and thymidine triphosphate, the enzyme polymerizes long poly(dT) tails, presumably at single-stranded nicks. Bender and Davidson[12] have successfully utilized these SV40–poly(dT) circles to map the polyadenylated RNAs of several C-type oncornaviruses by direct visualization in the electron microscope. In principle these molecules can be constructed using SV40 DNA radiolabeled with α-^{32}P-labeled deoxynucleoside triphosphates by the nick-translation reaction. Terminal deoxynucleotidyltransferase can then be used to catalyze the addition of several hundred thymidylic acid residues resulting in a molecule of high specific activity ($>10^8$ cpm/μg) for use in the sandwich hybridization experiments described above. Theoretically, circular DNA molecules much larger than SV40 can be used as a template for the addition of poly(dT). This would allow more radioactivity to be associated with the site of a specific DNA–RNA hybrid complex. In this case, the limiting factor will be the stability of the poly(A)–poly(dT) circles and this remains to be determined.

Radioactively Labeled Bacterial Plasmids Containing Tracts of Poly(dT). It should be possible to utilize as a hybridization probe any bacterial plasmid containing an insert of exogenous DNA which was constructed by A–T tailing. The radioactively labeled plasmid (labeled *in vitro* by nick translation using α-^{32}P-labeled deoxynucleoside triphosphates) in its denatured form can be used as a probe where the poly(dT) residues of the plasmid form duplexes with the poly(A) tract at the 3' end of the mRNA. Preliminary experiments carried out in our laboratory using this approach are encouraging and we are presently investigating the optimal conditions for sandwich hybridization using such a radioactive plasmid as a probe.

Addendum

Recently sandwich hybridization has been used to map a cDNA copy of a specific Ad2–SV40 hybrid mRNA whose 5' end includes the start of the adenovirus 2 fiber gene.[13] Unlabeled cDNA was synthesized using

[11] R. Roychoudhury, E. Jay, and R. Wu, *Nucleic Acids Res.* 3, 863–877 (1976).
[12] W. Bender and N. Davidson, *Cell* 7, 595–607 (1976).
[13] A. R. Dunn, M. B. Mathews, L. T. Chow, J. Sambrook, and W. Keller, *Cell* 15, 511–526 (1978).

oligo(dT) as a primer in the presence of RNA-dependent RNA polymerase. After hybridization to restriction enzyme fragments of adenovirus 2 DNA, a second round of hybridization was carried out using ^{32}P-labeled SV40 DNA in precisely the manner described using mRNA as the intermediary in the sandwich technique. The authors were able to show that the cDNA hybridized to adenovirus 2 DNA fragments, which included the four leader sequences known to be associated with adenovirus 2 fiber mRNA.[14,15]

By using a cDNA copy of a specific mRNA for sandwich hybridization, it is possible to overcome the problems associated with RNA degradation by trace amounts of ribonuclease.

[14] L. T. Chow, R. E. Gelinas, T. R. Broker, and R. J. Roberts, *Cell* **12**, 1–8 (1977).
[15] L. T. Chow and T. R. Broker, *Cell* **15**, 497–510 (1978).

[55] 5' Labeling and Poly(dA) Tailing

By P. G. BOSELEY, T. MOSS, and M. L. BIRNSTIEL

Here we describe two methods of ordering DNA restriction fragments. Both are based on terminal labeling and subsequent partial digestion by restriction enzymes. The first method utilizes radioactive terminal labeling and is suitable for the rapid preparation of a restriction map.[1,2] The second method utilizes poly(dA) terminal labeling for the preparation of an ordered set of overlapping fragments. Such ordered fragments are ideal for the base sequence analysis of a DNA molecule using the Maxam and Gilbert procedure.[3] The technique is especially useful for DNA containing repetitive sequences.[4]

Partial Digestion Mapping by Radioactive Terminal Labeling

The principle of the method is the partial digestion of a DNA fragment with a single labeled terminus and its subsequent electrophoresis on agarose gels. The partial digestion results in a complete spectrum of products, but a simple overlapping series, all with a common labeled terminus, can be visualized after gel electrophoresis and autoradiography (see Fig. 1). The relative mobility of each labeled fragment is compared with those of molecular weight standards to determine the restriction sites distance

[1] H. O. Smith and M. L. Birnstiel, *Nucleic Acids Res.* **3**, 2387 (1976).
[2] P. Botchan, R. H. Reeder, and I. B. Dawid, *Cell* **11**, 599 (1977).
[3] A. M. Maxam and W. Gilbert, *Proc. Natl. Acad. Sci. U.S.A.* **74**, 560 (1977).
[4] T. Moss, P. G. Boseley, and M. L. Birnstiel, in preparation.

METHODS IN ENZYMOLOGY, VOL. 65

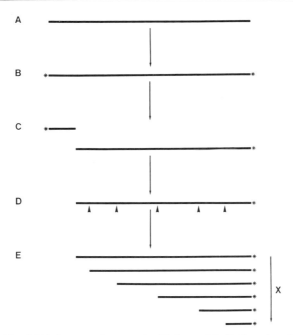

FIG. 1. (A) The DNA fragment is treated with bacterial alkaline phosphatase and then terminally labeled (*) using polynucleotide kinase. (B) The labeled fragment is cleaved asymmetrically with a restriction enzyme. (C) The two fragments are separated on an agarose gel and then eluted and purified. (D) The large fragment is partially digested with a restriction enzyme (the arrows show the restriction enzyme's sites). (E) The partial digestion products are electrophoresed and autoradiographed to reveal a discrete series of fragments whose lengths correspond directly to the location of the restriction sites (× indicates the mobility of these fragments on an agarose gel).

from the labeled terminus in base pairs. Thus, the order of the fragments and their lengths correspond directly to the location of the restriction sites along the DNA molecule. Using a series of restriction enzymes a detailed map is quickly obtained.

The chosen DNA fragment is terminally labeled, using normally [γ-^{32}P]ATP and T$_4$ polynucleotide kinase.[5] The fragment is then cleaved asymmetrically by a restriction enzyme, and the resulting DNA molecules are separated on preparative agarose gels. The separate uniquely labeled fragment is then recovered from the gel and a portion restricted by a selected enzyme under conditions resulting in a partial digestion. The partial digestion products are electrophoresed on an agarose gel in parallel with labeled DNA standards. After the gel has electrophoresed for an optimal length of time it is dried down and subjected to autoradiography.

[5] J. R. Lillehaug, R. K. Kleppe, and K. Kleppe, *Biochemistry* 15, 1858 (1976).

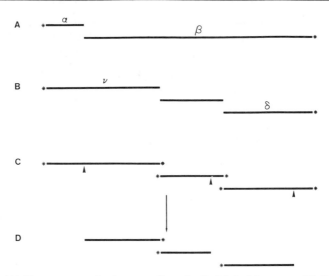

FIG. 2. (A) First asymmetric cleavage of terminally labeled fragment. (B) Second asymmetric cleavage of terminally labeled fragment. (C) Terminal labeling after total digestion of whole fragment using the second asymmetric cleavage (arrows indicate restriction enzymes giving asymmetric cleavage for each individual labeled fragment). (D) Electrophoretic separation, elution, and purification results in three fragments uniquely labeled toward the center of the whole fragment.

Fragment Selection and Terminal Labeling

A simple, initial map of the DNA fragment can be readily obtained by using infrequently cutting restriction enzymes, such as those with hexanucleotide recognition sites (see Danna [51], this volume). This information is used to select several enzymes that give asymmetric fragments with extensive overlaps (as illustrated by fragments β and ν in Fig. 2A and B). Restriction sites distal to the labeled termini, which are more difficult to assign, can be cross-correlated using such overlapping fragments. Furthermore, the additional fragments (e.g., α and δ of Fig. 2A and B) can be used to confirm the locations of sites close to their labeled termini.

An additional aid to the initial mapping can be used for cloned DNA fragments. Most vectors have well characterized restriction maps, and these can be used to produce the desired uniquely labeled fragments. For example, if the required site lies symmetrically within the cloned fragment it can still be terminally labeled, while the desired labeled asymmetric fragments can be obtained by using a restriction site lying within the vector DNA.

5′ Terminal labeling of DNA fragments requires free 5′ termini lacking the phosphate group. The chosen fragment (up to 25 pmoles DNA ends) is treated with bacterial alkaline phosphatase[6] (BAP) (Worthington BAPC) in 0.5 M Tris-HCl, pH 8 (200 μl final volume, BAP 1.5 units/ml) for 30 min at 37°. The BAP is removed by first diluting in 300 μl 0.5M NaCl in TE buffer (10 mM Tris-HCl, pH 8, 1 mM EDTA) and 25 μl of 2M sodium trichloroacetate, pH 7 (used for removing strongly bound proteins). The mixture is shaken twice with phenol–chloroform (1 : 1) and twice with chloroform–isoamyl alcohol (24 : 1). The DNA is ethanol precipitated, stored at −20° for more than 2 hr, and then centrifuged at 200,000 g for 30 min at 4°. The DNA pellet is washed with 80% ethanol and recentrifuged. The DNA is then dried, redissolved in 30 μl TE buffer, and 3.5 μl of 10× polynucleotide kinase buffer is then added to give a final concentration of 50 mM Tris-HCl, pH 7.6, 10 mM MgCl$_2$, 12 mM NaCl, 5 mM HTT, 1 mM spermidine.[5] Vacuum-dried [α-^{32}P]ATP (Amersham-Searle), with an equivalent molar ratio to DNA ends, is dissolved in the mixture and then 1 μl (2 units) of T$_4$ polynucleotide kinase is added (PL-Biochemicals). The reaction is incubated at 37° for 30 min and terminated by heat denaturing the enzyme at 65° for 5 min. The volume is increased by the addition of 100 μl of buffer A (0.15 M NaCl, 10 mM Tris-HCl, pH 8, 1 mM EDTA) and 40 μl of 80% glycerol–bromphenol blue mixture. The mixture is desalted using a Sephadex G-75 column (10 cm × 0.9 cm diameter) and the fractions are Cerenkov counted prior to pooling. The pooled fractions are brought to a final NaCl concentration of 0.3 M and then ethanol precipitated. After more than 2 hr at −20°, the labeled DNA is centrifuged at 200,000 g for 30 min at 4°, washed with 80% ethanol, and recentrifuged.

Labeling of DNA fragments with flush ends (e.g., *Alu*I fragments,[7] AG↓CT) or 3'-protruding ends (e.g., *Pst*I fragments[8] CTGCA↓G) is achieved by denaturation methods such as dimethyl sulfoxide (DMSO),[9] alkaline denaturation,[10] or boiling[3] (see Chaconas [10], this volume). Furthermore, should 5' terminal labeling not be possible the alternative 3' labeling could be used.[11]

Recovery of Uniquely Labeled Fragments

The labeled DNA fragment is cut asymmetrically with a restriction enzyme, using the recommended buffers of the suppliers (New England Bio-Labs, Beverly, Massachusetts). The fragments from the enzyme di-

[6] A. Torriani, this series, Vol. 12, Part B, p. 212.
[7] R. J. Roberts, P. A. Myers, A. Morrison, and K. Murray, *J. Mol. Biol.* **102**, 157 (1966).
[8] N. L. Brown and M. Smith, *FEBS Lett.* **65**, 284 (1976).
[9] P. G. Boseley, A. Tuyns, and M. L. Birnstiel, *Nucleic Acids Res.* **5**, 1121 (1978).
[10] P. G. Boseley, T. Moss, M. Mächler, R. Portman, and M. L. Birnstiel, *Cell* **17**, 19 (1979).
[11] R. Roychoudhury, E. Jay, and R. Wu, *Nucleic Acids Res.* **3**, 101 (1976).

gestion are separated on either 5 mm slab or tube agarose gels. For fragments [up to 15 kilobases (kb)] we have loaded up to 10 μg DNA per band per square centimeter of gel surface. The electrophoresis buffer used is 0.2 M glycine, 0.15 M sodium hydroxide pH 8.5,[1] while the "stop solution" for the digestion is 20 mM Na-EDTA, pH 7.5, 10 mM Tris-HCl, pH 7,8, 10% glycerol, 0.01% bromphenol blue (final concentrations).[12] The agarose concentration of the gels and electrophoresis times are optimized for the separation of the particular sized fragments (see Section VI, this volume, and below). After completion of electrophoresis the gels are stained for 30 min in 1 μg/ml ethidium bromide[13,14] in TE buffer and the bands visualized using long wave uv light (366 nm). The DNA bands are then cut out and forced from a sterile 2-ml syringe through a 22G needle into centrifuge tubes containing at least 10 volumes of buffer A (see above). The tubes are mixed by inverting and then left overnight at 4°. The gel particles are removed by centrifugation at 200,000 g for 30 min at 4°, and the supernatants loaded onto 0.3 ml Whatman DE52 columns previously equilibrated with buffer A. The columns are washed with 30 ml of buffer before eluting the bound DNA with 1 M NaCl. The DNA is ethanol precipitated and centrifuged, and the pellets are washed with 80% ethanol before vacuum drying.

Partial Digestion Mapping Procedure

Partial digestions of the selected DNA fragment with a single labeled terminus are optimized for a particular enzyme by diluting the restriction enzyme and/or by adding suitable amounts of sonicated calf thymus DNA to prevent complete digestion. A time course reaction is carried out to evaluate whether the dilution and quantity of calf thymus DNA sufficiently reduces the reaction rate. An example of this kinetic approach is shown in Fig. 3, a partial Hae III digestion of Xenopus laevis rDNA. It is clear that the largest fragments are well resolved after short digestions (1–5 min), and the smallest fragments are best resolved after 15–30 min reaction times. An alternative method of producing partial digestions is by using the drug distamycin A which competes with the restriction enzyme by binding to the DNA.[15]

All enzymes used for partial digestion mapping are diluted in 50 mM Tris-HCl, pH 7,6, 10 mM DTT, 100 μg/ml BSA, and 50% glycerol prior to use. The partial time course reactions employ a final 50-μl volume from which five 10-μl aliquots are taken (e.g. 1, 5, 10, 15, and 30 min). Each aliquot, or time point, is immediately mixed with 3 μl of the stop solution

[12] W. Schaffner, K. Gross, J. Telford, and M. Birnstiel, *Cell* **8**, 471 (1976).
[13] C. Aaij and P. Borst, *Biochim. Biophys. Acta* **269**, 192 (1972).
[14] P. A. Sharp, W. Sugden, and J. Sambrook, *Biochemistry* **12**, 3055 (1973).
[15] V. V. Nosikov and B. Sain, *Nucleic Acids Res.* **4**, 2263 (1977).

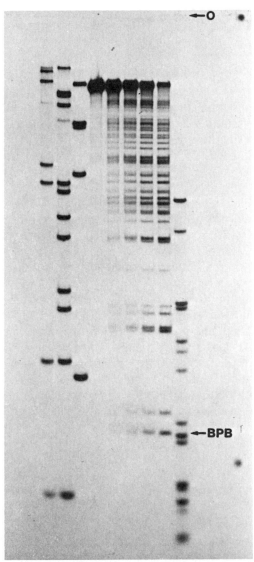

FIG. 3. Autoradiograph of a HaeIII partial digestion of a *Xenopus laevis* rDNA fragment. The central channels show time point aliquots (1–30 min) from this digestion. Three of the other channels show labeled digests of λ and PM2 DNAs while channel A represents undigested labeled rDNA fragments. BPB: bromphenol blue.

(see above) and heated at 65° for 5 min. After cooling on ice the samples are bench centrifuged to bring down the condensed water prior to loading on the analytical agarose gels. By analyzing only a portion of the sample from each time point on a preliminary gel, and retaining the remainder at $-20°$, the optimal time points for several enzymes can be later simultaneously compared on a "master gel."

Analysis of Partial Digests by Gel Electrophoresis and Autoradiography

For fragments of up to 5 kb in length we have found 1.5% agarose gels (40 × 17 × 0.2 cm) to be of most use. Lower percentage gels can be used for larger fragments, and these are limited only by the accurately known molecular weight standards available. Obviously, large viral DNAs with many restriction sites are best mapped by using smaller segments.[16]

The aliquots from the partial digestion time course reactions are loaded on the 1.5% agarose gel (using glycine–NaOH buffer) and initially electrophoresed into the gel at 50V, in parallel with the undigested fragment, and end-labeled markers, e.g., λ/HindIII-EcoRI[17] and *PM2–Hae*III.[18] The gel is then electrophoresed at 100 V until the bromphenol blue dye migrates 25 cm. The analytical gels are then dried onto Whatman DE81 paper[1] and autoradiographed. When necessary the gels are placed at $-70°$ with preflash-activated X-ray film and fast-tungstate screens (see Laskey [45], this volume).

On developing the autoradiograph, a discrete series of bands are obtained whose lengths correspond directly to the order of restriction sites along the DNA molecule (see Fig. 2). The mobility of the bands in the molecular weight standards are used to construct a semilogarithmic plot of DNA fragments length versus the relative mobility in centimeters. Thus the location of the fragments' restriction sites in base pairs from the labeled termini are obtained. Using such overlapping fragments as β and ν (Fig. 2A and B), a reliable map can be rapidly obtained for a series of enzymes.

Additional Comments

The use of overlapping fragments helps prevent the possible loss of bands while confirming others. However, a nonpreferred site of an enzyme can result in a very faint band. (see Fig. 3, which shows light and

[16] R. Wittek, A. Menna, H. K. Müller, D. Schümperli, P. G. Boseley, and R. Wyler, *J. Virol.* **28**, 171 (1978).
[17] K. Murray and N. E. Murray, *J. Mol. Biol.* **98**, 551 (1975).
[18] M. Noll, *Cell* **8**, 349 (1976).

dark bands). In such cases, the use of flash activated films and fast tungstate screens (see Laskey [45], this volume) help to enhance weak bands and eliminate possible band loss. On the other hand, care must be taken not to include light bands which might be due to contamination by another enzyme activity or to relaxed specificity of the enzyme.[1] Confirmation of the restriction map is most simply achieved by matching the fragment length predicted from partial digests with those actually found in complete digests of the whole molecule (Fig. 1A) or pieces of it (Fig. 2B). Such complete digests can be analyzed by either ethidium bromide staining or by autoradiography of terminally labeled DNA after gel electrophoresis. The number and lengths of the fragments are then used to decide whether a questionable site is present or not. However, this correlation is impossible in those cases where modification of the restriction site prevents complete digestion and may even make it difficult to detect all partial digestion products. For example, the methylation of the DNA by *E. coli*, even in modification defective mutants, can be sufficient to block such restriction enzymes as *Mbo*I[19] or *Xba*I.[20] Thus, with these enzymes the number of restriction sites is best regarded as a minimal estimate.

DNA fragments which include spacer elements can frequently contain repetitive areas, and those having a high GC content will contain many restriction sites.[10] Such closely spaced restriction sites are not easy to map with fragments over 2 kb in length. In our model fragment the second asymmetric cleavage (Fig. 2B) would be useful for mapping repetitious areas lying toward the center of the fragment. Thus, the complete fragment is cut with this enzyme and labeled at its 5' ends (Fig. 2C). Using the partial digestion mapping information previously obtained, other enzymes are chosen which give asymmetric cuts in each of the three fragments (Fig. 2D). The resulting uniquely labeled fragments are then digested with the frequently cutting endonucleases. As the size of the fragments are relatively small, an accurate measurement of the restriction sites distances relative to the DNA standards can be made. Such a methodology has been successfully applied to *Xenopus laevis* rDNA where the restriction mapping was confirmed by DNA sequencing.[10]

Many restriction enzymes having hexanucleotide recognition sequences also have their central tetranucleotides recognized by a second enzyme. For example, the central CCGG of the *Sma*I recognition site sequence CCCGGG[21] is recognized by *Hpa*II.[22] Therefore each *Sma*I site is in turn recognized by *Hpa*II and so a further cross-correlation on the

[19] R. E. Gelinas, P. A. Myers, and R. J. Roberts, *J. Mol. Biol.* **114**, 169 (1977).
[20] R. S. Zain and R. J. Roberts, *J. Mol. Biol.* **115**, 249 (1977).
[21] S. A. Endow and R. J. Roberts, unpublished observations.
[22] D. E. Garfin and H. M. Goodman, *Biochem. Biophys. Res. Commun.* **59**, 108 (1974).

location of these combined sites may be made. Using a series of enzymes, this can clearly remove anomalies in the locations of doubtful restriction sites.

In using the recommended gel system above, care must be taken in comparing different enzyme digests on a master gel. The different requirements of the various restriction enzymes consequently alters the salt concentrations of a given sample. This in turn can cause retardation in the mobility of the DNA in the gel. Should this become critical, desalting by ethanol precipitation and washing, or by using higher ionic strength buffers will eliminate such a problem.

Preparation of Ordered Partial Restriction Fragments by
    Poly(dA) Tailing

The partial digestion mapping procedure described above facilitates the observation of a set of terminally labeled overlapping fragments from among the many fragments produced by partial restriction. These overlapping fragments all have a common terminus but their other termini are staggered across the DNA molecule, located at the restriction sites for the particular enzyme used. By poly(dA) labeling the DNA at a single terminus or by extension of the DNA fragments by other homopolymers, it also is possible to purify such a set of overlapping fragments from a partial digest. The termini of these fragments distal to the homopolymer end can then be radioactively labeled and each fragment base sequence analyzed by the Maxam and Gilbert procedure.[3] This will yield stretches of sequence information in a common direction from every restriction site on the DNA molecule. By appropriate choice of restriction enzyme(s) for partial digestion, it is often possible to overlap these stretches of sequence, and thus obtain continuous sequence information over large segments of the DNA in a single experiment. The technique is equally applicable to DNA with repetitious or unique sequences. However, in many cases it may offer the only approach to the sequencing of repetitious DNA.

Our approach involves the 3' extension of duplex DNA with poly(dA), which is synthesized using terminal transferase.[23] The DNA is cleaved asymmetrically to provide two fragments with uniquely labeled termini. These fragments are then partially digested with restriction enzymes (having many sites) and the overlapping set of terminally labeled fragments

[23] F. S. Bollum, in "The Enzymes" (P. D. Boyer, ed.), 3rd ed., Vol. 10, pp. 145–171. Academic Press, New York.

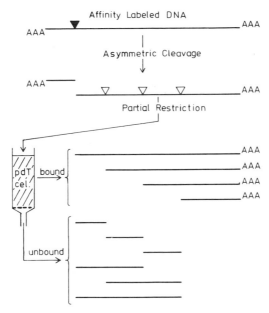

FIG. 4. Basic experimental scheme of the poly(dA) labeling technique.

selected by its affinity to poly(dT)–cellulose.[24] Such an experimental scheme is shown in Fig. 4.

Poly(dA) Tailing of DNA with Terminal Transferase

Terminal (deoxynucleotidyl)transferase is used to synthesize a poly(dA) affinity label on the 3'-hydroxyl termini of the DNA molecule to be studied. However, for terminal transferase to label duplex DNA efficiently and homogeneously, the 3' terminal nucleotide(s) must first be available to the enzyme.[23] Unpaired 5' terminal nucleotides and base-paired termini both strongly inhibit 3'-terminal labeling.[25,26] Since unpaired 5' terminal nucleotides are often generated when DNA fragments are excised by restriction from recombinant plasmids or phage, it is essential to be able to poly(dA) label such termini.

Roychoudhury et al.[27] have shown that replacement of the cofactor Mg^{2+} by Co^{2+} increases the synthetic activity of terminal transferase on duplex DNA having 5' unpaired terminal nucleotides. However, we have found that in the presence of Co^{2+} only a small percentage of the termini

[24] H. Aviv and P. Leder, Proc. Natl. Acad. Sci. U.S.A. 69, 1408 (1972).
[25] D. A. Jackson, R. H. Symons, and P. Berg, Proc. Natl. Acad. Sci. U.S.A. 69, 2904 (1972).
[26] P. E. Lobban and A. D. Kaiser, J. Mol. Biol. 78, 453 (1973).
[27] R. Roychoudhury, E. Jay, and R. Wu, Nucleic Acids Res. 3, 863 (1976).

are actually labeled. Jackson *et al.*[25] demonstrated that λ (5')-exonuclease digestion of duplex DNA with 5' unpaired terminal nucleotides to generate a free 3' terminus can be successfully used to enhance terminal transferase activity on the DNA. We have, therefore, applied this technique.

Since λ exonuclease is some 300 times more active on double than single-stranded DNA,[28] it is difficult to find conditions under which all molecules with 5' unpaired nucleotides will be digested to the same extent. λ-Exonuclease digestion of a fragment excised from a plasmid with endonuclease *Eco*RI was studied at four different temperatures (0°, 6°, 14°, and 20°). Gel electrophoresis of the digestion products showed that digestion at 14° yielded the most homogeneous products. Under the conditions given below approximately 30 nucleotides are removed from the DNA per 5' terminus, the number removed from individual termini at most ranging between about 10 and 50.

λ-*Exonuclease Digestion.* Before poly(dA) tailing with terminal transferase, DNA molecules having 5' unpaired nucleotides are routinely digested with λ-exonuclease under the following conditions. DNA at 4×10^{-8} to 14×10^{-8} M (5' termini) in 67 mM glycine-KOH, pH 9.4, 2.5 mM MgCl$_2$[28] is incubated with 600 units/ml of λ-exonuclease for 2.5 to 6 min at 14°, the shorter incubation time being used with the lower DNA concentration and vice versa. λ-exonuclease prepared essentially as described by Radding[29] at 60,000 units/ml was a gift from Dr. S. G. Clarkson. Digestion is terminated by the addition of ice-cold 0.5 M EDTA to a final concentration of 10 mM and then placing on ice. Cold 3 M NaCl is added to a final concentration of 0.3 M and the DNA extracted twice with phenol–chloroform and twice with chloroform–isoamyl alcohol. The DNA is finally ethanol precipitated and dried. The λ-exonuclease digestion is conveniently carried out in a 10–30-ml glass centrifuge tube, e.g., Corex 15-ml tube, in which the subsequent extractions to denature and remove the enzyme can also be made. The DNA is usually stored in TE buffer and added directly to the reaction solution. Here care is taken that not more than 0.5 mM of the Mg^{2+} in the reaction solution was chelated by EDTA in the TE buffer. The reaction solution is equilibrated to 14° for about 10 min in a water bath before addition of λ exonuclease (diluted to 6000 units/ml in reaction buffer). The solution is gently shaken for a few seconds to allow thorough mixing and the reaction allowed to proceed.

As yet we have not attempted to λ exonuclease digest DNA fragments having base-paired termini. However, Lobban and Kaiser[26] have described conditions under which base-paired termini may be homoge-

[28] J. W. Little, *J. Biol. Chem.* **242**, 679 (1967).
[29] C. M. Radding, this series, Vol. 21, p. 273.

neously digested with λ exonuclease. Termini with 3' unpaired nucleotides will probably not require digestion before terminal transferase labeling. It should be noted that DNA eluted from agarose may be partially resistant to λ-exonuclease digestion (personal observations). The DNA used in the above experiments is purified by CsCl density gradient centrifugation[9] and this gives reproducible results.

Terminal Transferase Labeling. The λ-exonuclease digested DNA is redissolved in ten times diluted TE buffer at a concentration between 10×10^{-7} and $30 \times 10^{-7} M$ (3' termini) and the reaction buffer components then added. The DNA at 5.8×10^{-7} to $17 \times 10^{-7} M$ (3' termini) in 100 mM Na-cacodylate, 8 mM MgCl$_2$, 1 mM dATP,[23,26] and 0.01 mCi/ml [α-^{32}P]dATP (Amersham) is incubated for 30 min at 37° with 730 units/ml (0.1 mg/ml) terminal transferase (Boehringer, Mannheim). [α-^{32}P]dATP is added to weakly radioactively mark the poly(dA) labels which can then be used to trace the DNA in subsequent experimental steps.

The reaction is terminated and the DNA extracted and precipitated, as described in discussion of the λ-exonuclease digestion. A 1.5-ml Eppendorf tube or a 3-ml siliconized glass Corex centrifuge tube are convenient for the terminal transferase reaction, since they allow subsequent extractions to be carried out in the same tube.

The extent of poly(dA) tailing of the DNA was usually assayed directly after termination of the reaction, as described below under affinity chromatography. Under the conditions described above ≥90% of the DNA could be reversibly bound to poly(dT)–cellulose. Incorporation of radioactive label into acid-precipitable DNA[26] indicated that about 14 nucleotides were polymerized per 3' terminus.

Production of Unique Poly(dA) Tailed DNA Fragments

Terminal transferase synthesizes poly(dA) at both 3' termini of the duplex DNA molecule. It is, therefore, necessary to produce uniquely labeled fragments from this molecule before the method of partial restriction and affinity chromatography can be applied. The most obvious approach to resolve this problem is to restrict the poly(dA) tailed DNA molecule in order to produce two large labeled fragments separable by gel electrophoresis (Fig. 4). This method has been described in detail in the section on partial mapping and so will not be further discussed. Several other possibilities may however exist depending on the DNA molecule under investigation.

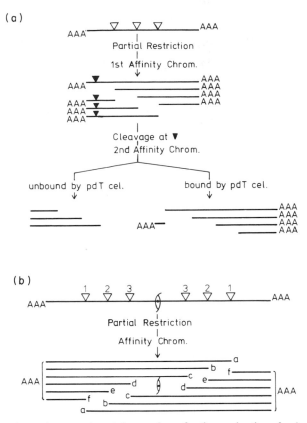

FIG. 5. Two alternative experimental procedures for the production of uniquely poly(dA) labeled fragments.

Figure 5a shows a possible approach when a single restriction site exists very near one terminus of the doubly labeled DNA molecule. Here two rounds of affinity chromatography are necessary. One separates internal partial restriction products from those with poly(dA) tailed termini. The second separates the two sets of overlapping fragments after total restriction to remove one labeled terminus.

Figure 5b shows an even simpler approach, possible if a dimer of the DNA molecule, with dyad symmetry, is available. Such a dimer may often be readily prepared by cloning. In this case the two poly(dA) tailed termini are one and the same and thus a unique terminus in fact already exists.

Partial Restriction

After the poly(dA) tailed fragment has been prepared, it is then partially digested with a restriction enzyme. The enzyme chosen is one giving suitably spaced sites which later allows the separation of a set of overlapping fragments. As this is a preparative partial digestion it is essential to optimize the yield of terminally labeled fragments. The number of units of enzyme required for optimal digestion is estimated approximately from

$$\text{units} = \frac{\mu\text{g of DNA}}{\text{No. of restriction sites}} \quad \text{for a 1 hr reaction}$$

where 1 unit of enzyme digests 1 μg of the fragment to completion in 1 hr. Analytical partial digests are carried out from which aliquots are taken at suitable time intervals and the digestion terminated by addition of EDTA to 10 mM. The optimal digest is one which leaves approximately 35% of the material undigested.[30] In order to evaluate this percentage, standard amounts of undigested DNA are electrophoresed in parallel with the time course aliquots on an agarose gel. Thus, after ethidium bromide staining,

---

[30] Using simple reaction kinetics it is possible to estimate the extent of digestion required to obtain the maximum yield of molecules cleaved at a single restriction site.

If n_0 is the initial number of molecules X, each with m restriction sites, n_m is the number of uncleaved molecules at any given time t, and n_{m-1} and n_{m-2} are the number of initial molecules cleaved once and twice, respectively, at any given time, then under ideal conditions:

$$n_m X_m \xrightarrow{cm} n_{m-1} X_{m-1} \xrightarrow{c(m-1)} n_{m-2} X_{m-2} \rightarrow \text{etc.}$$

where c is a reaction constant. At any time t, n_m will be given by

$$n_m = n_0 e^{-cmt} \tag{1}$$

Then by considering the fate of molecules initially cleaved in the time interval t to $t + \delta t$ and integrating over the range $t = 0$ to t', one obtains a formula for the number of singly cleaved molecules n_{m-1}

$$n_{m-1} = n_0 m (e^{-c(m-1)t'} - e^{-cmt'}) \tag{2}$$

Differentiating Eq. (2) with respect to t' and equating to zero, the conditions for n_{m-1} to be maximal are shown to be

$$c t' = \ln (m/(m/m - 1)) \tag{3}$$

Substituting back from Eq. (3) into Eqs. (2) and (1), one obtains both the maximum yield of singly cut molecules and the number of molecules remaining intact at this digestion time. For a molecule with 6 restriction sites ($m = 6$) the maximum yield of singly cleaved fragments is 40% of the input DNA. At $m = 10$ this yield is 39%, and at $m = 20$ it is 38%. These maxima all occur when about 35% of the input DNA remains uncleaved. Further consideration of Eq. (2) shows that the yield of singly cleaved molecules remains $\geq$75% of the maximum yield at any digestion time which leaves between 50 and 15% of the input DNA uncleaved.

the time point giving uncleaved DNA of equal intensity to the 35% standard is chosen to be optimal. A preparative digest is then carried out using these conditions and the reaction is terminated by adding EDTA to 10 mM and by cooling on ice.

Chromatography of Poly(dA) Tailed Restriction Fragments on Poly(dT)–Cellulose

After partial restriction the poly(dA) labeled and unlabeled DNA fragments are separated by affinity chromatography on poly(dT)–cellulose, essentially as described by Bantle et al.[31] A 0.4-ml column of poly(dT)–cellulose (Collaborative Research Inc., grade T3), packed in a 1-ml plastic syringe barrel, is washed with several milliliters of 0.3 M NaOH and then equilibrated with 0.4 M NaCl and TE buffer at room temperature. The terminated partial digest of poly(dA) labeled DNA is diluted five times with 0.4 M NaCl and TE buffer and applied to the column. The column is subsequently washed with several milliliters of 0.4 M NaCl and TE buffer to remove all traces of unlabeled DNA. Bound, poly(dA) labeled DNA is then eluted with TE buffer, and 100-μl fractions collected. Fractions containing DNA are identified by Cerenkov counting of the ^{32}P incorporated in the poly(dA) label and then pooled.

Assay of Transferase Reaction. About 1 μg of DNA is removed from the transferase labeling reaction directly after addition of EDTA. This is then diluted with 0.4 ml of 0.4 M NaCl and TE buffer and applied to a 0.1-ml poly(dT)–cellulose column as described above. The eluant containing unlabeled DNA is collected, the column washed with a further 0.4 ml of 0.4 M NaCl and TE buffer and the labeled DNA eluted with 0.4 ml of TE buffer. Twenty microliter aliquots of nonbinding and binding DNA are then electrophoresed on an agarose slab gel. The percentage of DNA poly(dA) labeled can be visually estimated from a photograph of the ethidium bromide stained gel.

Application of Partial Restriction Fragments to Sequencing

A particular example of the use of the poly(dA) affinity labeling technique to produce ordered fragments for DNA sequencing is shown in Fig. 6. In this experiment nine uniquely 5'-^{32}P-labeled fragments were obtained after fractionation of a DNA molecule containing highly repetitive se-

---

[31] J. A. Bantle, I. H. Maxwell, and W. E. Hahn, *Anal. Biochem.* **72**, 413 (1976).

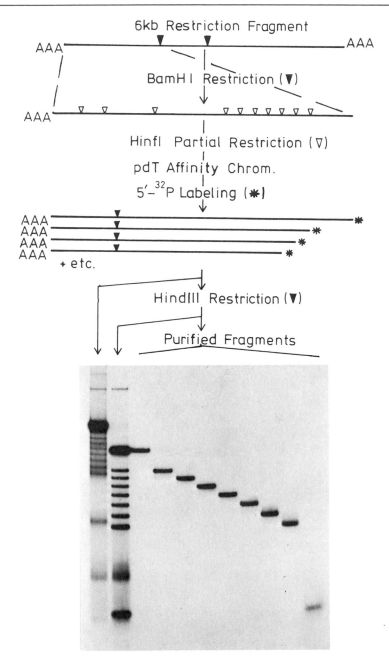

FIG. 6. An example of the application of the poly(dA) labeling technique to repetitious sequence DNA.

quences. Seven of these fragments were used to sequence the repetitive segment of this DNA by the Maxam and Gilbert procedure.[3,4]

In the experiment of Fig. 6, the poly(dA) affinity labeled termini were removed by secondary cleavage after 5'-^{32}P-labeling with T$_4$ polynucleotide kinase.[3] This improved resolution of the fragments in gel electrophoresis by removing the molecular weight heterogeneity of the poly(dA)-labeled terminus. It also produced uniquely ^{32}P-labeled fragments. However, 5' labeling with T$_4$ polynucleotide kinase of the strand immediately adjacent to the poly(dA) tail is extremely poor relative to an unpaired 5' terminus. Thus, it is often not necessary to remove this 5' label before sequencing.[4] Should it provide a problem, it may be removed by strand separation using the unique 3' poly(dA) tail to select the required strand. The DNA in TE buffer and at a concentration low enough to prevent rapid renaturation is heated to 100° for 5 min in a siliconized Eppendorf tube, then cooled on ice, NaCl is added to a final concentration of 0.4 M, and the solution is applied to a 0.2-ml poly(dT)–cellulose column, prepared as described above. The column is washed and eluted as above except that the eluant containing the required strand is collected and further handled in siliconized equipment to reduce losses.

In conclusion, although this method has been primarily developed for sequencing, the preparation of overlapping fragments certainly has other applications. For example, in conjunction with the Southern transfer technique,[32] genes lying within a particular DNA fragment can be readily located by hybridization.[33]

[32] E. M. Southern, *J. Mol. Biol.* **98**, 503 (1975).
[33] T. Moss and M. L. Birnstiel, *Nucleic Acids Res.* **6**, 3733 (1979).

Section VIII

Nucleotide Sequencing Techniques

[56] RNA Polymerase Nascent Product Analysis

By M. TAKANAMI

Synthesis of RNA with RNA polymerase is generally initiated with purine nucleoside triphosphates. Under a reduced NTP concentration, however, RNA synthesis can be initiated with oligonucleotides as primer at defined sites of single-stranded DNA. By applying this principle, the entire region of a DNA segment is able to be transcribed into RNA sequences from which the sequence of the template is deduced. This procedure was applied for the sequence determination of the *lac* operator[1] and of fd promoters.[2,3] Since new methods for direct DNA sequencing were developed,[4,5] the primed RNA synthesis method is rarely used because of complexity of the sequencing procedure. However, this method may still be effective for sequence analysis of DNA segments containing heterogeneous ends and no substrate sites of known restriction endonucleases.

Principle and Limitations

Kinetic studies on RNA synthesis reaction indicate that the apparent K_m estimated for nucleotides for polymerization is about 0.015 mM and about one-tenth the concentration for initiation.[6,7] Therefore, initiation of RNA synthesis is the rate-limiting step, and when the NTP concentration is lowered to 20 μM or less, the efficiency of RNA synthesis becomes very low. Under these conditions, the addition of appropriate oligonucleotides significantly stimulates RNA synthesis. With double-stranded DNA containing promoter, the primed synthesis still occurs within the promoter regions.[8,9] When RNA synthesis is performed on single-stranded DNA, the primed synthesis is now preferentially initiated at sites of which the sequence is complementary to the primer sequence. However, there are several limitations for application of this method, as described below.

[1] W. Gilbert and A. Maxam, *Proc. Natl. Acad. Sci. U.S.A.* **70**, 3581 (1973).
[2] K. Sugimoto, T. Okamoto, H. Sugisaki, and M. Takanami, *Nature (London)* **253**, 410 (1975).
[3] M. Takanami, K. Sugimoto, H. Sugisaki, and T. Okamoto, *Nature (London)* **260**, 297 (1976).
[4] F. Sanger and A. R. Coulson, *J. Mol. Biol.* **94**, 441 (1975).
[5] A. Maxam and W. Gilbert, *Proc. Natl. Acad. Sci. U.S.A.* **74**, 560 (1977).
[6] D. D. Anthony, C. W. Wu, and D. A. Goldthwait, *Biochemistry* **8**, 246 (1969).
[7] K. M. Downey and A. G. So, *Biochemistry* **9**, 2520 (1970).
[8] K. M. Downey, B. S. Jurmark, and A. G. So, *Biochemistry* **10**, 4970 (1971).
[9] E. G. Minkley and D. Pribnow, *J. Mol. Biol.* **77**, 255 (1973).

Double-stranded DNA can be used as template after denaturation, but if the primer sites are located on both the strands, analysis of the products becomes complicated. Unless appropriate primers are used, it is necessary to separate the strands of the template. RNA chains initiated on single-stranded DNA do not grow longer, compared with those on double-stranded DNA. For example, the size of RNA formed on single-stranded DNA phage fd is less than about 100 bases even at the optimal conditions for RNA synthesis. Therefore, the distance of transcription should be short enough to terminate RNA chains at the end of the template, otherwise the product size becomes heterogeneous even if initiation occurred at specific primer sites. The upper limit of the template size may be about 100 bases. In order to select appropriate primers, it is necessary to know at least part of the sequence of the template. This may be given by analysis of pyrimidine tracts. Another way is to find primers which stimulate RNA synthesis significantly by testing various oligonucleotides of known sequences. The priming ability of an oligonucleotide is not influenced by the presence or absence of phosphomonoester at the 5' end. However, only short oligonucleotides, less than pentamers, appear to work well as primer under the conditions used. The important step of this method is fractionation of the reaction products, because the products are generally very heterogeneous even with appropriate primers.

Conditions for Primed Synthesis

Reaction mixtures contain, in a final volume of 1 ml, 8 mM $MgCl_2$, 50 mM KCl, 40 mM Tris-HCl (pH 7.9), 0.1 mM dithiothreitol, about 10 pmole denatured DNA as template, about 30 pmole RNA polymerase, 20 μM each of one [α-^{32}P]NTP and three other NTPs, and about 1 nmole oligonucleotide as primer. The concentration of the primer should be high enough to force RNA polymerase to initiate synthesis from the sites at which the primer hybridizes with the template. The NTP concentration used by others[1] is 5 μM. In general, the lower the NTP concentration is, the more defined species of RNA appear to be synthesized. However, the yield of transcripts concomitantly decreases. RNA polymerase holoenzyme is prepared by the procedure described in Volume XXI, Section VII (articles [42] and [43]) in this series or by any other procedures. Although the effect of the sigma factor on the primed synthesis is not elucidated, the efficiency of RNA synthesis with the core enzyme alone is very low under the above conditions.

Incubation is done for 3 hr or more at 37°. At the end of incubation, 20 μg of DNase I are added and incubation is continued for an additional 20 min to destroy the template. The solution is treated with 80% phenol, followed by addition of 0.1 mg carrier tRNA, and passed through a

Sephadex G100 column (0.6 × 20 cm), equilibrated with 0.1 M NaCl, 0.02 M Tris-HCl(pH 7.6). The RNA fraction is collected and precipitated by adding 2 volumes of ethanol. The precipitate is dissolved in 50 μl of 10 mM Tris-HCl(pH 7.6), 0.1 mM EDTA, and 9 M urea. The solution is heated for 3 min at 95°, chilled, and fractionated by electrophoresis on polyacrylamide gel columns or slabs (8.5 to 15%) under denaturing conditions.[1-5] A general treatment of the use of polyacrylamide gel electrophoresis for fractionation of labeled RNA is given in Volume XXI, Section III (articles [7] and [8]) in this series. When an RNA band is reelectrophoresed to minimize the contamination with neighboring bands, the band region visualized by autoradiography is cut into a slice, directly mounted on the top of a gel column or slab of which the gel concentration is slightly higher than that for the first run, and electrophoresed. RNA is extracted from gel by the procedure described in Volume XXI, Section III (article [7]) in this series. The two-dimensional system which uses electrophoresis on cellulose acetate in the first dimension and homochromatography on DEAE–cellulose thin layer in the second dimension[10,11] is also useful for fractionation of transcripts up to about 50 bases in length.

Sequence Analysis of Transcripts

Nearest neighbor analysis based on ^{32}P transfer experiments[12] has been generally used for sequence analysis of transcripts. However, the new sequencing procedure which resolves partial nuclease digests of terminally labeled RNA by gel electrophoresis[13] would be more useful, since oligonucleotides with 5'-OH ends can be used as primer, and RNA chains primed with such oligonucleotides are selectively labeled with ^{32}P in the polynucleotide kinase reaction.

[10] G. G. Brownlee and F. Sanger, *Eur. J. Biochem.* **11**, 395 (1969).
[11] E. Jay, R. Padmanabhan, and R. Wu, *Nucleic Acids Res.* **1**, 331 (1974).
[12] G. G. Brownlee, "Determination of Sequences in RNA." Am. Elsevier, New York, 1972.
[13] H. Donis-Keller, A. M. Maxam, and W. Gilbert, *Nucleic Acids Res.* **4**, 2527 (1977).

[57] Sequencing End-Labeled DNA with Base-Specific Chemical Cleavages[1]

By ALLAN M. MAXAM and WALTER GILBERT

In the chemical DNA sequencing method,[2] one end-labels the DNA, partially cleaves it at each of the four bases in four reactions, orders the

[1] This work was supported by the U.S. National Institute of General Medical Sciences, Grant GM 09541.
[2] A. M. Maxam and W. Gilbert, *Proc. Natl. Acad. Sci. U.S.A.* **74**, 560 (1977).

(a)

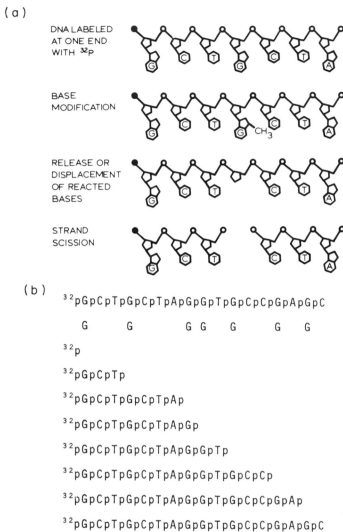

(b)

FIG. 1. Sequencing end-labeled DNA by limited, base-specific chemical cleavage. In (a) three consecutive reactions cleave one DNA molecule at one guanine, and when these reactions cleave at a guanine in all such molecules they generate the nested set of end-labeled fragments listed in (b). (a) Chemical cleavage begins (top) with a unique DNA fragment labeled at one 5′ or one 3′ end with ^{32}P. This DNA can be single- or double-stranded and of any length. First, in a limited reaction, some reagent specifically modifies one or two of the four DNA bases, generally by substituting a purine ring (above) or pyrimidine ring. Then, a second reaction removes this modified base from its sugar, and finally a third reaction eliminates both phosphates from that sugar to break the DNA. In these reactions the DNA breaks only at sugars without bases, and only appropriately modified bases can be removed from their sugars. (b) Partial chemical cleavage produces nested end-labeled prod-

products by size on a slab gel, and then reads the sequence from an autoradiogram by simply noting which base-specific agent cleaved at each successive nucleotide along the strand. This technique will sequence DNA made in and purified from cells. No enzymatic copying *in vitro* is required, and either single- or double-stranded DNA can be sequenced. Only one agent is needed to address it to a chosen stretch of DNA, a restriction endonuclease which cuts the DNA in that region. Given one such cut, this method can display the sequence of at least 250 bases in both directions from it in one experiment. It can, from the same cut and at the same time, check this sequence by deriving the corresponding 500 bases on the complementary strand. These distances over which the sequence can be read are determined by the gel electrophoresis technology.

Specific cleavage of the DNA is central to this method. Most chemical schemes which cleave at one or two of the four bases involve three consecutive steps: modification of a base, removal of the modified base from its sugar, and DNA strand scission at that sugar. Figure 1 illustrates these steps with methylation of guanine at N-7, release of the methylated base, and phosphate elimination which breaks the DNA. The second and third steps in such schemes have the important property that each is contingent on the one which preceded it. That is, the DNA will break only at a sugar without a base, and only an appropriately modified base can be removed from its sugar (Fig. 1). The specificity resides in the first reaction, which is with only a few percent of the bases; the subsequent strand-cleaving reactions must be quantitative. It follows from this that reagents which react with the DNA bases are most important, for they ultimately determine where the DNA will break. Dimethyl sulfate and hydrazine are two such reagents, and they have long been known to substitute and open base rings in ways which make the DNA vulnerable to cleavage.

In most of the base-specific cleavage schemes, one reaction disrupts the electronic structure of a DNA base, making it possible for a second reaction to break the bond between that base and its sugar. The reactions of dimethyl sulfate and then piperidine with the purine ring, and of hydrazine and then piperidine with the pyrimidine ring, illustrate how bases are removed from DNA in this manner (Figs. 2 and 3). Specifically, dimethyl sulfate methylates (among other positions) the 7-nitrogen of

---

ucts. If the ^{32}P-labeled DNA at the top undergoes the reactions illustrated in (a), every molecule will break at one of the seven guanines, leaving the end-labeled pieces below. Each radioactive product extends from the original ^{32}P-phosphate at the common end to the phosphate adjacent to a guanine which was methylated at the variable end. If this DNA fragment is cleaved in four such reactions, each specific for different bases, and the four different nested sets of products are electrophoresed in parallel on a gel and autoradiographed, its sequence can be read off the four meshing band patterns on the film (see Fig. 4).

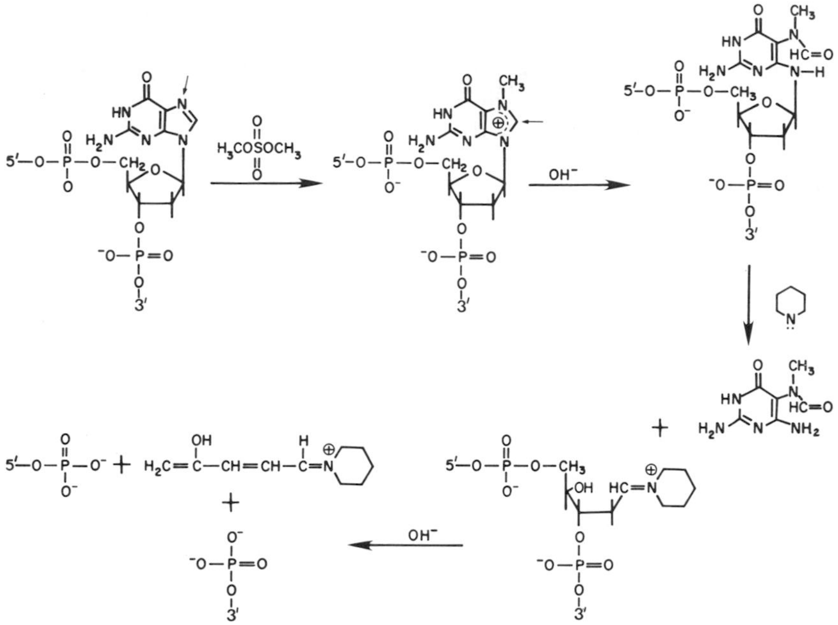

FIG. 2. Sequential reactions with dimethyl sulfate and piperidine which break DNA at guanines. Dimethyl sulfate methylates the N-7 of guanine in DNA (upper left), fixing a positive charge into the N-7–C-8–N-9 imidazole portion of the purine ring. When base attacks C-8, which shares this charge, its bond with N-9 breaks (upper right).[3] Piperidine then displaces the ring-opened 7-methylguanine (right) and catalyzes β-elimination of both phosphates from the sugar (lower left), possibly as the piperidone (lower middle). When one guanine N-7 in a DNA molecule becomes methylated in a limited dimethyl sulfate reaction, piperidine will break the DNA at that base. Some adenines also get methylated at N-3 in the same reaction, but they do not react with piperidine in ways which lead to strand scission. Lane G in Fig. 4 displays pieces of end-labeled DNA which arise when these reactions partially cleave it at guanines.

guanine, which opens between carbon 8 and nitrogen 9 in a base-catalyzed reaction,[3] and piperidine then displaces the ring-opened 7-methylguanine from its sugar (Fig. 2). The nucleophile hydrazine, on the other hand, splits a thymine or cytosine ring,[4-7] leaving ring fragments which can again be displaced from the sugar by piperidine (Fig. 3). If molar sodium chloride is present during this limited hydrazine reaction, only cytosine

[3] P. D. Lawley and P. Brookes, *Biochem. J.* **89,** 127 (1963).

[4] F. Baron and D. M. Brown, *J. Chem. Soc.* p. 2855 (1955).

[5] A. Temperli, H. Turler, P. Rust, A. Danon, and E. Chargaff, *Biochim. Biophys. Acta* **91,** 462 (1964).

[6] A. R. Cashmore and G. B. Petersen, *Nucleic Acids Res.* **5,** 2485 (1978).

[7] N. K. Kochetkov and E. I. Budowskii, "Organic Chemistry of Nucleic Acids," Part B, pp. 401–423. Plenum, New York, 1972.

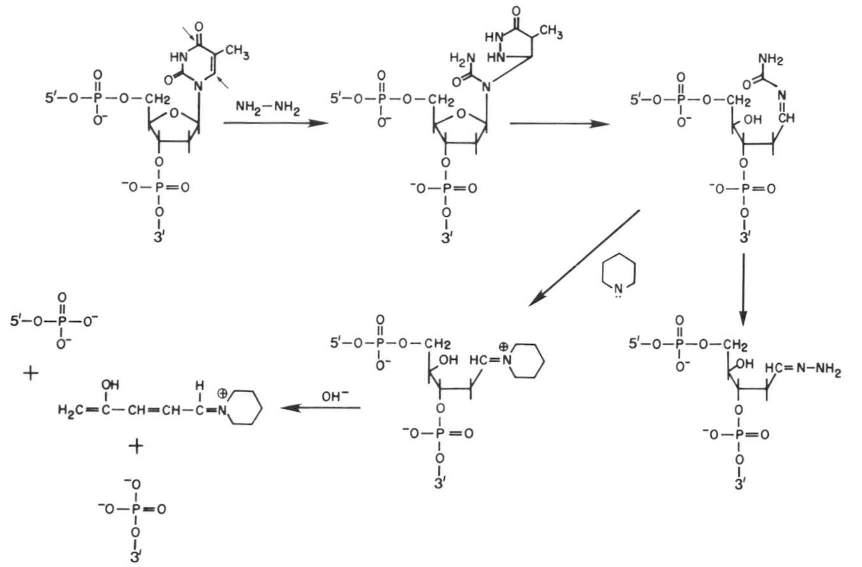

FIG. 3. Sequential reactions with hydrazine and piperidine which break DNA at thymines. Hydrazine attacks thymine in DNA (upper left) at C-4 and C-6, opens the pyrimidine ring, and cyclizes with C-4–C-5–C-6 to form a new five-membered ring. Extensive hydrazine reactions release this new pyrazolone ring (upper middle) and the N-2–C-2–N-3 urea fragment of the thymine base (upper right), leaving sugars in the DNA backbone as hydrazones (lower right) (reviewed by Kochetkov and Budowskii[7]). Piperidine reacts with all these glycosides and catalyzes β-elimination of both phosphates from the sugar (lower left), possibly as the piperidone (lower middle). When hydrazine splits one thymine or cytosine ring in one DNA molecule in a limited reaction, piperidine will break the DNA at that base. Lane C + T in Fig. 4 displays pieces of end-labeled DNA which arise when these reactions partially cleave it at thymines and cytosines.

reacts appreciably. In other reactions specific for two bases, acid weakens adenine and guanine glycosidic bonds by protonating (rather than methylating) purine ring nitrogens, and alkali opens adenine and cytosine rings,[8] which are again displaced by piperidine. In all cases, piperidine also goes on to catalyze β-elimination of phosphates from the empty sugar to finally break the DNA strand, leaving the sugar free, one phosphate on an unlabeled fragment, and the other phosphate on a fragment carrying the original ^{32}P end label.

Base-specific chemical cleavage is only one step in sequencing DNA by the method we have described. In sections which follow we present techniques for producing discrete DNA fragments, end-labeling DNA, segregating end-labeled fragments, extracting DNA from gels, and then

---

[8] N. K. Kochetkov and E. I. Budowskii, "Organic Chemistry of Nucleic Acids," Part B, pp. 381–397. Plenum, New York, 1972.

protocols for partially cleaving it at specific bases using the chemical reactions. These are in turn followed by a section on electrophoresis of the chemical cleavage products on long-distance sequencing gels, and a guide for troubleshooting problems in sequencing patterns. Finally, at the end, we integrate many of these techniques into efficient strategies for deriving and confirming the sequence of a plasmic, phage, or viral chromosome in 250-nucleotide blocks.

Unlabeled DNA Segments

A defined DNA region is easily mapped for restriction enzyme cleavage, sequenced, or analyzed for protein contacts[9-11] if it is purified away from the rest of the DNA as a unique restriction fragment. This is especially true for foreign DNA cloned in bacterial hosts. In a later section on strategies we consider sequencing and mapping objectives which merit isolation of unlabeled DNA segments. Given the need, up to several hundred picomoles of a large restriction fragment can be prepared from phage, viral, or plasmid DNA using the protocol of procedure 1. The first step is to digest several hundred picomoles of the chromosomal DNA with the appropriate restriction endonuclease. This is about 1 mg of 5000 base pair, plasmid-vector-size DNA. Although 1000 conventional units of restriction enzyme will cut up this 1000 μg of DNA in 1 hr at 37°, one-tenth this amount or less may suffice in a longer reaction. The extent of this digestion can easily be checked by electrophoresing a small aliquot of the enzyme–DNA mixture alongside a small-scale limit digest of the same DNA and staining. If all bands correspond, cleavage is complete. Terminal phosphates are then removed with alkaline phosphatase; the two enzymes are extracted into phenol; residual phenol is extracted into ether; and, after ethanol precipitation to desalt and concentrate the DNA, restriction fragments are resolved by polyacrylamide gel electrophoresis. DNA fragments 2000–3000 base pairs long resolve on the gel specified in procedure 1. After fragments have run their course, the band pattern in this slab gel is visualized by uv shadowing,[12] compared with a photograph of the analytic pattern above, and the relevant band identified and the DNA extracted and ethanol precipitated. This DNA segment will encompass the region of interest, and after end-labeling a small portion of it (Table I) and cutting with a second enzyme (procedure 7) or separating

---

[9] W. Gilbert, A. Maxam, and A. Mirzakekov, in "Control of Ribosome Synthesis" (N. O. Kjeldgaard and O. Maaloe, eds.), pp. 139–148. Munksgaard, Copenhagen, 1976.
[10] R. Ogata and W. Gilbert, Proc. Natl. Acad. Sci. U.S.A. 74, 4973 (1977).
[11] A. D. Johnson, C. O. Pabo, and R. T. Sauer, this volume, Article [76].
[12] S. M. Hassur and H. W. Whitlock, Anal. Biochem. 41, 51 (1971).

strands (procedure 8), it can be mapped for cleavage by various other restriction enzymes[13,14] or sequenced in part from each labeled end.

Procedure 1. Large-Scale Preparation of Unique DNA Segments with a Restriction Enzyme and a Polyacrylamide Gel

   1 mg DNA (phage, virus, plasmid)
 120 μl Distilled water
  20 μl 200 mM Tris-HCl, pH 7.4, 50 mM MgCl$_2$, 10 mM dithio-threitol, 100 μg/ml bovine serum albumin, $\pm$500 mM NaCl
  50 μl Restriction enzyme, 100 units

Combine the above in a 1.5-ml Eppendorf snap-cap tube and mix. Heat at 37° for 16 hr. Electrophorese 2 μl on a gel and stain with ethidium bromide. If digestion is complete, proceed; if not, add more restriction enzyme, heat for 2 hr, and analyze again.

  5 μl 2.0 M Tris-HCl, pH 8.0
  5 μl Alkaline phosphatase, 0.1–0.5 unit

Heat at 37° for 30 min.

    2 μl 1.0 M EDTA
 200 μl Redistilled phenol, saturated with 50 mM Tris-HCl, pH 8.0
     1 mM EDTA

Close the cap and mix well (Vortexer). Centrifuge 1 sec to separate the phases (Eppendorf). Remove the bottom phase with a Micropet.

 200 μl Phenol (same as above)

Mix well, centrifuge 1 sec, and remove the bottom phase.

 500 μl Diethyl ether

Mix, centrifuge 1 sec, and remove the top phase.

 500 μl diethyl ether

Mix, centrifuge 1 sec, and remove the top phase.

  25 μl 3 M Sodium acetate
 750 μl 95% Ethanol

Close the cap and mix (invert the tube four times). Chill at $-70°$ for 5 min (Dry Ice–ethanol bath). Centrifuge at 12,000 g for 5 min (Eppendorf, 4°).

---

[13] P. G. Boseley, T. Moss, and M. L. Birnstiel, this volume, Article [55].
[14] H. O. Smith and M. L. Birnstiel, *Nucleic Acids Res.* **3**, 2387 (1976).

Remove the supernatant with a Pasteur pipette. Add to the tube, but do not mix the DNA into

> 1 ml 70% (v/v) Ethanol

Centrifuge at 12,000 g 15 sec, and remove supernatant. Place the tube under vacuum for a few minutes (desiccator).

> 100 μl 10%(v/v) Glycerol, 50 mM Tris-borate, pH 8.3, 1 mM EDTA, 0.05%(w/v) xylene cyanol, 0.05%(w/v) bromphenol blue

Close cap and redissolve DNA (Vortexer and manual agitation). Centrifuge a few seconds (Eppendorf). Load on a long, thick, nondenaturing slab gel

> 5%(w/v) Acrylamide
> 0.17%(w/v) Methylene bisacrylamide
> 50 mM Tris-borate, pH 8.3, 1 mM EDTA
> Slab gel: 6 × 200 × 400 mm, or 3 × 200 × 400 mm
> Sample wells: two 6 × 30 mm, or four 3 × 30 mm
> Aged overnight to reduce uv absorption

Electrophorese until the smallest fragment wanted has traversed most of the gel. Remove glass plates, transferring the gel to a thin sheet of transparent plastic wrap (Saran wrap). Lay the gel on the Saran wrap on a fluorescent thin-layer chromatography plate in a dark room. Irradiate with short-wave uv, and observe dark bands (DNA) on a bright background (fluoresence from dye in plate). Excise gel segment containing desired DNA band with a razor blade. Drop gel piece into a siliconized 12-ml centrifuge tube. Crush the gel with an 8-mm-diameter siliconized glass rod. Hand grind the crushed gel to a paste with the rod. Add to the gel paste 8 ml gel elution solution:

> 8 ml 500 mM Ammonium acetate, 10 mM magnesium acetate, 1 mM EDTA, 0.1%(w/v) sodium dodecyl sulfate

Seal the tube and invert four times the mix. Heat at 37° for 10 hr (air incubator). Centrifuge at 10,000 g for 15 minutes. Remove the supernatant and save. Add to the pellet 4-ml gel elution solution:

> 4 ml 500 mM Ammonium acetate, 10 mM magnesium acetate, 1 mM EDTA, 0.1%(w/v) sodium dodecyl sulfate

Seal the tube and invert four times to mix. Centrifuge at 10,000 g for 15 min. Remove this second supernatant and combine with the first. Gravity filter DNA solution through Millipore Millex unit attached to a plastic syringe mounted on a ring stand (do not force liquid through membrane filter with syringe plunger). Collect filtrate (12 ml) in a siliconized 50-ml centrifuge tube. Add 3 volumes of ethanol, which just fills the tube

35 ml 95% Ethanol

Seal the tube and invert four times to mix. Chill at $-70°$ for 10 min (Dry Ice–ethanol bath). Centrifuge at 10,000 g for 15 min (angle rotor, 4°). Remove the supernatant with a pipette.

0.25 ml 300 mM Sodium acetate

Redissolve the DNA (roll the tube horizontally, and Vortex). Transfer DNA solution to a 1.5-ml Eppendorf snap-cap tube.

0.75 ml 95% Ethanol

Close the cap and invert the tube four times to mix. Chill at $-70°$ for 5 min (Dry Ice–ethanol bath). Centrifuge at 12,000 g for 5 min (Eppendorf, 4°). Remove the supernatant with a Pasteur pipette.

1 ml 95% Ethanol

Centrifuge 12,000 g for 15 sec and remove supernatant. Place the tube under vacuum for a few minutes (desiccator). The DNA residue appears as a pellet, a film, or both, and can be recovered in a small volume of water.

DNA Labeled at One End with [32]P

DNA Ends and End-Labeling Reactions. Sequencing DNA by limited internal cleavage requires that it have an atom or group on one end which can be detected later on a nested set of cleavage products. Most commonly this is ^{32}P-phosphate or a ^{32}P-labeled nucleotide enzymatically incorporated at the end of single or double-stranded DNA. This end should be unique, that is, most molecules in the sample should terminate with the same nucleotide in the sequence. Unique ends may originate *in vivo* or *in vitro,* and may be found on restriction fragments, a phased single-stranded reverse transcript of RNA, or double-stranded DNA with a singular gap, nick, or tail. This requirement for uniform ends, however, is not absolute. DNA with ragged ends, including S1-nuclease-resistant[15] or ligand-protected segments, can be sequenced if fragments of slightly different length can be resolved after labeling; when such fragments are less than 100 nucleotides long, a 1.5 mm, 8% sequencing gel (procedure 16) will often do this.

Three enzymes with three different ^{32}P-labeled precursors have been used to label DNA ends. They are polynucleotide kinase and [γ-^{32}P]ATP

[15] V. Vogt, this volume, Article [32].

TABLE I

³²P END-LABELING REACTIONS FOR DNA SEQUENCING*

End labeled	Enzyme	Precursor	³²P-Labeled groups incorporated	DNA	Literature references	Our protocol
5'	Phage T4 polynucleotide kinase	[γ-³²P]Adenosine triphosphate 1500 Ci/mmole[a,b] 1000–3000 Ci/mmole[c,d,e] >5500 Ci/mmole[b,c]	p	ss, ds	f–k	Procedures 2, 4, and 5
3'	Calf thymus terminal transferase	[α-³²P]Adenosine triphosphate 300 Ci/mmole[c,d] 2000–3000 Ci/mmole[d]	(pA)$_n$	ss, ds	l–n	Procedure 6
3'	Phage T4 DNA polymerase E. coli DNA polymerase I	[α-³²P]Deoxynucleoside triphosphate(s) 300 Ci/mmole[c,d] 1000–3000 Ci/mmole[c,d]	(pdN)$_n$	ds	o–u	

* Abbreviations: p, phosphate; N, nucleoside; dN, deoxynucleoside; ss, single-stranded; ds, double-stranded.

[a] I. M. Glynn and J. B. Chappell, *Biochem. J.* **90**, 147 (1964).

[b] P. F. Schendel and R. D. Wells, *J. Biol. Chem.* **248**, 8319 (1973).

[c] Source of ^{32}P precursors, New England Nuclear.

[d] Source of ^{32}P precursors, Amersham.

[e] Source of ^{32}P precursors, ICN.

[f] G. Chaconas and J. H. van de Sande, this volume, Article [10].

[g] C. C. Richardson, *Proc. Natl. Acad. Sci. U.S.A.* **54**, 158 (1965).

[h] J. H. van de Sande, K. Kleppe, and H. G. Khorana, *Biochemistry* **12**, 5050 (1973).

[i] G. Chaconas, J. H. van de Sande, and R. B. Church, *Biochem. Biophys. Res. Commun.* **66**, 962 (1975).

[j] J. R. Lillehaug, R. K. Kleppe, and K. Kleppe, *Biochemistry* **15**, 1858 (1976).

[k] K. L. Berkner and W. R. Folk, *J. Biol. Chem.* **252**, 3176 (1977).

[l] R. Roychoudhury and R. Wu, this volume, Article [7].

[m] H. Kossel and R. Roychoudhury, *Eur. J. Biochem.* **22**, 271 (1971).

[n] J. C. Chang, G. F. Temple, R. Poon, K. H. Neumann, and Y. W. Kan, *Proc. Natl. Acad. Sci. U.S.A.* **74**, 5145 (1977).

[o] M. D. Challberg and P. T. Englund, this volume, Article [6].

[p] P. T. Englund, *J. Mol. Biol.* **66**, 209 (1972).

[q] R. Wu and E. Taylor, *J. Mol. Biol.* **57**, 491 (1971).

[r] K. Kleppe, E. Ohtsuka, R. Kleppe, I. Molineux, and H. G. Khorana, *J. Mol. Biol.* **56**, 341 (1971).

[s] J. E. Donelson and R. Wu, *J. Biol. Chem.* **247**, 4654 (1972).

[t] E. Soeda, G. Kimura, and K. Miura, *Proc. Natl. Acad. Sci. U.S.A.* **75**, 162 (1978).

[u] E. Schwarz, G. Scherer, G. Hobom, and H. Kossel, *Nature (London)* **272**, 410 (1978).

for 5' ends, terminal transferase and an [α-^{32}P]ribonucleoside triphosphate for 3' ends, and DNA polymerase with [α-^{32}P]deoxynucleoside triphosphates complementary to 5' terminal nucleotides, also for 3' ends. Table I summarizes these reactions, provides references to their enzymology and use for end-labeling DNA, and notes commercial sources and methods of preparation for the relevant ^{32}P-labeled precursors. While some versions of these end-labeling reactions are not 100% efficient, the high specific activity of available ^{32}P-labeled nucleoside triphosphates (300–6000 Ci/mmole, Table I) permits one to sequence 1–2 strand picomoles of a DNA fragment without difficulty, and much less if intensifying screens[16–18] are used with X-ray films on sequencing gels.

[γ-^{32}P]ATP Synthesis. Polynucleotide kinase transfers the γ-phosphate of ATP to the 5' ends of DNA strands,[19] and when this phosphate contains ^{32}P an efficient reaction would label 5' terminal nucleotides to the same specific activity as the [γ-^{32}P]ATP. Two published procedures for synthesizing high specific activity [γ-^{32}P]ATP employ the same enzymes and phosphorylated intermediates, but different precursors: the phosphate–ATP exchange reaction of Glynn and Chappell[20] and the phosphate–ADP forward phosphorylation of Schendel and Wells.[21] Though the forward reaction should generate higher specific activity [γ-^{32}P]ATP, we have assumed that it must be chromatographically separated from excess ADP left in the reaction mixture, which would otherwise reverse the polynucleotide kinase reaction,[22] and to avoid exposure to millicurie amounts of ^{32}P in columns and fractions have used[2] the exchange reaction instead. (Unpurified Schendel–Wells [γ-^{32}P]ATP may, however, be quite compatible with exchange kinase reactions.) A modified Glynn–Chappell exchange synthesis for 1500–2000 Ci/mmole [γ-^{32}P]ATP that can be used without purification is in procedure 2.

Labeling 5' Ends with Polynucleotide Kinase.[23] High specific activity [γ-^{32}P]ATP, synthesized by one of the above methods or obtained commercially (Table I), can be used with polynucleotide kinase to end-label DNA by forward phosphorylation of 5'-hydroxyls[19,24] or by exchange with existing 5'-phosphates.[22,25,26] Forward phosphorylation of ends of restric-

[16] R. A. Laskey, this volume, Article [45].
[17] R. A. Laskey and A. D. Mills, FEBS Lett. 82, 314 (1977).
[18] R. Swanstrom and P. R. Shank, Anal. Biochem. 86, 184 (1978).
[19] C. C. Richardson, Proc. Natl. Acad. Sci. U.S.A. 54, 158 (1965).
[20] I. M. Glynn and J. B. Chappell, Biochem. J. 90, 147 (1964).
[21] P. F. Schendel and R. D. Wells, J. Biol. Chem. 248, 8319 (1973).
[22] J. H. van de Sande, K. Kleppe, and H. G. Khorana, Biochemistry 12, 5050 (1973).
[23] See also G. Chaconas and J. H. van de Sande, this volume, Article [10].
[24] J. R. Lillehaug, R. K. Kleppe, and K. Kleppe, Biochemistry 15, 1858 (1976).
[25] G. Chaconas, J. H. van de Sande, and R. B. Church, Biochem. Biophys. Res. Commun. 66, 962 (1975).
[26] K. L. Berkner and W. R. Folk, J. Biol. Chem. 252, 3176 (1977).

TABLE II
RESTRICTION ENDONUCLEASES WHICH CLEAVE DNA INTO FRAGMENTS READILY END-
LABELED WITH T4 POLYNUCLEOTIDE KINASE AND [γ-^{32}P]ATP OR WITH
DNA POLYMERASE AND [α-^{32}P]dNTP[a]

Restriction enzyme	Sequence recognized[b]	Fragment end structure after cleavage[b]	Mean fragment length (bp)[c]
*Hin*fI	GANTC	$_p$ANTCNN GNN	256
*Hpa*II	CCGG	$_p$CGGNN CNN	256[c]
*Sau*I	GATC	$_p$GATCNN NN	256
*Taq*I	TCGA	$_p$CGANN TNN	256[c]
*Eco*RII	CC^A_TGG	$_p$CCXGGNN NN	512
*Ava*II	GG^A_TCC	$_p$GXCCNN GNN	512
*Ava*I	CYCGRG	$_p$YCGRGNN CNN	1024[c]
*Bam*HI	GGATCC	$_p$GATCCNN GNN	4096
*Bgl*II	AGATCT	$_p$GATCTNN ANN	4096
*Eco*RI	GAATTC	$_p$AATTCNN GNN	4096
*Hin*dIII	AAGCTT	$_p$AGCTTNN ANN	4096
*Sal*I	GTCGAC	$_p$TCGACNN GNN	4096[c]
*Xba*I	TCTAGA	$_p$CTAGANN TNN	4096
*Xho*I,*Blu*I	CTCGAG	$_p$TCGAGNN CNN	4096[c]
*Xma*I	CCCGGG	$_p$CCGGGNN CNN	4096[c]

[a] When the above restriction enzymes cleave DNA, they make symmetric, staggered cuts which leave 5' ends extended. Double-stranded DNA with protruding 5' ends is more efficiently phosphorylated by polynucleotide kinase than DNA with flush or recessed ends [J. R. Killehaug, R. K. Kleppe, and K. Kleppe, *Biochemistry* **15**, 1858 (1976)]. The foreshortened 3' ends of the other strand can be labeled by repair with one or more [α-^{32}P]deoxynucleoside triphosphates complementary to bases in the template strand extension and a DNA polymerase (see text).

[b] R. J. Roberts, this volume, Article [1].

[c] Average cleavage frequencies are based on a probability equal to one-fourth of encountering each unique base in a recognition sequence; since eukaryotic DNA generally is deficient in the neighbor-pair CpG [G. J. Russell, P. M. B. Walker, R. A. Elton, and J. H. Subak-Sharpe, *J. Mol. Biol.* **108**, 1 (1976)], these restriction enzymes may cleave it less frequently.

tion fragments requires that their 5'-phosphates be removed beforehand. Procedure 4 describes how to dephosphorylate DNA with alkaline phosphatase and to then remove the phosphatase. After that procedure 5 gives two forward kinase protocols for this dephosphorylated DNA, Procedure 5A for flush or recessed 5' ends, and procedure 5B for DNA with protruding ends or no complementary strand. Fifteen restriction enzymes which leave protruding 5' ends after cleaving double-stranded DNA are listed in Table II. Procedure 5C is a version of the exchange kinase reaction for DNA with 5'-phosphates tested by Berkner and Folk[26] on ends produced by cleavage with EcoRI. They achieved 60–70% of theoretical ^{32}P-phosphate incorporation on these protruding 5' ends.

Labeling 3' Ends with Terminal Transferase.[27] Terminal deoxynucleotidyltransferase will polymerize ribonucleotides onto the 3' ends of DNA strands in a template-independent reaction, using ribonucleoside triphosphates as precursors.[28] When these triphosphates are α-labeled with ^{32}P, they label 3' ends of the DNA (Table I). If the reaction is extensive, and the product is subjected to alkaline hydrolysis or the appropriate ribonuclease, the DNA 3' ends will retain two ^{32}P-phosphates, one on either side of the initial ribonucleotide added. Procedure 6 is a modification of the reaction described by Roychoudhury et al.[27,29] for 3' labeling single- or double-stranded DNA. It employs the volatile base piperidine to hydrolyze off the DNA all but the first ^{32}P-ribonucleotide added by the terminal transferase, plus ^{32}P-phosphate from the second one. (If high specific activity [α-^{32}P]dideoxynucleoside triphosphates are available, strict single addition of the labeled nucleotide is possible with this enzyme, eliminating the need for alkaline hydrolysis[30].) Roychoudhury and Wu[27] and Chang et al.[31] show sequence patterns derived from DNA 3' end-labeled with terminal transferase and an [α-^{32}P]ribonucleoside triphosphate.

Labeling 3' Ends with DNA Polymerase.[32] The 3' ends of double-stranded DNA fragments can also be labeled with a DNA polymerase and [α-^{32}P]deoxynucleoside triphosphates. If restriction enzyme cleavage produces a DNA fragment with recessed 3' ends (Table II), E. coli DNA polymerase I or T4 DNA polymerase, plus labeled triphosphates chosen

[27] See also R. Roychoudhury and R. Wu, this volume, Article [7].
[28] H. Kossel and R. Roychoudhury, *Eur. J. Biochem.* **22,** 271 (1971).
[29] R. Roychoudhury, E. Jay, and R. Wu, *Nucleic Acids Res.* **3,** 863 (1976).
[30] K. Olson and C. Harvey, *Nucleic Acids Res.* **2,** 319 (1975).
[31] J. C. Chang, G. F. Temple, R. Poon, K. H. Neumann, and Y. W. Kan, *Proc. Natl. Acad. Sci. U.S.A.* **74,** 5145 (1977).
[32] See also M. D. Challberg and P. T. Englund, this volume, Article [6].

by reference to cleavage points in the recognition sequence (Table II), will label both its ends by filling them out with [5'-^{32}P]deoxynucleotides.[32–35] When restriction cleavage leaves 3' ends flush with 5' ends, on the other hand, these enzymes can also exchange the existing 3'-terminal nucleotide for a radioactive one, given the appropriate [α-^{32}P]deoxynucleoside triphosphate.[32,36,37] Uniform labeling with these enzymes, restricted to DNA ends, requires conditions under which synthesis exceeds exonuclease action, and full incorporation of the ultimate 3' nucleotide when repair or exchange of two or more consecutive nucleotides is possible. Gel sequencing patterns obtained from DNA 3'-^{32}P-labeled with these reactions have been exhibited.[38,39,40]

Segregating Two Labeled Ends or Complementary Strands. One or a whole collection of DNA fragments can be end-labeled using the kinase, transferase, or polymerase and appropriate ^{32}P-labeled precursor as just described. If the DNA has just one unique 5' or 3' end, it can be sequenced immediately after that end has been labeled. In all other cases, because one-dimensional gel sequencing methods cannot analyze two different sequences simultaneously, the various labeled fragments, fragment ends, or complementary strands have to be separated from one another. With single-stranded DNAs of different but discrete lengths, simply electrophoresing the products of an end-labeling reaction on the appropriate denaturing gel will do this. For a double-stranded DNA restriction fragment with ^{32}P on both 5' or both 3' ends, there are two choices. One is to cut it at one or more sites along its length with another restriction enzyme to release different-sized, singly end-labeled end-pieces, and then separate the latter by electrophoresing them on a native gel. Procedure 7 describes how to subdivide a doubly end-labeled DNA fragment in this way. The other choice is to melt the double-stranded fragment and separate its complementary strands by electrophoresing them through a gel.

Strand Separation. Agarose gel separation of dissociated DNA strands was first reported by Hayward with whole bacteriophage DNA,[41] and has since been achieved with smaller DNA restriction fragments in

[33] R. Wu, *J. Mol. Biol.* **51**, 501 (1970).

[34] R. Wu and E. Taylor, *J. Mol. Biol.* **57**, 491 (1971).

[35] K. Kleppe, E. Ohtsuka, R. Kleppe, I. Molineauz, and H. G. Khorana, *J. Mol. Biol.* **56**, 341 (1971).

[36] P. T. Englund, *J. Mol. Biol.* **66**, 209 (1972).

[37] J. Donelson and R. Wu, *J. Biol. Chem.* **247**, 4654 (1972).

[38] E. Soeda, G. Kimura, and K. Miura, *Proc. Natl. Acad. Sci. U.S.A.* **75**, 162 (1978).

[39] E. Schwarz, G. Scherer, G. Hobom, and H. Kossel, *Nature (London)* **272**, 410 (1978).

[40] R. Grosschedl and G. Hobom, *Nature (London)* **277**, 621 (1979).

agarose[42-44] and polyacrylamide[2,45] gels. (Electrophoresis on cellulose acetate at pH 3.5 with[46] and without[47] a second homochromatography dimension, and reversed-phase column chromatography,[48] also resolve the strands of small fragments.) Why complementary DNA strands separate in gel electrophoresis is not known, but it could be due to a difference in their size, charge, or shape. Since the purine-rich L strand of α DNA moves faster than the pyrimidine-rich H strand,[41] this difference is apparently not molecular weight. One view is that when duplex DNA is dissociated each strand folds on itself into one predominant semi-stable conformer prescribed by its primary structure. This interaction need not be comprised entirely of Watson–Crick base pairs. In this way complementary strands, the same in size but different in nucleotide sequence, might acquire different shapes and be differentially retarded by the gel matrix during electrophoresis. Aside from the mechanism, strand separation can be of great advantage in DNA sequence determination. When restriction fragments are labeled at both 5' or both 3' ends, separation of their strands obviates the need for secondary restriction enzyme cleavage prior to sequence analysis. One can then "sequence right through" what would have been the second cleavage site, and get confirming sequence from the other strand as well.

Originally we described a 1:30-cross-linked, 8% polyacrylamide gel for separating the strands of DNA restriction fragments dissociated and loaded in alkali.[2] While good for separation of strands of less than 200 nucleotides, longer ones often failed to resolve on such gels. Szalay *et al.*[45] have since reported good separation of larger strands on a 1:60-cross-linked 5% gel, and we find this lower gel percentage and cross-linkage of clear benefit. However, in comparing separation of the strands of microgram amounts of a 789 base pair fragment exposed to alkali and to heat in various solvents, we observed inadequate denaturation with "cracking buffer"[45] and superior strand separation after heat denaturation in dimethyl sulfoxide. The denaturation conditions and gel we currently use for resolving complementary strands hundreds of nucleotides long are described in procedure 8.

Elution of DNA from Polyacrylamide Gels.[49] Before it can be sequenced, a singly end-labeled DNA strand or end-piece resolved on one of

[42] C. Tibbets, U. Pettersson, K. Johansson, and L. Phillipson, *J. Virol.* **13**, 370 (1974).
[43] S. J. Flint, P. H. Gallimore, and P. A. Sharp, *J. Mol. Biol.* **96**, 47 (1975).
[44] D. Perlman and J. A. Huberman, *Anal. Biochem.* **83**, 666 (1977).
[45] A. A. Szalay, K. Grohmann, and R. L. Sinsheimer, *Nucleic Acids Res.* **4**, 1569 (1977).
[46] T. Maniatis, A. Jeffrey, and D. G. Kleid, *Proc. Natl. Acad. Sci. U.S.A.* **72**, 1184 (1975).
[47] F. G. Grosveld and J. H. Spencer, *Nucleic Acids Res.* **4**, 2235 (1977).
[48] H. Eshaghpour and D. M. Crothers, *Nucleic Acids Res.* **5**, 13 (1978).
[49] See also H. O. Smith, this volume, Article [46].

the above gels has to be extracted from the gel matrix. A technique which extracts DNA from gel should do so quantitatively and leave the DNA undegraded, concentrated, and free of gel buffer, enzyme inhibitors, and semisoluble polyacrylamide. It should also allow simultaneous processing of tens of gel slices quickly. Because it meets most of these criteria, we prefer the gel elution technique in procedure 9. Its essential steps are crushing the gel, soaking the crushed gel in a salt solution long enough for the DNA to diffuse out, filtering off gel fragments, and precipitating DNA out of the filtrate with ethanol. This crush–soak–filter–precipitate method efficiently recovers DNA fragments ten to several thousand base pairs long from small slices of polyacrylamide gel.

Elution of DNA from crushed gel takes place inside a plastic cone, plugged with siliconized glass wool (the plug retains polyacrylamide fragments in a later step). A 1000-μl (blue) pipettor tip is an appropriate vessel for the elution. Glass wool for the plug should be siliconized so that it does not bind DNA or RNA, and dimethyldichlorosilane-treated glass wool is available commercially from chromatography equipment suppliers (Alltech Associates), or the pipette tip may be plugged with untreated glass wool and then siliconized by centrifuging dimethyldichlorosilane [1% (v/v) in carbon tetrachloride] and several water rinses through it and drying.

For elution, the gel slice is first dropped into the pipette tip and pushed against the plug with a glass rod to crush it, and then mashed against the wall to reduce it to a paste. (Homogenizing the gel in phenol with a tissue grinder is not necessary and should not be done as it releases low molecular weight polyacrylamide which will then copurify with the DNA.) Ammonium acetate (salt) in the gel elution solution promotes diffusion of DNA out of the gel matrix and is more soluble in ethanol than sodium acetate in the following step. Sodium dodecyl sulfate is included to denature any contaminating DNase, and magnesium ions and carrier RNA promote precipitation of dilute DNA from aqueous ethanol. The 10-hr diffusion period is more than adequate and, depending on the gel pore size and the DNA molecular radius, much less time may be required. If the gel is at least 5%(w/v) polyacrylamide and the pipette tip is plugged tightly, the only residue noticeable after glass wool filtration and ethanol precipitation is DNA (RNA).

Procedure 2. [γ-³²P]ATP Synthesis

  10×-concentrated reaction mixture (store at −20° in 100 μl lots)
  500 mM Tris-HCl, pH 8.0
  50 mM Magnesium chloride

    20 mM Reduced glutathione
    10 mM 3-Phosphoglyceric acid
     1 mM EDTA
    0.1 mM NAD$^+$
    0.4 mM ATP

Enzyme mixture (prepare at 0°–4°C): Obtain ammonium sulfate suspensions of rabbit muscle glyceraldehyde-3-phosphate dehydrogenase (Calbiochem 35632, 0.85 units/μl) and yeast 3-phosphoglycerate kinase (Calbiochem 52518, 6.2 units/μl).

    50 μl Glyceraldehyde-3-phosphate dehydrogenase, 50 units
    25 μl 3-Phosphoglycerate kinase, 150 units

Combine the above in an Eppendorf snap-cap tube and mix gently. Centrifuge 5 min (Eppendorf, 4°), and remove supernatant.

    100 μl of 3.2 M ammonium sulfate, 100 mM Tris-HCl, pH 8, 10 mM
    mercaptoethanol, 1 mM EDTA, 0.1 mM NAD$^+$

Resuspend the pellet gently (stir with a disposable Micropet). Centrifuge 5 min (Eppendorf, 4°), and remove supernatant.

    100 μl 3.2 M ammonium sulfate, 100 mM Tris-HCl, pH 8, 10 mM
    mercaptoethanol, 1 mM EDTA, 0.1 mM NAD$^+$

Resuspend the pellet gently (stir with a disposable Micropet). Centrifuge 5 min (Eppendorf, 4°). Remove all the supernatant with a pointed glass capillary.

    25 μl Distilled water

Carefully rinse the walls of the tube and surface of the pellet using a pointed glass capillary, and then withdraw all the liquid.

    75 μl Distilled water

Gently dissolve the pellet (stir with a disposable Micropet), and use within 30 min.

The exchange reaction: Obtain 50 mCi carrier-free, HCl-free ^{32}P-phosphate in 500 μl water in a glass V-vial.

    500 μl ^{32}P-phosphate, 50 mCi, in water (5.5 nmole phosphate)

Place behind radiation shield and remove lead top (but not lead bottom), and the rubber seal.

    50 μl 10×-concentrated reaction mixture (20 nmole ATP)

Insert a 5-μl disposable Micropet to the bottom of the V-vial and stir and gently expel air bubbles to mix. Spot 0.1 μl on a polyethyleneimine (PEI)–cellulose plate (procedure 3).

  1 μl Enzyme mixture (above):

Mix again by expelling air from a 5-μl Micropet inserted to the bottom of the vial. Spot 0.1 μl on the PEI-cellulose plate at 1, 5, 10, 20, and 30 min, mixing the reaction mixture each time before sampling. Develop the plate in 0.75 M potassium phosphate, pH 3.5, and autoradiograph (procedure 3). If exchange of ^{32}P-phosphate into ATP is acceptable, proceed. (If not, spot 0.1 μl on a new PEI-cellulose plate, add another 1 μl enzyme mixture, mix, and again follow the reaction over time by taking aliquots onto the plate.)

  20 μl 100 mM EDTA

Withdraw contents of the vial (575 μl) into a 1-ml plastic pipette (using a mechanical pipetting device), expel to mix, withdraw again, and transfer to a clean conical polypropylene tube. Heat at 90° for 3 min (Temp-Block containing sodium chloride crystals), and cool to room temperature.

  1.4 ml 95% Ethanol

Mix by repeatedly withdrawing into and expelling out of a 1 ml plastic pipette (with mechanical pipetting device). Distribute into clean polycarbonate tubes and store at −20°. To prepare for use in the polynucleotide kinase reaction (procedure 5), transfer an aliquot containing the picomoles of ATP required to a siliconized tube, dry under vacuum (desiccator), and redissolve in 5 μl water.

  The preceding procedure describes a modified Glynn–Chappell exchange synthesis[20] for high specific activity [γ-^{32}P]ATP. In coupled reverse reactions, 3-phosphoglycerate is first phosphorylated with unlabeled ATP by 3-phosphoglycerate kinase, to 1,3-diphosphoglycerate, and this is reduced and dephosphorylated by glyceraldehyde-3-phosphate dehydrogenase to glyceraldehyde 3-phosphate and orthophosphate. Then in the forward reactions this phosphate and added ^{32}P-orthophosphate are reincorporated into ADP to yield [γ-^{32}P]ATP, which is favored by the equilibria. Since orthophosphate is an intermediate in these reactions, the proportion which is radioactive will determine the specific activity of the [γ-^{32}P]ATP.

  Exchangeable phosphate in the reaction mixture includes 20 nmole in the γ position of input ATP, 5.5 nmole in 50 mCi ^{32}P-phosphate, and any orthophosphate present as an impurity in all other components. If the latter is negligible, the theoretical yield at equilibrium is 78% exchange of ^{32}P-phosphate into the gamma position of ATP. The final preparation would then contain 20 μCi/10 pmole/μl [γ-^{32}P]ATP of specific activity 2000 Ci/mmole in 70% ethanol, 1 mM EDTA, and is stable at $-20°$. This [γ-^{32}P]ATP should be taken to dryness under vacuum, to remove the ethanol, and redissolved in water before introducing it into polynucleotide kinase reactions.

A common initial difficulty with the synthesis is low level of exchange of ^{32}P-phosphate into ATP, which decreases both the yield and specific activity of the [γ-^{32}P]ATP. Causes include inactive enzymes, enzyme inhibitors, the oxidation-reduction state of cofactors, insufficient input ATP, and exogenous nonradioactive phosphate. ATP, 3-phosphoglyceric acid, ^{32}P-phosphate, reagents, enzymes, water, and vessels are potential sources of nonradioactive orthophosphate which will decrease the specific activity. The concentration of ATP in a stock solution should be determined spectrophotometrically and the ATP checked for hydrolysis by chromatography on PEI-cellulose (procedure 3) and inspection of the plate under short-wave uv, and then dispensed accurately. Note that "carrier-free ^{32}P-phosphate" is commercially defined as that to which cold phosphate has not been *added*, and is not claimed to have the specific activity of ^{32}P at 100% isotopic enrichment. Consider trace amounts of phosphate in choosing sources for all other reaction mixture components as well, and then prepare them with deionized, distilled water in clean plastic vessels (avoid phosphate-containing detergents). The procedure suggested above includes two ammonium sulfate precipitations of the enzymes to free them of endogenous orthophosphate; as ammonium sulfate itself inhibits the reaction, residual amounts should be rinsed out of the tube after obtaining the final pellet. Although said to be tightly bound to muscle glyceraldehyde-3-phosphate dehydrogenase, NAD$^+$ stimulates reactions containing the Calbiochem enzyme we use and is therefore included in the 10×-concentrated reaction mixture. Reduced glutathione (G—SH) easily oxidizes (G—S—S—G), and the latter may inhibit glyceraldehyde-3-phosphate dehydrogenase[50] and should therefore be replaced rather than supplemented if suspect. In general, problems with enzymes, cofactors, reagents, inhibitors, and kinetics may be systematically evaluated in series of 500 μl mock reactions, each containing 20 nmole ATP, 5.5 nmole unlabeled orthophosphate, 10 μCi ^{32}P-phosphate as

[50] S. F. Velick, this series, Vol. 1 [60].
[51] K. Randerath and E. Randerath, Vol. 21, Part A [40].
[52] A. Bernardi, *Anal. Biochem.* 59, 501 (1974).
[53] R. A. Ludwig and W. C. Summers, *Virology* 68, 360 (1975).

tracer, and all other components the same as suggested for the full-scale preparation, followed by thin-layer chromatography on PEI-cellulose (procedure 3) and autoradiography.

Procedure 3. Identification of ^{32}P-*Labeled Nucleotides by PEI–Cellulose Thin-Layer Chromatography and Autoradiography*

Chromatography plate: Buy or spread a $0.1 \times 200 \times 200$ mm PEI–cellulose thin layer on a plastic support (Macherey-Nagel, Polygram CEL300PEI). Prechromatograph the plate in distilled water until the front reaches the top, air-dry, and cut into eight 50×100 mm rectangular pieces with scissors. Draw a line across one plate, 15 mm from one edge, on the plastic back with a black felt-tip marker. Place the plate on white paper with the cellulose layer up, and spot small aliquots ($0.1–5$ μl) of ^{32}P-labeled samples and uv-absorbing markers at the black origin line, which can be seen through the plate against the white background. Asymmetric spacing between spots aids sample identification later on the autoradiograph. Air dry.

Chromatography tank and solvents: Obtain a wide-mouth glass jar, about 80 mm in diameter and 100 mm high, with a screw-on cover. Cut a 3-mm glass rod to a length equal to the inside diameter of the jar and place it inside the jar, at the bottom. Introduce 15 ml chromatography solvent into the jar. (See tabulation.)

Chromatography solvent for PEI cellulose plate	Compounds separated	Literature references
0.40 M Sodium dihydrogen phosphate, titrated to pH 3.5 with phosphoric acid	DNA ($R_f = 0$) ATP ($R_f = 0.3$) Orthophosphate ($R_f = 0.6$)	This work, results not shown
1.0 M Lithium chloride	Mononucleotides	51
5% (v/v) Acetic acid, titrated to pH 3.7 with pyridine	Mononucleotides	52,53

Chromatography and autoradiography: Place the PEI-cellulose plate in the chromatography jar, with the bottom edge in the solvent, cellulose layer against glass rod and plastic back leaning against wall of jar. Screw on the cover, chromatograph until the solvent front reaches the top, and remove and dry the plate. Expose to inexpensive X-ray film and develop. If exposures of less than 5 sec are required ([γ-^{32}P]ATP synthesis, procedure 2), affix three strips of double-stick tape to the plastic back of plate, stick to palm of hand, and press against different sectors of film for several different second or subsecond exposures.

Procedure 4. Removing DNA Terminal Phosphates with Phosphatase

150 μl Distilled water
25 μl DNA with 5' terminal phosphates
5 μl [5'-^{32}P]DNA (the same fragment as tracer)
20 μl 500 mM Tris-HCl, pH 8.0

Combine the above in a 1.5-ml Eppendorf snap-cap tube and mix. Spot 1 μl on a PEI–cellulose chromatography plate (procedure 3).

1 μl alkaline phosphatase, 0.01–0.1 unit

Mix and heat at 37° (protruding or flush ends) or 60° (recessed ends) for 30 min, spotting 1 μl on PEI–cellulose at 1, 5, 10, 20, and 30 min. Chromotograph the PEI–cellulose plate in 0.75 M potassium phosphate, pH 3.5, and autoradiograph (see procedure 3). If all the ^{32}P has been released from the DNA (sticks at origin) as ^{32}P-phosphate (moves with solvent), proceed. (If not, spot 1 μl on a new PEI–cellulose plate, add more alkaline phosphatase, and again follow the reaction over time by taking aliquots onto the plate.)

2 μl 100 mM EDTA
200 μl Phenol, saturated with 50 mM Tris-HCl, pH 8, 1 mM EDTA

Close the cap and mix well (Vortexer). Centrifuge 1 sec (Eppendorf) to separate the phases. Remove the bottom phase with a Micropet.

200 μl Phenol (above)

Mix, centrifuge 1 sec, and remove the bottom phase, as above.

500 μl Diethyl ether

Mix, centrifuge 1 sec, and remove the top phase.

25 μl 3.0 M Sodium acetate
750 μl 95% Ethanol

Close the cap and mix (invert the tube four times). Chill at −70° for 5 min (Dry Ice–ethanol bath). Centrifuge at 12,000 g for 5 min (Eppendorf, 4°). Remove the supernatant with a Pasteur pipette.

1 ml 95% Ethanol

Centrifuge at 12,000 g 15 sec, and remove supernatant. Place the tube under vacuum for a few minutes (desiccator). The product is 5'-hydroxyl DNA ready for phosphorylation with [γ-^{32}P]ATP and polynucleotide kinase (procedure 5).

Terminal phosphates are removed from DNA restriction fragments with *E. coli* alkaline phosphatase at pH 8.0 prior to 5'-end-labeling with polynucleotide kinase and [γ-^{32}P]ATP. The appropriate amount of phosphatase can be quickly determined in the assay above (PEI–cellulose chromatography and autoradiography, procedure 3) if some of the fragment has already been end-labeled and is available as tracer. After phosphatasing, phenol extractions remove the phosphatase, ether extraction removes residual phenol, and ethanol precipitation removes the buffer and concentrates the DNA.

Procedure 5. Labeling DNA 5' Ends with Polynucleotide Kinase and [γ-^{32}P]ATP

A. Forward reaction for flush or recessed 5' ends: Combine the following in a siliconized 1.5-ml Eppendorf tube.

 5 μl Dephosphorylated DNA, 1-50 pmole 5' ends
 35 μl 20 mM Tris-HCl, pH 9.5, 1 mM spermidine, 0.1 mM EDTA

Mix, heat at 90° for 2 min, quick-chill in ice water, and immediately add the following.

 5 μl 500 mM Tris-HCl, pH 9.5, 100 mM MgCl$_2$, 50 mM dithiothreitol, 50%(v/v) glycerol
 5 μl [γ-^{32}P]ATP, >1000 Ci/mmole, 50 pmole or more
 1 μl T$_4$ polynucleotide kinase, 20 units

Mix and heat at 37° for 15 min.

B. Forward reaction for protruding 5' ends: Combine the following in a siliconized 1.5-ml Eppendorf tube.

 5 μl Dephosphorylated DNA, 1–50 pmole 5' ends
 35 μl Distilled water
 5 μl 500 mM Tris-HCl, pH 7.6, 100 mM MgCl$_2$, 50 mM dithiothreitol, 1 mM spermidine, 1 mM EDTA
 5 μl [γ-^{32}P]ATP, >1000 Ci/mmole, 50 pmole or more
 1 μl T$_4$ polynucleotide kinase, 20 units

Mix and heat at 37°C for 30 min.

C. Exchange reaction for protruding 5' ends[26]: Combine the following in a siliconized 1.5 ml Eppendorf tube.

 5 μl DNA with 5'-terminal phosphates, 1–50 pmole 5' ends
 25 μl distilled water

5 μl 500 mM Imidazole-HCl, pH 6.6, 100 mM MgCl$_2$, 50 mM dithiothreitol, 1 mM spermidine, 1 mM EDTA

3 μl 5 mM Adenosine diphosphate

10 μl [γ-^{32}P]ATP, >1000 Ci/mmole, 100 pmole (preferably more[26])

1 μl T$_4$ polynucleotide kinase, 20 units

Mix and heat at 37°C for 30 min.

A–C. Reaction stop and ethanol precipitation of the end-labeled DNA.

200 μl 2.5 M Ammonium acetate

1 μl tRNA, 1 mg/ml

750 μl 95% ethanol

Close the cap and mix (invert the tube four times). Chill at −70° for 5 min (Dry Ice–ethanol bath). Centrifuge at 12,000 g for 5 min (Eppendorf, 4°). Remove the supernatant with a Pasteur pipette.

250 μl 0.3 M Sodium acetate

Close the cap and redissolve the DNA (Vortexer).

750 μl 95% Ethanol

Invert to mix, chill, centrifuge, and remove supernatant.

1 ml 95% Ethanol

Centrifuge at 12,000 g 15 sec and remove supernatant. Place the tube under vacuum for a few minutes (desiccator). The product is 5′-^{32}P-labeled DNA plus carrier RNA, essentially free of salt and ready for secondary restriction enzyme cleavage (procedure 7) or strand separation (procedure 8).

In the procedure 5A forward reaction, double-stranded DNA with dephosphorylated flush or recessed 5′ ends is first heat-denatured and then rephosphorylated with polynucleotide kinase and [γ-^{32}P]ATP. Spermidine stimulates incorporation[54] and inhibits a nuclease in some polynucleotide kinase preparations.[55] The ATP concentration in the reaction mixture should be at least 1 μM; DNA fragments with recessed 5′ ends may require higher concentrations of ATP[24] if denaturation is not effective. A DNA restriction fragment which has been end-labeled and processed in this manner will almost always undergo complete digestion with a second restriction enzyme, but if it should not the denaturation may be omitted or a renaturation step introduced. The Procedure 5B forward reaction for dephosphorylated double stranded DNA with protruding 5′ ends (Table

---

[54] J. R. Lillehaug and K. Kleppe, *Biochemistry* 14, 1225 (1975).

[55] A. Panet, J. H. van de Sande, P. C. Loewen, H. G. Khorana, A. J. Raae, J. R. Lillehaug, and K. Kleppe, *Biochemistry* 12, 5045 (1973).

I), or single-stranded DNA, differs only in having no denaturation step and Tris-HCl at pH 7.6 instead of pH 9.5 with glycerol. The procedure 5C exchange reaction is essentially that of Berkner and Folk[26] which, by avoiding the phosphatase treatment, is faster for end-labeling restriction fragments, but may be less efficient. Adding ammonium acetate to 2 M stops these reactions, promotes DNA renaturation, and prevents precipitation of protein while enhancing DNA precipitation from ethanol. Carrier RNA coprecipitates the DNA and makes it adhere to the wall of the tube after centrifugation. The two ethanol precipitations remove most but not all of the unincorporated label.

Procedure 6. Labeling DNA 3' Ends with Terminal Transferase and [α-32P]ATP[27]

    5 μl DNA, 1–50 pmole 3' ends
   10 μl Distilled water

Combine the above in a siliconized 1.5-ml Eppendorf snap-cap tube. Mix, heat at 90° for 1 min, and quick-chill in ice water. Centrifuge for a few seconds (Eppendorf).

   2.5 μl 1.0 M Potassium cacodylate, pH 7.6, 10 mM CoCl$_2$, 2 mM dithiothreitol (prepared as described in Roychoudhury and Wu[27])
    5 μl [α-32P]NTP, 300–3000 Ci/mmole, 50 pmole (preferably more)
    1 μl Terminal deoxynucleotidyl transferase, several units

Mix and heat at 37° for 30 min.

    1 μl Terminal transferase, several units

Mix and heat at 37° for 30 min.

    1 μl 100 mM NTP

Mix, heat at 37° for 10 min, and chill in ice

  200 μl 2.5 M Ammonium acetate (0°)
    1 μl tRNA, 1 mg/ml
  750 μl 95% Ethanol

Close the cap and mix (invert the tube four times). Chill at −70° for 5 min (Dry Ice–ethanol bath). Centrifuge at 12,000 g for 5 min (Eppendorf, 4°). Remove the supernatant with a Pasteur pipette.

  250 μl 0.3 M Sodium acetate

Close the cap and redissolve the DNA (Vortexer).

  750 μl 95% Ethanol

Invert to mix, chill, centrifuge, and remove supernatant.

    1 ml 95% Ethanol

Centrifuge at 12,000 g 15 sec and remove supernatant. Place the tube under vacuum for a few minutes (desiccator).

    100 μl 1.0 M Piperdine (freshly diluted)

Close cap and redissolve DNA (Vortexer and manual agitation). Centrifuge for a few seconds (Eppendorf). Place stretchable tape under cap and close tightly. Heat at 90° for 30 min (under weight in water bath). Remove the tape and centrifuge for a few seconds (Eppendorf). Punch holes in the tube cap with a dissecting needle. Freeze the sample, and lyophilize. Redissolve the DNA in 10 μl water, freeze, and lyophilize. Redissolve DNA in 10 μl water, freeze, and lyophilize again. The product is 3′-^{32}P-labeled DNA, essentially free of salt and ready for strand separation (procedure 8). Secondary cleavage (procedure 7) with a restriction enzyme which cuts only double-stranded DNA requires that the DNA be renatured, though this may take place in the labeling and digestion mixtures.

Procedure 7. DNA Fragment Division by Secondary Restriction Enzyme Cleavage and Gel Electrophoresis

    1 μg DNA restriction fragment, ^{32}P labeled at both ends (salt-free pellet in Eppendorf tube)
    20 μl Distilled water

Close cap and redissolve DNA (Vortexer and manual agitation). Centrifuge a few seconds (Eppendorf).

    2 μl 200 mM Tris-HCl, pH 7.4, 50 mM MgCl$_2$, 10 mM dithiothreitol, ±500 mM NaCl
    0.5 μl Restriction enzyme, 1 unit

Mix and spot 0.2 μl on a PEI–cellulose chromatography plate. Heat at 37° for 60 min (water bath). Mix and spot 0.2 μl on the PEI–cellulose plate. Chromatograph plate in 0.75 M sodium phosphate, pH 3.5, and autoradiograph (procedure 3). If most of the label remains at the origin after digestion, proceed.

    5 μl 50%(w/v) Glycerol, 25 mM EDTA, 0.25%(w/v) xylene cyanol, 0.25%(w/v) bromphenol blue

Mix and load on preelectrophoresed (30 min) nondenaturing gel.

5–10%(w/v) Acrylamide
0.17–0.33%(w/v) Methylenebisacrylamide
50 mM Tris-borate, pH 8.3, 1 mM EDTA
Slab gel: 1.5 × 200 × 200 mm
Sample well: 1.5 × 20 mm

Electrophorese at 300–500 V (regulated). Stop electrophoresis when the tracking dyes have moved to positions which indicate adequate resolution of the end-labeled DNA fragments (predetermined empirically). Remove all liquid, one glass plate, and the two spacers from the slab gel, leaving it undisturbed on the other glass plate. Cover the gel with Saran wrap, smoothing it down with a paper towel and taping it tightly to the back of the plate. Spot 0.5 and 5 μl ^{32}P-phosphate-containing red ink (100 μCi/ml) on each corner of the covered slab, and 0.5 μl at the position of each tracking dye, dry, and cover with transparent tape. Expose the gel to two stacked sheets of X-ray film, and develop. Cut rectangles encompassing DNA band images out of one film. Match this template autoradiograph to the slab gel by superimposing red ink spots and their images on the film. Trace through holes in film onto Saran wrap covering the gel. Cut the slab gel along the tracings with a thin razor blade. Lift out gel slices with a spatula, and elute DNA (procedure 9).

A double-stranded DNA restriction fragment labeled at both 5' or both 3' ends (procedure 5 or 6) is cleaved by a second restriction enzyme into two or more subfragments, and the labeled end pieces are separated by electrophoresis on a polyacrylamide gel. Restriction enzyme preparations sometimes contain a phosphatase which may hydrolyze 5' terminal ^{32}P-phosphates or a 3'→5'-exonuclease which may remove 3'-terminal ^{32}P-nucleotides. Release of labeled phosphate or nucleotides from DNA can be detected by PEI–cellulose chromatography and autoradiography (procedure 3). If present, phosphatase can usually be inhibited by 1 mM orthophosphate in the restriction enzyme buffer.

*Procedure 8. DNA Strand Separation by Denaturation and Gel
   Electrophoresis*

   1 μg DNA restriction fragment, ^{32}P-labeled at both ends (salt-free
   pellet in Eppendorf tube)
   40 μl 30%(v/v) dimethyl sulfoxide, 1 mM EDTA, 0.05% (w/v) xylene
   cyanol, 0.05%(w/v) bromphenol blue

Close cap and redissolve DNA (Vortexer and manual agitation). Centrifuge a few seconds (Eppendorf). Heat at 90° for 2 min and quick-chill in ice water. Immediately load on a preelectrophoresed strand separation gel.

5%(w/v) Acrylamide
0.1%(w/v) Methylenebisacrylamide
50 mM Tris-borate, pH 8.3, 1 mM EDTA
Slab gel: 1.5 × 200 × 400 mm
Sample well: 1.5 × 20 mm

Electrophorese at 300 V (8 V/cm) or less (gel must not heat up). Stop electrophoresis when the tracking dyes have moved to positions which indicate adequate resolution of the complementary DNA strands (predetermined empirically). Remove all liquid, one glass plate, and the two spacers from the slab gel, leaving it undisturbed on the other glass plate. Cover the gel with Saran wrap, smoothing it down with a paper towel and taping it tightly to the back of the plate. Spot 0.5 and 5 μl ^{32}P-phosphate-containing red ink (100 μCi/ml) on each corner of the covered slab, and 0.5 μl at the position of each tracking dye, dry, and cover with transparent tape. Expose two stacked sheets of X-ray film to the gel and develop. Cut rectangles encompassing DNA band images out of one film. Match this template autoradiograph to the slab gel by superimposing red ink spots and their images on the film. Trace through holes in film onto Saran wrap covering the gel. Cut the slab gel along the tracings with a razor blade. Lift out gel slices with a spatula, and elute DNA (procedure 9).

Dilute, salt-free DNA is heated in a denaturant, rapidly cooled, immediately layered on a preelectrophoresed (30 min) large-pore gel, and then electrophoresed in such a way that the gel does not heat up. Thicker slab gels should be used for larger amounts of DNA, keeping in mind that DNA weight per DNA fragments per slot cross-sectional area is a relevant overloading parameter. Electrophoresis may be terminated when the faster-moving, double-stranded form of the DNA has traversed most of the gel (to reveal the extent of renaturation), or when it has run off and the slower-moving single strands have traversed most of the gel (to maximize the distance between them).

Procedure 9. Elution of DNA (or RNA) from Polyacrylamide Gel

Plug 1000-μl pipette tip tightly with siliconized glass wool. Seal the apex of the tip with a small flame. Drop a small gel slice into the plugged and sealed tip. Crush the gel with a 3-mm diameter siliconized glass rod. Hand grind the crushed gel to a paste with the rod.

0.6 ml 500 mM Ammonium acetate, 10 mM magnesium acetate, 1 mM EDTA, 0.1%(w/v) sodium dodecyl sulfate, 10 μg/ml tRNA (Omit the carrier RNA if labeling is to follow gel elution.)

Stir the paste into this solution with a Micropet. Seal the top of the tip with two layers of Parafilm. Heat at 37° for 10 hr (air incubator). Remove the Parafilm and cut off the flame-sealed point. Set the tip in a siliconized 10 × 75 mm centrifuge tube. Centrifuge at 3000 rpm for 2 min (clinical centrifuge). (If the DNA is ^{32}P-labeled, its movement from the tip to the tube below can be followed with a Geiger counter.)

    0.2 ml 500 mM Ammonium acetate, 10 mM magnesium acetate, 1 mM EDTA, 0.1%(w/v) sodium dodecyl sulfate, 10 μg/ml tRNA (Omit the carrier RNA if labeling is to follow gel elution.)

Centrifuge at 3000 rpm for 2 min (clinical centrifuge). Discard the tip and add to the eluate in the tube

    2 ml 95% Ethanol

Seal the tube with Parafilm and invert four times to mix. Chill at $-70°C$ for 5 min (Dry Ice–ethanol bath). Centrifuge at 10,000 g for 15 min (angle rotor, 4°). Remove the supernatant with a Pasteur pipette.

    0.5 ml 300 mM Sodium acetate

Redissolve the DNA (Vortexer).

    1.5 ml 95% Ethanol

Invert to mix, chill, centrifuge, and remove supernatant.

    25 μl Distilled water

Redissolve the DNA (roll tube horizontally, and Vortex).

    1 ml 95% Ethanol

Invert to mix, chill, centrifuge, and remove supernatant. Place the tube under vacuum for a few minutes (desiccator). The DNA–RNA residue appears as a pellet, a film, or both, is essentially salt-free, and can be recovered in a small volume of water (20 μl).

Base-Specific Chemical Cleavage Reactions

    Eight two- or three-step chemical procedures which will partially cleave DNA at one or two of the four bases are summarized in Table III.[2,56] Reactions R1–R4 are those we first found useful for sequencing DNA.[2] The dimethyl sulfate reactions R1(G > A) and R2(A > G) distinguish adenines from guanines by differences in their rates of methyla-

---

[56] T. Friedmann and D. M. Brown, *Nucleic Acids Res.* 5, 615 (1978).

TABLE III

BASE-SPECIFIC CLEAVAGE REACTIONS FOR SEQUENCING DNA[a]

Reaction	Cleavage	Base modification	Modified base displacement	Strand scission	Experimental details
R1	G > A	Dimethyl sulfate	Heat at pH 7	Sodium hydroxide	Reference 2
R2	A > G	Dimethyl sulfate	Acid	Sodium hydroxide	Reference 2
R3	C + T	Hydrazine	Piperidine	Piperidine	Reference 2
R4	C	Hydrazine + salt	Piperidine	Piperidine	Reference 2
R5	G	Dimethyl sulfate	Piperidine	Piperidine	Procedure 10
R6	G + A	Acid	Acid	Piperidine	Procedure 11
R7	C + T	Hydrazine	Piperidine	Piperidine	Procedure 12
R8	C	Hydrazine + salt	Piperidine	Piperidine	Procedure 13
R9	A > C	Sodium hydroxide	Piperidine	Piperidine	Procedure 14
R10	G > A	Dimethyl sulfate	Heat at pH 7	Piperidine	Procedure 15
R11	G	Methylene blue	Piperidine	Piperidine	Reference 56
R12	T	Osmium tetroxide	Piperidine	Piperidine	Reference 56

[a] This table is a summary of reactions which have been successfully used to partially cleave end-labeled DNA at one or two of the four bases. Reactions R1–R4 are those we originally recommended,[2] while R5–R8 are new versions, which are faster and provide salt-free cleavage products for thin sequencing gels. Reaction R9 is a viable alternative to R6 for adenine cleavage. Friedmann and Brown[56] have demonstrated piperidine cleavage at guanines after uv irradiation in the presence of methylene blue or rose bengal (R11), and at thymines after reaction with osmium tetroxide (R12).

TABLE IV

SUMMARY OF FOUR BASE-SPECIFIC REACTIONS FOR SEQUENCING END-LABELED DNA[a]

G	G + A	T + C	C
See procedure 10	See procedure 11	See procedure 12	See procedure 13
200 μl buffer	10 μl water	10 μl water	15 μl 5 M NaCl
5 μl [^{32}P]DNA	10 μl [^{32}P]DNA	10 μl [^{32}P]DNA	5 μl [^{32}P]DNA
1 μl DMS	2 μl pip for, pH 2	30 μl hydrazine	30 μl hydrazine
206 μl	22 μl	50 μl	50 μl
20°, 10 ± 5 min	20°, 60 ± 20 min	20°, 10 ± 5 min	20°, 10 ± 5 min
50 μl DMS Stop	Freeze	20 μl HZ Stop	200 μl HZ Stop
750 μl ethanol	Lyophilize	750 μl ethanol	750 μl ethanol
Chill 5 min	—	Chill 5 min	Chill 5 min
Centrifuge 5 min	20 μl water	Centrifuge 5 min	Centrifuge 5 min
Pellet	Lyophilize	Pellet	Pellet
250 μl 0.3 M NaAc	—	250 μl 0.3 M NaAc	250 μl 0.3 M NaAc
750 μl ethanol	—	750 μl ethanol	750 μl ethanol
Chill 5 min	—	Chill 5 min	Chill 5 min
Centrifuge 5 min	—	Centrifuge 5 min	Centrifuge 5 min
Pellet	—	Pellet	Pellet
Ethanol rinse	—	Ethanol rinse	Ethanol rinse
Vacuum dry	—	Vacuum dry	Vacuum dry
100 μl 1.0 M piperidine	100 μl 1.0 M piperidine	100 μl 1.0 M piperidine	100 μl 1.0 M piperidine
90°, 30 min	90°, 30 min	90°, 30 min	90°, 30 min
Lyophilize	Lyophilize	Lyophilize	Lyophilize
10 μl water	10 μl water	10 μl water	10 μl water
Lyophilize	Lyophilize	Lyophilize	Lyophilize
10 μl water	10 μl water	10 μl water	10 μl water
Lyophilize	Lyophilize	Lyophilize	Lyophilize
10 μl formamide–NaOH–dyes	10 μl formamide–NaOH–dyes	10 μl formamide–NaOH–dyes	10 μl formamide–NaOH–dyes
90°, 1 min	90°, 1 min	90°, 1 min	90°, 1 min
Quick chill	Quick chill	Quick chill	Quick chill
Load on gel	Load on gel	Load on gel	Load on gel

[a] The abbreviated protocols aligned above indicate how four base-specific cleavage reactions are coordinated to sequence one end-labeled DNA fragment. They are given only as a chronological guide, and initially the more detailed descriptions in Procedures 10–13 should be followed. DMS, dimethyl sulfate; pip for, piperidine formate; HZ, hydrazine.

tion and depurination, bands from adenine cleavage being stronger in the first and bands from guanine cleavage being stronger in the second. Experience has since revealed that adenines and guanines are best identified when bands from one are dependably present with or absent from those of the other. Hence reactions R5(G) and R6(G + A) (Table III) have replaced R1(G > A) and R2(A > G) for sequencing purines. These new reactions are also faster and easier to perform because they involve two rather than three chemical steps. The hydrazine reactions R7(C + T) and R8(C) are much like the original R3 and R4 for pyrimidines, but employ less carrier DNA and RNA and no magnesium acetate. Note that twice as much end-labeled DNA is introduced into two-base-specific reactions (G + A and C + T) as is put into one-base-specific reactions (G and C). Current sequencing practice based on our original account[2] should be compared in detail with the new procedures (10–15) and modified accordingly. The four contemporary cleavage schemes R5–R8 (G, G + A, C + T, C) all yield dry, salt-free cleavage products which can be dissolved in a small volume of formamide for loading on thin sequencing gels. We recommend them as a set of four DNA cleavages for sequencing the four DNA bases, provide detailed protocols for conducting the relevant reactions in procedures 10–13, and summarize all four in Table IV.

Reagents. Before partially cleaving end-labeled DNA with these base-specific reactions, give some consideration to the safe storage and safe handling of sequencing reagents. Both dimethyl sulfate and hydrazine are poisonous and volatile, and hydrazine is flammable. Both should therefore be stored and dispensed in a fume hood, using plastic or latex gloves. Discard waste (supernatants after ethanol precipitation of sequencing reactions) and disposable transfer pipettes into solutions of inactivating reagents, 5 *M* sodium hydroxide for dimethyl sulfate, and 3 *M* ferric chloride for hydrazine. (Large amounts of concentrated dimethyl sulfate or hydrazine should not be poured into these solutions.)

Some of the sequencing reagents are labile. Anhydrous dimethyl sulfate may absorb moisture from the air and hydrolyze to methanol and sulfuric acid; hydrazine can oxidize to diimine; and piperidine may also undergo oxidation. To avoid side reactions with decomposition and oxidation products which may produce artifacts in the cleavage patterns, reagents of high purity should be obtained and then kept sealed from the atmosphere. Each reagent should be stored in two lots: most of it in the original bottle, tightly capped, and some of it (1–2 ml) as a working solution in a small, screw-capped glass tube. When the latter portion works poorly, it can then be replaced immediately with a fresh aliquot. Hydrazine which has undergone prolonged or repeated exposure to air, especially, gives anomalous cleavage in that the reaction with thymine is not salt-suppressed and the cleavage products may blur. The hydrazine working solution should be replaced every day. Since the DNA is heated at 90° at various stages, other reagents and salts to which it is exposed prior to and during these high temperature reactions should be pure. These include water, piperdine, sodium acetate, magnesium acetate, EDTA, ethanol, sodium cacodylate, magnesium chloride, sodium chloride, and sodium hydroxide, and commercial preparations we have used are listed at the end of this article. Carrier DNA should be sheared, deproteinized, and adjusted to a concentration of 1 mg/ml. Dissolve the DNA (commercially available calf thymus DNA is suitable) in buffer–EDTA, sonicate extensively, extract three times with redistilled phenol, extract two times with ether, blow away residual ether with nitrogen, and dialyze against distilled water. Measure the DNA concentration (A_{260}) and dilute with water to 1 mg/ml.

Reaction Vessels. All the initial base-modification reactions are conveniently done in 1.5-ml Eppendorf conical polypropylene tubes with attached snap-caps. These plastic tubes should be treated with a 1%(v/v) solution of dimethyldichlorosilane in carbon tetrachloride (do not use benzene or toluene with plastic), and rinsed with distilled water and dried to provide a more hydrophobic inner surface. After stopping these reactions, the modified DNA can be ethanol-precipitated and sedimented in these same tubes in the Eppendorf microcentrifuge [which achieves 15,000 rpm (12,000 g) within 20 sec]. In general, the use of small volumes, a −70° Dry Ice–alcohol bath, and a high-speed microcentrifuge makes it possible to quickly remove excess reagent and to quantitatively recover polymeric products.

Base Modification Reactions. The DNA sequencing chemistry begins with a base modification reaction, the extent of which determines the degree of DNA cleavage in a subsequent phosphate-elimination reaction (Fig. 1). How many bases are modified on each fragment depends on the concentration of dimethyl sulfate (G), acid (G + A), hydrazine (C + T), or alkali (A > C) to which the DNA is exposed, and the temperature and duration of the reaction. Each such reaction could be controlled by heating the reagent and DNA separately to the appropriate reaction temperature, starting it by mixing the two, and then stopping it at the appropriate time by rapidly consuming or removing excess reagent. Manipulations of this kind are not practical when sequencing several DNA fragments simultaneously, using three different base-specific reagents per fragment. For speed and convenience we therefore use temperature shifts and dilution to control the base modification reactions. How long these tubes remain at 20°, the base modification reaction time, will ultimately determine the range of sizes of the DNA cleavage products after the final reaction.

There are two considerations in choosing a reaction time. (a) For a short fragment or a short-range sequencing objective, beginning from a labeled end, over how many specific partial degradation products should the label be distributed? The first 5? The first 10? (b) For the more common long range sequencing objective, this question becomes, what is the resolving power of the gel electrophoretic system, the limit of its ability to separate fragments of chain length n from $n + 1$ nucleotides? This is about 170 nucleotides for 1.5×400 mm 20% polyacrylamide slab gels and at least 250 nucleotides for 0.5×400 mm 10% gels.

That a given DNA fragment is very large is of no concern in choosing appropriate reaction kinetics. If it is, for instance, 5000 nucleotides long, all that matters is how much of it (a) needs to be or (b) can be sequenced from its one labeled end. Only one or the other of these criteria dictate the extent to which bases should be modified in sequencing reactions. Once the sequencing target is so defined, the goal is to react one base within that region in each labeled molecule. Since any one base in the DNA for which a given reagent is specific is as likely to react as any other, these one-hit reactions evenly distribute the radioactive label among cleavage products across the region of interest. The practice of conducting several identical reactions for different times and pooling the products, common with nucleases, is inappropriate and wholly unnecessary to achieve uniform cleavage distribution with these chemical reactions.

Times and temperatures for dimethyl sulfate, acid, hydrazine, and alkali reactions appropriate for sequencing 250 nucleotides from a labeled DNA end are given in the cleavage protocols, procedures 10–15. They can be adjusted if too much or no material is consistently left unreacted at the top of any of the four sequence ladders. Generally, to obtain greater or lesser DNA cleavage, increase or decrease the time or temperature of base modification, but not the reagent concentration. Hydrazine $(17–18\,M)$ and alkali $(1.2\,N)$ concentrations should not be varied because they have been established to maximize base specificity and minimize side reactions. Decreasing the concentration of hydrazine is to be especially discouraged; as a base it establishes an appropriate reaction pH in addition to reacting directly with pyrimidine rings in the DNA.

After a base modification reaction of the appropriate duration, cold (0°) sodium acetate and 3 volumes of cold (0°) ethanol are added to chill and dilute the reaction mixture and to precipitate the DNA. A short, high-speed spin (15,000 rpm for 5 min) quantitatively pellets the DNA, leaving most unreacted dimethyl sulfate or hydrazine in the supernatant, which is discarded. Then a second precipitation from sodium acetate–ethanol washes reagent residue out of the DNA, a final ethanol rinse removes sodium acetate, and evacuation for a few minutes evaporates residual ethanol.

Strand Scission Reactions. At this stage DNA which has been reacted with dimethyl sulfate contains a few methylated purine bases; with acid, missing purine bases; or with hydrazine, ring-opened pyridmidine bases. Piperidine will now open the 7-methylguanines (Fig. 2), displace all ring-opened bases from sugars (Figs. 2 and 3), and catalyze β-elimination of phosphates from the empty sugars to cleave the DNA. Aqueous piperidine for this reaction should be diluted from the concentrated reagent each time it is used, and this requires some attention as the amine is retained by pipette walls. For a 1.0 M solution, transfer 100 μl piperidine (free base) into 0.9 ml distilled water, rinse the Micropet repeatedly, and mix well by inverting the tube several times. All the above reactions can now be done at once by dissolving the DNA in diluted piperidine and heating at 90°. Two kinds of vessel have been used for these high-temperature reactions with the volatile amine base, a glass capillary and an Eppendorf microcentrifuge tube.

Originally[2] we performed all piperidine reactions in sealed capillaries as follows. The DNA, pelleted, rinsed, and dried in the original Eppendorf microcentrifuge tube, was dissolved in 25 μl 0.5 M piperidine and drawn up into a pointed glass capillary tube. The capillary was then closed at both ends by melting with a flame and submerged in 90° water for 30 min. After this reaction, the capillary was opened and its contents returned to the Eppendorf tube for the next step. Vaporized piperidine cannot escape from these flame-sealed capillaries, and they provide for a relatively small vapor space around the liquid. All ring-opening, displacement, and strand scission reactions go reliably to completion when done this way. Nonetheless, transferring many DNA samples into and out of capillaries involves manipulations and potential losses that might be avoided.

The high-temperature piperidine reactions can be done in the original Eppendorf microcentrifuge tubes, with certain modifications and precautions. Performing piperidine cleavage right in the Eppendorf tubes means the DNA never leaves these vessels from the time it goes in for base modification until it comes out for loading on a sequencing gel. This eliminates capillary manipulations. However, it can also introduce a new problem, and one should watch for its symptoms. Piperidine is volatile and can escape through leaks around the tube cap. When piperidine escapes, its concentration in the DNA solution dwindles; the reaction with hydrazine-modified nucleosides then may not go to completion, and bands in pyrimidine sequence ladders will be smeared. The following arrangement abates this problem. The DNA, pelleted, rinsed, and dried in an

Eppendorf tube, is first dissolved in 100 μl 1.0 M piperidine. A single or double layer of conformable tape (Teflon or polyvinyl) is stuck to the underside of the tube cap, which is then closed. The tape serves as a gasket between the cap and rim of the tube into which it seats. This sealed tube is finally set in a rack in a 90° water bath, and a heavy weight is placed on it so that pressure inside cannot pop the cap (that it will otherwise pop open indicates a good seal between the cap and rim). After 30 min at 90°, the top of the tube is punctured and three lyophilizations remove the piperidine, the last two of which require no more than 15 min each. This leaves four salt-free portions of the original end-labeled DNA, one cleaved at guanines (G), one at both guanines and adenines (G + A), another at cytosines and thymines (C + T), and the fourth at cytosines (C). When these four nested sets of cleavage products are dissolved in a formamide–NaOH–electrophoresis dyes solution and heat-denatured, they are ready to be electrophoresed through a denaturing polyacrylamide sequencing gel.

Procedure 10. Limited DNA Cleavage at Guanines (G)

> 200 μl 50 mM Sodium cacodylate, pH 8.0, 1 mM EDTA
> 5 μl End-labeled DNA, in water

Combine the above in a 1.5-ml Eppendorf snap-cap tube. Mix (Vortexer), chill to 0° in ice, and add

> 1 μl Dimethyl sulfate, reagent grade

Close the cap on the tube and mix (Vortexer). Heat at 20° for 10 ± 5 min.

> 50 μl 1.5 M Sodium acetate, pH 7.0, 1.0 M mercaptoethanol, 100
> μg/ml tRNA (0°)
> 750 μl 95% Ethanol (0°)

Close the cap and mix well (invert the tube four times). Chill at −70° for 5 min (Dry Ice–ethanol bath). Centrifuge at 12,000 g for 5 min (Eppendorf, 4°). Remove the supernatant with a Pasteur pipette and transfer to a dimethyl sulfate waste bottle containing 5 M sodium hydroxide.

> 250 μl 0.3 M Sodium acetate (0°)

Close the cap and redissolve the DNA (Vortexer).

> 750 μl 95% Ethanol (0°)

Invert to mix, chill, centrifuge, and remove supernatant.

1 ml 95% Ethanol

Centrifuge at 12,000 g 15 seconds and remove supernatant. Place the tube under vacuum for a few minutes (desiccator).

100 μl 1.0 M piperidine (freshly diluted)

Close cap and redissolve DNA (Vortexer and manual agitation). Centrifuge for a few seconds (Eppendorf). Place conformable tape under cap and close tightly. Heat at 90° for 30 min (under weight in water bath). Remove the tape and centrifuge for a few seconds (Eppendorf). Punch holes in the tube cap with a dissecting needle. Freeze the sample, and lyophilize. Redissolve the DNA in 10 μl water, freeze, and lyophilize. Redissolve DNA in 10 μl water, freeze, and lyophilize again.

10 μl 80%(v/v) formamide, 10 mM NaOH, 1 mM EDTA, 0.1%(w/v) xylene cyanol, 0.1%(w/v) bromphenol blue

Close cap and redissolve DNA (Vortexer and manual agitation). Centrifuge for a few seconds (Eppendorf). Heat at 90° for 1 min and quick chill in ice water. Load on sequencing gel(s) immediately (procedure 18).

Procedure 11. Limited DNA Cleavage at Guanines and Adenines (G + A)

10 μl Distilled water
10 μl End-labeled DNA, in water

Combine the above in a 1.5-ml Eppendorf snap-cap tube. Mix, and then add

2 μl 1.0 M piperidine formate, pH 2.0 [4%(v/v) formic acid, adjusted to pH 2.0 with piperidine]

Mix, and heat at 20° for 60 ± 20 min. Punch holes in the tube cap with a dissecting needle. Freeze the sample, and lyophilize. Redissolve the DNA in 20 μl water, freeze, and lyophilize again.

100 μl 1.0 M Piperidine (freshly diluted)

Close cap and redissolve DNA (Vortexer and manual agitation). Centrifuge for a few seconds (Eppendorf). Place conformable tape under cap and close tightly. Heat at 90° for 30 min (under weight in water bath). Remove the tape and centrifuge for a few seconds (Eppendorf). Freeze the sample, and lyophilize. Redissolve the DNA in 10 μl water, freeze, and lyophilize. Redissolve DNA in 10 μl water, freeze, and lyophilize again.

> 10 μl 80%(v/v) formamide, 10 mM NaOH, 1 mM EDTA, 0.1%(w/v) xylene cyanol, 0.1%(w/v) bromphenol blue

Close cap and redissolve DNA (Vortexer and manual agitation). Centrifuge for a few seconds (Eppendorf). Heat at 90° for 1 min and quick chill in ice water. Load on sequencing gel(s) immediately (procedure 18).

Procedures 12 and 13. Limited DNA Cleavage at Pyrimidines

12. Cytosines and thymines (C+T)	13. Cytosines (C)
10 μl Distilled water 10 μl End-labeled DNA	15 μl 5 M Sodium chloride 5 μl End-labeled DNA

Combine the above in a 1.5-ml Eppendorf snap-cap tube. Mix, and then add

> 30 μl Hydrazine, 95%, reagent grade

Close cap on tube and mix gently (manual agitation). Heat at 20° for 10 ± 5 min.

> 200 μl 0.3 M Sodium acetate, 0.1 mM EDTA, 25 μg/ml tRNA (0°)
> 750 μl 95% Ethanol (0°C)

Close the cap and mix well (invert the tube four times). Chill at −70° for 5 min (Dry Ice–ethanol bath). Centrifuge at 12,000 g for 5 min (Eppendorf, 4°). Remove the supernatant with a Pasteur pipette and transfer to a hydrazine waste bottle containing 2 M ferric chloride.

> 250 μl 0.3 M sodium acetate (0°)

Close the cap and redissolve the DNA (Vortexer).

> 750 μl 95% Ethanol (0°)

Invert to mix, chill, centrifuge, and remove supernatant.

> 1 ml 95% Ethanol

Centrifuge at 12,000 g 15 sec and remove supernatant. Place the tube under vacuum for a few minutes (desiccator).

100 μl 1.0 M Piperidine (freshly diluted)

Close cap and redissolve DNA (Vortexer and manual agitation). Place conformable tape under cap and close tightly. Centrifuge for a few seconds (Eppendorf). Heat at 90° for 30 min (under weight in water bath). Remove the tape and centrifuge for a few seconds (Eppendorf). Punch holes in the tube cap with a dissecting needle. Freeze the sample, and lyophilize. Redissolve the DNA in 10 μl water, freeze, and lyophilize. Redissolve DNA in 10 μl water, freeze, and lyophilize again.

10 μl 80%(v/v) formamide, 10 mM NaOH, 1 mM EDTA, 0.1%(w/v) xylene cyanol, 0.1%(w/v) bromphenol blue

Close cap and redissolve DNA (Vortexer and manual agitation). Centrifuge for a few seconds (Eppendorf). Heat at 90° for 1 min and quick chill in ice water. Load on sequencing gel(s) immediately (procedure 18).

Procedure 14. Limited DNA Cleavage at Adenines and Cytosines (A > C)

100 μl 1.2 N Sodium hydroxide, 1 mM EDTA
  1 μl Sonicated carrier DNA, 1 mg/ml
  5 μl End-labeled DNA, in water

Combine the above in a 1.5-ml Eppendorf snap-cap tube and mix. Place conformable tape under cap and close tightly. Heat at 90° for 10 ± 5 min (under weight in water bath).

150 μl 1 N Acetic acid
  5 μl tRNA, 1 mg/ml
750 μl 95% Ethanol

Close the cap and mix well (invert the tube four times). Chill at −70° for 5 min (Dry Ice–ethanol bath). Centrifuge at 12,000 g for 5 min (Eppendorf, 4°). Remove the supernatant with a Pasteur pipette.

1 ml 95% Ethanol

Centrifuge at 12,000 g 15 sec and remove supernatant. Place the tube under vacuum for a few minutes (desiccator).

100 μl 1.0 M Piperidine (freshly diluted)

Close cap and redissolve DNA (Vortexer and manual agitation). Centrifuge for a few seconds (Eppendorf). Place conformable tape under cap and close tightly. Heat at 90° for 30 min (under weight in water bath). Remove the tape and centrifuge for a few seconds (Eppendorf). Punch holes in the tube cap with a dissecting needle. Freeze the sample, and lyophilize. Redissolve the DNA in 10 μl water, freeze, and lyophilize.

Redissolve DNA in 10 μl water, freeze, and lyophilize again.

 10 μl 80%(v/v) Deionized formamide, 10 mM NaOH, 1 mM EDTA,
   0.1%(w/v) xylene cyanol, 0.1%(w/v) bromphenol blue

Close cap and redissolve DNA (Vortexer and manual agitation). Centrifuge for a few seconds (Eppendorf). Heat at 90° for 1 min and quick chill in ice water. Load on sequencing gel(s) immediately (procedure 18).

Procedure 15. Limited DNA Cleavage at Guanines and Adenines (G > A)

 200 μl 50 mM Sodium cacodylate, pH 8.0, 1 mM EDTA
  1 μl Sonicated carrier DNA, 1 mg/ml in water
  5 μl End-labeled DNA, in water

Combine the above in a 1.5-ml Eppendorf snap-cap tube. Mix (Vortexer), and then add

 1 μl Dimethyl sulfate, reagent grade

Close the cap on the tube and mix (Vortexer). Heat at 20° for 10 ± 5 min.

 50 μl 1.5 M Sodium acetate, pH 7.0, 1.0 M mercaptoethanol,
   100 μg/ml tRNA (0°)
 750 μl 95% Ethanol (0°)

Close the cap and mix well (invert the tube four times). Chill at −70° for 5 min (Dry Ice–ethanol bath). Centrifuge at 12,000 g for 5 min (Eppendorf, 4°). Remove the supernatant with a Pasteur pipette and transfer to a dimethyl sulfate waste bottle containing 5 M sodium hydroxide.

 250 μl 0.3 M Sodium acetate (0°)

Close the cap and redissolve the DNA (Vortexer)

 750 μl 95% Ethanol (0°)

Invert to mix, chill, centrifuge, and remove supernatant.

 1 ml 95% Ethanol

Centrifuge at 12,000 g 15 sec and remove supernatant. Place the tube under vacuum for a few minutes (desiccator).

 100 μl 20 mM Ammonium acetate, pH 7.0, 0.1 mM EDTA

Close cap and redissolve DNA (Vortexer and manual agitation). Centrifuge for a few seconds (Eppendorf). Place conformable tape under cap and close tightly. Heat at 90° for 15 min (under weight in water bath). Centrifuge for a few seconds (Eppendorf).

10 µl 10 M piperidine

Place conformable tape under cap, close tightly, and mix. Heat at 90° for 30 min (under weight in water bath). Remove the tape and centrifuge for a few seconds (Eppendorf). Punch holes in the tube cap with a dissecting needle. Freeze the sample, and lyophilize. Redissolve the DNA in 10 µl water, freeze, and lyophilize again.

10 µl 80%(v/v) formamide, 10 mM NaOH, 1 mM EDTA, 0.1%(w/v) xylene cyanol, 0.1%(w/v) bromphenol blue

Close cap and redissolve DNA (Vortexer and manual agitation). Centrifuge for a few seconds (Eppendorf). Heat at 90° for 1 min and quick-chill in ice water. Load on sequencing gel(s) immediately (procedure 18).

Sequencing Gels

In our limited cleavage sequencing technique[2] and the interrupted resynthesis techniques of Sanger, Nicklen, and Coulson,[57-59] gel electrophoresis assorts nested sets of small polynucleotides by size.[60] Labeled DNA fragments in both the partially degraded and partially synthesized sets have one end in common and another end which varies in length. Each fragment in the array contains all of the nucleotides in the next smaller fragment, plus one more at the variable end. Thus neighboring cleavages or synthetic interruptions produce two DNA fragments which differ only by a small and discrete increment of charge and mass, one of the four mononucleotides. When electrophoresed through polyacrylamide gel, the larger fragment moves slower than the smaller one because of its slightly greater noncovalent interaction with the gel matrix. Sequencing gels in current use resolve these nested DNA strands of n and $n + 1$ nucleotides over the range of one to several hundred nucleotides.

Before pouring, running, and autoradiographing sequencing gels, it is useful to know what influences their resolution of DNA strands which differ in length by only a nucleotide. As it is the images of these fragments on X-ray film which must be distinguishable, as bands, two parameters define the resolution: the band thickness and the center-to-center band distance. When DNA fragments stack before entering the gel, encounter

[57] A. J. H. Smith, this volume, Article [58].
[58] F. Sanger, S. Nicklen, and A. R. Coulson, *Proc. Natl. Acad. Sci. U.S.A.* **74**, 5463 (1977).
[59] F. Sanger and A. R. Coulson, *J. Mol. Biol.* **94**, 441 (1975).
[60] See also T. Maniatis and A. Efstratiadis, this volume, Article [38].

no ionic or matrix discontinuities while moving, do not diffuse much in the gel, and radioactive emission from them does not scatter much before striking the X-ray film, the bands on the film will be thin (sharp). As for the distance from one band to the next, it will be determined mostly by how far the two corresponding fragments have been driven through the retarding matrix of the gel. However, the distance between two RNA's in gels of different concentrations has been observed to pass through a maximum,[61] suggesting that polyacrylamide density may influence polynucleotide separation independently of distance moved.

Gel sequencing methods have traditionally employed versions of a pH 8.3 polyacrylamide gel described by Peacock and Dingman[62] and adapted for small single-stranded DNA molecules by Maniatis *et al.*[63] This gel is polymerized from a solution containing 5–20%(w/v) acrylamide, 5%(w/w) of which is N,N'-methylenebisacrylamide, 7 M urea, 90 mM Tris-borate, pH 8.3, 2.5 mM EDTA, 0.1%(w/v) ammonium persulfate, and TEMED catalyst. We originally suggested[2] a 20% polyacrylamide version of this gel, but now prefer an 8% gel for most sequencing runs. The polymerization mixture we use for 8% gels contains 7.6%(w/v) acrylamide, 0.4%(w/v) bisacrylamide, 50%(w/v) urea (8.3 M), 100 mM Tris-borate, pH 8.3, 2 mM EDTA, 0.07%(w/v) ammonium persulfate, and TEMED catalyst (procedures 16 and 17). This solution is injected into a 0.3 × 200 × 400 mm mold to form a gel slab one-third as thick as those we used previously.[2] Sanger and Coulson[64] innovated thin DNA sequencing gels, and we have adopted and routinely use them for chemical cleavage sequencing. Our experience with 0.3-mm thin gels indicates three advantages: they can be run at twice the voltage (faster) without generating heat sufficient to break the glass plates which enclose them, they produce sharper bands on films by decreasing scatter during autoradiography, and they reduce acrylamide consumption. Procedure 17 details how to prepare one thin 8% sequencing gel with just three vessels: a sidearm flask, a syringe, and the gel mold. The gel polymerizes in the mold in about 10 min, and can soon thereafter be set up in the electrophoresis apparatus and preelectrophoresed as described in procedure 18.

How much of the end-labeled DNA cleaved by the four chemical reactions [G, G + A, C + T, C (Table IV)] should be loaded on such a gel? If fewer than 100 contiguous bases are to be sequenced, all of the material should be loaded at once and electrophoresed until tracking dye positions indicate migration appropriate for the distance of that tract from the labeled end. On the 8% gel (procedures 16 and 17), xylene cyanol (green) runs with fragments 70 nucleotides long. If, on the other hand, 200

[61] E. G. Richards, J. A. Coll, and W. B. Gratzer, *Anal. Biochem.* **12,** 452 (1965).
[62] A. C. Peacock and C. W. Dingman, *Biochemistry* **6,** 1818 (1967).
[63] T. Maniatis, A. Jeffrey, and H. van de Sande, *Biochemistry* **14,** 3787 (1975).
[64] F. Sanger and A. R. Coulson, *FEBS Lett.* **87,** 107 (1978).

or more bases proximal to the labeled end are to be sequenced, we suggest fractionating portions of the cleaved DNA (G, G + A, C + T, C) on three gels as follows. For bases 25–100 (beginning from the labeled end), run one-third of the DNA on an 8% gel (procedure 17) until the bromphenol blue has moved to the bottom, leaving the xylene cyanol halfway down. For bases 100–250, run another third of the material on a second 8% gel (procedure 17) until long after both dyes have run off the gel, and the xylene cyanol would have moved, by extrapolation, about 1.5 times its length. These two runs could be done on one 8% gel by loading it twice. However, this is not encouraged if the gel employs a wick. The second loading operation may disturb the ion concentration (pH) gradient between the electrodes, and if this discontinuity overtakes DNA bands already in the gel, it may diminish their resolution.

The last third of the chemically cleaved DNA fragment remains. This can be used to get bases 1–30, close to its labeled end, by electrophoresing it on a thin 20% gel (procedure 16, gel B) until the bromphenol blue has moved only a third of the way down its length. Or this DNA may be used instead to get sequence very far from its labeled end, beyond that obtained from the two 8% gels above. This can be attempted by running it on a third 8% gel for an even longer time, or on a 6% gel.[64]

Hints for setting-up, electrophoresing, and autoradiographing these sequencing gels can be found in procedure 18. For autoradiography, the slab gel, still in place on one of its glass plates, is covered with plastic wrap and placed on X-ray film inside a light-tight, folding-cardboard exposure holder (Kodak 149-2719), shielded and stiffened with sheets of lead and aluminum, respectively, glued to its back panel. This packet can then be stacked with others in a freezer, aluminum side up, and compressed with a lead brick for the duration of the exposure. Exposing the gel to preflashed film juxtaposed with an intensifying screen at −70° can produce an image of the sequencing band pattern much faster,[16–18] but in our experience may result in some loss of resolution.

Procedure 16. General Method for Preparing Polyacrylamide Sequencing Gels

Gel A 8%, 1:20 cross-linked, for cleavage products 25–250	Gel B 20%, 1:20 cross-linked, for cleavage products 1–30
7.6%(w/v) Acrylamide 0.4%(w/v) Bisacrylamide 8.3 M Urea 100 mM Tris-borate, pH 8.3, 2 mM EDTA	19%(w/v) Acrylamide 1%(w/v) Bisacrylamide 8.3 M Urea 100 mM Tris-borate, pH 8.3, 2 mM EDTA

Assemble the gel mold, and push its bottom edge into a roll of warmed (37°C) Plasticene. Measure the mold, calculate its volume in milliliters, and inflate about 10% to some multiple of 10 ml; this will be the final volume (*gelvol*) of the polymerization mixture. Measure out the following components, based on which of the above gels (A or B) is being prepared and the final volume (*gelvol*)

   7.6%(A) 19%(B) (wt/gelvol) Acrylamide
   0.4%(A) 1%(B) (wt/gelvol) Bisacrylamide
   50%(A) 50%(B) (wt/gelvol) Urea
   30%(A) 30%(B) (vol/gelvol) Distilled water
   10%(A) 10%(B) (vol/gelvol) 1.0 *M* Tris-borate, pH 8.3, 20 m*M*
         EDTA

Dissolve solids with stirring and gentle warming (water bath). Add distilled water to 100% of final volume (*gelvol*). Vacuum filter through filter paper or nitrocellulose membrane. Heat at 37° for 10 min (water bath). Stopper flask, connect to vacuum, and degas (should boil).

   0.7%(vol/gelvol) 10%(w/v) Ammonium persulfate

Mix well by swirling liquid in flask. Remove a few milliliters into a test tube, add 5 μl TEMED, mix quickly, pour into gel mold along one side, and allow to polymerize at the bottom as a plug. Add 10–50 μl TEMED to the rest of the polymerization mixture, mix very well by swirling, pour into gel mold until overfull, and check for air bubbles trapped in the mold. If present, rap front and back plates with the plastic handle of a screw driver to dislodge. Rub slot former in acrylamide solution at top of mold to displace air bubbles from teeth, and insert into mold. Observe polymerization within 10 min (Schlieren patterns around slot-former teeth); if polymerization takes longer, use more TEMED for the next gel.

Procedure 17. Preparing One Thin[64] 8% Polyacrylamide Sequencing Gel

   Obtain two 5 × 200 × 400 mm glass plates and two 0.5 × 10 × 400 mm Teflon spacers. Apply a very thin trail of silicone grease from a syringe near the long edges of both plates, sandwich the spacers between them, and secure with ten high-tension steel binder clips on each side. Push a short edge vertically into a roll of warmed (37°) Plasticene, and support the gel mold in an upright position with a ring-strand clamp. Combine the following polymerization mixture components in a 125-ml sidearm flask (the final volume should be 30 ml)

   15 gm Urea
   10 ml Distilled water

6 ml 38%(w/v) Acrylamide, 2%(w/v) bisacrylamide
3 ml 1.0 M Tris-borate, pH 8.3, 20 mM EDTA

Dissolve the urea with swirling and warming (water bath). Stopper flask, connect to vacuum, and degas (should boil).

200 μl 10%(w/v) Ammonium persulfate

Swirl liquid in flask to mix. Remove 1 ml into a test tube, add 2 μl TEMED, mix, inject into gel mold next to spacers with a Pasteur pipette, and allow to polymerize at the bottom as a plug. Attach a short, 18-gauge, blunt-ended hypodermic needle to a membrane filter unit (Millipore Millex, 0.45 μm), and the latter to a 30-ml plastic syringe. Add to the polymerization mixture in the sidearm flask.

5–15 μl TEMED

Mix well by swirling liquid in flask. Immediately pour polymerization mixture into the syringe, replace the plunger, and inject into gel mold. Look for air bubbles trapped in the mold; if present, rap front and back plates to dislodge. Rub slot former in acrylamide solution at top of mold to displace air bubbles from teeth, and insert into mold.

Procedure 18. Loading and Electrophoresing Sequencing Gels

Gently pry between slot-former wings and top of mold to loosen. Move slot-former up and out vertically by alternately lifting on one wing and then on the other (do not rock it forward and backward). Fill slots and space above gel with electrophoresis buffer. Such liquid and gel debris out of the slots with a blunt-ended hypodermic needle inserted in a hose connected to an aspirator vacuum with trap, or invert the gel and shake it out. Install slab gel in electrophoresis apparatus containing electrophoresis buffer, removing any air bubbles trapped at the bottom edge. Fill slots and space above gel with electrophoresis buffer. Connect gel to top electrode compartment with a thick paper wick saturated with electrophoresis buffer, or directly with a buffer pool which is continuous from the top electrode compartment to the top of the gel. Apply about 1600 V (0.5 mm gel) or 800 V (1.5 mm gel), and electrophorese for 30 min. Disconnect gel and underlayer 3 μl (0.3 mm gel) or 10 μl (1.5 mm gel) sequencing samples in adjacent slots with a pointed glass capillary Micropet, using a slow side-to-side sweeping motion. Reconnect gel and electrophorese at about 2000 V (0.3 mm gel) or 1000 V (1.5 mm gel), regulating or periodically adjusting the power so that the surface of the gel mold is 50° ± 5° throughout the electrophoretic run. When the xylene cyanol

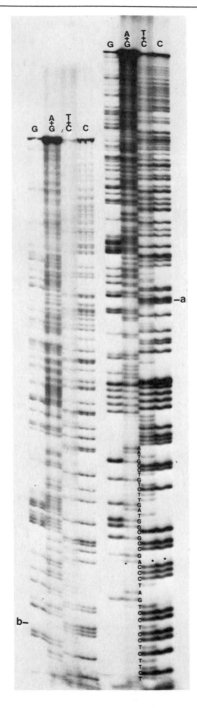

(green) or bromphenol blue marker dye has migrated to an appropriate position (see text), disconnect the gel and removed liquid from its slots with the aspirator needle, or invert the gel and shake it out. Remove one glass plate and cover the gel with thin, transparent plastic wrap (Saran wrap), smoothing it down with a paper towel and taping it tightly to the back of the remaining glass plate. Identify lanes and mark final electrophoresis dye positions with ^{14}C-containing ink (10 μCi/ml. Expose X-ray film to gel at $-20°$, or intensifying screen[16-18] to X-ray film to gel at $-70°$, in an aluminum-stiffened, light-tight exposure holder under 10–20 kg weight (lead brick).

Sequence Ladders

The autoradiograph of a sequencing gel (Fig. 4) exhibits four vertical ladders of staggered horizontal bands. Bands at the bottom are images of short fragments, produced by chemical cleavages close to the labeled end of the DNA, which moved fast through the gel. Bands higher in the pattern arise from cleavages progressively more distant from the labeled end, and the interval between them decreases logarithmically until a dark band abruptly terminates each ladder at the top. This band is an image of molecules which did not cleave at all in the sequencing reactions and are full length. Of the four ladders in each pattern (Fig. 4), bands in the first (G) derive from breaks at guanines, in the second (G + A) from breaks at both guanines and adenines, in the third (C + T) from breaks at cytosines and thymines, and in the fourth (C) from just cytosines. Note that, taken together, the two central ladders G + A and C + T contain all bands which arise from partial cleavage of the end-labeled DNA.

To read base sequence from these band patterns, go first to the one produced by electrophoresis for the shortest time (Fig. 4). Products of cleavages closest to the labeled end are at the bottom of this pattern. Beginning with the bottom most band, proceed upward along the juncture between the central G + A and C + T ladders, identifying each band in turn. If a band appears in the G + A ladder on the left it derives from

---

FIG. 4. Autoradiograph of a sequencing gel. Portions of double-stranded DNA, labeled with ^{32}P at one 5' end, were partially cleaved at guanines (G), guanines and adenines (G+A), cytosines and thymines (C+T), and cytosines (C), using the chemical reactions summarized in Table IV. The cleaved DNA was then electrophoresed on a 0.3 × 400 mm 8% sequencing gel (procedures 17 and 18) as follows: half of it was loaded (left) and electrophoresed until the xylene cyanol marker dye moved 300 mm, at which time the other half was loaded (right) and electrophoresis continued until the new xylene cyanol migrated 300 mm. To derive the sequence, begin at the bottom of the right pattern and read upward until bands are too closely spaced to continue (arrow a), find the corresponding position in the left pattern (arrow b), and again read upward until bands are not resolved.

cleavage at a purine, and if it also falls under G further to the left, that purine is a guanine; if not, it is an adenine. Likewise, a band in the C + T ladder on the right indicates a pyrimidine, and if it is also under C further to the right, that pyrimidine is a cytosine; if not on the far right, it is a thymine. Upon analyzing, in this manner, one band after another high up into the pattern, bands will eventually be found so closely spaced that perceiving their number and order becomes difficult. When this happens, switch to the bottom of the pattern produced by electrophoresis for a longer time (Fig. 4), and again read upward until successive bands are not resolved. When bands high in the four short-run ladders also appear low in the long-run ladders, the two patterns overlap and the two sequences read from them can be joined.

Sequencing patterns on X-ray films should be continuous and the bands in their ladders should be sharply resolved and consistent with one DNA sequence. Yet sometimes bands are doubled-up, retarded, extraneous, contradictory, too close together, too far apart, blurred, too dark, too light, or even missing, and the sequence reading breaks down. When such aberrations appear, one should apprehend that each is symptomatic of a specific physical irregularity in the DNA, infer what that is, and then root out the agent or manipulation which created it. Superficial observations such as "hydrazine cleaved the DNA at adenines" are often incorrect and almost always fail to uncover the cause. In this and similar sequencing techniques, the DNA undergoes a series of enzymatic and chemical reactions and is, at the end, electrophoresed in only one dimension. Thus faults in any of these reactions, not just the chemical ones which modify the bases, can put bands at wrong places in sequence ladders. DNA nicks, contaminating labeled DNA fragments, microheterogeneous labeled ends, incomplete base removal, incomplete strand scission, and intrastrand base-pairing will all do this, in addition to loss of specificity in a base modification reaction.

To correctly diagnose any aberration in a sequencing band pattern, we suggest first asking three questions about where it emerges.

1. Does it appear in patterns from all or just one of the gels on which the DNA was run?

2. Does it appear in all or just one or two of the four sequencing ladders (G, G + A, C + T, C)?

3. Does it appear at one position, in one locality, or throughout a given ladder?

With firm answers to these questions, go then to Table V and scan the fifteen entries in the Problem column for one which best describes the

[65] N. L. Brown and M. Smith, *J. Mol. Biol.* **116**, 1 (1977).
[66] H. Ohmori, J. Tomizawa, and A. M. Maxam, *Nucleic Acids Res.* **5**, 1479 (1978).

TABLE V
DIAGNOSIS AND CORRECTION OF ABERRATIONS IN SEQUENCING BAND PATTERNS

Problem	Probable causes	Suggested solutions
(1) Any or all of the following (a) Smearing of pyrimidine cleavage products (b) Loss of electrophoresis dye color in C and/or C+T samples (c) Bands in C and/or C+T ladders at positions where purines follow pyrimidine in the sequence (d) Large, water-insoluble pellets after the first ethanol precipitation (e) Poor suppression of T's in the C ladder	Fragments of DNA broken at pyrimidines contain one or more internal thymines and/or cytosines which reacted with hydrazine, but did not then cleave with piperidine. Each of these legitimate cleavage products blurs because its charge and/or mass is heterogeneous, due to the varying number and nature of internal hydrazine reaction products. Causes include an inadequate piperidine reaction, and secondary reactions with residual hydrazine during the 90° strand-scission and heat denaturation steps. Hydrazine can be carried through the procedure when ethanol precipitation and/or lyophilization fails to remove it. An insoluble complex of hydrazine, magnesium, and possibly other ions, appearing as a larger pellet, and poor vacuum during the lyophilization steps are common causes of this problem	If magnesium acetate is still used in the hydrazine stop solution as originally suggested,[2] delete it. If piperidine reactions are done in Eppendorf tubes, be sure to use at least 100 μl and to seal the cap tightly (otherwise use flame-sealed capillaries). After the piperidine reaction, employ a strong vacuum with rewetting of the residue to insure removal of any residual hydrazine during lyophilization
	Hydrazine has undergone oxidation	Replace the hydrazine with a fresh aliquot or new stock bottle
(2) T's are weak in the C+T ladder	Residual salt from ethanol precipitation of end-labeled DNA is carried over into the sequencing reactions and partially suppresses the reaction of hydrazine with thymine in the C+T sample	As a final purification step, dissolve end-labeled DNA in 25 μl water, and 1 ml ethanol, mix, chill, centrifuge, and dry the pellet. Then dissolve in water and distribute into sequencing reactions
(3) Bands are present in C and/or C+T ladders at *all*	A reaction between hydrazine and guanine at lower	Keep the samples at 4° or below during the ethanol

(*continued*)

TABLE V (*continued*)

Problem	Probable causes	Suggested solutions
guanine positions, not just those where guanine follows a pyrimidine in the sequence as in (lc) above and (4) below	pH (sodium acetate), after which piperidine breaks the DNA at guanines	precipitation steps. Use chilled (0–4°) hydrazine stop solution, sodium acetate, and ethanol
(4) Bands in C and/or C+T ladders at positions where purines follow pyrimidines in the sequence, but not at every purine in a run as in (3) above. Also see Problem (1c)	Residual sodium acetate from the second ethanol precipiton buffers piperidine, leading to incomplete phosphate elimination reactions, which require a high pH. The 3′ ends of 5′ labeled fragments then retain deoxyribosyl residues, have one less negative charge, and move one position slower in the gel	Rinse the last pellet well with 95% ethanol, or dissolve it in 25 μl water, add 1 ml 95% ethanol, mix, chill, and collect the reprecipitated DNA by centrifuging
	Some labeled DNA is not dissolved in the piperidine and therefore does not react with it at 90°	After adding piperidine to a tube containing dry DNA, close the cap and bat the tube with fingers to splash the liquid up the walls, and then collect it at the bottom with a quick spin.
(5) All DNA bands and both electrophoresis dyes are retarded in some ladders	Effect of residual piperidine on gel electrophoresis	Wet the DNA residue, freeze, and lyophilize with a strong vacuum as suggested in the chemical cleavage protocols
(6) Bands in A>G ladders produced by the older R2 reaction (Table III) are not sharp.	Overymethylation. Fragments cleaved at adenines carry too many ring-opened 7-methylguanines and are heterogeneous in charge	Employ the same dimethysulfate reaction conditions used for reaction R5 (Table III) guanine cleavage. (Weak A>G bands can be strengthened by mixing more often during the acid release, slightly increasing the acid concentration, or introducing extra labeled DNA into samples destined for A>G cleavage)
(7) All ladders exhibit uniform chemical cleavage of desired base-specificity, but the bands are not sharp enough to permit reading	Sample not evenly distributed in sample well	Layer samples with a pointed glass capillary, using a slow side-to-side sweeping motion.
	Difusion or secondary struc-	Turn the power up until the

TABLE V (*continued*)

Problem	Probable causes	Suggested solutions
very far into the sequence.	tural effects during electro-phoresis	gel runs at 50° (see fn. 65, p. 546)
	Parallax or scatter during autoradiography	Use a slab gel no more than 1.5 mm thick, preferably 0.5 mm, and compress the X-ray film against it with lead bricks during auto-radiography
(8) Sudden loss of expected bands, band spacing, or band order in all four ladders in the middle of a sequencing pattern	Nested end-labeled cleavage products become long enough to fold into stable hairpins and move anomalously in the gel (the so-called compression effect[65])	Be sure that the DNA is thoroughly denatured when loaded on the sequencing gel. Prerun and run the gel at a voltage which generates as much heat as the glass plates will tolerate (>50°)
(9) Loss of just one band in both pyrimidine ladders in the middle of a sequencing pattern	Presence of 5-methylcytosine, which reacts so slowly with hydrazine (relative to unmodified cytosine) that the DNA is not cleaved at that position[66]	Look for guanine at the corresponding position in the sequence of the complementary strand (see fn. 66, p. 546)
(10) Extraneous bands in all ladders throughout a sequencing pattern	Contaminating end-labeled fragment	Strive for better DNA resolution on the fragment division or strand separation gel, or use a different restriction enzyme for secondary cleavage
	Terminal heterogeneity generated before or during end-labeling ay a contaminating nuclease	Purify restriction endonuclease, phosphatase, kinase, or terminal transferase
	Unavoidable terminal heterogeneity originating *in vivo,* or as a consequence or DNase I or S1 nuclease treatment *in vitro*	After labeling ragged ends, cleave with restriction endonuclease(s) which will give end segments less than 100 base pairs long, denature, and fractionate on a 1.5 mm 8% sequencing gel (procedure 16). Strands differing in length by one or more nucleotides will resolve and be extracted for sequence analysis
(11) An extraneous band across all four ladders at	Some double-stranded DNA molecules are nicked by a	Attempt to derive the sequence surrounding the

(*continued*)

TABLE V (*continued*)

Problem	Probable causes	Suggested solutions
one position in a sequencing pattern	restriction enzyme at a sequence which closely resembles its true recognition site, or by a sequence-specific contaminating endonuclease. After end-labeling, a double stranded fragment isolated on a non-denaturing gel retains the hidden break, which is then revealed on the denaturing sequence gel. This can be confirmed by electrophoresing some unreacted end-labeled DNA in parallel with the sequencing samples. Unique endogenous nicks and alkali-labile nucleotides are unlikely causes but possible in rare cases and subject to similar tests	break on both strands, and compare it with the recognition sequences of all restriction enzymes used to generate that DNA fragment. In making this comparison recall that bands in the sequencing pattern arise from fragments which lack the nucleotide with which they are identified, while enzymatic cleavage products retain all nucleotides. If $\frac{3}{4}$, $\frac{4}{5}$, or $\frac{5}{8}$ bases are homologous, tentatively assume nicking at the variant site. Then avoid excesses of that restriction endonuclease in subsequent DNA digestions, ideally using just enough pure enzyme to cleave every legitimate site. If the problem persists, avoid that enzyme altogether and use another cleavage route or strand separation
(12) One or a few bands in two ladders produced by cleavage reactions which do not normally share base-specificity. Cleavage is consistent with base specificity everywhere else in the sequencing pattern	Sequence heterogeneity in the DNA at that position

Compression | Confirm by getting the complementary effect on the other strand.
See problem (8) above |
| (13) Difficulty in identifying the band corresponding to chemical cleavage of the ultimate 5'-nucleotide (^{32}P-labeled with polynucleotide kinase and [γ-^{32}P]ATP) | ^{32}P-Orthophosphate is the product of base-specific chemical cleavage of the ^{32}P-labeled 5' terminal nucleotide, and this migrates so rapidly that it is usually run off the gel and missed altogether. A diffuse hydrazinolysis artifact product which sometimes appears between the first | Put some ^{32}P-phosphate in formamide–buffer–dyes solution, and run as a marker adjacent to products of the four base-specific reactions on a 20% sequencing gel (procedure 16, gel B). Preelectrophorese the gel, load the five samples, electrophorese until the bromphenol blue marker has |

TABLE V (*continued*)

Problem	Probable Causes	Suggested solutions
	and third bands[66] further confuses the interpretation	proceeded one-third of the way down the gel, and autoradiograph. A band in one or two of the four usual lanes running even with the phosphate in the fifth indicates which base-specific reaction(s) cleaved the 5' terminal nucleotide
(14) Difficulty in identifying the band corresponding to chemical cleavage of the ultimate 3'-nucleotide (^{32}P labeled) and incorporated by terminal transferase or DNA polymerase)	With DNA 3' end-labeled by addition of more than one ^{32}P-nucleotide (procedure 6), chemical attack at the *ultimate*, added 3'-ribonucleotide releases ^{32}P-phosphate, which is usually run off the sequencing gel and missed altogether. Furthermore, be aware that none of the four recommended reactions (G, G+A, C+T, C) will cleave riboadenylic acid, since ribonucleotides do not depurinate readily. The three other ribonucleotides should cleave like deoxynucleotides in these reactions	Put some ^{32}P-phosphate in formamide–buffer–dyes solution, and run adjacent to products of the four base-specific reactions on a 20% sequencing gel as described under the solution to problem (13) above. This marker will indicate the product of a reaction with the *ultimate* 3' base
	With DNA 3'-end-labeled with DNA polymerase, chemical attack at the *penultimate* deoxynucleotide releases the ^{32}P-labeled ultimate 3' nucleotide intact	Digest some of the endlabeled DNA with DNAse and venom phosphodiesterase, lyophilize, dissolve in formamide–buffer–dyes, and run adjacent to products of the four basespecific reactions on a 20% sequencing gel as described under problem (13) above. This marker will indicate the product of a reaction with the *penultimate* 3' base
(15) More than one strong band appears at the position of uncleaved DNA at	One is single-stranded and the other double–stranded DNA	Increase dimethylsulfate or hydrazine reaction times, employ stronger denatura-

(continued)

TABLE V (*continued*)

Problem	Probable Causes	Suggested solutions
the top of all sequence ladders		tion before loading samples on the gel, or both
	Variant or incomplete restriction endonuclease cleavage	See problem (11)
	Contaminating end-labeled fragment	See problem (10)
	Fragment length heterogeneity	See problem (10)

artifact or difficulty you have. Finding one, consider whether the cause in the Table V second column is consistent with what you used, did, or did not do with that DNA fragment. If it is, try what the third column suggests to solve that problem the next time DNA is prepared, labeled, or sequenced. Table V does not necessarily list all the sequence reading problems one may encounter, their only causes, or the best solutions, but describes our experience to date which we hope will be useful.

Strategies for Sequencing Double-Stranded DNA

How one proceeds to sequence the DNA of a virus, plasmid, or other genetic element purified from cells will follow from the answer to a single question: Do I want to sequence the whole chromosome or just some defined stretch of it? Targeted sequencing suggests a scheme to create, identify, and further purify a corresponding segment of the DNA, while willfully foregoing this tactic will yield more sequence faster when an entire chromosome is the goal.

Sequencing a promoter, replication origin, insertion element, attachment site, structural gene, or other defined DNA region is usually preceded by mapping its physical and functional correlates onto a restriction endonuclease cleavage map of the DNA. Unique restriction fragments which hybridize with messenger RNA, are bound to a nitrocellulose filter by a site-specific protein ligand or are altered in length by recombination or mutation *in vivo* indicate in the beginning where these regions lie. Although sequence analysis could begin with such fragments, it will get done more efficiently if they are referred back to a comprehensive restriction map to identify a small fragment which contains all the relevant DNA and will resolve from all others in the digest. Then that one fragment may be prepared, in quantity sufficient for deriving and confirming the entire sequence, with one restriction enzyme and one large polyacrylamide gel (see the earlier section entitled Unlabeled DNA Segments). This strategy is commonly used to separate cloned eukaryotic chromosomal fragments from their bacterial vectors, but it is compelling whenever the sequencing target is small and mapped. We have often used it,[67,68] and the experimental details are in procedure 1.

Sequencing an entire virus, phage, or plasmid, on the other hand, requires no exhaustive restriction map nor stockpiles of isolated restriction fragments. One can begin simply by cleaving a small amount of chromosomal DNA into segments with ends which are easy to label. For both 5' and 3' labeling these will ideally have protruding 5' termini, and Table II lists fifteen restriction endonucleases which make staggered cuts to leave 5' ends extended. After the 5' or 3' ends of all these primary fragments have been heavily labeled, the whole collection can be denatured for strand separation or digested with a second restriction enzyme and electrophoresed on a gel. All fragments which strand separate or are cleaved will have ^{32}P on only one end and can be picked off the gel and sequenced immediately. If the two restriction enzymes are used in reverse order, they will yield the same set of fragments labeled at their opposite ends.

Procedure 19 details how to conduct this "shotgun" cutting, labeling, subcutting, and strand-separating of DNA. Because it employs superimposed enzymatic reactions and only one polyacrylamide gel, this procedure generates many singly end-labeled DNA segments with a minimum of labor. As an example of what this sequencing gambit can do, consider its application to the plasmid pBR322. Hinf cuts pBR322 at ten sites, AvaII cleaves it at eight, and the order and sizes of fragments produced by each enzyme are known.[69] Using Hinf–kinase–AvaII and AvaII–kinase–Hinf, in a double reciprocal application of procedure 19, about 40% of pBR322 (1700 nucleotides) can be converted into sequencable DNA in one day. Strand-separating en masse the doubly end-labeled Hinf and AvaII primary digests will set forth an even larger array of singly end-labeled fragments. If all the strands separated and were sequenced to their ends or, on longer ones, out to 200 bases, about 70% of the plasmid (3000 nucleotides) would be covered. Note that all the sequence derived from a 5' labeled set can be confirmed by sequencing complementary strands in the equivalent 3'-labeled set, and vice versa. In subsequent preparations, procedure 19 can employ restriction enzymes which cut up the chromo-

[67] A. M. Maxam, R. Tizard, K. G. Skryabin, and W. Gilbert, *Nature (London)* **267**, 643 (1977).
[68] S. Tonegawa, A. M. Maxam, R. Tizard, O. Bernard, and W. Gilbert, *Proc. Natl. Acad. Sci. U.S.A.* **75**, 1485 (1978).
[69] J. G. Sutcliffe, *Nucleic Acids Res.* **5**, 2721.

some in other ways. Each one will produce, very quickly and efficiently, unique sets of DNA fragments which are ready for sequencing.

After awhile this dispersed cutting and sequencing routine will begin to produce runs of nucleotides which overlap those already determined. Restriction enzyme recognition–cleavage sites[70] are excellent markers for finding these identities. To discover these overlaps, compare recognition sites appearing in new sequences with those of enzymes already used to create DNA ends for sequencing. Where the positions of cuts and recognition sequences seem to correspond, parallel and antiparallel sliding matches of bases on either side will reveal any overlap of new with old sequences. A given base sequence with a restriction enzyme site embedded in it can also be deliberately extended by cleaving the chromosome with that enzyme, end-labeling, and sequencing the appropriate fragment. In general, sequencing without a restriction map may perpetuate itself for some times by generating overlapping sequences and/or by revealing and then using one new restriction enzyme site after another. Sooner or later, however, avoiding redundant sequence determinations and orienting isolated sequences and gaps to be filled will necessitate an all-inclusive restriction map.

Eventually one must have criteria for deciding when a DNA base sequence is correct. Fortuitously, one gets sequence in blocks corresponding to restriction fragments, each having two complementary strands, one of which was analyzed from an end labeled with ^{32}P. Given the provisional sequence of one such strand, we know of no faster or more reliable way of verifying it than to sequence the complementary strand. This is easiest when the enzymatic reaction which labeled an end of one strand also labeled an end of the other, the two strands separated on a gel, and they are not overly long. Simultaneous base-specific partial cleavage and gel electrophoresis will then yield all the sequence on both strands at the same time. These two sequences can validate one another because each base-pair identified by a given chemical cleavage reaction on one strand is identified by a wholly different reaction on the other strand. Furthermore, if the break at a base-pair is distant from one labeled end, it will often be conveniently closer to the other end. The two sequences must be derived independently to be mutually confirming, though, for if good sequence from one strand forces reinterpretation of a poor pattern from the other, one then has only unconfirmed sequence from the one strand.

Sometimes sequencing all of both strands of a DNA restriction fragment in the manner just described may not be so easy. Its strands may fail to separate cleanly, its length may exceed the range of sequencing gels, or

[70] R. J. Roberts, this volume, Article [1].

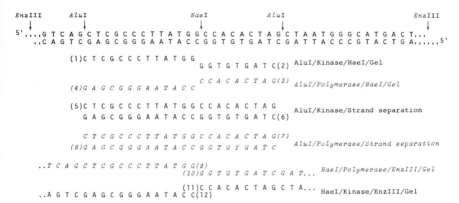

FIG. 5. Strategies for deriving and confirming the base sequence of double-stranded DNA. Along the top of the figure is a hypothetical DNA region containing recognition sequences for the restriction enzymes indicated. Generally, this DNA can be sequenced by cutting it with one restriction enzyme, labeling the ends with ^{32}P, cleaving with another enzyme or strand-separating, and then doing base-specific chemical cleavage on the segregated pieces. AluI will cut the DNA to release a double-stranded fragment which can be labeled with ^{32}P either at both 5′ ends with polynucleotide kinase, or at both 3′ ends with DNA polymerase or terminal transferase (Table I). HaeI will then cleave this AluI fragment asymmetrically, and after separation the two pieces can be sequenced from the labeled 5′ ends (1 and 2), or labeled 3′ ends (3 and 4). The uncut AluI fragment could be strand-separated instead and sequenced intact, again either from labeled 5′ ends (5 and 6) or 3′ ends (7 and 8). The original DNA (top) can also be cut with HaeI first, 5′ or 3′ end-labeled, digested with a third restriction enzyme which cleaves both halves (EnzIII), and sequenced outward in both directions beyond the AluI sites (9, 10, 11, and 12). Any odd-numbered sequence will confirm all or part of any complementary even-numbered sequence and vice versa.

its terminal nucleotides may be difficult to identify. One can still get the sequence of both its strands simply by sequencing the complementary strand from labeled 3′ ends instead of 5′ ends, or vice versa, or by splitting the fragment with another restriction enzyme and working from the new internal ends (Fig. 5). Rapid methods for identifying labeled 5′ and 3′ terminal nucleotides with nucleases, thin-layer chromatography, and autoradiography have been described.[27,32,36,52,53,71-74]

Eight different tetranucleotide-recognizing restriction endonucleases are now available,[70] each of which will cleave double-stranded DNA on average every 256 base pairs. Given these, strand separation techniques, 5′ end-labeling, 3′ end-labeling, and the ability to sequence at least 250 bases from any labeled end, it is difficult to avoid getting sequence from

[71] A. Bernardi and U. Bertazzoni, *Anal. Biochem.* **61**, 448 (1974).
[72] A. Bernardi and C. Gaillard, *Anal. Biochem.* **64**, 321 (1975).
[73] See also C. P. D. Tu and R. Wu, this volume, Article [62].
[74] K. Murray, *Biochem. J.* **131**, 569 (1973).

both strands. We advocate deliberate two-strand sequencing throughout the DNA for verification, and diagram strategies for doing it in Fig. 5. Results of all independent sequencing runs can then be displayed as left- or right-pointing arrows, each signifying which strand was sequenced and the restriction enzyme cleavage site from which it was done (for an example, see Tonegawa *et al.*[68]). We also advocate sequencing across all restriction fragment junctures, points at which the DNA was cut before and after end-labeling, for experience has revealed that it is in these regions that bases or even small restriction fragments are most apt to be missed. If, after having sequenced both strands and across all restriction fragment joints, we unconditionally find the sequence of one strand to be complementary at every position to the sequence of the other, we consider the primary structure of that DNA established.

Procedure 19. Three Consecutive Reactions and One Gel for Preparing 5'-^{32}P-Labeled DNA Fragments

Below is a fast procedure for producing, ready for sequence analysis, an array of 5' end-labeled segments of phage, viral, or plasmid DNA, or a large restriction fragment. In two successive enzymatic reactions, a small amount of DNA is first cut into pieces with a restriction enzyme (A), and the entire collection of fragments is then phosphorylated with ^{32}P. This is accomplished with the polynucleotide kinase exchange reaction,[22,25,26] which will work best if the restriction enzyme left protruding 5' ends (Table II). After both ends of all fragments have been so labeled, they are all denatured and electrophoresed to separate strands, or digested with a second restriction enzyme (B) and electrophoresed to sort out the secondary cleavage products. In either case, some undenatured or uncut material is run in parallel so that fragments which strand-separated or were split can be distinguished on the autoradiogram from those which remained intact. After elution from the gel, these singly end-labeled pieces of DNA can be sequenced immediately.

Primary restriction endonuclease cleavage (22 μl):

10 μg DNA (phage, virus, plasmid, or large restriction fragment)
15 μl Distilled water
2 μl 200 mM Tris-HCl, pH 7.4, 50 mM MgCl$_2$, 10 mM dithiothreitol (DTT), ±500 mM NaCl
5 μl restriction enzyme A, 10 units

Combine the above in a siliconized 1.5-ml Eppendorf snap-cap tube and mix. Heat at 37°C for 60 minutes.

End-labeling (51 μl):

10 μl Distilled water
5 μl 500 mM Imidazole-HCl, pH 6.6, 100 mM MgCl$_2$, 50 mM DTT,
   1 mM spermidine
3 μl 5 mM Adenosine diphosphate
10 μl [γ-^{32}P]ATP, >1000 Ci/mmole, 100 pmole (preferably more[26])
1 μl T$_4$ polynucleotide kinase, 20 units

Mix, heat at 37° for 30 min, and cool to 0° in ice.
   Precipitation of end-labeled DNA (1 ml):

200 μl 2.5 M Ammonium acetate
  1 μl tRNA, 1 mg/ml
750 μl 95% Ethanol

Close the cap and mix (invert the tube four times). Chill at −70°C for 5
minutes (Dry Ice–ethanol bath). Centrifuge at 12,000 g for 5 min (Eppendorf, 4°C). Remove the supernatant with a Pasteur pipette.

250 μl 0.3 M Sodium acetate

Close the cap and redissolve the DNA (Vortexer).

750 μl 95% Ethanol

Invert to mix, chill, centrifuge, and remove the supernatant.

1 ml 95% Ethanol

Centrifuge at 12,000 g for 15 sec and remove the supernatant. Place the
tube under vacuum for a few minutes (desiccator).

20 μl Distilled water

Close the cap and redissolve the DNA (Vortexer and manual agitation).
Centrifuge a few seconds (Eppendorf). Proceed with strand separation
(below) or secondary restriction endonuclease cleavage (further below),
or both.
   Strand separation (40 μl): Divide the end-labeled DNA restriction
fragments into two portions:

(A) 18 μl (90% for strand separation)
    20 μl 60%(v/v) dimethyl sulfoxide, 1 mM EDTA, 0.05% xylene
       cyanol, 0.05% bromphenol blue

Mix, heat at 90° for 2 min, and quick-chill in ice water.

(B) 2 μl (10% for undenatured marker)
    20 μl 60% (v/v) dimethyl sulfoxide, 1 mM EDTA, 0.05% xylene
       cyanol, 0.05% bromphenol blue
    15 μl Distilled water

Mix and keep at 0° in ice.

Immediately load both samples in adjacent slots of a strand separation gel (see procedure 8):

5%(w/v) Acrylamide
0.1%(w/v) methylenebisacrylamide
50 mM Tris-borate, pH 8.3, 1 mM EDTA
Slab gel: 1.5 × 200 × 400 mm
Sample well: 1.5 × 20 mm

Electrophorese, autoradiograph, and identify fragments which strand-separated. Elute 5′ end-labeled single-stranded fragments (procedure 9), and sequence.

Secondary restriction endonuclease cleavage (25 μl): Divide the end-labeled DNA restriction fragments into two portions:

(A) 18 μl (90% for secondary digestion)
2 μl 200 mM Tris-HCl, pH 7.4, 50 mM MgCl$_2$, 10 mM DTT, ±500 mM NaCl
5 μl Restriction enzyme B, 10 units
Mix and heat at 37° for 60 min.

(B) 2 μl (10% for undigested marker)
20 μl Distilled water
2 μl 200 mM Tris-HCl, pH 7.4, 50 mM MgCl$_2$, 10 mM DTT, ±500 mM NaCl
Mix and keep at 0° in ice.

Add concentrated gel loading solution to both samples:

5 μl 50%(v/v) glycerol, 25 mM EDTA, 0.25%(w/v) xylene cyanol, 0.25%(w/v) bromphenol blue

Mix and load the samples in adjacent slots of a nondenaturing gel (see procedure 7):

5–10%(w/v) Acrylamide
0.17–0.33%(w/v) Methylenebisacrylamide
50 mM Tris-borate, pH 8.3, 1 mM EDTA
Slab gel: 1.5 × 200 × 400 mm
Sample well: 1.5 × 20 mm

Electrophorese, autoradiography, and identify fragments which were cleaved. Elute double-stranded fragments labeled at one 5′ end (procedure 9), and sequence.

Commercial Sources of Reagents, Enzymes, and Equipment

Below are listed manufacturers or commercial suppliers of chemicals, enzymes, and equipment used in the DNA sequencing procedures in this paper. This does not imply that they are superior to others, and indicates only that they seem acceptable for end-labeling, chemically cleaving, and electrophoresing DNA. Following suppliers name, the product numbers are listed.

Acrylamide: Bio-Rad Laboratories, 161-0101
Adenosine diphosphate: P-L Biochemicals, 1426
Adenosine triphosphate: P-L Biochemicals, 100-A
Alkaline phosphatase: Worthington Biochemical Corp., BAPF
Ammonium acetate: R-Plus Laboratories, 01-2310-10
Bisacrylamide: Bio-Rad Laboratories, 161-0201
Bovine serum albumin: Pentex, Miles Laboratories, 81-001
Ammonium sulfate: R-Plus Laboratories, 01-0015-10
Bromphenol blue: BDH Chemicals, Ltd., 20015
Capillaries, glass, 1 × 150 mm: Kontes Glass Company, K744250-0023
Dimethyldichlorosilane: Ventron, Alfa Products, 69124
Dimethyl sulfate, 99%: Aldrich Chemical Company, D18,630-9
Dimethyl sulfoxide, 99%: Aldrich Chemical Company, 15,493-8
DNA, calf thymus: Worthington Biochemical Corp., DNA
EDTA, disodium: BDH Chemicals, 10093
Ethanol: Gold Shield
Exposure holders (light tight,
    for 35 × 43 cm X-ray film): Eastman Kodak Company, 149-2719
Film, X-ray, for autoradiography: Eastman Kodak Company, NS and RPR(XR)
Fluorescent thin-layer plates: Eastman Kodak Company, 13254
Gel plates, spacers, slot-formers, apparatus: Dan-Kar Plastic Products
Glass wool, siliconized: Alltech Associates, 4037
Glyceraldehyde-3-phosphate dehydrogenase: Calbiochem, 35632
Hydrazine, 95%: Eastman Organic Chemicals, 902
Magnesium acetate: BDH Chemicals, 10148
Magnesium chloride: BDH Chemicals, 10149
Membrane filter unit: Millipore Corp., Millex SLHA 025 OS or Swinnex
Microcentrifuge, 15,000 rpm: Eppendorf, 5412
Mixed bed resin (for deionizing formamide): Bio-Rad, AG501-X8(D)
3-Phosphoglycerate kinase: Calbiochem, 52518
PEI–cellulose thin-layer chromatography plates: Macherey-Nagel, CEL300PEI
Piperidine, 99%: Fisher Scientific, P-125
Power supplies, gel electrophoresis: E-C Apparatus, EC453 (1000 V), Isco, 493 (1000
    V), Dan-Kar Plastic Products, 203 (2000 volts)
Sodium acetate: BDH Chemicals, 10236
Sodium cacodylate: Fisher Scientific, S-257
Sodium chloride: BDH Chemicals, 10241
Sodium dodecyl sulfate: Pierce Chemical Company, 28364
Sodium hydroxide: BDH Chemicals, 45212
Spermidine trihydrochloride: Sigma, S2501

Tape, conformable (for Eppendorf tube gaskets): 3M Company, Plastic film tape No. 471
Thermometers, surface (stick onto gels): PTC, 310C (VWR Scientific, 61157-163)
Tris base: R-Plus Laboratories, 20-7209-10
Tubes, Microcentrifuge, 1.5 ml: Eppendorf, 2236411-1
Tubes, centrifuge, 10 × 75 mm: Rochester Scientific, R7045D
Urea: R-Plus Laboratories, 21-7601-10
Xylene cyanol: MC/B Manufacturing Chemists, XX0062

[58] DNA Sequence Analysis by Primed Synthesis

By ANDREW J. H. SMITH

Introduction

Several rapid methods are now available for DNA sequence analysis. All of these rely on high resolution electrophoresis on denaturing polyacrylamide gels to resolve oligonucleotides with one common end but varying in length at the other by a single nucleotide. The primed synthesis methods make use of the ability of DNA polymerases to synthesize accurately a complementary radioactive copy of a single-stranded DNA template using restriction enzyme generated DNA fragments as primers. The chain terminator sequencing procedure of Sanger et al.[1] is considered to be the most simple, rapid and accurate of this kind of sequencing method. It is now routinely used as the method of choice in place of the earlier rapid primed synthesis methods, the "plus and minus" procedure[2] and the partial ribosubstitution technique of Barnes.[3]

The chain terminator method makes use of the 2',3'-dideoxy and the β-D-arabinofuranosyl analogs of the deoxyribonucleoside triphosphates and their incorporation by DNA polymerase I onto the 3'-hydroxyl of an extending transcript.[4] Once incorporated, the 3' end is no longer a substrate for further extension. The dideoxy analogs simply lack the 3'-hydroxyl necessary for further chain growth, this being replaced by a hydrogen atom. In the case of a 3' end terminated in an arabinose analog, which is a 2' stereoisomer of the ribose sugar, the 2'-hydroxyl lying in trans to the 3'-hydroxyl, the mechanism of chain termination is less obvi-

[1] F. Sanger, S. Nicklen, and A. R. Coulson, *Proc. Natl. Acad. Sci. U.S.A.* **74**, No. 12, 5463 (1977).
[2] F. Sanger and A. R. Coulson, *J. Mol. Biol.* **94**, 442 (1975).
[3] W. M. Barnes, *J. Mol. Biol.* **119**, 83 (1978).
[4] M. R. Atkinson, M. P. Deutscher, A. Kornberg, A. F. Russell, and J. G. Moffat, *Biochemistry* **8**, 4897 (1969).

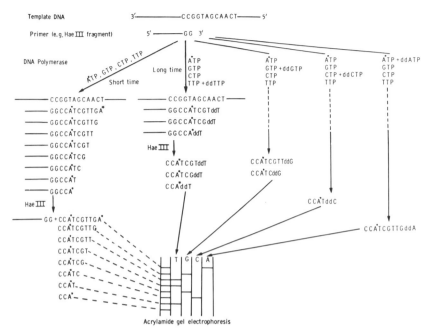

FIG. 1. A diagrammatic illustration of primed synthesis reactions carried out on a template of known sequence from the annealed strand of a *Hae* III restriction enzyme fragment using labeled dATP and, in turn, each dideoxynucleoside triphosphate. The diagram also illustrates the subsequent fractionation of the products of each reaction on a denaturing gel, from which the sequence can be deduced. The reaction on the left illustrates the products of a very short incubation in the absence of any chain terminator, which results in random termination at every base and therefore on gel electrophoresis in bands corresponding to oligonucleotides of every chain length.

ous. It is known, however, that the primer terminus binding site of the DNA polymerase I has a very low affinity for such a 3' end.[5] The low rate of incorporation of these analogs onto a 3' end, about 10^{-3} the rate of incorporation of a deoxyribonucleoside monophosphate, is also reflected in their correspondingly slower rate of removal by the 3' → 5' hydrolytic and pyrophosphorylytic activities of DNA polymerase I.[4] Effectively then, a 3' end terminated in a dideoxy or arabinose nucleoside monophosphate is inert to further extension even by removal of the terminating analogs by the proof reading function of the polymerase.

The chain terminator method involves synthesis by the Klenow subfragment of DNA polymerase I[6] (this lacks the 5' → 3'-exonuclease activity of the intact enzyme) of a complementary copy of the single-stranded

[5] J. A. Huberman and A. Kornberg, *J. Biol. Chem.* 245, 5326 (1970).
[6] H. Klenow, K. Overgaard-Hanesen, and S. A. Patkar, *Eur. J. Biochem.* 22, 371 (1971).

target sequence, primed with the directly adjacent annealed strand of a restriction fragment. The synthesis is carried out in the presence of the four deoxyribonucleoside triphosphates, one or more of which is α-^{32}P-labeled, and in turn each dideoxy- or arabinose nucleoside triphosphate in separate incubations. There is, therefore, in each reaction a base-specific partial incorporation of a terminating analog onto the 3' ends of the extending transcripts throughout the sequence. Parallel fractionation by gel electrophoresis of the size ranges of terminated labeled transcripts from each reaction, each with the common 5' end of the primer, allows a sequence to be deduced. The principle of the method is essentially outlined diagrammatically below in Fig. 1. As an alternative to Klenow DNA polymerase I, reverse transcriptase[7] can be used in this procedure. This similarly incorporates dideoxynucleoside triphosphates, leading to chain termination, although the reaction conditions required are slightly different.

The chain termination sequencing method has a number of advantages over the plus and minus and partial ribosubstitution procedures. It is more simple and rapid, since it involves just a single step reaction. There is a better incorporation of counts from the ^{32}P-labeled triphosphates since the reaction can be carried out for long enough to allow extension from every primer. Finally and most importantly, every nucleotide shows up as a band, even in runs of the same nucleotide.

General Application of Primed Synthesis Techniques. A requirement of the primed synthesis approach to DNA sequencing is a pure single-stranded template. In some cases the strands of a DNA molecule are different to separate in large enough quantities and unless, for example, the target sequence is inserted in a cloning vector, such as λ phage where strand separation may be possible or in a filamentous phage vector[8] in which pure single-stranded DNA can be isolated from the virion,[9] this kind of sequencing approach is impossible. Recently, however, it has proved possible to prepare pure single-stranded template from a linear duplex by treatment with *E. coli* exonuclease III,[9a] a 3' → 5'-exonuclease with a specificity for degrading duplex DNA.[10,11] A consequence of this specificity is that the product of a limit digest with exonuclease III is two noncomplementary single-stranded DNA molecules, as illustrated below.

[7] F. Sanger, personal communication.
[8] J. Messing, B. Gronenborn, B. Müller-Hill, and P. H. Hofschneider, *Proc. Natl. Acad. Sci. U.S.A.* **74**, 3642 (1977).
[9] P. H. Schreier and R. Cortese, *J. Mol. Biol.* **129**, 169 (1979).
[9a] A. J. H. Smith, *Nucleic Acids Res.,* **6**, 831 (1979).
[10] C. C. Richardson, I. R. Lehman, and A. Kornberg, *J. Biol. Chem.* **239**, 251 (1964).
[11] D. Brutlag and A. Kornberg, *J. Biol. Chem.* **274**, 241 (1972).

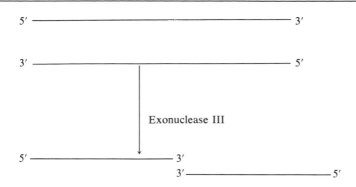

Single-stranded template prepared by this method has proved suitable as a template for primed synthesis with chain terminators, and the method should be applicable to most molecules. It is, of course, subject to the limitation that restriction fragments from the center of the linear duplex may prime on both strands. A further limitation of this approach is that for each half of the linear duplex, single-stranded DNA is only available from one strand. This is a disadvantage in DNA sequencing where it is desirable to be able to obtain sequences for the same region from both strands. However, in the case of a circular molecule with several well-distributed single-cutting restriction sites, cutting at each site followed by degradation with exonuclease III should, in turn, generate single-stranded template from both strands for some if not all of the molecule.

Application of Primed Synthesis Methods to RNA Sequencing.[11a] Both the plus and minus method and the chain termination method with dideoxy analogs have been successfully applied to the sequencing of mRNA in this laboratory using reverse transcriptase to copy the message.[12,13] Primers are usually chemically synthesized and specifically designed to hybridize to known sequences of the message, some of these sequences being common to a variety of mRNA molecules, for example, the poly(A) sequence. Alternatively restriction fragments can be used as primers. McReynolds *et al.*,[14] for example, successfully used a restriction fragment from a cDNA clone of an ovalbumin message to prime into and sequence the 5' end of the message which was not present in the clone.

[11a] See P. K. Ghosh *et al.*, this volume, Article [59].
[12] G. G. Brownlee and E. M. Cartwright, *J. Mol. Biol.* **114**, 93 (1977).
[13] P. H. Hamlyn, G. G. Brownlee, C.-C. Cheng, M. J. Gait, and C. Milstein, *Cell* **15**, 1067 (1978).
[14] L. McReynolds, B. W. O'Malley, A. D. Nisbet, J. E. Fothergill, D. Givol, S. Fields, M. Robertson, and G. G. Brownlee, *Nature (London)* **273**, 723 (1978).

Experimental Procedures

Materials. The sources and preparation of some of the materials routinely used in the experimental protocols described in the following sections are given here. The sources of material given here, however, should not be taken to imply that these are necessarily the only suitable ones.

A stock solution of 38% acrylamide/2% *N,N'*-methylene bisacrylamide in water is prepared for routine use in denaturing sequencing gels by using the specially purified grade of acrylamide available from BDH (Poole, Dorset, England). The solution is deionized for about 20 min by slowly stirring with 10 g/liter of BDH Amberlite MB1 monobed ion exchange resin, filtered and stored at 5°. It can be used for several months afterwards. The urea used is the ultra pure grade available from Schwarz Mann (Orangeburg, New York).

For the construction of the sequencing gels themselves glass plates 20 × 40 cm are used, one with a notch 16.5 × 2 cm cut out from the top. It has been found convenient to use vinyl insulation tape available from Universal Scientific (London, England) to tape the plates and side spacers together. This obviates the use of grease. The side spacers and well formers for use in the sequencing gels are 0.35 mm thick and are made from "Plastikard" obtainable from Slaters Ltd. (Matlock Bath, Derbyshire, England). Well formers are cut out by hand to mold slots 0.5 cm wide and 0.75 cm deep. The gels are run in a slab gel apparatus, a model of which (model RGA 505) is obtained from Raven Scientific (Haverhill, Suffolk, England). This apparatus allows direct contact between both buffer reservoirs and the gel, when the slab gel is inserted with the notched plate directly adjacent to the upper buffer reservoir.

The deoxyribonucleoside triphosphates, the 2',3'-dideoxynucleoside triphosphates and the adenine and cytosine β-D-arabinose nucleoside triphosphates are obtained from P.-L. Biochemicals (Milwaukee, Wisconsin). α-^{32}P-labeled deoxyribonucleoside triphosphates at a specific activity of 350 Ci/mmole, and a concentration of 1 mCi/ml are obtained from the Radiochemical Centre Ltd. (Amersham, Buckinghamshire, England).

The DNA polymerase I nach Klenow is obtained from Boehringer Ltd. (Mannheim, West Germany). Exonuclease III is obtained from New England Biolabs (Beverly, Massachusetts). Reverse transcriptase isolated from avian myeloblastosis virus is obtained from Dr. J. W. Beard, Life Sciences Incorporated (St. Petersburg, Florida).

Preparation of Restriction Fragments. Preparative digests of restriction fragments (normally 100 μg of DNA in total) are routinely fractionated on gradient polyacrylamide slab gets 40 × 20 cm × 3 mm thick as described by

Jeppesen.[14a,15] After electrophoresis the gel can be stained with ethidium bromide at a concentration of 1 μg/ml and the bands visualized under ultraviolet light. Alternatively if the DNA has been labeled before digestion, for example, by nick translation of a small amount of the total DNA with α-^{32}P-labeled deoxyribonucleoside triphosphates, the restriction fragments can be detected by autoradiography. The acrylamide containing the bands is excised and the DNA either eluted by the electroelution method used by Galibert et al.[16] or by soaking out the DNA from the gel. The soaking out is usually done in a buffer of 0.3 ammonium acetate 0.01 M magnesium acetate, 0.1% sodium dodecyl sulfate (SDS), 1 mM EDTA at 37° overnight, after which buffer is filtered from the gel fragments by centrifugation through glass wool. DNA eluted by either method is precipitated in 70% ethanol, 0.3 M ammonium acetate at $-70°$ for 10 min and centrifuged at about 100,000 g for 1 hr. The precipitate is washed in 95% ethanol and recentrifuged at 100,000 g for 15 min. The precipitated DNA is finally redissolved in water to a theoretical concentration of about 0.3 pmoles/μl. The fragments are then ready for use in priming. It is not necessary to extract the ethidium bromide from the DNA. Recovery of the DNA from the gel can, of course, be easily monitored if the DNA has been labeled. The soaking out procedure is found to be efficient for small restriction fragments of less than 200 base pairs, but the electroelution method gives a more quantitative recovery for fragments above this size. One of the problems with either of these elution methods is that sometimes soluble material from the polyacrylamide is eluted. This forms a heavy precipitation in 70% ethanol. Experience has shown that fragments contaminated with this are unsuitable for priming. However, further purification on a small DEAE column can remove the contaminating material. The restriction fragment solution is loaded onto a 0.2 cm³ DEAE column packed in a 1 cm³ disposable syringe and equilibrated in a 0.1 M NaCl, 0.01 M Tris-HCl, pH 7.5, buffer. The column is then washed with several column volumes of the same buffer and the DNA finally eluted by washing the column with 1.0 M NaCl, 0.01 M Tris-HCl and collected in as small a volume as possible. The NaCl concentration is adjusted to 0.3 M and the DNA precipitated in 70% ethanol and centrifuged as before.

Preparation of Template. Single-stranded templates for sequencing have been prepared from double-stranded DNA molecules with complementary strands of asymmetric base composition by alkaline cesium chloride ultracentrifugation. Strands separated by the poly(U,G) method

[14a] See P. G. N. Jeppesen, this volume, Article [39].
[15] P. G. N. Jeppesen, *Anal. Biochem.* **58,** 195 (1974).
[16] F. Galibert, J. Sedat, and E. B. Ziff, *J. Mol. Biol.* **87,** 377 (1974).

of Szybalski *et al.* [17] have also made suitable templates for primed synthesis methods. [3,18]

The exonuclease III method[9] offers an alternative approach. A linear duplex of DNA at a concentration of 5 pmoles in 10 μl of 1 × exonuclease III buffer [70 mM Tris-HCl, pH 8.0, 1.0 mM MgCl$_2$, 10 mM dithiothreitol (DTT)] is degraded at an approximate rate of 500 base pairs/hr/duplex end at 20° at an enzyme concentration of 0.5 units/μl of exonuclease III. One unit of enzyme activity is as defined by Richardson *et al.* [10] After digestion the DNA is extracted with phenol, the aqueous extracted with ether and DNA ethanol-precipitated in 70% ethanol 0.3 M sodium acetate, pH 6.0. The DNA is then redissolved in 10 μl 5 mM Tris-HCl, pH 7.6, and can then be used as a template. Digestion of the duplex can be carried out long enough to produce the limit digest product of two noncomplementary single strands or, alternatively, just long enough to expose the region of interest of the molecule as single-stranded DNA. For example, in a plasmid DNA molecule containing a cloned insert, the sequence of which is required, the plasmid can be cut at a single restriction site flanking the insert, and treatment with exonuclease III need only be carried out for long enough to convert the insert DNA to single strand for sequencing. However, in this case, a further reaction may be necessary to block the 3' ends of the partially degraded duplex with a dideoxynucleoside monophosphate before it can be used as a template. [9] The rationale for this is that because complementarity between the strands remains these will prime on one another, when a priming reaction is carried out, and will therefore incorporate label. Normally this is not a problem because this labeled material is of high molecular weight and fractionates well above the band pattern on the gel resulting from priming from the restriction fragment primer. However, if large restriction fragments are used as primers these have to be removed with a restriction enzyme cut prior to gel electrophoresis (see later). This is a problem since it is likely that the restriction enzyme used to cleave off the primer will also cleave labeled DNA from the priming of the two complementary strands of the template on one another. A result of this is that some of this labeled DNA will be of low molecular weight and become superimposed on the band pattern generated by the primer, making interpretation of the gel impossible. After addition of a dideoxynucleoside monophosphate to the 3' ends of the partially exonuclease III digested duplex and annealing of the restriction

[17] W. Szybalski, M. Kubinski, and W. C. Summers, Vol. 21, Part D, p. 383.
[18] R. Wayne Davies, P. H. Schreier, and D. E. Büchel, *Nature* (*London*) 270, 757 (1977).

fragment primer to this template, only the 3' end of the primer is then a substrate for chain extension by the polymerase. The 3' ends of the degraded duplex should be inert to extension due to the lack of any 3'-hydroxyl. The blocking reaction can be carried out as follows: 5 pmoles of exonuclease III treated DNA are terminated with a dideoxythymidine monophosphate by incubating in a solution of 20 μl containing 25 μM dATP, 10 μCi [α-^{32}P]dATP (350 Ci/mmole), 25 μM dCTP 25 μM dGTP, 500 μM ddTTP, 7 mM Tris-HCl, pH 7.6, 7 mM MgCl$_2$, 50 mM NaCl, 1 mM DTT, and 2 units of Klenow polymerase for 1 hr at 20°. The reaction is terminated by the addition of 1 μl of 0.1 M EDTA and the reaction mix heated at 70° for 15 min. The DNA is purified away from the unincorporated triphosphates by passing over a Sephadex G-100 column in a 1-ml disposal pipette run in a buffer of 5 mM Tris-HCl, pH 8.0, 0.01 mM EDTA. The DNA which should have incorporated a small amount of label can be followed down the column with a hand monitor and should be collected in as small a volume as possible, 100–200 μl. The DNA is then dried down, redissolved in about 10 μl of water, and is then ready for use as a template.

Annealing of Primer to Template. One strand of the restriction fragment which is complementary to the template is annealed as follows. One-half microliter of template usually at a concentration of 0.5 pmoles/μl is mixed with 2.5 μl of double-stranded restriction fragment at a theoretical concentration of 0.3 pmoles/μl and 0.5 μl of 10 × reaction buffer (0.5 M NaCl, 66 mM Tris-HCl, pH 7.6, 66 mM MgCl$_2$, 10 mM DTT). This gives a theoretical molar ratio of primer to template of 3 : 1. The mixture is sealed in a capillary heated at 100° in a boiling water bath for 3 min and the DNA annealed at 67° for 20 min. This is sufficient time to allow for a complete hybridization of the DNA. The DNA solution is then diluted to 10 μl with 1 × reaction buffer.

Primings. Only the chain termination procedure is described here. However, a detailed protocol for the plus and minus method can be found in a recent review by Barrell.[19]

Primings can be carried out using any size of restriction fragment, although with small fragments of less than 20 base pairs the concentration of primer may have to be increased to achieve the same degree of incorporation of label. This is probably due to the fact that at the temperatures of incubation used small primers melt off from the template easily. Larger primers, above 100 base pairs long, for example, are usually cleaved off

[19] B. G. Barrell, *Int. Rev. Biochem.* **17**, 125 (1978).

the copied material prior to electrophoresis. This is dealt with in a later section. The synthesis is carried out according to Sanger et al.[1] using the Klenow subfragment of DNA polymerase I. However, reverse transcriptase, which requires slightly modified reaction conditions, can be used, as mentioned previously, and does have some advantages (see later).

In a standard sequence analysis using DNA polymerase I, five reactions are usually carried out, four reactions with each dideoxy analog and one reaction with the arabinose CTP analog. The reason for including the arabinose CTP reaction is that while in the extension reactions with the A, G, and T dideoxy analogs every base gives rise to a band, there is a tendency for certain bands in the C extension to appear very faint, this invariably being the first C band in a run of C residues. The arabinose CTP analog, on the other hand, does not give rise to this problem, although there are other problems connected with its use which will be discussed later, and similarly with the arabinose ATP analog. In fact, there is no real necessity for using the ara-ATP analog, since the ddATP gives a perfectly adequate termination at each base.

In each extension the labeled nucleotide triphosphate used is by convention dATP. The $[\alpha\text{-}^{32}P]dATP$ is normally at a specific activity of 350 Ci/mmole and a concentration of 1 mCi/ml. For each reaction, 1 μl of the $\alpha\text{-}^{32}P$-labeled dATP in 50% ethanol is dried down in a siliconized tube under vacuum. A different reaction mix, containing deoxyribose nucleoside triphosphates and buffer, is used with each chain terminator. The composition of the A, G, C, and T reaction mixes which are, respectively, for use with the ddATP, ddGTP, ddCTP (and ara-CTP), and ddTTP analogs are given in the tabulation. In each mix one of the dNTPs, the competing dNTP (for example, the dTTP in the T mix) is at a low concentration compared to the other two dNTPs. In the A reaction mix the competing dNTP is the $[\alpha\text{-}^{32}P]dATP$ which is dried down separately.

	A mix (μl)	G mix (μl)	C mix (μl)	T mix (μl)
dTTP 0.5 mM	20	20	20	1
dGTP 0.5 mM	20	1	20	20
dCTP 0.5 mM	20	20	1	20
10 × reaction buffer	20	15	15	15

Two microliters of the annealed DNA solution are dispensed to five drawn out capillaries. To each is added 1 μl of a different reaction mix, although for the ara-C extension the C extension mix is used. To each reaction mix is added 1 μl of the appropriate dideoxynucleoside triphosphate solution. The concentration of the ddNTP is determined by the degree of extension required and by the base composition of the template DNA. Figure 2 shows a typical titration of ddATP concentration against chain extension. To obtain equivalent extensions for the C, G, and T reactions as in the A extension using the 0.5 mM stock of ddATP the following concentrations of stock solutions should be used on a template with an equimolar composition of the 4 bases: 2.0 mM ddTTP, 1.0 mM ddGTP, and 1.0 mM ddCTP. For the ara-C extension a stock concentration of 40 mM should be used. The 4 μl in each capillary are then used to take up the dried down [α-^{32}P]dATP by blowing and sucking up the solution in and out of the capillary in the bottom of each siliconized tube. Once the label is taken up into solution, 1 μl (0.2 units) of Klenow DNA polymerase I is added to each capillary and mixed into the reaction solution in a similar manner. The polymerase which is kept at a concentration of 1 unit/μl in 0.1 mM KH$_2$PO$_4$, pH 8.0, 50% glycerol buffer is diluted into the same buffer to a concentration of 0.2 units/μl before hand. The Klenow polymerase must be used since it lacks the 5′ → 3′-exonuclease activity of intact DNA polymerase I. If this activity were present, the 5′ terminus of the primer would be degraded and the gel analysis relies on all the oligonucleotides having a common 5′ end. The reactions are left to incubate in the capillaries for 15 min at 20°. After the 15-min incubation period, 1 μl of 0.5 mM dATP (unlabeled) is added to each reaction. This chase step is continued for a further 15 min period and each reaction is then terminated by the addition of 1 μl of 0.2 M EDTA. These conditions allow for a quantitative termination of the extending 3′ ends in a dideoxy or arabinose monophosphate.

In the final reaction volume the concentration of the competing dNTP's in the C, G, and T reactions is 2 μM. The concentration of the competing dNTP in the A reaction which is the concentration of the [α-^{32}P]dATP is about 0.5 μM. The molar ratios of deoxy- to dideoxynucleoside triphosphate, therefore, range from 1/200 for the A and T reactions to 1/100 for the C and G. The molar ratio of dCTP to ara-CTP is 1/4000.

Reverse transcriptase can be used as an effective alternative to DNA polymerase I with the four ddNTP analogs. The DNA is annealed as before and 2 μl of the annealed material aliquoted to each of four capillaries. To each of these is added 1 μl of reaction mix, which is the same for each of the four chain termination reactions. The composition of this is given in the tabulation. The 10 × reverse transcriptase reaction buffer is 0.5 mM Tris-HCl, pH 8.5, 0.05 M MgCl$_2$, 0.02 M DTT. One microliter of a different dideoxynucleotide triphosphate solution is then added to each capillary. The concentrations of the stock solutions of each ddNTP which should be used to give roughly equal extensions are 0.0001 mM ddATP,

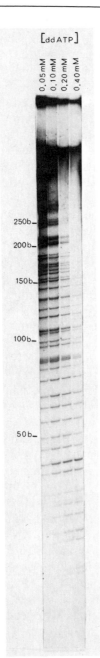

FIG. 2. An example of priming reactions carried out according to the text (with the ϕX AluI restriction fragment 13 annealed to the ϕX minus strand) using a range of ddATP concentrations that illustrate the relative degrees of extension which can be obtained. The concentrations given in the figure are the final concentrations of the ddATP in the reaction. Each reaction was carried out using 1 μCi of α-[32P]-labeled dATP (350 Ci/mmole). The distance in bases (b) from the 3' end of the primer is indicated in the figure against the band pattern.

0.5 mM dCTP	20 μl
0.5 mM dGTP	20 μl
0.5 mM dTTP	20 μl
10 × reverse transcriptase reaction buffer	20 μl

0.01 mM ddTTP, 0.02 mM ddCTP, and 0.01 mM ddGTP. The 4 μl in each capillary are then used to take up 1 μl of dried down [α-^{32}P]dATP as before. One microliter of avian myeloblastosis virus (AMV) reverse transcriptase, obtained at an activity of about 5 units/μl is diluted 1/5 in 1 × reverse transcriptase reaction buffer. One unit of AMV reverse transcriptase activity is as defined by Kacian and Spiegelman.[20] One microliter of the diluted enzyme is then added to each capillary and the contents mixed. The reaction carried out in the capillary is incubated at 37° for 15 min. After 15 min, 1 μl of 0.5 mM dATP (unlabeled) is added to each reaction and mixed, and the incubation continued for a further 15 min at 37°. The reaction is terminated by the addition of 1 μl of 0.1 M EDTA. If the chase step with cold dATP is omitted some termination occurs before A residues in the sequence due to the low concentration of [α-^{32}P]dATP.

In the final reaction volume the concentration of the competing dNTPs in the G, C, and T reactions is 25 μM. In the A reaction the concentration of the competing dNTP, which is the labeled dATP, is about 0.5 μM. The molar ratios of the deoxy- to dideoxy-NTP are, therefore, about 10 for the G and T reactions, 5 for the C reaction, and 20 for the A reaction.

Removal of the Primer. If the restriction fragment being used as a primer is large, it is necessary to cleave this from the radioactive copy of the template before gel electrophoresis. If this is not done, the length of time for electrophoresis has to be greatly increased in order to electrophorese the smallest of the labeled oligonucleotides to the bottom of the gel, and the proportion of sequence that can be resolved is decreased. Since the polymerase resynthesizes the restriction site used to generate the restriction fragment primer in the first place, the same restriction enzyme can be used to remove the primer. After the incubation period the addition of EDTA is omitted and 1 unit of restriction enzyme is added to each reaction, and the capillaries incubated for a further 5 min at 37°. The small amount of DNA present should be completely cut in this time. One microliter of 0.2 M EDTA is then added to terminate the reaction.

Alternatively cleavage by the single-site ribosubstitution method of Brown[21] can be used. This method involves an initial synthesis to incorpo-

[20] D. L. Kacian and S. Spiegelman, Vol. 29, Part E, p. 150.
[21] N. L. Brown, *FEBS Lett.* **93**, 10 (1978).

rate a ribonucleoside monophosphate onto the 3' end of the primer. The primer is then used for the chain termination reaction and it can be removed afterward by cleavage with ribonuclease or alkali at the position of ribosubstitution. The following procedure, for example, can be used to cleave off AluI restriction fragments. AluI recognizes and cuts the sequence

$$5'\ A\text{—}G\overset{\downarrow}{\underset{\uparrow}{\text{—}}}C\text{—}T\ 3'$$

The principle of this protocol is that in the presence of the annealed primer and template, $[\alpha\text{-}^{32}\text{P}]$dTTP and rCTP, and a buffer containing Mn^{2+}, DNA polymerase I will carry out a limited synthesis copying the adjacent sequence of the template containing the remainder of the AluI recognition sequence. The polymerase will, therefore, incorporate at least a single rCMP and dTMP residue onto the 3' end of each primer, and may incorporate more depending on the sequence further out from the AluI recognition site. The reaction is left long enough so that synthesis is extended as far as is possible from every primer terminus. After purification of the unincorporated deoxy- and ribonucleoside triphosphates away from the DNA, the chain termination reaction can be carried out as normal. Subsequent cleavage at each ribo-CMP residue adjacent to the primer, or in some cases at the several ribo-CMP residues adjacent to the primer produces oligonucleotides all with a common 5' terminus.

The annealing of template and primer is carried out as before but in a buffer of 100 mM Tris-HCl pH 7.6, 10 mM DTT, and 0.5 M NaCl. The annealed DNA is diluted to 7 μl with water and 1 μl 5 mM rCTP and 1 μl 10 mM MnCl$_2$ are added. Five microcuries of $\alpha\text{-}^{32}\text{P}$-labeled dTTP (350 Ci/mmole) are dried down and taken up into solution, and 1 unit of Klenow DNA polymerase I added. The reaction mixture is incubated for 30 min on ice. One microliter of 0.2 M EDTA is added before the mixture is loaded on a small Sephadex G-100 column in a 1-ml disposable pipette. The column is run in a buffer of 5 mM Tris-HCl, pH 7.6, 0.1 mM EDTA. The DNA which is labeled due to the incorporation of counts from the $[\alpha\text{-}^{32}\text{P}]$dTTP can be followed down the column with a monitor, collected in a minimum volume, and dried down to 10 μl. Two microliter aliquots of this are then taken for each chain termination reaction. After the chain termination has been carried out, 1 μl of 0.2 M EDTA and 1 μl of ribonuclease A at 10 mg/ml are added. The mixture is incubated for 60 min at 37°. An alternative is to cleave at the ribo-C residues with alkali by adding 1 μl of 2 M NaOH to each reaction. The samples are then sealed on a capillary and incubated at 100° for 20 min.

This is a useful method and is particularly advantageous when there is a second restriction enzyme site of the same kind used to generate the primer, in close proximity to and in the direction of chain extension from the primer. If this restriction enzyme is used to remove the primer and chain extension has also occurred past the second site, this second site will give rise to a separate pattern of bands superimposed on those due to extension from the primer. The single site ribosubstitution method obviously obviates this problem. In the case of *Alu*I restriction fragments this approach is a necessity since *Alu*I appears to be inhibited by single-stranded DNA. The method can be applied in principle to most restriction enzyme sites since, as mentioned, all four ribonucleoside monophosphates can be incorporated by DNA polymerase I in the presence of Mn^{2+}, although, it should be noted, to differing degrees of efficiency.

Gel electrophoresis. High resolution sequencing gels are made according to Sanger and Coulson.[22] Routinely these are 8% or 6% polyacrylamide, 7 M urea gels run in a continuous buffer system of 90 mM Tris-borate, pH 8.3, 1 mM EDTA. The gel plates, spacers, and well formers are as described in the Materials section. The inside of the notched plate is siliconized with Repelcote to facilitate easy removal of the plate from the gel after electrophoresis. The two plates and spacers are held together with waterproof vinyl electrophoresis tape. The volume of the gel is about 30 ml. Enough gel solution is made as follows:

Urea, 15 g
Stock 38% acrylamide–2% bisacrylamide, 6 ml
10 × Tris-borate, EDTA running buffer, 3 ml
1.6% Ammonium persulfate, 1 ml
Volume made up to 30 ml with water

A 10 × Tris-borate, EDTA running buffer stock solution is made with

Tris base, 108g
Boric acid, 55 g
Na_2 EDTA, 9.3 g
Dissolved in 1 liter of water; pH should be 8.3

After the urea is dissolved 30 μl of N,N,N',N'-tetramethylethylenediamine is added and the gel poured immediately, by holding the glass plates at an angle of 45° and pipetting the acrylamide solution in from a 25-ml pipette. The gels are allowed to set at an angle of 5° with the well former in position. After 1 hr the gel should have set and the well former can be removed. The wells should be immediately flushed out with buffer to remove any unpolymerized acrylamide. The electrophoresis tape on the bottom of the gel is removed and the slab gel inserted into the electrophoresis apparatus with the notched plate adjacent to the upper buffer tank. Both buffer tanks are filled with 1 × running buffer.

The reaction samples terminated by the addition of EDTA are then mixed with 5 to 10 μl of deionized formamide dye mix and heated at 100°

[22] F. Sanger and A. R. Coulson, *FEBS Lett.* **87**, 107 (1978).

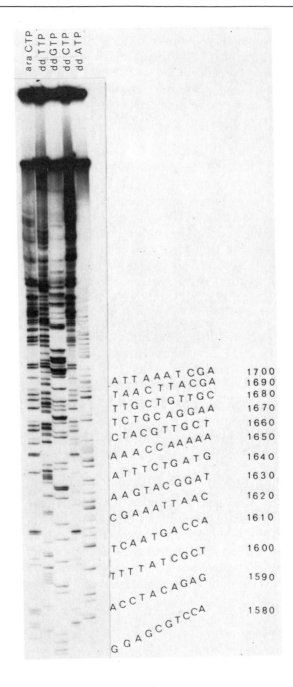

for 10 min. This denatures the radioactive transcripts from the complementary template. (This mix is prepared by stirring formamide with Amberlite MB1 monobed ion exchange resin to deionize the formamide, removing the resin by filtration and then adding bromphenol blue and xylene cyanol dyes to concentrations of 0.3%.) Prior to loading of the samples it is a good idea to flush the wells out again with buffer to remove any concentrations of urea. One to two microliters of the sample in formamide dye are loaded in each well using a finely drawn out capillary. The electrophoresis is carried out at a constant current of 30 mA. At this current the voltage is initially 1.7 kV, but drops rapidly to 1.3 kV and remains at this value for the rest of the electrophoresis. When the gels are run at this power they heat up to about 70°. It is important to maintain the gels at this high temperature, since the high urea concentration alone does not keep the oligonucleotides fully denatured. In particular, it does not prevent the formation of hairpin loop structures within a single-stranded oligonucleotide, which leads to a deviation from its expected mobility on the basis of chain length. (Even when the gels are run hot, very stable secondary structures are still present and cause problems when trying to deduce the sequence.) The bromphenol blue marker that corresponds on an 8% gel to oligonucleotides of a chain length of 30 nucleotides takes 90 min to migrate to the bottom of the gel; the xylene cyanol marker corresponding to a chain length of 80 nucleotides requires 180 min to migrate the same distance. There is enough sample denatured in the formamide to carry out several loadings on the same gel. Each set of samples should be loaded in adjacent wells preferably at intervals of 90 to 120 min apart. The samples should be briefly heated to 100° before loading each time. On an 8% gel sequences can easily be read accurately as far as bands corresponding to an oligonucleotide chain length of about 250 nucleotides, after $4\frac{1}{2}$ hr electrophoresis. Generally 6% gels are used to read sequences further out than this since the electrophoresis time required is shorter.

After electrophoresis the gel is fixed in 10% acetic acid for 10 min, washed, blotted, covered in Saran wrap, and autoradiographed overnight at room temperature. Using the quantities of DNA in this protocol and the same specific activity $[\alpha\text{-}^{32}P]dATP$, the gels should be easily readable after 12 hr of autoradiography.

Interpretation of Sequencing Gels

A typical result of a chain termination experiment carried out with DNA polymerase I is shown in Fig. 3. This is a priming on G4 phage plus

FIG. 3. Autoradiograph of a sequencing experiment carried out on G4 phage plus strand using a 60-nucleotide long *Hin*fI restriction fragment. The primer was not removed prior to the electrophoresis. Priming reactions were carried out with each dideoxynucleoside triphosphate and the cytosine arabinose nucleoside triphosphate. One microcurie of α-^{32}P-labeled dATP was used in each reaction. The derived sequence is given adjacent to the band pattern.

strand (isolated from the virion) with a 60 nucleotide long *Hin*fI restriction fragment. The primer was not split off before electrophoresis, and the oligonucleotides fractionated at the base of the gel are about 30 nucleotides from the primer site. The picture gives a fair impression of the kind of resolution that should be expected. The advantage of using the ara-C reaction can be clearly seen, since strong C bands show up where weak ones occur in the dideoxy extension. There is, however, one missing C band in the ara-C extension at nucleotide position 1586. This tends to be a problem with the ara-ATP and ara-CTP analogs. The disappearance of bands is also much more frequent when primers are split off with a restriction enzyme at 37°. Whether this is due to a more efficient proof reading of the arabinose residue or further extension from a terminated chain, at the higher temperature, is not known. The ara-CTP should, therefore, not be relied on alone. With the dideoxynucleoside triphosphate analogs no such loss of bands is seen at 37°.

In a typical priming, when the primer is not removed, the sequence can usually be read starting a few bases from the primer site. However, when the primer is removed with a restriction enzyme, labeled oligonucleotides of up to 20 residues are only present as very weak bands, if at all. This is thought to be due to small oligonucleotides melting off the template once the primer is removed and being rapidly degraded, by the exonuclease activity of the polymerase, or by some other contaminating nuclease, or by a combination of both.

Occasionally, although not seen in Fig. 2, regions of sequence are found where there are intense bands corresponding to oligonucleotides of the same chain length across all the tracks of the gel and in register with the band pattern of the gel. One explanation for these is that they are due to regions of the sequence which the polymerase finds difficult to copy and, therefore, represent premature points of termination common in each reaction. When reverse transcriptase is used as an alternative to DNA polymerase I, some of these "pile-ups" disappear, although new ones may be found in different parts of the sequence. In the case of reverse transcriptase, these are sometimes found to be sequences where the

TABLE I

Problem	Cause	Solution
Several band patterns superimposed upon one another, the bands of which are probably out of register with one another	(a) The restriction fragment may have been copurified with one or several other fragments that prime on the template	(a) Redigest the isolated fragment preparation with another restriction enzyme which only cuts the contaminating fragments, rerun the digest on a gel and reisolate the restriction fragment pure

TABLE I (*continued*)

Problem	Cause	Solution
	(b) If the primer has been generated by a restriction cut it may be a superimposed pattern generated by a second site in close proximity to the primer	(b) Use the single site ribosubstitution method or increase the dideoxy- to deoxy-NTP ratio
	(c) The preparation of single-stranded template may be contaminated with some single-stranded DNA from the complementary strand	(c) Repurify the template
	(d) If the template used has been generated by exonuclease III treatment and the primer has been removed with a restriction enzyme the extra bands are probably due to the restriction enzyme cutting the repaired template DNA	(d) Prepare the template inert to repair by blocking all the 3′ ends with a dideoxymonophosphate or digest to completion with exonuclease III
A band pattern in which every band appears as a doublet or triplet, etc.	(a) The restriction fragment has a heterogeneous 5′ end or if the primer is being split off with a restriction enzyme the enzyme has contaminating exonuclease activity	(a) Prepare a new preparation of restriction fragment with a purer enzyme or remove the primer with a different enzyme if a site is available
	(b) If the single site ribosubstitution method is being used this phenomenon may be due to incomplete ribosubstitution, i.e., in a *Hae*III site for example (5′ G–G$\downarrow$C–C 3′) this may be due to only partial addition of two ribo-CMPs to the primer	(b) Increase the time and enzyme concentration for the ribosubstitution reaction
A band or bands across all the tracks of the gel in a particular region of the sequence	(a) The band or bands may be due to the restriction fragment which has end-labeled during the priming. This will have a mobility similar to some of the chain terminated oligonucleotides if the copied material has been cleaved off the primer prior to electrophoresis, although the labeled restriction fragment will not necessarily	(a) The sequence should ideally be determined without cleaving off the primer, or by using a different primer in order to read the obscured region of sequence

(*continued*)

TABLE I (*continued*)

Problem	Cause	Solution
	be in register with the band spacing. Consequently, the reading of the sequence in this region may be obscured	
	(b) The band may be due to a pile-up of the polymerase at a particular sequence	(b) Try using reverse transcriptase instead of DNA polymerase I which will give pile-ups at different points
	(c) If the strand of the restriction fragment that is noncomplementary to the template is capable of forming some internal secondary structure that leaves a non-base-paired 5' extension, the strand will form a self-priming template and incorporate label during the priming reaction. An intense band will be seen at the same position in each gel track corresponding to the termination of some of the transcripts, from the self-priming reaction, at the 5' end of the strand. The sequence in this region will almost certainly be obscured. It may also be difficult to interpret the sequence below the position of this intense band since base-specific bands will be present that correspond not only to the template sequence but also to the sequence of the 5' extension	(c) The use of reverse transcriptase, which has no $3' \rightarrow 5'$ exonuclease activity, instead of the Klenow DNA polymerase I, may eliminate this problem. In order to self-prime using reverse transcriptase the strand must have a perfectly base-paired 3' terminus, which in most cases is unlikely
A high background of bands but of even intensity corresponding to every oligonucleotide chain length in each track. The background may be in register with the bands irrespective of chain length or only in register at low chain lengths forming a smear of	(a) If completely in register due to random nicking of the template (b) If only in register at low chain lengths this could be due to a number of possibilities. The most likely is that the primer and template have been prepared from preparations con-	(a) and (b) Short of preparing new DNA the sequence should be read by reading the strongest band at each chain length

TABLE I (*continued*)

Problem	Cause	Solution
background counts higher up the gel	taminated with some other DNA or RNA. When a priming reaction is carried out, the contaminating DNA or RNA gives rise to labeled oligonucleotides of heterogeneous sequence. At low chain length these oligonucleotides will have similar mobilities. At higher chain lengths these mobilities will diverge to give a smear of counts	
A sudden decrease in band spacing or jump size, sometimes to 0, so that bands in different tracks may line up, then a return to the normal band spacing	The relationship between oligonucleotide length and mobility is not necessarily a straightforward one. As oligonucleotide size increases by the further addition of residues to the 3' end, these may give rise to sequences capable of forming base paired loop structures during the gel electrophoresis. Such structures may give rise to an increased mobility for an oligonucleotide relative to what would be expected from its length. This is manifested as the sudden decrease in jump size between bands, known as a compression, as the oligonucleotide is increased in size by the addition of nucleotides involved in the formation of the secondary structure. Further addition of nucleotides not involved in the secondary structure results in a return to the normal band spacing	Some compressions can be removed by running the gels hotter; heat-resistant plates may be required in some cases. However, the only effective way of getting over problems of this kind is to derive the sequence from the other strand. Regions of rotationally symmetrical sequence in the duplex DNA give rise to compressions in different regions of sequence of the complementary strands

single-stranded template can assume internal secondary structure. In the case of DNA polymerase I, however, there is nothing which is similarly suggestive about the sequences around these pile-ups.

Apart from the general limitations of the method outlined above there are, as with any sequencing method, specific problems which may be encountered and which are peculiar to the particular primer or template being used and the sequences being analyzed. Table I summarizes some of the common problems which may be identified, their possible causes, and suggests some solutions. If problems such as these or others cause difficulty in determining an unambiguous sequence, the sequence should always be cross-checked by determining the same region of sequence from the complementary strand. In fact this should be adopted as a matter of routine even when there is no apparent difficulty in reading the gel.

[59] Determination of RNA Sequences by Primer Directed Synthesis and Sequencing of their cDNA Transcripts

By P. K. GHOSH, V. B. REDDY, M. PIATAK, P. LEBOWITZ, and S. M. WEISSMAN

I. Introduction

Until recently, RNAs have been sequenced primarily by methods which involve specific endo- and/or exonucleolytic degradation of *in vivo*[1] or *in vitro*[2] labeled, purified species of RNA followed by construction of sequences from analysis of degradation products. Although laborious, these methods have sufficed for analysis of RNAs which are present in cells in some abundance and which can be separated from other cellular RNAs. However, they cannot be used to sequence RNAs which are not obtainable in relatively pure form, and they are difficult to apply to RNAs of long chain length.

An alternative approach to analysis of RNA sequences involves synthesis of DNA copies of RNA, usually with the enzyme reverse transcriptase, followed by analysis of the resultant complementary DNAs (cDNAs). This approach has three important advantages over the more direct methods. First, it is not necessary to extensively purify RNAs before sequencing if primers are used which are radiolabeled and carry sequences complementary to a specific RNA. In practice DNA restriction fragments constitute the most easily obtainable specific primers. Second, the method developed by Maxam and Gilbert[3] for sequencing terminally

[1] G. G. Brownlee, "Determination of Sequences in RNA." Am. Elsevier, New York, 1972.
[2] R. G. Gupta and K. Randerath, *Nucleic Acids Res.* **4,** 3441–54 (1977); A. Simoncsits, G. G. Brownlee, P. S. Brown, J. R. Rubin, and K. Guitly, *Nature (London)* **269,** 833–836 (1976); H. Denis-Keller, A. M. Maxam, and W. Gilbert, *Nucleic Acids Res.* **4,** 2527–2538 (1977).
[3] A. Maxam and W. Gilbert, *Proc. Natl. Acad. Sci. U.S.A.* **74,** 560–564 (1977).

labeled DNAs is relatively easy to perform and allows determination of sequences of between 250–300 nucleotides in a single experiment. Third, it is possible to sequence RNAs of long chain length by sequentially binding a series of primers to adjacent regions of any given RNA and sequencing the cDNA products derived from overlapping regions.

Approaches of this type have recently been used successfully by McReynolds *et al.*[4], Szostak *et al.*,[5] and Hagenbuchle *et al.*[6] Our group has also used this approach to sequence specific regions of a number of simian virus 40 (SV40) mRNAs.[7–10] In our experiments we annealed specific radiolabeled SV40 restriction fragments of 20–100 nucleotides to polyadenylated RNA extracted from infected or transformed cells under rather stringent conditions, thus minimizing the likelihood of nonspecific binding to nonviral RNAs and synthesis of nonviral cDNAs. We also found it necessary to fractionate cDNA products, since cDNAs of different lengths and different internal sequences were frequently obtained with a single primer as a result of multiple alternative rearrangements of sequences within mRNA templates. The procedure we have used for RNA sequencing involves a multiplicity of steps, all of which must be performed with care to obtain optimal results. In brief, these steps are as follows: preparation of cellular or viral RNAs, isolation of restriction fragments complementary to specific regions of the RNA(s) being analyzed, ^{32}P labeling of the 5' termini of these fragments, hybridizations in solution of radiolabeled restriction fragments and RNA, isolation of the resultant RNA–DNA hybrids, synthesis of cDNA on RNA templates, destruction of the RNA templates, separation of cDNA molecules on polyacrylamide gels, and sequence analysis of individual cDNAs. The following sections deal with the details of this methodology.

II. Methods

A. Preparation of Cellular RNAs. Any RNA extraction procedure which completely removes endogenous DNA and proteins and reduces to

[4] L. McReynolds, B. W. O'Malley, A. D. Nisbet, J. E. Fothergill, D. Givol, S. Fields, M. Robertson, and G. G. Brownlee, *Nature (London)* 273, 723–728 (1978).
[5] J. W. Szostak, J. I. Stiles, C.-P. Bohl, and R. Wu, *Nature (London)* 265, 61–63 (1977).
[6] D. Hagenbuchle, M. Santer, J. Steitz, and R. T. Mans, *Cell* 13, 551–564 (1978).
[7] P. K. Ghosh, V. B. Reddy, J. Swinscoe, P. V. Choudary, P. Lebowitz, and S. M. Weissman, *J. Biol. Chem.* 253, 3643–3647 (1978).
[8] P. K. Ghosh, V. B. Reddy, J. Swinscoe, P. Lebowitz, and S. M. Weissman, *J. Mol. Biol.* 126, 813–846 (1978).
[9] V. B. Reddy, P. K. Ghosh, P. Lebowitz, and S. M. Weissman, *Nucleic Acids Res.* 5, 4195–4214 (1978).
[10] V. B. Reddy, P. K. Ghosh, P. Lebowitz, M. Piatak and S. M. Weissman, *J. Virol.* 30, 279–296 (1979).

a minimum enzymatic or chemical cleavage of RNA should provide an RNA preparation suitable for nucleic acid hybridization and cDNA synthesis. We have generally prepared RNA from cells grown in tissue culture by modification of the method of Penman[11] as follows. All steps are carried out at 4° unless otherwise noted. Precautions to avoid ribonuclease contamination include the use of gloves and baking glassware overnight at 200°. Although it is unnecessary for RNA to be labeled, light labeling with [³H]uridine is often helpful in the initial purification steps to follow recoveries and column chromatograms.

A cell pellet, generally 2–4 ml in volume and containing 10^9–10^{10} cells, is suspended in 20 ml of hypotonic buffer consisting of 0.01 M Tris-HCl, pH 7.4, 0.01 M NaCl, 0.003 M $MgCl_2$. After allowing 20 min for swelling, cells are lysed with ten strokes in a tight-fitting Dounce homogenizer followed by four passages through a 26-gauge hypodermic needle. This procedure results in complete lysis of cells without apparent lysis of nuclei. The cell lysate is then centrifuged for 10 min at 4000 rpm in a Sorvall SS34 rotor. The cytoplasmic supernatant is gently decanted, taking care not to disturb the nuclear pellet. Nuclei are resuspended in 10 ml of hypotonic buffer and recentrifuged. The nuclear pellet is again taken up in 10 ml of hypotonic buffer. Following addition of 1.5 ml of a mixture of 6.7% Tween 40 (v/v) and 3.3% sodium deoxycholate (w/v), the suspension is vortexed for 2–3 sec and nuclei sedimented a third time. The nuclear pellet is then resuspended one final time in 10 ml of hypotonic buffer and resedimented. This treatment with detergent and final wash step free nuclei of virtually all tabs of cytoplasmic material. All four cytoplasmic supernates are combined. Pelleted nuclei are taken up in 10 ml of 0.01 M Tris-HCl, pH 7.4, 0.5 M NaCl, 0.05 M $MgCl_2$ and incubated for 30 min at 37° in the presence of DNase (Worthington, electrophoretically purified), 50 μg/ml, resulting in degradation of DNA and loss of viscosity. From this point nuclear and cytoplasmic suspensions are treated identically. Both are brought to 0.5% in sodium dodecyl sulfate (SDS) and 1 mM in EDTA after which solid Proteinase-K (Merck) is added to a final concentration of 500 μg/ml and incubation carried out for 2 hr at 37°. Following incubation, equal volumes of redistilled phenol (water saturated and containing 0.2% 8-hydroxyquinoline) and chloroform : isoamyl alcohol (100 : 1) are added, and each suspension shaken on a wrist action shaker for 30 min at room temperature. Centrifugation is then carried out at 12,000–15,000 rpm for 20 min in the Sorvall SS34 rotor. If the aqueous phase in each extraction is well demarcated from a condensed interface, it is removed with a Pasteur pipette and reextracted with equal volumes of phenol and chloroform : isoamyl alcohol one or more additional times until no protein

11 S. Penman, *J. Mol. Biol.* **17**, 117–130 (1966).

appears at the interface. If the initial aqueous phase does not separate well from the lower phenol phase, it is removed along with protein at the interface and reextracted with an equal volume of phenol. The resultant aqueous phase usually separates satisfactorily from the phenol phase and can then be reextracted with equal volumes of phenol and chloroform: isoamyl alcohol until no protein remains at the interface. One-tenth volume of 20% potassium acetate, pH 5.4, and 2 volumes ethanol are then added to the clarified aqueous phases and precipitation of RNA carried out overnight at $-20°$.

Nuclear and cytoplasmic RNAs are next chromatographed on columns of oligo(dT)–cellulose to isolate polyadenylated RNA species.[12] The RNAs are first centrifuged at 1800 rpm for 20 min in the International RP-2 centrifuge, dried in a stream of N_2 and solubilized in 15–20 ml of 0.01 M Tris-HCl, pH 7.5, 0.5 M KCl. They are then loaded onto columns of 5–7 ml bed volume of oligo(dT)–cellulose: cellulose (1:1) made up in this same buffer. In order to maximize binding of poly(A) terminal RNAs, column effluents are cycled through columns five to ten times. Columns are then washed extensively with 0.01 M Tris-HCl, pH 7.5, 0.5 M KCl to remove all nonbinding nonpolyadenylated RNAs, after which polyadenylated RNAs are eluted with H_2O. Fractions containing poly(A) terminal RNAs are pooled and the RNAs precipitated by addition of potassium acetate and ethanol as above.

B. Isolation and Labeling of DNA Restriction Fragments. In order to obtain satisfactory nucleic acid sequencing results, it is necessary that DNA fragments used as reverse transcriptase primers be of relatively small size, i.e., from 20 to 150 nucleotides in length, and pure. Although it is often possible to obtain suitable fragments in pure form by cleavage of DNA with a single restriction enzyme, it is more frequently necessary to digest DNA with one restriction enzyme and then redigest certain of the resultant fragments with second restriction enzymes to obtain the desired fragments in highly purified form. In our experience cDNA sequencing has been most successful when >0.25 μg of a given DNA fragment is annealed to RNA. Therefore, large enough quantities of DNA must be digested initially to obtain adequate quantities of individual restriction fragments. For SV40 DNA with a chain length of 5226 nucleotides, digests are performed on a minimum of 150–200 μg DNA.

We have purchased restriction enzymes from either New England Biolabs, Bethesda Research Laboratories, Inc., or Boehringer Mannheim or have prepared them by published procedures. Digestions are carried out on unlabeled purified DNAs under conditions specified by these suppliers or described in the literature. Restricted fragments are usually

[12] H. Aviv and P. Leder, *Proc. Natl. Acad. Sci. U.S.A.* **69**, 1408–1412 (1972).

separated by electrophoresis on 4% polyacrylamide gels (set on 10% polyacrylamide traps) in 0.04 M Tris, 0.04 M sodium acetate, 1 mM EDTA, pH 7.8. Agarose gels may be used for separating DNA fragments too large to be resolved adequately on polyacrylamide gels. Fragments are localized by immersing gels briefly in ethidium bromide (0.1 μg/ml) and looking for fluorescent bands with a short wavelength uv light. The desired bands are excised from gels and homogenized in 2–3 ml of 0.1 × standard saline citrate [(SSC) : 0.15 M NaCl and 0.015 M sodium citrate, pH 7.0] for about 5 sec at low speed in a Tekmar homogenizer. The resultant acrylamide suspensions are then shaken for 12–15 hrs at room temperature on a mechanical wrist action shaker after which they are centrifuged at 20,000 rpm in the Sorvall SS34 rotor. The acrylamide pellets are resuspended in 2–3 ml of 0.1× SSC and after similar extraction centrifugation is again carried out. The supernates are then combined and any residual acrylamide pelleted. DNAs are precipitated from the final supernates by addition of 2 volumes of ethanol, overnight refrigeration at −20°, and centrifugation at top speeds for 1 hr at 4° in either the Spinco SW41 or SW50 rotors.

At this point, 5′ terminal labeling with ^{32}P is carried out on recovered DNA fragments (a) if they are of appropriate size and require no further restriction enzyme digestion or (b) if, after 5′ terminal phosphorylation, redigestion with second restriction enzymes would yield the desired fragments with a 5′ label on the correct DNA strand. If redigestion with second enzymes following phosphorylation would not yield the desired fragments with a 5′ label on the desired DNA strand, redigestion of the individual unlabeled parent fragments is carried out prior to labeling.

To remove 5′ terminal phosphate residues, DNA fragments (invisible on the bottom of centrifuge tubes) are taken up in 100 μl of 0.05 M Tris-HCl, pH 8.5, 0.01 M MgCl$_2$ and 5 mM β-mercaptoethanol containing 10–20 μg $E. coli$ alkaline phosphotase (Sigma) and incubated 40 min at 37°. Reactions are then extracted twice with phenol and DNA fragments precipitated by the addition of 0.1 volume of 3 M sodium acetate and 2 volumes of ethanol. Following centrifugation as above, DNA fragments are dissolved in 100 μl of 0.05 M Tris-HCl, pH 8.5, and 5 mM spermidine and heated at 100° for 3 min to denature the termini of the DNA strands. Following quick cooling in a Dry Ice–ethanol bath, dephosphorylated fragments are labeled at their 5′ termini by incubation for 50 min at 37° in reactions of 150 μl containing 0.05 M Tris-HCl, pH 8.5, 0.01 M MgCl$_2$, 5 mM β-mercaptoethanol, 5 mM spermidine, 0.15–0.2 mCi of [γ-^{32}P]ATP (New England Nuclear or ICN; 3000–5000 Ci/mmole), and 10–20 μg of T4 polynucleotide kinase prepared by the procedure of Richardson.[13] These

---

[13] C. Richardson, *Proced. Nucleic Acid Res.* **2**, 815–828 (1971); G. L. Cantoni and P. R. Davies, eds., Harper, New York, 1971.

reactions are also terminated by extraction with phenol after which DNA fragments are precipitated in the presence of 0.3 M sodium acetate and ethanol as above. Fragments which have already been cleaved to the desired size are then reelectrophoresed on 8–10% polyacrylamide gels in the aforementioned buffer system to separate them from all traces of labeled ATP. Fragments which require an additional cleavage with a second restriction endonuclease are redigested at this point and then electrophoresed on polyacrylamide gels. At the conclusion of these electrophoreses, wet gels are covered tautly with Saran wrap and autoradiographed, using Kodak RP X-omat film. Exposures of 1–2 min are usually satisfactory for obtaining good autoradiograms. Radiolabeled fragments are then covered from gels as described above.

C. *DNA–RNA Hybridizations.* Hybridizations of 5' terminally labeled DNA fragments and cellular RNA follow the procedure of Casey and Davidson.[14] Each DNA fragment (usually from 10 to 50 μCi, 0.25–1.5 μg DNA) is first solubilized in 150 μl of 80% formamide [(Eastman) deionized by stirring 2–4 hr with about 0.1 volume of Biorad Ag 501 $\times$ 8 mixed bed resin], 0.4 M NaCl, 0.01 M Pipes (piperazine, N,N-bis(2-ethanesulfonic acid), pH 6.4, and denatured by heating to 100° for 5 min. Polyadenylated RNA, usually from 0.2 to 1 mg in 150 μl of this same buffer, is then added to each denatured DNA fragment. Hybridization mixtures so constituted are brought to 85° for 5 min to denature RNA and then incubated 15–18 hr at 50°.

Following hybridization, it is advisable that double-stranded RNA–DNA hybrids be isolated from unhybridized radiolabeled DNA. This is easily accomplished by diluting hybridization reactions tenfold with 0.5 M KCl, 0.01 M Tris-HCl, pH 7.5, and passing them through columns of oligo(dT)–cellulose as above. Unhybridized DNA passes directly through these columns and duplex molecules composed of poly(A) terminal RNA and bound [³²P]DNA are eluted with H_2O. Fractions of the latter are pooled and hybrids precipitated by the addition of 0.1 volume of 3 M sodium acetate and 2 volumes of ethanol.

D. *cDNA Syntheses.* cDNA syntheses are next carried out on hybrid molecules with the enzyme reverse transcriptase serving as catalyst, the RNA components of hybrids as templates and the short hydrogen-bonded DNA fragments as enzyme primers. In these syntheses, primers are elongated in a 5' → 3' direction by the sequential addition of nonlabeled deoxynucleoside 5'-monophosphates. Thus, the final cDNA products are labeled with ³²P at their 5° termini, contain sequences complementary to internal and 5' terminal sequences of template RNAs, and may be used to determine the nucleotide sequences of these regions of the RNA templates (see Fig. 1). Reactions for synthesizing cDNAs contain 50 mM

[14] J. Casey and N. Davidson, *Nucleic Acids Res.* **4**, 1539–1552 (1977).

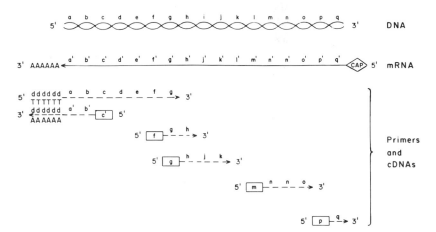

FIG. 1. Schematic diagram showing a hypothetical DNA, a processed mRNA transcript of this DNA containing both a splice and a tandem duplication of sequences, and the cDNAs obtained by reverse transcription of the mRNA using primers complementary to specific regions of the mRNA. Sequences a,b,c,d, etc., represent sequences on the coding strand of DNA, while sequences a',b',c',d', etc., represent the complementary sequences. Note that extension of the primer with the "f" sequence yields a cDNA with a sequence identical to that in template DNA; extension of the primer with the "g" sequence yields a cDNA lacking template DNA sequences, indicative of a splice in the mRNA; and extension of the primer with the "m" sequence yields a cDNA with a tandem duplication of template DNA sequences, indicative of a tandem duplication of sequences in the mRNA. Also, note that the cDNA synthesized with the fragment "p" as primer identifies the 5' terminal sequences of the mRNA while the two-step cDNA synthesis using oligo(dT) as the primer in the first step and fragment "c'" in the second step permits identification of the 3' terminal sequences of the mRNA. Polarities of the DNA, mRNA, and cDNAs are indicated.

Tris-HCl, pH 8.3; 6 mM magnesium acetate; 60 mM NaCl; 10 mM dithiothreitol; 1 mM each of dCTP, dTTP, dGTP, and dATP; 0.2 μg of a DNA–RNA hybrid; and 5 units of avian myeloblastosis virus reverse transcriptase in each 200 μl of reaction volume. Reactions are carried out for 3 hr at 41°. At the conclusion of the syntheses, reaction mixtures are made 0.2 N in NaOH and incubations are continued for an additional hour. Reactions are then neutralized with 1 N HCl and 0.5% SDS–phenol extraction and DNA precipitation carried out as above.

Most of the cDNA syntheses we have performed have made use of reverse transcriptase generously provided by Dr. J. Beard, Research Resources, Viral Oncology Program, National Cancer Institute. The efficiency of elongation of primers has varied somewhat from preparation to preparation of this enzyme. Furthermore, enzyme preparations have contained traces of ribonuclease which we have attempted to remove or neutralize by the following means.

(1) Chromatography on columns of BioGel P200 in 0.2 M KPO$_4$, pH 7.4, 2 mM dithiothreitol, 0.2% Triton X100, and 10% glycerol. (This purification method was kindly communicated to us by Drs. P. Berg and J. Wahl.)

(2) affinity chromatography on columns of Sepharose-bound 2'-3'-uridylic acid.

(3) inclusion of placental RNase inhibitor[15] in reaction mixtures.

(4) inclusion of 4 mM sodium pyrophosphate in reaction mixtures.[16]

The results of our cDNA syntheses using reverse transcriptase directly or following these treatments have been similar. Furthermore, control experiments with SV40 cRNA have suggested little or no degradation of this template during cDNA syntheses. Therefore, the amounts of RNase in reverse transcriptase appear to be too small to adversely affect cDNA syntheses by the method we have described. The lack of a major effect in our system may be related to the fact that the cDNAs we have studied, have generally been of relatively short length (several hundred nucleotides or smaller).

E. *Two-Step Hybridizations and cDNA Syntheses.* As mentioned above, cDNA syntheses carried out by the aforementioned one-step hybridization–elongation procedure are useful for establishing the internal and 5' terminal sequences of RNAs since RNA templates are transcribed by reverse transcriptase in a 3' → 5' direction. 3' terminal RNA sequences, however, cannot be determined by this procedure, and it has been necessary to utilize a two-step hybridization–elongation procedure to establish the 5' → 3' polarity with respect to RNA templates required for determination of 3' terminal sequences (see Fig. 1).

In the first step of this procedure, cDNAs are synthesized on polyadenylated RNA templates using oligo(dT) as the primer. The reaction mixture for this step is identical to the cDNA synthetic reaction described in the preceding section with three changes: (1) nonlabeled oligo(dT) (Collaborative Research) replaces a ^{32}P restriction fragment as the reverse transcriptase primer; to assure complete saturation of all poly(A) termini, oligo(dT) is added in approximately one-tenth the amount by weight as RNA. (2) One dXTP bears a ^{3}H label so that cDNA becomes labeled. (3) Actinomycin D is added to the reaction in a final concentration of 1 μg/ml to inhibit synthesis of double-stranded DNA. Incubation is carried out for 3 hr at 41° after which RNA templates are degraded with 0.2 N NaOH as described. cDNAs are then recovered by 0.5% SDS–phenol extraction of the reaction mixture, passage of the aqueous phase

[15] P. Blackburn, G. Wilson and S. Moore, *J. Biol. Chem.* **252**, 5904–5910 (1977).
[16] J. C. Myers, S. Spiegelman, and D. L. Kacian, *Proc. Natl. Acad. Sci. U.S.A.* **74**, 2840–2843 (1977).

through a column of Sephadex G100 in 0.01 M Tris-HCl, pH 7.5, 0.01 M MgCl$_2$, and 1 mM EDTA and precipitation of excluded DNA with Na acetate and ethanol as described.

In the second step, recovered cDNAs are annealed to a 5'-^{32}P-labeled restriction fragment and the latter extended in a 5' → 3' direction (see Fig. 1). If a DNA fragment derived from a site close to the 3' terminus of an RNA is chosen as a primer, cDNAs with a sequence identical to that at the 3' terminus of the RNA and terminating with a stretch of poly(dA) residues are obtained. In this step, a ^{32}P-labeled restriction fragment (approximately 0.25–0.5 μg) is first solubilized in 50 μl of 100% formamide and denatured at 95°–100° for 5 min. Precipitated poly(dT)–terminal cDNA (approximately 50–150 μg) is then taken up in an equal volume of 4× SSC and combined with the denatured restriction fragment. This mixture is aspirated into a capillary tube, maintained at 100° for 10 min to ensure denaturation of both DNA and cDNA and then incubated at 50° for 24 hr. DNA is recovered from the hybridization reaction by addition of 10 volumes of water, precipitation with ethanol, and centrifugation as described above. Following two washes of the pellet with ethanol, extension of the DNA primer bound to cDNA is carried out with reverse transcriptase in the basic cDNA synthetic reaction lacking actinomycin D. At the conclusion of this reaction, 0.25% SDS–phenol extraction and precipitation of cDNAs are carried out as noted. Precipitated cDNA–cDNA hybrids are then ready for denaturation and electrophoresis. *Escherichia coli* DNA polymerase I can also be used for these extensions, but total cDNA synthesis in our hands has been greater using reverse transcriptase.

F. Electrophoretic Separation of cDNAs. In order to obtain maximal information from cDNA syntheses, it is essential that the cDNAs be fractionated on polyacrylamide gels. The reasons are as follows.

(1) To prepare completely purified species for sequence analyses.

(2) To detect template heterogeneity with respect to internal splices and termini.

(3) To obtain information on the size of discrete cDNAs by coelectrophoresis with known size markers. As discussed below, sequences can usually be obtained on approximately 250–300 nucleotides at the 5' ends of cDNAs and the chain length of longer products can be approximated only from their electrophoretic mobility.

Fractionations of cDNAs are usually carried out on slab gels of 8% polyacrylamide (acrylamide : bisacrylamide of 19 : 1), 7 M urea in the 0.09 M Tris, 0.09 M boric acid, 0.025 M EDTA, pH 8.3 (1× TBE) buffer system of Peacock and Dingman.[17] The concentration of polyacrylamide can be

---

[17] A. C. Peacock and S. W. Dingman, *Biochemistry* 7, 668–674 (1968).

altered to achieve maximal resolution of cDNAs in specific size ranges. Gels measure 40 cm in length, 16 cm in width, and 2 mm in depth. Prior to electrophoresis, cDNA samples from either the single- or double-step primer elongations are taken up in 20 μl of 0.05 N NaOH and 5 M urea and incubated at 95° for 2 min to denature DNA. Bromphenol blue and xylene cyanol dyes are then added to final concentrations of 0.05% and electrophoresis is carried out at 400–800 V at room temperature. Electrophoresis may be stopped when the bromphenol dye reaches the bottom of the gel or may be continued for any desired length of time to obtain maximal resolution of specific gel bands.

The separation of cDNAs obtained with gels of uniform polyacrylamide concentrations is usually adequate for most purposes. Figure 2 demonstrates the separation of a multiplicity of cDNAs synthesized on a template of late polyadenylated cytoplasmic RNA from cells lytically infected with SV40 using an SV40 DNA restriction fragment 30 nucleotides in length as a primer. However, resolution of bands is sometimes not satisfactory on gels of this type. In such instances, we have used gradient polyacrylamide–7 M urea gels to enhance separation of cDNA products. In particular, linear gradients from 5% polyacrylamide–5% glycerol to 9% polyacrylamide–20% glycerol in 7 M urea and 1 × TBE have proved effective for such fractionations.

At the conclusion of electrophoreses, gels are autoradiographed and cDNAs recovered as described above. For heavily labeled cDNAs, exposure times of 5–10 min are satisfactory to obtain good autoradiograms; for lightly labeled cDNAs, exposure times of up to 18 hr may be necessary.

G. *Cleavage of cDNAs with Restriction Enzymes.* It is sometimes advantageous to cleave cDNAs with restriction enzymes prior to polyacrylamide gel electrophoresis. We have found this procedure useful in four specific situations.

(1) When cDNAs are longer than 350–400 nucleotides, their resolution on gels can often be enhanced by cleavage at an appropriate site. Little information is lost by this procedure since as noted above nucleotides beyond the first 250–300 nucleotides at the 5' end of a cDNA are not readily amenable to DNA sequence analysis.

(2) When a number of cDNAs are synthesized in quantities too small for sequencing, cleavage with a restriction enzyme sometimes results in their conversion to one or more species containing adequate radioactivity for sequencing of their 5' terminal nucleotides.

(3) cDNAs copies from spliced RNAs can often be identified by their failure to undergo cleavage with a restriction enzyme known to cleave at a specific site on the DNA which codes for a segment of RNA removed in the splicing reaction.

FIG. 2. Autoradiogram of the 8% polyacrylamide–7 M urea gel electrophoretic separation of 5'-^{32}P terminally labeled cDNAs obtained by reverse transcriptase catalyzed elongation of a 30 nucleotide restriction fragment of SV40 DNA bound near the 5' end of SV40 late mRNAs extracted from virus-infected cells. The multiplicity of cDNAs arises from the presence of a number of 5' termini and different splicing patterns within the SV40 19 S mRNAs.

(4) When a cDNA species is not abundant enough for sequencing or when it is over 250–300 nucleotides in length, the presence of a particular nucleotide sequence can be deduced from its susceptibility to digestion with an appropriate restriction enzyme. Furthermore, size analysis of the resultant radiolabeled fragment provides an approximate location of the site of enzymatic cleavage.

Cleavage of cDNAs can be carried out only with restriction enzymes which cleave single-stranded DNA. We have had good results with the *Hae*III enzyme, but other enzymes cleaving single-stranded DNA should be useful as well.[18] Since single-stranded DNAs are far more resistant to restriction endonuclease cleavage than double-stranded DNAs, vigorous digestion conditions must be employed. For cleavage with *Hae*III, [^{32}P]cDNAs are incubated in reaction mixtures containing 6 mM Tris-HCl, pH 7.5, 6 mM NaCl, 6.6 mM MgCl$_2$ 6 mM β-mercaptoethanol, and 5 units of enzyme in a total volume of 50 μl for 4–6 hr at 37°. Double-stranded cDNAs can be cleaved with any of the currently available restriction enzymes. Cleavages are carried out under conditions specified by suppliers or described in the literature. Digested cDNAs can be electrophoresed directly on 8–10% polyacrylamide–7 M urea gels without deproteinization.

H. Nucleotide Sequencing. Single-stranded 5'-^{32}P-labeled cDNAs recovered from 8% polyacrylamide–7 M urea gels are sequenced by the procedure of Maxam and Gilbert.[3] This procedure makes use of chemical methods to partially cleave terminally labeled DNAs at each of the four component bases. The products resulting from separate cleavage at each of the bases can then be electrophoresed on polyacrylamide gels and the DNA sequence read directly from the mobilities of the partial cleavage products. The method for cleaving DNAs has been well described in the original paper by Maxam and Gilbert and the reader is referred to it for methodological details. We note here three modifications of the procedure which we have found helpful.

(1) We have noted excessive degradation of cDNAs when methylation of guanylic acid residues is carried out with 0.5% dimethyl sulfate (DMS). We thus methylate G residues in the presence of 0.25% DMS. Methylation of adenylic acid residues has been satisfactory using 0.5% DMS.

(2) When the quantity of a radiolabeled cDNA is not great enough to carry out partial cleavages at all four bases and when the sequence of the relevant segment of template DNA is known, it is often advisable to carry out only two cleavage reactions, one at guanylic and to a lesser extent at

[18] G. N. Godson and R. J. Roberts, *Virology* **73**, 561–567 (1978).

adenylic acid residues and the second at thymidylic and cytidylic acid residues. With only these two cleavages, it may still be possible to discern if a cDNA sequence is or is not the same as the template sequence on DNA.

(3) The concentration of the polyacrylamide in gels used for DNA sequencing is determined by the length of DNA fragments being sequenced. In the original description of the method, a 20% poly-acrylamide–7 M urea gel of 1.5 mm thickness was recommended for sequencing DNAs of approximately 100 nucleotides. In our experiments we have found that 7 or 8% polyacrylamide–7 M urea gels of 0.3–0.5 mm thickness provide maximal resolution of cleavage products derived from DNAs of 250–300 nucleotides, confirming the earlier observations of Sanger and Coulson.[19]

The use of thin gels in DNA sequencing has been an especially important technical advance. These gels are superior to thicker gels for a number of reasons: (1) they require less time for prerunning, (2) they run faster at any given voltage setting, and (3) they do not overheat as readily when run at high voltages; hence, there is less disparity in the rate at which DNAs in the center and at the edges of the gel run. With gels of 0.3–0.5 mm thickness, 20 cm width and 40 cm length, electrophoreses can be carried out at 800 V for the first hour and 1000–1500 V thereafter.

III. Interpretation of cDNA Sequences

A. cDNA Sequences Complementary to Internal RNA Sequences. Since the sequence of SV40 DNA is known, it has been possible to evaluate the fidelity of reverse transcriptase-mediated copying of internal sequences of specific SV40 mRNAs in our transcriptional system. We have never noted insertion of nucleotides or changes in internal nucleotide sequences during reverse transcription. Thus, copying appears to be faithful under our transcription conditions, allowing internal RNA sequences to be directly deduced from sequences of cDNAs.

If the sequence of the DNA template for an RNA is not known, then the most that can be derived from cDNA sequencing is simply the sequence of its RNA template and any additional information, e.g., potential codons for initiation and termination of translation, that can be deduced from the RNA sequence. However, when the sequence of the DNA template for an RNA is known, then direct comparison of cDNA and DNA sequences localizes the genomic template for the RNA and may also provide information on steps that may be involved in its posttranscrip-

---

[19] F. Sanger and A. R. Coulson, *FEBS Lett.* **7**, 107–110 (1978).

tional modification. In our experiments with SV40 mRNAs, we have noted three different arrangements of internal RNA sequences and a fourth also seems plausible. All can be deduced from direct comparison of cDNA and DNA sequences as follows.

(1) Colinearity of RNA and DNA: When cDNA and template DNA sequences match, RNA is a colinear complementary copy of the DNA (see cDNA synthesized on primer fragment "f" in Fig. 1).

(2) Splices within mRNA: When the sequence of a cDNA coincides with the genomic DNA sequence to a point but then departs, a search must be made for the downstream (3' terminal) cDNA sequence on the sequence of template DNA. Most often this cDNA sequence can be detected at a distant site toward the 3' end of the template DNA strand (see cDNA synthesized on primer fragment "g" in Fig. 1). This spatial arrangement of cDNA sequences on template DNA defines a splice in an mRNA, the precise termini of which can be deduced by careful comparison of DNA and cDNA sequences.

(3) Insertion of RNA sequences, tandem duplications. When cDNA and genomic DNA sequences match to a point but then diverge, the search for the 3' terminal cDNA sequence on template DNA may lead to their identification at a position toward the 5' end of the template DNA strand (see cDNA synthesized on primer fragment "m" in Fig. 1). This occurrence defines the insertion of sequences within an RNA. In SV40, we have noted the insertion of viral sequences in one subgroup of 16 S late mRNA; in this case the inserted sequences are derived from the DNA region immediately adjacent to the sequences at which cDNA and template DNA sequences diverge, resulting in a tandem duplication of sequences in the mRNA. The mechanism for insertion of sequences in SV40 16 S mRNA is not known, but probably involves splicing within RNA molecules of greater than unit genomic length. Sequence duplications have also been noted in the 16 S mRNAs of polyoma (R. Kamen, personal communication).

(4) Linkage of host and viral mRNAs: In the investigation of viral mRNA structure by the method we have described, a cDNA and a viral DNA sequence may be identical to a point, but differ downstream, and the search for downstream cDNA sequences on template viral DNA may be unrevealing. In virus-infected or -transformed cells, this striking finding would suggest covalent linkage of viral and host RNA sequences. Such linkage could result from posttranscriptional recombination of viral and host mRNAs or from transcription of covalently linked viral and host DNA.

  B. *cDNA Sequences Complementary to the 3' Termini of mRNAs.* Determination of the 3' terminal sequences of mRNAs is greatly facilitated by the presence of contiguous "marker" tracts of poly(A). Thus one need only read the sequence on a cDNA transcript of the 3' terminal region of an mRNA up to the point of the long poly(dA) tract to obtain the 3' terminal sequence of that mRNA. Any sequence in cDNA not followed by poly(dA) cannot be transcribed from the 3' terminus of a polyadenylated mRNA but must represent an artifactual 3' terminal reverse transcriptase stop. In transcriptions at the 3' ends of early and late SV40 mRNAs we have not observed such artifactual stops.

C. *cDNA Sequences at the 5' Termini of RNAs.* Whereas no serious problems arise in determining the 3' termini of polyadenylated mRNAs from cDNA syntheses, two problems exist in determining the 5' termini of RNAs by this method. First, the absence of a transcribable marker at the 5' termini of RNAs analogous to poly(A) at 3' termini makes it very difficult to distinguish between reverse transcriptase stops at the 5' termini of *in vivo* RNAs and artifactual stops at sites short of *in vivo* termini. Artifactual stops may arise either as a result of creation of artifactual 5' termini by degradation of cellular RNAs during their isolation or at any subsequent step in the cDNA synthetic procedure or by premature terminations by reverse transcriptase. A number of causes for premature termination have been suggested. The presence of a modified base in an RNA may block elongation of the cDNA chain. In particular, N^6-dimethyladenosine blocks elongation by reverse transcriptase[6] since it cannot form a hydrogen bond between its exocyclic nitrogen and the base of an incoming nucleoside triphosphate. Similarly, N^1-methyladenosine has been found to interfere with reverse transcription of tRNA by DNA polymerase.[20] Secondary structure in an RNA chain has also been reported to inhibit distal cDNA chain elongation. It is also theoretically possible that certain nucleotide sequences may be read as transcriptional terminators by reverse transcriptase.

We have attempted to test for premature termination of reverse transcription in cDNA syntheses performed on high molecular weight SV40 cRNA. In this type of control experiment, we have found premature termination of transcription to be an infrequent event. Great caution must be exercised in attempting to deduce the 5' termini of RNAs solely on the basis of cDNA synthetic studies. Nevertheless, it may be possible to establish specific 5' termini: (1) if corroborative data from studies of *in vivo* labeled RNAs is available, (2) if the sequence of nucleotides adjacent to a 5' terminal capped structure is known from analysis of labeled mRNA, and (3) if a sequence in question is read through by reverse transcriptase during copying of any RNA template containing that sequence. Determination by polyacrylamide gel electrophoresis or sucrose

[20] B. Wittig and S. Wittig, *Nucleic Acids Res.* **5**, 1165–1178 (1978).

gradient sedimentation of the size of RNA extracted from cells and at various stages in the cDNA synthetic procedure may be helpful in ruling out extensive RNA degradation.

The second problem relevant to determination of 5' terminal RNA sequences is difficulty reading the last few 3' nucleotides of cDNA transcripts. This problem is due largely to the frequent presence of dark bands of undegraded cDNA at the top of all channels of DNA sequencing gels. In these instances, however, it is still possible to estimate 3' termini within 2–4 nucleotides.

[60] Computer Analysis of Nucleic Acids and Proteins

By CARY L. QUEEN and LAURENCE JAY KORN

The progress of molecular biology has been paced by the development of effective methods to determine the primary structure of macromolecules: proteins, RNA, and DNA. Recent techniques for sequencing DNA have become so rapid that analysis of the resultant data is a major undertaking. It is therefore natural to turn for aid to another achievement of contemporary science, the electronic computer.

In this article, we begin by outlining some general principles of computer utilization. Then we discuss the computer programs available for nucleic acid analysis and present in detail one comprehensive program that can be used for amino acid as well as nucleotide sequences. We conclude with a comment on the interpretation of computer-generated information.

General Principles

A computer analyzes data under the direction of a program. The quality of the program determines the range of functions the computer can perform, as well as the speed, and therefore the cost, of their performance. While writing a computer program is a painstaking project, using a well-designed program is easy once it is in operation. To facilitate setting the program up at a computer center, it is advisable to request as much help as needed from personnel such as user consultants.

Computer programs are conveniently stored on magnetic tape and will generally be received in that form. The tape, with appropriate identifying information specified by a consultant, should be deposited at the computer center. To use the program, the data is punched on IBM cards in accord with the instructions. Then the data cards, preceded and followed by several job control cards, are run through a card reader connected to the computer. The job control cards provide the computer with accounting information, set limits on the amount of computer resources which can be used, and direct the retrieval of the program on the tape. Control cards vary among computer centers and therefore must be made with the guidance of a user consultant. There is no tolerance for error in this task: even blank spaces on control cards are important. Fortunately, these cards do not have to be changed once they are made.

A computer program may come in one of two forms: a *source* program written in a language such as Fortran, or an *object* program produced by translation of the source program into a numeric language suitable for direct use by the computer. The translation is itself performed by another program called a compiler. If a source program is obtained originally, it is economical to have it translated and permanently stored as an object program, so the translation need not be performed by the compiler each time the program is run.

An alternative to the use of punched cards for communicating with a computer is to run the program interactively: the data can be typed directly on a teletype and the results received on the teletype or an attached cathode ray screen. Any program can be controlled in this fashion, and one of the programs we discuss later is specifically written for easy interactive use.

Problems and Costs

In theory using a program is straightforward, but in practice problems may arise—a situation not unfamiliar to experimental scientists. There are three possible sources of difficulty. The job control cards may be incorrect, particularly at first. This problem is generally symptomized by the computer's failure to produce anything intelligible. If it arises recheck the control cards, and be sure that the limits for resources such as computer storage are set high enough. Second, there may be an error on the data cards. The problem is often characterized by correct output to a point, followed by nonsense or termination. In this case refer to the data format instructions accompanying the program, and make sure they have been followed precisely. Try eliminating part of the data to isolate the error. Third, there may be a mistake in the program itself, sometimes indicated by an isolated error in the output. Such bugs can be most easily diagnosed and corrected by the original programmer.

Costs associated with computer data analysis vary depending on the amount of data and the nature and quality of the program. The specific programs we will discuss should not in general be too costly. Costs can often be further reduced by a preliminary consideration of what information is really desired. For example, in examining phage ϕX174 for dyad symmetries, probably only long dyad symmetries with a small central loop are of interest. The program we will describe in detail has the capacity to perform a search under such restrictions. It is also true that different computer centers charge at different rates. University computer centers, not being run for a profit, may be particularly *expensive*.

Computer Programs Available

The first applications of computers to nucleic acid analysis were directed toward the prediction of RNA secondary structure.[1,2] The most recent of these programs, described by Pipas and McMahon,[3] incorporates the thermodynamic suppositions of Gralla and Crothers.[4] The analysis of a tRNA by this program requires about a minute, and the time apparently increases as the square of the sequence length. Consequently, analyzing a sequence the length of phage Qβ could be a significant expense.

More recently, with the advent of rapid DNA sequencing technology, programs have been written to perform a variety of functions on long nucleotide sequences. The programs of Staden[5,6] are especially effective in keeping track of long nucleic acids, such as that of phage ϕX174, as they are progressively sequenced. Corrections, insertions, and deletions can be made to any part of the stored sequence. The programs also count nucleotide and codon frequency, translate using the genetic code, and locate restriction enzyme sites. They possess some ability to find repeated regions and dyad symmetries in DNA, and to compare proteins with the properties of the amino acids taken into consideration. The programs are meant to be used interactively: the researcher sits at a teletype and the computer prompts him with questions to state his desires. They are especially suitable for use on relatively small computers such as the PDP 11, which might be owned by a laboratory. The program of McCallum and Smith[7] shares some features with Staden's programs and runs rapidly on IBM machines.

The most comprehensive nucleic acid program currently available seems to be the one we described[8] and have since expanded by the inclu-

[1] I. Tinoco, Jr., O. C. Uhlenbeck, and M. D. Levine, *Nature (London)* **230**, 362 (1971).
[2] W. M. Fitch, *J. Mol. Evol.* **1**, 185 (1972).
[3] J. M. Pipas and J. E. McMahon, *Proc. Natl. Acad. Sci. U.S.A.* **72**, 2017 (1975).
[4] J. Gralla and D. M. Crothers, *J. Mol. Biol.* **73**, 497 (1973).
[5] R. Staden, *Nucleic Acids Res.* **4**, 4037 (1977).
[6] R. Staden, *Nucleic Acids Res.* **5**, 1013 (1978).
[7] D. McCallum and M. Smith, *J. Mol. Biol.* **116**, 29 (1977).
[8] L. J. Korn, C. L. Queen, and M. N. Wegman, *Proc. Natl. Acad. Sci. U.S.A.* **74**, 4401 (1977).

sion of more flexible data handling facilities and features for the analysis of proteins. It performs all the functions of the program discussed above, although not as conveniently for a sequence in progress, and has additional capabilities of two types. It performs other searching operations, notably for polypeptides coded by nucleic acids and oligonucleotides "reverse coded" by proteins. More importantly, it locates regions of possibly imperfect repetition and dyad symmetry in one or more nucleotide or amino acid sequences. The program is compatible with interactive use, but not specifically designed for it, and can be used on IBM computers. A complete analysis of a sequence 400 nucleotides long costs several dollars. The program has been used by a number of investigators and is available upon request.

A Comprehensive Program

Our program accepts sequences of nucleotides or amino acids as data. A nucleotide sequence is entered as a string of the letters A, C, G, T, and U. The letters T and U are not distinguished, so DNA sequences may be compared against RNA sequences. The letters P for purine, Q for pyrimidine, and N for nucleotide are accepted by procedures 4 and 14 below. An amino acid sequence is entered as a string of the standard three-letter abbreviations[9] preceded by an asterisk. Other symbols may be included in a nucleotide or amino acid sequence, but they only serve to hold a place and are not included in the analysis. Blank spaces may also be included to facilitate alterations on data cards, and are removed by the program before analyzing the sequence.

Any number of nucleic acid and protein sequences, each with a maximum length of 32,000 symbols, may be analyzed during a single computer run. One may indicate which of sixteen independent procedures is to be applied to each sequence or set of sequences (Table I). A variety of parameters may be set to guide the program (Table II). The analysis can be applied to a DNA sequence as entered, or to its opposite strand. An option exists to treat a nucleotide sequence as a circle. All procedures may be restricted to a specified part of a sequence. Conversely, several sequences can be concatenated by the program into a longer one. The simple input data format is described later.

The first procedure of the program (Table I) lists the entered sequence and numbers every tenth nucleotide or amino acid. For ease of comparison, numbers always refer to the original sequence, even when the opposite strand or only part of the sequence is displayed. Setting the parameter SHORT = 1 compresses the output to fit on a teletype. The second proce-

---

[9] A. L. Lehninger, "Biochemistry," p. 962. Worth Publ., New York, 1975.

TABLE I
PROCEDURES OF THE PROGRAM

1. Printing of sequence
2. Nucleotide and amino acid frequency
3. Dinucleotide frequency
4. AG- and CT-rich regions
5. AT- and GC-rich regions
6. AC- and GT-rich regions
7. Subsequence dictionary
8. Matching subsequence dictionary
9. Matching subsequence dictionary, simplified
10. Repeated regions
11. True symmetries (alphabetic palindromes)
12. Dyad symmetries (self-complementary regions)
13. Genetic code
     Translation of nucleotide sequences
     Reverse translation of amino acid sequences
14. Location of oligonucleotides and polypeptides
     Oligonucleotides in nucleic acids
     Polypeptides coded by nucleic acids
     Oligonucleotides reverse coded by proteins
     Polypeptides in proteins
15. Trinucleotide frequency
16. Codon frequency, separate reading frames

dure determines the number and percentage of each kind of nucleotide or amino acid. In this and other procedures for nucleic acids, percentages are based exclusively on the symbols A, C, G, T, and U.

The third procedure lists the frequency of the sixteen possible dinucleotides in a nucleic acid sequence, together with the expected frequencies calculated from the formula (expected proportion of XY) = (proportion of X) × (proportion of Y). These values are useful in detecting an anomalous distribution of dinucleotides. A χ^2 test for statistical significance should be employed in this context, since surprisingly large fluctuations can occur by chance alone.

Procedure 4 examines a nucleic acid for regions rich in the purines A and G or the pyrimidines C, T, and U. The criterion for richness is that six out of eight consecutive nucleotide bases be purines or pyrimidines. The program prints two copies of the nucleotide sequence and overlines the purine-rich tracts in one and the pyrimidine-rich tracts in the other. Similarly, procedure 5 locates GC- and AT-rich regions, which are postulated to have regulatory significance,[10] and procedure 6 finds AC- and GT- rich regions.

[10] W. Gilbert, in "RNA Polymerase" (R. Losick and M. Chamberlin, eds.), pp. 193–205. Cold Spring Harbor Lab., Cold Spring Harbor, New York, 1976.

TABLE II

PARAMETERS OF THE PROGRAM

Parameter	Range	Default[a]	Function
NUMSEQ	1–	10	Maximum number of entered sequences
NUMRES	0–	200	Maximum number of entered subsequences for procedure 14
MAXLEN	1–32,000	1000	Maximum length of entered sequences
MAXLENRES	1–32,000	20	Maximum length of entered subsequences for procedure 14
SHORT	0,1	0	SHORT = 1 compresses computer output
GAP	0–	0	Number of blanks inserted between concatenated sequences
NUMAGREE	3–	b	Number of matches required by procedures 8 and 9
PHASE	1–4	1	Coding frame in procedures 13 and 16 (4 = all frames)
MISSRES	0–	0	Number of mismatches allowed by procedure 14
MAXPROB	0.00001–0.02	0.002	Maximum probability of chance occurrence of homology
EXPECT	0–	c	Sets MAXPROB to expect this number of chance homologies
MINMATCH	3–	3	Minimum number of matches in a homology
MINRATIO	0.5–1.0	0.75	Minimum ratio of matches to length
LOOPLENGTH	0–	3	Maximum length of a loopout
DISTANCE	0–	d	Maximum distance between repeated regions
LOOPDIST	0–	20	Maximum length of central loop in dyad symmetry
DUBIOUS	0,1	0	DUBIOUS = 1 allows G–T matches in dyad symmetries

[a] The default value is the value chosen by the program if that parameter is not set.
[b] Chosen to produce a moderate amount of output.
[c] Chosen according to the value of MAXPROB.
[d] No maximum is placed on the distance.

The seventh procedure produces a list of all subsequences in a given sequence (Fig. 1). For this purpose, a sequence of n nucleotides is viewed as consisting of n nested "words," each of which begins at a different nucleotide and continues to the end of the sequence. The program arranges these words in lexicographical order, retaining only enough of each word to distinguish it from all others. The resulting dictionary provides a way of checking for the occurrence of a particular oligonucleotide. For example, a glance at Fig. 1 shows that TGATA occurs at positions 3 and 47 of the sequence, whereas TGC does not appear.

26	TGAC
53	TGAG
3	TGATAA
47	TGATAC
41	TGGCGGTGA
17	TGGCGGTGT
23	TGT
10	TTA
25	TTG

FIG. 1. Part of the dictionary of oligonucleotides produced by procedure 7 from the λ P_L sequence (Table IV). The number in the left-hand column indicates the starting position of the oligonucleotide within the sequence. All figures are reproduced from computer output.

A modification of the dictionary provides a way of finding perfectly repeated oligonucleotides. Our method is based on the fact that, in a dictionary, words in which the first few letters are the same occur close together. Procedure 8 creates a dictionary which lists only those oligonucleotides that agree with some other one to a specified number of places, set by the parameter NUMAGREE (Fig. 2). Inspection of this table reveals perfect repeats. Procedure 9 eliminates those oligonucleotides from the dictionary which are wholly contained within others, improving readability at the risk of some loss of information. Procedures 7–9 may also be used for amino acid sequences.

An advantage of this method for finding repeated regions is that a region which occurs more than twice will be readily apparent, because all the repetitions will be listed together. That fact may be utilized to search for tracts which appear in each of a set of sequences by concatenating the

2	30	ATAAATAC
1	2	ATAAATAT
1	44	CGGTGATAA
2	44	CGGTGATAC
3	42	CGGTGATAG
3	19	CGGTGTTA
2	20	CGGTGTTG

FIG. 2. Part of the dictionary of repeated oligonucleotides produced by procedure 8 from a concatenated sequence of three λ promoter regions (Table IV). There are 35 entries in the whole dictionary. The first numbers indicate the sequence in which the oligonucleotide appears ($1 = \lambda P_L$, $2 = \lambda P_R$, $3 = \lambda P_{RM}$), and the second numbers refer to the position within the sequence (position 60 = transcription initiation).

sequences and producing a dictionary from the resulting string. For example, from Fig. 2 it is obvious that the oligonucleotide CGGTGATA occurs in the phage λ right,[11] left,[12] and repressor maintenance[13] promoters. When using the concatenation feature of the program for this purpose, it is best to set the parameter GAP = 4 so that oligonucleotides will not continue from one sequence into the next. Of course, the dictionary method is primarily useful in detecting regions of perfect repetition.

A central part of our program consists of algorithms to find homologies which are not necessarily perfect. Procedures 10, 11, and 12 find repetitions, true symmetries, and dyad symmetries in a single sequence, while a similar procedure finds repetitions between two sequences (Fig. 3). The algorithms, which allow both mismatches and loopouts, have been previously described in detail.[8] The procedures to find repetitions, like the dictionary routines, may now be used for proteins as well as nucleic acids.

Several parameters may be set to provide the criteria for an acceptable homology of any type. MINMATCH establishes the minimum number of correct matches in a homology; MINRATIO sets the minimum ratio of matches to length. LOOPLENGTH is the maximum size of a loopout, so setting LOOPLENGTH = 0 prohibits loopouts entirely. DISTANCE gives the maximum difference in position of repeated regions in one or two sequences. Similarly, LOOPDIST defines the maximum number of nucleotides between the two parts of a symmetry or dyad symmetry. And setting DUBIOUS = 1 counts G-T pairs as matches in dyad symmetries.

For each homology, the computer determines the probability of finding at least as good a homology by chance alone. These probabilities should not be taken too literally, as they do not adequately reflect the base composition of the sequence and thus can be incorrect by a factor of 2 or more. However, they do provide a relative indication of how inherently improbable, and therefore how likely to be significant, each homology is. Moreover, only those homologies with a probability of chance occurrence less than the parameter MAXPROB are printed by the computer. Setting MAXPROB small will thus eliminate the inferior, more probable homologies, while setting MAXPROB larger will print more of them. If the parameter EXPECT is set instead of MAXPROB, then the computer itself will choose MAXPROB so that the number of homologies randomly expected to appear in the output is equal to the value chosen for EXPECT. Hence EXPECT keeps the amount of output relatively constant by applying a more stringent MAXPROB for longer sequences.

[11] A. Walz and V. Pirotta, *Nature* (*London*) **254**, 118 (1975).
[12] T. Maniatis, M. Ptashne, K. Backman, D. Kleid, S. Flashman, A. Jeffrey, and R. Maurer, *Cell* **5**, 109 (1975).
[13] A. Walz, V. Pirotta, and K. Ineichen, *Nature* (*London*) **262**, 665 (1976).

(A)

THE DYAD SYMMETRIES ARE:

```
      27        GA TATTT       33
      10        CTGATAAA        3
   0.875
   1E-03
   9E-01

      43        GGTGATAGAT     52
      38        CC CTATTTA     30
   0.800
   8E-04
   7E-01
```

THE NUMBER OF MATCHES IS 2   DUBIOUS = 0
MAXPROB = 1.99E-03   EXPECT = 1   MINRATIO = 7.50E-01
MINMATCH = 3   LOOPLENGTH = 3   LOOPDIST = 20

(B)

THE HOMOLOGOUS REGIONS ARE:

```
       1        GATAAAT ATCT       11
       4        GATAAATTATCT       15
   0.917
   2E-05
   1E-02

      20        GCG TGTTGAC        29
      19        GCGGTGTTGAC        29
   0.909
   3E-05
   2E-02

      35        TACCTCTGGCGGTGATAATG     54
      35        TACCACTGGCGGTGATACTG     54
   0.900
   0E+00
   0E+00
```

THE NUMBER OF MATCHES IS 3
MAXPROB = 1.99E-03   EXPECT = 1   MINRATIO = 7.50E-01
MINMATCH = 3   LOOPLENGTH = 3   DISTANCE = 10

FIG. 3. The dyad symmetries in λP_{RM} (A), and the repetitions between λP_R and λP_L (B), as printed by the computer. The upper lines in B are from λP_R and the lower lines from P_L. The numbers surrounding the regions indicate their position in the sequences. The numbers beneath each homology are, reading downward, the ratio of matches to length, the probability of random occurrence at any one position of a homology at least as good, and the number of homologies as good or better than that one expected by chance in the whole sequence. The values of the parameters used are printed beneath the regions (e.g., $1E-03 = 1 \times 10^{-3}$; $0E+00$ means less than 1×10^{-5} for the probability, the expected number is then printed as $0E+00$ also).

The various parameters may be selected for each run of the program to best suit the biological question under investigation. For example, the parameter DISTANCE may be used to facilitate a search for promoter homologies of the sort discovered by Pribnow,[14] which would be insig-

[14] D. Pribnow, *Proc. Natl. Acad. Sci. U.S.A.* **72**, 784 (1975).

```
                          10                                           20
PHE VAL ASN GLN HIS LEU CYS GLY SER HIS LEU VAL GLU ALA LEU TYR LEU VAL CYS GLY
UUQ GUN AAQ CAP CAQ CUN UGQ GGN UCN CAQ CUN GUN GAP GCN CUN UAQ CUN GUN UGQ GGN
                UUP             AGQ     UUP             UUP     UUP

                          30
GLU ARG GLY PHE PHE TYR THR PRO LYS ALA END
GAP CGN GGN UUQ UUQ UAQ ACN CCN AAP GCN UAP
    AGP                             UGA
```

FIG. 4. Reverse translation of the insulin B chain by procedure 13.

nificant were they not all located the same distance from a gene. By arranging the sequences so that transcription initiation occurs at the same position in each, and setting DISTANCE small, only such homologies will be found. Similarly, a moderate LOOPDIST may be appropriate in searching for dyad symmetries which mark the sites of protein–nucleic acid interaction. Moreover, a small DISTANCE, LOOPDIST, and MAX-PROB will reduce the output and expense of examining a long sequence, which otherwise can be substantial.

Procedure 13 uses the genetic code both to translate nucleotide sequences and to reverse translate amino acid sequences (Fig. 4). In this and other procedures, the program distinguishes the two kinds of sequence by the asterisk preceding proteins and automatically performs the correct function. The degeneracy of the genetic code means that amino acids cannot be reverse translated into unique codons. However, by using P for purine, Q for pyrimidine, and N for nucleotide, every amino acid can be represented by at most two nucleotide triplets. A nucleic acid sequence can be translated from the first, second, or third nucleotide by setting the parameter PHASE = 1, 2, or 3, while PHASE = 4 performs all three translations.

Procedure 14 searches for subsequences of various types in a given sequence (Fig. 5). It locates specified oligonucleotides in nucleic acids, polypeptides in proteins, polypeptides coded by nucleic acids, and oligonucleotides reverse coded by proteins. (An oligonucleotide reverse coded by a protein means an oligonucleotide occurring in any one of the nucleotide sequences that can code for the protein.) The locations of those subsequences found in the given sequence are printed, and those that do not appear are listed separately. In addition, the program prints another copy of the amino acid or nucleotide sequence and overlines the subsequences found. The oligonucleotides and sequence may contain the symbols P, Q, and N, which are matched, respectively, against any purine, any pyrimidine, and any nucleotide. The number of mismatches allowed in comparing a subsequence with the given sequence is set by the parameter MISSRES. In comparisons involving both amino acids and nucleotides, MISSRES refers to the amino acids.

Procedure 14 is especially useful in locating restriction enzyme recog-

	# OF SITES	SITES	FRAGMENTS		FRAGMENT ENDS	
HIND 2 (GTQPAC)	1	24	36	(60.0)	24	60
			24	(40.0)	1	24
PEP (*VAL LEU THR)	1	22				

THE FOLLOWING SITES DO NOT APPEAR:

HHA 2 (GANTC)   SAL 1 (GTCGAC)

```
       10        20            30           40          50          60
GGTGATAAAT TATCTCTGGC GGTGTTGACA TAAATACCAC TGGCGGTGAT ACTGAGCACA
```

FIG. 5. Subsequences found by procedure 14 in λ P_L.

nition sites, either directly from DNA or indirectly from coded protein. In addition to the sites themselves, the program lists in descending order the lengths of the DNA fragments produced. Next to each fragment is printed its percentage of total sequence length, start point, and end point. If the parameter REFLECT is set equal to 1, the program automatically searches both strands of the DNA sequence, and if the parameter FORM = 'C,' the sequence is treated as a circle. Note that if a restriction site is reverse coded by a protein, the site need not actually appear in the corresponding gene because of the degenerate code. Procedure 14 is also useful in aligning a protein with a nucleic acid thought to code for it. For this purpose it is only necessary to compare a small amino acid subsequence from the protein against the nucleic acid sequence.

Procedure 14 can sometimes have unexpected uses. For example, Sanger and his colleagues[15] noted similarities between eight ribosomal binding sites of phage ϕX174. Inspection of their table reveals that all the sites have an AUG initiation codon preceded by an AGG triplet at a distance of 6–9 nucleotides. We wondered whether there were any other such sites on ϕX174, so we used the program to search for the oligonucleotide AGGNNNNNNAUG and three oligonucleotides with additional N's. The program located nine more such sites, showing that either the observed homology is not sufficient for translation initiation or that ϕX174 produces undiscovered proteins. We note that one of the sites located is followed by 196 nucleotide triplets without an end codon.

Procedure 15 determines the total trinucleotide frequency. The final procedure 16 of the program lists the trinucleotide frequency in one reading frame and gives the corresponding amino acid distribution, with the frame selected by the parameter PHASE.

[15] F. Sanger, G. M. Air, B. G. Barrell, N. L. Brown, A. R. Coulson, J. C. Fiddes, C. A. Hutchison, III, P. M. Slocombe, and M. Smith, *Nature (London)* **265**, 687 (1977).

TABLE III
DATA FORMAT

Section	Format
Parameters	Parameter 1 = Number 1 Parameter 2 = Number 2 ;
Sequences	'Name of Sequence 1' 'Sequence 1' 'Name of Sequence 2' 'Sequence 2' 'End' ''
Subsequences	'Name of Subsequence 1' 'Subsequence 1' 'Name of Subsequence 2' 'Subsequence 2' 'End' ''
Procedures	'Name I' Procedure Numbers 0 'Name J' Procedure Numbers 0 'End' ''
Comparisons	'Name $I1$' 'Name $I2$' 'Name $J1$' 'Name $J2$' 'End' ''

Data Format

We now describe in detail the format required by the program for the input data cards. The data format is outlined in Table III, and, as an example, Table IV presents the data cards used to produce the figures in this paper. Distinct names, sequences, and numbers are always separated by one or more blank spaces. Names and sequences, but not numbers and parameters, must be enclosed in single quotes. The quote is the only symbol which must not be used *within* a name or sequence. The computer does not record when one data card ends and the next begins, so a sequence may extend from one card right onto the next. But an item ending in the last column of one card must not be followed by another item in the first column of the next card, or the computer will register one long item.

The input data consists of sets of five sections, which must appear in the order indicated by Table III. More than two items, any number in fact, may appear in each section. The first section is followed by a semicolon. The other sections are followed by cards reading 'END' '' where '' means two adjacent single quotes. Even if a section contains no data, its corresponding semicolon or end card must appear.

The first section contains the values of any parameters one may wish to set (Table II). In particular, the parameter NUMSEQ gives a maximum for the number of sequences to be entered in the next section. Its default value is 10, so if there are more than 10 sequences, it is necessary to set

TABLE IV
SAMPLE DATA CARDS[a]

SHORT = 1 GAP = 4 DISTANCE = 10 NUMAGREE = 7;
'LAMBDA PR'   'GATAAATATCTAACACCGTGCGTGTTGACTATTTTACCTCTG
                   GCGGTGATAATGGTTGCA'
'LAMBDA PL'   'GGTGATAAATTA TCTCTGGCGGTGTTGACATAAATACCACTG
                   GCGGTGATACTGAGCACA'
'LAMBDA PRM'   'TAAAATAGTCAACACGCACGGTGTTAGATATTTATCCCTTGC
                   GGTGATAGATT TAACGTA'
  'END' ''
'HIND 2' 'GTQPAC' 'HHA 2' 'GANTC'
'SAL 1' 'GTCGAC' 'PEP' '*VAL LEU THR'
  'END' ''
'LAMBDA PL' 7 14 0
'LAMBDA PRM' 12 0
'COMBINED' 8 0
  'END' ''
'LAMBDA PR' 'LAMBDA PL'
  'END' ''
SHORT = 1;
'INSULIN B' '*PHE VAL ASN GLN HIS LEU CYS GLY SER HIS LEU
VAL GLU ALA LEU TYR LEU VAL CYS GLY GLU ARG GLY PHE PHE
TYR THR PRO LYS ALA END'
  'END' ''
  'END' ''
'INSULIN B' 13 0
  'ENL' ''
  'END' ''

---

[a] Each line represents one data card.

the parameter NUMSEQ equal to at least the number of them. Conversely, when fewer than 10 sequences are entered, setting NUMSEQ small will reduce the amount of computer memory needed. If there is a sequence over 1000 symbols in length, the parameter MAXLEN must be set to its length or more. The same remarks apply to NUMRES if there are more than 200 subsequences and to MAXLENRES if any of them is over 20 symbols in length. Note that each nucleotide counts as one symbol and each amino acid as four.

The second section contains the nucleotide or amino acid sequences to be analyzed and their names. The third section contains the subsequences—oligonucleotides and polypeptides—for which procedure 14 is to search. These subsequences must also be named; no name in this or the previous section should contain more than 20 symbols. The fourth section directs the analysis of the individual sequences. Each entry consists of the name of a sequence followed by the numbers of the procedures (Table I) to be applied to it, then followed by the number 0. Each name must be spelled exactly as in the second section, including any blank spaces. The names may occur in any order, as may the numbers, and a name may appear more than once. The fifth data section contains the names of pairs of sequences to be compared for homologous regions.

Four special data features increase the flexibility of the program. The word 'ALL' may be used in place of the name of a sequence in sections 4 and 5. The indicated operations are then performed on all the sequences entered in section 2. For example, 'ALL' 1 0 performs procedure 1 on all the sequences, 'ALL' 'Name' compares the sequence called 'Name' against every other sequence, and 'ALL' 'ALL' compares all the sequences against each other. Similarly, the word 'COMBINED' may be used in sections 4 and 5 to perform the indicated operations on the sequence formed by concatenating all the sequences. The 'ALL' and 'COMBINED' features are designed in such a way that no sequence is compared against itself.

A card reading 'LIMITS' X Y, where X and Y are numbers, may be inserted immediately preceding a card 'Name' Numbers 0 in section 4. The numbered procedures are then applied to the part of the named sequence lying inclusively from position X to position Y (or to the end if Y is past it). Position of course refers to the number of nucleotides in nucleic acids and the number of amino acids in proteins. If $X > Y$ the sequence is treated as a circle and analyzed from X through the end to Y. A card 'LIMITS' 'LIMITS' X Y Z W preceding a card 'Name I' 'Name J' in section 5 restricts the comparison to the parts of the sequences between X and Y and between Z and W, respectively. Again, $X > Y$ or $Z > W$ invoke the circle option.

The final data feature, applicable only to nucleic acids, is applied by inserting a card 'BACKWARDS' in section 4. The procedures on the next card are then applied to the opposite strand, read in reverse direction, of the named nucleotide sequence. Cards reading 'BACKWARDS' 'BACKWARDS', 'BACKWARDS' 'FORWARDS', and 'FORWARDS' 'BACKWARDS' have the obvious effects in section 5. The 'ALL', 'COMBINED', 'LIMITS', and 'BACKWARDS' options can be used in any combination.

A complete collection of data cards for a run of the program consists of one or more successive sets of the five data sections, each set containing an entirely independent group of parameters, sequences and procedures. For example, Table IV needs two sets so the 'COMBINED' option used will not concatenate insulin B with the λ promoters.

Interpretation of Results

High-speed computer examination of nucleic acid sequences is a technique powerful enough to reveal many features of potential interest— repeated regions, dyad symmetries, AT-rich tracts, and others. However, it is a difficult problem to differentiate the features which are essentially random phenomena from those of biological significance. We can distinguish two aspects to the problem.

A minimum criterion for a feature to require attention is that it be statistically improbable, standing out from the background of random events. The program we have described prints, with each homology, the probability of *chance* occurrence at any one position of a homology at least as good. The program also prints the number of homologies, as good or better than that one, expected to occur by chance in the whole sequence. Homologies with low probability and expectation are of course most likely to be significant. However, even improbable homologies will occur by chance when many sequences or nucleotides are examined. For example, ϕX174 can be expected to contain a perfectly repeated dodecanucleotide. Moreover, an investigator searching a sequence for interesting features is likely to accept many types. Although the probability that any one type occurs by chance may be small, the probability that *some* pattern appears is much larger. Therefore, while intrinsic improbability may be a necessary criterion for biological significance, it is certainly not a sufficient one.

On a second level, even genuinely nonrandom features may have no biological interest. An improbable homology may be an evolutionary vestige of an ancient cross-over event rather than an effector of biological function. This is especially true when there is no *a priori* reason to expect a connection between the observed homology and the postulated function. Ultimately, the computer can only pinpoint features of possible interest; the establishment of function must rest, as always, with biological experimentation.

Acknowledgments

We would like to express our appreciation to M. Wegman for useful discussions and to D. Brutlag, L. Kedes, and J. Maizel for reporting on their experiences with our program. We thank our colleagues at the National Cancer Institute and Carnegie Institution for commenting on the manuscript. L. J. Korn is a Helen Hay Whitney Fellow.

[61] Chemical Synthesis of Deoxyoligonucleotides by the Modified Triester Method[1]

By S. A. NARANG, R. BROUSSEAU, H. M. HSIUNG, and J. J. MICHNIEWICZ

Introduction

The basic principle of the triester method is to mask each internucleotidic phosphodiester function by a suitable protecting group during the course of the building sequence. As uncharged molecules, the phosphotriester intermediates are soluble in organic solvents and amenable to such conventional purification techniques of organic chemistry as silica-gel column chromatography. After building a desired sequence, all the protecting groups can be removed at the final step to give a deoxyoligonucleotide containing natural 3' → 5'-phosphodiester bonds. The main advantages of this method are (1) opportunity for large-scale (10–20 g) synthesis, (2) significantly short time periods especially in the purification steps, and (3) high yields using almost stoichiometric amounts of the reactants. This is probably due to the absence of any endo-P-O' groups in the oligonucleotide chain thus avoiding chain scission and pyrophosphate formation.

Two approaches have been reported. The original triester approach[2] involved the phosphorylation of the 3'-hydroxyl group of a protected nucleoside or oligonucleotide with a substituted alkyl or aryl phosphate; the resulting phosphodiester was then condensed with the 5'-hydroxyl of another nucleoside or oligonucleotide. Subsequent improvements have been investigated and a modified triester approach has been developed[3] which has the following basic feature. Phosphorylation of the 3'-hydroxyl of a protected mononucleotide with bis(triazolyl)-*p*-chlorophenyl phosphate was followed by the addition of excess β-cyanoethanol to obtain a fully protected mononucleotide. The protected mononucleotide containing a fully masked 3'-phosphate was used as starting material to grow the

[1] NRCC No. 16779.

[2] A. M. Michelson and A. R. Todd, *J. Chem. Soc.* p. 2632 (1955); R. L. Letsinger and K. K. Ogilvie, *J. Am. Chem. Soc.* **91**, 3350 (1969); R. Arentzen and C. B. Reese, *J. Chem. Soc., Perkin Trans. 1* p. 445 (1977); F. Eckstein and I. Rizk, *Chem. Ber.* **102**, 2362 (1969); T. Neilson and E. S. Werstink, *J. Am. Chem. Soc.* **96**, 2295 (1974).

[3] K. Itakura, N. Katagiri, C. P. Bahl, R. H. Wightman, and S. A. Narang, *Can. J. Chem.* **51**, 3649 (1973); T. C. Catlin and F. Cramer, *J. Org. Chem.* **38**, 245 (1973).

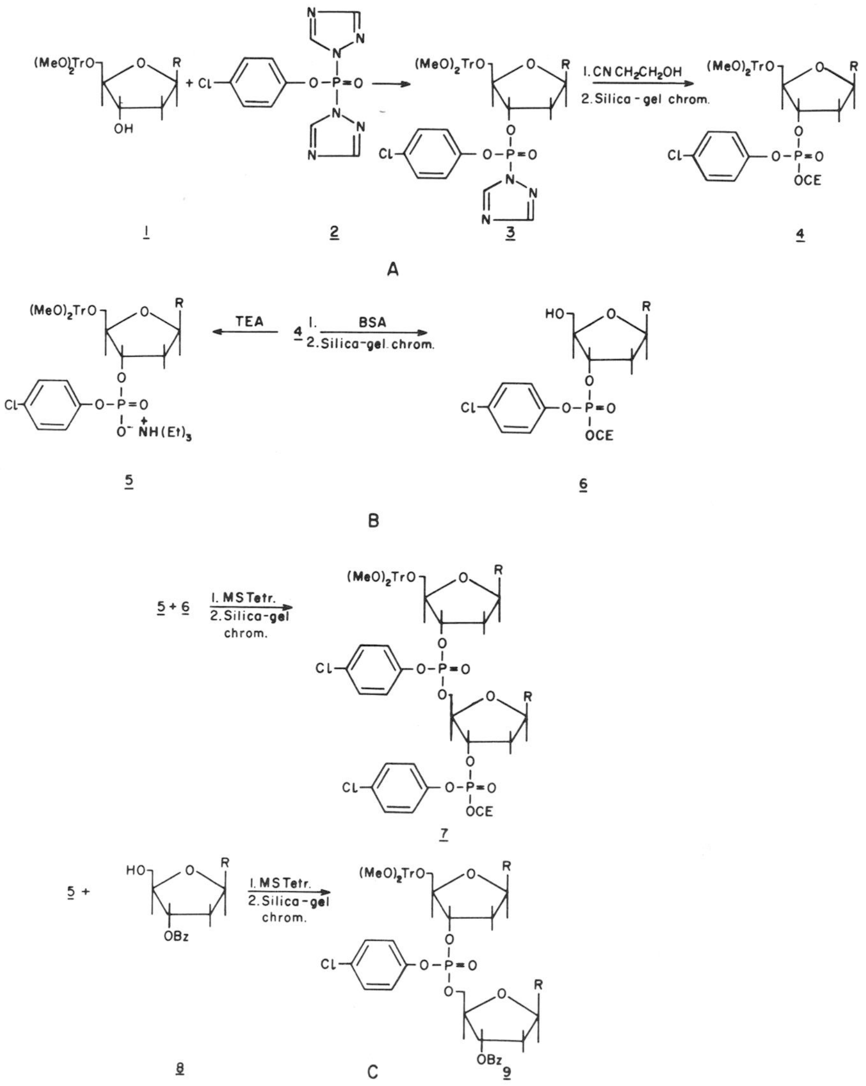

FIG. 1. Chemical synthesis of deoxyribooligonucleotides by the modified triester method. A, B, and C are the three stages of the synthesis procedure. MS tetr., mesitylene sulfonyl tetrazole; BSA, benzene sulfonic acid; TEA, triethylamine.

chain either from the $5' \rightarrow 3'$ or the $3' \rightarrow 5'$ direction, as shown in Fig. 1. Since any intermediate oligonucleotide so synthesized always contained a masked 3'-phosphate group, the original necessity of a phosphorylation step at each condensation stage was eliminated, thus simplifying the approach.

Preparation of Fully Protected Deoxymononucleotides

The preparation of protected deoxymononucleosides was carried out essentially by modification of the methods developed by Khorana.[4]

5'-O-Dimethoxytritylthymidine

Thymidine (18.1 g, 75 mM) first dried by evaporation with dry pyridine (2 × 50 ml) was then suspended in additional dry pyridine (200 ml) and treated with di-p-dimethoxytrityl chloride (30.4 g, 90 mM). This mixture was stirred for 4 hr at room temperature, during which it turned into a clear orange solution. The excess di-p-dimethoxytrityl chloride was decomposed by addition of methyl alcohol (30 ml) followed by 15 min of stirring, then the reaction mixture was poured into cold water (500 ml) where the product precipitated out as a gummy mass. This was dissolved in ethyl acetate (300 ml) and the aqueous layer further extracted with ethyl acetate (twice with 200 ml). The combined organics were washed with saturated aqueous sodium bicarbonate (100 ml), water (100 ml), and finally with saturated aqueous sodium chloride (100 ml). Evaporation of the solvent gave a gum which was redissolved in chloroform (30 ml) and applied to a silica gel GF-254 (type 60) short column (800 ml, dry volume) packed in chloroform. Elution was carried out with 3% methyl alcohol–chloroform (1 liter) followed by 5% methyl alcohol–chloroform (2 liters) which gave the desired product (37 g, 91%) (m.p. 116–118°) as a pale yellow solid [R_F = 0.8 in 10% methyl alcohol–chloroform on silica gel thin layer chromatography (tlc)].

N-Benzoyl-5'-O-dimethoxytrityldeoxycytidine

1. N-Benzoyldeoxycytidine. Deoxycytidine (20 g, 75 mM) was suspended in dry pyridine (300 ml), and benzoyl chloride (45 ml) was added dropwise and reaction mixture was maintained for 1 hr at room temperature. The reaction mixture was poured into ice water (800 ml) and the gummy precipitate extracted with ethyl acetate (three times with 500 ml). The ethyl acetate extract was dried over sodium sulfate, filtered, and evaporated to yield an oil which was next dissolved in a mixture of tetrahydrofuran (300 ml), methanol (150 ml), and water (60 ml) and the solution cooled to 0°. Sodium hydroxide solution (2 N) was added slowly with stirring to this solution until the pH rose to 12.0 as measured by a pH meter during addition. The reaction mixture was kept at 0° (ice bath) for

---

[4] H. Schaller, G. Weimann, B. Lerch, and H. G. Khorana, *J. Am. Chem. Soc.* **85**, 3821 (1963).

30–60 min until the complete disappearance of N^6, $O^{3'}$, $O^{5'}$-tribenzoyldeoxycytidine $(R_f \simeq 0.95)$ and the appearance of N^6-benzoyldeoxycytidine $(R_F \simeq 0.2)$ was observed by silica gel tlc (10% methyl alcohol–chloroform). The reaction mixture was neutralized by the addition of Dowex-50 (pyridinium form) ion-exchange resin, then filtered and the resin washed thoroughly with water. The total filtrate and washings were concentrated to 300 ml *in vacuo* and then extracted with ether (once with 300 ml) to remove benzoic acid and pyridine. The aqueous solution was heated (50°) to redissolve the desired product and filtered to remove any insoluble material. On cooling N-benzoyldeoxycytidine separated as crystalline material and some ether was added to complete the precipitation. The total yield in three crops (concentration after each crop) was about 80–90%.

N-*Benzoyl-5'-O-dimethoxytrityldeoxycytidine*. N-Benzoyldeoxycytidine (18.2 g, 55 mM) was dried by coevaporation with pyridine (30 ml) under vacuum. The dried material was dissolved in anhydrous pyridine (150 ml) and treated with di-p-methoxytrityl chloride (20 g, 60 mM) and kept for 2 hr at room temperature. The progress of the reaction was monitored by tlc on silica gel in methyl alcohol–chloroform (1 : 10, v/v) solvent. The excess of reagent was decomposed by the addition of methyl alcohol (100 ml) and the solution was evaporated to dryness *in vacuo*. The gummy material was dissolved in chloroform (200 ml) and washed with 5% sodium bicarbonate (twice with 200 ml) and water (twice with 200 ml). The chloroform layer, after drying with anhydrous sodium sulfate was concentrated to 40 ml volume and then chromatographed on a short column of silica gel (6 × 20 cm). The column was first eluted with chloroform (1 liter) followed by 3% methyl alcohol–chloroform (2 liters). Fractions of 10 ml were collected, and each fraction was then checked on tlc (10% methyl alcohol–chloroform). The pure N-benzoyl-5'-O-dimethoxytrityldeoxycytidine was isolated at 75–80% yield.

N-*Benzoyl-5'-O-dimethoxytrityldeoxyadenosine*

N-*Benzoyldeoxyadenosine*. Adenosine monohydrate (20.2 g, 75 mM) was dried by evaporation with anhydrous pyridine (twice with 30 ml) *in vacuo*. The material was slurried in dry pyridine (200 ml), cooled in an ice bath, and benzoyl chloride (44 ml) added slowly with stirring. The resulting clear pale yellow solution was stirred first at 0° for 30 min and then allowed to warm up to room temperature and stirred for a further 4 hr. The reaction mixture was poured into crushed ice water (800 ml) and the insoluble product was extracted with chloroform (four times with 300 ml). The combined chloroform extracts were washed with water (three times with 200 ml) and then evaporated to a semisolid material ($R_f = 0.9$ in 10%

methyl alcohol–chloroform). This material was dissolved in pyridine (300 ml) and ethyl alcohol (100 ml) and the solution was cooled in an ice bath. An ice-cold solution of equal volumes of ethyl alcohol and 2 N sodium hydroxide was added gradually with stirring until the pH rose to 12.0 as measured by pH meter. The reaction was monitored by tlc for the disappearance of the starting material and the appearance of product with $R_f = 0.2$. A total of 600 ml of ethyl alcohol–2 N sodium hydroxide was consumed in the time period of 45 min at 0°. An excess of Dowex-50 (pyridinium form) ion-exchange resin (500 ml) was then added; the slurry was stirred for 5 min (pH was checked to ensure neutralization to pH 6.3), filtered, the resin washed with water, and the combined filtrate and washings were evaporated *in vacuo* to 500 ml. This residual solution was washed with ethyl ether (three times with 200 ml); the aqueous layer was then further concentrated *in vacuo* to 200 ml, when the desired product started to precipitate. Cooling overnight at 4° in the refrigerator gave a very bulky precipitate which was filtered off and recrystallized in ethyl alcohol–water to give N-benzoyldeoxyadenosine (13.5 g, 51% yield, m.p. 113°–115°).

N-*Benzoyl-5'-dimethoxytrityldeoxyadenosine*.   N-Benzoyldeoxyadenosine (7.1 g, 20 mM) was dried by coevaporation with pyridine (10 ml) under vacuum. Its solution in anhydrous pyridine (50 ml) was treated with di-*p*-dimethoxytritylchloride (8.1 g, 24 mM) and the resulting solution was stirred at room temperature for 3 hr. The excess trityl chloride was decomposed by addition of methyl alcohol (30 ml) and after stirring for 15 min, the reaction mixture was poured onto crushed ice (500 ml) and extracted with ethyl acetate (three times 300 ml). The organic phases were washed once with saturated sodium bicarbonate (once with 200 ml) and water (twice with 100 ml). After removal of solvent, the gummy material was chromatographed on a short column (6 × 30 cm) of silica gel (800 ml, dry volume) packed in chloroform. The column was first eluted with 1 liter of 3% methyl alcohol–chloroform followed by 2 liters of 5% methyl alcohol–chloroform solvent. The desired product was isolated as brittle pale yellow foam (9.9 g, 77% yield).

N-*Isobutyryl-5'-O-dimethoxytrityldeoxyguanosine*

N-*Isobutyryldeoxyguanosine*.[5] Deoxyguanosine (21.0 g, 75 mM) was made anhydrous by repeated evaporation under *vacuo* of its suspension in anhydrous pyridine (100 ml). To the residue suspended in chloroform (300 ml) containing anhydrous pyridine (75 ml) was added isobutyryl chloride

[5] J. Stawinski, T. Hozumi, S. A. Narang, C. P. Bahl, and R. Wu, *Nucleic Acids Res.* **4**, 353 (1977).

(60 ml) in chloroform (200 ml) dropwise with stirring and cooling in ice bath. The reaction mixture became clear as the reaction proceeded, and after 2 hr at room temperature, it was decomposed with water (100 ml). The organic layer was washed twice with 5% sodium bicarbonate (200 ml) and water (once with 200 ml), dried over anhydrous sodium sulfate, and evaporated to an oily residue. This residue was redissolved in ethyl alcohol (150 ml) and kept at 0° while an ice-cold solution of 2 N sodium hydroxide (≈ 150 ml) was added gradually with stirring until the pH rose to 12.3. After completion of the reaction (30 min as monitored by tlc), the solution was neutralized by addition of excess Dowex-50 (pyridinium) ion-exchange resin which was removed by filtration and extensively washed with aqueous pyridine. The filtrate and washings were combined and concentrated to 300 ml, and this solution was washed with ether (200 ml) to remove excess isobutyric acid and pyridine. On cooling the aqueous solution to 4°, *N*-isobutyryldeoxyguanosine crystallized and the yield after three crops was about 80%.

N-Isobutyryl-5'-O'dimethoxytrityldeoxyguanosine.   *N*-Isobutyrylde-oxyguanosine (16.8 g, 50 mM) was dried by coevaporation with anhydrous pyridine (50 ml) *in vacuo.* The dried material was redissolved in anhydrous pyridine (150 ml) and treated with di-*p*-methoxytrityl chloride (18.6 g, 55 mM) at room temperature with stirring. After 3 hr, the reaction was complete as judged by tlc (silica gel) with 10% methyl alcohol–chloroform and the reaction mixture was treated with ice-cold water (100 ml), allowed to stand for 20 min, then extracted with ethyl acetate (three times 200 ml). The combined organic layers were further washed with water (three times with 100 ml), evaporated to dryness *in vacuo,* and redissolved in chloroform (40 ml). The final product was isolated in 75% yield with a short column of silica gel (6 × 30 cm) by eluting first with 1 liter of chloroform and then with 4% methyl alcohol–chloroform (2 liters).

Preparation of p-Chlorophenyl Phosphodichloridate

This was prepared by the method of Cramer and Winter[6] and also a more recent method of Reese *et al.*[7]

p-Chlorophenol (64.0 g, 0.5 M) was placed in a round-bottomed flask fitted with a reflux condenser and calcium chloride tube. Phosphorus oxychloride (250 ml, 2.7 M) and aluminum chloride (200 mg) was added and the mixture was gently refluxed for 48 hr. The reaction mixture was allowed to cool to room temperature, filtered through a glass wool plug,

[6] F. Cramer and M. Winter, *Chem. Ber.* **92**, 2761 (1959).
[7] G. R. Owen, C. B. Reese, and C. J. Ransom, *Synthesis* p. 704 (1974).

and the filtrate distilled under vacuum. The desired product was collected at 75°/0.3 mm Hg (lit. 140°/12 mm g) in 55% yield.

Phosphorylation of 5' Protected Nucleoside (Preparation of Fully Protected Mononucleotides[8] [4])

A mixture of p-chlorophenylphosphodichloridate (10.3 g, 42 mM), (1H)-1,2,4-triazole (6.0 g, 87 mM) and triethylamine (8.8 g, 87 mM) was stirred in anhydrous dioxane (300 ml), at 10°–15° for 30 min and then at room temperature for 1 hr. The mixture was then filtered and the filtrate was added directly to the N-acyl-5'-dimethoxytrityldeoxynucleoside (25 mM), which was previously made anhydrous by evaporation from anhydrous pyridine. The resulting solution was then concentrated to 100 ml under reduced pressure at ≤20°. The reaction mixture was kept at room temperature and monitored by tlc for the appearance of trityl-positive spot at the origin. After 2–3 hr β-cyanoethanol (55 mM) was added and the reaction maintained at room temperature for a further 2–4 hr and followed by tlc (disappearance) of the trityl-positive spot at origin and appearance of product at R_f 0.6–0.8. Aqueous pyridine (50%, 10 ml) was added and the reduction mixture stirred for 10 min, then evaporated to a sticky oil. This was dissolved in chloroform (300 ml) and washed with saturated aqueous sodium bicarbonate solution (150 ml) and water (150 ml). The organic layer was then evaporated to dryness, redissolved in chloroform (50 ml) and chromatographed on a short column (6 × 30 cm) of silica gel. The column was eluted with 1 liter of chloroform followed by 2 liters of 5% methyl alcohol–chloroform. Each fraction was checked on tlc and the fractions containing the desired product were pooled and evaporated to dryness. The average yield was 60–70%.

In the case of deoxycytidine and deoxyadenosine, the final residue was sometime found to be oily due to the presence of excess β-cyanoethanol. The residue was then dissolved in toluene (200 ml per 10 g) and washed extensively with water (three times with 100 ml). The organic layer was dried over sodium sulfate and filtered, and the filtrate was evaporated to dryness to yield a foamy residue.

Preparation of Arylsulfonyltetrazole Coupling Reagents[5]

To a solution of mesitylene sulfonyl or triisopropylbenzene sulfonyl chloride (10 g, 50 mM) and 1H-tetrazole (3.5 g, 50 mM) in redistilled dioxane (100 ml) at 12° was added dropwise a solution of triethylamine (5.02 g, 50 mM) in 50 ml dioxane over a period of 30 min. After the

---

[8] N. Katagiri, K. Itakura, and S. A. Narang, J. Am. Chem. Soc. 97, 7332 (1975).

addition, the reaction mixture was maintained for an additional 90 min at 12–15°, at which point the precipitate was filtered off and the filtrate evaporated to dryness under vacuum. The crude residue was dissolved in chloroform (100 ml) at room temperature and the solution washed quickly with ice-cold water, then dried over anhydrous sodium sulfate, filtered, and the filtrate evaporated to dryness *in vacuo*. The crude residue was recrystallized from toluene by heating to 55°–60°, and filtering to remove a small amount of insoluble material. On cooling the filtrate to 0°, the crystalline precipitate was removed and washed with a small amount of toluene giving the desired product in 40% yield (first crop): 1-mesitylene sulfonyl tetrazole, m.p. 108°–119°; 1-(2,4,6)-triisopropylbenzene sulfonyl tetrazole, m.p. 95°–97°.

Note: Recrystallization *must* be performed below 60° or rapid decomposition of the reagent occurred as evidenced by gas evolution and formation of an insoluble material.

Deblocking of Dimethoxytrityl Group from the Fully Protected Deoxynucleotides[5]

The fully protected compound (100 mg) was dissolved in methyl alcohol–chloroform (50 ml, 3 : 7, v/v) containing 2% benzene sulfonic acid and kept at 0° for 30 min by which time removal of the trityl group was complete as checked by tlc. The reaction mixture was washed successively with 5% sodium bicarbonate (20 ml) and then with water (20 ml). The chloroform layer was dried over anhydrous sodium sulfate, filtered and evaporated to dryness under reduced pressure. The detritylated product was isolated either by preparative tlc on silica gel plate or by medium-pressure liquid chromatography as described below. The yield was 80–90%.

Deblocking of Cyanoethyl Group from the Fully Protected Deoxynucleotides

Fully protected mono- or oligonucleotide was first dried by coevaporation of anhydrous pyridine (2 × 10 ml) and then dissolved in anhydrous pyridine (10–20 ml per mM) of the nucleotide component. Excess anhydrous triethylamine[9] (at least 10 M equivalent) was added and this reaction mixture was kept at room temperature and monitored by tlc (appearance of the trityl-positive spot at the origin). After 2–3 hr the reaction was complete and the mixture evaporated to a foam to remove excess

[9] A. K. Sood and S. A. Narang, *Nucleic Acids Res.* **4**, 2757 (1977); R. W. Adamiak, M. Z. Barciszewska, E. Biala, K. Grzeskowiak, R. Kierzek, A. Kraszewski, W. T. Markiewicz, and W. Wiewiorwski, *ibid.* **3**, 3397 (1976).

triethylamine and acrylonitrile liberated during the deblocking reaction. The foamy material was used as such in the coupling reaction as described below.

General Method for the Synthesis of Fully Protected Deoxyoligonucleotides

To the above reaction flask containing decyanoethylated mono- or oligodeoxynucleotidic material in anhydrous pyridine was added to 0.8 M equivalent of mono- or oligodeoxynucleotide containing a free 5'-hydroxyl group and the solution was evaporated in dryness in vacuo. The syrupy residue was redissolved in dry pyridine (10 ml per 1 mM of the nucleotidic components), followed by the addition of mesitylene sulfonyl tetrazole (3 to 5 M equivalents) or triisopropylbenzene sulfonyl tetrazole (5 to 10 M equivalents). The coupling reaction was over in 2 hr for lower-sized (up to hexamer) oligonucleotides and 6 to 10 hr for higher-sized oligomers (as judged by silica gel tlc for the complete disappearance of 5'-hydroxyl nucleotidic components on silica gel). The reaction mixture was decomposed with cold distilled water (5 ml), and the resultant solution was evaporated to a gum in vacuo. The gum was dissolved in ice-cold chloroform (100 ml) followed by washing with 5% sodium bicarbonate (twice with 50 ml) and water (once with 50 ml). The organic layer was dried over sodium sulfate or in the case of higher oligomer passed through Whatman phase separating filter paper to remove water. Since some higher oligomers were retained on the drying agent due to their lower solubility in chloroform, a considerable loss was observed. The dried crude mixture was either fractionated by preparative tlc on silica gel (20 × 20 cm, 2 mm thickness) in 10% methyl alcohol–chloroform or medium pressure liquid chromatography to obtain the desired product.

Medium-Pressure Liquid Chromatography for the Isolation of Product
Medium-Pressure Liquid Chromatography

The fractionation of each coupling reaction was carried out by medium-pressure liquid chromatography on prepacked silica gel columns[10] (Merck, LoBar, BorC) using a FMI laboratory pump at an operating pressure up to 100 psi. All solvents were thoroughly degassed prior to use. LKB Ultragrad was used to make a linear gradient of 5% methyl alcohol–chloroform to 20% methyl alcohol–chloroform for eluting the column. The eluent from the column was monitored for uv absorbance by an Altex detector. After checking each tube by tlc on silica gel plates, the

---

[10] Personal communication with Dr. D. J. Anderson, The Upjohn Co., Kalamazoo, Michigan.

fractions containing the desired compounds were pooled and concentrated to dryness. The advantages of this system are its reproducibility, improved resolution, and its capability of reusing columns after washing and equilibration.

Complete Deblocking of the Fully Protected Deoxyoligonucleotides

The dimethoxytrityl group was removed by treating the fully protected compound (10 mg) with a 2% solution of benzene sulfonic acid in chloroform: methanol (1 ml, 7:3) at 0° for 30 min. After neutralizing with 5% sodium bicarbonate solution, the reaction mixture was worked up as described previously. On removal of the solvent, the residue was purified by silica gel preparative tlc. The major band was eluted with chloroform–methanol (10:1, v/v) and the product was treated at 50° for 4-6 hr with excess of concentrated ammonium hydroxide ($\approx$5 ml) containing pyridine (0.5 ml). After removal of ammonia, the residue was washed with ether (twice with 1 ml) and the compounds containing phosphodiester bonds were isolated by tlc on polyethyleneimine (PEI–cellulose at 60° as described below.

Preparative Thin Layer Chromatography on Polyethyleneimine–Cellulose (PEI/uv$_{254}$) Plates[5]

Before applying the samples, the PEI plates were predeveloped with methanol, dried and redeveloped with distilled water by the ascending technique. The plates were then dried and stored in a refrigerator. This treatment removed a yellow discoloring material and apparently reactivated the ion-exchange capacity.

The samples of oligonucleotides ($\approx$100–150 A_{260} in 100 μl) were applied to the tlc plate as a narrow band along with a standard mixture of three dye markers[11] on each side of the sample. The plastic sheet was placed on a glass plate (20 × 20 cm) and a Whatman 3 MM paper was then clipped to the top of the plate and uniformly pressed against the PEI plate by a plastic strip (2 cm × width of plate). The plate was first developed in water for 3 cm and then in lithium chloride–7 M urea–0.025 M Tris (pH 8.0) at 60° until the blue dye marker was 2 cm from the top, or in LiOAc–7 M urea (pH 3.5) at room temperature.

The wet plates were washed in methanol (three times 50 ml) and then dried. The PEI–cellulose containing the desired band of product was removed and the product eluted with 2 M triethylammonium bicarbonate (pH 9.5). The eluent was collected in the centrifuge tube and concentrated

[11] F. Sanger, G. G. Brownlee, and B. G. Barrell, *J. Mol. Biol.* **13**, 373 (1965).

to dryness *in vacuo* in the presence of pyridine and finally with 0.2 *N* ammonium hydroxide.

The sequence of the synthetic oligomers[5,12] was determined and confirmed by using mobility shift analysis of Tu *et al.*[13]

[12] K. Itakura, N. Katagiri, S. A. Narang, C. P. Bahl, K. J. Marians, and R. Wu, *J. Biol. Chem.* **250**, 4592 (1975); C. P. Bahl, K. Marians, R. Wu, J. Stawinsky, and S. A. Narang, *Gene* **1**, 81 (1976); C. P. Bahl, R. Wu, J. Stawinsky, and S. A. Narang, *Proc. Natl. Acad. Sci. U.S.A.* **74**, 9666 (1977); J. W. Szostak, J. I. Stiles, C. P. Bahl, and R. Wu, *Nature (London)* **265**, 61 (1977).
[13] C. D. Tu, E. Jay, C. P. Bahl, and R. Wu, *Anal. Biochem.* **74**, 73 (1976).

[62] Sequence Analysis of Short DNA Fragments

By CHEN-PEI D. TU *and* RAY WU

Introduction

Rapid methods for sequencing long stretches of DNA have been developed,[1–3] and are being widely used. However, the first few nucleotides[2] or the first few dozen nucleotides[1,3] from the labeled ends are often excluded. In this article we describe a method for sequencing short DNA fragments from the labeled termini, up to 20 bases long. As shown in Fig. 1, this method consists of *in vitro* terminal labeling of the DNA fragments, followed by controlled *E. coli* exonuclease III digestion, and/or by strand separation on homochromatography or gel electrophoresis. Then, the sequences can be obtained after partial enzymatic digestions, two-dimensional fractionation by electrophoresis and homochromatography, and mobility-shift analyses.

The small DNA fragment (*Hin*d M) from SV40 DNA molecule produced by the restriction endonuclease *Hin*dII + III was used as an example.[4] The determination of the recognition sequence of a restriction endonuclease from *Haemophilus aegyptius* (*Hae*II) will also be described.

[1] R. Sanger and A. R. Coulson, *J. Mol. Biol.* **94**, 441–448 (1975).
[2] A. M. Maxam and W. Gilbert, *Proc. Natl. Acad. Sci. U.S.A.* **74**, 560–564 (1977).
[3] F. Sanger, S. Nicklen, and A. R. Coulson, *Proc. Natl. Acad. Sci. U.S.A.* **74**, 5463–5466 (1977).
[4] C.-P. D. Tu, R. Roychoudhury, and R. Wu, *Fed. Proc., Fed. Am. Soc. Exp. Biol.* **35**, 1595 (1976) (abstr.).

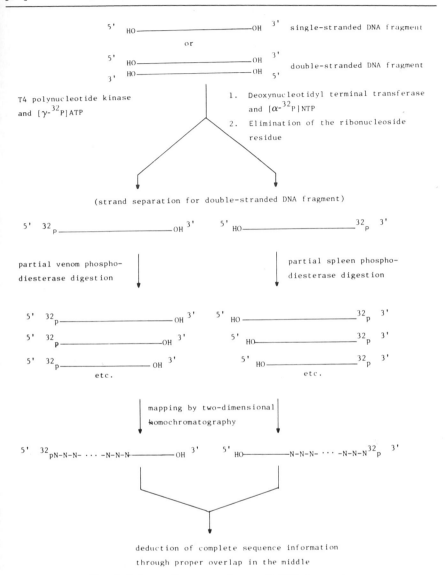

Fig. 1. Schemes for sequencing small DNA fragments.

Materials and Enzymes

All glassware used for the digestion of DNA with enzymes was siliconized.[5]

Pancreatic DNase (DNase I), highest grade, snake venom and spleen

[5] R. Roychoudhury and R. Wu, this volume, Article [7].

phosphodiesterase were purchased from Worthington Biochem. Co. (Freehold, NJ 07728). The venom phosphodiesterase was further purified to remove the contaminating phosphatase.[6] The spleen phosphodiesterase was further purified to remove endonuclease, nucleotidase and deaminase activity.[7]

*Hin*dII, *Hin*dIII, and *Hae*II were obtained from New England Biolabs (283 Cabot St., Beverly, Massachusetts).

Escherichia coli exonuclease III was purified according to Richardson and Kornberg[8] or Wu *et al.*[9]

DEAE–cellulose thin layer plates of 500 μm (7.5 : 1 of cellulose : DEAE–cellulose) were homemade[10,11]; the 250-μm-thick plates were purchased from Analtech Inc. (South Chapel St. Extension, Newark, Delaware).

Strand Separation of Short Duplex DNA by Homochromatography

Homochromatography separates oligonucleotides according to their size and base composition.[10,12] The component strands of a short duplex DNA with different composition should therefore be separable from each other on homochromatography under denaturing conditions (7 M urea, 65°).

The fragment *Hin*d M was produced by *Hin*dII cleavage and was therefore blunt-ended. It was terminally labeled at either of the 3' ends using deoxynucleotidyl terminal transferase and [α-^{32}P]CTP[5,13] or the 5' ends using T4 polynucleotide kinase and [γ-^{32}P]ATP[7] as previously described. The labeled *Hin*d M fragment was dissolved in a minimum volume of water (10 to 20 μl) and applied on a DEAE–cellulose thin layer chromatography (tlc) plate. Strand separation was achieved by developing the tlc plate in homomixture I of Jay *et al.*[10] at 65° until the yellow dye marker (orange G) has reached the top of the plate. Longer molecules of DNA can be displaced from the origin by increasing the concentration of RNA in the homomixture. As shown in Fig. 2, after development in homomixture I, the 5' end-labeled *Hin*d L and *Hin*d M fragments were each resolved into two species. These species corresponded very well with those from an unfractionated *Hin*cII digested SV40 DNA fragment.

[6] E. Sulkowski and M. Laskowski, Sr., *Biochim. Biophys. Acta* **240**, 443–447 (1971).
[7] R. Wu, E. Jay, and R. Roychoudhury, *Methods Cancer Res.* **12**, 87–176 (1976).
[8] C. C. Richardson and A. Kornberg, *J. Biol. Chem.* **239**, 240–250 (1964).
[9] R. Wu, G. Ruben, B. Siegel, E. Jay, P. Spielman, and C.-P. D. Tu, *Biochemistry* **15**, 734–740 (1976).
[10] E. Jay, R. A. Bambara, R. Padmanabhan, and R. Wu, *Nucleic Acids Res.* **1**, 331–353 (1974).
[11] C.-P. D. Tu, E. Jay, C. P. Bahl, and R. Wu, Anal. *Biochemistry* **74**, 73–93 (1976).
[12] G. G. Brownlee and F. Sanger, *Eur. J. Biochem.* **11**, 395–399 (1969).
[13] R. Roychoudhury, E. Jay, and R. Wu, *Nucleic Acids Res.* **3**, 863–878 (1976).

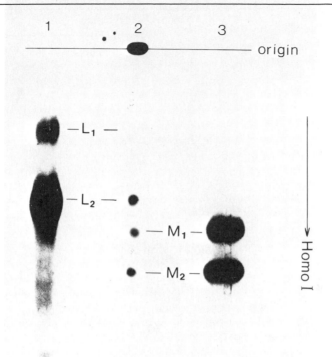

FIG. 2. Strand separation of *Hin*d L and *Hin*d M fragments on homochromatography. Lane 1, Isolated *Hin*d L fragment labeled at the 5′ ends. Lane 2, *Hin*cII-digested SV40 DNA fragments labeled at their 5′ ends. Lane 3, Isolated *Hin*d M fragment labeled at the 5′ ends. Homomixture used for fractionation was Homomix I of Jay *et al.*[10]

Exhaustive Digestion of 5′ End-Labeled DNA Fragments with E. coli Exonuclease III

The resolution of homochromatography has its limits. Oligonucleotides longer than 40 bases may not be well resolved by a homomixture containing 2% RNA solution. Higher concentration of RNA solution can displace longer oligonucleotides from the origin. However, for a tlc plate of fixed length (either 20 or 40 cm), increases in the RNA concentration used in the homomixture would decrease the separation between any two neighboring homologous oligonucleotides.

The scheme shown in Fig. 1 for sequencing small duplex DNA can be extended for determining longer duplex DNA sequences. When the

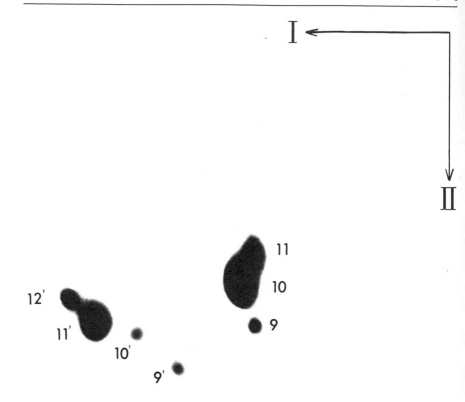

FIG. 3. Two-dimensional fractionation of exonuclease III-digested *Hin*d M fragment (5′ end labeled). I, electrophoresis on cellulose-acetate strip at pH 3.5; II, homochromatography in homomixture I.

strands of a duplex DNA fragment were longer than what can be resolved by a 2% homomixture I, strand separation can either be accomplished by gel electrophoresis or by shortening the strands using *E. coli* exonuclease III. Exonuclease III of *E. coli*[8] acts as a 3′ → 5′-exonuclease liberating 5′-mononucleotides from the 3′ ends of a double-stranded DNA. It was recently shown that the enzyme can hydrolyze a polynucleotide chain down to the last 5′ terminal dinucleotide.[14] When the temperature is raised stepwise from 25° to 60°, a gradation of longer 5′ terminal oligonucleotides of defined chain lengths is produced.

Approximately 5 pmoles of 5′ end-labeled *Hin*d M fragment was dissolved in 20 μl of 0.1 *M* glycylglycine buffer (pH 7.6), 10 m*M* MgCl$_2$, and 1 m*M* dithiothreitol containing 1.3 units of exonuclease III.[9] Incubation was carried out at room temperature (22° to 24°) for 2 hr. The reaction was

[14] R. Roychoudhury and R. Wu, *J. Biol. Chem.* **252**, 4786–4789 (1977).

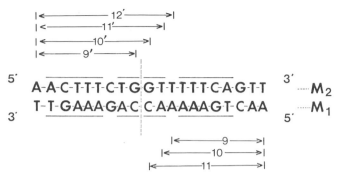

FIG. 4. Complete nucleotide sequence of *Hin*d M fragment of SV40 DNA. Dashed lines represented axis of twofold symmetry. Nucleotides involved in the symmetry were underlined. The nucleotide sequences of each spot in Fig. 3 was also indicated in the figure.

stopped by adding excess EDTA. The mixture was then passed through a small DE52 column (50 μl volume in an Eppendorf pipette tip) to remove proteins and salts.[7] The salt-free, exonuclease III treated *Hin*d M was then fractionated on two-dimensional electrophoresis–homochromatography. As shown in Fig. 3, two series of oligonucleotides were clearly resolved. Each of them was eluted and subjected to sequence analysis after partial digestion with venom phosphodiesterase as described in a later section. The sequence of each spot on Fig. 3 was shown in Fig. 4. By matching the longest oligonucleotides in each series, the complete sequence of *Hin*d M can be deduced. Large DNA duplex can also be sequenced after digestion with exonuclease III and strand separation.[15]

Generation of the Sequential Partial Degradation Products from an End-Labeled Oligonucleotide or DNA Fragment

1. Partial Venom Phosphodiesterase Digestion. Two digestion mixtures have been used.

a. The digestion mixture (10 μl) contained 2 mM Mg(OAc)$_2$, 20 mM Tris-acetate (pH 8.0 to 8.3, determined as a 1 M solution at room temperature), 30–50 μg carrier RNA,[10] 2 μg of venom phosphodiesterase, and the 5' end-labeled DNA sample to be digested (8000–40,000 cpm). Incubation was carried out at 37° and aliquots (1 μl) were pipetted into 10 μl of 2 mM EDTA in 0.1 M NH$_4$OH at 1, 2, 5, 10, 20, 40, 80, and 120 min. The mixture was boiled for 3 min, mixed with [^{14}C]dTMP carrier (approximately 4000–10,000 cpm), and dried in a desiccator. The partially digested DNA

[15] R. Wu, C.-P. D. Tu, and R. Padmanabhan, *Biochem. Biophys. Res. Commun.* **55**, 1092–1099 (1973).

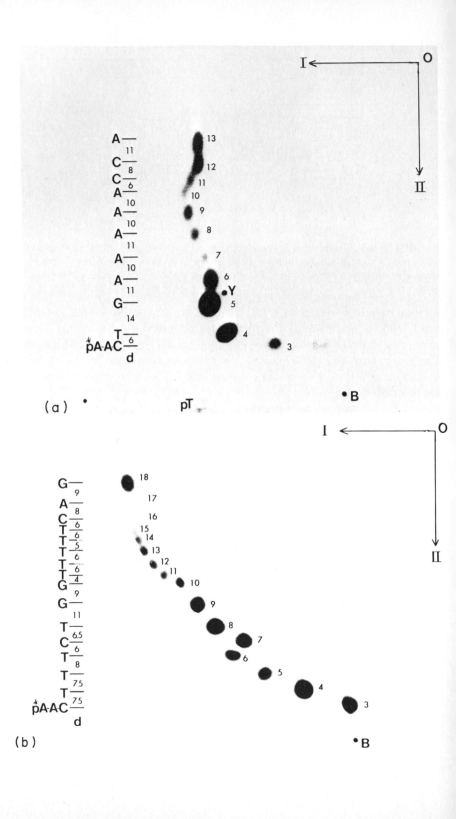

was dissolved in 6 μl of water, and 3 μl was applied onto cellogel for electrophoresis followed by homochromatography.

b. The reaction mixture (20 μl) contained 2 mM Mg(OAc)$_2$, 20 mM Tris-acetate (pH 8.0 to 8.3), 1.5 A_{260} units of 5'-dAMP, 0.5 μg venom phosphodiesterase, and the labeled DNA to be digested. Incubation was at room temperature, and eight aliquots (2 μl) were taken as described above. Because of the absence of carrier RNA in this procedure it is possible to label the 3' end of each digestion product for further end analysis.[15]

2. Partial Spleen Phosphodiesterase Digestion. Two digestion mixtures have been used.

a. The digestion mixture (10 μl) contained 3 mM KP$_i$ (pH 6.0), 0.3 mM EDTA, 0.02% Tween 80, 30 μg carrier RNA, approximately 0.1 units of spleen phosphodiesterase,[11] and 3' end-labeled oligonucleotides to be digested. Incubation was at 37°, and aliquots were pipetted into 20 μl of 0.1 M NH$_4$OH (4°) at 1, 2, 5, 10, 20, 40, 80, and 120 min. The mixture was boiled for 5 min before being dried in a desiccator.

b. The digestion mixture (25 μl) contained 3 mM KP$_i$ (pH 6.0), 0.3 mM EDTA, 0.02% Tween 80, 0.025 units of spleen phosphodiesterase and up to 5 μg of labeled oligonucleotides to be digested. Incubation was at room temperature, and 8 aliquots (3 μl) were taken as just described. This procedure allows further 5' end labeling of the digestion products for 5' end analysis.[15]

3. Partial Pancreatic DNase Digestion. Reactions were carried out at 37° in a volume of 20 μl containing 10 μg of calf thymus DNA, 50 mM triethylammonium bicarbonate (TEAB) buffer (pH 8.0), 3 mM CoCl$_2$, 1 mM CaCl$_2$, 2 μg of pancreatic DNase, and DNA samples to be digested. Aliquots were pipetted into excess EDTA (4°) at 1, 2, 5, and 30 min. The mixtures were boiled for 5 min before being dried in a desiccator. A single incubation at 37° for 5 min will usually produce all of the degradation products from dimer up to at least dodecanucleotides. However, the distribution of products is less even. Examples of partial pancreatic DNase digestions of 3' end-labeled and 5' end-labeled DNA fragments were given in Fig. 5. With 5' end-labeled DNA fragments, the distribution of partial pancreatic DNase digests can be made more even by further digestion (at 37°) of an aliquot of the digest with venom phosphodiesterase (0.5 μg per 50 μg carrier DNA) for 1 to 2 min in the presence of added 20 mM Tris-acetate (pH 8.0 to 8.3) and 2 mM Mg(OAc)$_2$.

FIG. 5. Two-dimensional electrophoresis-homochromatograms of partial pancreatic DNase digest of (a) 5' end-labeled M$_1$ (b) 5'-labeled M$_2$. The letters and numbers on these maps were as described in the legend to Fig. 7.

Fractionation of Sequential Partial Degradation Products on
Two-Dimensional Electrophoresis–Homochromatography

Brownlee and Sanger[12] introduced the two-dimensional fractionation procedure of oligodeoxynucleotides by electrophoresis on cellulose acetate at pH 3.5, followed by homochromatography on DEAE–cellulose thin-layer plates. This two-dimensional system has been used by Ling[16] as a sequencing method for the analysis of several long pyrimidine tracts from bacteriophage DNA. In his mobility-shift method, addition of a pdC to any pyrimidine tract always causes a mobility decrease (on the first dimension), which is easily distinguished from the addition of a pdT that always causes a mobility increase. However, for general DNA sequence analysis, distinction of all four nucleotides is required. Therefore, a more quantitative mobility-shift method is needed.

Recently, it has been shown that the addition of a purine nucleotide can be clearly distinguished from that of a pyrimidine nucleotide for both phosphorylated and dephosphorylated oligodeoxyribonucleotides on homochromatography using an improved homomixture.[10] This combined with electrophoresis on cellulose acetate at pH 3.5 in the first dimension has been studied and proposed as an independent method for sequencing oligodeoxyribonucleotides.[10,17] In this system, as shown in Fig. 6, the addition of a pdC (or dCp) causes a shift to the right. The addition of a pdA (or dAp) causes a more or less straight shift; and the addition of a pdT (or dTp) or a pdG (or dGp) causes a shift to the left. Thus, in principle, the addition of all four nucleotides can be distinguished. However, exceptions were observed, and unambiguous sequences were obtained only by confirmation from other methods. Therefore, the mobility method by visual inspection alone has not proved to be a complete method for sequence analysis.

The major difficulty arises for unambiguous distinctions between additions of pdA (or dAp) from pdG (or dGp), since the addition of pdA can cause a shift in either direction. When it causes a shift to the left, distinction between pdA and pdG is difficult.[11] Furthermore, with an oligonucleotide containing only A and C residues as in pA-C-A-A (Fig. 7), the observed shift with the addition of pdA is very much the same as that of a pdG, and the addition of a pdC causes an observed shift which cannot be predicted by visual inspection using the diagram shown in Fig. 6. However, when the same sequence, 5′ pA-C-A-A-T-TOH, was part of another oligonucleotide, 5′ pC-G-G-A-T-A-A-C-A-A-T-TOH (Fig. 7a), the observed shifts of A and C residues are different and follow those shifts

[16] V. Ling, *J. Mol. Biol.* **64,** 87–102 (1972).
[17] F. Sanger, J. E. Donelson, A. R. Coulson, H. Kössel, and D. Fischer, *Proc. Natl. Acad. Sci. U.S.A.* **70,** 1209–1213 (1973).

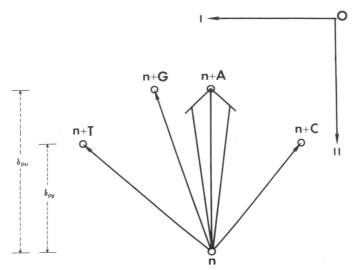

FIG. 6. The usual pattern of observed shift of an oligodeoxynucleotide (n) extended by a single nucleotide on two-dimensional homochromatography at pH 3.5. Dimensions and other descriptions are as in the legend to Fig. 7. δpu and δpy represent d values for addition of purine nucleotides and pyrimidine nucleotides, respectively.

described in Fig. 6. In an attempt to resolve these difficulties, thus making two-dimensional electrophoresis–homochromatography a more reliable sequencing method, we have developed the following mobility-shift analysis method for correct deduction of short DNA sequences.

Derivation of the Formula for the Calculation of the Mobility of Oligonucleotides[11]

A molecule in a fluid subjected to a voltage gradient E moves with a velocity $U = EQ'/K'$,[18] where Q' is the net charge on the molecule at a given pH and K' is a constant depending on the size and shape of the molecule. We express mobilities of all oligodeoxyribonucleotides relative to that of a reference marker, [14C]pdT, since pdT has a charge of (-1) at pH 2 to 5. For electrophoresis on cellogel at pH 3.5, the electrophoretic mobilities of oligodeoxynucleotides relative to pdT can thus be written as $U_T = Q/K$, where Q value is the algebraic sum of the mononucleotide charges q of a given oligonucleotide, and K is the resistance to motion relative to that of pdT.

The values of q for the four monodeoxynucleotides in two buffers at

[18] J. D. Smith, in "The Nucleic Acids" (E. Chargaff and J. N. Davidson, eds.), Vol. 1, pp. 267–284. Academic Press, New York, 1955.

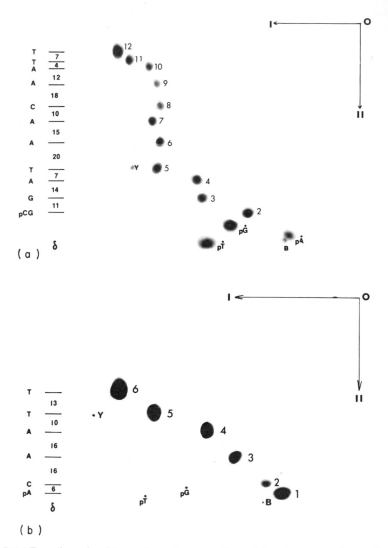

FIG. 7. (a) Two-dimensional homochromatograms of a partial snake venom phosphodiesterase digest of the dodecadeoxyribonucleotide d($\overset{*}{p}$C-G-G-A-T-A-A-C-A-A-T-T) labeled at the 5′ end. (b) Two-dimensional homochromatograms of a partial snake venom phosphodiesterase digest of the hexadeoxynucleotide d($\overset{*}{p}$A-C-A-A-T-T) labeled at the 5′ end. Dimension I, electrophoresis on cellulose acetate strip at pH 3.5, trimethylamineacetate buffer, 40-min electrophoresis run. Dimension II, homochromatography on a homemade 20 × 20 cm DEAE–cellulose thin-layer plate (7.5 : 1) with homomixture IV. B and Y represent the blue (xylene cyanol FF) and yellow (orange G) dye markers, and 0 represents the origin of electrophoresis. ^{14}C-Labeled 5′-mononucleotide markers are indicated by p$\overset{*}{N}$ on the homochromatograms. The d values (in millimeters) are labeled on the left side of a homochromatogram, and the mononucleotide added to the 3′ end of a previous oligomer is labeled to the left of the corresponding d values. The complete sequence of the original oligodeoxynucleotide can be read directly from the bottom mononucleotide (oligomer 1) to the top of the homochromatogram.

TABLE I
EXPERIMENTALLY DETERMINED q VALUES OF 5'-MONODEOXYRIBONUCLEOTIDE AT pH 2.8
AND pH 3.5

5'-dNMP	Pyridine-acetate[a] (pH 3.5)	Trimethylamine-acetate[b] (pH 3.5)	Pyridine-formate[c] (pH 2.8)	Trimethylamine-formate[d] (pH 2.8)	$pK_a'^{,e}$
dTMP	−1.0	−1.0	−1.0	−1.0	—
dGMP	−0.83	−0.83	−0.54	−0.57	2.9
dAMP	−0.40	−0.40	−0.10	−0.12	3.79
dCMP	−0.15	−0.14	−0.06	−0.06	4.6

[a] Five percent glacial acetic acid containing 5 mM EDTA titrated to pH 3.5 with pyridine.
[b] Five percent glacial acetic acid containing 5 mM EDTA titrated to pH 3.5 with 20% trimethylamine.
[c] Formic acid (1.25%) containing 5 mM EDTA titrated to pH 2.8 with pyridine.
[d] Formic acid (1.25%) containing 5 mM EDTA titrated to pH 2.8 with 20% trimethylamine.
[e] From Smith.[18]

two different pH's are given in Table I. The observed electrophoretic mobility U_T^{obs} on cellulose acetate of each partial product of the original oligonucleotide was calculated by comparison with the observed mobility of the [^{14}C]pdT marker by the following relationship

$$U_T^{obs} = \frac{\text{distance traveled by an oligomer in the first dimension}}{\text{distance traveled by [}^{14}\text{C]pdT in the first dimension}}$$

Since U_T and Q values for the oligonucleotides of known specific sequences can be obtained from the two-dimensional homochromatogram, K_n values ($n = 1, 2, 3, \ldots$, etc., the number of nucleotides) were obtained through the relation $K_n = Q/U_T$ and shown in Table II. By using these average K_n values ($\bar{K}_n$), the mobility of any oligonucleotide can be predicted. From these values, the predicted mobility shift between two adjacent nucleotides, n and $n + 1$, can be calculated by the relation

$$S^{calc} = U_{T,n+1}^{calc} - U_{T,n}^{calc}$$

Since these mobility shifts are characteristic of the mononucleotide unit added to oligomer n to give oligomer $n + 1$, comparison with the observed mobility-shift value

$$S^{obs} = U_{T,n+1}^{obs} - U_{T,n}^{obs}$$

for the addition of the four mononucleotides from $n = 1$ through the entire chain length of the oligomer, the sequence of the oligomer can be unambiguously derived.

TABLE II
EXPERIMENTALLY DETERMINED $\bar{K}_n$ VALUES AT pH 2.8 AND pH 3.5 FOR PHOSPHORYLATED OLIGODEOXYRIBONUCLEOTIDES

No.	Pyridine acetate (pH 3.5)	Trimethylamine acetate (pH 3.5)	Pyridine formate (pH 2.8)	Trimethylamine formate (pH 2.8)
1	1.0	1.0	1.0	1.0
2	1.31	1.30	1.30	1.28
3	1.75	1.72	1.61	1.58
4	2.10	2.02	1.91	1.89
5	2.54	2.40	2.22	2.18
6	2.91	2.76	2.52	2.49
7	3.29	3.06	2.82	2.80
8	3.60	3.37	3.13	3.10
9	3.90	3.65	3.43	3.40
10	4.23	3.99	3.74	3.70
11	4.58	4.31	4.04	4.0
12	4.90	4.68	4.34	4.30
13	5.29	5.09	4.65	4.61
14	5.57	5.34	4.95	4.91

Table III shows the calculation of mobility shifts and the deduction of nucleotide sequences for the two-dimensional electrophoresis–homochromatograms in Fig. 7.

Applications of the Mobility-Shift Analysis Method

The mobility-shift analysis method for sequencing short DNA fragments can be used as a simple and general method for the determination of recognition sequences for most of the type II restriction endonucleases regardless of the nature of the exposed end: 3' cohesive end, 5' cohesive end, or blunt-end. The determination of the recognition sequence of *Hae* II restriction endonuclease is described[19] as an example.

In the first step, we determined the nucleotide sequence of SV40 DNA at the single cleavage site of *Hae* II.[20] After cleaving SV40 DNA with *Hae* II, one aliquot of the sample was labeled at the 5' ends with ^{32}P, and another aliquot was labeled at the 3' ends with [^{32}P]CMP. The end-labeled DNA was digested with a different restriction enzyme (*Hin*d) followed by

[19] C.-P. D. Tu, R. Roychoudhury, and R. Wu, *Biochem. Biophys. Res. Commun.* **72**, 355–362 (1976).

[20] R. J. Roberts, J. B. Breitmeyer, N. F. Tabachnik, and P. A. Myers, *J. Mol. Biol.* **91**, 121–123 (1975).

TABLE III

OBSERVED AND CALCULATED U_T AND S VALUES FOR OLIGODEOXYRIBONUCLEOTIDES STUDIED IN TRIMETHYLAMINE ACETATE BUFFER AT pH 3.5 (FIG. 7)

N	Sequence	U_T^{obs}	U_T^{calc}	$U_T^{calc}(N)^a$	S^{obs}	S^{calc}	$S^{calc}(N)^a$
1	pA	0.36	0.40				
2	pA-C^b	0.44	0.42^b	1.08 (T)	0.08	0.02	0.58 (T)
3	pA-C-A	0.60	0.55	0.80 (G)	0.16	0.13	0.38 (G)
4	pA-C-A-A	0.74	0.66	0.88 (G)	0.14	0.11	0.33 (G)
5	pA-C-A-A-T	1.02	0.98		0.28	0.32	
6	pA-C-A-A-T-T	1.19	1.21		0.17	0.23	
1	pC	(0.14)	0.14				
2	pC-G	0.76	0.75	0.42 (A)	(0.62)	0.61	0.28 (A)
3	pC-G-G	1.04	1.05	0.80 (A)	0.28	0.30	0.05 (A)
4	pC-G-G-A	1.08	1.09	1.30 (G)	0.04	0.04	0.25 (G)
5	pC-G-G-A-T	1.33	1.33		0.25	0.24	
6	pC-G-G-A-T-A	1.30	1.30	1.46 (G)	−0.03	−0.03	0.13 (G)
7	pC-G-G-A-T-A-A	1.35	1.31	1.45 (G)	0.05	0.01	0.15 (G)
8	pC-G-G-A-T-A-A-C	1.30	1.23		−0.05	−0.08	
9	pC-G-G-A-T-A-A-C-A	1.32	1.24	1.36 (G)	0.02	0.01	0.13 (G)
10	pC-G-G-A-T-A-A-C-A-A	1.37	1.24	1.35 (G)	0.05	0	0.11 (G)
11	pC-G-G-A-T-A-A-C-A-A-T	1.48	1.38		0.11	0.14	
12	pC-G-G-A-T-A-A-C-A-A-T-T	1.58	1.48		0.10	0.10	

^{a} U_T^{calc} of S^{calc} for an alternative mononucleotide addition to the 3' end of a previous compound.

^{b} U_T^{calc} for pA-C was calculated as follows

$$\frac{q_A + q_C}{\bar{K}_n \text{ (for } n = 2)} = \frac{-(0.40 + 0.14)}{1.30} = -0.42$$

gel electrophoresis to separate the two labeled 5' ends or the labeled 3' ends. The nucleotide sequence around the *Hae*II cleavage site was then determined after partial pancreatic DNase digestion of the single-end labeled fragments. The sequence derived from the two-dimensional maps (Fig. 8) revealed that *Hae*II cleavage produced 3' cohesive ends, 4 nucleotides in length. The alignment of the two complementary staggered ends of the cleaved products yielded a complete duplex nucleotide sequence at the *Hae*II cleavage site in SV40 DNA as

5' G-C-A-A-C-A--G-C-G-C⌉ $\overset{*}{p}$T-C-A-C-A-C-C-A . . . 3'

3' . . . A-C-T-C-G-T-T-G-T$\overset{*}{p}$ ⌊C-G-G-G--A-G-T-G-T-G-G-T . . . 5'

The fact that there is only one cleavage site in the entire SV40 DNA molecule (about 5000 base pairs) is consistent with the number of probable cleavage sites from a hexanucleotide recognition sequence. Therefore, the hexanucleotide sequence with a twofold rotational symmetry

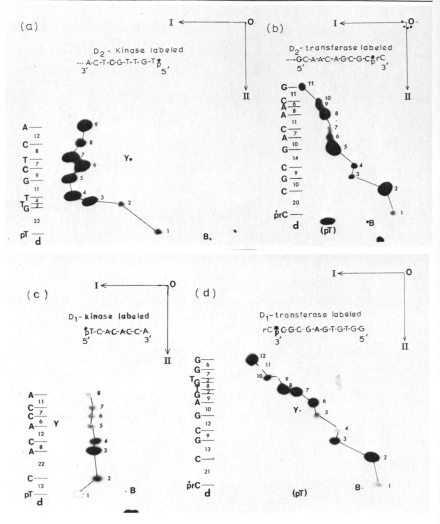

FIG. 8. Nucleotide sequence at the *Hae*II cleavage site of SV40 DNA. The figure shows two-dimensional maps of partial pancreatic DNase digests of (a) kinase labeled D_2, (b) transferase labeled D_2, (c) kinase labeled D_1, and (d) transferase labeled D_1. The letters on these maps are as described in the legend of Fig. 7.

$$5'\ \ A\text{-}G\text{-}C\text{-}G\text{-}C\text{-}T$$
$$3'\ \ T\text{-}C\text{-}G\text{-}C\text{-}G\text{-}A$$

is likely to be the recognition sequence in SV40 DNA.

In the second step, we analyzed the terminal sequences from a mixture of *Hae*II generated DNA fragments. Linear DNA, like λ DNA or adenovirus DNA, has naturally occurring ends which may interfere with

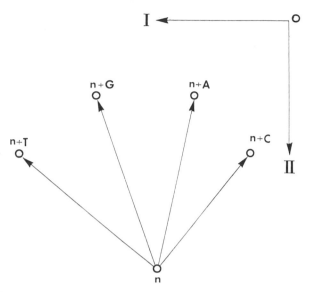

FIG. 9. The expected pattern of observed shifts of an oligodeoxyribonucleotide (n) extended by a single nucleotide on two-dimensional electrophoresis (at pH 2.8)–homochromatography.

the terminal nucleotide sequence analysis after *Hae*II cleavage. The plasmid DNA RSF2124[21] gave more than twelve *Hae*II-generated fragments on agarose gel electrophoresis (data not shown). Because *Hae*II produced 3' cohesive ends in SV40 DNA, the RSF2124 *Hae*II digested fragments were labeled at their 3' ends with terminal transferase,[5] and the terminal sequences were analyzed after partial pancreatic DNase digestion and mobility-shift analysis on two-dimensional electrophoresis–homochromatography. The result[19] showed a sequence 5' · · · G-C-G-Cp̈rC 3' common to all fragments, and the nucleotide at the 5' side of this common sequence is either a pdA or a pdG. Hence, the general recognition sequence at the *Hae*II cleavage site is

$$5' \; . \; . \; . \; \text{(A/G)-G-C-G-C} \big\uparrow \text{p(T/C)} \; . \; . \; . \; 3'$$
$$3' \; . \; . \; . \; \text{(T/C)p} \big\downarrow \text{C-G-C-G--(A/G)} \; . \; . \; . \; 5'$$

The mobility-shift analysis method for sequencing short DNA fragments is a reliable method. However, there are systematic deviations from the calculated mobility shifts found in several cases. These deviations happened wherever the sequences included a stretch of the same nucleotide. For example, there was a stretch of 5 T's in the sequence 5'

[21] M. So, R. Gill, and S. Falkow, *Mol. Gen. Genet.* **142**, 239–249 (1976).

pG-G-*T-T-T-T-T*-C-A-G-T-Tp$\overset{*}{r}$COH (not shown), and the observed mobility shifts (S^{obs}) for T's were consistently smaller than the calculated values (S^{calc}). On the other hand, whenever there was a stretch of A in a sequence such as 5' pA-A-C-T-G-*A-A-A-A-A-A*-C-C-A-GOH, the observed mobility shifts (S^{obs}) for the A's was consistently higher than that calculated (S^{calc}) (Fig. 5a). However, the deviation is small and a correct interpretation of the sequence can be made by using the mobility-shift analysis.

A Second System of Mobility-Shift Analysis

Electrophoresis at pH 3.5 was chosen probably because the four mononucleotides have the greatest separation among themselves. However, it may produce some ambiguities when used in a two-dimensional system for DNA sequence analysis.

This ambiguity between pdA and pdG shifts observed at pH 3.5 can be resolved by using a different pH for electrophoresis. We have chosen pH 2.8 to achieve the desired mobility shifts described in Fig. 9. When electrophoresis was carried out at pH 2.8 followed by homochromatography (C.-P. D. Tu, C. P. Bahl, E. Jay, and R. Wu, unpublished results), addition of a pdA (or dAp) residue to an oligonucleotide n always causes a backward shift (to the right) which can be distinguished from the addition of a pdG (or dGp) residue. The addition of a pdA can be distinguished from the addition of a pdC residue by a larger d value in the second dimension for pdA. The addition of a pdG (dGp) residue to an oligonucleotide n caused a forward shift (to the left) which can be distinguished from the addition of a pdT (dTp) by the smaller mobility increased due to pdG in the first dimension and a larger d value in the second dimension (Fig. 9).

Two new sets of $\bar{K}_n$ values for $n = 1$ to $n = 14$ (Table II) were obtained by running a number of two-dimensional electrophoresis-homochromatograms, with electrophoresis on cellulose–acetate strips carried out at pH 2.8 in pyridine–formate buffer or trimethylamine–formate buffer (Table I). The S^{obs} values were in very good agreement with S^{calc} values when the correct nucleotide addition was assigned (Fig. 10).

Conclusions

With the discoveries of increasing numbers of base-sequence specific restriction endonucleases, homogeneous populations of small DNA segments from 10 to 40 nucleotides in length from many regions of a large DNA molecule can be easily isolated. There is also increasing need to determine the recognition sequences of new restriction endonucleases.

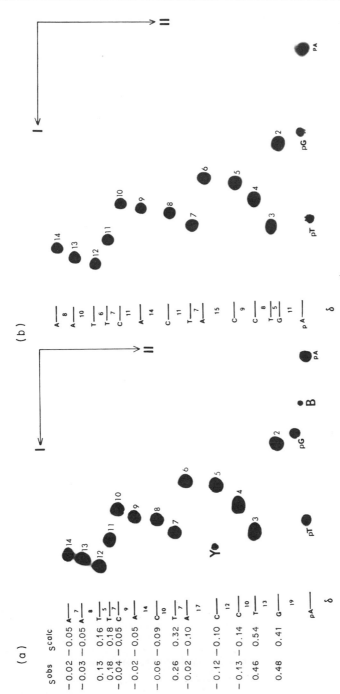

FIG. 10. Two-dimensional map of the partial venom phosphodiesterase digest of d(pA-G-T-C-C-A-T-C-A-C-T-T-A-AOH). The letters on the map indicated the following: I, Electrophoresis on cellulose-acetate at pH 2.8; II, homochromatography in homomix IV. The numbers at the left of each map between nucleotides indicated d values in millimeters; S^{calc} represented calculated mobility shifts and S^{obs} represented observed mobility shifts. (a) Trimethylamine–formate buffer at pH 2.8. (b) Pyridine–formate buffer at pH 2.8.

The methods described in this article are best suited for determining a short DNA sequence immediately adjacent to the labeled ends at the cleavage site. These methods are complementary to other new sequencing methods[1-3] that often miss the nucleotides immediately adjacent to the labeled ends. Furthermore, occasional abnormal electrophoretic mobility of some oligonucleotides on acrylamide gels (Sanger et al.[1-3] and unpublished observations) makes it desirable to check the sequence (usually within 15 nucleotides from the labeled end) by a second method, such as those described in this article.

Acknowledgments

We would like to acknowledge the collaboration of Drs. Ranajit Roychoudhury, Ernest Jay, and Chander P. Bahl for different portions of the work reported in this article.

[63] ³H and ³²P Derivative Methods for Base Composition and Sequence Analysis of RNA

By Kurt Randerath, Ramesh C. Gupta, and Erika Randerath

Tritium derivative methods for base composition[1-3] and sequence[3-10] analysis of small amounts of RNA developed in our laboratory will be described in this article. In addition, a ³²P derivative method recently developed in our laboratory for the sequence analysis of RNA[11,12] will also be presented. Distinct advantages of these methods are applicability to

[1] K. Randerath and E. Randerath, *Proced. Nucleic Acid Res.* **2**, 796 (1971).

[2] E. Randerath, C.-T. Yu, and K. Randerath, *Anal. Biochem.* **48**, 172 (1972).

[3] K. Randerath and E. Randerath, *Methods Cancer Res.* **9**, 3 (1973).

[4] K. Randerath, E. Randerath, L. S. Y. Chia, R. C. Gupta, and M. Sivarajan, *Nucleic Acids Res.* **1**, 1121 (1974).

[5] M. Sivarajan, R. C. Gupta, L. S. Y. Chia, E. Randerath, and K. Randerath, *Nucleic Acids Res.* **1**, 1329 (1974).

[6] R. C. Gupta, E. Randerath, and K. Randerath, *Nucleic Acids Res.* **3**, 2895 (1976).

[7] R. C. Gupta, E. Randerath, and K. Randerath, *Nucleic Acids Res.* **3**, 2915 (1976).

[8] R. C. Gupta and K. Randerath, *Nucleic Acids Res.* **4**, 1957 (1977).

[9] R. C. Gupta and K. Randerath, *Nucleic Acids Res.* **4**, 3441 (1977).

[10] E. Randerath, R. C. Gupta, L. S. Y. Chia, S. H. Chang, and K. Randerath, *Eur. J. Biochem.* **93**, 79 (1979).

[11] K. Randerath and R. C. Gupta, *Fed. Proc.* **38**, 499 (1979); abstr. 1420.

[12] R. C. Gupta and K. Randerath, *Nucleic Acids Res.* **6**, 3443 (1979).

nonradioactive RNA, high sensitivity, and the direct chromatographic identification of modified RNA components. Thus these methods make possible the analysis of small amounts of RNA of intact mammalian organisms, which is hard or impossible to achieve by conventional spectrophotometric or *in vivo* labeling methods.

The tritium derivative methods entail the chemical introduction of tritium label into the ribose moiety of nonradioactive ribonucleosides or ribonucleotides carrying free 2′- and 3′-OH groups. This is readily accomplished by oxidation with sodium metaperiodate, followed by reduction of the resulting dialdehydes with tritiated potassium or sodium borohydride to nucleoside trialcohols or nucleotide dialcohols, respectively [Eq. (1)].

Dialdehyde

Nucleoside trialcohol (R = H)
or
Nucleotide dialcohol
(R = phosphate, mono-
nucleotide, or poly-
nucleotide residue)

(1)

The tritium derivative methods for base composition analysis of RNA are particularly useful for RNAs containing modified bases such as tRNA. They have been applied in our laboratory (1) to comparative studies on tRNA from normal and neoplastic mammalian tissues[3,13-16] and tRNA from a virus and virally transformed cells,[17] (2) to the investigation of the effects of anticancer drugs[18-21] and a carcinogen[22] on tRNA base composition, and (3) to studies of the effects of anticancer drugs on tRNA modify-

[13] K. Randerath, *Cancer Res.* **31**, 658 (1971).
[14] K. Randerath, S. K. MacKinnon, and E. Randerath, *FEBS Lett.* **15**, 81 (1971).
[15] E. Randerath, L. S. Y. Chia, H. P. Morris, and K. Randerath, *Cancer Res.* **34**, 643 (1974).
[16] L. S. Y. Chia, H. P. Morris, K. Randerath, and E. Randerath, *Biochim. Biophys. Acta* **425**, 49 (1976).
[17] K. Randerath, L. J. Rosenthal, and P. C. Zamecnik, *Proc. Natl. Acad. Sci. U.S.A.* **68**, 3233 (1971).
[18] L. W. Lu, G. H. Chiang, D. Medina, and K. Randerath, *Biochem. Biophys. Res. Commun.* **68**, 1094 (1976).
[19] L. W. Lu, G. H. Chiang, W.-C. Tseng, and K. Randerath, *Biochem. Biophys. Res. Commun.* **73**, 1075 (1976).
[20] W.-C. Tseng, D. Medina, and K. Randerath, *Cancer Res.* **38**, 1250 (1978).
[21] L. W. Lu, W.-C. Tseng, and K. Randerath, *Biochem. Pharmacol.* **28**, 489 (1979).
[22] L. W. Lu, G. H. Chiang, and K. Randerath, *Nucleic Acids Res.* **3**, 2243 (1976).

ing enzymes.[20,21,23] A number of other laboratories[24-29] have also utilized this method. The tritium sequencing methods have been used by us and others to elucidate the primary structure of a number of tRNA species, i.e., tRNA$_{UAG}^{Leu}$ from yeast,[30] tRNA^{Phe} (Roe *et al.*[31]) and two tRNA^{Gly} species[32,32a] from human placenta, tRNA^{Phe} from a Morris hepatoma,[33] and tRNA^{Asn} from rat liver.[34]

The ^{32}P derivative method[11,12] entails chemical degradation of the RNA, ^{32}P-labeling of terminal 5'-hydroxyl groups, electrophoresis on sequencing gels, and thin-layer chromatographic identification of the labeled terminal components. This method has recently been applied in our laboratory to the structural analysis of tRNA^{Gly} species from human placenta and to tRNA^{Leu} and tRNA^{Ser} species from hepatoma.[32]

I. Tritium Derivative Method for Base Composition Analysis of RNA

A. Materials

N,N-Bis(2-hydroxyethyl)glycine (bicine) from Calbiochem (San Diego, California). Aqueous solutions can be stored up to 4 weeks at 4°. Do not freeze.[4]

Enzymes: RNase A (code R4875) and RNase T$_2$ (code R3751) from Sigma Chemical Co. (Saint Louis, Missouri), snake venom phosphodiesterase (code VPH) and *E. coli* alkaline phosphatase (electrophoretically purified, code BAPF) from Worthington Biochemical Corp. (Freehold, New Jersey). After dialysis against distilled water,[4] alkaline phosphatase is kept at −18°.

[23] L. W. Lu and K. Randerath, *Cancer Res.* **39**, 940 (1979).
[24] M. Simsek, J. Ziegenmeyer, J. Heckman, and U. L. RajBhandary, *Proc. Natl. Acad. Sci. U.S.A.* **70**, 1041 (1973).
[25] R. Reddy, T. S. Ro-Choi, D. Henning, H. Shibata, Y. C. Choi, and H. Busch, *J. Biol. Chem.* **247**, 7245 (1972).
[26] J. Horowitz, C.-N. Ou, M. Ishaq, J. Ofengand, and J. Bierbaum, *J. Mol. Biol.* **88**, 301 (1974).
[27] W. Schmidt, H. H. Arnold, and H. Kersten, *Nucleic Acids Res.* **2**, 1043 (1975).
[28] B. A. Roe, E. Y. Chen, and H. Y. Tsen, *Biochem. Biophys. Res. Commun.* **68**, 1339 (1976).
[29] K. Marcu, D. Marcu, and B. Dudock, *Nucleic Acids Res.* **5**, 1075 (1978).
[30] K. Randerath, L. S. Y. Chia, R. C. Gupta, E. Randerath, E. R. Hawkins, C. K. Brum, and S. H. Chang, *Biochem. Biophys. Res. Commun.* **63**, 157 (1975).
[31] B. A. Roe, M. P. J. S. Anandaraj, L. S. Y. Chia, E. Randerath, R. C. Gupta, and K. Randerath, *Biochem. Biophys. Res. Commun.* **66**, 1097 (1975).
[32] R. C. Gupta, B. A. Roe, and K. Randerath, *Nucleic Acids Res.*, in press (1979).
[32a] R. C. Gupta, B. A. Roe, and K. Randerath, in preparation (1979).
[33] A. S. Gopalakrishnan, E. Randerath, R. C. Gupta, and K. Randerath, to be published.
[34] E. Y. Chen and B. A. Roe, *Biochem. Biophys. Res. Commun.* **82**, 235 (1978).

Enzyme solution A contains (per microliter): 1.25 μg of RNase A, 1.25 μg of snake venom phosphodiesterase, and 1 μg of alkaline phosphatase. The solution is stored at $-18°$.

Enzyme solution B contains (per microliter): 0.20 μg of RNase A, 0.20 μg of snake venom phosphodiesterase, 0.16 μg of alkaline phosphatase, 10 nmole of $MgCl_2$, and 30 nmole of bicine-Na, pH 8.0. The solution is freshly prepared.

Enzyme solution C contains (per microliter): 0.10 μg of RNase T_2, 0.20 μg of snake venom phosphodiesterase, 0.10 μg of alkaline phosphatase, 10 nmole of $MgCl_2$, and 20 nmole of bicine-Na, pH 7.7. The solution is freshly prepared.

Potassium borohydride pellets from Alfa Inorganics (Beverly, Massachusetts).

Tritium-labeled potassium borohydride (code TRK.203; specific activity >3 Ci/mmole) and sodium borohydride (code TRK.45; specific activity >20 Ci/mmole) from Amersham Corp. (Arlington Heights, Illinois).

Borohydride solution A (2–5 Ci/mmole)[1]: Tritium-labeled potassium borohydride is dissolved in 0.1 N KOH (CO_2-free) at a concentration of 0.1 M. If necessary, the specific activity is adjusted by addition of unlabeled potassium borohydride. The solution is stored in portions of 20–50 μl at $-80°$.

Borohydride solution B (25–30 Ci/mmole)[6]: Sodium borohydride (specific activity > 20 Ci/mmole) is dissolved in 0.1 N KOH at a concentration of about 25 mM. If necessary, the specific activity is adjusted by the addition of unlabeled potassium borohydride. The solution is divided into 10-μl portions, lyophilized, and the dry residue stored at $-80°$. The residue of an individual portion is taken up in 10 μl of 0.01 N KOH immediately before use. Lyophilization serves to minimize losses of reducing capacity of high specific activity preparations.

Cellulose sheets (20 × 20 cm) from E. Merck (EM Laboratories, Elmsford, New York; #5502).

Solvent A[35]: Acetonitrile–4 N aqueous ammonia (3.4 : 1, by volume), freshly prepared.

Solvent B[2,35]: t-Amyl alcohol–methyl ethyl ketone–acetonitrile–ethyl acetate–water–formic acid, specific gravity 1.2 (4 : 2 : 1.5 : 2 : 1.5 : 0.18, by volume), freshly prepared.

Chromatographic markers: Solutions of labeled (0.05 mM, about 1 Ci/mmole) and unlabeled (2 mM) nucleoside trialcohols are prepared by periodate oxidation of nucleosides, followed by borohydride reduction, as described.[1–3]

[35] K. Randerath, E. Randerath, L. S. Y. Chia, and B. J. Nowak, *Anal. Biochem.* **59**, 263 (1974).

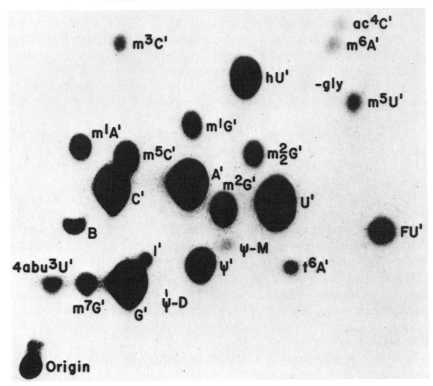

Fig. 1. Fluorogram of a nucleoside trialcohol map on a cellulose thin layer obtained by digestion to nucleosides and chemical tritium labeling of liver tRNA from fluorouridine-treated mice. First dimension from bottom to top; second dimension from left to right (for solvents, see text). For abbreviations of nucleoside trialcohols, see text. FU', trialcohol derivative of fluorouridine. ψ-D and ψ-M, traces of labeled products derived from pseudouridine (for the ^{3}H-labeled monoaldehyde derivative ψ-M, see text). B, background spot (not from RNA). Gly, glycerol. Adapted from Lu *et al.*[19]

X-Ray film: X-OmatR XR-1 or XR-5 from Eastman-Kodak.

2,5-Diphenyloxazole (PPO) and Omnifluor from New England Nuclear Corp. (Boston, Massachusetts).

Scintillation fluid: Solution of 0.3% Omnifluor (w/v) and 25% Triton X-100 (v/v) in xylene.[26,36]

[14]C-Ink from Schwarz/Mann (Orangeburg, New York).

B. Procedure

1. Outline of Procedure. The RNA is first degraded enzymatically to nucleosides. The RNA digest is then treated with periodate, followed by

[36] L. E. Anderson and W. O. McClure, *Anal. Biochem.* **51,** 173 (1973).

tritiated borohydride, to convert the nucleosides, via dialdehydes, to tritium-labeled nucleoside trialcohols [Eq. (1)]. The mixture of the nucleoside trialcohols is separated by two-dimensional thin-layer chromatography (tlc) on cellulose. Labeled compounds on chromatograms are detected by low-temperature solid scintillation fluorography,[37] eluted, and counted. Since the count rate of a nucleoside trialcohol has been shown to be directly proportional to the original concentration of the parent nucleoside,[1,2,38] the base composition can be directly calculated from the count rates of the nucleoside trialcohols. A map of a labeled digest is shown in Fig. 1.[39]

A standard procedure, as well as three scaled-down procedures are described below. While the standard procedure and its scaled-down version 1 have been applied mainly to the analysis of unfractionated tRNA and various RNA species thus far, the scaled-down versions 2 and 3 have been utilized for the analysis of oligonucleotides or oligonucleotide derivatives. In principle, however, the latter procedures can also be adapted to intact RNA.

2. *Enzymatic Digestion of RNA.*[1-3] To a solution of 50 μg of RNA (based on 1 A_{260} unit = 40 μg of RNA) in 40 μl of water are added: 0.5 μl of 1 M MgCl$_2$, 2.5 μl of 0.5 M bicine-Na, pH 8.0, and 8 μl of enzyme solution A. The mixture is incubated at 37° for 6 hr. The digest is processed further immediately or kept frozen at −80°. Repeated freezing and thawing should be avoided.[4,15] Frozen digest is reincubated for 5 min at 37° and vortexed briefly before further processing.

3. *³H-Labeling of Enzymatic Digest.*[1-3] To 15 μl of the above digest are added: 40 μl of water and 10 μl of 9 mM NaIO$_4$. The reaction mixture is incubated at 23° for 2 hr in the dark and then cooled briefly on ice. One microliter of 1 M potassium phosphate, pH 6.8, is added, followed immediately by 5 μl of borohydride solution A. It is important that the pH is kept between 7 and 8 during reduction. The reaction mixture is incubated at 23° for 2 hr in the dark. Excess borohydride is then destroyed by the addition of 100 μl of 1 N acetic acid. The solution is kept in the open tube for 10–15 min at room temperature. It is then evaporated in a stream of filtered air. Addition and decomposition of borohydride are carried out in a hood. The dry residue is taken up in 50 μl of 0.1 N formic acid. The specific activity of the nucleoside trialcohols is half that of the borotritide,

[37] K. Randerath, *Anal. Biochem.* **34**, 188 (1970).
[38] K. Randerath and E. Randerath, *Anal. Biochem.* **28**, 110 (1969).
[39] Abbreviations: A', C', U', G', etc., trialcohols of adenosine, cytidine, uridine, guanosine, etc.; 4abu³U, 3-(3-amino-3-carboxypropyl)uridine; N$_1$pN$_2$p . . . N'$_x$, oligonucleotide 3'-dialcohol; 1 A_{260} unit, amount of nucleotide material in 1 ml of solution giving an absorbance of 1 in a cell of 1 cm light path at 260 nm; tlc, thin-layer chromatography.

i.e., 1–2.5 Ci/mmole. The solution is stored at $-10°$ to $-20°$. Decomposition by self-radiolysis occurs on prolonged storage.

4. *Chromatography of Labeled Digest.*[1-3,35] Labeled digest corresponding to 2–5 μCi is applied to a cellulose thin-layer sheet at 2.5 cm from the left-hand and bottom edges. Development in the first dimension is with solvent A to 17 cm from the origin. After thorough drying, the sheet is trimmed and a Whatman 1 paper wick is attached to the original right-hand side of the sheet by stapling. Development in the second dimension is with solvent B to 4–5 cm on the wick. The sheet is then dried and the wick removed. Chromatography in either dimension is begun no later than 2–3 min after pouring the solvent into the tank (no saturation). The location of an individual nucleoside trialcohol on the map may be determined by cochromatographing an excess of a known radioactive reference compound with labeled digest.

5. *Detection and analysis of ³H-Labeled Nucleoside Trialcohols.*[1-3,37] Fourteen to 16 ml of a 7% (w/v) solution of PPO in diethyl ether is poured over the chromatogram and distributed evenly as rapidly as possible.[3,37] ¹⁴C-Ink is applied to a few points on the layer. This serves for later alignment of chromatogram and film. The layer is then placed in contact with Kodak XR Medical X-Ray Film in a darkroom. Exposure is carried out in the dark between glass plates held together with adhesive tape and wrapped in aluminum foil. If several films are to be exposed simultaneously, cardboard or aluminum sheets are placed in between to prevent penetration of light. The package is stored at $-70°$ to $-80°$ (Dry Ice box or deep freeze). Exposure time depends on the sample.[3,37] The sensitivity of the procedure is 1–5 nCi ³H per day for an average thin-layer spot.[37] Labeled digests of unfractionated tRNA usually require an exposure of about 2–4 days. The film is developed with Kodak liquid X-ray developer, fixed with Kodak liquid X-ray fixer, washed with water, and allowed to dry. Compound areas are marked by superimposing film and chromatogram and perforating the film around the spots with a pin. The darkened area usually does not completely coincide with the area actually occupied by the radioactive compound on the layer. While the four major nucleoside trialcohols may cause some overexposure of the film, the film may be underexposed at the sites of minor constituents. Thus, in the latter case a larger area has to be cut than indicated by the blackening of the film. This may be evaluated in preliminary experiments. The compound areas are cut from the sheet with scissors. The cutouts are eluted, layer side down, with 1.0 or 2.0 ml of 2 N aqueous ammonia for at least 1 hr at room temperature with shaking. Aliquots of 500 μl are transferred to vials, mixed with 10 ml of scintillation fluid, and counted in a scintillation spectrometer.

6. *Calculation of Base Composition.*[1-3] The base composition is calculated according to Eq. (2).

$$f_i = \left(cpm_i \Big/ \sum_{i=1}^{N} cpm_i \right) \times 100 \ [\%] \tag{2}$$

where f_i is the base composition value for individual nucleoside expressed as percentage of the total, N is the number of radioactive nucleoside derivatives, and cpm_i is the count rate of individual radioactive nucleoside derivative.

The base compositions of certain labile constituents have to be corrected for known losses.[2,3,10] Thus, recoveries under the conditions described are 67% for m^7G, 85% for m^3C, 95% for hU, and 57% for ac^4C. Twelve to 15% of m^1A is recovered as m^6A', the rest as m^1A'. Ninety-five percent of ψ is recovered as ψ, and about 5% as a radioactive decomposition product ψ–D. The base constituency of polynucleotides of known chain length is derived from Eq. (3).

$$f_i = \left(cpm_i \Big/ \sum_{i=1}^{N} cpm_i \right) \times \text{chain length [mole]} \tag{3}$$

7. *Scaled-Down Versions of the Procedure*

a. VERSION 1. Volumes of the standard procedure for RNA described above are reduced, but the final concentrations for the enzymatic digestion of the RNA (1 $\mu g/\mu l$ of RNA, 10 mM $MgCl_2$, 30 mM bicine-Na, 0.20 $\mu g/\mu l$ of RNase A, 0.20 $\mu g/\mu l$ of snake venom phosphodiesterase, and 0.16 $\mu g/\mu l$ of alkaline phosphatase) and the labeling reaction (0.7 mM total nucleoside, 1.4 mM $NaIO_4$, 15 mM potassium phosphate, 7 mM borohydride, 0.6 N acetic acid) are maintained. A minimum of about 2–5 μg of RNA, e.g., tRNA, can be analyzed this way.

b. VERSION 2.[4] About 0.1 nmole of an oligonucleotide (0.5–1 μl of solution) is incubated with 4 μl of enzyme solution B in a 25 × 6 mm glass tube for 6 hr at 37°. Labeling steps are analogous to those described above. Additions to the sample: 15 μl of water and 2 μl of 1.5 mM $NaIO_4$ (sufficient for oligonucleotides of chain length <10) for oxidation; 0.5 μl of 0.3 M potassium phosphate, pH 6.8, and 2 μl of borohydride solution A threefold diluted with 0.1 N KOH for reduction; 2 μl of 5 N acetic acid for decomposing residual borohydride. After drying the sample *in vacuo* over solid KOH and P_2O_5, the residue is taken up in 10 μl of water. Two to 3 μl of the labeled solution are chromatographed in the presence of 2–4 nmole each of the unlabeled major nucleoside trialcohols (uv-absorbing markers) and analyzed as described above.

c. VERSION 3.[6] Two to six pmoles of a ^3H-labeled oligonucleotide

3'-dialcohol (dried residue, see below) is incubated with 4 μl of enzyme solution C for 6 hr at 37°. Further processing is analogous to the previous procedure. Additions to the sample: 1 μl of 0.7 mM NaIO$_4$ (sufficient for oligonucleotides of chain length <10); 0.5 μl of 0.1 M potassium phosphate, pH 6.8, and 1 μl of borohydride solution B; 3μl of 1 N acetic acid. The dried residue is taken up in 4 μl of water and 1–2 μl of the labeled solution is chromatographed for base analysis.

8. *Comments.* A representative separation of nucleoside trialcohols obtained by digestion and labeling of tRNA from mouse liver exposed to 5-fluorouridine is shown in Fig. 1. The compounds indicated may be assayed reproducibly by the described procedure. In addition to these compounds, several other tRNA components may be determined, such as m^{2}A,[2] m$_2^6$A, uridine-5-oxyacetic acid,[2] 5-hydroxyuridine,[40] and 5-methylaminomethyl-2-thiouridine.[10,41]

It is important to adhere to the specified conditions as closely as possible to avoid the formation of nucleoside monoaldehydes and the loss of labile constituents:

a. FORMATION OF MONOALDEHYDES.[1,17] If the borohydride concentration during labeling is too low, nucleoside dialdehydes are reduced incompletely, resulting in the formation of tritium-labeled monoaldehydes. These compounds are located to the right of the corresponding nucleoside trialcohols (see Fig. 1 for the location of the monoaldehyde of ψ). A trace of the latter compound (<5% of ψ' radioactivity) is acceptable, as this indicates that the borohydride concentration was not too high (see below). In this case, the radioactivity of ψ-M is multiplied by 2 and added to the radioactivity of ψ'. Insufficient excess of borohydride leads to the appearance on the maps of monoaldehyde derivatives of additional nucleosides, particularly of A and G, the former partially overlapping m$_2^6$G', and the latter, ψ'. Such a chromatogram is not suitable for quantitative evaluation. The borohydride concentration may be too low for the following reasons: (a) Partial decomposition of the borohydride stock solution due to prolonged storage and/or repeated thawing and freezing. The higher the specific activity of the borohydride, the less stable is the preparation. (b) Variations in the amount of active borohydride in commercial batches. We have noted that commercial borohydride preparations sometimes contain up to 50% less or 30–40% more radioactivity than indicated on the label. (c) Partial decomposition of borohydride in the reaction mixture if pH is <7, and (d) presence in the sample of impurities, which are reduced by borohydride.

b. LOSS OF LABILE CONSTITUENTS. If the borohydride concentration or

[40] K. Murao, H. Ishikura, M. Albani, and H. Kersten, *Nucleic Acids Res.* **5,** 1273 (1978).
[41] K. Randerath, unpublished experiments (1972).

the pH of the reaction mixture is too high (for example, due to batch variations, see above), losses of labile constituents, particularly of hU, occur. Losses of alkali-labile constituents, particularly of m⁷G, occur also, if the digestion of the RNA is carried out for a longer period of time than specified.[2]

The reducing capacity of a borohydride solution may be assessed by labeling and analyzing a periodate-oxidized digest derived from a reference sample of tRNA at three or four borohydride concentrations. An appropriate adjustment of the amount of borohydride to be added may then be applied to the actual experiments.

The purity of the sample to be labeled is also important. Thus, a sample should not contain significant amounts of oxidizable or reducible material consuming reagents. For example, Tris buffer should not be present, because it reacts with periodate. For isolation and purification of oligonucleotides by PEI–cellulose tlc prior to tritium derivative analysis, see below.

The tritium derivative method is directly applicable to RNA's extracted from polyacrylamide gels.[42]

If two RNA preparations to be compared exhibit only small base composition differences, it is important that experiments are conducted in parallel.[15] In particular, the labeling reactions should be carried out simultaneously using the same borohydride preparation. It is also essential to cut the nucleoside trialcohol spots from the chromatograms in an identical manner. Several replicate chromatographic analyses (at least four) are carried out, and standard deviations are determined. Statistically significant differences between two samples are estimated on the basis of Student's t test.[3,15]

II. Tritium Derivative Methods for Sequence Analysis of RNA

A. Materials

Enzymes. RNase T_1 (No. 556785) and RNase U_2 (No. 556877 from Calbiochem, San Diego, California), nuclease S_1 (from Miles Laboratories, Elkhart, Indiana) polynucleotide kinase (from P-L Biochemicals Inc., Milwaukee, Wisconsin). Snake venom phosphodiesterase (Section I,A) is treated with acid to inactivate 5'-nucleotidase.[43] Spleen phosphodiesterase (from Boehringer Mannheim Biochemicals, Indianapolis, Indiana) is dialyzed against water.[8] Calf intestinal alkaline phosphatase

[42] L. S. Y. Chia, K. Randerath, and E. Randerath, *Anal. Biochem.* **55**, 102 (1973).
[43] E. Sulkowski and M. Laskowski, Sr., *Biochim. Biophys. Acta* **240**, 443 (1971).

(Grade I) from Boehringer Mannheim Biochemicals is dialyzed against water as described for bacterial alkaline phosphatase.[4] RNase Phy₁ is prepared as described by Pilly et al.[44] One unit of enzymatic activity is defined as the amount of enzyme that solubilized 35 μg of RNA per min at pH 4.5 and 40°.[44] The enzyme is stored in the presence of 40% glycerol at −18°. This enzyme has recently become available from P-L Biochemicals Inc. (No. 0924) and from Enzo Biochem. Research Products (New York, New York, No. ERN-Phy).

Enzyme solution D contains (per microliter): 0.14 unit of polynucleotide kinase, 0.1 nmole of [γ-^{32}P]ATP (10–20 Ci/mmole), 15 nmole of Tris-HCl, pH 8.0, 5 nmole of dithiothreitol, and 40 nmole of MgCl₂. The solution is freshly prepared.

Enzyme solution E contains (per microliter): 120 units of nuclease S₁, 120 nmole of sodium acetate, pH 4.5, 250 nmole of LiCl, and 0.2 nmole of ZnCl₂.

Tritium-labeled potassium borohydride from Amersham Corp. as above or sodium borohydride (5–15 Ci/mmole; code NET-023H) from New England Nuclear (Boston, Massachusetts). Borohydride solutions are prepared in 0.1 N KOH and stored lyophilized as above.

[γ-^{32}P]ATP (1000–2000 Ci/mmole) from ICN Pharmaceutical Inc. (Irvine, California) or New England Nuclear.

Yeast transfer RNA (Type I) from Sigma Chemical Co. (St. Louis, Missouri).

Polyethyleneimine (PEI)–cellulose sheets are prepared in the laboratory as described,[45] but a 0.5% solution of PEI-1000 (Dow Chemical Co., Midland, Michigan) is substituted for 1% Polymin P solution (see also Randerath et al.[4]). The size of the sheets is 20 × 25 cm or 20 × 22 cm.

For the elution of nucleotides from PEI–cellulose, sheets of Whatman 1 paper and Whatman P81 phosphate paper are thoroughly prewashed by descending flow-through development for about 24 hr with either 2 M LiCl or 4 M pyridinium formate, pH 4.0, depending on the eluent to be used for subsequent nucleotide extraction from PEI–cellulose. Small strips are cut from the prewashed papers as needed for the elution.

Silica gel sheets are Eastman No. 13181.

Button-type permanent Alnico magnets ($\frac{1}{2}$ inch diameter × $\frac{3}{8}$ inch) from General Hardware Manufacturing Co. (New York, New York).

Solvent C[4]: Acetonitrile–t-amyl alcohol–concentrated ammonia (2 : 1 : 1).

Solvent D[4]: t-Amyl alcohol–methyl ethyl ketone–water (3 : 6 : 1.2).

[44] D. Pilly, A. Niemeyer, M. Schmidt, and J.-P. Bargetzi, J. Biol. Chem. **253**, 437 (1978).
[45] K. Randerath and E. Randerath, J. Chromatogr. **2**, 110 (1966).

Phosphocellulose (Whatman, Pl Floc, cellulose phosphate) from Whatman Inc. (Clifton, New Jersey).

Calcium tungstate Lightning Plus Intensifying Screen from DuPont.

For additional materials, see Section I,A.

B. Analysis of Complete RNase T_1 and A Digests of RNA

1. Procedure A

a. OUTLINE OF PROCEDURE. A complete RNase T_1 or A digest of the RNA is separated by two-dimensional PEI–cellulose anion-exchange tlc.[10,30,31,46] Oligonucleotides are located under short-wave uv light and eluted.[10] Their base compositions are determined by tritium derivatization. Molar ratios of the oligonucleotides are determined in conjunction with base composition analysis. The oligonucleotides are sequenced by a tritium derivative method (formerly called procedure II[5]), which entails (1) controlled digestion of the oligonucleotide with snake venom phosphodiesterase–alkaline phosphatase and periodate oxidation of 3′ ends of partial digestion products, (2) borohydride reduction of the oligonucleotide dialdehyde intermediates obtained in step (1) to 3′ terminally ^{3}H-labeled oligonucleotide 3′-dialcohol derivatives [see Eq. (1)], and (3) deduction of the sequence of the parent oligonucleotide by separation of the ^{3}H-labeled oligonucleotide 3′-dialcohols according to chain length, in situ enzymatic liberation of ^{3}H-labeled 3′ terminal trialcohols, contact transfer, and identification of the trialcohols by two-dimensional tlc and scintillation counting. An example illustrating the separation of the oligonucleotide 3′-dialcohol intermediates according to chain length is shown in Fig. 2.

The same ^{3}H-labeled oligonucleotide 3′-dialcohol intermediates may also be obtained by "pseudoexonucleolytic" degradation of an oligonucleotide with alkaline phosphatase and periodate at pH 8, followed by borohydride reduction.[4] This procedure (formerly called procedure I) is not described in this article. For a comparison of both procedures, see Sivarajan et al.[5]

The 5′ termini of the oligonucleotides, which cannot be analyzed by these procedures, are determined separately as ^{3}H-labeled nucleoside trialcohols after venom phosphodiesterase treatment of the oligonucleotides and labeling.[4]

b. PREPARATION AND CHROMATOGRAPHY OF COMPLETE RNA DIGESTS.[10] For complete digestion by RNase T_1 or A, 120 μg ($3 A_{260}$ units) of low molecular weight RNA, e.g., tRNA, are incubated at 38° and at a

[46] K. Randerath and E. Randerath, this series, Vol. 12, Part A, p. 323.

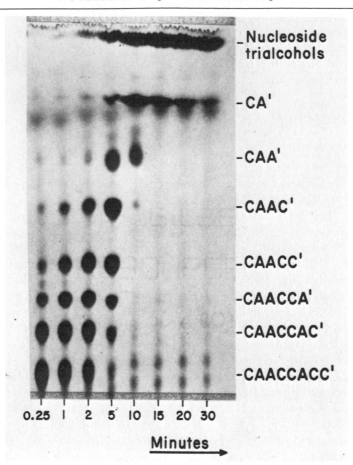

FIG. 2. Separation of ^{3}H-labeled oligonucleotide 3′-dialcohol intermediates indicating the time course of the degradation of CAACCACC-OH by venom phosphodiesterase–alkaline phosphatase. Detection of labeled compounds by fluorography. (Radioactive background is due to contaminants that were not removed in this experiment.)

concentration of 2 μg/μl in 0.02 M Tris-HCl, pH 7.6, containing 1 unit/μl of RNase T$_1$, for 2 hr or in 0.05 M Tris-HCl, pH 7.6, containing 1 μg/μl of RNase A, for 2.5 hr. To convert 2′,3′-cyclic phosphates to 2′- and 3′-phosphates, 1 N HCl is added to a final concentration of 0.14 N. After incubation at 22° for 35 min, HCl is neutralized by the addition of 0.5 N LiOH. RNase A digests are briefly heated to 100° and cooled rapidly on ice immediately before chromatography in order to dissociate G-rich oligonucleotides.

PEI–cellulose sheets (20 × 25 cm) are soaked in methanol–concentrated ammonia (1000 : 1, 250 ml/sheet) for 5 min and dried prior to application of digest. Digest is applied in 5-μl portions without intermediate drying at 2 cm from the left-hand and bottom edges. To remove salts, the sheet is soaked in 250 ml of methanol for 10 min after sample application and drying. After evaporation of methanol, a Whatman 1 paper wick is attached to the top of the first dimension and the origin area is wetted with 5 μl of water immediately before chromatography. The first dimension is developed parallel to the long side of the sheet as follows:

Water to the origin
7.5 M urea to 2 cm above the origin
0.3 M Tris-HCl–7.5 M urea, pH 7.9, to 5 cm
0.1 M LiCl–0.3 M Tris-HCl–7.5 M urea, pH 7.9, to 8 cm
0.2 M LiCl–0.3 M Tris-HCl–7.5 M urea, pH 7.9, to 11 cm
0.3 M LiCl–0.3 M Tris-HCl–7.5 M urea, pH 7.9, to 14.5 cm
0.4 M LiCl–0.3 M Tris-HCl–7.5 M urea, pH 7.9, to 20 cm
0.5 M LiCl–0.3 M Tris-HCl–7.5 M urea, pH 7.9, to 0.5 cm on the wick
0.6 M LiCl–0.3 M Tris-HCl–7.5 M urea, pH 7.9, to 3 cm on the wick

The wick is then cut off and the wet sheet soaked in two portions of 300 ml of methanol for 10 min each time. After drying, another wick is attached to the top of the second dimension and the sheet is developed as follows:

Water to the origin
7.5 M urea to 2 cm
0.4 M Li formate, 7.5 M urea, pH 3.5, to 5 cm
0.7 M Li formate, 7.5 M urea, pH 3.5, to 8 cm
1.2 M Li formate, 7.5 M urea, pH 3.5, to 13.5 cm
1.8 M Li formate, 7.5 M urea, pH 3.5, to 6 cm on the wick

After removal of the wick the sheet is soaked in methanol as above. The gradients described may have to be varied to some extent depending on the RNA to be analyzed. For example, digests containing oligonucleotides of chain length >10 require >0.8 M LiCl in Tris–urea and >2.5 M Li formate in urea as the final solvents of the gradients. Since the progress of the separation can be ascertained during chromatography by briefly removing the chromatogram from the tank and examining it under short-wave uv light, it is possible to select suitable chromatographic conditions during development. However, caution must be exercised if the digest contains U-rich oligonucleotides, which may decompose partially when exposed to uv light. The chromatography of G-rich oligonucleotides, which tend to streak, can be improved by development at elevated temperature (40°–50°).

c. ELUTION OF OLIGONUCLEOTIDES.[10] Nucleotide spots circled with pencil under short-wave uv light are cut from the chromatogram as indi-

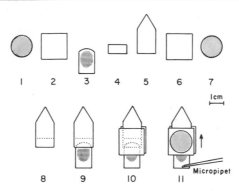

FIG. 3. Assembly of the elution device. 1 and 7, magnets for holding assembly together; 2 and 6, pieces of vinyl plastic; 3, cutout with nucleotide spot and a 3-mm long nucleotide-free zone at bottom end; 4, prewashed phosphate paper; 5, prewashed Whatman 1 paper. The width of 4 and 5 is the same and somewhat greater than that of the cutout. The elution device is assembled as indicated in 8–11. The phosphate paper is placed between layer and Whatman 1 paper. Eluent is applied to the nucleotide-free zone of the cutout as indicated in 11. The arrow indicates the direction of eluent flow.

cated in Fig. 3. The elution device is assembled in such a way (Fig. 3) that eluate passes from the layer through the phosphate paper into the Whatman 1 paper wick. Soluble PEI–cellulose coextracted with the nucleotide(s) is retained in the phosphate paper. If not removed, it interferes with nucleotide recoveries as well as subsequent ^{3}H-labeling reactions. Eluent is applied slowly in small portions close to the edge of the cutout (Fig. 3) and allowed to migrate into the tip. The assembly is then placed in a stream of warm air to dry the exposed parts. Elution into the tip is repeated two more times. Ten to twenty cutouts can be processed in parallel. The eluent is 2 M LiCl or 4 M pyridinium formate, pH 4.0. Lithium chloride samples are processed in the following way: The assembly is taken apart and the Whatman 1 paper cut about 2–3 mm below the tip. The piece containing the nucleotide is dried thoroughly in a stream of warm air, placed into 10 ml of ethanol–isopropanol (1 : 1, v/v) and agitated for 20 min to remove LiCl. This step is repeated once. After drying, the nucleotide is eluted from the paper with 300 μl of water for 1–3 hr in a covered plastic weigh-boat. The eluate is removed and the paper re-extracted briefly with 50–100 μl of water. The combined eluates are lyophilized and the residue taken up in 20 μl of water. For elution with pyridinium formate, the Whatman 1 paper, after cutting, is placed directly, without drying, in 300 μl of water. Further processing is as above. Pyridinium formate is removed by lyophilization.

Recoveries of oligonucleotides of chain length up to 10–12 are 60–70%. They are usually identical for the two eluents. However, better recoveries of G-rich oligonucleotides are obtained with pyridinium formate, while LiCl elution is better suited for large oligonucleotides (chain length > 10–12). Recoveries of larger oligonucleotides can be improved if all elution steps are carried out in the presence of 7–8 M urea, which may be removed by dialysis. Large oligonucleotides may also be digested endonucleolytically *in situ* to smaller fragments[4,47,48] and then eluted or further chromatographed directly after contact transfer (for contact transfer, see below). For example, a RNase T_1 digestion product may be digested *in situ* by wetting the spot with 15 μl/cm^2 of 0.1 M Tris-HCl, pH 7.6, drying, and application of 15 μl/cm^2 of RNase A (2 μg/μl). The spot is covered with Parafilm and sandwiched between glass plates. Incubation is at 38° for about 3 hr. *In situ* digestion with RNase T_1 may be performed analogously.

d. BASE ANALYSIS AND MOLAR RATIOS. The base composition of aliquots (0.1 nmole) of the eluted oligonucleotides is analyzed by the scaled-down version 2 of the tritium derivative method for base analysis (see above). The count rates of individual trialcohols may also be utilized to calculate the molar ratios of the oligonucleotides in a given digest.[10] For this purpose, the count rates derived from the 3' terminal positions of each individual oligonucleotide are compared. As most oligonucleotides in a pure RNA are present at a molar ratio of one per chain, the most frequent count rates of the 3' terminal trialcohols are normalized to 1.0. To obtain reproducible molar ratios, it is important to avoid loss of material when isolating the oligonucleotides from the PEI–cellulose chromatogram and to analyze the oligonucleotides from a particular map in parallel.

e. CONVERSION OF OLIGONUCLEOTIDES TO ^{3}H-LABELED OLIGONUCLEOTIDE 3'-DIALCOHOL INTERMEDIATES.[5] Aliquots of the eluted oligonucleotides (0.5 nmole) are incubated at 23° and at a concentration of 0.005 mM with 10 mM Mg acetate, 20 mM bicine-Na, pH 8.0, 0.1 μg/μl of bacterial alkaline phosphatase, and 0.01 μg/μl of snake venom phosphodiesterase. Six to eight 10-μl aliquots are taken at suitable time intervals up to about 30 min and immediately mixed with 5 μl of 0.75 mM NaIO$_4$. Oxidation is allowed to proceed for 45–60 min at 23° in the dark. The oxidized samples are kept frozen until borohydride treatment. Samples are reduced by adding 1 μl of 40 mM KB^{3}H$_4$ or NaB^{3}H$_4$ (6–10 Ci/ mmole). After 1 hr at 23° in the dark, 7 μl of 1 N acetic acid is added and the solutions are evaporated in a stream of air. The residues are taken up in 15 μl of water.

f. SEPARATION OF ^{3}H-LABELED OLIGONUCLEOTIDE 3'-DIALCOHOL INTERMEDIATES.[4] Aliquots of the labeled solutions (see preceding section)

---

[47] K. Randerath and E. Randerath, *Angew. Chem., Int. Ed. Engl.* **3**, 442 (1964).
[48] A. S. Gopalakrishnan, R. C. Gupta, E. Randerath, and K. Randerath, to be published.

corresponding to 0.02–0.03 nmole of original oligonucleotide are applied in 2-μl portions to a PEI–cellulose sheet (25 cm long). To remove radioactive contaminants present in some batches of borohydride, which interfere with the detection particularly of the larger oligonucleotide 3'-dialcohols, the sheet is predeveloped with 4 M lithium formate, 7 M urea, pH 3.5, to 3 cm above the origin.[32] The wet sheet is then soaked in 300 ml of methanol–concentrated ammonia (1000 : 1, v/v) for 10 min, dried, and cut at 1 cm above the origin. The lower part of the sheet containing most of the contaminants is discarded. After attaching a Whatman 1 paper wick, the upper part of the sheet is developed as follows:

Water to 2 cm above the lower edge of the chromatogram
0.05 M Tris-HCl, pH 8.0, to 4 cm
0.15 M Tris-HCl, 8.5 M urea, pH 8.0, to 8 cm
0.40 M Tris-HCl, 8.5 M urea, pH 8.0, to 11 cm
0.60 M Tris-HCl, 8.5 M urea, pH 8.0, to 16 cm
0.80 M Tris-HCl, 8.5 M urea, pH 8.0, to 1–2 cm on the wick

After removal of the wick, the sheet is dried thoroughly and finally soaked with agitation in 500 ml of methanol for 15 min.

g. ANALYSIS OF 3' TERMINI OF OLIGONUCLEOTIDE 3'-DIALCOHOL INTERMEDIATES.[4] Oligonucleotide 3'-dialcohol intermediates are detected on the chromatogram by fluorography[37] similarly as previously described. One intermediate of each chain length (see Fig. 2) is selected for 3' terminal analysis. To release the labeled terminus as a nucleoside trialcohol, 2–4 μl of a RNase T$_2$ solution (1 unit/μl) is applied to the center of each spot for *in situ* digestion. Spots are covered immediately with Parafilm, and the assembly is kept between heavy glass plates so as to keep the spots moist. Incubation is at 23° for 3 hr or overnight. After drying, 1 μl of a mixture of nonradioactive nucleoside trialcohols, 2–4 mM with respect to each component, is applied to the treated areas. After evaporation, each area is cut out and placed in contact with a silica gel layer close to one corner of a 10 × 20 cm sheet. The cutout–silica gel assembly is then sandwiched between magnets. Subsequent ascending chromatography quantitatively transfers nucleoside trialcohols from the cutout to the silica gel layer. Development is with solvent C to 15 cm above the origin (first dimension), followed by solvent D to 7.5 cm (second dimension). Compounds are located under uv light, cut from the chromatogram and assayed by direct counting[4]: Cutouts are placed, layer side up, in scintillation vials at 0°, and 10 or 20 μl of scintillation fluid, depending on spot size, is applied evenly to the layer. The samples are then counted in a liquid scintillation counter at 4–8°. Alternatively, the nucleoside trialcohols can be analyzed by two-dimensional tlc on cellulose following contact transfer, using the same solvents as for base analysis (see above). The labeled nucleoside trialcohols may also be rendered visible by fluorography on either cellulose or silica gel sheets.

h. ANALYSIS OF 5' TERMINI.[4] To release the 5' terminal nucleosides of oligonucleotides, 0.075–0.15 nmole of oligonucleotide is incubated at a concentration of 3–6 μM with 20 mM bicine-Na, pH 9.0, 10 mM Mg acetate, and 0.1 $\mu g/\mu l$ of snake venom phosphodiesterase at 38° for 60 min. To 15 μl of the incubation mixture an equal volume of 0.3 mM NaIO$_4$ is added. After incubation for 1 hr at 23° in the dark, 1 μl of 50 mM KB^{3}H$_4$ or NaB^{3}H$_4$ (6–10 Ci/mmole) is added and the solution is kept for 1 hr at 23° in the dark. Ten microliters of 1 N acetic acid are added, and the solution is evaporated in a stream of air. The residue is taken up in 15 μl of water. Three microliters of the labeled solutions are applied to a cellulose sheet together with suitable nucleoside trialcohol markers. Chromatography and analysis of the nucleoside trialcohols is as described above for 3' terminal analysis.

i. ANALYSIS OF OLIGONUCLEOTIDES CONTAINING 2'-O-METHYLATED NUCLEOSIDES.[49] Although 2'-O-methylated nucleosides cannot be analyzed by chemical tritium derivative methods, their presence in an oligonucleotide is readily detectable. When digested by the snake venom phosphodiesterase–alkaline phosphatase procedure described here, an oligonucleotide carrying a 2'-O-methylated nucleoside in an internal position of the chain does not give rise to an oligonucleotide 3'-dialcohol of chain length n, thus causing a gap in the series of oligonucleotide 3'-dialcohol intermediates separated by size. If the ribose-methylated nucleoside occupies the 5' terminal position, 5' terminal analysis (see above) will not yield a ^{3}H-labeled nucleoside trialcohol.

Ribose-methylated nucleosides are analyzed directly by a radioactive derivative procedure, which entails ^{32}P-labeling by the [α-^{32}P]ATP–polynucleotide kinase reaction[24,50] of the 5' terminus of ribonuclease T$_2$-stable 2'-O-methylated dinucleotides derived from the oligonucleotide, conversion of the labeled dinucleotide to the ^{32}P-labeled 2'-O-methylated nucleoside 5'-monophosphate, and identification of the monophosphate by its chromatographic properties on a PEI–cellulose thin layer.[49]

2. Procedure B

a. OUTLINE OF PROCEDURE. Oligonucleotides in a complete RNase T$_1$–alkaline phosphatase or RNase A–alkaline phosphatase digest of the RNA are converted to 3' terminally ^{3}H-labeled oligonucleotide 3'-dialcohols [Eq. (1)] and mapped on PEI–cellulose.[3,32] Labeled derivatives

[49] R. C. Gupta, K. Randerath, and E. Randerath, *Anal. Biochem.* **76,** 269 (1976).
[50] C. C. Richardson, *Proced. Nucleic Acid Res.* **2,** 815 (1971).

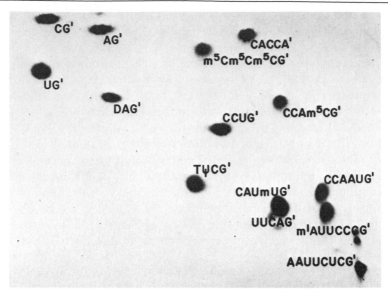

FIG. 4. PEI–cellulose map of a ^{3}H-labeled RNase T_1 digest of human placenta tRNA$_{GCC}$[Gly] (Gupta *et al.*[32]). First dimension (LiCl gradient), from right to left; second dimension (ammonium formate gradient), from bottom to top. Detection by fluorography. For experimental details, see text. CG′, UG′, etc., oligonucleotide 3′-dialcohols of CpG, UpG, etc. The overlapping compounds CAUmUG′ and UUCAG′ may be resolved by rechromatography. The spot marked CAUmUG′ contains some CAUmUm²G′.[32]

are located by fluorography[37] and eluted. A fluorogram of a map of a labeled digest derived from human placenta tRNA$_{GCC}$[Gly] is shown in Fig. 4. Base compositions of the oligonucleotide 3′-dialcohols are determined by the tritium derivative method for base analysis. Molar ratios of the oligonucleotides are calculated from the count rates of the ^{3}H-labeled oligonucleotide dialcohols.

The sequences of the oligonucleotide 3′-dialcohols are deduced by a readout procedure based on partial digestion of these compounds with specific endonucleases and resolution of the cleavage products by size on PEI–cellulose.[8,9] If the oligonucleotide 3′-dialcohols are derived from a RNase T_1 digest, they are partially degraded with RNases U_2, A, and Phy$_1$, while RNases T_1 and U_2 are used for RNase A digests. These partial digests are chromatographed alongside a controlled nuclease S_1–alkaline phosphatase digest, which contains all possible cleavage products. The sequence of the four major constituents can then be read directly from the chromatogram. An example illustrating this procedure is shown in Fig. 5. As the smallest labeled cleavage product obtained in this system is the dinucleotide from the 3′ end of the oligonucleotide 3′-dialcohol, positions 1 and 2 from the 3′ end have to be determined separately[8] (see below).

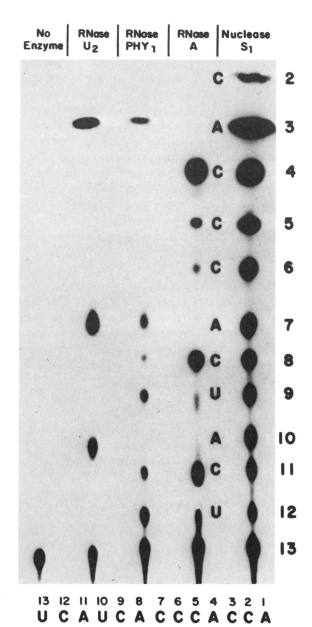

FIG. 5. Resolution by size on a PEI–cellulose thin layer of products obtained by controlled RNase U$_2$, RNase Phy$_1$, RNase A, and nuclease S$_1$ digestion of UCAUCACCCACCA′. Detection by fluorography. The numbers indicate both the chain lengths of the compounds and the positions of individual residues within the sequence. The vertical column of letters refers to individual residues identified on the basis of the cleavage patterns. Taken from Gupta and Randerath.[9]

The positions of modified constituents may be inferred from base composition data or, if necessary, determined in a separate experiment. Thus, internal positions of modified constituents in the oligonucleotide chain may be determined as follows[6,7]: The oligonucleotide 3'-dialcohol is partially digested with nuclease S_1–alkaline phosphatase and the labeled fragments separated by size by PEI–cellulose tlc and located by fluorography. Undesired nonradioactive partial digestion products are eliminated by periodate oxidation. The sequence of the parent compound is then deduced by enzymatic incorporation of ^{32}P-label into the 5' termini of the labeled oligonucleotide 3'-dialcohol intermediates, enzymatic release of ^{32}P-labeled nucleoside 5'-monophosphates, two-dimensional PEI–cellulose chromatography, and autoradiography.

b. PREPARATION AND ^{3}H-LABELING OF COMPLETE DIGESTS.[3,32] For complete RNase T_1 or RNase A digestion, 4–12 μg (0.1–0.3 A_{260} unit) of a low molecular weight RNA, e.g., tRNA, is incubated at a concentration of 1 μg/μl with 0.1 μg/μl of RNase T_1 or RNase A, 0.1 μg/μl of alkaline phosphatase, and 50 mM bicine, pH 7.8, at 38° for 1½ hr. One microliter of 30 mM NaIO$_4$ is added to 4 μl of the digest and the solution is incubated at 23° for 2 hr in the dark. After cooling on ice, 1 μl of 0.1 M potassium phosphate, pH 6.8, is added, followed immediately by the addition of 2 μl of 0.3 M NaB^{3}H$_4$ (40 Ci/mmole) and further incubation at 23° for 2 hr in the dark. After addition of 6 μl of 5 N acetic acid, the solution is dried in a stream of air. The residue is taken up in 20 μl of water.

c. CHROMATOGRAPHY OF ^{3}H-LABELED DIGESTS.[3,32] Labeled digest corresponding to 4–12 μg of RNA is applied in 5-μl portions to a PEI–cellulose thin-layer sheet (20 × 25 cm) at 2 cm from the left-hand and bottom edges. After the origin has been dried, radioactive contaminants are removed by predevelopment[32] as described (see Section II,B,1,f). A Whatman 1 paper wick is attached to the part of the sheet containing the purified digest and the chromatogram is developed as follows:

Water to 2 cm above the lower edge of the chromatogram
0.05 M LiCl to 5 cm
0.20 M LiCl to 11 cm
0.40 M LiCl to 21 cm
0.50 M LiCl to 6–7 cm on the wick

The wick is then cut off and the sheet dried and soaked in two portions of 300 ml of methanol 10 min each time. Another wick is attached to the top of the second dimension and the sheet developed as follows:

Water to the origin
0.2 M ammonium formate, pH 2.6, to 4 cm

1.0 M ammonium formate, pH 2.6, to 9 cm
2.5 M ammonium formate, pH 2.6, to 16 cm
4.0 M ammonium formate, pH 2.6, to 5–6 cm on the wick

After drying with warm air, the sheet is soaked in 300 ml of methanol for 10 min and dried. The oligonucleotide 3'-dialcohols are located by fluorography. The gradients described here may have to be varied to some extent depending on the RNA to be analyzed. If the digest contains oligonucleotides of chain length 10–14, an additional solvent of higher ionic strength (0.6–0.8 M LiCl and 4.5–6 M ammonium formate) has to be included. A suitable gradient can be established by preliminary experiments with small portions of the labeled digest corresponding to about 0.2–0.3 μg of RNA. If necessary, compounds may be rechromatographed directly by contact transfer (see Section II,B,1,g) to a fresh PEI–cellulose sheet and development with a gradient of Tris-HCl, 8.5 M urea, pH 8.0, as described in Section II,B,1,f).

The excellent resolution of a labeled RNA digest in the above system is illustrated by the chromatogram in Fig. 4. The chromatogram is essentially free of radioactive background spots, which is largely due to the effective purification by predevelopment at pH 3.5.

d. ELUTION OF OLIGONUCLEOTIDE 3'-DIALCOHOLS.[6,32] To remove PPO, the chromatogram of the labeled digest is soaked in 500 ml of methanol 15 min prior to elution. Phosphocellulose and glass wool are prewashed extensively with 4 M pyridinium formate, pH 4.2. Phosphocellulose (3–5 mm) is then layered on top of a small glass wool plug placed above the narrow part of a short Pasteur pipette, drawn out into a capillary. Oligonucleotide 3'-dialcohol spots are scraped from the chromatogram and layered on top of the phosphocellulose by using suction. A series of columns can be processed in parallel. Eluent is added from a 1-ml syringe. The columns are eluted twice with 400–600 μl of 4 M pyridinium formate, pH 4.2 (elution time, 10–15 min) and the remaining eluent is removed in a stream of air. The eluates are lyophilized. To remove residual pyridinium formate, residues are taken up in 200–400 μl of water, the solutions lyophilized, and residues finally extracted with three 500-μl portions of ether. The dry residues are taken up in 40–80 μl of water.

While this elution procedure is less time-consuming than the alternative procedure described in Section II,B,1, compounds isolated by the column technique may contain traces of phosphocellulose, which may occasionally interfere with enzymatic reactions requiring divalent cations. To prevent this, the concentration of divalent cation(s) in the incubation mixture may have to be raised.

e. BASE ANALYSIS OF OLIGONUCLEOTIDE 3'-DIALCOHOLS. The base

composition of the eluted oligonucleotide 3'-dialcohols (4–6 pmole) is analyzed by the scaled-down version 3 of the tritium derivative method for base analysis (see above).

f. MOLAR RATIOS. A small portion of the labeled RNA digest corresponding to about 0.2–0.3 μg of the RNA is chromatographed as described above. Oligonucleotide 3'-dialcohol spots located by fluorography are cut from the chromatogram. The cutouts are placed in vials and counted directly after addition of a minimum of 2 ml of scintillation fluid. The molar ratios of the original oligonucleotides are calculated by normalizing the most frequent count rates to 1.0.

g. PARTIAL ENDONUCLEOLYTIC DIGESTION OF OLIGONUCLEOTIDE 3'-DIALCOHOLS.[8,9] For controlled RNase T_1, A, U_2, and Phy_1 digestion, each reaction mixture contains, in a total volume of 20–50 μl, 1–10 pmole (4×10^4–4×10^5 dpm) of oligonucleotide 3'-dialcohol and 0.1 μg/μl of yeast tRNA as carrier. Incubation mixtures for RNase T_1 or RNase A digestion contain 20 mM bicine, pH 7.8, and 0.01 ng/μl of RNase T_1[51] or 0.1–0.2 ng/μl of RNase A.[51] Incubation mixtures for RNase U_2 or Phy_1 digestion contain 20 mM sodium acetate, pH 4.5, 2 mM EDTA, and 0.1–0.2 munit/μl of RNase U_2[51] or 0.34 munit/μl of RNase Phy_1. Incubation is at 23° for 15 min (RNase Phy_1) or 30 min (RNase T_1, A, or U_2). Aliquots (5–10 μl) are withdrawn at suitable intervals, e.g., at 2, 8, and 15 min (RNase Phy_1) or at 2, 8, 15, and 30 min (RNase T_1, A, or U_2), and applied to the same point at the origin of a PEI–cellulose thin-layer sheet (pretreated by soaking 5 min in 300 ml of methanol–concentrated ammonia (1000 : 1, v/v)). After each application, the spot is dried immediately with warm air.

For controlled nuclease S_1 digestion,[6] 20–50 pmole (0.8–2 $\times$ 10^6 dpm) of oligonucleotide 3'-dialcohol in 20–50 μl is incubated with 1 unit/μl of nuclease S_1, 0.125 M LiCl, 0.1 mM $ZnCl_2$, and 20 mM sodium acetate, pH 4.5, at 23° for 30 min. Aliquots (6–15 μl) are withdrawn at suitable time intervals, e.g., at 1, 8, and 15 min, and added successively to a tube containing 100 mM bicine-Na, pH 8.5, and 0.2 μg/μl of alkaline phosphatase. The volume of the bicine–phosphatase solution approximately equals that of the combined aliquots. Dephosphorylation is allowed to proceed for 30–45 min at 23°. To produce the intermediate of chain length ($n - 1$) that is sometimes obtained in low yield upon partial nuclease S_1 digestion, the parent oligonucleotide 3'-dialcohol (about 2 pmole) is partially digested in 10 μl with spleen phosphodiesterase (0.05 μg/μl), 20 mM bicine, pH 7.8, 2 mM EDTA, for 10–30 min at 38°. This digest is cochromatographed with a 10-pmole aliquot of the nuclease S_1 digest alongside the other partial digests.

---

[51] For short fragments (chain length <5), 2-3-fold higher concentrations of the enzymes are preferable.

h. CHROMATOGRAPHY OF PARTIAL DIGESTS AND READOUT OF SE-
QUENCE. The partial digests (Section II,B,2,g) are resolved on a PEI–
cellulose thin layer by size in the gradient described in Section II,B,1,f.
Predevelopment with lithium formate–urea is omitted. After drying, the
sheet is soaked in 300 ml of methanol for 10 min. Compounds are located
by fluorography.

Partial digestion of a 3' end-labeled oligonucleotide in combination
with resolution by size as described above enables the determination of
the location of the four major constituents relative to the labeled terminal
position.[8,9] While the positions of G, A, and pyrimidine residues in the
chain can be derived from the cleavage patterns for G, A + G, and C + U
produced by RNase T_1, RNase U_2, and RNase A, respectively,[8] C and U
residues can be distinguished on the basis of the cleavage patterns produced by partial RNase Phy_1 digestion.[9] Rules for the cleavage of dinucleotides[44] and polynucleotides[9] by RNase Phy_1 under partial digestion
conditions may be summarized as follows:

(i) C–N is resistant with the exception of C–A, which may be slightly
digested under certain conditions.
(ii) U–N is always cleaved.
(iii) N–C is cleaved with the exception of C–C.
(iv) N–U is resistant with the exception of U–U.

In cases where RNase Phy_1 does not cleave next to U or U derivatives,
e.g., U–Um or T–ψ, these may be distinguished from C by a previously
described chromatographic mobility-shift procedure.[8]

The derivation of the sequence of a 3' end-labeled oligonucleotide by
this procedure is shown in Fig. 5. Specific cleavage between the nth and
$(n + 1)$th position from the 3' terminus of a 3' end-labeled oligonucleotide
3'-dialcohol makes possible the identification of the $(n + 1)$th position.
Thus, partial RNase U_2 digestion indicates positions 4, 8, and 11 to be A
residues and partial RNase A digestion shows positions 3, 5, 6, 7, 9, 10,
12, and 13 to be pyrimidines. (The dinucleotide spot of the RNase A digest
was weak on the original film and therefore does not show in Fig. 5.) A
preliminary sequence can then be written as $(Pyp)_2Ap(Pyp)_2Ap(Pyp)_3$-
ApPypCpA', positions 1 and 2 having been determined in separate experiments[8] (for the identification of positions 1 and 2, see also below). The
remaining individual pyrimidines are identified by partial RNase Phy_1
digestion as follows. As shown in Fig. 5, intermediates of chain lengths 3,
7, 8, 9, 11, and 12 are present in the RNase Phy_1 digest. The fact that
cleavage products of chain lengths 2, 4, 5, and 6 are present in the RNase
A but not in the RNase Phy_1 digest indicates positions 3, 5, 6, and 7 to be
C, in agreement with the resistance of C–N bonds to RNase Phy_1 under

partial digestion conditions [rule (i)]. As shown by a comparison with the RNase A digest, chain lengths 9 and 12 must have arisen from Py–Py breaks. Since C–Py is resistant to RNase Phy_1[rule (i)], the presence of chain lengths 9 and 12 indicates positions 10 and 13 to be U. The absence of chain length 10 in the RNase Phy_1 digest (but its presence in the RNase U_2 digest) provides further evidence for U in position 10, as A–U bonds are virtually resistant to the action of RNase Phy_1[rule (iv)], while A–C is readily cleaved [rule (iii)]; note, for example, chain lengths 3 and 7 in the RNase Phy_1 digest. Thus, the sequence of the compound is

$$U_{13}-Py_{12}-A_{11}-U_{10}-Py_9-A_8-C_7-C_6-C_5-A_4-C_3-C_2-A_1{}'$$

As to position 12, note the weakness of spot 11 relative to spot 12. In agreement with the distinct preference of the enzyme for U–C over C–A bonds,[44] the relative intensities of spots 12 and 11 indicate that spot 12 is due to U–C cleavage and spot 11 to C–A cleavage. (In the sequence –U–U–A–, a U–U break would result in a much weaker spot than a U–A break.) The 5′ proximal sequence therefore must be U–C–A–U–. The relative intensities of spots 9 and 8 indicate similarly that these spots result from U–C and C–A breaks, respectively. These results extend the 5′-proximal sequence to U–C–A–U–C–A–C. . . . , thus establishing the sequence of the entire compound.

    i. IDENTIFICATION OF 3′ TERMINAL POSITIONS. Positions 1 and 2 (the 3′-terminus) may be identified as follows[8]:

    (i) If the phosphodiester bond between positions 1 and 2 is susceptible to cleavage by RNase T_1 or RNase A, 2–4 μl of the partial digest (see above) are further incubated for 1–2 hr at 38° after adding more of the respective RNase (enzyme–substrate ratio, 1 : 10, by weight). The ^{3}H-labeled nucleoside trialcohol released is then analyzed by silica gel thin-layer chromatography (Section II,B,1,g). If the 3′ terminal trialcohol is released by RNase T_1, position 2 must be G or a derivative of G. If it is released by RNase A, it must be a pyrimidine. RNase U_2 does not release the 3′ terminus as a trialcohol, if position 2 is A.[8]

    (ii) If position 2 is C, U, or A, it is identified most conveniently by cochromatography of the ^{3}H-labeled dinucleotide 3′-dialcohol with authentic nonradioactive reference compounds[8]: A solution containing 5–10 nmole of the latter is applied to the dinucleotide intermediate on a PEI–cellulose chromatogram of a partial S_1 nuclease–phosphatase digest. Following contact transfer, the compounds are separated on a second PEI–cellulose layer in water to the origin, 0.1 N acetic acid to 7 cm, 0.2 N formic acid to 17 cm.[8] This system discriminates between A, C, and U in position 2.

(iii) ^{3}H-Labeled nucleoside trialcohol is released from the oligonucleotide 3'-dialcohol by *in situ* treatment with RNase T_2 and analyzed as described (Section II,B,1,g).

(iv) If position 2 is a ribose-methylated nucleoside, nucleoside trialcohol may be released from position 1 by nuclease P_1–alkaline phosphatase digestion of the labeled dialcohol.[8]

(v) If position 2 is modified, it may be identified by a procedure[7] entailing 5' end labeling of the isolated dinucleotide 3'-dialcohol with ^{32}P by treatment with [α-^{32}P]ATP and polynucleotide kinase (see also the following section).

j. DETERMINATION OF THE POSITIONS OF MODIFIED NUCLEOSIDES.[6,7] A controlled nuclease S_1–alkaline phosphatase digest of the ^{3}H-labeled oligonucleotide 3'-dialcohol to be analyzed is prepared as described (Section II,B,2,g). After dephosphorylation, 3 μl of 10 mM $NaIO_4$ is added to the mixture, which is then incubated for 1 hr at 23° in the dark. This treatment converts undesired unlabeled fragments to nucleotide dialdehydes, which are being retained at the origin during subsequent chromatography. Oxidized nuclease S_1–alkaline phosphatase digest corresponding to 20 pmole of starting material is applied to a PEI–cellulose sheet, which has been pretreated with methanol–concentrated ammonia (1000 : 1). The intermediates are separated by size in Tris-HCl, 8.5 M urea, pH 8.0 (Section II,B,2,h) and located by fluorography. After removal of PPO with methanol, spots of oligonucleotide-3' dialcohol intermediates are cut out and eluted as described (Section II,B,2,d). If necessary, eluates of minor intermediates are pooled from several replicate sample lanes.

For 5' terminal labeling, 3 μl of enzyme solution D is added to a dried portion of each eluate containing 1–4 pmole of ^{3}H-labeled intermediate and the solutions are incubated at 38° for 30 min. To release ^{32}P-labeled nucleoside 5'-monophosphate from the 5' termini, 3 μl of enzyme solution E is added to each sample and the mixtures are incubated for 20–50 min at 38°.

For the analysis of the ^{32}P-labeled nucleoside 5'-monophosphates,[7] aliquots of these digests (2–4 μl) are applied together with suitable reference compounds (see Gupta *et al.*)[7] to 20 × 20 cm PEI–cellulose sheets. After the sheets have been soaked in 300 ml of methanol for 5 min, a wick is attached at 17 cm above the origin and the chromatogram is developed as follows: water to the origin, 0.4N acetic acid to 9 cm, 0.8N formic acid to 4 cm on the wick. After removal of the wick, the sheet is dried thoroughly and then soaked in a solution of 600 mg of Tris (free base) in 500 ml of methanol, followed by 500 ml of methanol (10 min each). For the second dimension, another wick is attached and the sheets are developed

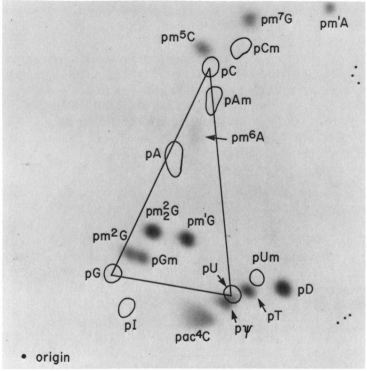

FIG. 6. Two-dimensional separation of a model mixture of nonradioactive and ^{32}P-labeled nucleoside 5'-monophosphates on a PEI–cellulose thin layer. First dimension from bottom to top, second dimension from left to right (for solvents, see text). Taken from Gupta et al.[7]

as follows: water to the origin, 0.22 M Tris-HCl, pH 8.0, to 5 cm on the wick. The ^{32}P-labeled compounds are detected by autoradiography. Sensitivity of detection may be considerably enhanced by sandwiching the film between the chromatogram and a calcium tungstate intensifying screen and carrying out the exposure at $-70°$.[52,53] A nucleoside 5'-monophosphate map[7] of the major and many minor constituents known to occur in tRNA is shown in Fig. 6.

C. Derivation of the Complete Sequence of a Ribonucleic Acid

1. Tritium Sequencing of Partial Fragments. Partial RNase T$_1$ or RNase A fragments of a ribonucleic acid may be prepared by described procedures (e.g., Brownlee[54]). Instead of conventional column chromatography, PEI–cellulose tlc may also be used for the isolation of

[52] R. A. Laskey and A. D. Mills, *FEBS Lett.* **82**, 314 (1977).
[53] R. Swanstrom and P. R. Shank, *Anal. Biochem.* **88**, 184 (1978).

partial fragments. For this purpose, solvent systems and elution techniques are similar to those described in this article. Partial fragments of a chain length of up to 15–20 can be sequenced directly by the tritium derivative methods. Large oligonucleotides are degraded to smaller complete or partial fragments with RNases T_1[54], A,[54] or U_2[10,55] and the smaller fragments are then purified by PEI–cellulose tlc and analyzed by tritium derivative methods. Spots of large oligonucleotides may be digested *in situ*, as described in Section II,B,1,c, and the products further analyzed after isolation (or contact transfer) and separation.

2. Readout Sequencing of Whole RNA or Large Partial Fragments

a. GENERAL ASPECTS. The basic features of procedure B, such as terminal labeling, specific enzymatic digestion, separation by size, and deduction of the sequence on the basis of the cleavage patterns,[8,9] are retained in procedures for sequencing whole RNA or large RNA fragments.[56–58] These procedures utilize the [γ-^{32}P]ATP-polynucleotide kinase reaction for 5' terminal labeling of RNA rather than chemical ^{3}H-labeling and polyacrylamide gel electrophoresis rather than chromatography for resolving the partial products. Highly base-paired regions of RNA may be resistant to enzymatic cleavage even under denaturing conditions, as indicated by the absence of the corresponding bands on the gel, and therefore are difficult to analyze by this method. As this procedure is detailed elsewhere in this volume, we shall mainly focus on conditions worked out recently in our laboratory[59] for the preparation of partial digests of 5' terminally ^{32}P-labeled RNA with RNase Phy$_1$.

b. 5' TERMINAL LABELING. This is carried out under conditions similar to those described by Simoncsits *et al.*,[57] except that calf intestinal phosphatase is used instead of bacterial alkaline phosphatase for removing the 5' terminal phosphate prior to labeling.[58] The labeled RNA is purified on a polyacrylamide gel and extracted[42] after adding 10–20 μg of yeast tRNA carrier. The RNA is precipitated at $-18°$ overnight after adding 3 volumes of acetonitrile–ethanol (4 : 1, v/v).[42]

c. DIGESTION OF TERMINALLY LABELED RNA WITH RNASE PHY$_1$.

i. Digestion in the absence of urea is performed in similar conditions as described in Section II,B,2,g. A solution of 0.3–1.0 × 10^6 dpm of end-

54 G. G. Brownlee, Determination of Sequences in RNA, *in* "Laboratory Techniques in Biochemistry and Molecular Biology" (T. S. Work and E. Work, eds.). Am. Elsevier, New York, 1972.

55 S. Takemura, H. Kasai, and M. Goto, *J. Biochem. (Tokyo)* **75**, 1169 (1974).

56 H. Donis-Keller, A. M. Maxam, and W. Gilbert, *Nucleic Acids Res.* **4**, 2527 (1977).

57 A. Simoncsits, G. G. Brownlee, R. S. Brown, J. R. Rubin, and H. Guilley, *Nature (London)* **269**, 833 (1977).

58 M. Silberklang, A. M. Gillum, and U. L. RajBhandary, *Nucleic Acids Res.* **4**, 4091 (1977).

59 R. C. Gupta, A. S. Gopalakrishnan, E. Randerath, and K. Randerath, unpublished (1978).

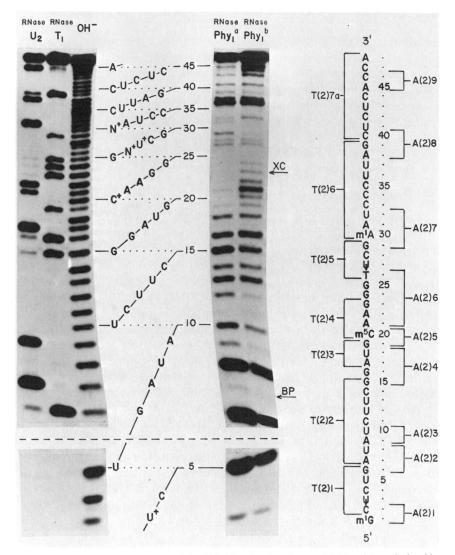

Fig. 7. The primary structure of the 3′ half-molecule of yeast tRNA$_{UAG}^{Leu}$ as derived by a combination of gel readout sequencing and tritium derivative analysis of complete RNase T$_1$ and A digests. The left part shows polyacrylamide gel electrophoresis patterns of partial digests of the 5′ terminally ^{32}P-labeled 3′ half-molecule. Phy$_1^a$, RNase Phy$_1$ digest prepared in the absence of urea at room temperature; Phy$_1^b$, digest prepared in the presence of 7 M urea at elevated temperature (see text). The part of the autoradiogram below the dashed line represents an additional gel run for a shorter period of time to display the short fragments. The positions of the modified nucleosides, indicated by the superscript plus, were derived from the complete RNase T$_1$ and A digests (indicated in the right-hand part of the figure). N, no cleavage obtained by partial enzymatic digestion: BP, bromphenol blue marker; XC, xylene cyanole FF marker. The right-hand part of the figure shows the complete sequence of the 3′ half-molecule derived by combining the readout data with the sequence data for complete RNase T$_1$ and A fragments obtained by procedure A. From E. Randerath, R. C. Gupta, L. S. Y. Chia, S. H. Chang, and K. Randerath, *Eur. J. Biochem.* **93**, 79 (1979).

labeled RNA and about 5 μg of yeast tRNA carrier is dried and the residue dissolved in 20 μl of 20 mM sodium acetate, 2 mM EDTA, pH 4.5. After adding 1 μl of tracking dye mixture (4 μg/μl each of bromphenol blue and xylene cyanol FF), the sample is incubated 5 min at 50° and briefly cooled at 23°. RNase Phy$_1$ (4 μl, 3.4 munits/μl) is added and the sample kept at 23°. Aliquots (6–8 μl) are withdrawn at 2, 8, and 15 min and successively added to solid urea (final concentration, about 7 M) in a tube kept in a Dry Ice–acetone bath. The pooled solutions are finally heated 1 min at 100°.

ii. Digestion in the presence of urea is performed as follows. A solution of 0.3–1.0 × 10^6 dpm of end-labeled RNA and about 1 μg of yeast tRNA carrier is dried and the residue dissolved in 20 μl of 20 mM sodium citrate, 1 mM EDTA, 7.5 M urea, pH 5.0. One microliter of tracking dye mixture is added (see above) and the solution is dried. The residue is taken up in 8 μl of water. After preincubation at 50° for 5 min, 12 μl of RNase Phy$_1$ (3.4 munits/μl)[60] is added and the sample is divided into two equal portions, which are incubated 10 min at 38° and 50°, respectively. The solutions are then combined and heated 1 min at 100°.

d. GEL ELECTROPHORESIS OF PARTIAL RNASE PHY$_1$ DIGESTS. Freshly prepared digests or digests that had been kept at −70° (5–8 μl each) are applied to a thin stacked polyacrylamide slab gel[56,61] (0.3–0.6 mm thick, 30 or 50 cm long to resolve chain lengths below or above 30, respectively, upper part 10% acrylamide, 3–4 cm long, lower part 20% acrylamide) containing 7 M urea and resolved alongside other partial digests. Electrophoresis is at pH 8.3 (90 mM Tris, 90 mM boric acid, 1 mM EDTA) and 1200 V for 4–6 hr to resolve chain lengths up to about 30, for 20–24 hr to resolve chain lengths up to about 80, and for 40–48 hr to resolve chain lengths up to 120. Digestion products are located by autoradiography for 12–48 hr at −20°, amplified by apposition of an intensifying screen (see Section II,B,2,j).

The central portion of Fig. 7 depicts cleavage patterns of the 3′ half-molecule obtained by digestion of yeast tRNA$_{UAG}$^{Leu} with RNase Phy$_1$ in the absence and the presence of 7 M urea, respectively.[10] A comparison of these two digests shows that a number of bonds (positions 19–24) that are cleaved under denaturing conditions are either not or only slightly cleaved under nondenaturing conditions, indicating the presence of a base-paired region in the 3′ half-molecule, which is resistant to cleavage by RNase Phy$_1$ under nondenaturing conditions. Thus, the secondary structure of the T arm has been preserved in the 3′ half-molecule. Similar observations have been made with RNase Phy$_1$ digests of whole tRNAs and other RNAs.[59]

---

[60] Because of the instability of RNase Phy$_1$ in 7.5 M urea at elevated temperature, a relatively high enzyme concentration is required under these denaturing conditions.

[61] F. Sanger and A. R. Coulson, *FEBS Lett.* **87**, 107 (1978).

The example given in Fig. 7 demonstrates that the cleavage rules deduced for 3′ terminally ^{3}H-labeled oligonucleotides apply also to 5′ terminally ^{32}P-labeled polynucleotides, with the following additions for certain modified nucleosides: (i) m^5C behaves like C; (ii) m^1A- and T- are not cleaved; (iii) ψ behaves like U. The resistance of m^1A and T to RNase Phy[1] has also been observed by Brown *et al.*[62] The behavior of other modified nucleosides needs to be further investigated.

3. Combination of Data from Complete and Partial Digests. Sequence information from the analysis of partial digests (Sections II,C,1 or II,C,2) provides the necessary data to establish overlaps for aligning the complete digestion products. While RNAs consisting only of the four major constituents may be sequenced solely by gel readout procedures, complementary methods are necessary for the identification and location of modified constituents in a RNA chain. The tritium derivative methods described in this article are ideally suited for this purpose, since they are highly sensitive without requiring biological labeling.

The example given in Fig. 7 illustrates the combination of the gel sequencing and tritium derivative procedures. Inspection of the figure shows that the two complementary procedures provide all the information necessary to deduce the sequence of the RNA.

III. [32]P Derivative Procedure (Procedure C) for Sequence Analysis of RNA Containing Modified Nucleotides

A. Materials

Enzymes: Nuclease P_1 from Yamasa Shoyu Co. (Tokyo), potato apyrase (code A 6132) from Sigma Chemical Co. (Saint Louis, Missouri).

Ultrapure urea from Schwarz/Mann (Orangeburg, New York), acrylamide and methylene bisacrylamide from BioRad (Richmond, California).

Macherey and Nagel PEI–cellulose sheets (40 × 20 cm and 20 × 20 cm) without indicator from Brinkmann Instruments (Westbury, New York). Prior to use, the smaller sheets are predeveloped in water.[46] After the front has migrated to the top of the sheets, the cover of the chromatographic tank is slightly moved back so as to allow water to evaporate from the upper portion of the sheets. Development is for 8–15 hr. The sheets are allowed to dry at room temperature for several hours before use. If not used the same day they are stored at −20°. The 40 × 20 cm sheets are used for the print step (see the following) without predevelopment.

[62] R. S. Brown, J. R. Rubin, D. Rhodes, H. Guilley, A. Simoncsits, and G. G. Brownlee, *Nucleic Acids Res.* **5**, 23 (1978).

^{99}Tc-Ink is prepared by diluting [^{99}Tc]NH$_4$TcO$_4$ (New England Nuclear, Boston, Massachusetts) with regular ink.

For additional materials, see Sections I,A and II,A.

B. Procedure

1. *Outline of Procedure.* Procedure C[11,12] is in part based on the observation recently reported by Stanley and Vassilenko[63] that under sufficiently mild conditions of chemical degradation of RNA the cleavage products arise mostly from random single "hits," leading to the formation of two sets of fragments, the first consisting of chains extending from the phosphorylated 5' terminus of the RNA (e.g., tRNA) to internal residues carrying 3' terminal phosphate groups. The second set of fragments contains chains extending from internal positions carrying a free 5'-hydroxyl group to the 3' terminus of the RNA. In this procedure, RNA is partially hydrolyzed by brief heating in water to generate the two sets of fragments. The free 5'-hydroxyl groups of the second set of fragments are then labeled with ^{32}P by the polynucleotide kinase reaction. The fragments are separated on a denaturing polyacrylamide gel into a series of 5' terminally ^{32}P-labeled fragments, each differing from its neighbor by the addition or the removal of a single 5' terminal nucleotide. The radioactive fragments are then contact-transferred to a PEI–cellulose anion-exchange thin-layer sheet ("print" step). The fragments are digested *in situ* with RNase T$_2$,[8,9] and the 5'-^{32}P-labeled nucleoside-3',5' diphosphates released in the previous step are resolved by thin-layer chromatography. After autoradiography, the nucleotide sequence can be read directly from the spot pattern displayed on the X-ray film. This is the only method presently known that allows one to display and identify directly the positions in the nucleic acid chain of modified nucleotides along with those of the major nucleotides. It also circumvents the difficulties arising from the resistance of highly base-paired regions of the RNA to enzymatic digestion (see Section II,C,2).

Escherichia coli tRNA^{Tyr} will be used as an example to illustrate the method.

2. *Controlled Hydrolysis of RNA.* A sealed tube containing 2–10 μg of RNA in 5–10 μl of water is placed in a water bath at 80° for 6 min. The solution is evaporated in a current of cool air.

3. *^{32}P-Labeling of 5' Ends.* The residue is dissolved in 5–10 μl of 40 mM Tris–HCl, pH 8.7, 13 mM MgCl$_2$, 10 mM dithiothreitol, 25 μM (γ-^{32}P)ATP (500–800 Ci/mmole). Polynucleotide kinase, 0.5 μl (3–5 units), is added and the solution kept at 38° for 25 min. One microliter of apyrase solution (4 munits/μl) is added per 5 μl of incubation mixture and the

[63] J. Stanley and S. Vassilenko, *Nature (London)* 274, 87 (1978).

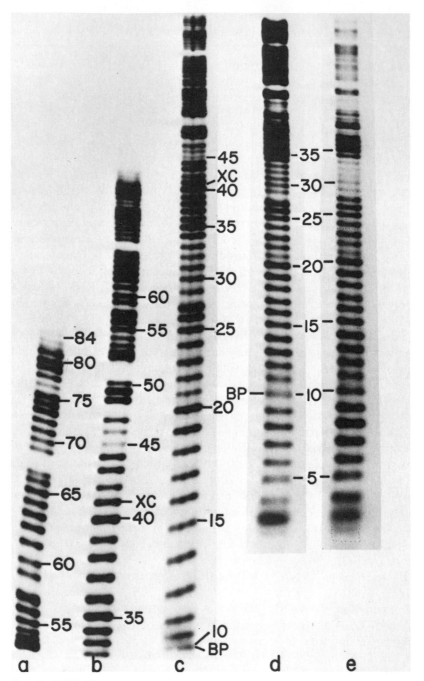

FIG. 8. Gel ladders (a–d) and a print ladder (e) obtained by heating 5 μg of *E. coli* tRNA^{Tyr} in 5 μl of water at 80° for 6 min, followed by 5′ terminal ^{32}P-labeling. Specific activity of $(\gamma\text{-}^{32}\text{P})$ATP was 300 Ci/mmole. Four identical aliquots were loaded on a 12% polyacrylamide slab gel (50 × 30 × 0.04 cm) at 0 hr (a), 4 hr (b), 8 hr (c), and 11 hr (d), and

solution kept at 38° for 20 min. The solution is evaporated in a current of cool air.

4. *Gel Electrophoresis.* The residue is dissolved in 20–40 µl of 90 mM Tris base, 90 mM boric acid, 1 mM EDTA, and 7 M urea. Two to four microliters each of bromphenol blue (BP) and xylene cyanol FF (XC) solutions (8 µg/µl) are added. Five microliter aliquots of the sample are fractionated on a slab of polyacrylamide (50 × 30 × 0.04 cm; 14 wells of dimensions 1 × 1 × 0.04 cm) prepared from 12% acrylamide, 0.4% methylene bisacrylamide, 7 M urea, 90 mM Tris base, 90 mM boric acid, 1 mM EDTA, 0.03% ammonium persulfate, and 0.03% N,N,N',N'-tetramethylethylenediamine. Prior to electrophoresis, the gel is allowed to age at least 8 hr and then preelectrophoresed 8–15 hr at 900 V, using 90 mM Tris base, 90 mM boric acid, and 1 mM EDTA as the electrophoresis buffer. The sample is loaded in alternate wells at 0, 4, 8, and 11 hr and electrophoresed at 1200 V (constant), about 20 mA for a total of 14 hr.

5. *Print Step.* A PEI–cellulose sheet (40 × 20 cm) is soaked in 500 ml of deionized water for 10 min. One glass plate is removed from the gel by gentle prying with a spatula. The gel surface is wetted slightly with deionized water. The drained PEI–cellulose layer is placed on top of the gel. Trapped air and water are extruded by gently stroking the plastic backing of the thin layer. The gel/PEI–cellulose assembly is wrapped in thin plastic foil such as Saran wrap. With the layer on top of the gel, the assembly is covered with a plain glass plate and 3 or 4 glass thin-layer chromatography tanks (or similar weight) are placed on top. After 15–24 hr at room temperature, the PEI–cellulose sheet is removed and dried in a current of warm air. Before autoradiography, the layer is marked with ^{99}Tc-ink dots for alignment after exposure. The film is sandwiched between the layer and the sensitive side of an intensifying screen. Exposure is at −20° or −70° for 20 min to a few hours, depending on the amount of radioactivity to be detected. Sensitivity is 5–10 times greater at −70° than at −20°.

Figure 8 depicts polyacrylamide gel ladders (a–d) and a PEI–cellulose print ladder (e) obtained by applying this procedure to *E. coli* tRNA[Tyr]. Chain lengths of up to 25–30 nucleotides are being transferred almost quantitatively by the printing technique. Transfer of larger fragments becomes progressively less efficient, but enough is transferred for terminal analysis of chain lengths of up to 120. (Larger RNA fragments have not been investigated thus far.) It should be noted that only the prints are

---

electrophoresed at 1200 V for 14 hr. (e) PEI–cellulose print ladder obtained from (d). Ladders (d) and (e) were lowered to facilitate reproduction. Autoradiography with intensifying screen for 30 min at −20°. BP and XC, positions of tracking dyes, bromphenol blue and xylene cyanol FF, respectively. Positions are numbered in the 3' to 5' direction.

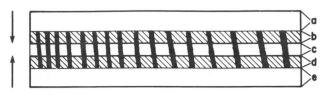

FIG. 9. Preparation of the two donor strips from the print ladder (schematic). (a, e) Nonradioactive regions of the layer. (b, d) Regions of the ladder treated with RNase T$_2$. (c) Untreated region of the ladder. After RNase T$_2$ treatment the strip is cut along the bold lines to obtain two donor strips for subsequent contact transfer of released ^{32}P-labeled digestion products to a PEI–cellulose sheet (see text). The arrows indicate the direction of solvent migration through the donor strip during contact transfer/chromatography.

autoradiographed; the gel autoradiogram is presented mainly to show that the procedure affords a reasonably regular spectrum of terminally labeled products. After contact transfer, ladders a, b, c, and d are suitable for the determination of positions 51–84, 32–62, 10–47, and 2–35, respectively, encompassing almost the entire sequence of the RNA.

Note the large gaps between bands 47 and 48, 50 and 51, and 67 and 69. Fragments 48 and 51 are retarded on the gel because of the presence at their 5′ ends of the hypermodified nucleosides ms^{2}i^{6}A and Q, respectively. Band 68 is missing due to the resistance to hydrolysis of the phosphodiester bond linking Gm69 and G68. In contrast, bands 27 and 28, 57 and 58, and 75 and 76 are incompletely resolved. This appears to be a secondary structure effect which occurs when pG or pC is added at the 5′ end of a stem region, resulting in the formation of a G≡C base pair.

6. *5′ Terminal Analysis and Reading of the Sequence.* This involves the following manipulations: (i) excision of the ladder from the print; (ii) treatment of the ladder with RNase T$_2$; (iii) preparation of the treated ladder for contact transfer; (iv) chromatography of RNase T$_2$ digestion products after contact transfer to a fresh PEI–cellulose acceptor sheet; (v) autoradiographic detection of products and reading of the sequence.

(i) Film and print are aligned and the positions of the ladders are marked by punching holes through the film into the layer. The print sheet is cut midway between individual ladders of interest, leaving 6–10 mm wide nonradioactive margins parallel to the ladder (Fig. 9a and e). Strips are marked appropriately with pencil.

(ii) Up to 5 ladder strips are soaked in 200 ml of 0.15 M sodium acetate, pH 4.5, for 10 min with agitation. Without drying, the strips are soaked in 200 ml of methanol for 10 min and dried. A solution of RNase T$_2$ is prepared containing 0.2 unit/μl in water. (This stock solution can be kept at −20° for many months.) Twenty microliter portions of this enzyme solution are applied evenly by streaking with a disposable micropipette to a strip of the ladder containing about one-third of each gel band (Fig. 9b) adjacent to the nonradioactive part of the layer (Fig. 9a). One 20-μl portion is sufficient for 5–6 cm. Leaving the center of the ladder (Fig. 9c) untreated, one proceeds with the other part (Fig. 9d) until the entire length of the ladder to be analyzed has been moistened with enzyme solution. The treated strip is covered with Teflon tape to retard evaporation. The strips are sandwiched between glass plates and kept at 38° for 1.5–2 hr or overnight.

(iii) The strips are soaked in 200 ml of methanol for 7 min, dried, and then cut lengthwise into two narrower strips, each comprising the RNase T$_2$-treated zone, as well as an about 6 mm wide strip from the adjacent nonradioactive area (total width, about 10 mm; length, up to 18 cm). (See Fig. 9.) The untreated center strip is discarded.

(iv) Following contact transfer, one of the two strips is analyzed by chromatography on PEI–cellulose in an ammonium sulfate system,[64] the other (from the same part of the ladder) in an ammonium formate system, pH 3.5. For contact transfer and chromatography of digestion products in ammonium sulfate, a Whatman 1 wick is attached at the top of a PEI–cellulose sheet (20 × 20 cm) by stapling and the layer of the donor strip is then placed on the acceptor sheet in such a way that its nonradioactive portion is below the origin line (about 2.5 cm from the bottom). The donor strip and the acceptor sheet are sandwiched between two strong magnetic bars. The preparation of such bars from individual magnets is described by Gupta and Randerath.[8,65] The chromatogram is developed at 4° with water to the origin line, then with 0.55 M ammonium sulfate to 3–4 cm on the wick. After removal of the magnetic bar and the donor strip, the layer is dried in a current of warm air and the wick is cut off.

For the analysis of the released 5′ terminal nucleotides in the ammonium formate system, donor strip and acceptor steet are pretreated with pH 3.5 buffer prior to the development in ammonium formate. Several (up to 5) ladder strips are soaked in 100 ml of 0.1 M ammonium formate, pH 3.5, for 7 min and dried with warm air. In this case, contact transfer and chromatographic separation are performed as two separate operations. For transfer, the layer of the donor strip is placed on the PEI–cellulose acceptor sheet so that its nonradioactive portion is below the origin line (1.5 cm from bottom) and held in place with magnets as previously described. The chromatogram is developed at room tempera-

[64] K. Randerath, *Experientia* **20**, 406 (1964).
[65] Magnetic bars prepared in this way ensure the quantitative transfer of the labeled mono- and dinucleotides to the acceptor sheet without streaking. Elongated spots are obtained if the magnets are too weak which may cause insufficient contact between donor layer and acceptor layer. For instance, magnetic bars assembled from standard bulletin board magnets or strips of magnetic rubber are not suitable for contact transfer.

AMMONIUM SULFATE

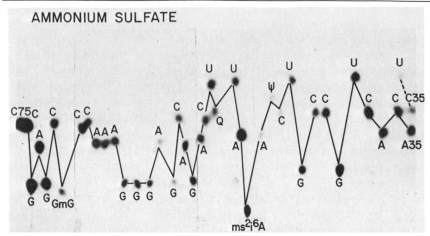

FIG. 10. 5' Terminal analysis of fragments 35–75 from the prints of *E. coli* tRNA^{Tyr} ladders a and b (Fig. 8). After RNase T$_2$ treatment *in situ*, the released 5' terminal nucleotides were contact-transferred to a PEI–cellulose sheet for further chromatographic analysis in 0.55 M ammonium sulfate. Autoradiography, as shown, enables one to read the sequence from position 35 (from 3' end) to position 75. Two termini each were found for positions 35 and 36, indicating that the tRNA used for the analysis was a mixture of tRNA$_1^{Tyr}$ and tRNA$_2^{Tyr}$.[66]

ture with water to the origin line, then with 4 M lithium formate, 7 M urea, pH 3.5, to 5 cm from the bottom of the acceptor sheet. The nucleotides will migrate onto the acceptor sheet in this solvent. After removal of the magnets and the donor strip, the wet sheet is soaked in 200 ml of methanol for 7 min, dried, and then soaked in 200 ml of 0.1 M ammonium formate, pH 3.5, for 7 min. After drying (5 min in warm air), a Whatman 1 wick is attached to the top of the sheet, which is developed in 1.75 M ammonium formate, pH 3.5, to 4 cm on the wick. After drying in a current of warm air, the wick is cut off.

(v) For autoradiography, the film is sandwiched between the layer and the sensitive side of an intensifying screen. Exposure is at −20° or −70° for up to several hours or overnight as required. If the chromatogram has weakly and strongly labeled nucleotides exposure may be repeated for a different length of time.

Figure 10 illustrates the resolution in the ammonium sulfate system of 5' terminal ³²P-labeled nucleotides released by *in situ* RNase T$_2$ treatment of print ladders from *E. coli* tRNA^{Tyr}. The sequence shown extends from the 5' side of the D stem of the RNA to the apex of the large variable loop. Note that pψp and pGm-Gp as well as the hypermodified nucleotides pQp

[66] H. M. Goodman, J. Abelson, A. Landy, S. Zadrazil, and J. D. Smith, *Eur. J. Biochem.* **13**, 461 (1970).

AMMONIUM FORMATE, pH 3.5

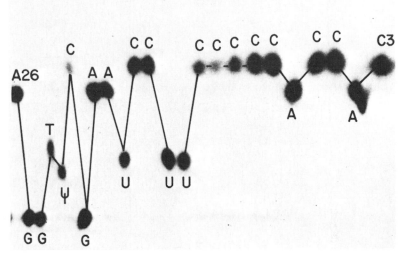

FIG. 11. 5′ Terminal analysis of fragments 3–26 from a print of ladder d (Fig. 8e). Chromatography in 1.75 M ammonium formate, pH 3.5. For further details, see text and Fig. 10.

and pms²i⁶Ap are well resolved from each other and the major nucleoside 3′,5-diphosphates.

Figure 11 exemplifies a print readout in the ammonium formate system extending from the 5′ side of the T stem to the acceptor stem. The modified nucleotides are seen again to be well separated from each other and the major nucleotides. pAp travels in a second front in this system, sometimes trailing slightly.

7. *Analysis of Nucleoside 3′,5′-Diphosphates.* Among several systems investigated, unbuffered ammonium sulfate and ammonium formate, pH 3.5, were found to give the most satisfactory results. Solubility effects have been shown to play a major role in the chromatography of nucleotides on PEI–cellulose in ammonium sulfate.[64] In accord with this observation, 0.55 M ammonium sulfate was found to be a particularly powerful solvent for the separation of methylated and other modified nucleoside 3′,5′-diphosphates from the parent nucleotides. Results obtained in the two systems complement each other so that most modified tRNA constituents can be identified on the basis of their relative R_F values in ammonium sulfate and ammonium formate, pH 3.5 (Fig. 12). The ammonium sulfate system may be used at room temperature, but resolution of nucleotides is somewhat better at 4°.

For identifying ³²P-labeled nucleoside 3′,5′-diphosphates that do not

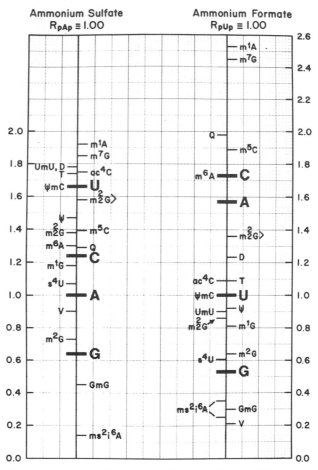

FIG. 12. Diagram of relative R_F values of nucleoside 3′,5′-diphosphates and ribose-methylated dinucleoside triphosphates in ammonium sulfate at 4° ($R_{pAp} \equiv 1.00$) and in ammonium formate, pH 3.5 at 23° ($R_{pUp} \equiv 1.00$). V, Q, and ms^2i^6A denote nucleoside 3′,5′-diphosphates of uridine-5-oxyacetic acid, 7-(4,5-*cis*-dihydroxy-1-cyclopenten-3-yl-amino-methyl)-7-deazaguanosine, and 2-methylthio-N^6-isopentenyladenosine, respectively.

separate well in both systems (such as pac^{4}Cp/pTp and pm^{6}Ap/pCp), one converts these compounds to the corresponding ^{32}P-labeled nucleoside 5′-monophosphates by treatment of the spots with nuclease P_1, followed by contact transfer and chromatography. Individual nucleoside 3′,5′-diphosphate spots to be analyzed are cut from the ammonium formate chromatogram. The cutout(s) are soaked in 20–50 ml of methanol for 5 min and dried. Four to six microliters of a solution containing 0.2 μg/μl of nuclease P_1 in 50 mM Tris-HCl, pH 7.3, are applied to the center of the cutout. After the cutouts have been covered with Parafilm, they are sandwiched between two glass plates and kept at 38° overnight. The treated cutouts are soaked in methanol and dried. One microliter of a mixture of nonradioactive major nucleoside 5'-monophosphates (about 5 nmoles each) is applied to the center of the cutout. Using button-type magnets, the compounds are contact-transferred[4] to a PEI–cellulose layer (20 × 20 cm) by development in either acetic acid/formic acid or Tris-HCl, pH 8.0, as described previously for the separation of nucleoside 5'-monophosphates.[7] After removing the magnets and cutouts, the chromatogram is dried in a current of warm air. Nonradioactive nucleotides are located in uv light and radioactive nucleotides are located by autoradiography.

8. *Comments.* The starting material should be at least 90% pure and undegraded; for example, RNAs or RNA fragments isolated from denaturing polyacrylamide gels were found to meet the requirements of the procedure. A control experiment is recommended in which a sample of the RNA is subjected to 5'-^{32}P-labeling and gel electrophoresis without prior heating in order to check for breaks. The parent molecule should have a phosphorylated or otherwise blocked 5' terminal hydroxyl group to prevent labeling of fragments encompassing the 5' end of the RNA, and chemical degradation of the RNA should be performed in such a way as to avoid overdigestion resulting in double "hits." These requirements hold regardless of whether cleavage is achieved by heating in formamide[63] or water. An acceptable level of digestion was obtained by heating the intact tRNA in water at 80° for 6–10 min. Digests obtained by heating at 90° for 2–3 min or at 100° for 0.5–1 min appear to be still suitable for sequence analysis, but heating at 100° for 5 min leads to overdigestion and substantial background contamination within and between gel bands. Simple hydrolysis in water offers advantages over formamide digestion because the sample is ready for the labeling reaction without additional steps, whereas formamide interferes with the polynucleotide kinase reaction and thus needs to be removed by ethanol precipitation of the RNA or by freeze-drying.[63]

In spite of nonuniform labeling of 5' ends in the polynucleotide kinase reaction, the extent of labeling of all modified nucleotides appears sufficient for their subsequent detection on the readout chromatograms. The apyrase digestion step serves to remove unreacted (γ-^{32}P)ATP and other unidentified ^{32}P-labeled contaminants which interfere with the detection of some shorter fragments (chain length <10) of the ladder. At the dilution used, the commercial apyrase preparation was found not to contain phosphatase and nuclease activities interfering with subsequent analysis.

Fifty centimeter long thin[61] (0.04 cm) 12% gels appear best suited for resolving the labeled constituents in tRNA and 5 S RNA hydrolysates; longer gels may be required for larger RNAs. The lack of diffusion on the print (Fig. 8e) is probably due to binding of the transferred polynucleotides to polyethyleneimine. As a rule of thumb, the distance between two adjacent gel bands should be about 3 mm or larger to enable their subsequent analysis.

In situ digestion on PEI–cellulose thin layers of 5' terminally ^{32}P-labeled polyribonucleotides with RNase T_2 results in the extensive (60–90%) release of the end groups as $5'$-^{32}P-labeled nucleoside $3',5'$-diphosphates, except that ribose-methylated nucleotides are obtained as dinucleotide derivatives and m_2^2G forms predominantly the corresponding $2',3'$-cyclic nucleotide derivative. Hypermodified compounds may also give rise to partial formation of cyclic nucleotides.

The pairs pac^{4}Cp/pTp and pm^{6}Ap/pCp are not sufficiently resolved in the two chromatographic systems described. Since the modified nucleosides in the RNA to be sequenced have usually been identified independently by base analysis of the intact RNA and ac^{4}C, rT, and m^{6}A, respectively, occur usually in specific positions of the RNA chain, the chance of misidentifying these nucleotides is small. The simple expedient of digesting the nucleoside $3',5'$-diphosphate to a nucleoside $5'$-monophosphate by *in situ* treatment of the spot with nuclease P_1, followed by contact transfer, allows one to resolve these compounds as ^{32}P-labeled nucleoside $5'$-monophosphates. Separation of pac^{4}C from pT and pm^{6}A from pC on PEI–cellulose thin layers has been described previously.[7] Yields of nucleoside $5'$-monophosphates after *in situ* nuclease P_1 treatment of nucleoside $3',5'$-diphosphate spots at pH 7.3 depend on the base, amounting to 20–50% for pyrimidine derivatives and 60–80% for purine derivatives (see also Fujimoto *et al.*[67]).

The procedure described is accurate, technically simple, and highly sensitive. A total of 1–2 μg of tRNA is sufficient to generate 4–6 ladders. At a specific activity of (γ-^{32}P)ATP of 400–500 Ci/mmole, 10^3–10^4 dpm are obtained in individual bands of the print ladder. This is ample radioactivity for subsequent digestion and analysis. The procedure is also extremely rapid: hydrolysis, labeling, gel electrophoresis, and starting the print can be done on one day; autoradiography of the print, excising the ladders, and RNase T_2 treatments require another half day; PEI–cellulose thin-layer chromatography, autoradiography, and reading of the sequence may take another day or perhaps 2 days. Thus, a single worker can read most of the sequence of a tRNA within a few days.

[67] M. Fujimoto, A. Kuninaka, and H. Yoshino, *Agr. Biol. Chem.* **38**, 1555 (1974).

TABLE I
COMPARISON OF PROCEDURES A, B, AND C

	Procedure A	Procedure B	Procedure C
Amount of tRNA or complete tRNA digest needed	2–3 A_{260} units of digest (80–120 μg)	0.1–0.3 A_{260} unit of digest (4–12 μg)	0.05–0.2 A_{260} unit of tRNA (2–10 μg)
Base analysis of oligonucleotides	Tritium derivative method; shows presence or absence of modified constituents	Tritium derivative method; shows presence or absence of modified constituents	Not applicable
Determination of molar ratios	Counting after digestion and labeling of nucleosides	Direct counting of ^{3}H–labeled oligonucleotides on fingerprint	Not applicable
Determination of positions of modified nucleosides[a]	As ^{3}H–trialcohols; RNase T_2 in situ; no isolation of oligonucleotide intermediates	As ^{32}P–monophosphates; nuclease S_1 in solution; isolation of intermediates required	As ^{32}P–nucleoside 3′,5′–diphosphates after in situ digestion of 5′ terminally ^{32}P-labeled fragments with RNase T_2
Identification of major nucleosides	Same as for modified nucleosides	Direct identification by specific enzymatic cleavage and readout of sequence	Same as for modified nucleosides

[a] May frequently be deduced from base composition or terminal analysis and readout sequence data.

IV. Comparison of Procedures

The choice of a suitable sequencing procedure depends on various factors, such as the amount of RNA available, the presence or absence of modified nucleosides, the secondary structure of the RNA (which may offer resistance to partial enzymatic digestion), the size of the RNA, etc. While the direct gel readout sequencing procedure is an excellent tool to sequence minute amounts (<0.5 μg)[10,32] of RNAs that contain no modified nucleosides, the presence of modified nucleosides in many RNAs makes the use of additional procedures, such as the ones detailed in this article, necessary. The most significant feature of these procedures is the direct chromatographic identification, as a radioactive derivative, of each position in the polynucleotide chain. Using tRNA as an example, we have shown that this applies to the four major and many modified constituents known to occur in tRNA. The salient features of procedures A, B, and C are summarized in Table I. It is evident that procedures B and C afford a considerably higher sensitivity than does procedure A, and are therefore applicable whenever only minute amounts of RNA are available for structural analysis. Procedures B and C complement each other owing to the completely different techniques used for the identification of the major and modified RNA constituents. A combination of these two procedures affords the highest overall sensitivity for sequence analysis of nonradioactive RNAs containing modified nucleosides.

Acknowledgments

Development and application of these procedures was supported by United States Public Health Service Grants CA-13591, CA-10893-P8 (Center Grant), and CA-16840. K.R. was a recipient of a Faculty Research Award from the American Cancer Society (PRA-108), while part of this work was carried out. We are grateful to Professor Harris Busch for support and encouragement.

[64] A Micromethod for Detailed Characterization of High Molecular Weight RNA

By FINN SKOU PEDERSEN and WILLIAM A. HASELTINE

Introduction

The methods presented in this article were developed for detailed studies of high molecular weight RNA available in small quantities.

The procedure involves specific cleavage of a small amount of nonradioactive RNA after guanosine residues by ribonuclease T_1 followed by $5'$-^{32}P-labeling of the T_1-resistant oligonucleotides.[1-3] The mixture of $5'$-^{32}P-labeled oligonucleotides is then fractionated by two-dimensional gel electrophoresis. The unique oligonucleotides can be eluted and their nucleotide sequence determined using recently developed methods for RNA sequencing.

When the procedures described here are applied to 200–400 ng of an RNA species about 10,000 nucleotides long, each T_1-resistant oligonucleotide will be labeled with ^{32}P at about 100,000 dpm. The high resolution gel electrophoresis system described makes it possible to isolate 50–80 pure unique T_1-resistant oligonucleotides in this case. The amount of radioactivity in each nucleotide is sufficient for complete nucleotide se-

[1] M. Szekeley and F. Sanger, *J. Mol. Biol.* **43**, 607 (1969).

[2] M. Simsek, J. Ziegenmeyer, J. Heckman, and U. L. RajBhandary, *Proc. Natl. Acad. Sci. U.S.A.* **70**, 1041 (1973).

[3] D. Frisby, *Nucleic Acids Res.* **4**, 2975 (1977).

quence determination. The total sequence information thus obtained from less than 1 μg of RNA corresponds to 10–15% of the entire RNA molecule.

We have used these techniques for extensive nucleotide sequence studies of retrovirus genomes. We have also applied them successfully in the characterization of genomes of influenzavirus,[4] vesicular stomatitis virus, and measles virus.

Materials and Solutions

Ribonuclease T_1 (Sankyo, Calbiochem) is dissolved at approximately 10 units per microliter in 10 mM Tris-HCl, pH 8.0, 2 mM EDTA and stored frozen at $-20°$. The ribonuclease activity is determined in the following assay[5]: Aliquots of the ribonuclease T_1 stock solution are diluted into 75 μl of 0.067 M Tris-HCl, pH 7.5, 2.7 mM EDTA, 3.3 mg/ml of yeast RNA. The mixtures are incubated for 15 min at 37° and cooled to 0°. Twenty five microliters of cold 25% trichloroacetic acid, 0.75% uranyl acetate is added and the mixture is centrifuged at 4° for 5 min at 10,000 g in a microcentrifuge (Eppendorf); 25 μl of the supernatant is mixed with 725 μl H_2O and the absorbance is measured at 260 nm. An absorbance of 0.1 above background corresponds to a total amount 1.2 $\times$ 10^{-2} units of RNase T_1 in the incubation mixture.

Bacterial alkaline phosphatase is purchased as a suspension in 3.2 M ammonium sulfate (Boehringer Mannheim). The precipitate is spun down and dissolved in 0.03 M sodium cacodylate, pH 8.0. The salt is removed by passage through a Sephadex G-25 column. To inactivate contaminating ribonucleases the phosphatase solution is made 0.1% in diethylpyrocarbonate (Eastman) and stirred for 1 hr at 0°. The diethylpyrocarbonate is removed by extraction four times with ether. The phosphatase activity is determined spectrophotometrically as the rate of release of p-nitrophenol from a 0.001 M solution of p-nitrophenylphosphate in 1.0 M Tris-HCl, pH 8.0. One unit is the activity liberating 1 μmole of p-nitrophenol per minute at 25°.[6] The molar absorbance index for p-nitrophenol in 1.0 M Tris-HCl, pH 8.0 is 1.62 $\times$ 10^4.

Polynucleotide kinase from T_4-infected $E. coli$[7] is purchased from New England Biolabs or from P-L Biochemicals and used as the amount of enzyme units specified by the manufacturer.

[4] U. Desselberger, K. Nakajima, P. Alfino, F. S. Pedersen, W. A. Haseltine, C. Hannoun, and P. Palese, *Proc. Natl. Acad. Sci. U.S.A.* (in press).
[5] K. Takahashi, *J. Biochem. (Tokyo)* **49**, 1 (1961).
[6] "Worthington Enzymes," p. 73. Freehold, New Jersey, 1972.
[7] C. C. Richardson, *Proceed. Nucleic Acid Res.* **2**, 815 (1971).

γ-^{32}P-Labeled ATP with a specific activity of 1500–2000 Ci/mmole is prepared by the following procedure adapted from published methods[8,9]: 50 μl of 3-phosphoglycate kinase/glyceraldehyde-3-phosphate dehydrogenase as a suspension at 6 mg/ml in $3.2 M$ ammonium sulfate (Boehringer Mannheim) is centrifuged for 5 min at 10,000 g in a microcentrifuge. The supernatant is removed, the pellet dissolved in 50 μl of 10 mM Tris-HCl, pH 8.0, 2 mM reduced glutathione (Sigma), 0.1 mM NAD[+] (Sigma), and dialyzed against the same buffer for 15–30 min at 0°. The main function of the dialysis is to remove ammonium ions, which are inhibitory for polynucleotide kinase. This is carried out immediately before the isotope exchange reaction. The reaction mixture (100 μl) [100 mM Tris-HCl, pH 8.0, 7 mM Mg(OAc)$_2$, 0.5 mM EDTA, 2 mM glutathione, 0.3 mM 3-phosphoglycerate (Sigma), 0.4 mM ATP (P-L Biochemicals)] is added to a tube containing 140 mCi of lyophilized H$_3$[32]PO$_4$ (NEN, carrier free, shipped in $0.02 M$ HCl; $N.B.$ do not transfer through a metal needle). The tube is shaken, 10 μl of the dialyzed enzyme solution is added and the mixture is left for 20–30 min at room temperature. A small aliquot is taken into a pulled capillary tube and applied to a PEI–cellulose plate (Macherey-Nagel, Brinkman), 6 × 2 cm. ATP is separated from unreacted phosphate by ascending chromatography in $0.75 M$ KH$_2$PO$_4$, pH 3.5. An X-ray film is exposed to the PEI plate for 1 sec. The film is developed and the degree of incorporation of ^{32}P into ATP can be determined by visual inspection (ATP is found halfway between the origin and the unreacted phosphate, which migrates near the front). If the film shows more than 50% conversion of ^{32}P-phosphate to [^{32}P]ATP, 300 μl of 7 mM EDTA is added to stop the reaction. The mixture is extracted with 400 μl of phenol (redistilled and saturated with 10 mM Tris-HCl, pH 8.0). The aqueous phase is extracted four times with ether, then the volume is adjusted to 1.6 ml with water. The solution (25 μM in ATP) is stored in aliquots at −20° and used in the polynucleotide kinase reactions without further treatment.

The stock solution of 40% (w/v) acrylamide, 1.3% N,N'-methylenebisacrylamide is made using practical grades of both compounds (Eastman). The solution is cleaned by stirring with 5 g/liter of charcoal and 15 g/liter of mixed bed resin [AG 501-X8 (D), 20-50 mesh, Bio-Rad] followed by filtration through several layers of filter paper (Whatman No. 1) and one layer of nitrocellulose filter (pore size 0.45 μm, S & S).

All reactions are carried out in siliconized polypropylene microcentrifuge tubes (capacity 1.5 ml, Eppendorf)

[8] I. M. Glynn and J. B. Chappell, $Biochem.$ $J.$ **90**, 147 (1964).
[9] A. M. Maxam and W. Gilbert, $Proc.$ $Natl.$ $Acad.$ $Sci.$ $U.S.A.$ **74**, 560 (1977).

TNE: 50 mM Tris-HCl, pH 7.5, 100 mM NaCl, 1 mM EDTA
SSC: 150 mM NaCl, 15 mM sodium citrate

Ribonuclease T_1 Digestion of RNA Followed by $5'$-^{32}P-Labeling of the Cleavage Products

Procedure. One microliter containing 200–400 ng of RNA in 20 mM TrisHCl, 2 mM EDTA, pH 8.0, is mixed with 1 μl containing 1 unit of RNase T_1 and 5 × 10^{-4} units of bacterial alkaline phosphatase in the same buffer solution and incubated for 60 min at 37°. Fifty microliters of polynucleotide kinase reaction mixture [10 mM KH$_2$PO$_4$/K$_3$PO$_4$, pH 9.5, 10 mM Mg(OAc)$_2$, 5 mM dithiothreitol, 15 μM γ-^{32}P-labeled ATP and 40 units/ml of polynucleotide kinase] is added to the digested RNA and the incubation is continued for 1–16 hr at 37°. The reaction is terminated by the addition of 50 μl of 0.6 M NH$_4$(OAc), 2 mg/ml yeast carrier RNA. The mixture is extracted with 100 μl phenol (redistilled and saturated with 10 mM Tris-HCl, pH 8.0) at room temperature. The aqueous phase is mixed with 300 μl of 95% ethanol and chilled for 20 min at −70°. The RNA is pelleted by centrifugation at about 10,000 g for 10 min in a microcentrifuge (Eppendorf). The pellet is lyophilized and dissolved in 5–10 μl of 10 mM sodium citrate, pH 5.0, 1 mM EDTA, 7 M urea, 0.1% bromphenol blue, 0.1% xylene cyanol FF, and 200 μg/ml of yeast carrier RNA. The sample is now ready for layering onto the first dimension gel.

Comments. If less than 200 ng of RNA is available, 100 ng of polyguanylic acid (Miles) may be included to ensure proper cutting specificity of the ribonuclease.

The inclusion of 10 mM phosphate in the polynucleotide kinase reaction mixture provides a simple way of inhibiting the phosphatase.[10] The polynucleotide kinase is the only partially inhibited at this concentration of phosphate.

Prolongation of the incubation time for the polynucleotide kinase reaction up to 16 hr results in a more uniform efficiency of labeling of the oligonucleotides.

Separation of Ribonuclease T_1-Resistant Oligonucleotides

The system described here is modified from the procedure of de Wachter and Fiers.[11] A number of changes has been made to increase the resolution of the gels and thereby the number of oligonucleotides available

[10] A. Efstratiadis, J. N. Vournakis, H. Donis-Keller, G. Chaconas, D. K. Dougall, and F. Kafatos, *Nucleic Acids Res.* **4,** 4165 (1977).
[11] R. de Wachter and W. Fiers, *Anal. Biochem.* **49,** 184 (1972).

for sequence analysis. These changes include an expansion of the two dimensions, a change of buffer system for the second dimension, and the use of thinner gels. The thin gels run for shorter time at high voltage because of the lower current requirement, and this results in improved resolution of the individual oligonucleotides. Moreover, the use of thin gels improves the photographic resolution by autoradiography, due to the proximity of the radiation source and the film.

The gel solution for the first dimension has the composition[11]: 10% acrylamide, 0.3% N,N'-methylenebisacrylamide, 0.025 M citric acid, 6 M urea (Schwarz Mann, Ultra pure), pH 3.5. The gel solution is filtered through a nitrocellulose filter (Millipore, pore size 0.45 μm) and the polymerization is started by addition of (per 100 ml of gel solution); 0.4 ml $FeSO_4 \cdot 7H_2O$ (2.5 mg/ml), 0.4 ml of ascorbic acid (100 mg/ml), and 0.04 ml of 30% H_2O_2.

The mold for the first dimension gel is made of two glass plates (400 × 20 × 0.45 cm), held together with paper clips around two Teflon strips (40 × 1 × 0.75 mm) covered with vacuum grease. The bottom of the mold is sealed with plasticine. The gel solution is poured into the cell and a Teflon comb with eight teeth, 5 mm wide and 1 cm deep is placed in the top of the solution. Two paper clips placed across the top of the cell keep the glass plates firmly pressed against the comb.

The polymerization of the gel is complete after 30 min at room temperature. The plasticine, the paper clips, and the comb are removed and the gel is placed vertically in a tank (38 × 24 × 7 cm) at room temperature containing 6 liters of 0.025 M citric acid leaving only about 1 cm of the gel plates uncovered. The slots are filled with 0.025 M citric acid and the gel is connected through a paper wick (Whatman 3MM) to another tank with 4 liters of the same solution. The two tanks are connected to a power supply and preelectrophoresis is carried out for 20 min at 1000 V. The slots are emptied and cleaned with a strip of filter paper and covered again with 0.025 M citric acid. The samples of 5' labeled T_1-resistant oligonucleotides are layered onto the gel through pulled capillary tubes. Electrophoresis is at 1700 V (the current is about 30 mA) until the most rapidly migrating dye (Bromphenol blue) has moved 18 cm (running time about 2 hr).

The procedure can be interrupted at this point and the gel stored frozen at $-70°$. The gel is then thawed at room temperature before the next step. One of the glass plates of the first dimension gel is removed and the gel covered with Saran wrap. A gel strip 1.5 cm wide, covering from 4 to 35 cm from the origin, is cut out with a razor blade and transferred to the glass plate for the second dimension gel. We find it convenient to keep the Saran wrap on the gel strip during handling. The second dimension glass plate has the dimensions 40 × 33 × 0.45 cm. Two Teflon strips (40 cm

$\times$ 1 cm $\times$ 0.38 mm) covered with a thin layer of vacuum grease are placed along the edges. The strip from the first dimension gel is placed perpendicular to the Teflon strips, 4 cm from the edge of the glass plate. The length of the gel strip is adjusted to leave a 2–3 mm distance to the Teflon strips on both sides. The Saran wrap is removed from the gel strip and another glass plate is placed on top. The glass plates are held together with paper clips and the opening closest to the gel strip is sealed with plasticine.

The second dimension gel solution has the composition: 50 mM Tris-borate, pH 8.3, 1 mM EDTA, 22.8% acrylamide, 0.8% N,N'-methylenebisacrylamide, 0.7 g/liter ammonium persulfate. The gel solution is filtered through a nitrocellulose filter (Millipore, pore size 0.45 μm) and 100 μl/liter of N,N,N',N'-tetramethylethylenediamine (TEMED) is added to catalyze polymerization. The gel solution is poured between the glass plates to 1 cm from the top, filling the space below the first dimension gel strip. The distance between the glass plates for the second dimension is smaller than the thickness of the first dimension gel strip. This helps to keep the strip in place and to press out air pockets between the gel strip and the glass plates.

The gel apparatus for the second dimension (made by Dan-Kar plastic products, Reading, Massachusetts) accommodates 4 gels. It has two lower reservoirs of 1-liter capacity each and one upper reservoir of 2 liters. The reservoirs are filled with 50 mM Tris-borate, pH 8.3, 1 mM EDTA. After complete polymerization of the gel (1–2 hr) the plasticine is removed and the gel is placed in the lower reservoir. The top of the gel is covered with 50 mM Tris-borate, pH 8.3, 1 mM EDTA and connected through a paper wick (Whatman No. 1) to the upper reservoir. Electrophoresis is carried out at room temperature until the fast migrating dye (bromphenol blue) has moved 25 cm from the top of the first dimension gel strip. The running time is about 5 hr at a constant current of 25 mA/gel or about 9 hr at a constant voltage of 750 V.

After electrophoresis, one of the glass plates is removed and the gel is covered with Saran wrap. For autoradiography we use Kodak XR-2 film, and expose the film for 10–60 min. An example of an autoradiogram is shown in Fig. 1.[12]

For nucleotide sequence analysis, gel pieces containing 5' labeled oligonucleotides are cut out with a 5 mm metal corkborer and transferred to tubes containing 200 μl of 1 mM EDTA and 200 μg/ml of carrier RNA. The tubes are left at 37° for 2 hr for elution. The liquid is transferred to

---

[12] E. Chan, W. P. Peters, R. W. Sweet, T. Ohno, D. W. Kufe, S. Spiegelman, R. Gallo, and R. E. Gallagher, *Nature (London)* **260**, 266 (1976).

FIG. 1. Autoradiogram showing 5′ labeled T_1-resistant oligonucleotides separated by two-dimensional gel electrophoresis. The direction of the electrophoresis for the first dimension is from left to right and for the second dimension from bottom to top. The RNA for this experiment was isolated from the retrovirus HL23(E3) (Chan et al.[12]), as text describes.

another tube; 20 μl of 3 M NaOAc, 0.1 M Mg(OAc)$_2$, 1 mM EDTA, pH 6.0, is added followed by 700 μl of chilled ethanol (95%). The tube is left for 20 min at $-70°$ and the precipitated RNA is pelleted. The supernatant is taken off and the RNA pellet is washed with 200 μl of 95% ethanol, lyophilized, and dissolved in 10 μl of 20 mM Tris-HCl, 2 mM EDTA, pH 7.5.

For nucleotide sequence determination the sample is divided into aliquots. The aliquots are partially digested with different sequence-specific ribonucleases, and the cleavage products are fractionated by electrophoresis in polyacrylamide gels under denaturing conditions. The fragmentation pattern is visualized by autoradiography for a few days at $-70°$ using intensifying screens, and the nucleotide sequence can be read off the autoradiogram.[13,14]

[13] H. Donis-Keller, A. Maxam, and W. Gilbert, Nucleic Acids Res. 4, 2527 (1977).
[14] A. Simoncsits, G. G. Brownlee, R.S. Brown, J.P. Rubin, and H. Gvilley, Nature (London) 269, 833 (1977).

Procedure for Isolation of Small Amounts of RNA from Retroviruses

In our studies of retrovirus RNA we use the following procedure for isolation of small amounts of viral RNA in high yields: The viruses are sedimented from tissue culture fluids for 3 hr at 17,000 rpm in a Beckman type 19 rotor at 4°. The pellets are resuspended in 1 ml of TNE and the viruses are sedimented through 30% sucrose–TNE in a Beckman SW 50.1 rotor for 2 hr at 50,000 rpm at 4°. The pellet is dissolved in 1 ml of 10 mM Tris–HCl, pH 7.4, 3 × SSC, 1 mM EDTA, 1% SDS, and 0.2 mg/ml of proteinase K (E. Merck) and the tube is incubated for 2 hr at 37°. The solution is layered onto a linear gradient of 15–30% sucrose in TNE, 0.2% SDS. The tube is centrifuged at 38,000 rpm for 2 hr at 25° in a Beckman SW 41 rotor. Fractions of the gradient are collected through a flow cuvette in a spectrophotometer reading at 260 nm. The fractions containing the 70 S RNA are pooled, mixed with an equal volume of 4 × SSC and passed through a 100-μl column (3 × 15 mm) of oligo(dT)–cellulose (Colloborative Research, T-3). The column is washed with 5 ml of 10 mM Tris-HCl, pH 7.4, 2 × SSC and the poly(A)-containing RNA is eluted in 0.4 ml of 10 mM Tris-HCl, pH 7.4. 40 μl of 3 M NaOAc, 0.1 M Mg(OAc)$_2$, 1 mM EDTA is added, followed by 1.1 ml of 95% ethanol and the tube is chilled for 20 min at −70°. The precipitated RNA is pelleted by centrifugation for 20 min at about 10,000 g in a microcentrifuge (Eppendorf). The pellet is lyophilized, dissolved in 0.4 ml of H$_2$O, reprecipitated under the same conditions. The pellet is washed with 200 μl of 95% ethanol, and the purified RNA is stored at −20° in 5–10 μl of 20 mM Tris-HCl, pH 8.0, 2 mM EDTA.

Acknowledgments

This work was supported by grants from National Institute of Health, RO1CA19341, American Cancer Society, VC 260, National Cancer Institute, 71057, and the Danish Natural Science Research Council. W.H. is a recipient of an American Cancer Society Faculty Research Award. F.S.P. is a fellow of the Leukemia Society of America, Inc. We thank Meredith Simon for expert technical help.

[65] Use of *E. coli* Polynucleotide Phosphorylase for the Synthesis of Oligodeoxyribonucleotides of Defined Sequence

By SHIRLEY GILLAM and MICHAEL SMITH

Oligodeoxyribonucleotides of defined sequence are finding an increasing number of applications in molecular biology. For example, they can be used as specific probes for the characterization and cloning of a DNA fragment.[1] Oligodeoxyribonucleotides can be specific mutagens for inducing defined point mutations in DNA at high efficiency.[2] In DNA sequencing, using the terminator method,[3,4] they provide specific primers for DNA polymerase and obviate the need for restriction fragments and strand separation of template DNA.[5] Analogously, oligodeoxyribonucleotides have been used as specific primers for DNA polymerase and reverse transcriptase using RNA as template to obtain a pure radioactive probe[6] or for sequence determinations.[7-9] Because of this wide variety of applications and the large number of possible oligodeoxyribonucleotides, it is desirable to have a simple, rapid synthetic method which will reliably provide a specific, pure oligodeoxyribonucleotide in the hands of an experimentalist who has no previous experience with this type of synthesis. The procedure we have developed fulfills these requirements and involves the stepwise addition of deoxyribonucleotide residues to an oligodeoxyribonucleotide primer catalyzed by *E. coli* polynucleotide phosphorylase in the presence of NaCl and $MnCl_2$ with the appropriate deoxyribonucleoside 5'-diphosphate as substrate[10-13]

$$d(pN_n) + ppdN' \rightarrow d(pN_n\text{-}N') + P_i$$

This reaction is irreversible, there being no detectable phosphorolysis.[10,11] The reaction has a requirement for oligodeoxyribonucleotide primers with a minimum of three phosphate residues for efficient synthesis.[12] The primers are commercially available or can be obtained by simple chemical synthesis.[14] Reactions are carried out on a nanomole to micromole scale. An alternate strategy involves the use of a diribonucleoside phosphate,

[1] D. L. Montgomery, B. D. Hall, S. Gillam, and M. Smith, *Cell* **14**, 673–680 (1978).
[2] C. A. Hutchison, III, S. Phillips, M. H. Edgell, S. Gillam, P. Jahnke, and M. Smith, *J. Biol. Chem.* **253**, 6551–6560 (1978).
[3] F. Sanger, S. Nicklen, and A. R. Coulson, *Proc. Natl. Acad. Sci. U.S.A.* **74**, 560–564 (1974).
[4] F. Sanger and A. R. Coulson, *FEBS Lett.* **87**, 107–110 (1978).
[5] M. Smith, D. W. Leung, S. Gillam, C. R. Astell, D. L. Montgomery, and B. D. Hall, *Cell* **16**, 753–761 (1979).
[6] T. H. Rabbitts, A. Forster, M. Smith, and S. Gillam, *Eur. J. Immunol.* **7**, 43–48 (1977).
[7] C. C. Cheng, G. G. Brownlee, N. H. Carey, M. T. Doel, S. Gillam, and M. Smith, *J. Mol. Biol.* **107**, 527–547 (1976).
[8] P. H. Hamlyn, S. Gillam, M. Smith, and C. Milstein, *Nucleic Acids Res.* **4**, 1123–1134 (1977).
[9] N. J. Proudfoot, S. Gillam, M. Smith, and J. I. Longley, *Cell* **10**, 807–818 (1977).
[10] S. Gillam and M. Smith, *Nature (London), New Biol.* **238**, 233–234 (1972).
[11] S. Gillam and M. Smith, *Nucleic Acids Res.* **1**, 1631–1648 (1974).
[12] S. Gillam, K. Waterman, M. Doel, and M. Smith, *Nucleic Acids Res.* **1**, 1649–1664 (1974).
[13] S. Gillam, P. Jahnke, and M. Smith, *J. Biol. Chem.* **253**, 2532–2539 (1978).
[14] M. Smith and H. G. Khorana, this series, Vol. 6 [94].

e.g., rAprA, as a primer which can be removed at the end of the synthesis.[12]

Because the course of a particular synthetic step is influenced by the incoming mononucleotide, the terminus of the growing oligonucleotide, and additional more subtle structural factors, the prediction of yields in any method of oligonucleotide synthesis is notoriously difficult. The present method is no exception; however, many of the important factors have been evaluated.[13] In general dADP and dCDP are more reactive than dGDP and dTDP. A deoxycytidylate residue at the 3' terminus of a primer is the best receptor and a deoxythymidylate is the poorest. The result is that yields of desired products range from 10 to 70%. When this is a consequence of a slow reaction, the yield can be increased by use of a larger amount of enzyme and/or a longer reaction time.[13,15] More difficult to deal with is the case where the initial (desired) product is consumed as a primer for a subsequent, more rapid reaction. Here, the reaction is controlled by increasing the concentration of NaCl and/or decreasing that of $MnCl_2$. In addition, recycling of unused primer results in acceptable yields.[13,15]

At each step of the synthesis, product isolation and characterization is necessary. Isolation is reliably achieved by chromatography on DEAE–cellulose in the presence of $7 M$ urea.[16] Characterization is by quantitative nucleoside analysis using high-pressure cation-exchange chromatography (see below) or sequence determination using the wandering spot method.[17] If a limited synthesis is planned (up to three synthetic steps), the desired product can be obtained by using the appropriate conditions selected from the examples described in this article. It is important to realize that a small amount of an oligodeoxyribonucleotide, e.g., 10 μg, is sufficient for many hundreds of experiments involving interactions with natural nucleic acids on the usual picomole scale. Thus a synthesis starting with 2 or 3 mg of a simple primer, and involving up to three steps working at low efficiency, is quite satisfactory. The isolation of trace amounts of an oligodeoxyribonucleotide can be facilitated by using a ^{3}H-labeled deoxyribonucleoside 5'-diphosphate as precursor.

When a synthesis involves more than three steps, it is essential to optimize the yield at each step. The best conditions are established using high-pressure anion-exchange chromatography (see below) on the anion-exchanger RPC-5.[18] In addition, unreacted primer is recycled. Using these

[15] E. Trip and M. Smith, *Nucleic Acids Res.* **5,** 1529–1538 (1978).
[16] G. M. Tener, this series, Vol. 12, Part A [47].
[17] G. G. Brownlee, and F. Sanger, *Eur. J. Biochem.* **11,** 395–399 (1969).
[18] A. D. Kelmers, D. E. Heatherly, and B. Z. Egan, this series, Vol. 29 [37].

strategies, it has been possible to synthesize oligodeoxyribonucleotides containing up to 13 nucleotides starting from a tetranucleotide primer.[2,19]

This article describes a convenient isolation of polynucleotide phosphorylase from commercially available *E. coli* B. The syntheses described include examples of possible next-neighbor combinations in the series of twelve oligodeoxyribonucleotides d(pT$_8$-*N*-*N'*),[20] where *N* is A, G, or C and *N'* is A, G, C, or T. These oligodeoxyribonucleotides are useful primers for DNA polymerase or reverse transcriptase on RNA templates which contain a natural 3'-poly(rA) tract[7-9] or to which the tract has been added *in vitro*.[21] They can also be used as primers on tracts of dA residues often found in DNA.[5] The oligodeoxyribonucleotides, d(pA$_5$-*N*-A$_3$), where *N* is G, C, or T are equally useful primers on DNA templates.[5] Their syntheses are an example of the use of a ^{3}H-deoxynucleoside 5'-diphosphate as substrate to allow the monitoring and isolation of a small amount of oligodeoxyribonucleotide. A new restriction endonuclease recognition sequence can be introduced by ligation at a preexisting site in DNA using a synthetic oligodeoxyribonucleotide. Thus, an *Xma*I (*Sma*I) sequence can be introduced into an *Eco*RI site using d(pA$_2$-T$_2$-C$_3$-G$_3$) or d(pA$_2$-T$_2$-C$_4$-G$_4$) whose syntheses are described below.[20] The dodecamers d(pG-T-A-T-C-C-T-A-C-A-A-A) and d(pG-T-A-T-C-C-C-A-C-A-A-A) are examples of more complex syntheses; they are designed to program specific transition mutations in bacteriophage ϕX174 DNA.[2]

Isolation of *E. coli* Polynucleotide Phosphorylase[11]

Assay Method

Principle. The polymerization of a deoxyribonucleoside 5'-diphosphate is measured by the incorporation of radioactive deoxyribonucleotide residues, in the presence of MnCl$_2$, into acid-insoluble polynucleotide. Because this reaction is readily inhibited by NaCl,[11,13] the enzyme also is assayed by the NaCl-insensitive incorporation of radioactive ribonucleotide residues, in the presence of MgCl$_2$, into acid-insoluble polynucleotide.

Procedure

(i) dADP polymerization
1 *M* Tris-HCl, pH 8.5   20 μl

[19] S. Gillam, F. Rottman, P. Jahnke, and M. Smith, *Proc. Natl. Acad. Sci. U.S.A.* **74,** 96–100 (1977).
[20] S. Gillam and M. Smith, unpublished results.
[21] O. Hagenbuchle, M. Santer, J. A. Steitz, and R. J. Mans, *Cell* **13,** 551–563.

$5 \times 10^{-2} M$ dADP	$5 \mu l$
[^{3}H]dADP (10 Ci/mmole, 50 μCi/ml)	$10 \mu l$
$10^{-1} M$ 2-mercaptoethanol	$20 \mu l$
H_2O	
Enzyme	
$10^{-1} M$ MnCl$_2$	$20 \mu l$
Total volume: 200 μl	

The components are mixed at 0°, MnCl$_2$ being added last to start the reaction, and incubation is at 37° for 20 minutes. Ice-cold 5% trichloroacetic acid (TCA), (3 ml), is added followed by bovine serum albumin (0.1 ml of 10 mg/ml). The precipitate is collected by centrifugation (5000 g, 10 min, 5°) and washed twice with ice-cold 5% TCA (3 ml). The acid-insoluble material is dissolved in NCS solubilizer (0.4 ml, Amersham) and then transferred to a scintillation vial containing toluene-Ominfluor (10 ml, New England Nuclear) prior to radioactive counting.

(ii) ADP polymerization

$1 M$ Tris-HCl, pH 8.5	$20 \mu l$
$5 \times 10^{-2} M$ ADP	$5 \mu l$
[^{3}H]ADP (5 Ci/mmole; 25 μCi/ml)	$5 \mu l$
$10^{-1} M$ 2-mercaptoethanol	$20 \mu l$
H_2O	
Enzyme	
$10^{-1} M$ MgCl$_2$	$10 \mu l$
Total volume: 200 μl	

The procedure is as for dADP polymerization except that MgCl$_2$ is used to start the reaction which is incubated at 37° for 10 min.

Units. One unit of enzyme catalyzes the incorporation of 1 nmole of dAMP into acid-insoluble polynucleotide at 37° in 1 min. The same amount of protein catalyzes the incorporation of 20–25 nmoles of AMP at 37° in 1 min.

Purification Procedure

All steps are carried out at 5° unless specified. Centrifugation is at 15,000 g for 10 min. Flow rates through ion-exchange columns are approximately 1 bed volume per hour, fractions of 0.1 bed volume being collected.

Step 1: Extract. Frozen *E. coli* B cells (300 g, late log phase, Grain Processing, Muscatine, Iowa) are divided into six 50-g portions each of which is mixed with buffer A (50 ml, 50 mM Tris-HCl, pH 7.8, 10 mM 2-mercaptoethanol) made up to 0.1 mM EDTA and thawed at room temperature. The suspension, mixed with 150 g of HCl-washed glass beads (149–210 μm, Potter Industries, Carlstadt, New Jersey), is disrupted in a Virtis homogenizer at half-maximum speed for 20 min. The supernatant solution obtained after centrifugation is adjusted to an absorbance at 280 nm of 50 by addition of buffer A (final volume, 1400 ml).

Step 2: Streptomycin Sulfate, Sodium Acetate, and Ammonium Sulfate Precipitations. The extract is adjusted to 0.5% streptomycin sulfate by addition of a 5% solution and the precipitate removed by centrifugation. The solution is acidified to pH 5.3 by addition of 1 *M* sodium acetate (pH 4.8) and the precipitate removed by centrifigation. The pH of the solution is quickly raised to 7.4 with 1 *M* NaOH. An equal volume of saturated (NH₄)₂SO₄, pH 7.4, is added dropwise. The precipitate is collected by centrifugation, dissolved in buffer A (100 ml), and dialyzed against the same buffer (2 liters) overnight.

Step 3: DEAE–Cellulose Chromatography. The dialysate is applied to a column (50 mm diameter × 200 mm) of DEAE–cellulose equilibrated with buffer A. The column is washed with buffer A containing 0.1 *M* NaCl until the absorbance at 280 nm is less than 0.1. The enzyme is eluted with buffer A containing 0.4 *M* NaCl and concentrated by precipitation with (NH₄)₂SO₄ as described in step 2. The precipitate is dissolved in buffer A (50 ml) and dialyzed overnight against buffer A containing 5% glycerol (1 liter).

Step 4: Phosphocellulose Chromatography. The dialysate is adjusted to pH 5.6 with 1 *M* sodium acetate, pH 4.8. Precipitated protein is removed by centrifugation and the solution applied to a column (25 mm diameter × 150 mm) of phosphocellulose equilibrated with buffer B (50 m*M* sodium acetate, pH 5.6, 10 m*M* 2-mercaptoethanol, 5% glycerol). The enzyme is eluted with buffer B and the pH of the eluted fractions immediately adjusted to 7.0 with 1 *M* Tris-HCl, pH 8.0. Careful control of pH during chromatography on phosphocellulose is critical. It is recommended that the phosphocellulose be suspended in buffer B and titrated to pH 5.6 before packing the column. The enzyme is precipitated with (NH₄)₂SO₄ as described in step 2 and collected by centrifugation. The precipitate is dissolved in buffer A (20 ml) and dialyzed overnight against buffer A containing 0.15 *M* NaCl and 5% glycerol (1 liter).

Step 5: DEAE–Sephadex A-50 Chromatography. The dialysate is applied to a column (25 mm diameter × 250 mm) of DEAE–Sephadex A-50 equilibrated with buffer A containing 0.15 *M* NaCl and 5% glycerol. The column is eluted with a linear gradient of NaCl in buffer A containing 5% glycerol (0.18 *M* to 0.4 *M*, total volume, 2 liters). This step is monitored using the polymerization of ADP. Fractions containing enzyme are combined, diluted with buffer A containing 5% glycerol to a NaCl concen-

TABLE I

PURIFICATION OF POLYNUCLEOTIDE PHOSPHORYLATION FROM *E. coli* B (300 g)

	Fractions	Volume (ml)	Protein (mg/ml)	Specific activity (unit/mg protein)	Total activity (units)
Step 1	Extract	1320	18.5	0.078	1905
Step 2	(a) Streptomycin sulfate	1450	14	0.095	1928
	(b) Sodium acetate	1370	10.5	0.11	1582
	(c) $(NH_4)_2SO_4$ precipitation	117	50	0.34	1989
Step 3	DEAE–cellulose	56	50	0.68	1904
Step 4	Phosphocellulose	16	50	1.8	1440
Step 5	DEAE–Sephadex A-50	11	12	8.0	1056
Step 6	Sephadex G-200	5	10	17.0	850

tration of 0.1 M and applied to a column (20 mm diameter × 50 mm) of DEAE–cellulose equilibrated with buffer A containing 5% glycerol. The concentrated enzyme is eluted with buffer A containing 0.5 M NaCl and 5% glycerol. Fractions with absorbance at 280 nm >0.1 are combined and the enzyme precipitated with $(NH_4)_2SO_4$ as in step 2. The precipitate is dissolved in buffer A containing 5% glycerol (10 ml) and applied directly to the Sephadex G-200 column.

Step 6: Sephadex G-200 Chromatography. The sample is applied to a column (25 mm diameter × 900 mm) of Sephadex G-200 equilibrated with buffer A containing 0.05 M NaCl and 5% glycerol. The column is eluted with the same buffer, fractions (3 ml) being collected at 15-min intervals. Those fractions containing enzyme are combined and concentrated by dialysis overnight against buffer A containing 50% glycerol. The enzyme (5 ml) is stored at −20°. The isolation is summarized in Table I.

The isolation has steps in common with other preparations.[22] Particular features of the present procedure are the good recovery of enzyme activity, achieved by minimizing exposure to acidic pH, and the absence of DNases from the product, which is stable for several years.

Stepwise Synthesis of Oligodeoxyribonucleotides of Defined Sequence

Procedure. The reaction mixture consists of the following
Oligonucleotide primer, 0.1–0.2 mM
Deoxyribonucleoside 5′-diphosphate, 1.5–3.0 mM
Tris-HCl, pH 8.5, 20 mM

[22] Y. Kimhi and U. Z. Littauer, this series, Vol. 12, Part B [34].

2-Mercaptoethanol, 10 mM
NaCl, 300–600 mM
E. coli polynucleotide phosphorylase, 2–10 units/ml·
MnCl$_2$, 5–20 mM

The reaction is started by addition of MnCl$_2$ and incubation continues for 2 to 16 hr at 37°. The reaction is stopped by addition of 1 M EDTA, pH 7.0, to twice the concentration of MnCl$_2$. After dilution to a NaCl concentration of 0.05 M, the product is loaded onto a column (10 mm diameter × 200 mm for 1 to 2 μmoles of oligonucleotide) and eluted with a linear gradient (50 bed volumes) of NaCl (0.05 to 0.2 M) in 50 mM Tris-HCl, pH 7.5, 7 M urea at ambient temperature. The flow rate is 30 ml/hr, fractions being collected at 10-min intervals and the oligonucleotides are monitored at 260 nm. Fractions containing a desired oligonucleotide are combined, diluted with an equal volume of H$_2$O, and applied to a column (10 mm diameter × 20 mm) of DEAE–cellulose. After washing the column with 0.05 M NH$_4$HCO$_3$ to remove urea and NaCl, the oligonucleotide is eluted with 0.5 M NH$_4$HCO$_3$, pH 8.0, fractions (1 ml) being monitored at 260 nm. The NH$_4$HCO$_3$ and water are removed by evaporation under reduced pressure in a rotary evaporator and the residual oligodeoxyribonucleotide (1 mM, in H$_2$O) stored at −20°.

High-Pressure Ion-Exchange Chromatography. This provides a rapid and sensitive tool for monitoring the progress of the enzymatic synthesis, for establishing the purity of isolated oligonucleotides and for quantitative nucleoside analysis. While a number of expensive machines designed for high-pressure chromatography are commercially available, a more economical approach is to assemble a system from commercially available components.[23] A suitable system consists of a two-chamber linear gradient maker (200 ml total volume) connected to a Milton-Roy minipump which propels buffer through a column (2 mm diameter × 500 mm) equipped with a septum injection port. Elution is monitored continuously using an Altex model 152 analytical uv detector equipped with a Linear Instruments model 385 dual channel recorder operating at a chart speed of 200 mm/hr. Anion-exchange separation of oligonucleotides is carried out at ambient temperature on a column packed with RPC-5,[18] using a linear gradient of ammonium acetate, pH 4.5 (0.5 to 3 M, total volume 100 ml)[15,24] at a flow rate of 20 ml/hr and a pressure of 70–100 kg/cm². Samples (1–10 μl of reaction mixtures) are applied directly to the column, without pretreatment, using a Hamilton microsyringe and elution monitored at 254 nm.

---

[23] Altex Scientific Inc., Berkeley, California.
[24] E. Trip and M. Smith, *Nucleic Acids Res.* **5**, 1539–1549 (1978).

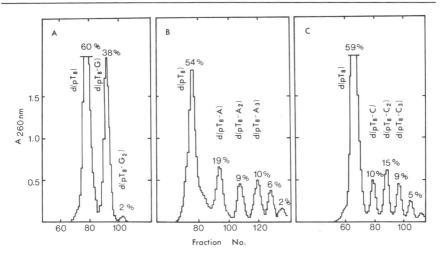

FIG. 1. Syntheses of d(pT$_8$-G), d(pT$_8$-A) and d(pT$_8$-C). (A) d(pT$_8$-C). The reaction (5 ml), containing d(pT$_8$) (1.8 μmole), dGDP (7.5 μmoles), 10 mM Tris-HCl, pH 8.5, 10 mM 2-mercaptoethanol, 300 mM NaCl, enzyme (45 units), and 10 mM MnCl$_2$, is incubated for 5 hr at 37°. (B) d(pT$_8$-A). The reaction (3 ml), containing d(pT$_8$) (1.4 μmole), dADP (4.5 μmoles), 10 mM Tris-HCl, pH 8.5, 10 mM 2-mercaptoethanol, 600 mM NaCl, enzyme (12 units), and 5 mM MnCl$_2$ is incubated for 15 hr at 37°. (C) d(pT$_8$-C). The reaction (6 ml), containing d(pT$_8$) (1.58 μmole), dCDP (9 μmoles), 10 mM Tris-HCl, pH 8.5, 10 mM 2-mercaptoethanol, 300 mM NaCl, enzyme (12 units), and 5 mM MnCl$_2$, is incubated for 10 hr at 37°. Each product is fractionated on a DEAE–cellulose column (10 mm diameter × 200 mm) as described in the text.

Cation-exchange chromatography, for nucleoside analysis is carried out on a column packed with Aminex A-5 (BioRad) at 55°,[24] eluted with a constant concentration (0.4 M) of ammonium formate, pH 4.5, and monitored at 254 and 280 nm. Samples are prepared for nucleoside analysis as follows: oligonucleotide (10 μg), $E.\ coli$ alkaline phosphatase (5 μg), snake venom phosphodiesterase (5 μg), 50 mM ammonium formate, pH 9.0, and 2 mM magnesium acetate (total volume, 50 μl). The mixture is incubated at 37° for 2 hr and an aliquot (1–10 μl) injected onto the column without pretreatment.

Synthesis of d(pT$_8$-A), d(pT$_8$-G), and d(pT$_8$-C).[20] The oligodeoxyribonucleotide d(pT$_8$) is an example of a slowly reacting primer.[13,15] Thus, the reactions of dADP and dCDP require increased NaCl and/or decreased MnCl$_2$ concentrations for optimal yields (Fig. 1b and 1c). Both reactions produce useful amounts of further addition products, e.g., d(pT$_8$-A$_2$) and d(pT$_8$-C$_2$). The slow reacting dGDP requires incubation with an increased amount of enzyme for an optimal yield of d(pT$_8$-G) (Fig. 1a).

TABLE II
Synthesis of d(pT$_8$-N-N')[a]

Primer (nmoles)[b]	Nucleotide residue added	Reaction conditions						Primer recovered (%)	Adducts (%)			
		Volume (ml)	NaCl (mM)	MnCl$_2$ (mM)	dNDP (mM)	Enzyme (units/ml)	Time (hr)		mono-	di-	tri-	tetra-
d(pT$_8$–A) (60)	pdC	0.5	600	5	1.5	4	5	70	15	8	4	
d(pT$_8$–A) (60)	pdG	0.7	600	10	1.5	3	2	48	22	24	3	3
d(pT$_8$–A) (100)	pdT	1.0	300	10	3	8	16	45	30	18	6	
d(pT$_8$–G) (100)	pdC	1.0	600	5	1.5	6	5	62	10	10	14	4
d(pT$_8$–G) (70)	pdA	0.5	600	4	1.5	3	4	75	8	7	9	
d(pT$_8$–G) (100)	pdT	1.2	300	10	3	11	11	76	8	6	9	3
d(pT$_8$–C) (75)	pdA	0.7	600	4	1.5	2	2	62	34	2		
d(pT$_8$–C) (75)	pdG	0.65	300	5	1.5	3	3	64	30	2	2	
d(pT$_8$–C) (60)	pdT	0.65	300	10	1.5	4	4	42	48	10		

[a] All reactions contain 10 mM Tris-HCl, pH 8.5, and 10 mM 2-mercaptoethanol.
[b] Numbers in parentheses are nanomoles of primer.

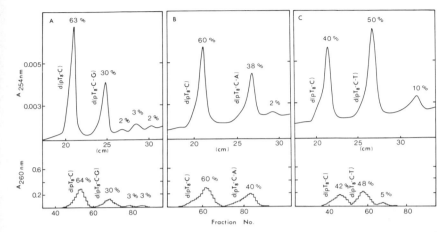

FIG. 2. Syntheses of (A) d(pT$_8$-C-G), (B) d(pT$_8$-C-A), and (C) d(pT$_8$-C-T). The reaction conditions are summarized in Table II, and the products are fractionated on a DEAE–cellulose column (8 mm diameter × 150 mm) and isolated as described in the text (lower diagrams). The upper diagrams illustrate the analysis of products by high-pressure anion-exchanger chromatography on RPC-5 (cm refers to the distance moved by the recorder paper after sample injection).

Synthesis of d(pT$_8$-N-N').[20] The conditions required for completion of the twelve members of this series are listed in Table II, and the isolation of d(pT$_8$-C-A), d(pT$_8$-C-G) and d(pT$_8$-C-T) is illustrated in Fig. 2, which also shows the results of high-pressure anion-exchange analysis of the reaction products. Notable are the great reactivity of d(pT$_8$-C) and the low reactivity of d(pT$_8$-G). The latter is analogous in behavior to d(pT$_8$) and requires a high concentration of NaCl and a low concentration of MnCl$_2$ for optimal yields of the desired products.

Synthesis of the series d(pA$_5$-N-A$_3$).[25] The hexanucleotides d(pA$_5$-G), d(pA$_5$-C), and d(pA$_5$-T) are obtained using standard conditions, i.e., 300 mM NaCl, 10 mM MnCl$_2$, 5 units/ml enzyme for 4 hr at 37°. Their elongation to produce d(pA$_5$-G-A$_3$), d(pA$_5$-C-A$_3$), and d(pA$_5$-T-A$_3$) illustrates the use of [3H]-labeled dADP in the last step of the syntheses to facilitate isolation of trace amounts of oligonucleotides. The reaction (0.4 ml), containing oligonucleotide (0.15 μmoles), [3H]dADP (10 μCi/μmole, 1 μmole), 10 mM Tris-HCl, pH 8.5, 10 mM 2-mercaptoethanol, 300 mM NaCl, *E. coli* polynucleotide phosphorylase (5 units), and 10 mM MnCl$_2$, is incubated for 6 hr at 37°. The isolation of d(pA$_5$-C-A$_3$) by monitoring the 3H-label is illustrated in Fig. 3.

Synthesis of d(pA$_2$-T$_2$-C$_3$-G$_3$) and d(pA$_2$-T$_2$-C$_4$-G$_4$).[13] These syntheses illustrate a basic strength of *E. coli* polynucleotide phosphorylase

[25] S. Gillam, K. Waterman, and M. Smith, *Nucleic Acids Res.* **2**, 613–624 (1975).

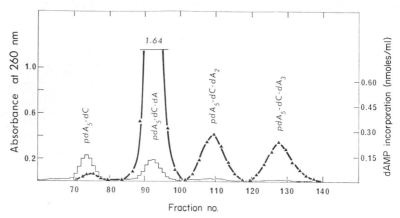

FIG. 3. Synthesis of d(pA₅-C-A₃). The reaction conditions and isolation procedure using DEAE–cellulose are described in the text.

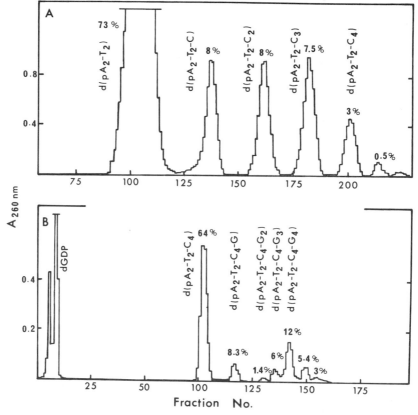

FIG. 4. Syntheses of (A) d(pA₂-T₂-C₃) and d(pA₂-T₂-C₄) and (B) d(pA₂-T₂-C₄-G₄). The reaction conditions and isolation procedure using DEAE–cellulose are described in the text.

TABLE III

CONDITIONS FOR ENZYMATIC STEPWISE SYNTHESIS OF $d(pG-T-A-T-C_2-T-A-C-A_3)$ AND $d(pG-T-A-T-C_3-A-C-A_3)$ CATALYZED BY *E. coli* POLYNUCLEOTIDE PHOSPHORYLASE AT 37°[a]

Primer	(μmoles)[b]	Volume (ml)	dNDP (μmoles)[b]	NaCl (mM)	MnCl₂ (mM)	Enzyme (units/ml)	Time (hr)	Product	Yield (%)
$d(pG-T-A-T)$	(3.90)	8	dCDP (24)	300	20	7	5	$d(pG-T-A-T)$	34
								$d(pG-T-A-T-C)$	15
								$d(pG-T-A-T-C_2)$	30
								$d(pG-T-A-T-C_2-C)$	12
$d(pG-T-A-T-C)$	(0.81)	2	dCDP (6)	600	5	9	7	$d(pG-T-A-T-C)$	36
								$d(pG-T-A-T-C_2)$	55
								$d(pG-T-A-T-C_3)$	9
$d(pG-T-A-T-C_2)$	(1.73)	5	dTDP (20)	100	20	12	16	$d(pG-T-A-T-C_2)$	40
								$d(pG-T-A-T-C_2-T)$	51
$d(pG-T-A-T-C_2-T)$	(1.40)	5	dADP (5)	600	20	12	7.5	$d(pG-T-A-T-C_2-T)$	81
								$d(pG-T-A-T-C_2-T-A)$	14
$d(pG-T-A-T-C_2-T-A)$	(0.31)	1.2	dCDP (1.8)	600	5	16	6	$d(pG-T-A-T-C_2-T-A)$	70
								$d(pG-T-A-T-C_2-T-A-C)$	15
$d(pG-T-A-T-C_2-T-A-C)$	(0.054)	0.25	[³H]dADP (0.4) (2 μCi/μmole)	600	5	8	6.5	$d(pG-T-A-T-C_2-T-A-C)$	20
								$d(pG-T-A-T-C_2-T-A-C-A)$	31
								$d(pG-T-A-T-C_2-T-A-C-A_2)$	15
								$d(pG-T-A-T-C_2-T-A-C-A_3)$	19
$d(pG-T-A-T-C_3)$	(0.62)	2	dADP (6)	600	20	12	16	$d(pG-T-A-T-C_3)$	58
								$d(pG-T-A-T-C_3-A)$	40
$d(pG-T-A-T-C_3-A)$	(0.30)	1.2	dCDP (3.6)	600	20	15	6	$d(pG-T-A-T-C_3-A)$	87
								$d(pG-T-A-T-C_3-A-C)$	8
$d(pG-T-A-T-C_3-A-C)$	(0.028)	0.1	[³H]dADP (0.15) (2 μCi/μmole)	600	5	7	6	$d(pG-T-A-T-C_3-A-C)$	35
								$d(pG-T-A-T-C_3-A-C-A)$	29
								$d(pG-T-A-T-C_3-A-C-A_2)$	6
								$d(pG-T-A-T-C_3-A-C-A_3)$	12

[a] All reactions contain 10 mM Tris-HCl, pH 8.5, and 10 mM 2-mercaptoethanol.

TABLE IV
OVERALL CONVERSIONS AT EACH STEP OF THE ENZYMATIC SYNTHESIS OF
d(pG-T-A-T-C$_2$-T-A-C-A$_3$) AND d(pG-T-A-T-C$_3$-A-C-A$_3$)[a]

Primer	(μmoles)[b]	Cycles	Product	(μmoles)[b]	Yield (%)
d(pG-T-A-T)	(3.90)	5	d(pG-T-A-T)		
			d(pG-T-A-T-C)	(0.81)	21
			d(pG-T-A-T-C$_2$)	(1.53)	40
			d(pG-T-A-T-C$_3$)	(0.62)	16
d(pG-T-A-T-C)	(0.81)	2	d(pG-T-A-T-C$_2$)	(0.56)	70
			d(pG-T-A-T-C$_3$)	(0.09)	11
d(pG-T-A-T-C$_2$)	(1.73)	5	d(pG-T-A-T-C$_2$-T)	(1.40)	81
d(pG-T-A-T-C$_2$-T)	(1.40)	6	d(pG-T-A-T-C$_2$-T-A)	(0.31)	22
d(pG-T-A-T-C$_2$-T-A)	(0.31)	4	d(pG-T-A-T-C$_2$-T-A-C)	(0.07)	22
d(pG-T-A-T-C$_2$-T-A-C)	(0.054)	1	d(pG-T-A-T-C$_2$-T-A-C-A$_3$)	(0.01)	19
d(pG-T-A-T-C$_3$)	(0.62)	3	d(pG-T-A-T-C$_3$-A)	(0.35)	56
d(pG-T-A-T-C$_3$-A)	(0.30)	5	d(pG-T-A-T-C$_3$-A-C)	(0.028)	9
d(pG-T-A-T-C$_3$-A-C)	(0.028)	1	d(pG-T-A-T-C$_3$-A-C$_3$)	(0.003)	12

[a] The amounts of product oligonucleotides are the sums of the material obtained after recycling unreacted primer at each step.
[b] Numbers in parentheses are quantity in micromoles.

catalyzed synthesis of oligodeoxyribonucleotides, i.e., the preparation of relatively long oligonucleotides in a few steps by taking advantage of the presence of tracts of the same nucleotide residue. The syntheses also illustrate two technical points: (i) the use of an organic solvent (dimethyl sulfoxide) in the enzyme reaction to promote successive additions of the same nucleotide and (ii) the use of high temperature (60°) during DEAE–cellulose–7 M urea purification of an oligonucleotide which aggregates at ambient temperature.

The first step is the synthesis of d(pA$_2$-T$_2$-C$_3$) and d(pA$_2$-T$_2$-C$_4$). The reaction (10 ml), containing d(pA$_2$-T$_2$) (7 μmoles), dCDP (40 μmoles), 25 mM Tris-HCl, pH 8.5, 10 mM 2-mercaptoethanol, glycerol (0.5 ml), dimethyl sulfoxide (2.0 ml), $E.$ $coli$ polynucleotide phosphorylase (45 units), and 20 mM MnCl$_2$, is incubated for 16 hr at 37°. The fractionation of the products on DEAE–cellulose–7 M urea is shown in Fig. 4a. The yield of d(A$_2$-T$_2$-C$_3$) is 0.5 μmole and of d(A$_2$-T$_2$-C$_4$) is 0.2 μmole.

For the synthesis of d(pA$_2$-T$_2$-C$_4$-G$_4$), the reaction (1 ml), containing d(pA$_2$-T$_2$-C$_4$) (0.14 μmole), dGDP (1.5 μmoles), 25 mM Tris-HCl, pH 8.5, 300 mM NaCl, 10 mM 2-mercaptoethanol, $E.$ $coli$ polynucleotide phosphorylase (3 units), and 10 mM MnCl$_2$, is incubated for 6 hr at 37°. The isolation of d(pA$_2$-T$_2$-C$_4$-G$_4$) (0.017 μmole) by chromatography on DEAE–cellulose–7 M urea at 60° is shown in Fig. 4b. The synthesis of d(pA$_2$-T$_2$-C$_3$-G$_3$) from d(pA$_2$-T$_2$-C$_3$) follows the same procedure.

Synthesis of d(pG-T-A-T-C-C-T-A-C-A-A-A) and d(pG-T-A-T-C-C-C-A-C-A-A-A).[2] The syntheses described to this point have involved two steps and have not required optimization of yields at each step. The same principle applies to three-step syntheses, e.g., d(pA$_4$) to d(pA$_4$-G$_3$-T-G$_2$) and d(pC$_3$) to d(pC$_3$-A$_3$-G-A$_3$), each of which can be accomplished in 1 week.[20] More extensive syntheses do require optimization of yields by recycling of unreacted primers. The procedures for two such syntheses are summarized in Tables III and IV. Another example of this type of synthesis is that of d(pT$_2$-A-G-C-A-G-A$_2$-C$_2$-G$_2$) starting with the primer d(pT$_2$-A-G).[19] All three of these oligonucleotides have proved to be very pure and capable of forming stable, specific hydrogen-bonded structures with the appropriate DNA.[1,2]

Conclusion

Primed synthesis using *E. coli* polynucleotide phosphorylase provides a simple and convenient route to a wide variety of oligodeoxyribonucleotides of defined sequence including oligonucleotides containing base analogs.[24]

Acknowledgments

We are grateful to Patricia Jahnke for the chemical synthesis of a number of oligodeoxyribonucleotide primers. The research is supported by the Medical Research Council of Canada of which M.S. is a Research Associate.

Section IX

Localization of Functional Sites on Chromosomes

A. Methods for Identifying Replication Origin and
Termination Sites
Articles 66 and 67

B. Restriction Enzyme Fragment: RNA Transcript Mapping
Articles 68 through 72

C. Marker Rescue of Mutants
Article 73

D. Assay of Biologically Active Fragments
Articles 74 and 75

E. DNA Fragment: Protein Interactions
Articles 76 through 79

[66] Mapping the Origin and Terminus of Replication of Simian Virus 40 DNA by Pulse Labeling

By CHING-JUH LAI

Principle

The chromosome of simian virus 40 (SV40) represents a unique replicon in the infected permissive cells. The replication process of the viral chromosome involves initiation, elongation, and termination. One approach to localization of functional sites in the viral genome that are involved in the DNA replication steps is through the use of a pulse-labeling technique. As adapted from the experiment performed originally to determine the direction of biosynthesis of proteins,[1] a brief pulse of ^{3}H-thymidine administered to cells which are actively synthesizing SV40 DNA, results in incorporation of the radiolabel at the growing points of the viral DNA. Molecules that are completed during the pulse (for a pulse time approximately equal to the time required for one round of replication) contain highest labeled radioactivity at or near the terminus, whereas the specific label is lowest at or near the origin of replication. ^{3}H-Thymidine pulse-labeled form I DNA is mixed with uniformly labeled [^{32}P]DNA and analyzed by restriction enzyme digestion. From the distribution of radiolabel (^{3}H/^{32}P ratio) in fragments of known position in the genome, one can deduce the origin, the terminus, and the direction of replication of the viral DNA.[2]

Materials

Virus strains: wild-type SV40 (strain 776) and derived deletion mutants, *dl*-1002 and *dl*-1007.

Medium: tissue culture growth medium

Radioactive chemicals: ^{3}H-thymidine (specific activity 24 Ci/mM or greater) and ^{32}P-orthophosphate

Enzymes: restriction enzyme *Hin*dII+III; pancreatic DNase I, and snake venom phosphodiesterase

Others: polyacrylamide gel or agarose gel and electrophoresis buffer

[1] H. M. Dintzis, *Proc. Natl. Acad. Sci. U.S.A.* **47**, 247 (1961).
[2] K. J. Danna and D. Nathans, *Proc. Natl. Acad. Sci. U.S.A.* **69**, 3097 (1972).

Methods

Pulse-Labeling of SV40 DNA. Initially a few pulses of different time periods are needed in order to establish a suitable gradient of label throughout the genomic fragments. These pulse periods range from less than one cycle to slightly over one cycle of replication. Confluent BSC-1 cells, infected with SV40 at a multiplicity of ~1–5 plaque-forming units (PFU) per cell for 48 hr at 37°, are exposed for 5, 10, or 15 min to a growth medium containing 100 μCi/ml ^{3}H-thymidine (specific activity 24 Ci/mmole) prewarmed to 37°. After removal of the medium, cells are lysed according to the procedure of Hirt.[3] The covalently closed SV40 [^{3}H]DNA (form I) is prepared from ethidium bromide–CsCl gradient centrifugation[4] and purified by a neutral sucrose gradient centrifugation or by direct electrophoresis in 1.4% agarose gel.[5]

Uniform Labeling of SV40 [^{32}P]DNA. Uniformly labeled, SV40 [^{32}P]DNA serves as a standard in the analysis of pulse-labeled DNA. To obtain uniformly labeled [^{32}P]DNA, SV40-infected BSC-1 cells are grown in low phosphate medium containing ^{32}P-orthophosphate (10 μCi/ml) from 24 to 60 hr postinfection. The ^{32}P-labeled form I SV40 DNA is prepared by the Hirt procedure and purified by ethidium bromide–CsCl isopycnic centrifugation or by electrophoresis as described earlier.

Restriction Enzyme Digestion of Labeled DNA and Gel Electrophoresis. To analyze the distribution of pulse label in the newly replicated molecules, pulse-labeled SV40 [^{3}H]DNA and uniformly labeled SV40 [^{32}P]DNA are mixed at a ^{3}H/^{32}P disintegrations/minute ratio of 5 or greater and cleaved with an enzyme that makes 10–20 scissions in SV40 DNA, for example, endo R · HindII + III. The endo R · Hind digest is applied to a slab gel (15 × 40 cm) of 4% polyacrylamide (acrylamide–bisacryalmide, 20 : 1) in 0.04 M Tris-HCl, 0.02 M sodium acetate, 2 mM EDTA.[6] Separation of the DNA fragments is performed by electrophoresis in the same buffer at 140 V for 16–20 hr. To analyze the radioactivity present in each fragment, [^{32}P]DNA bands are located by autoradiography of the wet gel, excised, and placed in scintillation vials. Each gel segment, dissolved in 0.4 ml 30% H_2O_2 by heating 68° for 24 hr,[7] is added with 10 ml scintillation fluid containing Triton X-100 and counted to separately determine the ^{3}H and ^{32}P radioactivity. The ratio of ^{3}H/^{32}P in each fragment is then determined.

Correction for Base Composition of DNA Fragments. The specific label in each DNA fragment is determined from the ^{3}H/^{32}P ratio measured and

[3] B. Hirt, *J. Mol. Biol.* **26**, 365 (1967).

[4] R. Radloff, W. Bauer, and J. Vinograd, *Proc. Natl. Acad. Sci. U.S.A.* **57**, 1514 (1965).

[5] P. A. Sharp, B. Sugden, and J. Sambrook, *Biochemistry* **12**, 3055 (1973).

[6] K. J. Danna and D. Nathans, *Proc. Natl. Acad. Sci. U.S.A.* **68**, 2913 (1971).

[7] R. W. Young and H. W. Fulhorst, *Anal. Biochem.* **11**, 389 (1965).

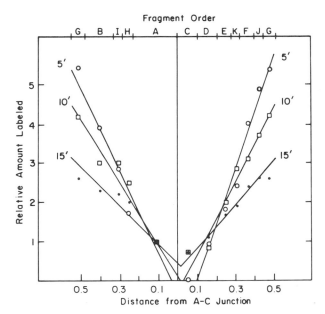

FIG. 1. Comparison of the position of each fragment in SV40 DNA and the relative amount of ³H-labeled fragment obtained from pulse-labeled newly completed molecules. Pulse-labeled, newly completed molecules of SV40 DNA were isolated after infected cells were exposed to ³H-thymidine for 5, 10, or 15 min. Each [³H]DNA sample was then mixed with uniformly labeled SV40 [³²P]DNA and digested with *H. influenzae* enzyme. Individual DNA fragments were isolated by electrophoresis and counted. On the oridinate is plotted the relative amount of each fragment labeled (³H/³²P ratio, corrected for thymidine content and normalized to the value for fragment A). On the *lower* abscissa is the distance of the midpoint of each fragment from the A–C junction. O, 5-min pulse time; □, 10-min pulse time; ●, 15-min pulse time. (From Danna and Nathans[2].)

the base composition of that fragment. To analyze the base composition, uniformly labeled [³²P]DNA is digested with the restriction enzyme, and each resulting DNA fragment is hydrolyzed to 5'-monophosphate deoxyribonucleotides by pancreatic deoxyribonuclease and snake venom phosphodiesterase.[8] The four monophosphate nucleotides are separated by high-voltage paper electrophoresis, and the radioactivity of each component is measured.[9] From such analyses the A + T% of the endo R DNA fragments can be determined and the ³H/³²P ratios corrected for variations in content. A less direct, but simpler way, to correct for variations in thymine content of different restriction fragments is to use for normalization the ³H/³²P ratios of fragments from DNA *uniformly* labeled with both ³H-thymidine and ³²P.[10]

[8] J. R. Lehman, M. Bessman, E. Simms, and A. Kornberg, *J. Biol. Chem.* **233**, 163 (1958).
[9] J. D. Smith, this series, Vol. 12, Part A, p. 180.
[10] S. Eisenberg, B. Harbers, C. Hours, and D. T. Denhardt, *J. Mol. Biol.* **99**, 107 (1975).

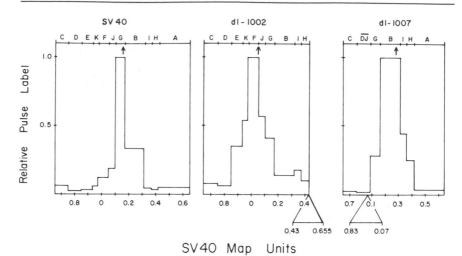

SV40 Map Units

Fig. 2. Pulse-label distribution in wild-type SV40 DNA and deletion mutant DNA. A mixture of 7-min pulse labeled [³H]DNA plus the same viral DNA uniformly labeled with ³²P was digested by endo R-HindII/III and the digest products separated by 4% polyacrylamide gel electrophoresis. The radioactivity in each DNA segment was determined after dissolving the gel in H_2O_2. The relative label in each case was the value obtained by normalizing ³H/³²P ratio of a given fragment to the fragment of highest ³H/³²P ratio. Which was set at 1.0. The arrow indicates the point on the map opposite R_1 for each genome. Δ, site and map units of deletion. (Data in part from Lai and Nathans.[11])

The Initiation Site and Termination Site of the SV40 Replicon. A typical comparison of the physical order of fragments with their pulse-labeled radioactivities is presented in Fig. 1. Two sets of gradients from the specific labels for each labeling time are evident: one on each side of the endo R · HindA–C junction indicating that SV40 replication is bidirectional. The slopes of both gradients are approximately equal, suggesting that both replication forks, one on each arm, are elongating at the same rate.

The origin of replication, as defined by the position of least label, can be located at a site in the genome corresponding to the intersection of the two gradient extrapolations. As is shown in Fig. 1, the origin of SV40 DNA replication maps within HindC at or near the HindA–C junction. The termination site of DNA replication, on the other hand, is located within or near the physical locations of the fragment containing the highest specific label. In the case of the SV40 replicon, the terminus for replication is within or near HindG segment at a position approximately opposite to the origin in the SV40 genome.

Pulse-label experiments have also been used to demonstrate that the terminus of replication of SV40 DNA is not a specific sequence of nu-

[11] C.-J. Lai and D. Nathans, *J. Mol. Biol.* **97**, 113 (1975).

cleotides.[11] If termination of SV40 DNA replication does not require specific sequences, mutants containing a deletion in one replication arm would terminate at a new site opposite the unique origin. On the other hand, no such change of terminus would be found in the mutant if termination is site-specific. As shown in Fig. 2, a mutant with a large deletion clockwise to the replication origin (*dl*-1007) terminates clockwise to the wild-type terminus, whereas a mutant with a large deletion counterclockwise to the origin (*dl*-1002) terminates counterclockwise to the wild-type terminus. Thus using the pulse-labeling technique it is possible to demonstrate that termination, unlike initiation, of SV40 DNA replication involved no site-specific sequences, rather the terminus is where two replicating forks join.

[67] Electron Microscopic Methods for Locating the Origin and Termination Points for DNA Replication

By GEORGE C. FAREED and HARUMI KASAMATSU

Electron microscopy has provided an important tool in the study of the molecular biology of DNA. It has been particularly useful in studies of the size and configuration of DNA molecules; however, prior to the use of site-specific bacterial restriction endonucleases, the ability to orient a linear DNA or identify a position on circular DNA was formidable. Schnös and Inman[1] in 1970 used partial denaturation mapping to provide the first visual evidence for a specific origin and bidirectional replication of bacteriophage λ DNA. The type II bacterial restriction endonucleases have subsequently simplified the identification and orientation of functional sites on DNA molecules. These enzymes recognize and generally produce scissions at or near specific oligonucleotide sequences in double-stranded DNA. The termini created by these specific cleavages serve as particularly useful reference points for the orientation of sites on circular and large linear DNA molecules in the electron microscope. In this article we will present a method employing restriction endonucleases and electron microscopy to locate the origin and termination of points for a variety of circular replicating genomes.

Most replicating, double-stranded, circular genomes possess characteristic Θ (theta) or "Cairns" forms which arise after initiation of replication at a unique site. These molecules when seen in the electron microscope contain two forks with three branches at each fork and no free ends. Frequently, two of the branches have identical contour lengths and corre-

[1] M. Schnös and R. B. Inman, *J. Mol. Biol.* **51,** 61 (1970).

METHODS IN ENZYMOLOGY, VOL. 65

spond to the newly replicated portion of the molecule. A physical reference point is created by the cleavage of the replicating DNA with a restriction endonuclease which cuts the genome at a single, unique site. One can then orient the position of the branch points or forks with respect to the cleavage site and follow their movement in molecules at different stages of replication. By analysis of such a spectrum of replicating molecules, one can accurately identify both the origin and termination sites as well as the direction of fork movement.

Experimental Design

In order to perform this analysis, it is preferable to select a purified bacterial restriction endonuclease whose cleavage site is present at a single, unique site on the replicon. Such a nuclease must be free of contaminating (exo- or endo-) nucleolytic activities which might alter the expected morphology of replicating DNA molecules. Next, it is essential that replicative intermediates at various stages of replication are purified from other DNA structures. This may involve a combination of ultracentrifugation, gel electrophoretic, and column chromatographic steps which remove nonreplicating (parental or progeny) genomes. These procedures also may allow for the replicative intermediates to be fractionated grossly into "young" (less than 50% replicated) and "mature" (greater than 50% replicated) intermediates. Such a fractionation simplifies the assignment of the origin and termination points. Finally, to carry out the analysis relatively small quantities (10 to 50 ng) of the cleaved replicating DNA are needed. The cleaved DNA is then examined in the electron microscope, and contour length measurements are determined for the linear branches.

When the cleavage site is some fixed distance away from the initiation site, there will exist a fraction of replicative intermediates which have not yet duplicated that site. After cleavage of these molecules, linear structures will be generated which contain a bubble (the replicated region) attached to two linear branches or tails created by cleavage in the unreplicated region of the molecule. When replication is unidirectional, the distance of one of the linear branches from one branch point to the end of the molecule will be constant, regardless of the extent of replication. In contrast, when replication is bidirectional, both linear branches will decrease in length as replication proceeds. Furthermore, if the two replication forks are moving at the same rate, there should be a parallel decrease in the lengths of both linear branches during bidirectional replication. Those molecules which have replicated the cleavage site will yield linear structures with two terminal branches at each end. The linear segment connecting the two branch points corresponds to the unreplicated portion of the molecule. In bidirectional replication, the lengths of the terminal branches will increase as the forks converge on the termination point, whereas in unidirectional replication, the lengths of only one of the terminal branches will increase. With the aid of a linear plot of the lengths of the linear branches or tails from one hundred cleaved molecules, versus the extent of replication, one may extrapolate to 0% replication for assignment of initiation site. A variety of aberrant structures which do not conform with those predicted may be observed. These may arise by random breakage of the replicating DNA (the forks which are often partially single-stranded are particularly vulnerable) or be due to the presence of other contaminating DNA molecules. They generally should be excluded from consideration or analyzed separately to determine whether secondary initiation sites or alternate replication modes exist.

Since this method was first employed for the analysis of SV40 DNA replication,[2] we will describe the procedure for the isolation of fractionated, SV40 replicative intermediates, their cleavage by endo R · EcoRI, and subsequent electron microscopic analyses. The method has also been used to analyze a variety of other replicating DNAs[3-13] and to locate the interaction of specific proteins with functional sites on certain DNA molecules.[14-16]

Procedure

Cells and Virus. Either primary or secondary African green monkey kidney cells or a continuous line of green monkey kidney cells, such as CV-1 or BSC-1 cells, are cultivated in 6–150 mm dishes using Eagle's minimal essential medium (MEM) supplemented with 5% fetal calf serum.

[2] G. C. Fareed, C. F. Garon, and N. P. Salzman, *J. Virol.* **10**, 484 (1972).
[3] J.-I. Tomizawa, Y. Sakakibara, and T. Kakefude, *Proc. Natl. Acad. Sci. U.S.A.* **71**, 2260 (1974).
[4] M. A. Lovett, L. Katz, and D. R. Helinski, *Nature (London)* **251**, 337 (1974).
[5] D. Perlman and R. H. Rownd, *Nature (London)* **259**, 281 (1976).
[6] F. Cabello, K. Timmis, and S. N. Cohen, *Nature (London)* **259**, 285 (1976).
[7] C.-J. Lai and D. Nathans, *J. Mol. Biol.* **97**, 113 (1975).
[8] W. W. Brockman, M. W. Gutai, and D. Nathans, *Virology* **66**, 36 (1975).
[9] T. Shenk, *Cell* **13**, 791 (1978).
[10] L. V. Crawford, C. Syrett, and A. Wilde, *J. Gen. Virol.* **21**, 515 (1973).
[11] L. V. Crawford, A. K. Robbins, and P. M. Nicklin, *J. Gen. Virol.* **25**, 133 (1974).
[12] D. L. Robberson, L. V. Crawford, C. Syrett, and A. James, *J. Gen. Virol.* **26**, 59 (1975).
[13] H. F. Tabak, J. Griffith, K. Geider, H. Schaller, and A. Kornberg, *J. Biol. Chem.* **249**, 3049 (1974).
[14] J. Griffith, M. Dieckmann, and P. Berg, *J. Virol.* **15**, 167 (1975).
[15] H. Kasamatsu and M. Wu, *Biochem. Biophys. Res. Commun.* **68**, 927 (1976).
[16] S. I. Reed, J. Ferguson, R. W. Davis, and G. R. Stark, *Proc. Natl. Acad. Sci. U.S.A.* **72**, 1605 (1975).

When the cells reach confluency, the medium is removed and SV40 (strain 776) is added [50 plaque forming units (PFU) per cell] in 1.5 ml of MEM. After a 2-hr absorption period at 37°, 25 ml of MEM with 2% fetal calf serum is added and the infected cells are maintained at 37°.

Preparation of Replicating SV40 DNA. At 32 to 40 hr after infection, the medium is removed and 10 ml of fresh medium containing ^{3}H-thymidine (50 μCi/ml, 25 mCi/μmole, New England Nuclear Corp.) is added to each dish. Pulse labeling is carried out at 37° for 10 min in order to label the replicating SV40 DNA (one cycle of SV40 replication is completed in 15 to 20 min at 37°). At the end of the pulse-labeling period, the medium is removed, the cells are rinsed with phosphate-buffered saline, and then lysates are prepared by the addition of 0.6% sodium dodecyl sulfate, 0.02 M Tris-HCl, pH 7.4, and 0.02 M EDTA (4 ml/dish). After 10 min at 25°, the lysate is gently transferred to a centrifuge tube and 0.25 volume of 5 M NaCl is added. The lysate is rinsed by inverting gently 8 to 10 times and then placed at 4° for 8 hr. This Hirt[17] extract is centrifuged at 10,000 rpm for 40 min in the SS34 rotor for a Sorvall RC-5 centrifuge. The supernatant is removed and precipitated with 2 volumes of absolute ethanol at −20°. The precipitated DNA is collected by centrifugation and dissolved in 6.25 ml of 0.01 M Tris-HCl, pH 7.4, 0.001 M EDTA, and 0.01 M NaCl. To this is added 6.93 gm CsCl and 0.9 ml ethidium bromide (2 mg/ml solution in H_2O). The final density of the dissolved CsCl is 1.5644 gm/cm³. The sample is subjected to equilibrium centrifugation in a Beckman fixed angle 50 Ti rotor for 40 hr at 38,000 rpm and 18°. The 9-ml gradient is fractionated into 0.3-ml aliquots, and tritium counts from trichloroacetic acid precipitates of 10 μl from each fraction are determined by liquid scintillation spectrometry. The fractions from the heavy shoulder of the tritium count profile are pooled as ''young'' replicating DNA, and those fractions on the light side of the peak of tritium counts are pooled as ''mature'' replicating DNA. The replicating SV40 DNA molecules are fractionated in ethidium bromide–CsCl density gradients due, in part, to the covalently closed template strands. The replicated region consists of two DNA segments that are able to bind dye in an unrestricted fashion as linear DNA, whereas the unreplicated region retains the more restricted dye-binding properties of closed circular duplex DNA.[18] The pooled fractions are extracted twice with 3 volumes of CsCl-saturated isoamyl alcohol, and the aqueous phase is dialyzed against 0.02 M Tris-HCl, pH 7.5, 0.01 M NaCl, and 0.001 M EDTA at 4°.

The replicating DNA molecules are further purified by neutral sucrose velocity gradient centrifugation. Samples (0.3 ml) are layered onto 12 ml

---

[17] B. Hirt, *J. Mol. Biol.* **40**, 141 (1969).
[18] E. D. Sebring, T. J. Kelly, Jr., M. M. Thoren, and N. P. Salzman, *J. Virol.* **8**, 478 (1971).

of 5–30% sucrose gradients in the above dialysis buffer. Sedimentation is carried out in the Beckman SW41 rotor at 40,000 rpm and 10° for 7 hr. The radiolabeled DNA sedimenting between 22 S and 27 S is pooled, dialyzed as above, and concentrated by vacuum dialysis. It should be noted that alternative fractionation schemes such as agarose gel electrophoresis or chromatography on benzoylated, naphthoylated DEAE–cellulose may also be useful.

Restriction Endonuclease Cleavage of Purified Replicating DNA. A number of restriction endonucleases which cleave SV40 DNA at single, unique sites[19] are now known and available from commercial sources (New England Biolabs, Bethesda Research Laboratories, Miles, Boehringer-Mannheim, Worthington) or can be purified according to published protocols. These enzymes (and their sites of cleavage on SV40 DNA) are *Eco*RI (0.0 map units), *Hpa*II (0.72), *Hae*II (0.83), Bam I (0.14), *Taq*I (0.54) and *Bgl*I (0.67). The desired restriction endonuclease should be shown to lack nonspecific endo- and exonucleolytic activities. Since *Eco*RI has been used to orient the replicating DNA for SV40, polyomavirus,[10,11] and colicin E1,[3,4] its use will be described. The fractionated replicating DNA (1 to 10 μg/ml) is incubated at 37° in 0.01 M Tris-HCl, pH 7.6, 0.005 M MgCl$_2$ and 0.05 M NaCl with 10 units of *Eco*RI for 60 min. The reaction is terminated by the addition of EDTA to a final concentration of 20 mM and an aliquot is taken for electron microscopy.

Preparation of DNA for Electron Microscopy. Various modifications and improvements have been achieved since the original basic protein monolayer technique of Kleinschmidt was introduced for the electron microscopic visualization of DNA.[20–22] These methods have been described in detail in previous volumes and should be referred to accordingly.

Two basic protein monolayer DNA spreading techniques have been used for the examination of replicating DNA molecules.[20,21,23,24] These are the aqueous spreading technique and the formamide spreading technique. In the aqueous spreading technique,[20,21] DNA samples are suspended in buffered salt solutions at neutral pH containing cytochrome *c*. The aqueous DNA spreading solutions are spread on to hypophase solutions. The cytochrome *c* monolayer film which contains the DNA is picked up with a supporting film on a copper grid. The DNA molecules may be stained prior to shadowing, and the specimen is shadowed with heavy metals such as platinum. The preparation is then ready for the examination under the

[19] G. C. Fareed and D. Davoli, *Annu. Rev. Biochem.* **46,** 471 (1977).
[20] A. K. Kleinschmidt, this series, Vol. 12, Part B, p. 361.
[21] R. W. Davis, M. Simon, and N. Davidson, this series, Vol. 21, Part D, p. 413.
[22] R. B. Inman, Vol. 29, this series, Part E, p. 451.
[23] A. K. Kleinschmidt and R. K. Zahn, *Z. Naturforsch. Teil/B* **14,** 770 (1959).
[24] B. C. Westmoreland, W. Szybalski, and H. Ris, *Science* **163,** 1343 (1969).

electron microscope. With the aqueous spreading technique, DNA molecules with good contrast are visualized and can be photographed for the length analyses.

In the formamide spreading technique,[21,24] DNA samples are suspended in buffered solutions containing formamide and cytochrome c. These DNA spreading solutions are layered on to hypophase solutions. Subsequent processing of the preparations prior to electron microscopic examination is similar to that for the aqueous spreading technique. In the presence of formamide, single-stranded DNA is extended and can be distinguished from double-stranded portions of DNA; whereas in the aqueous technique, the single-stranded DNA collapses and appears as a bushed structure. The replicating DNA molecules may contain single-stranded regions at each growing fork,[1] or they may contain large single-stranded regions if there is a time lag in the synthesis of the two antiparallel progeny strands.[25] Such replicating DNA molecules should be extended with the use of the formamide spreading technique for length analyses. Although commercially available formamide is, in general, satisfactory, it may be necessary to purify the formamide further by crystallization or distillation. In general, the concentrations of formamide and salt are selected such that the spreading is done at 25°, below the melting temperature for the DNA. The choice of the concentration of formamide in the DNA spreading and in the hypophase solutions depends on the structure of the replicating DNA, and these conditions should be selected accordingly.[21,26]

Schnös and Inman introduced a valuable denaturation mapping technique.[1,27] Under suboptimal denaturation conditions, a region in DNA with A+T-rich sequences will denature, while other segments remain double-stranded. After formaldehyde fixation of single-stranded DNA, these partially denatured DNA molecules are prepared for electron microscopy. For DNAs with sequences very rich in A+T interrupted by G+C-rich sequences, the visualization of denaturation loops leads readily to the orientation of the DNA molecules. The technique has been used to locate the origin and the terminus of DNA replication and to demonstrate the directionality of DNA replication. The method has been described in a previous volume.[22]

Various modifications of the above techniques have been introduced. These include the use of different buffers with various pH ranges. Different spreading conditions influence the length of DNA molecules, since the

[25] D. L. Robberson, H. Kasamatsu, and J. Vinograd, *Proc. Natl. Acad. Sci. U.S.A.* 69, 737 (1972).

[26] R. W. Hyman, *J. Mol. Biol.* 61, 369 (1971).

[27] M. Schnos and R. B. Inman, *J. Mol. Biol.* 55, 31 (1971).

mass-to-length ratio of DNA depends on many physical and chemical parameters. The use of a standard DNA of known size and configuration in the spreading mixture may be essential for the length analyses in electron microscopy.

The amount of DNA required for each spreading varies considerably with different techniques. From 2.5 to 25 ng DNA per spreading sample have been described.[20–22] With a very concentrated DNA solution after restriction enzyme treatment, a small amount of the concentrated DNA solution may be diluted into the spreading solution directly without affecting the results of the spreading. However, with a less concentrated DNA solution, the solution is converted into the spreading solution without cytochrome c by dialysis. The removal of detergents, alcohol, and sucrose are essential for the DNA spreading, and for the best result the DNA samples are stored in buffered solutions.[20–22]

When DNA samples are prepared for electron microscopic analysis after a restriction nuclease incubation, significant artifacts may be observed due to constituents in the nuclease preparation. This high background on the grid may be prevented by removal of proteins with phenol extraction after the nuclease reaction. The DNA may then be concentrated by ethanol precipitation. In such occasions, the organic solvent should be carefully removed by dialysis or by gel filtration.

Length Analysis; Localization of the Origin and the Terminus for DNA Replication. The specific enzyme cleavage site, which occurs at some fixed distance from the initiation site, will serve as a reference point on cleaved replicating forms. These molecules are photographed, and the lengths of individual linear segments in a replicative intermediate are measured. Alignment of the intact measured molecules according to the degree of replication should elucidate the direction, the origin, and the terminus of replication. It is important to decide the direction of DNA replication from the viewed molecules. After the restriction enzyme treatment, the Θ form of replicating molecules is converted to a linear structure containing a bubble or a linear structure with forks at the end. When replication forks pass the restriction site, the terminally branched structure as seen in late or mature replicating molecules is generated. Schematic diagrams of the cleaved, replicating molecules are presented in Fig. 1A and B. Since the two linear ends of the DNA molecules cannot be distinguished in electron micrographs, careful consideration of the aligned molecules is necessary in order to orient the branch points and linear ends.

If replicated segments, *L2* and *L3, L5* and *L6,* or *L7* and *L8* (see Fig. 1), do not contain appreciable single-stranded portions, the length of *L2* should be equal to that of *L3,* similarly *L5* = *L6, L7* = *L8.* The length of the unit or parental DNA molecule is computed from the length measurements as

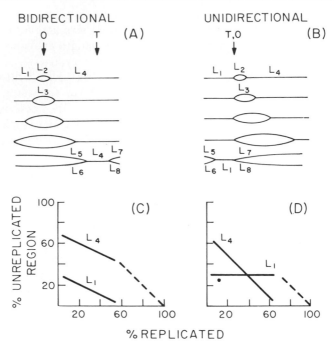

FIG. 1. Schematic and graphic representations for the analysis of bidirectional and unidirectional replication modes. The letters O and T locate the origin and termination points, respectively.

$$L = L1 + \frac{L2 + L3}{2} + L4 \quad \text{or} \quad L1 + \frac{L5 + L6}{2} + \frac{L7 + L8}{2}$$

Generally, the unit length of the DNA molecule is determined by measuring the contour lengths of purified, intact nonreplicating molecules. When replicated segments contain appreciable single-stranded regions, the length of two replicated segments may be different or obvious bushed structures may be observed with aqueous spreading technique. Such replicating DNA molecules are extended with the formamide spreading technique. Care should be taken for the computation of the length measurement, since the mass-to-length ratio of single-stranded DNA varies under different spreading conditions.[21] Random breakage of DNA molecules may occur during these manipulations generating small L numbers. These molecules should be discarded. Individual lengths are expressed as a fraction or percent of the mean value of the unit DNA length, and the extent of replication is expressed as the fraction corresponding to the replicated region.

If replication is bidirectional and the two growing forks move with equal rates, the lengths of two linear segments (*L1* and *L4*) intercepted by a bubble decrease as replication proceeds (Fig. 1A). The percent lengths for the shorter unreplicated segment, *L1*, or the longer unreplicated segment, *L4*, of individual replication DNA molecules are plotted as a function of the extent of replication (Fig. 1C). The collection of actual data points is expressed as an idealized single line *L1* or *L4*. Extrapolation to 0% replication locates the initiation site for the replication. Generally, one uses the result obtained from *L1* to determine the origin for DNA replication. In the example in Fig. 1C, at 0% replication *L1* is 30% of the genome in length. This position is the cleavage site at either 0.3 or 0.7 map units from the origin on a circular genome with arbitrary map units from 0.0 to 1.0. The termination site for the bidirectional replication is located opposite the origin on the circular genome either 80% or 20% of the genome length from the restriction nuclease cleavage site.

If replication is unidirectional, the length of one of two linear segments (*L1*) intercepted by a bubble will be constant, while the other segment, *L4*, decreases in length (Fig. 1B). The origin of DNA replication lies at the fork proximal to the constant linear segment. Thus, extrapolation to 0% replication locates the initiation site *and* the termination site for replication to be 30% or 70% of the genome length away from the cleavage site in the diagram (Fig. 1D).

From this analysis, the initiation site and the termination site for SV40 DNA replication were 33 and 83% of the genome length from the one end of *Eco*RI cleavage site, respectively, or 67 and 17% of the genome length from the other end. The direction of the DNA replication was bidirectional.

There may be intermediates in DNA replication which arise through alternate or different modes of replication. Some of the replicative intermediates of bacteriophage λ DNA showed a unidirectional mode, while most of the others conformed with a bidirectional replication mode.[1] Most restriction endonucleases do not cleave single-stranded DNA; therefore, replicating molecules with large single-stranded regions (e.g., displacement loops in replicating mitochondrial DNA) may yield partially single-stranded circular molecules with two linear, duplex branches.[28] These and other unexpected or alternate configurations of replicating DNA must be taken into consideration for classification of molecules prior to computation of length measurements. The mapping of functional sites on DNA by the electron microscopic methods described in this article does not resolve those sites at the level of specific nucleotide sequences. This is in part due to the inherent length fluctuations in DNA molecules spread by the basic protein film method.[21]

[28] W. M. Brown and J. Vinograd, *Proc. Natl. Acad. Sci. U.S.A.* **71**, 4617 (1974).

[68] Transcription Maps of Polyoma Virus-Specific RNA: Analysis by Two-Dimensional Nuclease S1 Gel Mapping

By Jennifer Favaloro, Richard Treisman, and Robert Kamen

This article describes the methodology used in our laboratory to map polyoma virus transcripts on the physical map of the viral genome.[1] The ultimate characterization of viral RNA is the determination of the complete nucleotide sequence of individual RNA species. Although the techniques allowing such a detailed analysis are now available (for example, cDNA copies of the RNAs can be prepared and then sequenced[2,3]), we here restrict our attention to the more modest goal of localizing the 5′ ends, internal "splice" points, and the 3′ ends of viral RNA molecules with respect to the template DNA sequence. The biology of the small DNA tumor viruses has historically presented a serious obstacle to the study of their transcription; these viruses stimulate rather than inhibit host metabolism, and, thus even under the most favorable circumstances, polyoma virus-specific RNA comprises only a small proportion of the RNA synthesized in an infected or virally transformed cell.[4] During the early phase of the productive virus infection of permissive mouse cells, which is defined as the period preceding the initiation of viral DNA replication, only 0.0001 to 0.001% of the cellular cytoplasmic RNA (10–100 molecules/cell) is virus specified.[5-7] The proportion of viral RNA in transformed cells can be even lower. After the initiation of viral DNA replication (which defines the beginning of the late phase of productive infection), the level of viral transcription increases markedly as template DNA accumulates, reaching a maximum of about 0.13% of the cytoplasmic RNA (5000–10,000 molecules per cell[6]). This quantitative difficulty has been overcome by a variety of RNA–DNA molecular hybridization techniques. A convenient method for the purification of radioisotopically pure viral RNA by preparative hybridization to polyoma virus DNA covalently

---

[1] For a review of polyoma virus genome structure, see M. Fried and B. E. Griffin, *Adv. Cancer Res.* **24,** 67 (1977).

[2] P. K. Ghosh, V. B. Reddy, J. Swinscoe, P. V. Choudary, P. Lebowitz, and S. M. Weissman, *J. Biol. Chem.* **253,** 3643 (1978).

[3] V. B. Reddy, P. K. Ghosh, P. Lebowitz, and S. M. Weissman, *Nucleic Acids Res.* **5,** 4195 (1978).

[4] N. Acheson, *Cell* **8,** 1 (1976).

[5] R. Kamen, D. M. Lindstrom, H. Shure, and R. Old, *Cold Spring Harbor Symp. Quant. Biol.* **39,** 187 (1974).

[6] L. T. Bacheler, *J. Virol.* **22,** 54 (1977).

[7] B. Cogen, *Virology* **85,** 222 (1978).

bound to cellulose is described elsewhere.[8] A rapid procedure for the analysis of labeled viral RNA using an adaptation of the Southern transfer technique[9] ("mini-blot" hybridization) has also been described.[10] We here concentrate on the powerful approach recently introduced by Berk and Sharp.[11] In our adaptation of this technique, hybrids formed between polyoma virus RNA and DNA are digested with single-strand-specific endonuclease S1[12] and the resulting duplexes or their component DNA strands are sized by gel electrophoresis. The two-dimensional variation of this basic procedure, presented below, is a general "diagonal" assay for splicing within RNA molecules. We shall first describe in some detail the standard procedures used in our laboratory for the preparation of viral nucleic acids and hybridization probes. While the bulk of these procedures are not novel, we include them so that all of the techniques used in the study of polyoma virus transcripts, which may be generally applicable to other systems, are available in one place.

Preparation of Virus and Viral Nucleic Acids

Materials and Buffers

Culture medium: Dulbecco's modified Eagle's medium (DMEM)

Cells: Secondary mouse embryo fibroblasts or BALB/c 3T6, as specified

Serum: Fetal calf or horse (Gibco Bio-cult) as specified

Proteinase K: Merck, purchased from BDH Chemicals, Ltd.

Ribonuclease A: Sigma grade X-A, heated to 100° before use

Deoxyribonuclease: Worthington RNase-free (DPFF), further purified by passage through 5'-(4-aminophenylphosphoryl)uridine-2'(3')-phosphate agarose (Miles-Yeda code 34-203)

Bovine serum albumin: Boehringer (electrophoretically pure)

DNA polymerase I: Boehringer grade I, 4900 units/mg

Escherichia coli RNA polymerase: holoenzyme purified from E. coli MRE 600 (Microbiological Research Establishment, Porton) as described in this series[13]

Oligo(dT)–cellulose: oligo(dpT)$_n$–cellulose–T1 (Boehringer)

Avian myeloblastosis virus (AMV) reverse transcriptase: specific activity 54,000. Obtained from Life Sciences, Inc.

[8] A. J. Flavell, S. Legon, A. Cowie, and R. Kamen, Cell 16, 357 (1979).
[9] E. M. Southern, J. Mol. Biol. 98, 503 (1975).
[10] F. Birg, J. Favaloro, and R. Kamen, Proc. Natl. Aead. Sci. U.S.A. 74, 3138 (1977).
[11] A. J. Berk and P. A. Sharp, Cell 12, 721 (1977).
[12] V. M. Vogt, Eur. J. Biochem. 33, 192 (1973).
[13] P. Berg, K. Barrett, and M. Chamberlain, this series, Vol. 21, p. 506.

Hirt mix: 0.6% (w/v) SDS (sodium lauryl sulfate, BDH), 10 mM EDTA, 10 mM Tris-HCl, pH 7.5

TNE: 150 mM NaCl, 50 mM Tris-HCl, pH 7.5, 1 mM EDTA

TSE: 100 mM NaCl, 50 mM Tris-HCl, pH 7.5, 1 mM EDTA

TE: 10 mM Tris-HCl, pH 7.5, 1 mM EDTA

Phenol: redistilled and stored at $-20°$ under N_2. Equilibrated with an equal volume of TNE before use

Phenol mixture: redistilled phenol, chloroform, and isoamyl alcohol (50 : 50 : 1) containing 0.1% (w/v) 8-hydroxyquinoline. Equilibrated with TNE and stored at $-20°$

Nick-translation buffer (10×): 500 mM Tris-HCl, pH 7.5, 50 mM MgCl$_2$, 10 mM 2-mercaptoethanol

cRNA mix (5×): 200 mM Tris-HCl, pH 7.5, 50 mM MgCl$_2$, 5 mM EDTA, 5 mM dithiothreitol

Tris–saline: 8 g/liter NaCl, 0.38 g/liter KCl, 1 g/liter Na$_2$HPO$_4$, 1 g/liter dextrose, 3 g/liter Tris base, 0.015 g/liter phenol red; adjusted to pH 7.4 with HCl

Lysis buffer[14]: 140 mM NaCl, 1.5 mM MgCl$_2$, 10 mM Tris-HCl, pH 8.6, 0.5% Nonidet P-40 (BDH), 0.015% (w/v) Macaloid (American Baroid Corp.)

2 × PK buffer − 200 mM Tris-HCl, pH 7.5, 25 mM EDTA, 300 mM NaCl, 2% (w/v) SDS, 400 μg/ml proteinase K

dT binding buffer: 10 mM Tris-HCl, pH 7.5, 500 mM NaCl, 0.1% (w/v) Sarcosyl (Ceiba-Geigy NL35)

dT midwash buffer: 10 mM Tris-HCl, pH 7.5, 100 mM NaCl, 0.1% Sarcosyl

dT elution buffer: 5 mM Tris-HCl, pH 7.5, 1 mM EDTA, 0.1% SDS

Polyoma Virus Stocks

Polyoma virus is produced by the low multiplicity infection of secondary whole mouse embryo cells. Cells at 5×10^6 cells per 90 mm culture dish are infected with 0.05 plaque-forming units (PFU) per cell of plaque purified polyoma virus. The A2 strain[1] of polyoma virus is our standard wild-type. The medium, 10 ml per dish of Dulbecco's modified Eagle's medium (DMEM) supplemented with 5% fetal calf serum, is changed after 3 and 6 days of incubation at 37°, and the virus is harvested at 10–14 days by freezing and thawing the cells plus medium three times. This lysate is sonicated in a Dawes Soniclean Automatic bath for 1 hr, and virus titer is

---

[14] U. Lindberg and J. Darnell, *Proc. Natl. Acad. Sci. U.S.A.* **65,** 1089 (1970).

determined by hemagglutination and plaque assay. Titers between 1–4×10^9 PFU/ml are routinely obtained.

The virus stocks are checked for lack of production of defective DNA in a subsequent cycle of infection by analyzing the polyoma virus DNA produced 48 hr after infection of 3T6 cells (two cultures, 5×10^6 cells per 90 mm dish) at 10 PFU/cell. The Hirt supernatant fraction is prepared and processed as described below (see discussion of polyoma virus DNA, ^{32}P-labeled *in vivo*) except the Sepharose 4B step is omitted, and aliquots representing 10% of the total preparation are digested with a variety of restriction endonucleases such as *Hpa*II,[15] *Sst*I,[16] and *Mbo*II[17] under conditions appropriate for each enzyme. The digests are fractionated by electrophoresis through a 1.4% agarose gel containing 0.5 μg/ml ethidium bromide. The gel patterns produced are visualized under long wavelength uv illumination, and compared with those of standard nondefective A2 polyoma DNA. All of our virus stocks are produced from a common plaque isolate originally produced by B. E. Griffin, or from a second virus stock derived by one low multiplicity passage using this isolate. The generation of defectives has never been detected in the many parallel low multiplicity virus stocks we have subsequently produced. It is critical that virus stocks which do not produce defectives during one cycle at relatively high multiplicity are used for transcription studies.

Polyoma Virus DNA

Polyoma virus DNA is prepared by selective SDS extraction[18] of 3T6 cells infected with polyoma virus. Subconfluent 3T6 cells (5×10^6 cells per 90 mm dish) are infected with A2 polyoma virus at 10 PFU per cell in 1 ml of DMEM. After 1 hr of incubation at 37° with occasional shaking, 10 ml of DMEM medium containing 5% fetal calf serum are added. Harvesting is at 48–60 hr by removal of the medium followed by the addition of 1 ml/dish Hirt mix and after 5 min, 0.25 ml of 5 M NaCl. The cell lysate is collected into 25 ml Beckman polycarbonate centrifuge tubes with the aid of a silicone rubber policeman and left at 0° for 4–15 hr before centrifugation at 25,000 rpm for 1 hr at 4° in the 30 rotor of a Beckman L2 65B centrifuge. The supernatant is vortexed with an equal volume of phenol, and the phases are separated by centrifugation at 10,000 rpm for 5 min, 18°, in the HB4 rotor of a Sorvall RC2 centrifuge. Nucleic acid in the aqueous phase is precipitated by adding 2 volumes of ethanol; after at

[15] P. A. Sharp, B. Sugden, and J. Sambrook, *Biochemistry* **12**, 3055 (1973).
[16] Originally isolated by S. Goff and A. Rambach, personal communication.
[17] R. E. Gelinas, P. A. Myers, and R. J. Roberts, *J. Mol. Biol.* **114**, 169 (1977).
[18] B. Hirt, *J. Mol. Biol.* **26**, 365 (1967).

least 2 hr at −20°, it is recovered by centrifugation and redissolved in TE buffer. Superhelical polyoma virus DNA is then purified from contaminating cell DNA or RNA by equilibrium centrifugation on CsCl–ethidium bromide gradients,[19] each containing up to 100 μg DNA (the Hirt extract from 12 dishes) in 5.75 ml TE buffer, 6 gm CsCl and 0.25 ml of 5 mg/ml ethidium bromide. The initial refractive index is 1.392. The tubes are topped with liquid Paraffin, capped, and centrifuged at 45,000 rpm, 20°, for at least 16 hr in the 65 angle rotor of a Beckman L2 65B centrifuge. Form I DNA (the lower band, visible in uv light) is collected, and the ethidium bromide is removed by repeated extraction with isoamyl alcohol. Two volumes of distilled water are added to the aqueous phase to dilute the CsCl, and the DNA is precipitated by adding a further 2 volumes of ethanol. The form I polyoma virus DNA is subsequently further purified by velocity sedimentation through sucrose gradients to remove mitochondrial DNA and low molecular weight components. Up to 250 μg of viral DNA (in 0.5 ml of TE buffer) is processed on each linear gradient (38.5 ml) containing 5–30% (w/v) sucrose in TNE buffer. The gradients are centrifuged in the SW27 rotor at 23,000 rpm and 18° for 21 hr. The fractions containing polyoma virus DNA are located by reading the optical density at 260 nm. These fractions are pooled, and the DNA is precipitated by adding NaCl to 0.5 M and 2 volumes of ethanol. The DNA is stored in TE buffer at −20°. The yield of form I DNA is 5–15 μg per 90 mm dish of infected 3T6 cells.

Polyoma Virus DNA, [32]*P-Labeled in Vivo*

In vivo ^{32}P-labeled polyoma DNA[20] is prepared by a modified version of routine DNA production. Four dishes of 3T6 cells (5 × 10⁶ cells per 90 mm dish) are washed with DMEM minus phosphate and infected with polyoma A2 virus diluted in the 1 ml of the same medium at 10 PFU per cell. DMEM minus phosphate supplemented with 3% horse serum (10 ml) is added to each dish after 1 hr. At 24–27 hr postinfection, the medium is replaced by 10 ml of fresh DMEM minus phosphate containing 2.5 mCi ^{32}P-phosphate per dish (Radiochemical Centre, PBS-1). The cells are harvested at 48 hr as described above into a siliconized 30 ml Corex tube, and the cell lysate is left at −20° for 1 hr before centrifugation at 12,000 rpm (4°) for 1 hr in the HB-4 rotor of a Sorvall RC2 centrifuge. The Hirt supernatant is incubated with 200 μg/ml proteinase K for 30 min at 37° and then extracted with an equal volume of phenol. After centrifugation at 10,000 rpm (20°) for 10 min, nucleic acids are precipitated from the aque-

[19] F. Radloff, W. Bauer, and J. Vinograd, *Proc. Natl. Acad. Sci. U.S.A.* **57**, 1514 (1967).
[20] R. Kamen and H. Shure, *Cell* **7**, 361 (1976).

ous phase by addition of 2 volumes of ethanol and stored at $-20°$ for 1 hr. The pellet obtained after centrifugation at 10,000 rpm (4°) for 20 min is washed with 75% ethanol to remove residual SDS and phenol prior to dissolution in 200 μl of TE buffer. RNA is digested with ribonuclease A (20 μg/ml) for 15 min at room temperature, and the DNA is purified, after addition of SDS to 0.1% (w/v), by passage through a 5-ml Sepharose 4B column equilibrated in TE buffer. The excluded peak, containing the viral DNA, is pooled and digested with the desired restriction endonuclease after addition of the appropriate buffer components. The restriction digestion is terminated by addition of EDTA to 15 mM, SDS to 0.1% (w/v), and proteinase K to 100 μg/ml. After a further 15 min at 37°, the solution is extracted with phenol and the DNA fragments in the aqueous phase are precipitated with ethanol after adding NH$_4$OAc to 0.5 M and yeast RNA carrier (25 μg/ml). The restriction digest is subsequently fractionated by agarose or polyacrylamide gel electrophoresis. The selective labeling conditions used here allow the preparation of ^{32}P-labeled restriction fragments of polyoma virus DNA sufficiently pure for use in hybridization analysis without prior isolation of the viral DNA present in the Hirt supernatant. The procedure must be carried out rapidly until the bulk of the radioactivity is removed by the Sepharose 4B column in order to avoid radiochemical fragmentation of the DNA. We estimate the initial specific activity of the viral DNA prepared by this method to be 5–10 × 10^6 cpm/μg.

Polyoma Virus DNA, ^{32}P-Labeled in Vitro

High specific activity ^{32}P-labeled polyoma virus DNA is prepared *in vitro* by the "nick translation"[21,22] activity of *E. coli* DNA polymerase I using very high specific activity [α-^{32}P]deoxyribonucleoside triphosphates. In a typical reaction, 200 μCi each of [α-^{32}P]TTP and dCTP (Radiochemical Centre, 2000–3000 Ci/mmole) are evaporated to dryness in a siliconized glass test tube (3 × ½ inches) and redissolved in 80 μl nick translation buffer (1×) containing 50 μg/ml bovine serum albumin, 2 μM dATP and dGTP, 1.5 μg form I polyoma virus DNA, and 9 units of *E. coli* DNA polymerase I. With the several different batches of Boehringer *pol*I that we have used the addition of DNase was not required. Incubation is at 11°–14° for 2–3 hr in a 1.5-ml plastic Microfuge tube (Eppendorf). The reaction is terminated by addition of EDTA to 10 mM. When labeled restriction fragments are required, the enzymes are then inactivated by heating to 70° for 1 min, and the sample is diluted to 500 μl with the appropriate restriction enzyme buffer. After incubation with the selected

[21] P. W. J. Rigby, M. Dieckmann, C. Rhodes, and P. Berg, *J. Mol. Biol.* **113**, 237 (1977).
[22] M. Botchan, W. Topp, and J. Sambrook, *Cell* **9**, 269 (1976).

endonuclease, the reaction is terminated by addition of EDTA to 15 mM, NH$_4$OAc to 0.5 M, carrier yeast RNA to 25 μg/ml, and SDS to 0.1% (w/v) before precipitation of the nucleic acids by addition of 2 volumes of ethanol or 1 volume of isopropanol. When total polyoma virus DNA probe is wanted, the nick translation reaction is directly terminated in the manner described for the endonuclease digestion. In both cases, unincorporated triphosphates are removed by passing the sample (redissolved in 100 μl of TE buffer) through a 1 × 18 cm column of Sephadex G-75 equilibrated with TNE buffer containing 0.1% SDS. The labeled DNA recovered in the excluded peak is stored at −20° after addition of carrier yeast RNA (25 μg/ml) and 2 volumes of ethanol. Typically, 25–40% of the radioactivity is incorporated into DNA resulting in a specific activity of 1–2 × 10^8 cpm/μg. Higher specific activities are obtained by using all four [α-^{32}P]deoxyribonucleotide triphosphate and omitting the nonradioactive dATP and dGTP. We find that highly radioactive DNA should be used within 5 days to obtain maximum sensitivity. Since, on the other hand, the α-labeled deoxyribonucleotide triphosphates currently provided by the Radiochemical Centre work well for at least 1 month, it is most economical to prepared small batches of nick translated DNA weekly rather than large batches less frequently.

Restriction Fragments of High Specific Activity ^{32}P-*Labeled Polyoma Virus DNA*

RNA mapping by the endonuclease S1 gel electrophoresis method[11] requires DNA restriction fragments free of single-strand nicks. All restriction enzymes that we use are checked for nicking activity by alkaline agarose gel electrophoresis[23] of digested DNA; we have not observed significant contaminating endonucleolytic activity among many preparations of different enzymes, whether purified in our laboratory or obtained from commercial sources, but still recommend this simple assay as a routine precaution. Unlabeled restriction fragments and low specific activity ^{32}P-labeled fragments are fractionated on neutral agarose or polyacrylamide gels and eluted by standard procedures described elsewhere in this volume. High specific-activity DNA labeled *in vitro* by nick translation, however, requires somewhat different treatment. Using the nick translation protocol described above, we are able to synthesize DNA at greater than 5 × 10^7 cpm/μg (1 hr incubation at 11°–14°) which contains about one nick per molecule. Intact single strands of restriction fragments less than 1 kb in length can be obtained from such DNA by two-

[23] M. W. McDonnell, M. N. Simon, and F. W. Studier, *J. Mol. Biol.* **110**, 119 (1977).

dimensional fractionation on agarose gels as will be described in detail below. Briefly, the restriction digest is first fractionated along one edge of a square agarose gel under neutral conditions. The entire gel is equilibrated, by soaking, with an alkaline buffer and electrophoresis is continued in a direction perpendicular to the first dimension. Intact single-stranded fragments, visualized by autoradiography, are cut out of the gel and eluted electrophoretically using neutral buffer. Nicked derivatives of each fragment occur as a smear migrating faster in the alkaline dimension. The two-dimensional fractionation avoids contamination of smaller fragments with nicked derivatives of larger fragments.

Synthesis of Asymmetric cRNA

Complementary RNA (cRNA) is defined as the RNA synthesized *in vitro* by *Escherichia coli* RNA polymerase using form I polyoma virus DNA as the template. At the ionic strengths (50–150 mM KCl) normally used for *in vitro* RNA synthesis, *E. coli* RNA polymerase copies both strands of polyoma virus DNA. However, transcription of this template proceeds efficiently at elevated ionic strengths, and under such conditions (450 mM KCl) transcription of one DNA strand (called the L strand) predominates.[5] A typical reaction mixture contains[24] 40 mM Tris-HCl (pH 7.5); 10 mM MgCl$_2$; 1 mM EDTA; 1 mM dithiothreitol; 1 mM each ATP, GTP, CTP, and UTP; 20 μg/ml superhelical (form I) polyoma virus DNA; 25 μg *E. coli* RNA polymerase holoenzyme; 10 μCi/ml [^{3}H]UTP (10–30 Ci/mmole, Radiochemical Centre TRK.289); and 450 mM KCl. The KCl is added last. Incubation is at 39.5° for 3–4 hr until incorporation ceases, which normally occurs when 25–35% of the triphosphates have been incorporated into RNA. A marked turbidity signals a successful synthesis. A 15 to 20-fold net synthesis of asymmetric cRNA is obtained. The reaction is terminated by addition of SDS to 0.1% (potassium dodecyl sulfate will precipitate, heat to 37° to dissolve) and EDTA to 20 mM. Protein is removed by extraction with an equal volume of phenol mixture, and the cRNA is precipitated from the aqueous phase by addition of 2 volumes of ethanol. The ethanol pellet is redissolved in TE buffer (0.1 ml for each 1 ml of initial reaction volume) and then MgCl$_2$ (to 10 mM) and pancreatic DNase (to 20 μg/ml) are added. After 30 min at 37°, SDS (0.1%) and EDTA (15 mM) are added, and the solution is extracted with an equal amount of phenol mixture. Nucleoside triphosphates are removed by passage of the aqueous phase through a 5-ml column of Sephadex G-75 (equilibrated in TNE buffer containing 0.1% SDS), and the cRNA is re-

---

[24] R. Kamen, J. Sedat, and E. Ziff, *J. Virol.* **17**, 212 (1976).

covered from the excluded peak by ethanol precipitation. The cRNA is self-annealed at a concentration greater than 100 μg/ml in TSE buffer containing 0.1% SDS at 68° for 1 hr, and then contaminating double-stranded RNA is removed by chromatography on a 5-ml CF-11 (Whatman) cellulose–ethanol column[25] as follows: ethanol is added to a final concentration of 35% (v/v), the sample is loaded onto the column (formed in a 10-ml plastic syringe, and equilibrated with TSE–35% ethanol), which is then washed with 20 ml of TSE–35% ethanol. Single-stranded RNA is eluted with TSE buffer containing 15% ethanol, collecting 1-ml fractions; double-stranded RNA is subsequently eluted with 1 mM EDTA, 0.1% SDS. Approximately 80% of the cRNA should be recovered in the single-stranded peak.

Preparation of ³²P-Labeled Asymmetric cRNA

Asymmetric cRNA labeled with ³²P at high specific activity is prepared by *in vitro* phosphorylation with polynucleotide kinase and [γ-³²P]ATP (Radiochemical Centre, 2000 Ci/mmole, PB168). The cRNA is cleaved by mild alkaline hydrolysis (50 mM Na$_2$CO$_3$, 1 mM EDTA; 15 min at 52°[26]) to generate fragments about 400 nucleotides in length and is then labeled *in vitro* using the procedure described in this volume.[27] This method is used for the preparation of ³²P-labeled cRNA rather than direct synthesis because we find that commercial preparations of α-³²P-labeled ribonucleoside triphosphates do not incorporate at the high substrate concentrations required for asymmetric transcription at high ionic strength.

Synthesis of cDNA to Polyoma Virus cRNA

To prepare hybridization probe complementary to the E strand of polyoma virus DNA, complementary DNA[28] is synthesized *in vitro* using asymmetric cRNA as template. A typical reaction (100 μl) contains 100 μCi [α-³²P]TTP (2000–3000 Ci/mmole, Radiochemical Centre, PB207, evaporated to dryness in a siliconized glass tube), 10 μM nonradioactive TTP, 200 μM each dGTP, dATP, and dCTP, 3.0 μg cRNA, 50 mM Tris-HCl, pH 8.1, 8 mM MgCl$_2$, 2 mM dithiothreitol, 40 mM KCl, 100 μg/ml actinomycin D, 4 mg/ml calf thymus DNA oligonucleotides (prepared as described in Taylor *et al.*[28]), and 40 units of AMV reverse transcriptase. After incubation at 37° for 2–3 hr, EDTA (10 mM) and NaOH (0.6 M) are added and the RNA is hydrolyzed by further incubation at 37° for 2 hr.

[25] R. M. Franklin, *Proc. Natl. Acad. Sci. U.S.A.* **55,** 1504 (1966).
[26] J. M. Coffin and M. A. Billeter, *J. Mol. Biol.* **100,** 293 (1976).
[27] G. Chaconas and J. H. van de Sande, this volume, Article [10].
[28] J. M. Taylor, R. Illmensee, and J. Summers, *Biochim. Biophys. Acta* **442,** 324 (1976).

The solution is neutralized with acetic acid and then passed through a 4-ml column of Sephadex G-75 equilibrated with TNE buffer containing 0.1% SDS to remove nonincorporated radioactivity. Normal yields are approximately 0.5 μg of cDNA at 1–2 × 10^8 cpm/μg. Prior to use as probe, the cDNA is self-annealed at a concentration of about 1 μg/ml in TNE buffer containing 1 M NaCl for 2 hr at 68°.

Preparation of RNA from Polyoma Virus-Infected Cells

3T6 cells (5 × 10^6 cells per 90 mm dish) are infected with A2 polyoma virus at 50 PFU per cell. The culture medium is DMEM supplemented with 5% fetal calf serum. For preparations containing late viral RNA[29], the cells are harvested 30 hr after infection by removing the culture medium and scraping off the cell sheet (with the aid of a rubber policeman and a minimal volume of ice-cold sterile Tris–saline) into siliconized 30-ml Corex tubes. The cells are washed three times with 10 volumes of Tris–saline by repeated sedimentation (5000 rpm in the HB-4 Sorvall rotor) and resuspension with a wide-mouth pipette. The washed cell pellet is resuspended in ice-cold lysis buffer using 5 ml for each 20 dishes, vortexed, and underlayered with an equal volume of 24% (w/v) sucrose, 1% NP40, in lysis buffer. After 5 min at 0° the lysate is centrifuged at 8000 rpm, 4°, for 20 min in the HB-4 rotor of an RC2B Sorvall centrifuge. The cytoplasmic (upper) phase is removed with a wide-mouthed pipette and added to an equal volume of 2 × PK buffer. The clear sucrose phase is discarded, the nuclear pellet is resuspended in lysis buffer (5 ml per 20 dishes), and an equal volume of 2 × PK buffer added. Complete disruption of the nuclei as well as reduction of the DNA viscosity is achieved by repeating shearing through a 19-gauge syringe needle. In large-scale preparations, this is alternatively accomplished by blending in a Sorvall Omnimixer. Both the nuclear and cytoplasmic fractions are incubated at 37° for 30 min, shaken briefly with an equal volume of phenol mixture in screw-capped sterile bottles, and then centrifuged at 10,000 rpm, 18°, for 10 min. The proteinase K digestion virtually eliminates interfacial denatured protein. Nucleic acids in the aqueous phases are precipitated by the addition of 2.5 volumes of ethanol. After at least 2 hr at −20° the precipitates are recovered by centrifugation for 10 min at 10,000 rpm in the HB4 rotor of an R2 Sorvall. The pellets are washed with a solution containing 75% (v/v) ethanol in 0.1 M NaOAc (pH 5.3) by resuspension and resedimentation. DNA is then digested with pancreatic deoxyribonuclease I (2 μg/ml) in a buffer containing 50 mM Tris-HCl, pH 7.5, and 10 mM MgCl$_2$ for 30 min at 37°. Digestion is terminated by adding EDTA and SDS to final concen-

[29] A. Flavell and R. Kamen, *J. Mol. Biol.* **115**, 237 (1977).

trations of 10 mM and 0.2%, respectively; protein is extracted with phenol mixture, and the RNA is precipitated with 2 volumes of ethanol after addition of NaOAc (pH 5.3) to 0.2 M. Oligodeoxyribonucleotides still contaminating the nuclear RNA preparations are removed by gel filtration chromatography on Sepharose 4B (large-scale preparations) or by extraction of the ethanol pellet with 20% NaOAc followed by reprecipitation from ethanol (small-scale preparations).

Polyadenylated RNA is purified by oligo(dT)–cellulose column chromatography.[30] The RNA sample, in dT-binding buffer, is loaded onto an oligo(dT)–cellulose column (formed in a 5-ml plastic syringe, up to 10 mg of RNA can be processed per milliliter of dT–cellulose) which has been washed with dT-binding buffer. The flow-through is reapplied to the column to ensure complete binding of the polyadenylated RNA. The column is washed sequentially with 2 ml per milliliter of column volume of dT-binding buffer, three times with 2 ml dT midwash buffer, and three times with 1 ml of dT-elution buffer. The fractions are assayed by reading the optical density at 260 nm. Normal yields of polyadenylated RNA are 2–3% of the input. Nonpolyadenylated RNA (the flow-through and the first binding buffer fractions) and polyadenylated RNA (the first two dT-elution buffer fractions) are stored as suspensions in 70% ethanol at $-20°$. NaOAc (0.2 M, pH 5.3) and carrier yeast RNA (25 μg/ml; unnecessary if the concentration of polyadenylated RNA is greater than 10 μg/ml) are added to the polyadenylated RNA fractions before the ethanol.

The same basic purification scheme is used to prepare RNA from other cells. Polyoma virus-transformed cell lines are usually grown to confluence in mass culture vessels (Sterilin), harvested, and then treated exactly as described above. Study of "early" polyoma virus RNA is facilitated by use of temperature-sensitive mutants which overproduce this RNA at late times of infection.[7] We infect 3T6 cells with the tsA mutant[31] at approximately 10 PFU per cell, incubate at 32° for 36 hr and then at 42° for 4–6 hr before harvesting. Under these conditions, early mRNAs are present at the same high level as late mRNAs. The overproduction is attributed to the inactivation of the polyoma virus large-T protein, which may be a repressor of early RNA synthesis.[7]

RNA Mapping by Gel Electrophoresis of Endonuclease S1-Resistant RNA–DNA Hybrids

The RNA mapping technique introduced by Berk and Sharp[11] is based on the gel electrophoretic analysis of single-strand-specific endonuclease

[30] See H. Nakazato and M. Edmonds, this series, Vol. 29, p. 431.
[31] M. A. Fried, *Virology* **40**, 605 (1970).

S1-resistant hybrids formed between the RNA and specific restriction fragments of the template DNA. The data obtained are used to position the ends and splice points of transcripts with respect to the physical map of the DNA. Complex families of related RNAs which may comprise only a small proportion of the total RNA sample can be characterized in detail.

Preparation of S1 Nuclease-Resistant RNA–DNA Hybrids

Materials and Buffers

Micropipettes: adjustable mechanical pipettes (Gilson Pipetman) with disposable plastic tips

Microcentrifuge tubes: conical plastic microtubes 1.5 ml capacity (Eppendorf 3810)

Centrifuge: Eppendorf model 5412

Carrier RNA: avian (chicken or turkey) nonpolyadenylated cytoplasmic RNA (prepared from embryo fibroblasts by the procedure described above) or *E. coli* rRNA are used for the annealing step. Yeast RNA (BDH) is used for the precipitation of nuclease-resistant hybrids

S1 nuclease: Sigma type III, 100–200,000 units/mg

Formamide Analar grade (BDH), twice recrystallized by slow stirring at 0°

Hybridization buffer: 40 mM PIPES buffer (Sigma), pH 6.4, 400 mM NaCl, 1 mM EDTA, in 80% (v/v) recrystallized formamide. Stored at −20°

S1 digestion buffer[32]: 280 mM NaCl, 30 mM NaOAc (pH 4.4), 4.5 mM Zn(OAc)$_2$, 20 μg/ml denatured calf thymus DNA (Sigma type I), containing 200 units/ml S1 nuclease added immediately before use

Termination mix: 2.5 M NH$_4$OAc, 50 mM EDTA

Procedure. Each hybridization contains viral DNA (or a restriction fragment of viral DNA), an appropriate aliquot of the RNA sample to be analyzed, and carrier nonhomologous rRNA, all dissolved in 10 μl of hybridization buffer. The normal amount of double-stranded polyoma virus DNA used is 5–10 ng (or the equivalent weight of restriction fragment); the quantity of RNA of course varies with the experiment, but as a rough guide we use 50–200 ng of total cytoplasmic polyadenylated RNA to map late mRNAs present in RNA extracted from cells late during infection or to map "early" mRNAs in RNA extracted from cells infected with

[32] T. E. Shenk, C. Rhodes, P. W. J. Rigby, and P. Berg, *Proc. Natl. Acad. Sci. U.S.A.* **72**, 989 (1975).

the mutant tsA. Viral RNA in polyoma virus-transformed cells is studied using 2–10 μg of cytoplasmic polyadenylated RNA per annealing. It is critical that the DNA be in excess, and we thus routinely hybridize three different quantities of a new RNA preparation to assure that the final gel band intensity is proportional to RNA concentration. Carrier non-homologous RNA is added so that each annealing contains a total of 25 μg RNA. All nucleic acids are stored as ethanolic suspensions to facilitate setting up the annealings. Calculated volumes of these suspensions are combined in 1.5-ml conical Microfuge tubes (WARNING: use only Eppendorf tubes, others may leak!) and the DNA and RNA is pelleted by centrifugation in an Eppendorf angle Microfuge for 15 min. All of the supernatant is carefully removed (the small volume remaining in the tube after decantation and any droplets on the walls are removed by capillarity with sterile glass capillary pipettes), and residual solvent is evaporated in an evacuated dessicator for 5 min. The pellets are dissolved in 10 μl of hybridization buffer by repeated pipetting up and down in the yellow plastic disposable tip of a Gilson micropipettor. Incomplete dissolution is a common cause of failure; when [^{32}P]DNA is used it is helpful to monitor the tube and pipette tip during this step to ensure that the entire pellet is being taken up. After careful closure of the Microfuge tube caps, the tubes are *completely submerged* in an 85° water bath for 15 min to denature the DNA. Although this temperature is well above that theoretically required, better results are achieved and we detect no degradation of the RNA caused by the high temperature. The tubes are then rapidly transferred to 52° water bath where they are incubated submerged for 3 hr. Under these conditions, RNA–DNA hybridization goes to completion, but the DNA does not reassociate. The tubes are opened with their tips still immersed in the water bath, and 0.3 ml of S1 buffer (0°) are added. After rapid mixing with a Vortex, the tubes are stored on ice. When S1 buffer has been added to all the tubes in the experiment, they are incubated at the desired temperature for 30 min before addition of 75 μl termination mix, 2 μl of 10 mg/ml yeast RNA, and 400 μl of isopropanol to precipitate the hybrids. For analysis on alkaline gels, the S1 incubation temperature is 37°, but for neutral or two-dimensional gel analysis, an empirically established temperature of 0°–20° is used to minimize nuclease cleavage of the RNA strand opposite an excised single-stranded DNA loop (see below). After 15 min in a Dry Ice–methanol bath or 2 hr at −20°, the hybrids are recovered by centrifugation and complete removal of the supernatant is as described above. If ^{32}P-labeled DNA has been used, the radioactivity in the dried pellets may be quantitated by Cerenkov counting in a liquid scintillation spectrophotometer.

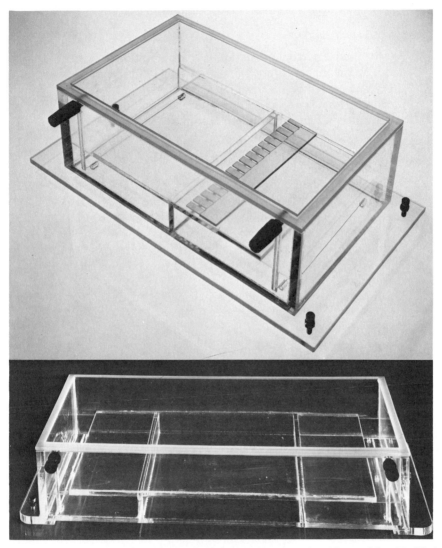

FIG. 1. Apparatus for horizontal agarose gel electrophoresis without wicks. Modified from the original design of W. Shaffner. The upper gel box (dimensions are 10 cm high × 16.5 cm wide × 28.5 cm long) is for one-dimensional gels, and is shown in place on its leveling table. The lower gel box (dimensions are 10 × 22 × 48 cm) is for two-dimensional gels and has leveling screws built in. The boxes were constructed by J. Kitau of the Imperial Cancer Research Fund workshops.

Analysis of S1 Nuclease-Resistant Hybrids by Gel Electrophoresis

Materials and Buffers

Agarose gel electrophoresis apparatus: horizontal agarose gels are run in the apparatus shown in Fig. 1 which is a modification of the original design of W. Schaffner

Urea–polyacrylamide gel electrophoresis apparatus: vertical slab gels are run in an apparatus similar to that used for DNA sequencing

Agarose: Sigma type II, medium EEO.

Alkaline gel loading buffer: 50 mM NaOH, 1 mM EDTA, 2% (w/v) Ficoll (Pharmacia), 0.015% bromocresol green (Gurr code 6410)

Neutral gel loading buffer: 10 mM Tris-HCl (pH 7.5), 1 mM EDTA, 2% (w/v) Ficoll, 0.015% bromocresol green

Urea-polyacrylamide loading buffer: 10 M urea, 1 mM EDTA, 0.1 N NaOH, 0.015% bromcresol green

Neutral agarose electrophoresis buffer[15]: 40 mM Tris, 5 mM NaOAc, 1 mM EDTA, adjusted to pH 7.8 as a 10× concentrate with glacial acetic acid

Alkaline agarose electrophoresis buffer[23]: 30 mM NaOH, 1 mM EDTA

X-ray film: Fuji Rx

X-ray intensifying screens: Fuji Mach II

Nitrocellulose: Sartorius 0.1 μm pore size (SM 11309) nitrocellulose filters, purchased (from V. A. Howe in Britain) as 16 × 16 or 16 × 20 cm sheets

Polythene sleeving: 5 inch diameter, 1000 gauge

Denaturation buffer: 0.2 N NaOH, 0.5 M NaCl

Neutralization buffer: 0.5 M NH$_4$OAc, 0.5 M NaCl

20 × SSC: 3.0 M NaCl, 0.3 M sodium citrate, filtered through paper

Blot presoak solution: 0.2% (w/v) each bovine serum albumin, Ficoll, and polyvinyl pyrrolidone in 6 × SSC

Blot hybridization solution: presoak solution additionally containing 0.1% SDS and 400 μg/ml denatured calf thymus DNA

Neutral agarose gels: Gels containing 2% (w/v) agarose in neutral electrophoresis buffer are prepared as follows: the agarose is dissolved in the buffer (150 ml per gel) by boiling on a stirring hot plate, using a reflux condenser to avoid evaporation loss, and cooled to about 50°. The glass gel plate (14.9 × 20 × 0.4 cm) is surrounded with a Sellotape collar to retain the agarose and placed on a Perspex platform (a flat plastic sheet with three adjustable feet) made horizontal with the aid of a carpenter's spirit level. The slot formers (fashioned from 1 mm Perspex and containing twenty-two 1 × 45 mm teeth) are held in place with plasticine sup-

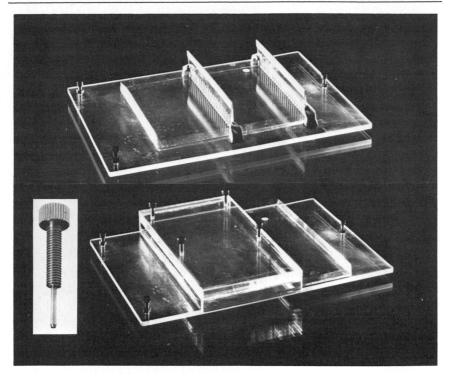

FIG. 2. Methods for casting horizontal agarose gels. The upper photograph shows a one-dimensional gel (20 × 14.9 cm) with two slot formers in place. The lower photograph shows a two-dimensional gel, with the template for forming four wells in a 20 × 20 cm gel in place. The insert shows one of the four well-formers.

ports; two registers, one 1 cm from one edge of the plate and a second 11 cm from the edge may be used if required (Fig. 2). The warm agarose is poured onto the plate and allowed to set. After removal of the slot former(s) and the Sellotape collar, the gel on its glass plate is placed on the horizontal support of the gel box illustrated in Fig. 1. Buffer is added to a level which submerges the gel by 0.3–0.5 cm. The samples, dissolved in 10 μl of neutral gel loading buffer by pipetting up and down with a disposable plastic pipette tip, are loaded into the slots under the buffer. A marker mix, containing a distribution of double-stranded DNA fragments of various known sizes is loaded into the first and last gel slots. Electrophoresis is at 20–40 V until the dye has migrated 8 cm (15 hr at 1 V/cm).

Alkaline agarose gels: Alkaline agarose gels[23] are prepared and run in an analogous manner. The gels (2% agarose) are formed in a neutral unbuffered solution (50 mM NaCl, 1 mM EDTA) and then run in alkaline gel electrophoresis buffer. We have observed that addition of NaOH to hot agarose before pouring the gel causes partial hydrolysis resulting in uneven band migration (the DNA nearer the glass bottom plate moves slower than the DNA near the gel surface). The nuclease S1-resistant hybrids are dissolved in alkaline gel loading buffer (10 μl) and loaded as described above. Electrophoresis is either overnight (15 hr) at 1 V/cm or at higher voltage (up to 7.5 V/cm) until the dye has migrated 8 cm. Since the dye diffuses rapidly on the alkaline gels, it is advisable to cover the gel with a second glass plate after loading.

Urea–polyacrylamide slab gels: Vertical slab gels (23 × 32 cm) are prepared as described by Maxam and Gilbert.[33] Samples are dissolved in 10 μl of urea–polyacrylamide gel loading buffer and heated at 52° for 5 min prior to loading. Electrophoresis at 300 V is continued until the dye reaches the bottom of the slab (about 12 hr for a 5% gel). For sizing DNA fragments of 100–800 nucleotides, 5% gels are used. Smaller fragments are sized on 10% gels.

Two-dimensional agarose gels: Agarose gels for two-dimensional analysis are prepared in a manner similar to that used for one-dimensional neutral gels except that 200 ml of solution is used for each 20 × 20 cm square gel. Sample wells are formed using a Perspex template with retractable threaded stainless steel rods (1.5 mm diameter) as illustrated in Fig. 2. The nuclease S1-resistant hybrids are dissolved in 10 μl of neutral agarose gel loading buffer. The gel is placed in the larger gel box shown in Fig. 1, and the samples are loaded into the wells under neutral agarose gel electrophoresis buffer. Electrophoresis in the first dimension is at 1 V/cm for 12–14 hr. The gel slab is then removed to a plastic tray, equilibrated with alkaline agarose gel buffer (two changes of 5 volumes each for 30 min), and returned to the gel box now containing alkaline gel buffer. The sample wells are filled with 10 μl of alkaline gel loading buffer containing DNA size markers, and electrophoresis is continued in a direction perpendicular to the first dimension at 1–3 V/cm until the marker dye has moved 8 cm. The gel is covered with a second glass plate during electrophoresis to minimize dye diffusion. Up to four 20 × 20 cm gels (16 samples) can be run in two layers in the apparatus illustrated in Fig. 2.

Detection of Gel Bands by Direct Autoradiography

When RNA has been annealed to ³²P-labeled DNA, the gel bands are detected directly by autoradiography or by fluorography with intensifying screens. The agarose gels (previously soaked in 7% trichloroacetic acid in the case of alkaline gels) are mounted on glass plates and dried overnight under several layers of 3MM paper (Whatman) weighted with another

[33] A. Maxam and W. Gilbert, this volume, Article [57].

glass plate. Drying is completed by placing the glass plate on top of an 85° water bath (CAUTION—the gel will crack if left too long). The dried gel on the glass plate is then covered with Saran wrap and autoradiographed at room temperature or fluorographed with an intensifying screen and pre-flashed film at −70°.

Detection of Gel Bands by Transfer to Nitrocellulose and Annealing to High Specific Activity Probes

Greater sensitivity is achieved by annealing RNA to unlabeled DNA, transfer of the resulting gel bands to nitrocellulose,[9] and hybridization to high specific activity ^{32}P-labeled probes prior to autoradiography.

The RNA–DNA hybrids in neutral gels are denatured by gently rocking the gel in 5 volumes of denaturation buffer for 45 min (unnecessary for alkaline gels). The gel is neutralized for 45 min in 5 volumes of neutralization buffer and then placed on a sheet of double-thickness 3MM paper saturated with 20 × SSC. The paper is draped over a glass plate supported on a plastic tray so that the ends of the paper are immersed in a reservoir of 20 × SSC. Areas of the 3MM paper not covered by the gel are masked with Saran wrap. A sheet of nitrocellulose the same size as the gel (wetted with distilled water and then soaked in 20 × SSC) is placed on top of it taking care to exclude air bubbles, and the nitrocellulose is covered with an piece of Whatman No. 1 paper of the same dimensions (presoaked in 20 × SSC). A stack of folded paper towels, a glass plate, and a weight (about 1 kg) are then piled on. After approximately 24 hr, the nitrocellulose sheet is removed, placed between two sheets of dry 3MM paper, baked for 2 hr at 80° in a vacuum oven, and stored in a dessicator. Prior to hybridization, the nitrocellulose sheet is incubated at 68° in a large volume (about 100 ml) of blot presoak solution. The presoaked blot is placed in a length of polythene sleeving sealed along three edges with a plastic bag sealer (Calor model 24.03). Hybridization solution (degassed under vacuum prior to addition of SDS) containing 5 ng/ml of denatured, in vitro ^{32}P-labeled, probe polyoma virus DNA (or the equivalent amount of kinase-treated cRNA or ^{32}P-labeled cDNA to cRNA) is then added and the bag is sealed excluding air bubbles. Approximately 20 ml of hybridization solution is used for one to six 8 × 15 cm blots contained in a single bag. After annealing for 24 hr in a shaking water bath at 68°, the nitrocellulose sheets are removed and washed with 0.3 × SSC, 0.1% SDS (several changes at room temperature, one wash at 68° for 10 min, then several further changes at room temperature), mounted on glass plates, covered with Saran wrap, and exposed under autoradiographic or fluorographic conditions.

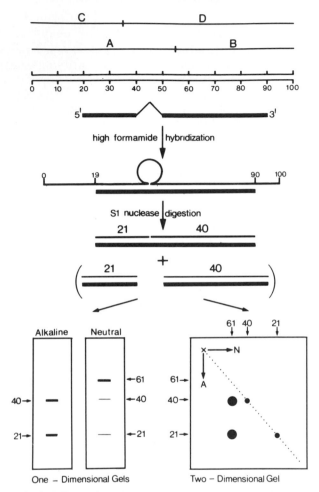

FIG. 3. Hypothetical illustration of nuclease S1 gel electrophoresis RNA mapping. See text for explanation.

Interpretation of One-Dimensional Gels

Rationale

Consider the simple hypothetical example diagrammed in Fig. 3—the mapping of a single "spliced" RNA transcribed from one strand of a linear duplex DNA. We assume that detailed restriction enzyme analysis of this DNA has been used to construct a physical map divided into 100 arbitrary map units (Fig. 3). The 5′ end of the RNA lies at map unit 19, there is a "splice" deleting those sequences represented in the DNA between 40 and 50, and the 3' end lies at 90 map units. In the first step of the analysis used to establish these positions, an excess of purified DNA, ^{32}P-labeled and lacking single-strand nicks, is combined with a sample of unlabeled RNA containing a small proportion of the hypothetical RNA species in a solvent containing 80% formamide. After heating to denature the DNA, the nucleic acids are annealed at a temperature allowing DNA–RNA hybridization but excluding DNA reassociation. Because RNA–DNA hybrids are considerably more stable than the corresponding DNA–DNA duplexes in high concentrations of formamide, it is normally possible to find annealing conditions which are above the T_m of the DNA–DNA double-strand but below that of the DNA–RNA hybrid.[34] The hybrid formed (Fig. 3) will have single-stranded "tails" representing the DNA sequences before the 5' end and after the 3' end of the RNA, as well as an internal loop of single-stranded DNA corresponding to the sequences from map unit 40–50 which are missing in the RNA. Digestion with endonuclease S1 is then used to remove all of the single-stranded sequences without nicking either strand within duplex regions. The nuclease digestion simultaneously hydrolyzes nonhomologous RNA and the noncomplementary DNA strand. The S1 nuclease-resistant structure is a hybrid containing a continuous RNA strand hydrogen bonded to a DNA strand with a specific nick at the site where the loop was excised. In a proportion of the hybrids, the nuclease will have also cleaved the RNA strand opposite the loop in the DNA strand, producing two separate hybrid molecules. The extent of this S1 nuclease "cut-through" varies with the length of the loop and the temperature of S1 digesion; we assume in our hypothetical example the common result that 10–20% of the hybrids have been cut through. Procedures for minimizing cut-through are described in a subsequent section.

The next step in the mapping technique is to fractionate the S1-resistant hybrids by electrophoresis through agarose or polyacrylamide gels. The gel patterns are visualized by direct autoradiography or fluorography. Marker DNA fragments of known size are run in parallel, and the lengths of the S1-resistant structures are estimated from the resulting semilogarithmic standard curves. In our hypothetical example, electrophoresis on a neutral gel would yield a major hybrid band 61 map units long as well as two minor bands of 40 and 21 map units resulting from nuclease cleavage of the RNA strand. Electrophoresis on an alkaline agarose gel to examine only the component single-stranded DNA would yield bands of 40 and 21 map units. Alternatively, both gel analyses can be combined in a two-dimensional fractionation, using electrophoresis in a

[34] J. Casey and N. Davidson, *Nucleic Acids Res.* **4**, 1539 (1977).

neutral buffer as the first dimension and in an alkaline buffer as the second. This alternative procedure is discussed in detail later.

The hypothetical results discussed this far establish that the RNA is spliced and have determined the lengths of its colinear portions. Controls are of course necessary to demonstrate that the "splice" does not correspond to an S1 nuclease-sensitive site within a complete duplex caused by, for example, a long AT tract. If an RNA complementary to the entire DNA is available (such as cRNA in the case of polyoma virus DNA) this control is easily done. In our experience, with a variety of DNA molecules, there are no specific sites within hybrids which are efficiently cleaved by S1. Berk and Sharp[35] added a powerful further method to confirm the existence of splices. Exonuclease VII[36] is used instead of endonuclease S1 to digest the hybrids. This enzyme specifically removes single-stranded DNA tails with either 5' or 3' ends, but does not nick internal loops. Thus, exo VII treatment of our hypothetical hybrid would generate on alkaline gels a fragment 71 map units long corresponding to the distance between the actual 5' and 3' ends of the RNA. Unfortunately, we have had little success with preparations of this enzyme purified in our laboratory and will thus not discuss its use further.

Separate annealing experiments with an appropriate overlapping set of restriction fragments of the DNA are next used to map the ends of the colinear segments of the RNA. For example, with the restriction fragments indicated in our illustration (Fig. 3), annealing to fragment A would generate the DNA fragment of 21 map unit and a new 5 map unit single-stranded DNA; annealing to fragment C would yield only one S1-resistant DNA 16 map units long. These data suggest that the 21-map unit long DNA lies within fragment A with one end 16 map units before the C/D junction at map unit 19 and its other end 5 map units after the C/D junction at map unit 40. Further measurements using small restriction fragments and polyacrylamide gel electrophoresis can be used to locate the ends more precisely. If a very accurate restriction map is available (namely, the DNA sequence) S1 mapping can locate RNA ends with a precision of ±10 nucleotides.

The 5' to 3' orientation of our hypothetical RNA remains to be established. This can be done directly in the case of polyadenylated RNA simply by cleaving the RNA with mild alkali, selecting small polyadenylated RNA fragments and then assaying which end of the RNA is still present by S1 gel analysis. If the chemical polarity of the DNA strands with respect to the restriction map is known, the issue is resolved by determining to which strand the RNA anneals. Hybridization with intact

[35] A. J. Berk and P. A. Sharp, *Proc. Natl. Acad. Sci. U.S.A.* **75**, 1274 (1978).
[36] J. W. Chase and C. C. Richardson, *J. Biol. Chem.* **249**, 4553 (1974).

single strands of DNA restriction fragments[37] can be used for this purpose. This variation has an additional advantage. Certain short DNA–RNA hybrids, such as small leader sequences, may be unstable in the high formamide annealing conditions but may form efficiently under less stringent annealing conditions which are possible when separated DNA strands are used. This has been particularly useful in mapping the leader sequences on late SV40 mRNAs.[37] A second method of DNA strand assignment is possible when the DNA bands on the agarose gels are detected by the Southern transfer[9] procedure rather than by direct autoradiography of radioactive DNA. In this case, highly radioactive strand-specific probes can be hybridized to the DNA immobilized on nitrocellulose.

Mapping Late Polyoma Virus mRNAs by One-Dimensional Electrophoresis

A more concrete and somewhat more complex example of S1 gel mapping is shown in Fig. 4. The agarose gel patterns illustrated are selected from our work mapping the mRNAs transcribed from the L DNA strand of the late region (extending counterclockwise from 70 to 25 map units on the standard physical map) of polyoma virus DNA. These polyadenylated mRNAs accumulate during the late phase of productive infection. Previous work[20] had suggested that there were two late mRNAs, a larger one sedimenting at 19 S with a 5' end at map unit 67 and its 3' end at map unit 26, as well as a more abundant smaller one sedimenting at 16 S which comprised the 3' terminal portion of the 19 S species from 48 map units to 26. One-dimensional alkaline agarose gel electrophoresis of the S1-resistant hybrids formed between total late cytoplasmic polyadenylated RNA and a large fragment of viral DNA which includes the entire late region (*Bgl*I + *Eco*RI 1; 0–72.5 map units) resulted in the pattern of three bands seen in Fig. 4, tracks a–d and f. These bands correspond to DNA fragments approximately 2.30, 1.88, and 1.25 kilobases (kb) long. Various amounts of RNA were hybridized to a fixed quantity of DNA in this experiment. With the lowest amount of RNA (Fig. 4, track a), where the DNA is in excess, the intensity of the bands reflects the relative abundances of the different RNA species. The colinear RNA which produced the 1.25 kb band is clearly the most abundant; its size is that expected for the 16 S mRNA. The 2.3 kb band was generated by the least abundant mRNA; its size is that expected for the 19 S mRNA. The 1.88 kb band, which has an intermediate abundance, was unexpected. The faint band at 1.4 kb seen in tracks d and f is produced by early mRNA and will not be discussed here. The 5' and 3' ends (the polarity had already been determined) of the three colinear late transcripts were mapped by

[37] C-J. Lai, R. Dhar, and G. Khoury, *Cell* 14, 971 (1978).

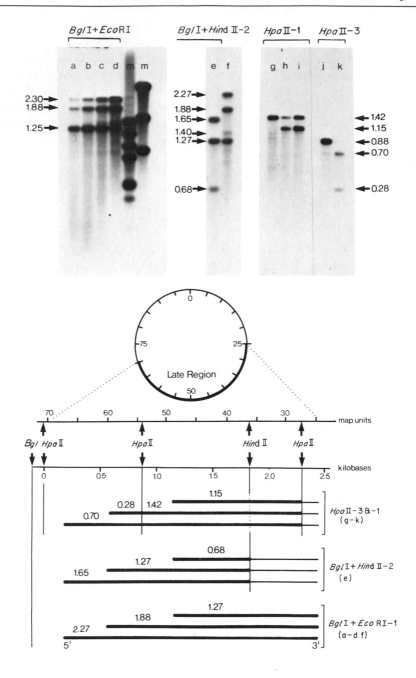

hybridization to a variety of smaller restriction fragments. For example, hybridization to a fragment extending from 36.2–72.5 map units (*BglI* + *HindII* 2; Fig. 4, track e) produced three bands 1.65, 1.27, and 0.68 kb long. As shown in the diagram (Fig. 4), these lengths can be used to determine the location of the 5' ends of the three mRNAs. The calculated positions for the 5' ends were then confirmed by hybridization with two additional fragments (*HpaII* 1, 27.2 to 54.0 map units and *HpaII* 3, 54.0 to 70.6 map units) which further divide the late region. The unique map assignments deduced from these data and from further experiments using other restriction fragments are shown schematically in Fig. 4. Polyoma virus determines three rather than two late mRNAs which have a common polyadenylated 3' end at map unit 24.5 and 5' ends at map units 48.7, 59.9, and 67.0, respectively. This conclusion has been confirmed by electron microscopy of DNA–RNA hybrids[38] and is consistent with cell-free translation studies showing that the activities encoding the three viral capsid proteins can be partially resolved on sucrose gradients as 19 S, 18 S, and 16 S components.[39,40]

We have chosen the example of late polyoma virus in RNAs to illustrate one limitation of the S1 gel mapping technique. Direct analysis of late viral RNA by examination of 5' terminal cap sequences[8] and T1 oligonucleotides,[41] as well as the electron micrographic studies,[38] showed that the

[38] R. Kamen and J. Parker, unpublished results; H. Manor, M. Wu, and N. Davidson, personal communication; M. Horowitz, S. Bratosin, and Y. Aloni, personal communication.
[39] S. G. Siddell and A. E. Smith, *J. Virol.* **27**, 427 (1978).
[40] T. Hunter and W. Gibson, *J. Virol.* **28**, 240 (1978).
[41] S. Legon, A. Flavell, A. Cowie, and R. Kamen, *Cell* **16**, 373 (1979).

FIG. 4. Mapping late polyomavirus mRNAs by one-dimensional nuclease S1 gel electrophoresis. The upper part of the figure shows autoradiographs of alkaline agarose gels fractionating S1 nuclease-resistant hybrids formed between various ^{32}P-labeled restriction fragments of viral DNA and either polyomavirus cRNA or cytoplasmic polyadenylated RNA extracted from mouse cells at a late time during productive infection by wild-type virus or the tsA mutant. The numbers indicate the sizes of DNA fragments in kilobases. Tracks a–d and f are hybrids with a restriction fragment which includes the entire late region (*BglI* + *EcoRI* − 1, 0–72.5 map units). (a) 0.058 μg WT mRNA. (b) 0.145 μg WT mRNA. (c) 0.290 μg WT mRNA. (d) 0.580 μWT mRNA. (f) 0.25 μg WT mRNA. Tracks labeled m are size markers. Track e, hybrid between 0.25 μg WT mRNA and the restriction fragment from 36.2–72.5 map units (*BglI* + *HindII* − 2). Tracks g–i are hybrids with the restriction fragment from 27.2–54.0 map units (*HpaII*-1). (g) 0.08 μg cRNA, (h) 1.0 μg tsA mRNA, (i) 1.0 μg WT mRNA. Tracks j and k are hybrids with the restriction fragment extending from 54.0–70.6 map units (*HpaII*-3). (j) 0.08 μg cRNA, (k) 1.0 μg WT mRNA. The circle shows the standard physical map of polyomavirus DNA with the late region expanded in linear form below. The diagram indicates the deduced topography of the three late mRNAs, with the lengths of the different restriction fragments which were protected from S1 nuclease digestion shown as emboldened lines.

three colinear late mRNAs are, in fact, spliced to a common family of leader sequences derived from the DNA in the vicinity of 67 map units. We initially missed these leaders in the S1 analysis because they are very heterogeneous and few form hybrids more than 40–50 nucleotides in length. The unusual structure of these leaders is discussed in Legon *et al.*[41] Subsequent S1 gel analysis using 10% polyacrylamide gels detected the expected heterogeneous distribution of small S1-resistant DNA fragments, but it is doubtful whether they would have been detected independent of the results obtained with different techniques.

Interpretation of Two-Dimensional Gels

Rationale

Two-dimensional agarose gel electrophoresis of endonuclease S1-resistant hybrids was developed to overcome two limitations encountered in the one-dimensional approach. The mapping of different internal splices within a family of similar RNA molecules by the Berk and Sharp procedure involves the detection of DNA–RNA hybrids on neutral gels and the subsequent reelectrophoresis of these hybrids on denaturing gels to determine the size of the component single-stranded DNA fragments. This is difficult with minor species which can only be detected on the neutral gels after long exposure. We further wanted to study viral transcripts in certain polyoma virus-transformed cells. The low abundance of these transcripts made their detection with *in vivo* labeled DNA (less than 10^7 cpm/μg) marginal, and eliminated the possibility of sequential electrophoresis on a second gel. On the other hand, we were able to prepare labeled viral DNA at very high specific activity (3×10^8 cpm/μg), but this DNA was too fragmented for direct use in S1 gel mapping experiments except when small restriction fragments were required. We, therefore, decided to combine the Berk and Sharp technique with the very sensitive Southern transfer procedure commonly used in gene mapping. The annealings were thus done with unlabeled DNA, and the final pattern of DNA fragments separated on agarose gels was transferred to nitrocellulose and subsequently detected by hybridization to highly radioactive probes. Since this technique did not allow sequential analysis of hybrids on a second alkaline gel, we further developed a two-dimensional technique which allowed the entire analysis to be done on one gel. The S1 nuclease-resistant DNA–RNA hybrids are loaded into a circular well in one corner of a neutral agarose gel and separated by electrophoresis. The entire gel is then equilibrated with an alkaline buffer and electrophoresis is continued under

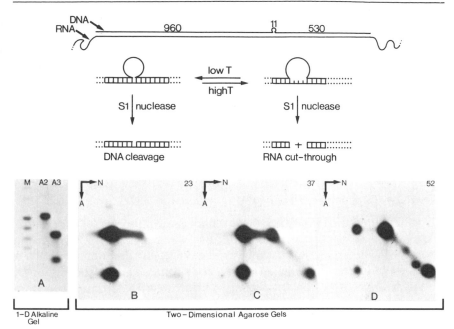

FIG. 5. Control experiment to examine the effect of nuclease S1 digestion temperature on the cleavage of the RNA strand opposite a small DNA loop in an RNA–DNA hybrid. Hybrids containing an eleven nucleotide long DNA loop were constructed by annealing cRNA transcribed from the DNA of a viable deletion mutant (polyomavirus strain A3) to an appropriate restriction fragment (Sst-2; 52.9–82 map units) of strain A2DNA; the numbers show the position of the deletion in nucleotides from the ends of the fragment. The hybrid in the vicinity of the DNA loop is expected to have two alternative structures, as shown, either with all possible base pairs formed or with a short single-stranded RNA bridge. Low temperature would favor the maximally base-paired structure. The hybrids were digested with S1 nuclease at various temperatures and were then fractionated either on a one-dimensional alkaline agarose gel or by two-dimensional electrophoresis as described in the text. The DNA in the gels was transferred to nitrocellulose and then annealed to nick-translated polyomavirus DNA. (A) One-dimensional alkaline agarose gel of hybrids digested at 37°. Track M, polyoma DNA size marker. Track A2, control hybridization of A2cRNA to A2DNA showing complete protection of the DNA fragment. Track A3, A3cRNA–A2DNA hybrids showing quantitative cleavage at the site of the DNA loop. (B) Two-dimensional fractionation of A3–A2 hybrids digested at 23°. (C) Same as B but digested at 37°. (D) Same as B but digested at 52°. N and A indicate the neutral and alkaline dimensions.

denaturing conditions in a directional perpendicular to the initial separation. After transfer to nitrocellulose, the "spot" pattern of single-stranded DNA is visualized by annealing with ^{32}P-labeled probes and subsequent autoradiography. The continuous DNA strands in hybrids between unspliced RNAs and their template DNA will have the same relative mobility in the neutral and the alkaline dimensions and thus they will fall along a diagonal line (see Fig. 3). In practice, this diagonal is visible as a faint line corresponding to the general gray background (caused by DNA or RNA nicking) always seen in one-dimensional gel tracks. By contrast, the component DNA strands of an S1-digested hybrid formed between a spliced RNA and its template DNA will migrate together in the neutral dimension (held together by the RNA strand) but will resolve in the alkaline dimension as two or more discrete "spots" vertically aligned below the diagonal (Fig. 3). A line drawn between these spots should extrapolate to a point on the diagonal corresponding to the sum of the lengths of the component DNA single strands.

Sensitivity of Splice Detection and Minimization of RNA Strand Cleavage

We illustrate the two-dimensional technique with a control experiment designed to ask whether very small loops can be efficiently excised from RNA–DNA heteroduplexes without cleavage of the opposing RNA strand. Hybrids containing an 11-nucleotide long single-stranded loop in the DNA strand were constructed by hybridizing cRNA transcribed from DNA of the A3 strain of polyoma virus to a restriction fragment of strain A2 DNA. Comparison of the DNA sequences of these two DNA molecules had demonstrated that A3 DNA has an 11-base pair deletion with respect to A2 DNA.[42] Figure 5 shows schematically the structure of the heteroduplex expected between A3 cRNA and a 1500-nucleotide fragment of A2 DNA including the sequences deleted in A3 DNA. Digestion of this hybrid under our standard conditions (excess S1 nuclease, 37°) followed by analysis on a one-dimensional alkaline agarose gel (Fig. 5A) showed complete cleavage of the A2 DNA strand at the site of the deletion in A3 cRNA to generate a 960 nucleotide and a 530 nucleotide fragment; in a control annealing with homologous A2 cRNA, the entire length of the fragment was protected from nuclease digestion. Two-dimensional gel analysis of A2 DNA–A3 cRNA hybrids digested with S1 nuclease under the same conditions resulted in the pattern shown in Fig. 5C; four dark "spots" were seen after annealing the nitrocellulose blot to nick-translated polyoma virus DNA. The two darkest spots were vertically aligned below the diagonal in the alkaline dimension and had mobilities in this dimension corresponding to the 960-nucleotide and the 530-nucleotide single-stranded DNA fragments seen on the one-dimensional alkaline gel. The location of these fragments clearly indicates that they were the collinear components of a hybrid species comprising two DNA fragments held together by a continuous RNA strand. A further pair of rather less dark spots were detected which had the same alkaline dimension mobilities as the major pair but which lay along the diagonal. These result

---

[42] E. Soeda, J. R. Arrand, N. Smolar, and B. E. Griffin, submitted for publication (1979).

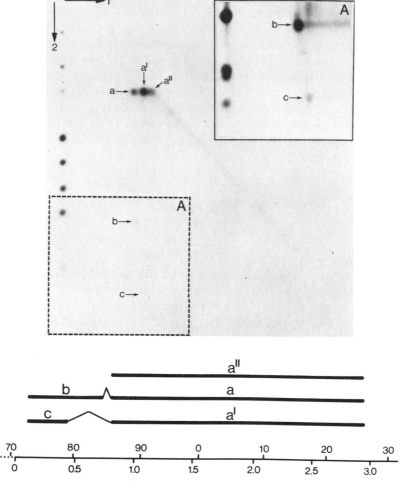

FIG. 6. Two-dimensional nuclease S1 gel mapping of early polyomavirus mRNAs. Cytoplasmic polyadenylated mRNA from cells infected with the tsA mutant was hybridized to a restriction fragment spanning the early region (map unit 72.5–26.6) of polyomavirus DNA. The S1-resistant hybrids were fractionated on a 20 × 20 cm two-dimensional gel, transferred to nitrocellulose, and annealed to nick-translated DNA as described in the text. Dimension 1, neutral; dimension 2, alkaline. The series of spots along the left edge of the gel are polyomavirus DNA size markers. The diagram shows the deduced mRNA structures with respect to a linearized map of the early region. The upper numbers are map units; the lower numbers kilobases.

from S1 nuclease cleavage of the cRNA strand opposite the 11-nucleotide single-stranded DNA loop. We reasoned that cleavage of the RNA strand occurred because a proportion of the hybrids had DNA loops longer than 11 nucleotides in which all possible base pairs were not formed, resulting in a short single-stranded RNA bridge across the bottom of the loop (Fig. 5). To investigate this possibility, we varied the temperature of S1 nuclease digestion. At 9° (not shown), the RNA strand in essentially all molecules remained intact, but the DNA loop was still cleaved so that only the pair of spots below the diagonal were seen. The pattern obtained at 23° was similar (Fig. 5B), whereas at 52° (Fig. 5D) most of the RNA strands were nicked and therefore the DNA fragments occurred predominantly as the two spots along the diagonal. We now routinely digest hybrids for two-dimensional analysis at 18°–20° to minimize RNA strand cleavage, but in certain instances (very small splices) lower temperatures (0°–9°) are necessary to minimize cut-through.

Mapping Early Polyoma Virus RNAs by Two-Dimensional Gel Electrophoresis

The "early" polyoma virus mRNAs are transcribed from the E DNA strand of the region extending clockwise from about 70 to 26 map units[5,6] (see physical map in Fig. 4). Figure 6 shows the two-dimensional analysis of S1 nuclease-resistant hybrids formed between early mRNAs and a fragment of polyoma virus DNA which includes the entire early region. The three darkest "spots" (a, a', and a'') are horizontally aligned—a and a' are below the diagonal, while spot a'' lies on it. This is the pattern produced by DNA fragments which had different mobilities in the neutral dimensional but the same mobility in the alkaline dimension. The minor spot, b, lies directly below spot a, and thus a and b were the two continuous DNA components of a hybrid molecule held together by a spliced mRNA. Similarly, spot c (best visible on the darker exposure shown in Fig. 6A) lies directly below spot a', and thus a' and c were originally present in a slightly smaller DNA–RNA hybrid. The diagram in Fig. 6 shows the RNA structures deduced from this two-dimensional gel. The precise sizes and map positions of the continuous DNA fragments are from one-dimensional alkaline agarose and urea–polyacrylamide gel experiments using a variety of different restriction fragments. We concluded that there are two major early viral mRNAs which share common 5' and 3' ends but which differ in the size of an internal splice. The spot on the diagonal, a'', is produced by S1 cleavage of the RNA strand opposite the smaller of the two splices and can be eliminated by lower temperatures of nuclease digestion.

The experiment shown in Fig. 7 illustrates further complexities in the structure of minor polyoma virus early RNAs which were revealed by examination of two-dimensional gels in greater detail. The hybrids here were formed between unit length viral DNA cleaved once within the noncoding sequence between the early and late regions (map unit 72.5; see

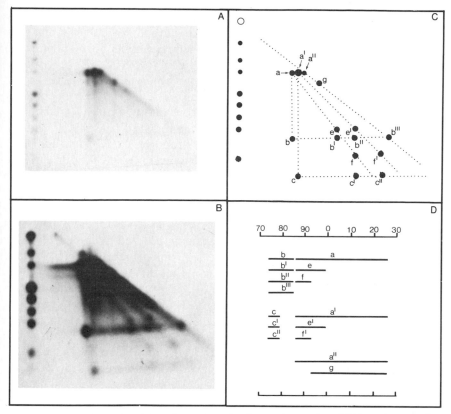

FIG. 7. Two-dimensional S1 gel mapping of polyomavirus early mRNAs, minor species. Cytoplasmic polyadenylated RNA from cells infected with the tsA mutant was hybridized to linear polyomavirus DNA cleaved with *Bgl*I at 72.5 map units. After nuclease S1 digestion at 18°, the hybrids were fractionated by two-dimensional agarose gel electrophoresis. The DNA was transferred to nitrocellulose and annealed with ^{32}P-labeled cDNA to cRNA. (A) Short autoradioagraphic exposure (14 hr, no screen). (B) Long fluorographic exposure (48 hr with intensifying screen). (C) Schematic diagram of the gel pattern. (D) Deduced structures for major and minor RNAs with respect to the linearized physical map of the early region.

Fig. 4); the Southern transfer of the gel was annealed with a strand-specific probe (cDNA to cRNA) so that only the DNA strands protected from S1 nuclease digestion by early mRNAs were detected. Figure 7A and B are short and long autoradiographic exposures of the same nitrocellulose sheet so that both major spots may be seen. Figure 7C is a schematic representation of the positions of the different lines and spots detected. The pattern initially appears to be rather complex, but the logic of the two-dimensional array allows one to deduce the RNA structures shown in Fig. 7D. To help explain this logic, we introduce some nomenclature. The vertical lines below spots we call "vertical tie-lines"; these result from nonspecific nicking of the DNA strand within a continuous DNA–RNA hybrid. Although these vertical tie-lines are methodological artifacts, they are useful interpretative aids, since they connect spots corresponding to the component single-stranded DNA fragments of DNA–RNA hybrids. For example, spots a and b, which are the colinear components of the major early mRNA with the smaller splice (see Fig. 7D) are connected by a vertical tie-line. There are in addition "horizontal tie-lines," such as the line b b' b'' b''' or c c' c''. These are also the result of nonspecific nicking, but of a different sort. If either the DNA strand or the RNA strand were nicked before hybridization, a random length DNA–RNA hybrid would result. Consider the consequence of a random nick within the "a" portion of the major early RNA which generates spots a and b. The S1 nuclease-resistant hybrid would have a random size in the neutral dimension, but in the alkaline dimension it would generate a full-length fragment b as well as a random subfragment of a. The b fragments derived from such hybrids account for the horizontal tie-line b b' b''; similarly, nonspecific nicking within the a' portion of major RNA containing the smaller splice (a' + c) accounts for the horizontal tie-line c c' c''. The random degradation products of a and a' in the same family of hybrids occur as the two "secondary diagonal" lines (a'e'f' and aef) which extend from the full-length spots a and a'. The major diagonal comprises DNA fragments originating from nicked hybrids with a continuous DNA strand. The first secondary diagonal (a'e'f') comprises DNA fragments which were attached to fragment c by an RNA strand. The other secondary diagonal (aef) comprises DNA fragments which were attached to fragment b.

Once the origin of the tie-lines and secondary diagnoals is understood, the interpretation of the minor spots in the two-dimensional array becomes rather easy. Each spot occurs on a tie-line or a secondary diagonal. Thus, for example, the spots e and e' have the same alkaline dimension mobility and each occurs on a secondary diagonal. Spot b' on its horizontal tie-line is directly below e, and spot c' on its horizontal tie-line occurs below e'. Fragment e had already been mapped on the viral genome by one-dimensional experiments, and with the additional information provided by the two-dimensional gel, the structures shown in Fig. 7D could be deduced. Using similar arguments, one can deduce that spots f anf f' correspond to DNA fragments originally present in hybrids also containing b'' and c'', respectively (Fig. 7D).

Mapping Symmetric RNA Transcripts by Two-Dimensional Gel Electrophoresis

All of the examples discussed thus far concern only asymmetric transcripts of duplex DNA. When transcripts of both strands of the DNA occur (symmetric RNA), as in the case of late polyoma virus nuclear RNA,[5, 43] the effect of competing RNA–RNA self-annealing must be considered when interpreting nuclease S1 gel electrophoresis mapping data. Under high formamide hybridization conditions, RNA–RNA annealing is most favored and occurs even in the presence of excess of amounts of homologous DNA.[44] If the RNA molecules which self-anneal are only complementary for part of their lengths, the single-stranded regions of the partial RNA–RNA duplexes remain available for hybridization to DNA. Thus, the S1 nuclease-resistant structures formed will contain regions of both RNA–RNA and RNA–DNA duplex. We refer to nuclease-resistant DNA fragments generated by hybridization to single-stranded regions of RNA–RNA partial duplexes as "shadows." The shadows can be used to map the 5' or 3' ends of the symmetric RNA species. Such experiments are simplified either by using strand-separated DNA in the initial hybridization or strand-specific probes when analyzing Southern transfers of the agarose gels. The two-dimensional system is particularly informative in this instance because the nuclease-resistant duplex structure, comprised largely of RNA, migrates far slower in the neutral dimension than its component single-stranded DNA in the alkaline dimension. A DNA spot well below the diagonal without a corresponding vertically aligned spot, the length of which would sum to the extrapolated position along the diagonal, is characteristic of a shadow fragment.

Other Applications of Two-Dimensional Nuclease S1 Gel Mapping

The two-dimensional technique can be applied in general to ask whether the transcript(s) of a purified DNA is spliced. The characteristic location of spots either on or below the major diagonal immediately yield the answer. Thus far the method has been used to study rabbit globin mRNA and its nuclear precursors by hybridizing bone marrow RNA to a plasmid DNA containing the rabbit β-globin gene.[45] Similar analyses can be done with any other cloned gene. In the case of reiterated genes, it has been possible to generate hybrids directly by hybridizing cellular RNA to genomic DNA.[46] The possibility of detecting splicing in transcripts of single-copy genes by the same procedure is under investigation.

[43] Y. Aloni and H. Locker, *Virology* **54**, 495 (1973).
[44] R. Treisman and R. Kamen, unpublished observations.
[45] R. A. Flavell, personal communication.
[46] D. M. Glover, personal communication.

[69] Transcription Maps of Adenovirus

By PHILLIP A. SHARP, ARNOLD J. BERK, and SUSAN M. BERGET

The lytic cycle of adenovirus 2 can be divided into an early and late phase. In general, the early phase precedes viral DNA replication, while the late phase consists of events following the onset of DNA synthesis. All true early mRNAs and proteins can be found in Ad2 transformed cell lines. However, only a subset of these early gene products are required to establish a transformed state: these map within the left terminal 8% of the genome (see Fig. 6 for conventional genome orientation).[1,2] With the transition from early to late phase of infection several new mRNAs appear in the cytoplasm which are transcribed from the same regions of the viral DNA as the early mRNAs but which have novel sizes and sequence content.[3] These mRNAs will be referred to as intermediate RNAs in Fig. 6. The polypeptides coded for by these mRNAs have not been established nor has a critical function been assigned to any of the viral products produced during this intermediate period of infection. During the late stage of infection most of the biosynthetic capacity of the infected cell is directed toward making virion components. There are at least nine structural components in the virion; many of which have been positioned in the virus particle relative to the 12 vertices and the DNA core.[4] The only known example of a late polypeptide which is not a virion component is the 100K polypeptide. This protein seems to be involved in assembly and transport of the hexon capsid component.[5]

Most, if not all, Ad2 mRNAs have a spliced structure, i.e., RNA sequences from two or more noncontagious regions of the genome are joined together in the mature mRNA. Figure 1 is a schematic of an RNA–DNA hybrid between a spliced mRNA and a strand of DNA. The intervening sequence c forms a loop at the splice point between a and b. When the spliced segment adjacent to the 5′ terminus of a mRNA is short, it is commonly referred to as a leader sequence. The term *leader* probably should be reserved for sequences spliced on the 5′ end of a mRNA which are not translated.

[1] P. H. Gallimore, P. A. Sharp, and J. Sambrook, *J. Mol. Biol.* **89**, 49–72 (1974).
[2] F. L. Graham, A. J. van der Eb, and H. L. Heijneker, *Nature (London)* **251**, 687–691 (1974).
[3] D. J. Spector, M. McGrogan, and H. J. Raskas, *J. Mol. Biol.* **126**, 395–414 (1978).
[4] L. Philipson, U. Pettersson, and U. Lindberg, (1975). "Molecular Biology of Adenovirus." Springer-Verlag, Berlin and New York.
[5] H. S. Ginsberg, M. J. Ensinger, R. S. Kauffman, A. J. Mayer, and U. Lundholm, *Cold Spring Harbor Symp. Quant. Biol.* **39**, 419–426 (1974).

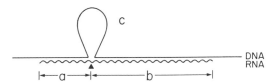

FIG. 1. Diagram of a hybrid between a spliced RNA molecule and the coding strand of genome DNA. The spliced RNA (wavy line) is composed of two colinear transcripts (a and b) joined at a splice point (▲). An intervening sequence (c) occurs in the DNA between the sequences complementary to a and b.

Two methods have been recently developed to map spliced Ad2 mRNAs: the nuclease gel electrophoresis procedure and electron microscopy of RNA–DNA hybrids. The strengths and limitations of these two techniques will be briefly reviewed. A summary transcription map of Ad2 is presented as the conclusion.

Nuclease Procedures for RNA Mapping

Radioisotope Labeled Viral DNA and Unlabeled RNA

The nuclease gel electrophoresis method is based on digestion of RNA–DNA hybrids similar to that diagrammed in Fig. 1 with a single-strand-specific nuclease. The length of the nuclease-resistant DNA segments is determined by electrophoresis in either neutral or denaturing gels.[6-8] Typically the viral DNA is ^{32}P-labeled and gel electrophoresis bands of protected labeled DNA are detected by autoradiography. Two single-stranded-specific nucleases have been used: S1 endonuclease and exonuclease VII. S1 endonuclease will degrade all single-stranded DNA portions of the RNA–DNA hybrid yielding colinear segments of viral DNA.[9,10] If the S1 digestion products are resolved in a neutral gel, RNA–DNA hybrids the length of a, b, and a + b are usually observed. The a + b length hybrid occurs because S1 endonuclease cleavage of the ribophosphodiester bond opposite a nick is inefficient. If the S1 digestion products are displayed by alkaline gel electrophoresis,[11] single-stranded DNA segments the length of a and b are resolved. mRNAs are mapped by

[6] A. J. Berk and P. A. Sharp, *Cell* **12**, 721–732 (1977).
[7] A. J. Berk and P. A. Sharp, *Proc. Natl. Acad. Sci. U.S.A.* **75**, 1274–1278 (1978).
[8] A. J. Berk and P. A. Sharp, *Cell* **14**, 695–711 (1978).
[9] T. Ando, *Biochim. Biophys. Acta* **114**, 158–168 (1966).
[10] V. M. Vogt, *Eur. J. Biochem.* **33**, 192–200 (1973).
[11] M. W. McDonell, M. N. Simon, and F. W. Studier, (1977). *J. Mol. Biol.* **110**, 119–146 (1977).

using fragments generated by restriction endonucleases that have cleavage sites within a colinear segment.

Exonuclease VII[12,13] will degrade single-stranded DNA from either 3' or 5' termini. Digestion of an RNA–DNA hybrid such as that shown in Fig. 1 with exonuclease VII yields a DNA strand with length $a + b + c$; the total length along the genome spanned by the sequences encoding the spliced RNA. Exonuclease VII digestion products are generally resolved on an alkaline agarose gel. Again the segment of DNA protected by mRNA from nuclease degradation can be mapped by using restriction endonuclease fragments that have termini within the colinear portions of the hybrid.

Radioisotope-labeled viral DNA can either be prepared by extraction of virus from cells cultured in the presence of the radioisotope or by using nick translation to incorporate radioactive nucleotides into viral DNA.[14,15] Viral DNA labeled with ^{32}P *in vivo* to specific activities of 10^6 to 5×10^6 cpm/μg can be prepared at reasonable yields using 20 mCi of ^{32}P-phosphate per liter of suspension cell HeLa culture.[16] The critical step in preparing this DNA is to extensively wash infected cells with phosphate-free media before suspending the cells in this media and 2% horse serum for labeling. Viral DNA prepared from such a culture 48 hr postinfection will have 25–50% intact strands using conventional virus purification procedures. It should be noted that the rate of damage to DNA in an aqueous sample is chiefly a function of the total level of radiation per volume of solution and not the specific activity of the DNA. Thus, storing the DNA at dilute concentrations (≤ 5 μg/ml) is advised.

Radioisotope labeling by nick translation must be critically controlled to avoid extensive nicking of the product.[14,15] Typically, nick translation mixtures which result in a 30% substitution of labeled nucleotides give a final product approximately 300 nucleotides in length. Obviously single-stranded DNA this short cannot be used in the nuclease gel technique to map mRNAs of 500–2000 nucleotides. By reducing the concentration of endonucleases added to the nick translation mixture, products with single-strand lengths of 2000–5000 nucleotides can be prepared, which have specific activities of $1–5 \times 10^7$ cpm/μg.

Restriction endonuclease cleavage fragments of labeled viral DNA are purified by electrophoresis of digested viral DNA in agarose gels. Specific

[12] J. W. Chase and C. C. Richardson, *J. Biol. Chem.* **249**, 4545–4552 (1974).
[13] J. W. Chase and C. C. Richardson, *J. Biol. Chem.* **249**, 4553–4561 (1974).
[14] P. Rigby, D. Rhodes, M. Dieckmann, and P. Berg, *J. Mol. Biol.* **113**, 237–251 (1977).
[15] T. Maniatis, A. Jeffrey, and D. G. Kleid, (1975). *Proc. Natl. Acad. Sci. U.S.A.* **72**, 1484–1489 (1975).
[16] S. J. Flint, P. H. Gallimore, and P. A. Sharp, *J. Mol. Biol.* **96**, 47–68 (1975).

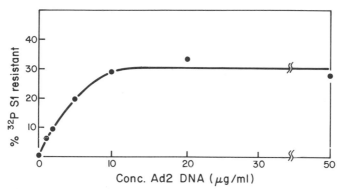

FIG. 2. Kinetics of Ad2 DNA/RNA hybridization in 80% formamide. Small amounts of ^{32}P-labeled poly(A)(+) cytoplasmic RNA was prepared from cells 30 hr postinfection. About 0.02 μg of this RNA (2.5 × 10⁴ cpm) was dissolved in a standard hybridization mixture as described in the text and incubated for 3 hr at 60°. The percentage of S1-resistant ^{32}P-labeled RNA was determined. An Ad2 concentration of 10 μg/ml is saturating under these conditions.

bands are detected by staining with 0.5 μg/ml of ethidium bromide and visualized with long wavelength uv.[17] A fluorescence 16 W long wavelength uv lamp must be used because short wavelength uv induces pyrimidine dimers which create substrates for S1 in duplex nucleic acid.

RNA–DNA hybrids are formed as follows. A mixture of a labeled viral DNA fragment and cell RNA are ethanol precipitated in a small plastic vial. The precipitate is dissolved in hybridization buffer, 80% formamide, 0.4 M NaCl, 0.05 M PIPES, pH 6.4, 0.001 M EDTA.[18] After dissolving the precipitate, the hybridization mixture is incubated at 68° for 10 min to denature the Ad2 DNA. Hybridization is carried out at 60° for Ad2 DNA which is 58% (G + C); the melting temperature (T_{m_i}) of viral DNA under these conditions is 56°. Casey and Davidson[18] have studied the effect of formamide concentration and temperature on the rate of RNA–DNA hybridization and DNA–DNA renaturation kinetics. In 80% formamide there is a 15°–20° difference between the melting temperature of 58% (G + C) DNA and the corresponding RNA–DNA hybrid. Thus Ad2 DNA–RNA hybridizations are done at temperatures above the melting temperature of duplex DNA, ensuring that no nuclease resistant duplex DNA is formed. The kinetics of RNA–DNA hybridization under these conditions is a factor of 12 slower than the kinetics of hybridization at optimal temperature in aqueous solution with equivalent salt concentrations.[18] As shown in Fig. 2, a total Ad2 DNA concentration of 10 μg/ml is required to drive a trace amount of labeled viral RNA into hybrid in a 3-hr

[17] P. A. Sharp, W. Sugden, and J. Sambrook, *Biochemistry* **12**, 3055–3063 (1973).
[18] J. Casey and N. Davidson, *Nucleic Acids Res.* **4**, 1539–1552 (1977).

incubation. In most cases, hybridization is done in great DNA excess, and thus the rate of hybrid formation is proportional to the concentration of DNA; thus an Ad2 concentration of 20 μg/ml would be saturating in a 1.5-hr incubation. RNA chains of 3000 bases can be incubated at 60° in the high formamide hybridization mix for up to 3 hr without noticeable strand scission. When carrying out hybridizations with high molecular weight RNA chains of 10,000 to 20,000 nucleotides, incubations have been shortened to 0.5 hr.

In 80% formamide solutions the window between the temperature of melting of RNA–DNA hybrid and the corresponding DNA–DNA duplex decreases with the percent (G + C) of the duplex segment.[18] The optimal hybridization temperature for 40% (G + C) DNA is 48°, in this case the temperature of melting duplex DNA and corresponding RNA–DNA hybrid are 46° and 52°, respectively. In general, the temperature of hybridization should be adjusted to be slightly higher than the melting temperature of duplex DNA.

After hybridization of ^{32}P-labeled DNA and cell RNA, the hybridization mixture is diluted tenfold into nuclease reaction mix at 0°. For S1 digestion, the final buffer is 0.29 M NaCl, 0.03 M NaCH$_3$CO$_2$ (pH 4.6), 0.001 M ZnSO$_4$. S1 nuclease is added immediately after dilution of the hybridization mix. The final reaction mix contains approximately 20 μg/ml of nucleic acid to which 5 units of nuclease activity are added per 0.1 ml volume. Sufficient S1 nuclease activity is added to digest 20 μg/ml of denatured ^{32}P-labeled DNA to over 95% TCA solubility under these conditions. The concentration of S1 added is not critical as a tenfold increase does not effect either the mobility or quantity of hybrid found after gel electrophoresis. After S1 digestion, the sample is ethanol precipitated and loaded onto a neutral[17] or alkaline[11] agarose gel in 0.01 M Tris, 0.001 M EDTA, pH 7.5. Neutral gels are directly dried before exposure to film for autoradiography, while alkaline gels are first neutralized by washing in 0.05 M Tris, pH 7.5, and then handled in a similar fashion. The use of enhancement screens during autoradiography at −70° permits detection of 10–20 cpm of DNA in a band with an overnight exposure.

The chief advantage of the nuclease mapping method is the rapidity and precision with which transcripts can be mapped. The precision of this procedure for defining the termini of the colinear segments of an RNA is limited by the specificity of the S1 nuclease.

Wiegand et al.[19] estimated that purified S1 nuclease in 0.1 M NaCl had 75,000-fold more activity on single-stranded as compared to double-stranded DNA. These investigators also showed that S1 nuclease would cleave a deoxyphosphodiester bond opposite an interruption of one such

[19] R. C. Wiegand, G. N. Godson, and C. M. Radding, J. Biol. Chem. 250, 8848–8855 (1975).

bond in an otherwise duplex DNA. This activity at nicks was also suggested by earlier work where it was found that the product of digestion of superhelical DNA with S1 was primarily a linear structure and not a nicked circle.[20] Ghangas and Wu[21] have shown that S1 digestion will remove a single unpaired nucleotide from the end of a duplex segment of DNA without appreciable nibbling off of paired bases. Also Shenk et al.[22] have used digestion with S1 to detect single base mismatches in heteroduplexes. These experiments suggest that purified S1 nuclease has little or no activity on duplex DNA, and therefore it should be possible to define the ends of colinear sequences of mRNA to within one or two nucleotides by appropriate digestion with this nuclease.

The activity of S1 nuclease on RNA and RNA–DNA hybrids has not been as well characterized. Single-stranded RNA is degraded at about one-seventh the rate of single-stranded DNA.[10] Again the nuclease shows little activity on a substrate of duplex RNA or RNA–DNA hybrid. Perhaps a major distinction between S1 nuclease activity on RNA is its inability to cleave completely a ribophosphodiester bond opposite a nick in the DNA strand of a RNA–DNA hybrid.[20] The inability of S1 to completely cleave this RNA bond can be exploited by resolving the RNA–DNA hybrid on a neutral gel, and after recovery of a band from the neutral gel, the length of its colinear segments can be determined by electrophoresis on alkaline gels.[7]

The most frequent problem in mapping RNAs by the nuclease gel procedure is the renaturation of labeled viral DNA. When detecting small amounts of RNA, less than 1% of the total DNA added to the hybridization mixture enters hybrid. The remaining 99% must be solubilized during nuclease digestion. Slight amounts of renaturation typically generate a band of intact fragment DNA and a diffuse background of smaller fragments. However, in some cases, when high DNA concentrations are used, artifactual distinct bands (smaller than the intact fragment) of viral DNA are observed (A. J. Berk and P. A. Sharp, personal communication). These are probably kinetic intermediates in the process of DNA renaturation. In both cases, a control hybridization mixture lacking RNA will display these bands after nuclease digestion. If DNA renaturation does occur during the process of nuclease digestion, artifactual bands of duplex DNA specific for the presence of mRNA can be generated. These bands arise from digestion of R loop type structure (see below). To prevent DNA renaturation the hybridization mixture must be rapidly quenched with cold S1 buffer and the nuclease immediately added.

[20] P. Beard, J. F. Morrow, and P. Berg, J. Virol. 12, 1303–1313 (1973).
[21] G. S. Ghangas and R. Wu, J. Biol. Chem. 250, 4601–4606 (1975).
[22] T. E. Shenk, C. Rhodes, P. W. J. Rigby, and P. Berg, Proc. Natl. Acad. Sci. U.S.A. 72, 989–993 (1975).

For mapping with exo VII, hybridization is carried out in 80% form-amide and at high temperature as described previously. However, for exo VII digestion, the hybridization mix is diluted with a tenfold volume of 0.03 M KCl, 0.01 M Tris (pH 7.4), 0.01 M NaEDTA (pH 7.4). Exo VII digestion is done at 45° for 60 min. To a standard reaction mixture of 0.33 ml containing 0.1 μg of Ad2 DNA and 500–1000 μg of RNA, sufficient exo VII activity is added to digest twice this quantity of denatured DNA to 95% TCA solubility in 60 min at 45°. The reaction mix is terminated by the addition of NaCl to 0.1 M and 2 volumes of ethanol.

Chase and Richardson[12] identified and purified exo VII from an exonuclease I and DNA polymerase I negative mutant of *E. coli*, HMS137. Exo VII, which is active in the presence of EDTA, degrades single-stranded DNA and has little or no activity on RNA. The presence or absence of a 5′-phosphoryl terminus has no effect on the activity nor is the nuclease inhibited by the presence of duplex DNA or total cell RNA in the reaction mix. The oligonucleotides generated by exo VII range be-tween 2 and 25 residues long and have 5′-phosphoryl and 3′-hydroxyl termini.[13] Ghangas and Wu[21] have tested the precision of exo VII activity in removing single-stranded tails from otherwise perfect duplex DNA. Extensive exo VII digestion left unpaired tails to 2–3 nucleotides in length. Thus exo VII digestion is less precise than S1 nuclease in defining the terminus of a duplex segment. However, unless DNA segments of less than 25–50 base pairs are being analyzed, the effect of a length heteroge-neity of a single nucleotide would not be observed in most gel systems.

DNA renaturation is a greater problem during exo VII digestion than S1 digestion as processional exo VII digestion leaves long regions of single-stranded DNA which can nucleate renaturation during incubation. The second concern for generating artifactual bands by exo VII digestion is the possible presence of short inverted repeats or pallindromic se-quences which may form intramolecular duplex hairpins and block degra-dation. In Ad2 DNA, the two components of an inverted repeated se-quence of less than 50 base pairs in length are known to map at 72 and 74 on the genome. Single-stranded DNA from this region is often seen in a loop structure in the electron microscope.[23] Not surprisingly, exo VII digestion of single-stranded Ad2 DNA is blocked by intramolecular rena-turation at this position. A control to detect these barriers to exo VII digestion, is to renature ^{32}P-labeled single-stranded DNA from a restric-tion endonuclease cleavage fragment with an excess of unlabeled DNA from a second fragment which has a fraction of its sequences complemen-tary to the labeled DNA.[7,8] The renatured mixture is digested with exo VII, and the resistant material resolved on an alkaline gel. If the labeled

[23] M. Wu, R. J. Roberts, and N. Davidson, *J. Virol.* **21,** 766–777 (1977).

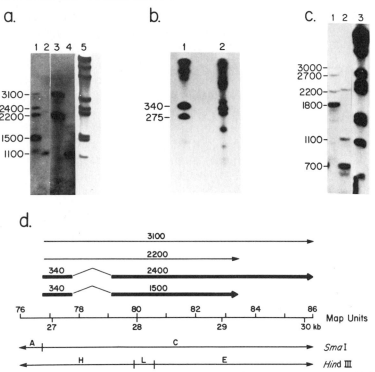

FIG. 3. Autoradiograms of gels of S1- and exo VII-digested early RNA–DNA hybrids from early region 3, and the deduced structures of the RNAs. (a) Alkaline agarose gel of the S1 products of early RNA hybridized to *Sma*I C (track 1) and *Hin*dIII H (track 2); the exo VII products of early RNA hybridized to *Sma*I C (track 3) and *Hin*dIII H (track 4). Track 5 is a marker *Sma*I digest of Ad2. There is a faint band in track 1 migrating at ~1200 nucleotides which we have not interpreted because a corresponding band has not been observed in hybridizations to other restriction fragments. (b) Urea gradient (8 *M*) polyacrylamide gel of the S1 products of early RNA hybridized to *Hin*dIII H (track 1) and a marker *Hae*III digest of SV40 DNA (track 2). (c) Neutral agarose gel of the S1 products of early RNA hybridized to *Sma*I C (track 1) and *Hin*dIII H (track 2), and of a marker *Sma*IA digest of Ad2 DNA. The faint band migrating at 2.21 kb in track 2 is due to a small fraction of renatured *Hin*dIII H DNA. (d) Diagram of the deduced structure of the region 3 cytoplasmic RNAs. The relevant restriction maps are shown below the genome map. Colinear transcripts comprising the cytoplasmic RNAs are presented by lines above genome sequences included in the RNA. Bold lines represent abundant cytoplasmic RNAs, and thin lines represent less abundant cytoplasmic RNAs. Caret symbols connecting two colinear transcripts indicate that they are joined at a splice point into one covalently continuous RNA chain, and arrowheads indicate the 3' ends of the RNAs. The lengths in nucleotides of the colinear transcripts are indicated by numbers above the bold and narrow lines. From Berk and Sharp.[8] Copyright © MIT, published by the MIT Press.

single-stranded DNA is totally susceptible to exo VII digestion, a band corresponding to the common set of sequences among the two fragments should be observed. This one region from 72 to 74 is the only segment of

Ad2 DNA where we have found a secondary structure that blocks exo VII digestion.[8] Likewise all of the DNA sequences in SV40 are susceptible to exo VII digestion.[7]

An example of an analysis of RNAs from the early region 3 of Ad2 is shown in Fig. 3. The figure's legend contains the details as to what uniformly ^{32}P-labeled restriction endonuclease cleavage fragments were used as probes in an experiment. The structure of the early RNAs deduced from this and other data are schematically presented below the gels. Data used to map colinear segments of RNA from this region are not presented here. An example of experiments mapping colinear segments can be seen in Fig. 3a, track 1, and Fig. 3b, track 1. After resolution of S1 digestion products in an alkaline agarose gel with the *Sma*I C fragment as a probe, a prominent 1500-nucleotide long band was seen as well as faint 2200, 2400, and 3100 bands (Fig. 3a, track 1). (A 340-nucleotide long band observed in other experiments with *Sma*I C has migrated off the end of this gel.) When a probe of the *Hin*dIII H fragment was used in similar experiments, prominent bands of 275 and 340 were observed (Fig. 3b, track 1). As the right terminus of the *Hin*dIII H fragment maps at 79.9, the 275 nucleotide band positions the 5' end of the 1500 colinear RNA at 79.1. Many other fragments from this region were also used as probes to map these RNAs. A direct experiment demonstrating that the 340 nucleotide RNA was joined to the 1500 segment was performed by first resolving the S1 digestion products on a neutral agarose gel. A prominent 1800 nucleotide band (Fig. 3c, track 1) was observed which was shown to be composed of the 1500 + 340 segment by subsequential electrophoresis of the ^{32}P-labeled DNA from this band in an alkaline agarose gel.

Electron Microscope Mapping of Adenovirus mRNAs

Mapping of mRNAs by R Loop Formation

Few mRNA mapping techniques can yield as much information as rapidly as electron microscopy. The power of the electron microscope in characterization of nucleic acids has been recognized for several years.[24] With the recognition that RNA–DNA hybrids were more stable in high concentrations of formamide than the homologous set of DNA strands,[25] there quickly developed procedures for visualizing RNA–DNA hybrid in the midst of a length of duplex DNA.[26] Such a structure, where a complementary RNA segment has annealed to one strand of a longer duplex

[24] R. W. Davis, M. Simon, and N. Davidson, this series, Vol. 21, Part D, pp. 413–428.

[25] R. L. White and D. S. Hogness, *Cell* 10, 177–192 (1977).

[26] M. Thomas, R. L. White, and R. W. Davis, *Proc. Natl. Acad. Sci. U.S.A.* 73, 2294–2298 (1976).

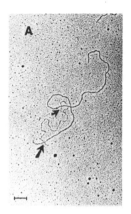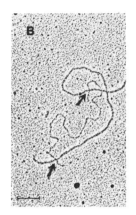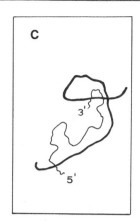

FIG. 4. Electron micrographs of R loop polypeptide II (hexon) mRNA and *Hin*dIIIA fragment. (A) Micrograph of a R loop situated on the total *Hin*dIIIA fragment. (B) Enlargement of just the hybrid part of this structure. (C) Schematic of the hybrid portion. Note the 5' and 3' tails extending from the R loop structure; these are denoted by arrows in (A) and (B). Scale, 0.1 μm.

DNA segment displacing a strand of DNA during the formation of the RNA–DNA hybrid is referred to as an R loop. An example of an R loop formed by the Ad2 mRNA for polypeptide II and the restriction endonuclease cleavage fragment *Hin*dIII A is shown in Fig. 4.[27] The *Hin*dIII A fragment spans from 50.1 to 73.6 units on the genome, and the ends of the R loop map the RNA–DNA portion of this hybrid at 51.7 and 61.3 units on the genome. To obtain significant map coordinates for the hybrid termini, data are collected from measurements of a large number of such structures and are treated statistically as described by Davis *et al.*[24]

The short single-stranded RNA segment at the 5' end of the hexon mRNA R loop shown in Fig. 4 is due to the presence of an RNA sequence which is not complementary to the adjacent DNA sequence (see below). However, even when R loops are formed with RNA which is completely colinear with the DNA sequence, segments of single-stranded RNA are frequently observed at the termini of the loops. In these cases, the RNA sequences are displaced by branch migration, the formation of DNA–DNA base pairs at the expense of RNA–DNA base pairs.[28] During the process of mounting the nucleic acid mixture for electron microscopy, the formamide concentration and temperature are decreased compared to conditions used for R loop formation. Under these conditions, the DNA duplex is almost as stable as the RNA–DNA hybrid, and fluctuations in the breaking and forming of base pairs at the end of the hybrid segment

[27] S. M. Berget, C. Moore, and P. A. Sharp, *Proc. Natl. Acad. Sci. U.S.A.* **74,** 3171–3175 (1977).

[28] C. S. Lee, R. W. Davis, and N. Davidson, *J. Mol. Biol.* **48,** 1–22 (1970).

lead to displacement of the RNA sequences. The rate of such branch migration and conditions which increase the stability of the RNA–DNA hybrid have been studied by Thomas et al.[26]

The kinetics of formation of R loops have also been studied by Thomas et al.[26] The general mechanism for R loop formation appears to be (a) generating of a partially denatured DNA segment in the region homologous to RNA and (b) hybridization of complementary RNA to single-stranded DNA in the partially denatured segment. This hybridization step follows the kinetics and concentration dependence expected for RNA–DNA hybridization in solution.[29] Under most reaction conditions, the rate-limiting step in the formation of R loops is the generation of partially denatured regions of DNA which are available for hybridization.[26] At a given temperature and in the presence of formamide, the probability of a given region being so denatured is dependent on its $(G + C)$ content. Thus it is not possible to use the observed frequency of R loop formation in two regions of a genome as a relative measure of the concentration of complementary RNA sequences unless an independent control is done to show that equal concentrations of the RNAs will give rise to equal numbers of R loops.

The difficulty of forming RNA–DNA hybrids in the $(G + C)$-rich portion of a DNA can be partially alleviated by performing the hybridization with denatured DNA in formamide at high temperatures (see above and Casey and Davidson[18]). After formation of the RNA–DNA hybrids, the temperature can be lowered to permit renaturation of the two DNA strands and the resulting molecules can be subsequently mounted for visualization of R loops. However, even with this protocol, the kinetics of formation of R loops in different regions of the genome which vary in fraction $(G + C)$ may be different.

Both early and late adenovirus 2 mRNA populations have been mapped by formation of R loops. An example of each type of experiment will be briefly discussed in terms of RNA and DNA concentrations and hybridization conditions.

Meyer et al.[30] formed R loops with late Ad2 mRNA and DNA in a buffer of 70% (v/v) formamide, 0.5 M NaCl, 0.01 M EDTA, 0.1 M N-[Tris(hydroxymethyl)methyl]glycine adjusted to pH 8.0 with NaOH. In this buffer, the T_m of Ad2 DNA was 53.4° as determined by hyperchromicity. Poly(A)(+) late mRNA isolated from cells 24 hr postinfection and Ad2 DNA were mixed at concentrations of 100 and 10 μg/ml, respectively, and incubated at 52°. The hybridizations were done in sealed micropipettes in approximately 10 μl volumes. Under these conditions

[29] J. G. Wetmur and N. Davidson, J. Mol. Biol. **31**, 349–370 (1970).
[30] J. Meyer, P. D. Neuwald, S. P. Lai, J. V. Maizel, Jr., and H. Westphal, J. Virol. **21**, 1010–1018 (1977).

a molecule of viral DNA contained an average of two and four R loops after incubation of 1 and 10 hr, respectively. About 4% of this RNA preparation should be viral RNA, and thus the concentration of complementary RNA was about 4 μg/ml. These kinetics agree reasonably well with the more defined kinetic experiments of Thomas et al.[26] using ribosomal RNA of Saccharomyces cerevisiae and a clone of complementary cellular sequences. A comparison of the histograms of Meyer et al.,[30] showing the map positions of observed R loops as a function of incubation time and other data on the relative abundance of mRNA from different regions of the genome,[31] suggest that the efficiency of R loop formation varies by as much as 50-fold between various regions of the viral DNA.

Chow et al.[32] have combined an extensive R loop mapping of late Ad2 mRNAs with the earlier data on the map position of sequences which code for particular polypeptides to assign polypeptides to abundant mRNAs. Map positions for the 5' and 3' ends of the bodies of mRNAs for Ad2 polypeptides II (hexon), III (penton), IV (fiber), V (major core), pVI (precursor hexon associated), pVII (precursor major core), pVIII (precursor hexon associated), IX (hexon associated), IV_{a2} (virion component), and 100K (nonvirion) were proposed. The coordinates of the ends of the R loops were measured relative to either an end of the viral DNA or to the ends of restriction endonuclease cleavage fragments. A sufficient number of examples of R loops for the most abundant RNAs were observed to allow the assignment of quite precise confidence limits for the coordinates of the intact mRNAs.

Early adenovirus 2 mRNAs are present in infected cells at one-tenth or less the abundance of late mRNAs. There are about 250–500 copies/cell of most early mRNAs by 8 hr postinfection.[31] When cytoplasmic poly(A)(+) RNA (100 μg/ml) isolated from cells 8 hr postinfection was incubated with Ad2 DNA (10 μg/ml) for 16 hr at 52°, over 90% of the viral DNA molecules contained R loops in the early region adjacent to the right termini.[33] The lowest frequency of R loops, 10%, was observed in early region 2. Thus it would be possible to map the low abundance early Ad2 mRNAs by R loop formation. However, these studies did not establish coordinates for individual early mRNAs.

Electron Microscopic Mapping of Spliced mRNA Molecules

In 1962, the electron microscope was used to map the position and size of deletions in heteroduplexes of viral DNAs.[34,35] A deletion loop, created

[31] S. J. Flint and P. A. Sharp, J. Mol. Biol. 106, 749–771 (1976).
[32] L. T. Chow, J. M. Roberts, J. B. Lewis, and T. R. Broker, Cell 11, 819–836 (1977).
[33] P. D. Neuwald, J. Meyer, J. V. Maizel, Jr., and H. Westphal, J. Virol. 21, 1019–1030 (1977).
[34] R. W. Davis and N. Davidson, Proc. Natl. Acad. Sci. U.S.A. 60, 243 (1968).
[35] B. C. Westmoreland, W. Szybalski, and H. Ris, Science 163, 1343 (1969).

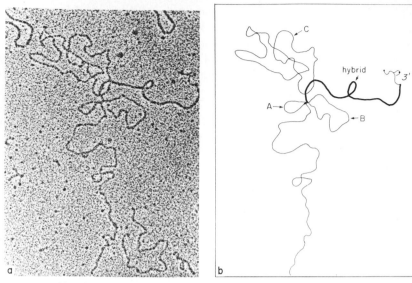

FIG. 5. (a) Electron micrograph of a hybrid of polypeptide II (hexon) mRNA and Ad2 EcoRIA DNA. Purified hexon (II) mRNA was hybridized to single-stranded EcoRIA DNA, and the resultant hybrids were spread for visualization in the electron microscope as described.[27] In (b) duplex regions are indicated by heavy lines and single-stranded DNA by light lines. Loops A, B, and C are regions of EcoRIA DNA separating those sequences which hybridize to hexon (II) mRNA. From S. M. Berget and P. A. Sharp, Brookhaven Symp. Biol. **29**, 332–344 (1977).

when one strand of a heteroduplex contains a segment of DNA which the second strand does not contain, is easily visualized and mapped by electron microscopy. When spliced mRNAs are hybridized to DNA, the RNA–DNA hybrid contains deletion type loops of DNA at the spliced points (see Fig. 1). An electron micrograph of such a hybrid is shown in Fig. 5. Here a purified mRNA fraction which codes polypeptide (II) (hexon) of Ad2 has been hybridized to a single strand of the EcoRI A fragment.[27] From R loop mapping results, it was known that the 3' end of this mRNA extended beyond the right terminus of the EcoRI A fragment. The partially extended strand of RNA can be seen protruding from the RNA–DNA duplex. At the other end of the RNA–DNA hybrid are three deletion type loops of single-stranded DNA. The lengths of these loops show that RNA segments from positions 16.6, 19.3, and 26 have been spliced to the body of RNA mapping at 51.7 on the conventional genome map. Measurements of the distance between loops A and B, and B and C suggest these colinear segments are of lengths $80 + 20N$ and $110 + 10N$, respectively.[27] Thus electron microscopy of such RNA–DNA hybrids not only maps the body of a mRNA but also positions all the leader segments present in the mRNA which are complementary to the single-stranded probe. Obviously, splices which remove very short intervening sequences (less than 100 bases) and thus generate small loops in the RNA–DNA hybrid could go undetected in these hybrids. In addition, short leader segments less than 20 bases in length might not form stable RNA–DNA hybrids under the mounting conditions for electron microscopy and would not be scored.

The map position of the 5' proximal leader of a given mRNA defines the minimal span of DNA which constitutes a transcriptional unit for that mRNA. This follows immediately if most spliced mRNAs are processed by intramolecular splicing from a larger precursor.[27,36]

Hybrids of single-stranded viral DNA and mRNA can be formed in two ways: first, separated strands of viral DNA can be hybridized in excess to enriched mRNA fractions; second, total Ad2 DNA can be denatured and incubated in formamide at high temperatures where DNA renaturation does not occur but RNA–DNA hybridization does occur. For the latter method, the conditions of incubation, DNA concentration, and time of incubation can be deduced from the previous discussion on nuclease mapping. Because each DNA molecule contains only one set of sequences complementary to leader segments, the molar concentration of viral DNA strands should be greater than that of total mRNA chains in the reaction mix. Ideally each single strand of DNA should have approximately one mRNA hybridized to it. To ensure complete hybridization of leader segments, the incubation mixture can be diluted to 0.5 μg/ml of viral DNA in 40–50% formamide and further incubated at room temperature. This solution can then be used directly for mounting for electron microscopy as described by Davis et al.[24]

The chief advantage of using hybridization in high formamide to create RNA–DNA hybrids is that the necessity of separating the strands of Ad2 DNA is eliminated. Intact strands of Ad2 DNA can be separated by sedimenting denatured DNA to equilibrium in a CsCl gradient containing the polyribohomopolymer (UG).[37,38] Typically the yield of separated strands is not good, nor do all preparations of the polyribohomopolymer give equivalent results. However, hybridizations with separated strands of viral DNA can be performed at lower temperatures and for shorter periods of incubation. Under these less stringent conditions, there should be less variation in the efficiency of hybridization to mRNAs from different regions of the genome.

[36] D. F. Klessig, Cell 12, 9–21 (1977).
[37] M. Landgraf-Leurs and M. Green, J. Mol. Biol. 60, 185 (1971).
[38] C. Tibbetts, U. Pettersson, K. Johansson, and L. Philipson, J. Virol. 13, 370 (1974).

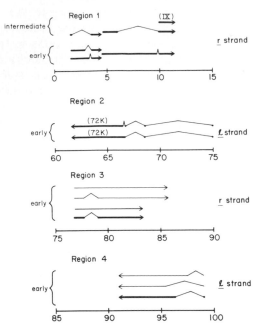

FIG. 6. Transcription map of early Ad2 mRNAs. The map and symbols are discussed in the text. Lines with arrows pointing to the right and left indicate transcripts from the *r* and *l* strands, respectively, of the genome. In region 1, the proposed structure of mRNA synthesized during an intermediate stage of infection are shown. There may also be different RNAs expressed from the other three early regions during the intermediate stage of infection.

Ad2 Early Transcription Map

Figure 6 is a transcription map of early Ad2 mRNAs.[8,39,40] The four regions of the genome complementary to early mRNAs have been designated regions 1 → 4 in a left to right order on the conventional map. The coordinates of each region are given in 1% units; the total Ad2 genome is 35,000 base pairs in length, and thus each unit represents 350 base pairs. Three early mRNA species hybridize to region 1. The two smaller RNAs have common 5' and 3' termini which map at 1.5 and 4.4, respectively, but differ by having two different-size internal segments excised by RNA splicing. In Fig. 6 the spliced segments are connected by caret symbols. The third early mRNA from region 1 spans from 4.5 to 11.2. This mRNA has a short segment of intervening sequences near map position 10 removed by RNA splicing. These three mRNAs can also be found in Ad2-

[39] G. R. Kitchingman, S. Lai, and H. Westphal, *Proc. Natl. Acad. Sci. U.S.A.* **74**, 4392–4395 (1977).

[40] S. J. Flint, *Cell* **10**, 153–166 (1977).

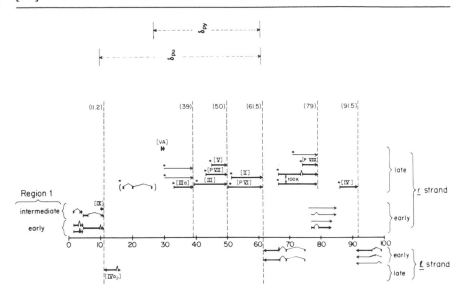

FIG. 7. Transcription map of the total Ad2 genome. See text for discussion of the map.

transformed cell lines. Graham et al.[2] have reported that an 8% fragment of viral DNA contains sufficient information for cell transformation. Only the two small early mRNAs mapping between 1.5 and 4.4 are completely encompassed within this region.

After 8 hr postinfection three other mRNAs appear from region I. These have been designated intermediate mRNAs in Figs. 6 and 7.[3] Data are not available to accurately map these three RNAs; the map positions given in Fig. 6 represent an interpretation of electron microscope R loop mapping and the sizing of RNAs by gel electrophoresis.[39,41] A comparison of the spliced structure of the intermediate and early mRNAs suggests that expression of this region of viral DNA may be regulated at the level of RNA splicing. The short mRNA mapping between 9.7 and 11.2 codes for polypeptide IX which is hexon-associated in the virion. This mRNA has been purified and mapped on the genome.[42]

Early region 2 is thought to code for one polypeptide; a single-stranded DNA-binding protein of molecular weight 72,000.[43,44] There are two forms of early mRNA from region 2 which differ by the presence or absence of a splice near 66.7 (L. Chow and T. Broker, personal communi-

[41] L. T. Chow, R. E. Gelinas, T. R. Broker, and R. J. Roberts, Cell 12, 1–8 (1977).

[42] M. B. Mathews and U. Pettersson, J. Mol. Biol. 119, 293–298 (1978).

[43] J. B. Lewis, J. F. Atkins, P. R. Baum, R. Solem, R. F. Gesteland, and C. W. Anderson, Cell 7, 141–151 (1976).

[44] P. C. vander Vliet, A. J. Levine, M. J. Ensinger, and H. S. Ginsberg, J. Virol. 15, 348–354 (1975).

cation). Preparations of early mRNA vary in the proportion of these two forms. All the early mRNAs from this region have leader segments from 68.6 and 74.9.

Four early RNAs are complementary to region 3 (Fig. 6). The RNA represented schematically by a thick line is abundant while the forms represented by thin lines are less abundant. It is not clear whether the unspliced RNAs observed in cytoplasmic RNA from this region are due to leakage of precursors from the nucleus to the cytoplasm or whether these are functional mRNAs.

Three RNAs have been detected in early region 4; all of which have common 5' and 3' termini but differ in having various segments deleted by RNA splicing. It is unlikely that these RNAs are intermediates in a processing pathway because all three have similar kinetics of appearance in the cytoplasm during pulse labeling experiments (A. J. Berk, personal communication).

It is likely that RNAs with different structures are also synthesized from regions 2–4 during an intermediate stage of infection (L. Chow and T. Broker, personal communication). In the case of region 2, the intermediate and late forms of the 72K mRNA appear to have a 5' proximal leader mapping at 71.5, which may signify the presence of a new exclusively late promoter for transcription of this mRNA.

Total Transcription Map of Ad2

A current total transcription map of Ad2 is presented in Fig. 7. The early mRNAs from regions 1–4 are also shown here for the purpose of orientation.

Most of the map positions for late mRNAs are taken from the electron microscopy data of Chow *et al.*[32] In addition, some unpublished data of Susan Berget collected by gel nuclease mapping of late mRNAs are included. The only late mRNA complementary to the l strand of Ad2 maps between 16.1 and 11.2. This mRNA probably codes for the IV_{a2} polypeptide.[45] The other known late mRNAs are complementary to the r strand and map in the region of the viral DNA between 29 and 92.[46] These mRNAs can be divided into five families with 3' ends at either 39, 50, 61.5, 79, or 91.5.[47–50] All the mRNAs within a family share a common 3' termi-

[45] J. B. Lewis, J. F. Atkins, C. W. Anderson, P. R. Baum, and R. F. Gesteland, *Proc. Natl. Acad. Sci. U.S.A.* **72**, 1344–1348 (1975).

[46] P. A. Sharp, P. H. Gallimore, and S. J. Flint, *Cold Spring Harbor Symp. Quant. Biol.* **39**, 457–474 (1974).

[47] L. Philipson, U. Pettersson, U. Lindberg, C. Tibbetts, B. Venström, and T. Persson, *Cold Spring Harbor Symp. Quant. Biol.* **39**, 447–456 (1974).

[48] M. McGrogan and H. J. Raskas, *J. Virol.* **23**, 240–249 (1977).

[49] E. Ziff and N. Fraser, *J. Virol.* **25**, 897–906 (1978).

[50] S. M. Berget, personal communication.

nal map position. The 5' ends of the bodies of mRNAs within a family vary, and since sequences adjacent to the 5' termini are probably translated, these mRNAs are presumed to code for different polypeptides. In Fig. 7, the tripartite leader set mapping at 16.6, 19.3, and <26> is denoted by the asterisk. All mRNA bodies marked with an asterisk have been shown to have the tripartite leader set spliced to their 5' termini.[27,41] Recent data have shown that many cytoplasmic RNAs can have other RNA segments in addition to a tripartite leader set spliced to their 5' termini (L. Chow and T. Broker, personal communication). Dunn et al.[51] have purified two forms of late mRNA coding for a hybrid protein which is specified by both Ad2 and SV40 sequences. The amino terminal sequence of this protein is identical to that of polypeptide IV (fiber). Accordingly, there are probably two forms of mRNA for the polypeptide IV (fiber). One of these two mRNAs simply has the tripartite leader set spliced to its 5' terminus. The second form of this mRNA has an additional segment from 79.0 spliced between the tripartite leader set and the body of the mRNA. These multiple forms of leaders may reflect alternative pathways for synthesis of these mRNAs or intermediate forms in the processing of these mRNAs.

Viral polypeptides have been assigned (Fig. 7) to many of the late mRNAs mapping between 29 and 92. These assignments have been made on the basis of the results of translation of RNAs selected by hybridization to different restriction endonuclease cleavage fragments[45] and on the known sizes of mRNAs coding for particular polypeptides.[52] The sizes of mRNAs coding for particular viral polypeptides have also been determined by translation of RNA resolved by sedimentation in denaturing sucrose gradients,[52] or by electrophoresis in methyl mercury–agarose gels (R. P. Riccardi and B. E. Roberts, personal communication). R. P. Riccardi and B. E. Roberts (personal communication) have also mapped mRNA sequences coding for viral polypeptides by the hybridization arrest of translation (HART) procedure.[53] In this procedure the ability of a fragment to hybridize to a given mRNA and block its in vitro translation is assayed. In general the HART results agree with and refine the previous translation mapping data.

The combination of experiments used to make assignments of mRNAs to polypeptides give definite map positions for sequences coding for a particular polypeptide. These map positions for polypeptides were then correlated with the map positions of mRNAs. In cases where several mRNAs overlap the coordinates for a polypeptide coding sequence, the

[51] A. R. Dunn, M. B. Mathew, L. T. Chow, J. Sambrook, and W. Keller, Cell (in press).

[52] J. B. Lewis, C. W. Anderson, J. F. Atkins, and R. F. Gesteland, Cold Spring Harbor Symp. Quant. Biol. 39, 581–590 (1974).

[53] B. M. Paterson, B. E. Roberts, and E. L. Kuff, Proc. Natl. Acad. Sci. U.S.A. 74, 4370–4374 (1977).

correctly sized mRNA was selected. Thus, in all cases the assigned map positions and sizes of mRNAs agree with that expected of the RNA coding for the given polypeptide. A more extensive discussion of the use of similar data for assigning polypeptides to mRNAs can be found in Chow *et al.*[32]

Acknowledgments

Supported by an American Cancer Society Personal Faculty Award grant, an American Cancer Society research grant (No. VC-151C) and a National Science Foundation Grant (No. PCM76-20603) to Phillip A. Sharp.

[70] Definition and Mapping of Adenovirus 2 Nuclear Transcription

By JOSEPH R. NEVINS

Introduction

Unlike the situation in prokaryotes, the immediate products of genes in mammalian cells are not used directly as mRNA but undergo extensive posttranscriptional modifications.[1-3] In addition to regulation at the level of transcription, the control of gene expression in mammalian cells may also involve these posttranscriptional modifications. Therefore any step concerned with processing of a nuclear precursor molecule can potentially be involved in the regulation of gene expression. It thus becomes essential to identify and define the nuclear primary RNA transcript (the immediate gene product) and determine the exact details by which a mature mRNA is produced from a nuclear precursor.

Since the faithful transcription of a pure eukaryotic DNA template with a purified RNA polymerase cannot at present be carried out, then an *in vivo* system is required to identify and define a primary transcript that is an mRNA precursor. Additionally, the particular transcript must be produced in sufficient amounts to allow very short periods of isotope labeling which must be used to identify new gene products which are unmodified (not yet processed). Adenovirus-infected HeLa cells provide an ideal system for studying the formation of primary RNA transcripts. Late during

[1] A. J. Shatkin, *Cell* **9**, 645 (1976).
[2] J. E. Darnell, W. Jelinek, and G. Molloy, *Science* **181**, 1215 (1973).
[3] J. E. Darnell, *Prog. Nucleic Acid Res. Mol. Biol.* **22**, 327 (1978).

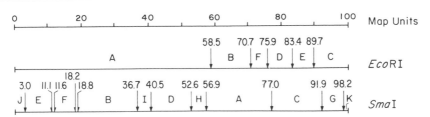

FIG. 1. Cleavage maps of adenovirus 2 DNA generated by restriction endonucleases *Eco*RI and *Sma*I. One map unit is equal to 355 base pairs. Cleavage sites are taken from data collected by Dr. Marc Zabeau and distributed at the 1976 Cold Spring Harbor Meeting on DNA tumor viruses.

the infection cycle, most virus-specific transcription occurs from a single large transcription unit encompassing 30,000 nucleotides on the adenovirus DNA genome (right-hand 84% of the genome).[4-7] In addition, at least 20% of the total RNA synthesis occurring in virus-infected nuclei is from this single transcription unit.[8] Therefore, the major prerequisites cited above are met.

Finally what makes adenovirus an excellent model system for the study of the formation of primary transcripts is the availability of highly refined physical maps of the adenovirus 2 genome which have been obtained using restriction endonucleases.[9] This then allows a relationship to be established between DNA structure, mRNA location, and definition of transcription units responsible for mRNA production (Fig. 1).

General Procedures

Solutions

MEM[10] (Joklik-modified): supplemented with 5% fetal calf serum
PBS (20×): 160 g/liter NaCl; 4 g/liter KCl; 43.2 g/liter $Na_2HPO_4 \cdot 7H_2O$; 4 g/liter KH_2PO_4; $MgCl_2$ added to 1 mM when diluted to 1×
Tris-HCl, 0.01 M, pH 8.1

[4] S. Bachenheimer and J. E. Darnell, *Proc. Natl. Acad. Sci. U.S.A.* **72,** 4445 (1975).
[5] R. Evans, N. W. Fraser, E. Ziff, J. Weber, M. Wilson, and J. E. Darnell, *Cell* **12,** 733 (1977).
[6] S. Goldberg, J. Weber, and J. E. Darnell, *Cell* **10,** 617 (1977).
[6a] E. B. Ziff and R. M. Evans, *Cell* **15,** 1463 (1978).
[7] J. Weber, W. Jelinek, and J. E. Darnell, *Cell* **10,** 612 (1977).
[8] L. Philipson, U. Pettersson, U. Lindberg, C. Tibbetts, B. Venström, and T. Persson, *Cold Spring Harbor Symp. Quant. Biol.* **39,** 447 (1974).
[9] C. Mulder, J. R. Arrand, H. Delius, W. Keller, U. Pettersson, R. J. Roberts, and P. A. Sharp, *Cold Spring Harbor Symp. Quant. Biol.* **39,** 397 (1974).
[10] The following abbreviations are used: MEM, minimal essential media; PBS, phosphate-buffered saline; HSB, high salt buffer; SDS, sodium dodecyl sulfate; kb, kilobases.

CsCl (density 1.2 g/cm): 22.4% (w/w) in 0.05 M Tris-HCl, pH 7.4

CsCl (density 1.45 g/cm³): 42.2% (w/w) in 0.05 M Tris-HCl, pH 7.4

Iso-hi-pH: 0.01 M Tris-HCl, pH 8.4; 0.14 M NaCl; 0.0015 M MgCl₂ NP-40, 5% (v/v)

HSB: 0.01 M Tris-HCl, pH 7.4; 0.5 M NaCl; 0.05 M MgCl₂; 0.002 M CaCl₂

DNase: 1 mg/ml in 0.1 M sodium acetate, 0.15 M iodoacetic acid, pH 5.3 (heated to 55° for 40 min)

Tween-DOC ("Magik") solution[11]: 3.3% (w/w) sodium deoxylcholoate; 6.6% (v/v) Tween 40; stored frozen at −20°

SDS, 20% (w/v)

EDTA, 0.5 M

1 M Sodium acetate, pH 5.1

LiCl, 10 M

ETS: 0.01 M Tris-HCl, pH 7.4; 0.01 M EDTA; 0.2% (w/v) SDS

NETS: ETS with NaCl; molarity (0.1, 0.2, etc., refers to molarity of the NaCl)

Phenol–acetate: liquefied phenol saturated with 0.05 M sodium acetate, pH 5.1

Pronase: 1 mg/ml in 0.05 M Tris-HCl, pH 7.5; 0.001 M EDTA; 0.5% (w/v) SDS

Chloroform–1% (v/v) isoamyl alcohol

Phenol-SSC: liquefied phenol saturated with 1× SSC

SSC (20×): 3 M NaCl; 0.3 M Na citrate; pH 7.0

TESS: 0.02 M TES, pH 7.4; 0.01 M EDTA; 0.3 M NaCl; 0.2% (w/v) SDS

Pancreatic RNase (200×): 500 μg/ml in 0.05 M Na acetate, pH 5.1; heated to 80° for 10 min

T1 RNase (20×): 100 units/ml in 0.05 M Na acetate, pH 5.1; heated to 80° for 10 min

E buffer: 0.04 M Tris-HCl; 0.005 M Na acetate; 0.001 M EDTA, pH 7.8

Agarose, 1.4% (w/v): 140 g/liter agarose (Bio-Rad, Richmond, California) dissolved by heating in E buffer

Ethidium bromide: 10 mg/ml; diluted 1 : 10,000 for staining

Glycerol, 50% (v/v)–bromphenol blue, 0.2% (w/v)

NaOH, 1 N

HEPES, 2 M

Poly(U)–Sepharose: prepared according to procedure of Wagner et al.[12] or can be obtained commercially (Pharmacia, Uppsala, Sweden)

[11] S. Penman, J. Mol. Biol. **17,** 117 (1966).
[12] A. F. Wagner, R. L. Bugianesi, and T. Y. Shen, Biochem. Biophys. Res. Commun. **45,** 184 (1971).

Purification of Virus. Purified adenovirus 2 is added to a culture of growing HeLa cells at a multiplicity of 1000 particles/cell. Cells are maintained at 37° for 40–48 hr, at which time they are collected by centrifugation (600 g, 5 min) and washed twice with PBS. The cells are then resuspended in 0.01 M Tris-HCl, pH 8.1, at a density of 2 × 10⁷ cells/ml, and are subjected to three cycles of freeze–thawing. Virus may be kept stored frozen at this stage without loss of infectivity. They are then further disrupted by sonication for 30 sec with a Branson Sonicator (Model W185, Heat Systems-Ultrasonics, Inc.). The suspension is then extracted with an equal volume of Freon 113 (Freon TF MS 182 Degreaser, Miller-Stephenson Chemical Co., Danbury, Connecticut), the phases separated by centrifugation and the aqueous phase reextracted with Freon. The second aqueous phase is then layered onto 20 ml preformed linear CsCl density gradients (density 1.2–1.45) in SW 27 tubes. The gradients are spun at 20,000 rpm for 2 hr at 4°. The virus bands are collected by puncturing the bottom of the tubes and diluted with an equal volume of 0.01 M Tris, pH 8.1. The diluted virus is then layered onto 30 ml preformed linear CsCl density gradients (density 1.2–1.45) in SW27 tubes. The gradients are centrifuged at 20,000 rpm for about 18 hr at 4°. The virus bands are collected and diluted for storage with 0.01 M Tris-HCl, pH 8.1, 0.1 M NaCl, 0.1% BSA (bovine serum albumin).

Quantitation of the virus in the purified sample, before dilution in BSA, is made by the relation: 1 A_{260} = 1.1 × 10¹² virus particles/ml.[13] Virus stored in this manner retains infectivity for about 4–6 weeks.

Growth of Cells and Virus Infections for Labeling. HeLa cells are maintained in suspension cultures in MEM, at concentrations between 3 × 10⁵ and 8 × 10⁵ cells/ml. For virus infection, an appropriate number of cells (usually about 10⁸ cells) are centrifuged (600 g, 5 min) and resuspended in MEM containing no serum at a concentration of 10⁷ cells/ml. Purified adenovirus is then added at a multiplicity of 2000 particles per cell and allowed to adsorb for 30 min. The infected cells are then diluted 20-fold with warm MEM (containing 5% fetal calf serum) and maintained at 37°.

³H-*Nucleoside Labeling and RNA Extractions.* To prepare labeled RNA late in infection (16–25 hr postinfection), cells are centrifuged and resuspended in warm MEM containing 5% serum at a concentration of 3 × 10⁶ cells/ml. ³H-Uridine (30 Ci/mmole) or ³H-adenosine (35 Ci/mmole) (New England Nuclear) is then added at concentrations up to 300 μCi/ml. Also 20× PBS or powdered medium in appropriate amounts is added to bring the isotope solution to isotonic conditions.

After an appropriate labeling period, incorporation is stopped by pouring the cells onto an equal volume of crushed, frozen PBS. After centrifugation (600 g, 5 min) the cells are washed with PBS and recentrifuged.

[13] J. V. Maizel, D. O. White, and M. D. Scharff, *Virology* **36**, 115 (1968).

This is repeated once more. The nuclear RNA is then extracted by the hot phenol procedure.[14] The washed cells are resuspended in cold iso-hi-pH (approximately 5×10^7 cells/ml), made 0.5% to NP-40, and lysed by gentle vortexing for 5 min. Nuclei are removed by centrifugation ($1000\,g$, 3 min) and resuspended in iso-hi-pH. Tween-DOC solution[11] is added ($\frac{1}{10}$ volume), the suspension quickly mixed, and the nuclei are again pelleted. The detergent-washed nuclei are resuspended in HSB (5×10^7 cells/ml) to which has been added DNase ($\frac{1}{10}$ volume) causing the nuclei to lyse as is evident by greatly increased viscosity. The tube is now warmed slightly and the solution pipetted up and down to aid in DNase digestion. When the solution loses viscosity ($\sim$1–2 min), 1/20 volume of $1\,M$ Na acetate is added followed by an equal volume of $0.05\,M$ Na acetate, pH 5.1, in order to reduce the NaCl concentration and lower the pH. SDS and EDTA (1/50 volumes each) are added followed by 2 volumes of phenol–acetate. Extraction is performed at 60° for 5 min, with periodic shaking or vortexing. The phases are separated by centrifugation ($1000\,g$, 5 min), and the phenol phase is removed leaving the aqueous phase and the interphase. The extraction is then repeated using an equal volume of phenol–acetate and an equal volume of chloroform, this time at room temperature. A final extraction is performed with 2 volumes of chloroform. The final resulting aqueous phase is removed and the RNA precipitated at $-20°$ with 2 volumes of ethanol. Recovery of acid precipitable radioactivity in RNA averages 75% with this procedure.

Poly(U)–Sepharose Chromatography. A column of 0.3 ml bed volume of poly(U)–Sepharose in a pasteur pipette is sufficient to collect the poly(A)-containing nuclear RNA from 10^8 cells. After packing, the column is washed with 10 volumes of 70% formamide in ETS and then equilibrated with $0.3\,M$ NETS. The RNA sample (in $0.3\,M$ NETS) is slowly passed through the column. It has been found that large nuclear poly(A)-containing RNA is often not quantitatively retained on poly(U)–Sepharose. Prior denaturation in 50% formamide (65°, 2 min) followed by tenfold dilution in $0.3\,M$ NETS and then loading onto the column may improve the binding. After loading the sample, the column is washed successively with 10 column volumes of $0.3\,M$ NETS, ETS, and 10% formamide in ETS. The poly(A)-containing RNA is then eluted with 10 volumes of 70% formamide in ETS. The RNA is then precipitated with 2 volumes of ethanol after adding yeast RNA to 50 μg/ml and LiCl to $0.1\,M$. (It is advisable to use Li as the cation rather than Na for the ethanol precipitation due to the formation of an ethanol-insoluble Na precipitate with the formamide.)

Preparation of Viral DNA. Purified virus is diluted sixfold into the

[14] R. Soeiro and J. E. Darnell, *J. Mol. Biol.* **44**, 551 (1969).

Pronase solution (self-digested for 30 min prior to use) and incubated at 37° for 60 min. One-quarter volume of chloroform–isoamyl alcohol and an equal volume of SSC–phenol are added and the mixture shaken. The phases are separated by centrifugation and the aqueous phase reextracted as above. The final aqueous phase is made 0.15 M to NaCl, and the DNA is precipitated twice with 2 volumes of ethanol at −20°.

Restriction Endonuclease Digestion of Viral DNA and Purification of Fragments. Extracted viral DNA (usually 1–2 mg) is centrifuged, dried, and dissolved in an appropriate buffer. Digestion is carried out with the desired restriction endonuclease under appropriate conditions for the particular enzyme. After digestion, a small aliquot (about 0.5 μg DNA) is removed, glycerol is added to 5% and bromphenol blue to 0.02%, and the sample layered onto a small cylindrical gel (0.5 × 6 cm) of 1.4% agarose containing 1 μg/ml of ethidium bromide. Electrophoresis is carried out at 70 V for about 2 hr or until the dye is near the bottom. Bands are visualized under uv light, and the pattern examined for completeness of digestion. If the digestion has gone to completion, then the remaining major part of the sample is made 0.1% to SDS, 5% to glycerol, and 0.02% to bromphenol blue and loaded onto a 1.4% agarose slab gel.[15]

For up to 1 mg of DNA a 20 × 40 × 0.5 cm gel is employed. For 1–2 mg of DNA a slab of 40 × 40 × 0.5 cm is used. Electrophoresis is carried out for the length of time necessary to separate the fragments. Bands are stained in ethidium bromide (1 μg-ml in E buffer), visualized under uv light, cut out of the gel with a scalpel. The gel slice is placed in a dialysis bag with 1× E buffer and sealed at each end by tying a knot. The bags are then laid in a flat rectangular tray with electrodes at either end and containing 1× E buffer of a sufficient depth to cover the bags. The DNA is then eluted out of the gel slice into the buffer in the dialysis bag by electrophoresis (current at approximately 100 mA). Small fragments (300–1000 nucleotides) elute in approximately 1–3 hr; large fragments may require up to 12 hr for complete elution. Elution may be followed by visualization of the ethidium-stained DNA under uv light. The eluted DNA is then recovered from the bags and precipitated with ethanol.

Preparation of DNA Filters. Either intact viral DNA or purified DNA restriction fragments are loaded onto nitrocellulose filters according to the procedures described by Gillespie[16] in a previous volume of this series. DNA in 0.1 × SSC is denatured by the addition of NaOH to 0.1 N. After 15 min, the solution is neutralized with monobasic sodium phosphate and diluted 10- to 20-fold with cold 5 × SSC. Appropriate aliquots are then

[15] J. B. Lewis, J. F. Atkins, C. F. Anderson, P. R. Baum, and R. F. Gesteland, *Proc. Natl. Acad. Sci. U.S.A.* **72**, 1344 (1975).
[16] D. Gillespie, this series, Vol. 12, Part B, p. 641.

passed slowly through Millipore filters previously washed with 5 × SSC. The filters are then washed with 10–20 ml of 5 × SSC, dried *in vacuo* at room temperature for 2–4 hr, and then baked at 80° *in vacuo* for 4 hr. It has been found that it is convenient to prepare filters containing four times the final desired amount of DNA, with the filters then being cut into quarter sections with a razor blade prior to hybridization.

One difficulty which may be encountered concerns the ability of very small DNA fragments to be bound by nitrocellulose filters. Adenovirus DNA fragments as small as 630 nucleotides are retained, but one of 250 nucleotides is not. Therefore, it is advisable when attempting to bind DNA of this size range to filters to first check for binding with labeled DNA.

Preparation of RNA for Hybridization. Most preparations of nuclear RNA will contain sufficient amounts of contaminating viral DNA to interfere in subsequent hybridizations. This can readily be removed by precipitating the RNA from 2 M LiCl. The ethanol-precipitated RNA is centrifuged (10,000 g, 20 min) and dissolved in ET (ETS without SDS). LiCl is then added to a final concentration of 2 M and the solution left for 16–24 hr at 4°. The RNA is centrifuged, dissolved in NETS, and precipitated with ethanol.

In general, most hybridizations involve viral DNA restriction fragments which are smaller than the RNA molecules. It has been observed that when an RNA molecule completely hybridizes the DNA on a filter, the hybrid has a tendency to come off the filter.[17] Prior to hybridization, it is thus important to break the RNA to small fragments, which can be achieved by controlled partial alkaline hydrolysis. The ethanol-precipitated RNA is pelleted, dissolved in 0.1 M NETS, and chilled to 4° on ice. NaOH is added to 0.2 N final concentration, and after 20–30 min, the solution is neutralized with an equal volume (to that of the NaOH) of 2 M HEPES, and the RNA is precipitated with two volumes of ethanol. These conditions are sufficient to reduce the RNA size to about 200–500 nucleotides.[18] The RNA precipitate is then centrifuged and dissolved in TESS.

DNA–RNA Hybridizations. Appropriate DNA filters are wetted in TESS solution and then marked with a pencil as desired. At this time filters can be cut with a razor blade into small sections containing the amount of DNA required in the hybridization. The cut filters are placed in glass scintillation vials (Wheaton R2550-4 with polyseal cap) with the appropriate RNA sample. Hybridizations are carried out at 65° for 24–48

[17] M. Haas, M. Vogt, and R. Dulbecco, *Proc. Natl. Acad. Sci. U.S.A.* **69,** 2160 (1972).
[18] W. Jelinek, G. Molloy, R. Fernandez-Munoz, M. Salditt, and J. E. Darnell, *J. Mol. Biol.* **82,** 361 (1974).

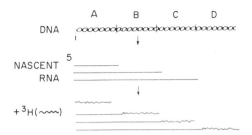

Fig. 2. Schematic representation of nascent RNA chain labeling. In this case transcription begins at the left end of DNA fragment A and continues in a rightward direction. Nascent RNA molecules in various stages of transcription then become labeled with ³H-nucleotides at their growing 3′ ends. Therefore, after a brief exposure to radioisotope, the shortest labeled molecules will be specific to fragment A, and the longest labeled molecules specific to fragment D.

hr, depending on the volume (and thus the RNA concentration). Filters are then removed to a plastic disposable beaker (Falcon, No. 4020) containing about 50 ml of 2× SSC and incubated at 65° for 15–20 min. The wash procedure is repeated once. The filters are then incubated with T1 RNase (5 units/ml) and pancreatic RNase (2.5 μg/ml) in 2× SSC at 37° for 30 min. The filters are washed several times with 2× SSC at room temperature, removed, dried, and counted in an appropriate scintillation fluid. The completeness of the initial hybridization can be checked by adding a fresh DNA filter to the hybridization fluid followed by an additional incubation at 65°.

Definition of Transcription Units by Nascent Chain Analysis

If a radioactive nucleotide is added to a culture for a short period of time such that the time of labeling is less than that required for the synthesis of a complete RNA chain, then a distribution of labeled molecules will be obtained that represent nascent chains in which label has been deposited only at the growing ends (Fig. 2). A particular transcription unit will thus give rise to a series of labeled molecules which represent growing chains originating from a single promoter site. The origin of a transcription unit as well as the dimensions of the transcription unit can be determined by such an analysis, since the shortest molecules will hybridize to DNA fragments nearest the promoter, and the termination site will be reflected by the limit in size of labeled molecules. In the diagram pictured in Fig. 2, the promoter would thus be in fragment A as it would hybridize the shortest labeled molecules, and the terminator would be in fragment D, since it would detect the longest terminally labeled molecules. In addition, since an ordered array of DNA fragments may be used to hybridize the nascent RNA, it then becomes possible to determine the direction of transcription. That is, are the molecules getting larger in a leftward or rightward direction in relation to map locations of the DNA fragments employed.

Solutions

> Dimethyl sulfoxide (DMSO), 99% (v/v): containing 0.01 M Tris-HCl, pH 7.4, 0.001 M EDTA
> Sucrose, 5% (w/v): in 99% DMSO
> Sucrose, 20% (w/v): in 99% DMSO
> Yeast RNA: 10 mg/ml in ETS

Labeling of Nascent Chains. Infected cells are labeled with ^{3}H-uridine for 2 min, as described previously. The time of labeling will depend upon the nature of the transcription unit. In the case of the late adenovirus transcription unit, 5–10 min is probably required for transcription, allowing a synthesis rate of 50–100 nucleotides per second.[19,20] Thus, 2 min of ^{3}H-uridine would ensure that the time to complete a chain had not been exceeded. However, if the transcription unit were smaller, then a shorter period of labeling would be required. For instance, in HeLa cells, where the average transcription unit is ~5 kb, a labeling time of 30 sec is sufficiently short to produce a nascent chain profile.[21]

Analysis of Nascent Chains in DMSO–Sucrose Gradients. After nascent chains have been terminally labeled, the second aspect of the experiment is to separate the chains according to size. This can be achieved in most cases by either gel electrophoresis or sucrose gradient sedimentation. Sucrose gradients offer the advantages of being able to accept very large molecules as well as a greater capacity for mass of RNA. The second point is most important when one is dealing with very briefly labeled preparations, since to obtain a sufficient amount of radioactivity, the sample size must often be large (2 × 10⁸ cells or more).

Since the very large nuclear RNA molecules can possibly aggregate in aqueous sucrose gradients, and various conformational states of RNA molecules can affect their sedimentation behavior, then sedimentation under completely denaturing conditions provides the best and least ambiguous analysis. This can be achieved in sucrose gradients prepared in 99% DMSO.[22] Linear gradients of 38 ml are formed in SW 27 (Spinco) polyallomer tubes using 5 and 20% sucrose solutions in DMSO (CAUTION:

[19] H. Greenberg and S. Penman, *J. Mol. Biol.* **21**, 527 (1966).
[20] J. E. Darnell, M. Girard, D. Baltimore, D. F. Summers, and J. V. Maizel, *in* "The Molecular Biology of Viruses" (J. Colter, ed.), p. 375. Academic Press, New York, 1967.
[21] E. Derman, S. Goldberg, and J. E. Darnell, *Cell* **9**, 465 (1976).
[22] J. H. Straus, R. B. Kelley, and R. L. Sinsheimer, *Biopolymers* **6**, 793 (1968).

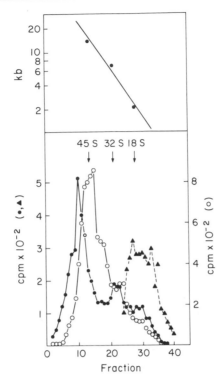

Fig. 3. Analysis of RNA chains labeled for 2 min with ³H-uridine in late adenovirus infected cells by sedimentation in a dimethyl sulfoxide–sucrose density gradient. Centrifugation was for 64 hr in a SW 27 rotor. Fractions were hybridized to filters containing DNA fragments SmaF (▲), SmaH (○), and EcoF (●). Sedimentation was from right to left.

DMSO will dissolve cellulose nitrate centrifuge tubes). Precipitated RNA samples are pelleted out of ethanol and dissolved in 0.1 ml of ETS. DMSO (0.4 ml) is added, and the samples are heated to 65° for 3 min. The RNA is then layered on the gradients and centrifuged at 27,000 rpm at 29°. The nuclear RNA from 1×10^8 to 2×10^8 cells can be sedimented through such a gradient with good resolution of RNA molecules. Centrifugation for 64 hr will sediment 32 S preribosomal RNA approximately halfway down the gradient. The gradients are then collected by pumping from the bottom of the gradient using a peristaltic pump while continuously monitoring the A_{260} with a recording spectrophotometer equipped with a flow cell. In this way preribosomal 45 S (14 kb) and 32 S (6.7 kb) RNA molecules provide convenient sedimentation markers.

Since the DMSO will dissolve nitrocellulose filters used in the subsequent hybridizations, the fractions must be ethanol precipitated. To each fraction is added 3 volumes of 0.2 M NETS and yeast RNA as carrier (50 μg/ml final concentration) and then 2 volumes of ethanol. The RNA is then pelleted and prepared for hybridization as described. Hybridizations are carried out to DNA filters as discussed in the section on general procedures.

An analysis of nascent chains produced in late adenovirus-infected cells is shown in Fig. 3. As can be seen, when RNA is hybridized with DNA fragments going from left to right on the genome (refer to Fig. 1), the size of the nascent molecules is greater. This would thus indicate that transcription is proceeding in a left to right direction. When the average size of the molecules hybridizing to the various fragments is determined, it is found that RNA hybridizing to SmaF (11.6–18.2) is approximately 1900 nucleotides in length, RNA hybridizing to SmaH (52.6–56.9) is approximately 15,000 nucleotides in length and RNA hybridizing to EcoF (70.7–75.9) is about 24,000 nucleotides in length. Thus the nascent molecules are larger when going from left to right on the genome by values corresponding to the distance between the fragments. This then suggests a common origin (promoter) for the RNA, and extrapolation from the sizes of the nascent molecules indicates a promoter at approximately map position 16. (Nascent chains from 16 to 18 which grow ~1000–1500 nucleotides during the pulse label give a sedimentation distribution peaking at about 1900 nucleotides.)

uv Transcription Mapping to Define Transcription Units

Ultraviolet irradiation of cells results in the introduction of thymine dimers in the DNA which become terminators of transcription.[23] Since it appears that the initiation event is not affected,[23] then the probability of transcription of a particular sequence after uv irradiation will be the probability that no uv lesion has been formed between the sequence being assayed and its promoter. Thus, the rate of uv inactivation of transcription of a particular sequence is a measure of its distance from its promoter. It has been possible to use the method of uv transcription mapping to determine the order of genes within a particular transcription unit, since the most promoter–distal sequences will have the greatest sensitivity while the promoter–proximal sequences will have the least sensitivity.[24–28] In addition, if the order of genes or sequences (such as restriction fragments)

[23] W. Sauerbier, R. L. Millette, and P. B. Hackett, *Biochim. Biophys. Acta* **209**, 368 (1970).
[24] P. B. Hackett and W. Sauerbier, *Nature (London)* **251**, 639 (1974).
[25] P. B. Hackett and W. Sauerbier, *J. Mol. Biol.* **91**, 235 (1975).
[26] L. A. Ball and C. N. White, *Proc. Natl. Acad. Sci. U.S.A.* **73**, 442 (1976).
[27] G. Abraham and A. K. Banerjee, *Proc. Natl. Acad. Sci. U.S.A.* **73**, 1504 (1976).
[28] K. Glazier, R. Raghow, and D. W. Kingsbury, *J. Virol.* **21**, 863 (1977).

of a DNA template is known, then the number of transcription units can be determined.[6,20,29-31]

Procedure. Cells infected for 16 hr are pelleted and washed once with MEM containing no phenol red (phenol red absorbs uv and thus is omitted) and then resuspended in MEM lacking phenol red and with 5% fetal calf serum at a concentration of 3×10^6 cells/ml. Samples of approximately 25 ml are placed in a 25×150 mm plastic petri dish, irradiated with a General Electric germicidal lamp at an incident fluence of 40 erg/mm²/sec for various lengths of time, and then transferred to spinner vessels maintaining the cells in a light-protected environment. After allowing 20 min for equilibration of transcription on the uv-damaged templates, ^{3}H-uridine (50–100 μCi/ml) is added. Labeling is carried out for 5 min and then terminated, and the nuclear RNA is extracted from the samples. The RNA is precipitated with LiCl, broken with NaOH, and then hybridized to desired DNA filters.

Analysis of Data. The effect of uv irradiation on the synthesis of RNA from a particular sequence can be described by

$$R/R_0 = e^{-Kd} \tag{1}$$

where R is the rate of transcription at a particular uv dose, R_0 is the rate of transcription in the control sample, d is the uv dose, and K is the uv inactivation rate constant which is proportional to the distance of the sequence from its promoter.[32] Since the effect of increasing uv dose on RNA synthesis is exponential, then a plot of uv dose versus log survival of synthesis should yield a straight line for a particular sequence being transcribed at a given distance from a promoter. The slope of this curve (K value) is then a measure of the relative sensitivity of the sequence in question to uv irradiation and therefore its relative distance from the promoter. Thus, if three genes (or DNA restriction fragments) are known to be part of a single transcription unit, then their relative sensitivities to uv (slopes of the dose curves) will give their order within the transcription unit. Alternatively, if the order of genes (or DNA sequences) along a DNA molecule is known, then the uv sensitivities of labeling RNA specific to the genes or DNA sequences will indicate the number of transcription units, that is, the number of promoters and the order of sequences within a transcription unit.

Such an approach has been used to demonstrate that the region from 16 to 100 on the adenovirus genome is part of a single large transcription

[29] A. R. Brautigam and W. Sauerbier, *J. Virol.* **13**, 1110 (1974).
[30] K. Hercules and W. Sauerbier, *J. Virol.* **12**, 872 (1973).
[31] K. Hercules and W. Sauerbier, *J. Virol.* **13**, 341 (1974).
[32] A. R. Brautigam and W. Sauerbier, *J. Virol.* **12**, 882 (1973).

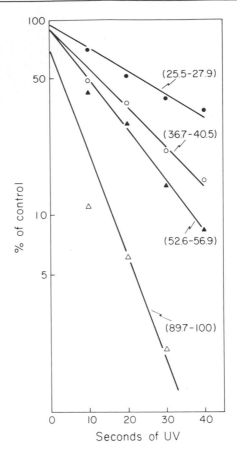

FIG. 4. Ultraviolet inactivation of transcription of adenovirus nuclear RNA. RNA from a control culture and cultures irradiated with various doses of uv was labeled for 40 min with ^{3}H-uridine and then extracted and hybridized to filters containing the DNA fragments *Hpa*F (●), *Sma*I (○), *Sma*H (▲), and *Eco*C (△). The lines were drawn by linear regression analysis of the data.

unit, utilizing a single promoter at approximately map position 16.[6] The results of an experiment demonstrating this type of analysis are shown in Figs. 4 and 5. Figure 4 depicts the effect of increasing uv dose on the rate of labeling RNA specified by four regions of the adenovirus genome. As can be seen, when DNA fragments going from left to right on the genome are used, the sensitivity to uv (slope of the curves) of transcription increases.

When the slopes of these dose curves are then plotted as a function of their map position on the genome (Fig. 5), a straight line is obtained with an origin near map position 10. Thus, this indicates that these sequences are part of a single transcription unit having a promoter near map position 10 on the genome (the actual location of the promoter being 16.45[6a]).

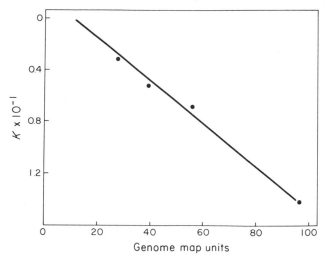

FIG. 5. Ultraviolet sensitivity of synthesis of late adenovirus nuclear RNA plotted versus map positions of DNA fragments used in the assays. The slopes of the curves in Fig. 4 were calculated from the relationship of $K = (\ln R_0/R)/d$ [see Eq. (1)]. These values were then plotted as a function of map positions of the particular DNA fragment.

Mapping of Products of Transcription Units

If a region of transcription identified by nascent chain analysis or by uv transcription mapping yields nuclear RNAs of varying sizes, such as intermediates or products in a processing pathway, then the origin of these molecules may be mapped by first size fractionating the RNA followed by hybridization to a series of restriction fragments. Not only is the resolution increased by first separating the different sizes of molecules and thus simplifying the analysis, but in addition the gel analysis will yield information regarding the lengths of the RNAs. If the length of an RNA molecule is known, and it exhibits hybridization across a boundary between two DNA fragments, then the positions of the ends of the RNA can be estimated by the percentage of hybridized radioactivity in the adjacent fragments. In addition, if an RNA were found to span three contiguous fragments (fragments B, C, D) thus completely covering the middle fragment (fragment C) with no hybridization to fragments on either side of these three (A and E), then an independent measure of its length would be obtained; that is, the RNase-resistant radioactivity hybridized to the middle fragment (C) would be proportional to the length of that particular DNA fragment. Then the amount of hybridized radioactivity in each of the adjacent fragments (B and D) could be converted to number of nucleotides with the relation established from the middle fragment (C). In this way, the positions of the termini of the RNA could be established as well as the length of the molecule. This type of analysis has been employed to identify products of nuclear RNA processing from early and late adenovirus-infected cells[33,34] as well as late adenovirus cytoplasmic mRNAs.[35,36] The existence of spliced RNAs will obviously complicate such an analysis making it necessary to employ a large number of small DNA fragments for hybridization. In this way, sequences in the DNA which were spliced out in the formation of the RNA product would be indicated by a lack of hybridization of a specific RNA to an internal fragment in a series of contiguous fragments, the remainder of which did hybridize to the RNA.

Solutions

Acrylamide (30%, w/v)-N,N'-methylenebisacrylamide (1.5%, w/v)

Loening's buffer[37] (10×): 43.6 g/liter Tris base; 41.4 g/liter NaH_2PO_4; 2.92 g/liter EDTA, pH 7.8

N,N,N',N'-Tetramethylethylenediamine (TEMED)

Ammonium persulfate, 10% (w/v)

SDS, 20% (w/v)

Formamide: deionized before use by stirring with a mixed bed resin [AG 501-X8 (D), Bio-Rad, Richmond, California] with subsequent filtration

Glycerol, 50% (v/v)–bromphenol blue, 0.2% (w/v)

Preparation of Gels and Electrophoresis. Since the RNA is to be hybridized to a variety of DNA fragments after electrophoresis, cylindrical tube gels which can be easily fractionated are most suitable. Gels prepared according to a modification of the procedure of Loening[37] and consisting of 2.4% acrylamide and 0.5% agarose are used for resolution of RNA species 5000 nucleotides (28 S) or less in length. The gel solution consists of 4.0 ml acrylamide–bisacrylamide, 5.0 ml 10× Loening's buffer, 0.5 ml 20% SDS, 0.05 ml TEMED, 39.5 ml H_2O, 0.25 g agarose, 0.5 ml ammonium persulfate. The agarose is first dissolved in 20 ml of the water by heating in a boiling water bath. When the agarose has cooled to about 70°, the remaining ingredients are separately mixed (persulfate added last) and then mixed with the agarose. Gels are cast in 0.7 × 10 cm Perspex tubes sealed at the bottom with Parafilm. Solidification of the agarose and polymerization of the acrylamide requires about 1 hr. A piece of dialysis tubing is secured to the open top end of the tube and punctured with a needle. The Parafilm is removed and the gel tube is inverted, allowing the

---

[33] E. A. Craig and H. J. Raskas, *Cell* **8**, 205 (1976).
[34] J. R. Nevins, *J. Mol. Biol.* **130**, 493 (1979).
[35] M. McGrogan and H. Raskas, *Proc. Natl. Acad. Sci. U.S.A.* **75**, 625 (1978).
[36] J. R. Nevins and J. E. Darnell, *J. Virol.* **25**, 811 (1978).
[37] U. E. Loening, *Biochem. J.* **102**, 251 (1967).

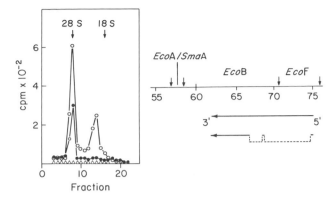

FIG. 6. Electropherogram of early adenovirus poly(A)-containing nuclear RNA labeled with ³H-uridine from 2.5 hr to 4 hr after infection. The RNA was subjected to electrophoresis in a 2.5% acrylamide gel, eluted, and hybridized to DNA fragments $EcoF$ (●), $EcoB$ (○), and $EcoA/SmaA$ (△). Each fraction represents a 2-mm gel slice. Electrophoresis was from left to right. The map positions of the RNA were derived from the hybridization data as described in the text. Positions of 5′ terminal spliced segments are those as have been described.[40,41] Adapted from Blanchard et al.[43]

gel to slide down against the dialysis membrane. The flat surface formed against the Parafilm now becomes the top of the gel. The gels are placed in a suitable apparatus with 1× Loening's buffer containing 0.2% SDS as the electrophoresis buffer.

Samples of ethanol-precipitated nuclear RNA are pelleted, briefly dried, and dissolved in a small volume of ETS (20–50 μl). An equal volume of formamide is added, and the sample is heated to 65° for 2 min to denature the RNA. Glycerol is added to 5% final concentration along with bromphenol blue (0.02%), and the sample is layered onto the gel. Electrophoresis is at 70 V (approximately 5 mA/gel) for 3–5 hr, at which time the bromphenol blue should be just running out of the gel. The gel is easily removed from the tube and is placed in an apparatus for slicing. A convenient design is an aluminum trough designed with slots cut along each side of the trough 1 mm apart through which a razor blade can be inserted for slicing. The slices are placed in glass scintillation vials with 0.5 ml of TESS solution, tightly capped, and incubated at 65° for 24 hr. These conditions are sufficient to elute 80–90% of 28 S and 18 S rRNA from gel slices. The slices are left in the vials, and filters are added for hybridization. Usually, the RNA is sufficiently broken during the elution procedure such that alkaline cleavage is not necessary. However, if a small DNA fragment contained on a filter were being used for the hybridization, then it would probably be advisable to further break the RNA with alkali.

Analysis of Size Fractionated RNA by Overlapping Hybridizations. An example of such an analysis is depicted in Fig. 6. In this case, nuclear RNA from early adenovirus-infected cells was labeled with ^{3}H-uridine and then extracted and selected on a poly(U)–Sepharose column. The RNA was then subjected to electrophoresis as described above, and the gel fractions were hybridized to DNA fragments EcoB, EcoF, and EcoA/ SmaA, three contiguous fragments from the center of the genome. The EcoB filters hybridized a 28 S species and a 22 S species of RNA, while the EcoF filters only hybridized the 28 S RNA. No hybridization was detected to the EcoA/SmaA filters. Since transcription in this region is known to proceed in the leftward direction,[38,39] then it follows that the 28 S RNA has its 3' end in EcoB and a 5' end in EcoF, while the 22 S RNA must lie completely within EcoB. From the ratio of hybridized 28 S RNA cpm in EcoB versus EcoF (1.94 : 1) and taking the DNA-coded size of the RNA to be about 4900 nucleotides [measured length of 5100 nucleotides minus 200 for the nontranscribed poly(A)], it would place 1660 nucleotides in EcoF and 3230 nucleotides in EcoB. This would then position the 5' terminus at 75.4 and the 3' terminus at 61.6 on the genome. Since the 22 S RNA did not exhibit hybridization across two fragments, then its exact coordinates cannot be determined. It must, however, lie entirely within EcoB and thus, must share sequences with the 28 S RNA. Its precise location could be resolved by performing the hybridization of the 22 S RNA to DNA fragments representing the left and right halves of EcoB such that the RNA would give overlapping hybridization.

A map arrangement of the 28 S RNA consistent with the hybridization data is shown in Fig. 6. Independent measurements have placed a 20 S mRNA as shown in the diagram, with noncontiguous leader segments as shown.[40,41]

Recent experiments employing these techniques have in fact demonstrated that the 22 S RNA possesses a 3' terminus in common with the 28 S RNA (at 61.6) but has lost sequences between 68.8 and 74.6, retaining a small leader segment at 75.[42] Finally, the processing of the 28 S precursor has also been demonstrated to occur in isolated nuclei using the techniques previously described.[43]

[38] P. A. Sharp, P. H. Gallimore, and S. J. Flint, Cold Spring Harbor Symp. Quant. Biol. 39, 457 (1974).

[39] U. Pettersson, C. Tibbetts, and L. Philipson, J. Mol. Biol. 101, 479 (1976).

[40] A. J. Berk and P. A. Sharp, Cell 12, 721 (1977).

[41] G. R. Kitchingman, S. P. Lai, and H. Westphal, Proc. Natl. Acad. Sci. U.S.A. 74, 4392 (1977).

[42] C. J. Goldenberg and H. J. Raskas, Cell 16, 131 (1979).

[43] J.-M. Blanchard, J. Weber, W. Jelinek, and J. E. Darnell, Proc. Nat'l. Acad. Sci. U.S.A. 75, 5344 (1978).

Acknowledgments

This work was supported by grants from the National Institutes of Health (CA16006) and the American Cancer Society (NP2136) to Dr. J. E. Darnell. The author was a Fellow of the Jane Coffin Childs Memorial Fund for Medical Research. The author would also like to thank Dr. James E. Darnell for critically reading the manuscript.

[71] Restriction Fragments from *Chlamydomonas* Chloroplast DNA

By J. D. ROCHAIX

The chloroplast DNA of *Chlamydomonas reinhardii* represents an attractive model system for studying the cooperation between the nucleocytoplasmic and organellar protein synthesizing apparatus in eukaryotic cells. This is largely due to the fact that this unicellular alga can be manipulated with ease both at the biochemical and genetic level. Since Sager's discovery of the first uniparental mutant in *C. reinhardii*,[1] numerous mutations of this type have been mapped, and there is now convincing evidence that these mutations are located in the chloroplast DNA.[2,3] Electron microscopic[4] and restriction enzyme analyses[5,6] of this DNA have shown that it is circular with a molecular weight of 125×10^6 to 130×10^6. A restriction map of this DNA has been established recently.[7]

The availability of pure restriction fragments of the chloroplast DNA allows detailed studies on the organization and function of this DNA. This article describes the methods used for obtaining chloroplast DNA restriction fragments, and some of their properties.

Chloroplast DNA preparation

Media. TAP (Tris–acetate–phosphate) medium contains the same constituents as TMP medium[8] except that 1 ml of glacial acetic acid is added per liter instead of concentrated hydrochloric acid and the $MgSO_4$ can be omitted.

[1] R. Sager, *Proc. Natl. Acad. Sci. U.S.A.* **40**, 356 (1954).
[2] R. Sager, "Cytoplasmic Genes and Organelles." Academic Press, New York, 1972.
[3] N. W. Gillham, *Annu. Rev. Genet.* **8**, 347 (1976).
[4] W. Behn and R. G. Herrmann, *Mol. Gen. Genet.* **157**, 25 (1977).
[5] S. H. Howell, P. Heizmann, and S. Gelvin, *in* "Acides nucléiques et synthèse de protéines chez les végétaux" (L. Bogorad and J. H. Weil, eds.), p. 313. CNRS, Paris, 1977.
[6] J. D. Rochaix, *in* "Acides nucléiques et synthèse de protéines chez les végétaux" (L. Bogorad and J. H. Weil, eds.), p. 77. CNRS, Paris, 1977.
[7] J. D. Rochaix, *J. Mol. Biol.* **126**, 597 (1978).
[8] S. J. Surzycki, this series, Vol. 23, p. 4.

Solutions

A Buffer: 0.1 M NaCl, 50 mM EDTA, 20 mM Tris-HCl, pH 8.0.
Pronase solution: 10 mg/ml Pronase in 0.01 M Nacitrate, pH 5.0, predigested for 2 hr at 37° and stored frozen
TE: 10 mM Tris-HCl, pH 8.0, 1 mM EDTA

In order to obtain preparative amounts of high molecular weight chloroplast DNA, it is advisable to use the cell wall-deficient CW15+ (Davies and Plaskitt[9]) strain. This strain is grown on TAP medium at room temperature, and usually not more than 3 l of culture are processed at a time. It is our experience that the yield and the quality of the chloroplast DNA decreases when large amounts of cells are processed together. When the cell density reaches 3×10^6 to 5×10^6 cells/ml, the cells are collected by low-speed centrifugation (4 min at 2500 g) in the cold. The pellet is resuspended gently with cold TAP medium, the cells are distributed equally into four 30-ml Sorvall Corex tubes, and centrifuged at 2000 g for 5 min. Each of the four pellets is resuspended separately with 8 ml of cold A buffer. After gentle resuspension of the pellet, the cells are transferred into a 50-ml Erlenmeyer flask and 0.5 ml of Pronase solution is added. The cells are transferred into a 50° water bath after addition of 0.5 ml of 20% SDS. This procedure is repeated with the three other pellets. It is important not to pool the cell lysates before or during the incubation at 50° because difficulties with proper solubilization of the lysate may occur. Pronase solution is added after 45 min and again after 90 min of incubation. After 2 to 2.25 hr the color of the lysate starts turning from green to brown. At this point the lysates are cooled on ice, pooled together, and 2 volumes of distilled phenol (saturated with 0.1 M Na borate) are added. The mixture is shaken gently by hand, and after 20 min it is centrifuged in a Sorvall swingout rotor at 7000 g for 15 min. The aqueous phases are pooled and nucleic acids are precipitated with 2 volumes of EtOH. The DNA is spooled on a glass rod, rinsed in 70% EtOH, dried, and resuspended in TE buffer overnight at 4°. It is important to spool the DNA because centrifugation of the EtOH precipitate coprecipitates an unidentified component which will inhibit the restriction endonucleases later on. The resuspended nucleic acids are treated with 30 μg/ml pancreatic RNase for 45 min at 37°. After phenol extraction the DNA is spooled, rinsed, dried, and resuspended in about 10 ml TE overnight at 4°.

Four to six CsCl gradients are prepared by adding 3.35 ml of DNA solution to 4.27 g of solid CsCl. The gradients are centrifuged to equilibrium for at least 60 hr at 35,000 rpm in the Ti50 rotor at 18° to 20°. Gradients can be fractionated either with an ISCO fraction collector or

[9] D. R. Davies and A. Plaskitt, *Genet. Res.* **17,** 33 (1971).

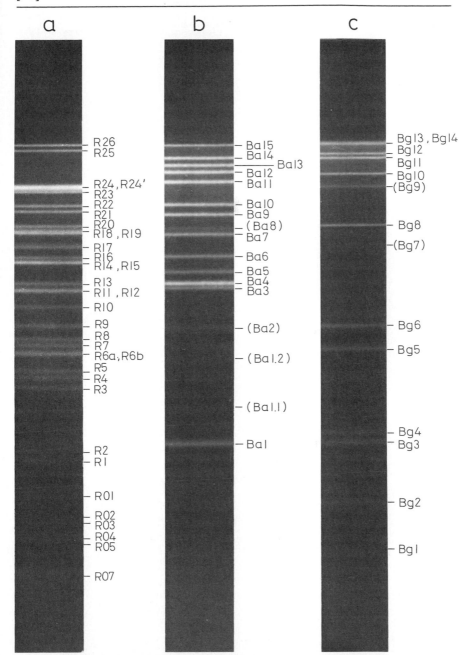

FIG. 1. Agarose gel electrophoretic pattern of chloroplast DNA restriction fragments of *C. reinhardii* produced by *Eco*RI (a), *Bam*HI (b), and *Bgl*II (c). Electrophoresis was performed in 0.8% agarose gels of 40 V for 14 hr at 4°.

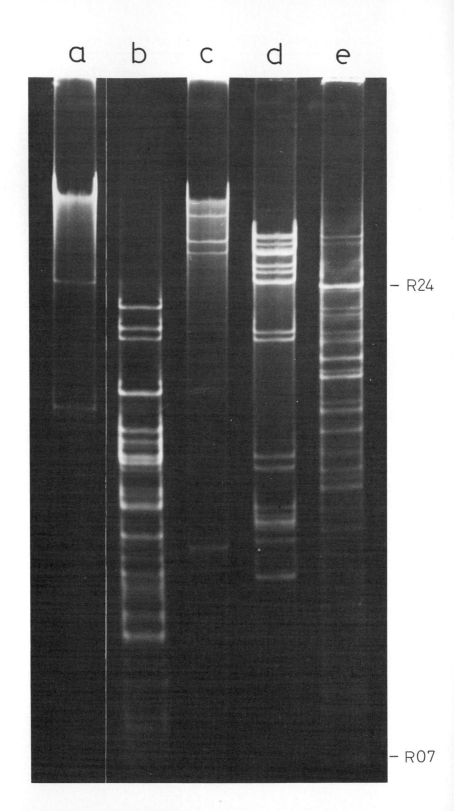

TABLE I

SIZE OF CHLOROPLAST DNA RESTRICTION FRAGMENTS[a]

Fragment no.	EcoRI	BamHI	BglII
07	0.55R		
05	0.72		
04	0.76		
03	0.81		
02	0.85	0.5	
01	0.93	0.75	
1	1.16	2.00R	0.80
2	1.25	(3.40)	1.10
3	1.76	4.20	1.50
4	1.90	4.25R	1.60
5	1.97	4.66	2.55
6a	2.14	5.20	2.96
6b	2.14		
7	2.23	6.00	(5.28)
8	2.30	(6.10)	6.20
9	2.50	7.10	(9.5)
10	2.78	8.10	11.95
11	3.10	10.60	16.75
12	3.10	12.30	18.5
13	3.20	13.90	27.8R
14	3.70	16.20	33.85R
15	3.70	23.75	
16	3.83		
17	4.10		
18	4.54		
19	4.60R		
20	4.83		
21	5.38R		
22	5.55		
23	6.35		
24'	6.65		
24	6.70R		
25	10.80		
26	11.50		

[a] The EcoRI, BamHI and BglII fragments are indicated with their numbers in Fig. 1. Sizes are given in 10^{-6} daltons. Fragments containing ribosomal RNA gene sequences are indicated by R. Fragments whose molecular weight is in parentheses are non-chloroplast DNA fragments (cf. text).

FIG. 2. Agarose gel electrophoretic pattern of chloroplast DNA fragments obtained after digestion with SalI (a), HindIII (b), SmaI (c), PstI (d), and EcoRI (e).

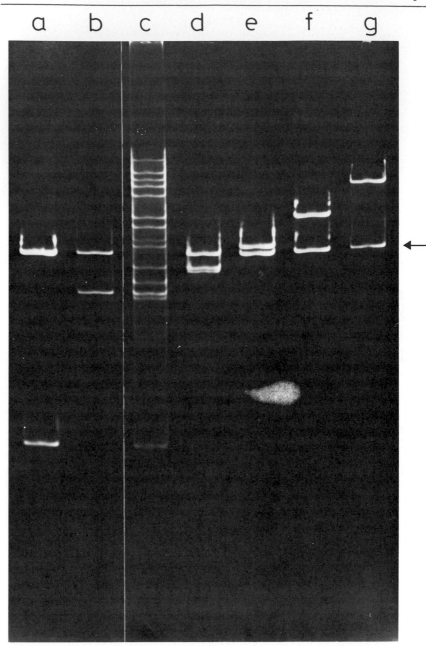

FIG. 3. Agarose gel electrophoretic patterns of chloroplast DNA plasmid hybrids digested with *Bam*HI (slots a, b, d, e, f, g). Slot c represents a *Bam*HI chloroplast DNA digest. The arrow shows the linear pBR313 vector plasmid.

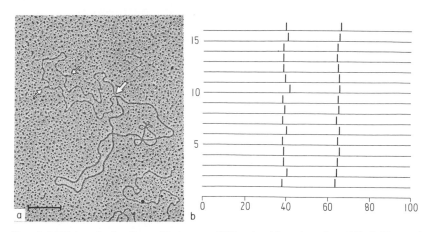

FIG. 4. (a) Heteroduplex formed between pCRI and a chloroplast plasmid hybrid containing *Eco*RI fragment R23. The thick arrow indicates the insertion site of the chloroplast DNA fragment into pCRI. The two thin arrows indicate the hairpin structures on the single-stranded chloroplast DNA which has a kinky appearance. Scale, 1 kb. (b) Drawing of 16 single-stranded R23 fragments with secondary structure features.

manually by puncturing the bottom of the tube. The chloroplast DNA peak ($\delta = 1.696$ g/ml) is readily visible on the light side of the nuclear DNA peak ($\delta = 1.724$ g/ml). A second CsCl equilibrium density gradient is required in order to obtain pure chloroplast DNA with a final yield of 200 to 300 μg per 3 liters of initial culture.

Digestion of the Chloroplast DNA of *Chlamydomonas reinhardii* with Restriction Endonucleases

In order to achieve a complete digestion of the chloroplast DNA it is recommended to use two to three times the amount of restriction enzyme required to digest the same quantity of λ DNA.

Figure 1 shows an agarose gel pattern of chloroplast DNA fragments obtained after digesting the DNA with the restriction endonucleases *Eco*RI (a), *Bam*HI (b), and *Bgl*II (c). The fragments have been numbered with increasing size from R07 to R26 for *Eco*RI, from Ba1 to Ba15 for *Bam*HI, and from Bg1 to Bg14 for BglII. The size of these fragments has been measured by electron microscopy[6] and by comparative agarose gel electrophoresis using DNA fragments of known length as standards such as the λ DNA *Eco*RI fragments[10] or the PM2 DNA *Hae*III fragments.[11] The results are shown in Table I. The sum of the molecular weights of the

[10] M. Thomas and R. W. Davis, *J. Mol. Biol.* **91,** 315 (1974).
[11] K. S. Schmitz and B. R. Shaw, *Science* **197,** 661 (1977).

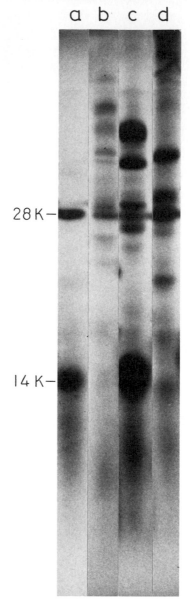

FIG. 5. Autoradiogram of *in vitro* synthesized polypeptides labeled with ^{35}S-methionine and electrophoresed on a 12.5% SDS–polyacrylamide gel. The templates used are pCRI (slot a) and hybrid plasmids containing *Eco*RI fragment R15 (b), R12 (c), and R3 (d). (Courtesy of P. Malnoe.)

fragments produced by each of the three restriction enzymes is close to 126×10^6.

It is apparent in Fig. 1 that several restriction fragments (whose numbers are in parentheses) are present in submolar amounts relative to the other chloroplast DNA fragments. We have shown that these fragments are contaminating nonchloroplast DNA fragments which originate from nuclear ribosomal DNA[12] (fragments Ba1.1, Ba1.2, Bg7) and from mitochondrial DNA[7] (Fragments Ba2, Ba8, Bg9; the corresponding EcoRI fragments comigrate with the fragments R18, R17, and R03).

Figure 1 shows further that several restriction fragments are present in more than one molar amounts. Most of these fragments (i.e., fragments R07, R24, Ba1, and Ba4) are repeated twice in the chloroplast genome and contain gene sequences of the chloroplast ribosomal RNA.[6,7] In contrast fragments R6a and R6b contain different sequences, although they cannot be separated one from another under the conditions used. Further restriction endonuclease digestion patterns of the chloroplast DNA of C. reinhardii are shown in Fig. 2. Slots a, b, c, d, and e display SalI, HindIII, SmaI, PstI, and EcoRI chloroplast DNA fragments, respectively, fractionated by agarose gel electrophoresis. The size of the restriction fragments can be estimated from the known size of the EcoRI fragments (cf. Table I).

Isolation of Chloroplast Restriction Fragments

Chloroplast DNA restriction fragments can be prepared using two different approaches.

Preparative Agarose Gel Electrophoresis. Several EcoRI fragments and most BamHI and BglII fragments can be obtained in pure form by cutting out the corresponding bands from preparative gels. The large chloroplast DNA fragments can be recovered efficiently from agarose gels by dissolving the agarose gel pieces with KI and by subsequent KI density centrifugation.[13] However, this approach is tedious.

Molecular Cloning. Because several chloroplast DNA restriction fragments have almost the same size (e.g., fragments R18 and R19) it is not possible to purify them individually by agarose gel electrophoresis. These fragments can be inserted, however, in appropriate plasmids or bacteriophages. Over 80% of the chloroplast DNA of C. reinhardii has been cloned,[14] and most of the hybrid plasmids appear to be stable and faith-

[12] Y. Marco and J. D. Rochaix, unpublished (1978).
[13] H. Blin, A. V. Gabain, and H. Bujard, *FEBS Lett.* **53**, 84 (1975).
[14] J. D. Rochaix, *in* "Genetics and Biogenesis of Chloroplasts and Mitochondria" (T. Bücher *et al.*, eds.), p. 375. Elsevier/North Holland, Amsterdam, 1976.

fully replicated in the *E. coli* host. It is, therefore, possible to obtain large amounts of pure chloroplast DNA fragments. Figure 3 displays the pattern obtained when *Bam* chloroplast DNA hybrids are digested with *Bam*HI and electrophoresed on an agarose gel. The arrow indicates the linear vector plasmid pBR313,[15] and the other fragments represent the cloned *Bam* restriction fragments which can be seen to correspond to the native *Bam* restriction fragments (slot c).

Use of Chloroplast Restriction Fragments

The availability of pure chloroplast DNA fragments, either purified by biochemical means or cloning, allows the study of several aspects of chloroplast DNA organization and function.

a. Secondary structure of single-stranded chloroplast DNA. Figure 4 shows a heteroduplex formed between a hybrid plasmid containing *Eco*RI fragment R23 and the vector plasmid pCRI. Under these conditions only the chloroplast DNA appears to be single stranded. The thick arrow indicates the insertion site of the chloroplast DNA into the plasmid and the two thin arrows indicate two small hairpin structures, 100 to 200 nucleotides in length, which appear reproducibly in all the 16 molecules examined as shown in Fig. 4. These hairpin structures are also present in other chloroplast restriction fragments.

b. Chloroplast restriction fragments allow the construction of a fine structure map of particular chloroplast genes.

c. Use of chloroplast restriction fragments as templates for *in vitro* protein synthesis. Cloned chloroplast DNA restriction fragments of *C. reinhardii* have been used as templates[14] in the *in vitro* coupled transcription translation system of Zubay.[16] Figure 5 shows an autoradiogram of ^{35}S-methionine-labeled *in vitro* synthesized polypeptides electrophoresed on an SDS–polyacrylamide gel. It can be seen that the vector plasmid pCRI (slot a) produces two major polypeptides of 28,000 and 14,000 daltons in this system. Several of the chloroplast hybrids (slots b, c, and d) synthesize in addition polypeptides ranging in size between 10,000 and 60,000 daltons. It is possible to identify specific chloroplast genes contained in these hybrid molecules by immunoprecipitation of the *in vitro* products with specific chloroplast protein antisera.[17]

d. Chloroplast restriction fragments which contain known chloro-

[15] F. Bolivar, R. L. Rodriguez, M. C. Betlach, and H. W. Boyer *Gene* **2**, 75 (1977).

[16] G. Zubay, D. A. Chambers, and L. C. Cheong, *in* "The Lactose Operon" (J. R. Beckwith and D. Zipser, eds.), p. 375. Cold Spring Harbor Lab., Cold Spring Harbor, New York, 1970.

[17] J. D. Rochaix and P. Malnoe, *in* "Chloroplast Development" (G. Akoyunoglou *et al.*, eds.), p. 581. Elsevier/North Holland, Amsterdam, 1978.

plast genes can be used for isolating specific chloroplast mRNAs, and they can be used as hybridization probes for measuring the concentration of individual mRNAs.

e. Restriction enzyme analysis of the chloroplast DNA of uniparental mutants of *C. reinhardii*[2,3] should be particularly helpful for correlating the genetic and physical map of this chloroplast DNA.

Acknowledgment

This work was supported by the Swiss National Science Foundation no. 3.740.76.

[72] Template Function of Restriction Enzyme Fragments of Phage M13 Replicative Form DNA

By RUUD N. H. KONINGS

I. Introduction

The F-specific filamentous coliphages, among which are included M13, fl, fd, and ZJ/2, are unique among *E. coli* phages in that they do not kill or lyse their host cells (for a recent review, see Denhardt et al.[1]). Their genome consists of a circular covalently closed single-stranded DNA molecule (MW about 2×10^6) which upon infection is converted into a double-stranded replicative form molecule (RF-I). Both *in vivo* as well as *in vitro* only the nonviral strand of this duplex DNA molecule functions as a template for transcription.[2] The DNA genome codes for at least nine polypeptides whose corresponding genes have been ordered into a circular genetic map.[3,4]

By using restriction enzyme fragments of replicative form DNA as a template, the technique of "DNA-dependent *in vitro* protein synthesis" has proved to be a very powerful tool both for the elucidation of the genetic organization as well as for the identification and localization of transcription–initiation (promoter-sites) and –termination signals on the

[1] D. T. Denhardt, D. Dressler, and D. S. Ray, "The Single Stranded DNA Phages." Cold Spring Harbor Lab., Cold Spring Harbor, New York, 1979.

[2] R. N. H. Konings and J. G. G. Schoenmakers, *in* "The Single-Stranded DNA Phages" (D. T. Denhardt, D. Dressler, and D. S. Ray, eds.). p. 507–530. Cold Spring Harbor Lab., Cold Spring Harbor, New York, 1979.

[3] C. A. van den Hondel, A. Weijers, R. N. H. Konings, and J. G. G. Schoenmakers, *Eur. J. Biochem.* **53**, 559–567 (1975).

[4] R. N. H. Konings, T. Hulsebos, and C. A. Van den Hondel, *J. Virol.* **15**, 570–584 (1975).

filamentous phage genome.[4-15] In addition these studies have made it possible to establish, first, the direction of transcription[4]; second, whether genes overlap each other[4,15]; and, third, whether particular phage encoded proteins are synthesized as precursor molecules.[4]

In this article, the techniques we have found most useful for the expression of restriction enzyme fragments of phage M13 replicative form DNA in a coupled transcription and translation system will be described. In addition, the methods used for the identification and characterization of the *in vitro* synthesized polypeptides will be reported.

The identity (gene–product relationship) of the polypeptide made under the direction of a particular restriction enzyme fragment is established with the aid of the following procedures: (a) The polypeptide should not be made, and consequently no polypeptide band should be visible after autoradiographic analysis, in case the coupled system is not programmed by the "wild-type fragment" but by its homologous fragment carrying an amber mutation in the gene coding for the polypeptide whose synthesis is under investigation. The polypeptide band, however, should reappear on the autoradiograph, in case besides the "amber fragment" also (amber) suppressor-tRNA is added to the cell-free system. (b) The polypeptide should not be labeled, and consequently no polypeptide band should be visible after autoradiographic analysis, in case polypeptide synthesis is studied in the presence of a radioactively labeled amino acid(s) which does not occur in the amino acid sequence of the polypeptide whose synthesis is under investigation (differential labeling technique). (c) By analyzing the immunoprecipitates formed between the *in vitro* products and the antibodies raised against a particular phage specific protein on polyacrylamide gels.

[5] R. N. H. Konings, *FEBS Lett.* **35**, 155–160 (1973).
[6] R. N. H. Konings and J. G. G. Schoenmakers, *Mol. Biol. Rep* **1**, 251–256 (1974).
[7] R. N. H. Konings, J. Jansen, T. Cuypers, and J. G. G. Schoenmakers, *J. Virol.* **12**, 1466–1472 (1973).
[8] G. F. Vovis, K. Horiuchi, and N. D. Zinder, *J. Virol.* **16**, 674–684 (1975).
[9] G. Pieczenik, K. Horiuchi, P. Model, C. McGill, B. J. Mazur, G. F. Vovis, and N. D. Zinder, *Nature (London)* **253**, 131–132 (1975).
[10] P. Model, K. Horiuchi, C. McGill, and N. D. Zinder, *Nature (London)* **253**, 132–134 (1975).
[11] C. A. van den Hondel, R. N. H. Konings, and J. G. G. Schoenmakers, *Virology* **67**, 487–497 (1975).
[12] L. Edens, R. N. H. Konings, and J. G. G. Schoenmakers, *Nucleic Acids Res.* **2**, 1811–1820 (1975).
[13] L. Edens, P. van Wezenbeek, R. N. H. Konings, and J. G. G. Schoenmakers, *Eur. J. Biochem.* **70**, 577–587 (1976).
[14] L. Edens, R. N. H. Konings, and J. G. G. Schoenmakers, *Virology* **86**, 354–367 (1978).
[15] C. A. van den Hondel, L. Pennings, and J. G. G. Schoenmakers, *Eur. J. Biochem.* **68**, 55–70 (1976).

The observation that a particular restriction enzyme fragment directs the synthesis of a phage-specific polypeptide does not only give information about its genetic content but also answers the question of whether a promoter (RNA-initiation site) is located on the fragment. Information about the exact position of this promoter can be obtained by comparing the size of the restriction enzyme fragment and the size of the gene of its encoded polypeptide. In addition promoter sites (and termination signals) can be mapped by studying coupled transcription and translation under the direction of restriction enzyme fragments with overlapping nucleotide sequences.[4,12,14,16]

Preliminary Remarks

All reagents are made from sterilized stock solutions. All glassware and related equipment is either heated at 200° for 2 hr, autoclaved, or treated with sulfochromic acid. All manipulations are carried out at low temperature (0°–4°) unless otherwise specified. All *in vitro* reactions are carried out in sterile plastic tubes (Eppendorf), and all solutions are handled with sterile pipettes and/or sterile pipette tips (Oxford Laboratories). Dialysis tubing is boiled in 5% sodium bicarbonate containing 0.1 M EDTA and washed in distilled water. The pH of all buffers is measured at 20°.

II. Preparation of Restriction Enzyme Fragments

Reagents and Buffer Solutions

The pH of all buffers is measured at 20°.
Lysozyme (egg white), from Sigma
Sodium dodecyl sulfate (SDS), from Sigma
RNase A (bovine pancreas), from Boehringer
Sephadex G-100, from Pharmacia
Cesium chloride (analytic grade) from Merck
Ethidium bromide, from Sigma
Bromphenol blue from Sigma
Acrylamide, from Eastman Organic Chemicals
N,N'-Methylenebisacrylamide, from Eastman Organic Chemicals
N,N,N',N'-Tetraethylenediamine (TEMED), from Eastman Organic Chemicals
M-9 medium (per liter): $Na_2HPO_4 \cdot 2H_2O$ (7 g), $KH_2PO_4 \cdot 2H_2O$ (3 g),

[16] L. Edens, R. N. H. Konings, and J. G. G. Schoenmakers, *J. Virol.* **28**, 835–842 (1978).

NH$_4$Cl (1 g), and NaCl (0.5 g). After autoclaving are added 10 ml 0.01 M CaCl$_2$, 10 ml 0.1 M MgSO$_4$, 3 ml 1 mM FeCl$_3$, 1 ml 20% glucose, and 25 ml 4% casamino acids (Difco).

TE buffer: 10 mM Tris-HCl, pH 7.6; 1 mM EDTA

STE buffer: 10 mM Tris-HCl, pH 7.6; 100 mM NaCl; 1 mM EDTA

E buffer: 40 mM Tris-HAc, pH 7.8; 20 mM sodium acetate; 2 mM EDTA

Phenol solution: 500 ml of freshly distilled phenol (Merck) and 0.5 g of 8-hydroxyquinoline (Merck) saturated with STE buffer

Preparation of Replicative Form DNA

Replicative form DNA (RF) is prepared from cells (*E. coli* K37 *sup* D (S26 R1E[17]) infected either with bacteriophage M13 wild type or with M13 phages carrying an amber mutation in a particular gene. Cells are grown under vigorous aeration (P. E. C. Fermentor; Chemap AG., Switzerland) in 10 liters of M-9 medium. At a density of about 4×10^8 cells/ml, aeration is stopped and the culture is infected with the appropriate phage at a multiplicity of 20. After 5 min of phage adsorption and penetration, aeration is restarted, and incubation is continued for 90 min at 37°. The cells are harvested by centrifugation (15 min, 10,000 rpm, 2°; Beckmann, rotor JA 10) washed once with 500 ml of STE buffer (pH 8.0), and resuspended in 80 ml of the same buffer. Then freshly dissolved lysozyme (final concentration 0.1 mg/ml) is added, and the mixture is held for 15 min at 25°. Cells are lysed by the addition, under gently stirring (at this step shearing forces should be avoided) of 0.25 volume of 20% SDS, followed by incubation for 5 min at 60°. Subsequently 4 volumes of STE buffer and 2.5 volumes of 5 M NaCl are added and mixed with the lysate by slow inversion and rolling of the tube. The lysate then is chilled at 0° for 4 hr and centrifuged for 30 min at 14,000 rpm (2°; Beckman, rotor JA 14). Under these conditions, DNA of the host cell is sedimented at the bottom, forming a white granular pellet, whereas RF DNA remains in the supernatant fluid. In case after the centrifugation step still clumps are present in the supernatant, the treatment with 5 M NaCl should be repeated.

The supernatant is extracted twice with an equal volume of chloroform–isoamyl alcohol (24 : 1, v/v) saturated with STE buffer. After the second extraction the nucleic acids are precipitated from the aqueous layer by the addition of 0.1 volume of 3 M sodium acetate (pH 5.5) and 2.5 volumes of ethanol ($-20°$). After standing overnight at $-20°$, the precipitate is collected by centrifugation (30 min, 14,000 rpm, 2°), dissolved in STE buffer and reprecipitated. To destroy contaminating RNA, the RF preparation is dissolved in 50 ml of STE buffer and treated for 1 hr at 37°

[17] A. Garen, S. Garen, and R. C. Wilhelm, *J. Mol. Biol.* **14,** 167–178 (1965).

with RNase A (50 μg/ml, previously heated for 15 min at 80° to destroy contaminating DNase activity). RNase is removed from the incubation mixture by two extractions at room temperature with phenol saturated with STE buffer. RF DNA is recovered from the aqueous layer by ethanol precipitation as described and dissolved in 10 ml of STE buffer. The degradation products of RNA and other low molecular weight material are removed by passing the solution through a Sephadex G-100 column (5 × 50 cm) which is equilibrated in STE buffer. The material eluting at the void volume is collected, and the nucleic acids are precipitated with sodium acetate and ethanol as described. After centrifugation, the precipitate is redissolved in 10 ml of STE buffer and centrifuged to equilibrium in cesium chloride–ethidium bromide. (To minimize photodegradation of dye-complexed DNA, all operations on stained DNA are performed as much as possible in the dark.)

To a centrifuge tube of Spinco rotor 65, 2.4 ml of the RF solution (containing a maximum of 500 μg DNA), 4.24 g of cesium chloride, and 2.0 ml of ethidium bromide (2 mg/ml) are added. After dissolution of the cesium chloride, the solution is overlaid with mineral oil and centrifuged at 20° for 30 hr at 38,000 rpm. Following centrifugation the DNA bands are visualized with the aid of a long-wave uv lamp (366 nm; Desaga, Heidelberg). The lower band contains exclusively RFI (covalently closed, circular double-stranded DNA), while the upper band contains RFII (double-stranded circular DNA containing at least one single-strand break). The DNA bands are removed from the tube with the aid of a hypodermic syringe by piercing the side wall just below the DNA band. Ethidium bromide is removed from the DNA by five extractions with isoamyl alcohol saturated with cesium chloride. After dialysis of the RF solutions against three changes of 300 volumes of STE buffer (16 hr at 2°), the DNA is recovered by ethanol precipitation as described. After standing for 2 hr at −80°, the precipitate is collected by centrifugation (30 min, 25,000 rpm, 2°; Spinco, rotor SW-41), washed once with 0.3 M sodium acetate in 70% ethanol (0°C), dried in a stream of nitrogen gas, and finally dissolved at a concentration of 0.5 mg/ml in 0.1 M Tris-HCl (pH 7.8). As such the DNA solution can be stored for months at −20°. The yield of replicative form DNA is approximately 1.5 mg.

Isolation of Restriction Endonucleases

The restriction endonucleases endo R·HapII, endo R·HindII, endo R·HaeII, endo R·HaeII, endo R·AluI, endo R·BamI, and endo R·HinfI, are isolated according to standard procedures previously described.[15,18,19]

[18] C. A. Van den Hondel and J. G. G. Schoenmakers, *J. Virol.* **18**, 1024–1039 (1976).
[19] C. A. Van den Hondel and J. G. G. Schoenmakers, *Mol. Biol. Rep.* **1**, 41–45 (1973).

Almost all of these purification procedures are described in this volume. A great number of these enzymes now also can be purchased from commercial sources (Bio-Labs, Beverly, Massachusetts; Boehringer, Mannheim, Germany; Porton, Witshire United Kingdom; Sigma, St. Louis, Missouri).

Cleavage of Replicative Form DNA with Restriction Endonucleases

Digestion of replicative form DNA (usually 10 μg) with the appropriate amount of restriction endonuclease (this should be determined in pilot experiments) is carried out for 3 to 5 hr at 37° in 0.25 ml of 10 mM Tris-HCl (pH 7.6), 7 mM magnesium chloride, 7 mM 2-mercaptoethanol, 0.1 mM EDTA, and 0.25 μg/ml gelatin (Oxford). After incubation, the reactions are terminated by the addition of EDTA to a final concentration of 20 mM. Subsequently the mixtures are extracted twice at room temperature with an equal volume of phenol saturated with a STE buffer containing 0.1% SDS. The DNA fragments in the aqueous layer are concentrated by the addition of 0.1 volume of 3 M sodium acetate, 0.1 volume of 0.1 M magnesium acetate, and 2.5 volumes of cold ethanol (−20°). After thoroughly mixing, the mixture is chilled for 1–2 hr at −80°. Subsequently the precipitate is collected by centrifugation (10 min, 12,000 rpm; Eppendorf centrifuge, model 5412), dried in a stream of air, and dissolved in 100 μl of electrophoresis buffer (E buffer) supplemented with 15% sucrose, 0.2% bromphenol blue and 0.02 M EDTA.

Fractionation of the Restriction Enzyme Fragments on Polyacrylamide Gels

The separation of the restriction enzyme fragments is achieved by electrophoresis on vertical slab gels (40 × 18 × 0.2 cm).[3] Routinely we use polyacrylamide as supporting medium, but for this purpose agarose can be used as well.[20,21] The discontinuous polyacrylamide gel consists of a layer of 25 cm of 3% polyacrylamide on top of a layer of 12 cm of 10% polyacrylamide in E buffer. The 3% gel contains 3% acrylamide, 0.15% N,N'-methylenebisacrylamide, 40 mM Tris-HAc (pH 7.8), 20 mM sodium acetate, and 20 mM EDTA. The solution is polymerized by the addition of TEMED and ammonium persulfate (final concentrations: 0.001% (v/v) and 0.05% (w/v), respectively). The 10% gel contains 10% acrylamide (w/v), 0.5% N,N'-methylenebisacrylamide (w/v), 0.00075% TEMED (v/v), and 0.05% ammonium persulfate (w/v) in E buffer.

[20] J. Messing, B. Gronenborn, B. Muller-Hill, and P. H. Hofschneider, *Proc. Natl. Acad. Sci. U.S.A.* **74,** 3642–3646 (1977).
[21] R. N. H. Konings and C. A. Van den Hondel, unpublished results.

For electrophoresis a gel chamber similar to that described by De-Wachter and Fiers is used.[22] After the acrylamide is polymerized, the sample comb is removed and the wells are filled with E buffer, 30–50 μl of the DNA sample is loaded into a 1 cm sample well. Electrophoresis is performed at room temperature with a constant current of 40 mA until the bromphenol blue marker has reached the bottom of the gel (about 20 to 24 hr). After the run, the cell holding the polyacrylamide gel is taken out of the buffer compartment and is carefully disassembled by gently prying apart the glass plates of the mold. The DNA bands are stained by soaking the gel for 30 min in a solution of 2 μg/ml ethidium bromide in E buffer. Subsequently the bands are visualized with a short-wave uv S68 lamp (Ultra-Violet Products Inc., San Gabriel, California) and photographed with a Polaroid MP-4 Land Camera (CU 5) adapted with a Kodak 23A red filter.

Extraction of Restriction Enzyme Fragments from the Polyacrylamide Gel

For successful expression of restriction enzyme fragments in the DNA-dependent cell-free system, it is essential that the fragments are not contaminated with gel fragments and/or gel impurities (both contaminants interfere very strongly with the amino acid incorporation). For these reasons the purification of the restriction enzyme fragments should be carried out with great care. In our hands, two extraction procedures have proved to come up to these requirements quite satisfactory.

Method A: Extraction by Electrophoresis. After visualization of the DNA bands on the polyacrylamide gel, the appropriate gel segments are excised with a scalpel. Subsequently they are cut in pieces and placed in the lower portion of a disposable 10-ml pipette fitted at the tapered end with dialysis tubing. The tubing is tied so that it holds about 1 ml of E buffer. A small cotton plug is inserted in the pipette, and the gel fragments are packed on top of this plug. Finally the gel is compacted with a cotton plug on top. The pipette is then placed in a standard cylindrical gel apparatus (Pleuger), and the DNA is eluted at room temperature by electrophoresis in E buffer. After electrophoresis (18 hr at 100 V) the DNA solution in the lower compartment (dialysis bag) is removed with a hyperdermic syringe. Subsequently the DNA solution, together with the dialysis membrane (during electrophoresis significant amounts of the DNA fragments adhere reversible to the membrane), is extracted twice with an equal volume of phenol saturated with STE buffer containing 0.1% SDS. After phenol extraction the DNA is concentrated by repeated ethanol precipitation at $-80°$. Finally the DNA fragments are dissolved in

[22] R. DeWachter and W. Fiers, this series, Vol. 21, pp. 167–178.

0.1 M Tris-HCl, pH 7.8 (final concentration 0.125 μM; equivalent to 0.5 mg/ml RF). Following this purification procedure, the recovery of the DNA fragments varies between 60 and 90%. The recovery of the larger fragments is higher than that at the smaller fragments throughout.

Method B: Extraction by Diffusion. After excision of the gel segments, they are minced by pushing through a hyperdermic syringe without a needle. The fine minced gel beads are resuspended in 5–10 volumes of STE buffer containing 0.1% SDS, and shaken for several hours at room temperature. The extract is separated from the beads by centrifugation (10 min, 5,000 rpm, 2°; Heraeus-Christ minifuge) and the gel beads are reextracted twice. Finally the extracts are recombined, filtered through nitrocellulose filters (Millipore, 0.45 μm), and extracted twice with phenol saturated with STE buffer containing 0.1% SDS, as described in the previous paragraph. After ethanol precipitation ($-80°$), the fragments are dissolved in STE buffer and further purified by means of centrifugation (16 hr, 32,000 rpm, 4°; I.E.C., rotor SB-60) through a 4-ml 5–30% glycerol gradient in TE buffer. After centrifugation, the bottoms of the tubes are punctured with a needle and 3-drop fractions are collected. The fractions containing the fragments are pooled, and the fragments are recovered by repeated ethanol precipitation at $-80°$. Finally they are dissolved in 0.1 M Tris-HCl (pH 7.8) at a concentration of 0.125 μM (equivalent to 0.5 mg RF per milliliter). Following this purification procedure the recoveries are about 80%.

III. The DNA-Dependent Cell-Free Protein-Synthesizing System

Two coupled transcription and translation systems have found the widest application for DNA-directed cell-free protein synthesis. One is the so-called "preincubated S-30 system" and the other the "DEAE system."[23,24] In our hands, both systems have proved to be useful for the elucidation of the genetic organization as well as for the localization of functional sites on the M13 genome.[4–7,11–16] The main advantage of the DEAE system is that it is more dependent on the addition of exogenous tRNA (suppressor tRNA and/or formylmethionyl-tRNA) than the S-30 system. In addition, the endogenous amino acid incorporation of the DEAE system is lower than that of the S-30 system. For these reasons, currently we favor the use of the DEAE system for our studies on the genetic organization of the M13 genome.

[23] G. Zubay, D. A. Chambers, and L. C. Cheong, *in* "The Lactose Operon" (J. R. Beckwith and D. Zipser, eds.), pp. 375–393. Cold Spring Harbor Lab., Cold Spring Harbor, New York, 1970.

[24] L. M. Gold and M. Schweiger, this series, Vol. 20, pp. 537–542.

Reagents and Buffer Solutions

Antifoam (Anti-Foam-Spray), from Sigma

DEAE-cellulose (type Cellex D, catalogue No. 7480530), from Bio-Rad Laboratories

Micrococcal nuclease (specific activity 9.7 kunits/mg), from PL-Biochemicals

Ribonucleoside triphosphates (neutralized before use to pH 7.0 with NH_4OH), from Boehringer

Phosphoenolpyruvate (crystallized tricyclohexylammonium salt; neutralized before use to pH 7.0 with NH_4OH), from Boehringer

Pyruvate kinase (rabbit muscle; crystalline suspension in ammonium sulfate solution), from Boehringer

tRNA (*E. coli* B-stripped), from General Biochemicals

Folic acid (calcium salt), from Serva

L-[^{35}S]Methionine (specific activity 145 Ci/mmole), from the Radiochemical Centre, Amersham

[^{3}H]-Labeled L-amino acids (specific activity higher than 15 Ci/mmole), from The Radiochemical Centre, Amersham

Ethyleneglycolbis(2-aminoethylether) N,N'-tetraacetic acid (EGTA), from Sigma

DNase I (electrophoretically purified), from Worthington Biochemical Corp.

Hershey broth (per liter): Bacto-nutrient-broth (Difco) 8 g, Bacto-peptone (Difco) 5 g, and NaCl 5 g. Add 5 ml of a 20% solution of glucose after autoclaving

Buffer A: 10 mM Tris-HCl, pH 7.5; 10 mM magnesium acetate; 22 mM ammonium acetate; 1 mM dithiothreitol

Consult Section II for other reagents and buffer solutions.

Bacterial Strains

In principle any strain of *Escherichia coli* K 12 or *E. coli* B is suitable for the preparation of cell-free DNA-dependent protein synthesizing systems. Most popular, however, are strains which lack RNase I. Routinely we prepare our cell-free extracts from the RNase I-deficient strain: *E. coli* K12 AB301 met$^-$, his$^-$, RNase I$^-$ (λ).

Growth of Cells

Normally, bacteria are grown at 37° under vigorous aeration in a rich medium such as Hershey broth.[25] Growth is conducted in a P.E.C. fer-

---

[25] M. H. Adams, "Bacteriophages." Wiley (Interscience), New York.

mentor (Chemap AG., Switzerland) which provides vigorous aeration and a generation time of about 20 min. Since excessive foaming may occur under these conditions, the use of an appropriate sterile antifoam is indicated (minimal amounts should be used). To start growth, the preheated medium is inoculated with 0.01 volume of a fresh stationary phase culture (a few days old culture can also be used if it has been kept refrigerated). When the culture reaches a density of 1×10^8 to 2×10^8 cells/ml, the cells are harvested by pouring quickly onto ice. Cells are collected by centrifugation in the cold (2°) at 10,000 rpm for 10 min (Beckmann, rotor JA10). Immediately thereafter the cells are washed at 0° by resuspension in approximately 250 ml of buffer A and collected by centrifugation in the cold (2°) at 14,000 rpm for 10 min (Beckmann, rotor JA14). Unless the cells are going to be used immediately, they should be quickly frozen ($-70°$, Dry Ice–acetone). The frozen cells are stored either at $-80°$ or in liquid nitrogen. Cells stored for months at these temperatures yield extracts as active as those prepared from fresh cells. Storage at $-20°$ for a substantial period of time results in extracts with decreased activity.

Preparation of Suppressor tRNA

Amber suppressor tRNA is prepared from the permissive *E. coli* strain K37 (S26R1E) carrying the *sup D* (serine) gene.[17] 10 g of freshly harvested cells (wet weight) are washed twice with STE buffer and finally suspended in 20 ml of this buffer. An equal volume of phenol saturated with STE buffer is added, and the mixture is shaken for 1 hr at 25°. The mixture then is centrifuged for 10 min at 5000 rpm (Heraeus-Christ minifuge) and the aqueous layer is removed. The phenol layer is reextracted with 20 ml of STE buffer. To the combined aqueous layers 0.1 volume of 3 M sodium acetate and 2.5 volumes of ethanol ($-20°$) are added and after mixing allowed to stand overnight at $-20°$. The resulting precipitate is collected by centrifugation (10 min, 20,000 rpm, 2°; Beckmann, rotor JA20), extracted three times with 10-ml portions of 1 M NaCl. To the combined fractions 2.5 volumes of ethanol ($-20°$) are added and allowed to stand at $-20°$ for 30 min. After centrifugation the precipitate is washed with 75% ethanol and then extracted three times with 2-ml portions of 0.2 M Tris-HCl (pH 8.8). The combined fractions are incubated at 37° for 5 hr in order to strip the tRNA of its amino acids. After cooling, 8 ml of 0.1 M sodium acetate (pH 5.0), 1.2 ml of 5 M NaCl, and 16 ml of ethanol ($-20°$) are added, and allowed to stand at $-20°$ for 30 min. After centrifugation the precipitate is washed with 0.5 M NaCl in 70% ethanol and dissolved in distilled water to obtain a concentration of approximately 5 mg RNA/ml. To free it from remaining ribosomal RNA, DNA, oligonucleotides, polyphosphates, and polysaccharides, the RNA solution is passed through a DEAE–cellulose column. The solution is applied to the column in 0.01 M Tris-HCl (pH 7.6), and washed with 2 volumes of this buffer, followed by 3 volumes of 0.2 M NaCl, 0.01 M Tris-HCl (pH 7.6). The RNA is eluted from the column with 2 M NaCl in 0.01 M Tris-HCl (pH 7.6) and subsequently dialyzed against three changes of 300 volumes of 0.3 M sodium acetate (pH 5.5). After dialysis the tRNA is concentrated by the addition of 2.5 volumes of ethanol ($-20°$). After standing overnight in the freezer ($-20°$) the precipitate is collected by centrifugation (10 min, 20,000 rpm, 2°; Beckmann, rotor JA20), dried in a stream of nitrogen gas, and dissolved in distilled water at a concentration of 25 mg/ml. Finally the tRNA is stored in aliquots of 20 μl at $-20°$. The yield of tRNA is approximately 3 mg/g of cells.

The Coupled Cell-Free System

The DNA-dependent transcription-translation system is essentially that described by Gold and Schweiger,[24] with particular attention to details described by O'Farrell and Gold.[26] In addition in the preparation of the cell-free system the following modifications are incorporated.

a. After the first low-speed centrifugation of disrupted cells the pellet, which contains cell debris, unbroken cells, and glass beads, is resuspended in a small amount of buffer A and homogenized. The homogenate is centrifuged for 5 min in a Beckmann JA20 rotor (2°, 10,000 rpm) and the supernatant is combined with the supernatant from the first centrifugation for further purification.

b. The postribosomal supernatant (S-100) is freed of nucleic acids by passage through a DEAE–cellulose column (1 × 5 cm) of Bio-Rad Cellex D. The protein fraction is eluted from the column with buffer A + 0.25 M ammonium chloride. The fractions with the highest A_{280}/A_{260} ratio are combined and stored in aliquots of 200 μl (protein concentration 8–10 mg/ml) in liquid nitrogen.

c. If, for the analysis of the *in vitro* products, a cell-free system with a very low endogenous protein synthetic activity is required, immediately before the addition of the ribosomes and supernatant proteins to the cell-free system they are preincubated (15 min at 20°) with 10 μg/ml micrococcal nuclease in the presence of 1 mM calcium chloride.[27] After the preincubation 2 mM EDTA is added to chelate the calcium ions and to destroy the RNase activity. The reaction mixtures we use as standard for phage

[26] P. Z. O'Farrell and L. M. Gold, *J. Biol. Chem.* **248**, 5512–5519 (1973).
[27] R. B. Pelham and R. J. Jackson, *Eur. J. Biochem.* **67**, 247–256 (1976).

M13 duplex DNA directed transcription and translation include (per milliliter): Tris-HCl (pH 7.8), 50 μmole; magnesium acetate, 11.6 μmole; ammonium acetate, 110 μmole; potassium acetate, 50 μmole; dithiothreitol, 2,4 μmole; phosphoenolpyruvate, 20 μmole; ATP, 2 μmole; CTP, UTP, and GTP, 0.5 μmole each; leucovorin (folic acid, calcium salt), 0.3 μmole; the nonlabeled amino acids, 0.25 μmole each; pyruvate kinase, 25 μg; $E.$ $coli$ wild-type or amber-suppressor tRNA (stripped), 1 mg; ribosomes 1.92 mg; and supernatant protein 2 mg. Phage M13 replicative form DNA (or equimolar amounts of restriction enzyme fragments) is normally added at a concentration of 20 μg/ml (5 pM). Usually the polypeptides are labeled with L-[^{35}S]-methionine (50 μCi/ml; specific activity, 145 Ci/mmole). In case of differential labelings studies they are, however, labeled with tritium-labeled amino acids of high specific activity (60 μCi/ml; specific activity >15 Ci/mmole).

After mixing the reagents at 0°, the reaction mixtures are incubated at 37°.

The synthesis of phage M13 specific proteins proceeds via endogenous $E.$ $coli$ RNA polymerase. We never have observed that it is necessary to add extra amounts of RNA polymerase to the cell-free system. After appropriate incubation for transcription and translation (in general 30 min) the reactions are terminated by chilling of the incubation tubes (Eppendorf) in ice water. The synthetic activity of the coupled system then is destroyed by the addition of 1 μl of a solution containing DNase I and RNase A (1 mg/ml each) per 50 μl reaction mixture. The reaction mixtures then are reincubated at 37° for 10 min, chilled again, and processed as described in the subsequent paragraphs.

Determination of Amino Acid Incorporation into Acid-Insoluble Material

Total protein synthesis is determined by spotting 10-μl samples of the reaction mixtures on Whatman 3 MM filter disks and placing the disks in a beaker containing approximately 150 ml of 10% trichloroacetic acid (TCA). The disks then are washed with two 150-ml portions of 10%TCA, followed by boiling for 30 min in 150 ml of 5% TCA. After cooling the disks are washed at room temperature with 150 ml of ethanol–ether (1 : 1, v/v) and ether. After drying the radioactivity present on the filters is measured in a Packard liquid scintillation counter. The stimulation of amino acid incorporation is dependent on the amino acid used to label the in $vitro$ synthesized polypeptides as well as on the DNA template added to the cell-free system. In case replicative form DNA is added, we routinely obtain a tenfold stimulation of the ^{35}S-methionine incorporation above the endogenous level. This corresponds to about 30 μg of newly synthesized bacteriophage proteins per milliliter. The absence of a stimulation of the amino acid incorporation does not necessarily mean that under the direction of the exogenous template (e.g., restriction enzyme fragment) no (phage specific) proteins are made. Frequently we have observed that, due to the addition of the exogenous template, the synthesis of the endogenous proteins is suppressed in favor of the synthesis of the phage M13 DNA encoded polypeptides. For these reasons it is advisable always to analyze the *in vitro* products on polyacrylamide gels (either before or after immunoprecipitation).

IV. Radioimmunological Identification of the *in Vitro* Synthesized
    Polypeptides

Reagents and Buffer Solutions

   Sepharose-protein A (Sepharose CL-4B), from Pharmacia
   PBS buffer: 10 mM phosphate, pH 7.2; 0.9% NaCl
   Immunoprecipitation buffer (I buffer): 10 mM phosphate, pH 7.2;
     0.9% NaCl; 1% Triton X-100; 0.1% sodium deoxycholate; and
     0.1% SDS
   Cracking buffer: 0.125 M Tris-HCl, pH 6.8; 2% SDS;
     2-mercaptoethanol; 20% glycerol; 0.002% bromphenol blue; 8 M
     urea

   Consult previous sections for other reagents and buffer solutions.

Preparation of Antisera

   White New Zealand rabbits are used for the preparation of the antisera. Antisera against phage-specific proteins are prepared by intraperitonial injection of an emulsion of 2 mg of protein (in 1 ml PBS buffer containing 0.1% SDS) and an equal volume of Freund's complete adjuvant. Approximately 4 weeks later this injection is repeated twice at 2-week intervals with the same amount of protein emulsified in 0.2 ml Freund's incomplete adjuvant. Blood is collected by heart puncture 2 weeks after the last injection. The blood is incubated at 37° for 1 hr and at 4°C overnight to allow the clot to retract. The serum is collected, and the remaining blood cells are removed by centrifugation (20 min; 20,000 rpm, 2°; Beckmann, rotor JA20). The sera are stored at −20° in aliquots of 1 ml until use.

Immunoprecipitation of in Vitro Synthesized Phage-Specific Polypeptides

   In the cell-free system only very small amounts of phage-specific proteins are made. For this reason the radioimmunological identification of the *in vitro* synthesized polypeptides can be performed with the aid of the so-called "indirect immunoprecipitation technique." After *in vitro* protein synthesis to every 75 μl of reaction mixture 5 μl of a five times concentrated I buffer is added. After homogenization this mixture is diluted with 125 μl of H_2O, followed by the addition of 2 ml of I buffer. The resulting solution is clarified by centrifugation for 10 min at 58,000 rpm (2°; I.E.C., rotor B-60). After centrifugation to each 150 μl of supernatant 2 μl of antiserum is added, and after mixing the mixture is incubated for 16 hr at 4°C. Subsequently the complexes are precipitated by the addition of either 40 μl of a 10% suspension of *Staphylococcus aureus* cells[28] or by the addition of 40 μl of a 10% suspension of Sepharose–protein A in I buffer. Following incubation for 2 hr at 4°, the precipitate is collected by centrifugation (5 min, 12,000 g, 2°) through a 1.5 cm layer of 10% (w/v) sucrose in I buffer. Subsequently the precipitate is washed three times with I buffer and finally dissolved in 25 μl of cracking buffer and heated for 3 min in a boiling water bath. After cooling the polypeptides are analyzed by electrophoresis on polyacrylamide gels as described.

V. Analysis of the *in Vitro* Synthesized Polypeptides on Polyacrylamide Gels

To establish the gene–product relationship of the *in vitro* synthesized polypeptides after their synthesis (or after their precipitation with antisera) they are routinely fractionated on vertical slab gels of polyacrylamide. The apparatus we use for the preparation of the gel (12 × 12 × 0.12 cm) is that described by Studier.[29] This apparatus is also commercially available (Aquebogue Machine and Repair Shop, Aquebogue, New York).

Reagents and Buffer Solutions

    Coomassie brilliant blue, from Gurr Ltd.
    Dimethyl sulfoxide (DMSO), from Merck
    2,5-Diphenyloxazole (PPO), from Merck

Consult previous sections for other reagents and buffer solutions.

Preparation of Samples for Polyacrylamide Gel Electrophoresis

After reincubation of the *in vitro* samples with DNase I and RNase A, to every 50 μl of reaction mixture 75 μl of cracking buffer is added and

[28] S. W. Kessler, *J. Immunol.* **115**, 1616–1624 (1975).
[29] W. Studier, *J. Mol. Biol.* **79**, 237–248 (1973).

incubated for 3 min in a boiling water bath. After cooling 10–20 μl of the sample is placed in the sample well of the polyacrylamide gel. The immunoprecipitates obtained after the addition of protein A–Sepharose are dissolved in the same buffer. Usually the immunoprecipitate derived from 5 μl of reaction mixture is dissolved in 25 μl of cracking buffer.

Polyacrylamide Gel Electrophoresis

Routinely two different gel systems are used for the analysis of the *in vitro* synthesized polypeptides.

Gel System A: SDS-Phosphate System. The gels contain 10% acrylamide, 0.25% N,N'-methylenebisacrylamide, 0.05% TEMED (v/v), 0.4 mg/ml ammonium persulfate, 0.1% SDS, 0.1 M sodium phosphate, pH 7.2, and 8 M urea. The electrode buffer consists of 0.1 M sodium phosphate, pH 7.2, 0.1% SDS, 1% 2-mercaptoethanol (v/v), and 8 M urea. Electrophoresis is carried out at 20 mA/gel until the bromphenol blue marker has reached the bottom of the gel (about 16 hr). Usually 20 μl of the protein sample is loaded per 1 cm sample well.

Gel System B: SDS-Tris-Glycine System. This system consists of the discontinuous buffer system described by Laemmli.[30] The gels have an acrylamide-bisacrylamide ratio of 15 : 0.4. The resolving gel (11 cm) contains 15% acrylamide, 0.375 M Tris-HCl (pH 8.8), 0.1% SDS, 0.002 M EDTA, and 8 M urea. The spacer gel (1 cm) contains 5% acrylamide in 0.125 M Tris-HCl (pH 6.8), 0.1% SDS, 0.002 M EDTA, and 6 M urea. The electrode buffer contains 0.025 M Tris, 0.192 M glycine, 0.1% SDS, 0.002 M EDTA, and 6 M urea (pH about 8.3). Usually 20 μl of the protein sample are layered under the electrode buffer into the 1 cm sample well of the slab gel. Electrophoresis is carried out at room temperature for 4 hr at 200 V.

The difference in resolving power between both gel systems is that with gel system A a better separation is obtained for polypeptides of low molecular weight (MW less than 10,000). In addition, estimations of molecular weights with the aid of gel system A are more reliable. Besides the fact that separation of the *in vitro* synthesized polypeptides on gel system B is less time-consuming than the separation on gel system A, the former has the additional advantage that, due to the band-sharpening effect of the discontinuous buffer system, polypeptides that are not resolved with the SDS–phosphate system (gel system A) become visible.[4]

Recently we have observed that with the SDS–Tris–glycine system (gel system B) a reproducible separation of polypeptides with molecular weights ranging between 3,000 and 10,000 can be obtained in cases where

[30] U. K. Laemmli, *Nature (London)* **227**, 680–685 (1970).

the spacer gel is omitted and the concentration of Tris-HCl in the resolving gel is raised to 0.5 M (pH 8.0). In addition, the concentrations of glycine and Tris-HCl in the electrode buffer should be lowered to 0.077 M and 0.1 M (pH 8.3), respectively.

The urea in the gels has a dual function: first, it causes the phage-encoded proteins (particularly the major capsid protein encoded by gene VIII) to migrate as monomers; and second, it causes a resolution of polypeptides of low molecular weight.

Autoradiographic Analysis of the Polyacrylamide Gel

After electrophoresis, the slab gels are carefully removed from the gel chamber and fixed for autoradiography by staining for 2 hr in 5 volumes of 0.2% Coomassie brilliant blue in 50% methanol and 7.5% acetic acid. The gels are destained by immersing for several hours in a mixture containing 5% methanol and 7.5% acetic acid. This procedure washes out the urea and any soluble radioactive material, and the stained bands (which represent *E. coli* extract proteins) serve as convenient standards by which to identify any ambiguous autoradiographic bands.

Following destaining, the gel slab is transferred to a piece of prewetted Whatman 3 MM filter paper and dried under vacuum essentially as described by Fairbanks et al.[31] (A similar apparatus is commercially available from Bio-Rad Laboratories.) After drying, the gels on the paper backing are pressed together with X-ray film (Kodak RP/R 54) for varying time periods (dependent on the amount of radioactivity loaded onto the gel) before the film is developed according to standard procedures. Whenever it is necessary to compare autoradiographs, they are scanned with a Kipp microdensitometer equipped with an integrator.

In case the polypeptides have been labeled *in vitro* with tritium-labeled amino acids (differential labeling), after electrophoresis the polypeptide bands are visualized by scintillation autography (fluorography[32]). For this purpose after destaining the gel is dehydrated by immersing twice for 0.5 hour in five times its volume of dimethyl sulfoxide (DMSO). Subsequently the gels are immersed for 3 hr in 4 volumes of 20% (w/v) 2,5-diphenyloxazole (PPO) in DMSO, followed by immersion for 1 hr in 20 volumes of water. Thereafter the gels are dried as described and after drying X-ray film is placed in direct contact with the gel and exposed for various times at $-80°$.

---

[31] G. Fairbanks, Jr., C. Levinthal, and R. H. Reeder, *Biochem. Biophys. Res. Commun.* **20**, 393–399 (1965).

[32] W. M. Bonner and R. A. Laskey, *Eur. J. Biochem.* **46**, 83–88 (1974).

Acknowledgments

The present investigations have been carried out in part under the auspices of the Netherlands Foundation for Chemical Research (SON) and with financial aid from the Netherlands Organization for the Advancement of Pure Research (ZWO).

The author wishes to express his sincere appreciation for helpful criticism and comments to John G. G. Schoenmakers and Mari A. Smits.

[73] Mapping Simian Virus 40 Mutants by Marker Rescue[1]

By CHING-JUH LAI

I. Principle

The technique of marker rescue for mapping simian virus 40 (SV40) mutants was adapted from experiments originally described by Hutchison and Edgell[2] for mapping mutants of bacteriophage ϕX174. The procedure is presented schematically in Fig. 1. Specific wild-type DNA fragments obtained from digestion with restriction endonucleases (endo R) are separately hybridized to singly nicked form II DNA from an SV40 mutant [for example, a temperature-sensitive (ts) mutant], forming a partial heteroduplex molecule. Upon infection of cells, the partial heteroduplex molecules are completed by cellular DNA repair enzymes. Following incubation under conditions where wild-type virus but not the mutant can form plaques, one finds plaques if the DNA fragment used for annealing covers the mutational site. For each ts mutant only one endo R DNA fragment in a given complete digest "rescues" the mutant, from which one can infer that the mutational site falls between the map coordinates of the ends of the active fragment. Further localization of the mutational site can be achieved by rescuing with subfragments or overlapping fragments from several consecutive restriction enzyme digestions.

II. Materials

Cell Culture. Primary African green monkey kidney cells, or established cell lines derived from monkey kidney such as BSC-1, CV-1 cells.

Medium Components

Minimum essential medium (Eagle's)
Fetal calf serum

[1] C.-J. Lai and D. Nathans, *Virology* **60**, 466 (1974).
[2] C. A. Hutchison III and M. H. Edgell, *J. Virol.* **8**, 181 (1971).

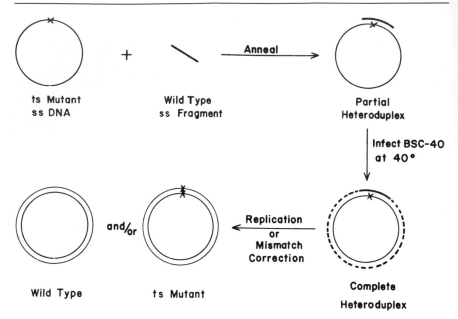

FIG. 1. Scheme of partial heteroduplex formation between single-strand (ss) circles of mutant DNA and a wild-type fragment, and subsequent intracellular events. × indicates the mutant site. From Lai and Nathans.[1]

Penicillin and streptomycin
Glutamine
Agar, Difco
Neutral red

Virus Strains. Temperature-sensitive mutants and their parental wild-type virus (kindly provided by Drs. P. Tegtmeyer and R. G. Martin).

Restriction Enzymes

*Hpa*I
*Hin*dII+III

III. Methods

Preparation of Singly Nicked Form II DNA from ts Mutants. Several enzymatic as well as physicochemical methods have been described to obtain singly nicked DNA from the covalently closed cirlce.[3] It is also possible to obtain form II DNA nicked at a specific location using certain single-cut restriction enzymes in the presence of ethidium bromide.[4,5] A

[3] H. T. Tai, C. A. Smith, P. A. Sharp, and J. Vinograd, *J. Virol.* **9**, 317 (1972).
[4] R. C. Parker, R. M. Watson, and J. Vinograd, *Proc. Natl. Acad. Sci. U.S.A.* **74**, 851 (1977).
[5] D. Shortle and D. Nathans, *Proc. Natl. Acad. Sci. U.S.A.* **75**, 2170 (1978).

convenient physicochemical method involving only visible light irradiation of form I DNA in the presence of ethidium bromide is described. The reaction mixture containing approximately 40 μg/ml form I DNA, 1 M CsCl, 70 μg/ml ethidium bromide, 1 mM EDTA and 10 mM Tris-HCl, pH 8, is placed 10 cm in front of a slide projector lens. The irradiated sample is measured for the conversion of form I to form II DNA by taking aliquots after several time periods and analyzing either by neutral sucrose gradient centrifugation or, more conveniently, by agarose gel electrophoresis.[6] The extent of irradiation is considered optimal for singly nicked DNA when 30% of the covalently closed circle is converted to form II. It is found that a 20–40 min period is required under the conditions described. The nicked form II DNA can be separated from the remaining form I DNA by CsCl–ethidium bromide gradient and by a neutral sucrose gradient centrifugation. The form II DNA fraction after removal of ethidium bromide by chromatography on a Dowex-50 column is dialyzed against 0.1 SSC buffer (containing 0.015 M NaCl and 0.015 M sodium citrate) and stored frozen. Conditions for producing a single nick in SV40 form I DNA by means of endonuclease cleavage in the presence of ethidium bromide are given in Shortle and Nathans;[5] namely, to a reaction mixture containing form I DNA (100 μg/ml) and 120 μg/ml of ethidium bromide endo R·BglI (100 units/ml) is added, and the mixture is incubated at 25° for 2 hr. Other restriction enzymes as well as pancreatic DNase I can also be employed under appropriate conditions. With this procedure over 90% of the DNA can be converted to singly nicked form II, which can be used directly after dialysis.

Preparation of DNA Fragments. Since SV40 DNA is highly infectious in permissive monkey kidney cell lines, wild-type virus plaques would form even if a trace of an active fragment is present in the fragment preparation used for annealing (enhanced rescue to be described). A successful marker rescue experiment, therefore, requires pure DNA fragments for hybridization with the mutant DNA. A useful strategy to obtain pure DNA fragments is stepwise digestion with restriction enzymes and purification of the resulting fragments through polyacrylamide gel electrophoresis. The stepwise digestion and purification has been valuable to obtain unambiguous results for mapping mutational sites of SV40 ts mutants. As an example, wild-type DNA is first cleaved with endo R·HpaI giving three large fragments designated as HpaI A, HpaI B, and HpaI C in the order of decreasing size. These separated HpaI fragments, obtained by electrophoresis on 3.5% polyacrylamide gel [acrylamide : bisacrylamide (20 : 1)], are subsequently digested with endo R·HindII + III, yielding their respective subfragments upon further electrophoretic sep-

[6] P. A. Sharp, B. Sugden, and J. Sambrook, *Biochemistry* **12**, 3055 (1973).

TABLE I

MARKER RESCUE WITH *Hin* FRAGMENTS (PFU/dish)[a]

ts mutants	*Hin* B	*Hin* I	*Hin* H	*Hin* A	*Hin* C	None
A7	42,54	0,0				0,0
A58	0,0	24,31				0,0
A207	0,0	40,37				0,0
A209	0,0	23,10				0,0
A40			20,20	0,0	0,0	0,0
A47			16,17	0,0	1,0	0,0

ts mutant	*Hin* D	*Hin* E	*Hin* K	*Hin* FJG	*Hin* F	*Hin* J	*Hin* G	None
B201	0,0	0,0	0,0	51,52	32,46	0,0	0,0	0,0
B204	0,0	0,0	0,0	9,27	16,23	0,0	0,0	0,0
B218	0,0	0,0	28,37	0,0				0,0
C219	0,0	0,0	0,0	24,20	0,0	25,28	0,0	0,0
C240	0,0	0,0	0,0	>100,>100	0,1	42,55	1,0	0,0
BC214	0,0	0,0	0,0	34,38	3,4	0,0	43,66	0,0
BC216	0,0	0,0	0,0	65,58	2,7	0,0	60,66	0,0
D202	0,0	20,23	0,0	0,0				0,0
D222	0,0	27,24	0,0	1,1				0,0

[a] Data from Lai and Nathans.[11] BSC-40 cells were infected with heteroduplexes formed between form II DNA from a ts mutant and endo R · *Hin*d fragments of wild type DNA. Only the *Hin*d subfragments derived from the active larger *Hpa*I fragment determined from a preliminary experiment were tested. Infection was carried out in duplicate dishes (0.6 cm diameter), each received 0.020–0.030 µg mutant DNA. PFU are plaque forming units scored as the number of plaques in each dish.

aration on 4% polyacrylamide gel.[7] The two-step digestion cleaves the SV40 genome into a total of eleven fragments ranging from 1100 to 200 nucleotides. Using this set of digest products for marker rescue mapping, most ts mutants of SV40 have been localized in fragments of size between 200 and 400 nucleotides.

Formation of Partial Heteroduplexes. As diagrammed earlier, the marker rescue procedure utilizes infection of permissive cells with partial heteroduplex DNA molecules. *In vitro* annealing of the DNA components to generate partial heteroduplexes is carried out as follows. Singly nicked form II DNA (0.04 µg) from a ts mutant is mixed with a tenfold molar excess of a purified wild-type virus DNA fragment in 150 µl of 0.1 SSC, 20 µl of 1 *M* NaOH is added, and the mixture is held for 10 min at room

[7] K. J. Danna and D. Nathans, *Proc. Natl. Acad. Sci. U.S.A.* **68**, 2913 (1971).

temperature, after which the solution is neutralized by adding 30 μl of buffer containing 1 M Tris pH 7.0 : 1 M HCl (2 : 1, v/v). DNA reannealing is next carried out at 68° for 30 min. Under these conditions, less than 20% of the mutant DNA reassociates, whereas over 50% of the DNA forms heteroduplexes with the DNA fragments present, as determined by hydroxyapatite column chromatography.[8] The hybridization mixture containing partial heteroduplex DNA is used within 3 hr for infectivity assay by plaquing on a suitable cell monolayer at the nonpermissive temperature.

DNA Infectivity Assay.[9] Cells from primary African green monkey kidney and several established lines of similar origin have all been used for SV40 plaque assay. For incubation at a nonpermissive temperature for SV40 ts mutants. BSC-40 cells, a derived line of BSC-1, adapted for growth at 40°[10] are satisfactory. Confluent BSC-40 cells grown in minimum essential medium (MEM) supplemented with 10% fetal calf serum in a tissue culture petri dish (6 cm diameter) are rinsed once with 3 ml of phosphate-buffered saline. The reassociated DNA solution (0.2 ml), mixed with 25 μl of 10× MEM containing DEAE–dextran (2.5 mg/ml) is applied to the center of the confluent cell monolayer in duplicate of 0.11 ml per dish. After 30-min incubation at room temperature, the DEAE–dextran mixture is removed, the cells are rinsed again with 3 ml of MEM with 2% serum supplement, and overlayed with 5 ml of 0.9% agar in MEM containing 5% fetal calf serum. The cell plates are placed at 40° in a CO_2 incubator. The infected cells are refed with 2.5 ml of the agar overlay at 5 days postinfection and stained with 2.5 ml overlay containing 0.005% neutral red at 9 to 10 days after infection. Wild-type virus plaques appear in one of a set of DNA fragments used for heteroduplex formation, whereas other fragments are not active (see Table I). From these experiments the mutational site of a defective mutant is physically localized within the active fragment.

IV. Discussion

Enhanced Marker Rescue. Using endo R fragments for mapping SV40 ts mutants by this procedure, it has been observed that some DNA fragment preparations give ambiguous results in which a ts mutant can be rescued by two or more fragments. These results can generally be explained if a trace of the active fragment is present in the other fragment preparation. Subsequent experiments have demonstrated that the efficiency of rescue can be greatly increased if one or more inactive frag-

[8] W. W. Brockman, T. N. H. Lee, and D. Nathans, *Virology* 54, 384 (1973).

[9] J. H. McCutchan and J. S. Pagano, *J. Natl. Cancer Inst.* 41, 351 (1968).

[10] W. W. Bockman and D. Nathans, *Proc. Natl. Acad. Sci. U.S.A.* 71, 942 (1974).

ments are also present during heteroduplex formation.[2] Presumably the enhanced effect is a result of protection of the single-stranded DNA in the heteroduplexes which would otherwise be more susceptible to nucleolytic enzymes. The enhanced rescue technique is particularly valuable for finer mapping where shorter DNA segments are to be used.

Proximity Analysis.[11] Mapping mutants by the marker rescue procedure thus far described localizes the mutational sites to the nucleotide sequences contained within the active fragment. Often two or more mutants may be rescued with the same endo R fragment. In order to provide additional information on defective sites of these mutants, a test for proximity of the two mutations can be further performed. This analysis is based on a possible rescue mechanism through correction of mismatched base pairs by cellular enzymes. Mismatched regions represented by two mutational sites in a heteroduplex molecule would not be corrected if they are identical or located in close proximity. The procedure for proximity analysis, essentially similar to marker rescue, involves infection of cells with DNA heteroduplexes formed between form II DNAs of two ts mutants. Wild-type virus plaques develop if the two sites are repaired independently to generate wild-type viral sequences at both sites. In the case of SV40 ts mutants, DNA from many heteroduplex pairs are not infectious, indicating that the two mutational sites in the pair are identical or very close. The minimal distance required for independent repair of two mismatched base pairs is not known. However, since some heteroduplex pairs that generate wild-type virus are known to map within a 214 base pair segment, the minimal distance must be less than this length.

Acknowledgments

I wish to thank Dr. D. Nathans for his critical reading of the manuscript and Drs. P. Tegtmeyer and R. G. Martin for making this work possible by generously supplying their ts mutants of SV40 while I was in Dr. Nathans' laboratory.

[11] C.-J. Lai and D. Nathans, *Virology* **66,** 70 (1975).

[74] Microinjection of Early SV40 DNA Fragments and T Antigen

By A. GRAESSMANN, M. GRAESSMANN, and C. MUELLER

The microinjection technique is a valuable tool for the assay of biological material at the cellular level. Originally designed for the transplanta-

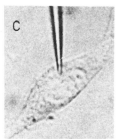
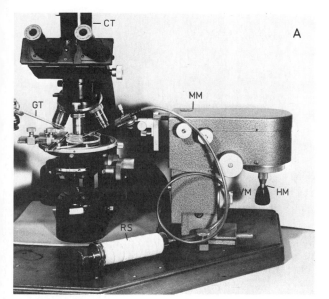
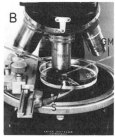

FIG. 1. Assembled instruments for microinjection. (A,B), Slide with recipient cells (S); glass tube (GT) connected to CO_2 cylinder; camera tubus (CT) for attachment of TV or photographic camera; micromanipulator (MM); glass microcapillary (GM); 50 ml Rekord syringe (RS) protected by adhesive textile tape; lever for horizontal micromovements (HM); adjustment knob for vertical micromovements (VM). (C), Capillary tip inside a recipient cell.

tion of mammalian cell nuclei into foreign cells,[1] this method was further miniaturized in our laboratory some years ago, and since then has mainly been used for studies on SV40 and polyoma virus gene expression. Some of the results obtained so far have been reviewed recently.[2] Since the technique can also be used to assay the biological activity of other viral and nonviral nucleic acids as well as proteins in mammalian tissue culture cells,[3–6] it will be described here in some detail. A summary of our results on mapping of early SV40 gene functions will serve as an example for the employment of the microinjection technique.

[1] A. Graessmann, *Exp. Cell Res.* **60**, 373 (1970).
[2] A. Graessmann, M. Graessmann, and C. Mueller, *Curr. Top. Microbiol. Immunol.* **87**, in press (1979).
[3] D. W. Stacey and V. G. Allfrey, *Cell* **9**, 729 (1976).
[4] D. W. Stacey, V. G. Allfrey, and H. Hanafusa, *Proc. Natl. Acad. Sci. U.S.A.* **74**, 1614 (1977).
[5] A. Graessmann and M. Graessmann, *Hoppe-Seyler's Z. Physiol. Chem.* **352**, 527 (1971).
[6] R. Tjian, G. Fey, and A. Graessmann, *Proc. Natl. Acad. Sci. U.S.A.* **75**, 1279 (1978).

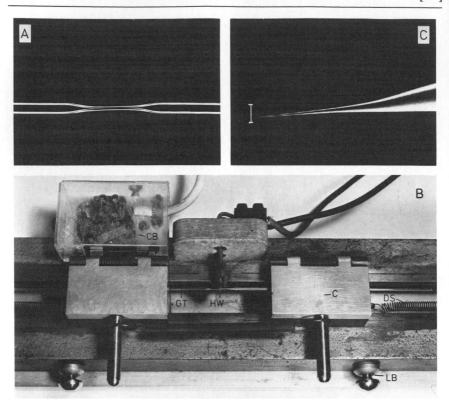

FIG. 2. Pulling of glass microcapillaries. (A), Glass tube with constriction. (B), Partial view of the drawing apparatus with platinum heating wire (HW), glass tube (GT), carriages (C) connected to drawing springs with adjustable tension (DS), locking bolts (LB) to hold carriages in place before the glass tube is inserted, and circuit-breaker (CB) to turn off the current when carriage moves. (C), Microcapillary tip. Bar: 10 μm.

Procedure

General. The microinjection is performed under a phase-contrast microscope (Fig. 1): a glass microcapillary, prefilled from the tip with the biological sample, is directed into the cells to be injected with the aid of a micromanipulator, and an appropriate sample volume is transferred by gentle air pressure exerted by a syringe connected to the capillary. Recipient cells are grown on glass slides imprinted with numbered squares for convenient localization of the cells injected. Events are followed in individual cells, e.g., by immunofluorescence and autoradiography.

Preparation of Glass Microcapillaries. Glass tubes of 1.2 mm inner and 1.5 mm outer diameter (SGA Scientific, Inc., Bloomfield, New Jersey) are broken into pieces of about 30 cm in length and cleaned by treatment with

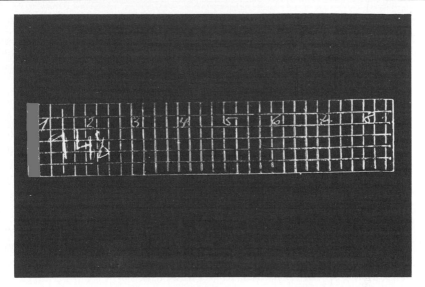

FIG. 3. Scored and numbered glass slide.

a mixture of one-third concentrated H_2SO_4 and two-thirds concentrated HNO_3 (v/v) for 24 hr and extensive washings with tap water, double-distilled water, and ethanol (p.a.). Glass tubes pretreated in this way are air-dried at 120° for 24 hr and kept in a closed container until use.

Before pulling of the capillary a constriction has to be introduced at the middle of the glass tube. This is done by hand using the small flame of a bunsen burner: the glass is softened in the flame and the ends pulled apart outside, resulting in a constriction 8 to 15 mm in length and 0.3 to 0.5 mm in diameter (Fig. 2A). The glass tube is then clamped into a capillary puller, the middle of the constriction being surrounded by the heating wire (Fig. 2B). The shape of the capillary to be obtained is shown in Fig. 2C; the tip should be rigid and open with an outer diameter of about 0.5 μm. Conditions for the mechanical pulling must be optimized by varying the temperature of the heating wire and the pulling forces at the carriages. The puller shown in Fig. 2B was built in our workshop, but commercial pullers suited for this purpose are available from E. Leitz, Wetzlar, West Germany, or from D. Kopf Instruments, Tujunga, California.

Usually, capillary tips are further treated by connecting the capillary to a 50 ml syringe, dipping the tip into 50% HF for 1 sec and then serially washing it by suction and pressure exerted by the syringe with double-distilled water, ethanol (p.a.), tetrahydrofuran (p.a.), 0.5% dichloro-dimethylsilane in tetrahydrofuran (v/v), and again with tetrahydro-furan and ethanol. This treatment produces a smooth capillary tip, but for many purposes it may be omitted without affecting the microinjection procedure. Capillaries are conveniently stored in an upright position (tips upside) by inserting them into fitting holes pierced into a Perspex block. The block is placed into a petri dish and covered with a beaker. Capillaries are then air-dried at 120° for 4 hr.

Preparation of Glass Slides. Glass slides (5 × 1 cm) are coated on both sides with a melted mixture of one-third beeswax and two-thirds stearin (w/w) using a cotton plug, resulting in a thin, supple film. This film is then scored (squares of 1 or 4 mm²) and numbered (Fig. 3) with a steel needle and the glass etched with a paste made from CaF_2 (precipitated) and 40% HF, which is spread over the slide and left there for 10 to 15 min. The slides are kept under running tap water overnight to split off the wax. Finally the slides are treated with a mixture of one-third concentrated H_2SO_4 and two-thirds concentrated HNO_3 (v/v), washed with running tap water for 24 hr, followed by washings with double-distilled water and ethanol (p.a.). They are air-dried and wrapped in aluminum foil, sterilized in an oven at 200° for 24 hr and stored until use.

Cells on Slides. Cells are grown on slides under standard conditions for cell culture. For the investigations described here, TC7 cells (a subline of CV1 monkey cells) and primary mouse kidney cells, prepared from 10-day-old mice (NMRI),[7] were grown in Dulbecco's medium containing 10% fetal calf serum.

Sample. Whenever feasible, the material to be injected is dissolved in an isotonic injection buffer (0.048 M K_2HPO_4, 0.014 M NaH_2PO_4, 0.0045 M KH_2PO_4, pH 7.2), but other compositions such as 0.01 to 0.1 M Tris-HCl, pH 7.2, will be tolerated by the cells. Concentrations up to 1 mg DNA, 5–10 mg RNA, or 10–20 mg protein per milliliter injection solution can be handled. The sample is centrifuged at 10,000 to 15,000 g for 10 min directly before microinjection and then transferred as a small drop to a 100-mm plastic petri dish furnished with a moistened filter paper. The drop may be placed either directly on the plastic surface or on a small sheet of Parafilm, a volume of 2 μl being sufficient. The sample dish is kept on ice.

The preparation of the materials dealt with here, superhelical SV40 DNA I, SV40 specific early complementary RNA (cRNA), early SV40 DNA fragments, and SV40 T antigen (purified from SV80 cells, a human SV40 transformed cell line, by Dr. R. Tjian)[8] has already been described in detail.

Microinjection. A Leitz Ortholux microscope with phase-contrast equipment (Phaco 10/0.25 objective lens, Periplan GF 16× oculars) and a

---

[7] E. Winocour, *Virology* **19**, 158 (1963).

[8] R. Tjian, *Cell* **13**, 165 (1978).

Leitz micromanipulator (Fig. 1) are employed for microinjection in our laboratory. These instruments are placed vibration-free in a room reserved for the purpose which can be uv-sterilized for long-term experiments. The microcapillary is fixed to the instrument holder of the manipulator, and the tip focused under the microscope. Next the sample drop is brought into plane, and the needle is filled by capillary attraction forces supported by negative pressure exerted by the syringe. The capillary tip is slightly raised by turning the vertical adjustment knob of the manipulator, and the cells (immersed in medium in an open 60 mm petri dish) are placed on the microscope stage. A gentle stream of CO_2 maintains the pH optimum. For longer experiments, the stage must be heated. Cells are brought into focus, and a distinct field is chosen for injection. The capillary is now lowered until it is nearly in focus, and an individual cell is approached by operating the manipulator lever for horizontal movements. This cell is injected by further lowering the capillary tip once it is directly above the cell. Both movements of the capillary are controlled by one hand (the right one), while the other hand exerts a gentle pressure on the syringe. A dent is seen on the cell surface when the capillary touches the cell. Microinjection itself is marked by a slight enlargement (swelling) of the cell. Moreover, the nucleus gains in contrast. Either the nucleus or the cytoplasm can be injected. After transfer of the sample, the capillary tip is brought just above the plane of the cell layer and moved to the next cell. About 150 to 200 cells can be injected within 10 min with some practice. After microinjection, cells are maintained as usual, checked at appropriate intervals for possible cytotoxic effects of the sample, and finally processed for evaluation.

The volume injected per cell can be estimated by measuring the distance covered by the meniscus within the pipette after microinjection of a certain number of cells and by determining the inner diameter of the capillary. By this method, we determined a mean injection volume of 1×10^{-11} to 2×10^{-11} ml per fibroblast culture cell, but up to 10^{-10} ml can be transferred. Another approach is to microinject a radiolabeled compound of high activity. A microinjection volume of 1×10^{-10} ml per HeLa cell was also measured using ^{125}I-labeled bovine serum albumin and a slightly modified microinjection system.[3]

Immunofluorescence. The immunofluorescence technique is ideally suited for evaluation of events after microinjection. Cells on slides are washed in phosphate buffered saline (PBS), air-dried, fixed in a mixture of two-thirds acetone and one-third methanol (v/v) at $-20°$ for 10 min, and air-dried again. The injected field is then covered with 5 μl of the appropriately diluted antiserum, and the slide incubated in a humid chamber at $37°$ for 30 to 45 min. Slides are extensively washed with PBS after each staining step. Antisera used for the experiments summarized here were hamster anti-SV40 T, rabbit anti-adeno 2 fiber (both from Cold Spring Harbor Laboratory) and monkey anti-SV40 U (from A. Lewis, Jr.) γ-globulins as first antibodies, and fluorescein-conjugated goat anti-hamster, goat anti-rabbit (Cappel Laboratories), and rabbit anti-monkey (Gibco) γ-globulins as second antibodies. SV40 capsid protein was demonstrated by direct immunofluorescence using fluorescein-conjugated globulin prepared in our laboratory from guinea pigs immunized with purified empty SV40 capsids. Immunofluorescence was observed with a Zeiss standard microscope equipped for reflected light excitation and Zeiss filter combination 48 77 10.

Detection of DNA Synthesis. Stimulation of cellular DNA synthesis upon transfer of SV40 nucleic acids or SV40-related proteins is tested in confluent cultures of TC7 cells or primary mouse kidney cells continuously held in medium with 1% fetal calf serum beginning 24 hr before microinjection. One-tenth microcurie of [methyl-³H]thymidine (Amersham Buchler, 50 Ci/mmole) is added per milliliter of incubation medium directly after microinjection or, if the time course of DNA synthesis is to be followed, at defined time intervals.

After incubation cells are washed, fixed, and stained for viral antigens as described; briefly inspected for the staining quality; and processed for autoradiography. Kodak nuclear track emulsion type NTB is diluted with an equal volume of water, heated to 46°, and stirred. Slides held with a pair of tweezers are dipped into the emulsion and air-dried in an upright position at room temperature for about 10 min. They are then transferred to a slide-holder, wrapped air- and light-tight, and stored over silica gel at 4° in the refrigerator. Three to seven days later, the slides are developed with Kodak D-19 developer for 5 min, washed shortly with water, fixed with Kodak Unifix 1/2/1 for 10 min, and finally washed under running tap water for 30 min. Slides are evaluated under a Zeiss standard microscope by alternately inspecting them at fluorescence excitation or with visible light, whereby stimulation of DNA synthesis can be correlated to the presence of viral antigens. A field of mock injected cells is monitored for incorporation of [methyl-³H]thymidine. The background of DNA-synthesizing cells should be less than 5%.

Microinjection of T Antigen and SV40 DNA Fragments

SV40 T antigen (big-T) is a multifunctional protein[9] synthesized in productively and abortively infected cells.

---

[9] R. Weil, C. Salomon, E. May, and P. May, *Cold Spring Harbor Symp. Quant. Biol.* **39,** 381 (1975).

1. This protein is a virus-coded protein.[10-14] Monkey cells or mouse cells microinjected with early SV40 cRNA efficiently synthesized T antigen, even when cellular RNA synthesis was blocked chemically.[10,11]

2. This protein stimulates cellular DNA synthesis, as shown in cells microinjected with either SV40 cRNA[10] or purified T antigen.[6]

3. This protein initiates late SV40 gene expression when present in appropriate quality[15] and quantity[16] (at least 1×10^6 to 2×10^6 molecules/cell).

4. This protein provides helper function for the growth of adenovirus 2 in monkey cells. TC7 cells infected with adenovirus 2 do not synthesize viral fiber protein (in less than 0.01% of the infected cells), but do so in about 90% of cells upon microinjection of either cRNA,[17] T antigen,[6] or a 23K fusion protein that is partly coded by the C terminal early SV40 coding region.[6]

5. This protein contains the antigenic sites for both T and U antigens. Intranuclear SV40 T and U antigen accumulation is demonstrable in monkey cells, microinjected with either cRNA[17] or T antigen[6] (A. Graessmann, unpublished results).

To map these T antigen functions, we prepared different early SV40 DNA fragments by digestion of DNA I with appropriate restriction endonucleases (Fig. 4) and tested their biological activity in microinjected TC7 and primary mouse kidney cells. The results of these experiments are summarized in Table I.

Summary

All early SV40 functions tested so far (Table I) are also detectable in cells microinjected with the SV40 DNA fragment HpaII/BamHI A (map coordinates 0.16–0.735, Fig. 4). By contrast, fragments cut upstream of map position 0.16 have lost the ability to induce late gene expression

[10] M. Graessmann and A. Graessmann, *Proc. Natl. Acad. Sci. U.S.A.* **73**, 366 (1976).

[11] A. Graessmann, M. Graessmann, H. Hoffmann, J. Niebel, G. Brandner, and N. Mueller, *FEBS Lett.* **39**, 249 (1974).

[12] C. Ahmad-Fadeh, B. Allet, J. Greenblatt, and R. Weil, *Proc. Natl. Acad. Sci. U.S.A.* **73**, 1097 (1976).

[13] C. Prives, E. Gilboa, M. Revel, and E. Winocour, *Proc. Natl. Acad. Sci. U.S.A.* **74**, 457 (1977).

[14] E. Paucha, R. Harvey, R. Smith, and A. Smith, *INSERM* **69**, 789 (1977).

[15] K. Cowan, P. Tegtmeyer, and D. D. Anthony, *Proc. Natl. Acad. Sci. U.S.A.* **70**, 1927 (1973).

[16] A. Graessmann, M. Graessmann, E. Guhl, and C. Mueller, *J. Cell Biol.* **77**, 1 (1978).

[17] C. Mueller, A. Graessmann, and M. Graessmann, *Cell* **15**, 579 (1979).

TABLE I

EARLY SV40-SPECIFIC FUNCTIONS IN MONKEY CELLS MICROINJECTED WITH SV40 NUCLEIC ACIDS OR T ANTIGEN

Sample injected	Percentage of early SV40 genome region	Presence of intranuclear antigens			Cell DNA synthesis in T-positive cells[a]	Initiation of late SV40 gene expression[b] (complementation of tsA viruses at 41.5°)	Helper function for adenovirus 2[c]
		T	U	Capsid			
SV40 DNA I	100	+	+	+	+	ND[d]	+
SV40 cRNA		+	+	−	+	+	+
T antigen (from SV80 cells)		+	+	−	+	ND	+
DNA fragments							
HpaII/BamHI A	100	+	+	−	+	+	+
HpaII/PstI A	80	+	−	−	+	−	−
BamI A	69	+	−	−	+	−	−
HpaI/HpaI B	58	+	−	−	−	−	−
BglI/BamHI A	100	−	−	−	−	−	−
cRNA (from BglI/BamHI A)	100	+	+	−	+	ND	ND

[a] Identical results were obtained in primary mouse kidney cells.

[b] Tested in TC7 cells preinfected with SV40 tsA7 or tsA58 virus and constantly held at the nonpermissive temperature of 41.5°. Intranuclear capsid protein accumulation is the indicator of late viral gene expression in these cells.

[c] Tested in TC7 cells preinfected with adenovirus 2. Fiber protein synthesis is the indicator of adenovirus 2 growth in these cells.

[d] ND, not done.

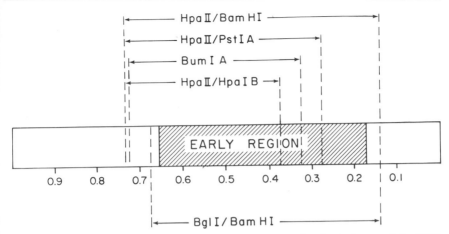

FIG. 4. Assignment of microinjected DNA fragments to the physical map of the SV40 viral genome. Linear SV40 DNA III as produced by cleavage with restriction enzyme *Eco*RI serves as reference.

(complementation of tsA virus at 41.5°), U antigen synthesis, or helper function for adenovirus 2. However, fragment *Bum*I A, containing about 70% of the early coding region, still induces cellular DNA synthesis, whereas of all the functions tested, only intranuclear T antigen could be detected upon microinjection of the *Hpa*II/*Hpa*I B fragment.[17,18]

DNA sequences located at, or clockwise beyond the *Bgl*I cut (map position 0.67) seem to be required for correct intranuclear transcription, since the fragment *Bgl*I/*Bam*HI failed to induce early SV40 functions in microinjected cells (Table I).[17] However, cRNA transcribed from this fragment is translated into a protein(s) with SV40 T and U antigen specificity and the capacity to stimulate cell DNA synthesis following microinjection.

NOTE ADDED IN PROOF:

The injection volume per cell can be enlarged by a factor of 10^5 using as recipients multinucleated giant cells generated by fusion of tissue culture cells with appropriate agents, e.g., polyethylene glycol.[19]

[18] M. Graessmann, A. Graessmann, and C. Mueller, *INSERM* **69**, 233 (1977).

[19] A. Graessmann, M. Graessmann, and C. Mueller, *Biochem. Biophys. Res. Comm.* **88**, 428 (1979).

[75] Assay of Transforming Activity of Tumor Virus DNA

By A. J. VAN DER EB and F. L. GRAHAM*

Introduction

Several techniques have been developed in the past two decades for assaying biological activity of DNA extracted from mammalian viruses.[1] These methods were intended to measure infectivity of viral nucleic acid preparations in cultured cells. Surprisingly, although high specific infectivities could be obtained with a number of different DNAs, especially with the DEAE–dextran technique,[2,3] most of these procedures appeared to be unsuitable for the transformation of nonpermissive cells with viral DNA or for the transformation of permissive cells with noninfectious DNA fragments. Reproducible transformation with viral DNA, however, became possible after the development of the calcium technique in 1973.[4,5] This technique, which was originally developed as an assay for infectivity of adenovirus DNA, appeared suitable also for demonstrating transformation with several viral DNAs, including DNA of adenoviruses, SV40, BK virus, and herpesvirus.[5-9] The differences between the biological effects brought about by the DEAE–dextran and the calcium technique have never been explained adequately, although both methods have been used extensively in virus research. A possible reason for the failure to obtain transformation with the DEAE–dextran technique may be the relative toxicity of this compound for many cultured cells.

It is clear from results obtained by Farber and co-workers[10,11] that uptake of DNA into the cellular cytoplasm or nucleus alone is not necessarily correlated with biological expression. For instance, a facilitator, such as polyornithine, which was very effective in stimulating uptake of DNA into cells, appeared to be unsuitable for demonstrating infectivity.

* Research scholar of the National Cancer Institute of Canada.

[1] F. L. Graham, *Adv. Cancer Res.* **25**, 1 (1977).
[2] J. H. McCutchan and J. S. Pagano, *J. Natl. Cancer Inst.* **41**, 351 (1968).
[3] J. S. Pagano, *Prog. Med. Virol.* **12**, 1 (1970).
[4] F. L. Graham and A. J. van der Eb, *Virology* **52**, 456 (1973).
[5] F. L. Graham and A. J. van der Eb, *Virology* **54**, 536 (1973).
[6] P. J. Abrahams and A. J. van der Eb, *J. Virol.* **16**, 206 (1975).
[7] J. Van der Noordaa, *J. Gen. Virol.* **30**, 371 (1976).
[8] S. Bacchetti and F. L. Graham, *Proc. Natl. Acad. Sci. U.S.A.* **74**, 1590 (1977).
[9] N. J. Maitland and J. K. McDougall, *Cell* **11**, 233 (1977).
[10] F. E. Farber, J. L. Melnick, and J. S. Butel, *Biochim. Biophys. Acta* **390**, 298 (1975).
[11] F. E. Farber, *Biochim. Biophys. Acta* **454**, 410 (1976).

In addition, comparison of the results obtained with the DEAE–dextran and calcium technique indicates that assays for infectivity may not always be suitable for the detection of transforming activity.

In this article we describe the procedures for assaying transformation of cultured cells by viral DNA or DNA fragments, in particular of adenovirus and SV40 DNA. Since the only method found so far to be consistently suitable for the demonstration of transforming activity with these viral DNAs is the calcium technique, we shall limit our discussions to this method and the modifications developed later to enhance its effectiveness.

Principle of the Assay Method

Transforming activity of adenovirus and SV40 DNA is assayed essentially according to the "calcium technique" or "calcium phosphate technique," that was originally developed to measure infectivity of adenovirus DNA.[4]

Briefly, the method is as follows: the viral DNA to be studied is diluted into an isotonic HEPES-buffered saline (pH 7.05), which contains 10 μg/ml of a carrier DNA, usually salmon sperm DNA. To this mixture, $CaCl_2$ is added to a final concentration of 125 mM, which causes the formation of a calcium phosphate precipitate due to the presence of phosphate ions in the HEPES-buffered saline. The viral DNA and carrier DNA molecules coprecipitate with the calcium phosphate, a process requiring approximately 5 to 20 min. Double-stranded DNA in the form of a calcium phosphate complex appears to be more resistant to shear than uncomplexed DNA, as was demonstrated by P. Sheldrick and M. Laithier (personal communication) for native herpes simplex virus DNA.

The calcium phosphate–DNA mixture is inoculated onto monolayer cultures in petri dishes or bottles, by pipetting an aliquot of the mixture onto the cultures. The cultures are incubated for 4–5 hr at 37°, and the medium is subsequently changed with fresh growth medium. If required, the cells can be trypsinized after a few hours and replated at lower cell densities or reseeded in soft agarose medium, depending on which procedure is used for the selection of transformed cell colonies. Foci of transformed cells can usually be scored and isolated after 2–4 weeks.

Preparation of Primary Cultures for Transformation Assay

Reagents

Phosphate buffered saline (PBS)
Trypsin (Difco, 1 : 250), 0.5% solution in phosphate buffered saline (PBS), without calcium and magnesium

Tissue culture growth medium, e.g., Eagle's minimal essential medium (MEM), plus antibiotics

Calf serum or fetal calf serum

Procedure. Transforming activity of adenovirus DNA or DNA fragments can be assayed on monolayer cultures of a variety of cell types. Commonly, nonpermissive or semipermissive cultures of rat or hamster cells are used, but fully permissive human embryonic kidney cells can be transformed also, provided the lytic reaction of the virus is prevented by using fragmented DNA.[12,13]

Primary baby rat kidney (BRK) cells have been most frequently used in transformation studies because such cultures appear to be particularly suitable for studies of viral transformation. The procedure for the preparation of primary BRK cell cultures is described here briefly, since the quality of the cultures is important for obtaining reproducible results. The cultures are prepared as follows: kidneys of 6 to 7-day-old rats are removed aseptically immediately after killing the animals (e.g., by breaking the neck). The kidneys are obtained most conveniently from the dorsal side of the animals, after removal of the skin and cutting through the spine with scissors. The kidneys are washed in PBS containing antibiotics, and are then freed as completely as possible from the enveloping membranes and traces of the urinary duct and blood vessels. The cleaned kidneys are transferred to a suitable vessel, e.g., to a 50-ml centrifuge tube, and are minced with scissors into small pieces. The tissue clumps are then trypsinized in a solution of 0.5% trypsin in phosphate buffered saline according to standard procedures. After an initial 15–20 min of agitation at 37° the suspension is pipetted up and down vigorously, and the individual cells and small cell clumps are harvested. The remaining tissue clumps are further digested by one or two additional incubations with trypsin. Trypsin action is stopped by pipetting the cell suspension in trypsin into an ice-cold mixture of PBS plus calf serum (final serum concentration approximately 5–10%). The cells are collected by low speed centrifugation, and the cell pellet is suspended into 20–30 ml of Eagle's minimal essential medium (MEM) containing antibiotics and 10% newborn or fetal calf serum. After a few minutes at room temperature, the cell suspension is filtered through three layers of sterile cotton gauze to remove precipitated fibrous material, and growth medium is added to the filtered suspension to reach the desired cell concentration. In our experience, best results are obtained if the cells from one pair of baby rat kidneys are suspended in

---

[12] F. L. Graham, P. J. Abrahams, C. Mulder, H. L. Heijneker, S. O. Warnaar, F. A. J. de Vries, W. Fiers, and A. J. van der Eb, *Cold Spring Harbor Symp. Quant. Biol.* **39**, 637 (1974).

[13] F. L. Graham, J. Smiley, W. C. Russell, and R. Nairn, *J. Gen. Virol.* **36**, 59 (1977).

50–60 ml growth medium and are distributed among 10–12 plastic petri dishes with a diameter of 5 cm, or into dishes or flasks having an equivalent total surface area (200–240 cm²). After an incubation for 1 day at 37°, the cultures are washed once with PBS and are incubated for a second day in 5 ml of fresh growth medium. Two days after seeding, the monolayers usually will have reached a stage of about 70–80% confluency, and then are ready for infection with DNA. Highest transformation frequencies are obtained if the cultures are subconfluent at the time of infection. Occasionally, the cultures grow faster and reach about 50% confluency as early as 24 hr after seeding. In this case, it is advisable to infect the cells sooner, i.e., after about 1½ days or, alternatively, to shift the cultures to a lower temperature of 33° during the second day. This shift-down of the temperature slows the growth rate of the cells, but does not seem to affect susceptibility of the cells to transformation.

The initial concentration of cells used for seeding the dishes appears to be an important factor for the transformation assay. We have obtained reproducible results if the cells from a pair of baby rat kidneys are distributed over 10–12 dishes of 5 cm diameter. If the same amount of cells is used to prepare larger numbers of dishes (e.g., 15–20), the cultures usually grow sufficiently fast to still reach a stage of about 70% confluency in 2 days. We have observed repeatedly, however, that such cultures are less sensitive to transformation by adenovirus DNA, and that colonies of spontaneously transformed (fibroblastic) cells may appear after an incubation of 3–4 weeks.

Other types of primary cultures, e.g., derived from baby hamster kidney,[12] human embryonic kidney,[12,13] and rat[5] or hamster embryo (A. J. van der Eb, unpublished results) have also been used in transformation studies with adenovirus DNA. Kidney cultures are normally infected at subconfluency, whereas embryo cultures can be used also when they have just reached confluency.

Transformation by adenovirus DNA has also been assayed on permanent rat cell lines, rather than on primary cells,[14,15] and for some purposes lines may have the advantage of convenience. However, it should be pointed out that the choice of cell could have an important effect on the results obtained (cf. Graham et al.[16]), and generally we have found primary rat kidney cells to be most reproducibly susceptible to transformation.

[14] S. Yano, S. Ojima, K. Fujinaga, K. Shiroki, and H. Shimojo, *Virology* **82**, 214 (1977).
[15] K. Shiroki, H. Handa, H. Shimojo, S. Yano, S. Ojima, and K. Fujinaga, *Virology* **82**, 462 (1977).
[16] F. L. Graham, T. Harrison, and J. Williams, *Virology* **86**, 10 (1978).

Transformation Assay Using the Calcium Technique

Reagents

For the calcium technique:
HEPES-buffered saline (HeBS); 10× concentrated HeBS contains

NaCl	80	g
KCl	3.7	g
Na$_2$HPO$_4$ · 2H$_2$O	1.25	g
Dextrose	10	g
N-2-Hydroxyethylpiperazine N'-2-ethanesulfonic acid (HEPES) (e.g., Calbiochem A grade or Merck)	50	g
H$_2$O	to 1000	ml

Store at −20° in 20–25 ml aliquots.
Salmon sperm DNA (highly polymerized; e.g. Sigma Chemical Company). Salmon sperm DNA stock solution contains 1 mg of DNA/ml 0.01 M NaCl, 0.01 M Tris-HCl pH 7.5, 0.001 M EDTA. Store at 0°.
CaCl$_2$, 2.5 M stock solution, sterilize by autoclaving or by filtration and store in plastic containers.
Viral DNA prepared according to standard procedures, or specific viral DNA fragments obtained by digestion with restriction endonucleases. The concentration of viral DNA should be ⩾ 30 µg/ml. The restriction endonuclease fragments should have a corresponding concentration of ⩾ 30 µg "genome equivalent"/ml.[16a]

Assay of Transforming Activity of Adenovirus DNA

The transformation assay to be described below is based on the use of the calcium technique with monolayer cell cultures in 5–6 cm (plastic) petri dishes. One day prior to infection, the medium of the dishes is changed with 5 ml of fresh growth medium. Subconfluent cultures, e.g., baby rat kidney cells, are infected with a solution containing viral DNA (or DNA fragments), salmon sperm DNA (5 µg), and calcium chloride (125 mM) in HEPES-buffered saline (pH 7.05). This mixture in 0.5 ml is pipetted into dishes containing the cell cultures without removal of the growth medium.

The inoculum containing the viral DNA is prepared as follows. A solution of HEPES-buffered saline (HeBS) is prepared prior to use, from a 10× concentrated stock solution. The 1× HeBS buffer is adjusted with

---

[16a] "Microgram genome equivalent" is obtained by dividing the amount of DNA fragment by the fraction of the genome the fragment represents.

0.3–1 M NaOH to pH 7.05 using a pH meter, at room temperature, and the solution is sterilized by membrane filtration or by autoclaving. To prepare the DNA inoculum, the 1× HeBS buffer (pH 7.05) is distributed into sterile tubes. Viral DNA is added, usually to final concentrations of 0.5–10 μg/ml buffer. If specific DNA fragments are used rather than intact viral DNA, the amount of fragment DNA should be equivalent to 0.5–10 μg of intact DNA per ml buffer (0.5–10 μg genome equivalent). Carrier DNA, usually salmon sperm DNA (highly polymerized) is added from a stock solution of 1 mg/ml, to a final concentration of 10 μg/ml, and the solutions containing viral and carrier DNAs are thoroughly mixed. The final step in the preparation of the DNA inoculum consists of the addition of CaCl$_2$ from a 2.5 M stock solution, to a final concentration of 125 mM. The solution is immediately mixed again, and left at room temperature for about 20 min. Addition of calcium ions to the HeBS causes the formation of a calcium phosphate precipitate, which becomes visible as a light turbidity after a few minutes. The formation of this precipitate represents the crucial step, since the viral and carrier DNA molecules are coprecipitated in this reaction with the calcium phosphate, a process that appears essential for DNA uptake.[4]

After about 20 min incubation at room temperature, 0.5 ml of the calcium phosphate–DNA suspension is pipetted dropwise into the cell cultures in 5 cm petri dishes, containing 5 ml of growth medium. In early studies the medium was removed from the cultures prior to inoculation of the DNA–calcium phosphate suspension. In later work[17] it was found that equally satisfactory results were obtained if the medium were left on the cultures at this stage, and this is now the most common procedure. The cultures are then transferred to 37° and the precipitate is allowed to adsorb to (and be taken up by) the cells for 4–5 hr. Microscopic examination of the dishes reveals the presence of a fine granular precipitate, which usually aggregates to larger clumps in the course of the 4 hr. After the adsorption–uptake period, the medium is replaced with fresh MEM plus 8 or 10% newborn calf serum, and the medium is changed every 3–4 days thereafter for a period of at least 3 weeks. For transformation with adenovirus DNA, the growth medium can be changed 3–4 days after infection to medium containing a low concentration of calcium (0.1 mM CaCl$_2$) or to calcium-free medium. Adenovirus-transformed cells are known for their ability to grow in calcium-free media,[18] whereas the proliferation of untransformed fibroblastic cells, in particular, is somewhat inhibited in the absence of calcium ions. The use of calcium-free or low calcium media is recommended, therefore, particularly when hamster or

[17] F. L. Graham, G. Veldhuisen, and N. M. Wilkie, *Nature (London), New Biol.* **245**, 265 (1973).

[18] A. E. Freeman, P. H. Black, R. Wolford, and R. J. Huebner, *J. Virol.* **1**, 362 (1967).

rat embryo cultures are used, since these may contain large numbers of rapidly growing fibroblastic cells. An alternative or additional method of suppressing proliferation of fibroblasts is to use medium containing 5% horse serum rather than calf serum.

The volume of viral DNA added to the HeBS should not exceed approximately 20% of the buffer volume. If this is not possible because the viral DNA solution is too dilute, the DNA preparation should be either concentrated, or adjusted to 1× HeBS (pH 7.05), e.g., by dialysis or by the addition of concentrated HeBS.

Transformed colonies become visible as early as 8–10 days after the infection of primary BRK cells with adenovirus 5 (Ad5) DNA. The number of colonies increases during the following 7 days and remains approximately constant afterward. Longer incubation periods are required to obtain maximum numbers of transformed foci with DNA of the oncogenic adenoviruses [Ad3 and Ad7 (3–4 weeks), and Ad12 (4–5 weeks)]. The dose–response curve for transformation with Ad5 DNA or Ad5 DNA fragments is approximately linear, at least up to about 1.5 μg/dish.[19] The specific transforming activity of intact Ad5 DNA on primary BRK cells is approximately five transformed foci per microgram of DNA. However, efficiencies may vary from one experiment to another, for reasons that are not yet understood. Fragments of viral DNA, prepared by shearing or by incubation with restriction endonucleases, usually have a transforming activity similar to that of intact DNA, provided the fragments are not too small[19] or the transforming region is not cleaved by the restriction endonuclease.[20] Lower specific transforming activities are observed frequently if restriction endonuclease-generated fragments are used at concentrations of 3 μg/dish or higher.[20] This decline in transforming activity is probably caused by impurities originating from the agarose or acrylamide gel, which copurify with the DNA fragments.

As mentioned earlier, primary baby rat kidney cells are particularly suitable for transformation studies, since the cultures have a very limited life span and cannot be transferred for more than two to three passages. In addition, the cells remain highly contact inhibited, and gradually disappear from the culture in the course of 2–4 weeks, leaving clearly visible the colonies of transformed cells. The latter property is essential for the identification and isolation of cells transformed by the Ad5 *Hpa*I E fragment (4.5% of the genome) which represents only part of the genetic information normally involved in adenovirus transformation.[21]Baby rat

[19] F. L. Graham, A. J. van der Eb, and H. L. Heijneker, *Nature* (*London*) **251**, 607 (1974).
[20] A. J. van der Eb, C. Mulder, F. L. Graham, and A. Houweling, *Gene* **2**, 115 (1977).
[21] A. J. van der Eb, H. van Ormondt, P. I. Schrier, J. H. Lupker, H. Jochemsen, P. J. van den Elsen, R. J. de Leys, J. Maat, C. P. van Beveren, R. Dÿkema, and A. de Waard, *Cold Spring Harb. Symp. Quant. Biol.* **44**, in press (1979).

kidney cells have also been found to be essential for the detection of transformation by certain "transformation defective" mutants of Ad5.[16]

We have observed generally lower specific transforming activities with DNA from weakly oncogenic Ad3 or Ad7 (approximately one focus per microgram of DNA),[21a] and with DNA from highly oncogenic Ad12 (approximately 0.5 foci/μg DNA or less; H. Jochemsen, unpublished results; S. Mak *et al.*[21b]). However, others have reported transforming efficiencies for Ad12 DNA in permanent rat cell lines which are more similar to that obtained with Ad5 DNA,[14,15] so that it is not clear whether a real difference exists between these adeno serotypes.

Transformed colonies can be isolated from the dishes, e.g., by trypsinization, using a small glass or stainless steel cylinder ± 1 cm in length and 4–5 mm inner diameter, to separate the colony from the rest of the culture. The cylinders can be stuck to the bottom of the dish with sterilized silicon grease. Alternatively, cultures can be isolated by scraping the cells of a transformed colony from the dish by means of a Pasteur pipette. In both procedures, the monolayers should be rinsed first with warm PBS, and most of the PBS should be removed. The isolated cells are transferred to culture vials with a small surface area (e.g., tissue culture tubes, small petri dishes, or multiwell dishes), and the cells are incubated in a small volume of growth medium containing 10% fetal calf serum. Permanent lines can usually be established with a success rate of 60–80%.

For quantitative assays of transforming activity, colonies are visualized and counted most conveniently by fixing and staining the cultures at the end of the transformation assay. Immediately before fixation, the medium should be removed and the cultures washed with two changes of PBS. The cells can be fixed, e.g., with a mixture of 3 volumes of methanol and 1 volume of glacial acetic acid. The fixation time is 20 min. The cultures are then dried, stained with Giemsa for 20–30 min, and finally washed in running tap water and air-dried.

Assay of Transforming Activity of DNA from Viruses of the SV40 Polyoma Group and of Herpesviruses

SV40. Transforming activity of DNA of SV40 virus is assayed essentially as described for Ad5 DNA. Foci of densely packed, multilayered cells appear as early as 9 days after infection of primary BRK cells and reach a maximum number after about 3 weeks. The specific transforming activity of SV40 DNA is 20–25 foci/μg DNA. The dose–response curve with form I circular DNA is approximately linear from 1 μg up to about 10 μg/dish, but the curve does not seem to extrapolate to zero.[6,12] This phe-

[21a] R. Dÿkema, B. M. M. Dekker, M. J. M. van der Feltz, and A. J. van der Eb, *J. Virol.*, in press (1979).

[21b] S. Mak, I. Mak, J. R. Smiley, and F. L. Graham, *Virology*, in press (1979).

nomenon appears to be caused by an increase in the specific transforming activity at DNA concentrations lower than ± 1 μg/dish. Linear SV40 DNA molecules of genome size and SV40 DNA fragments, obtained by digestion with restriction endonucleases that do not introduce breaks in the early region, have approximately the same transforming activity as circular DNA.[22]

Other cell types that have been transformed with SV40 DNA or DNA fragments include baby hamster kidney (P. J. Abrahams, unpublished results), mouse 3T3 cells,[6] and human diploid fibroblasts. Relatively high transformation frequencies are obtained with 3T3 cells, if the cultures are trypsinized and replated at lower cell densities 1 day after infection. Successful transformation of human diploid fibroblasts by SV40 DNA, using the DEAE–dextran technique, has been reported.[23-25] Human diploid fibroblasts, however, are semipermissive for SV40 replication, hence it is possible that these transformations were caused by intact virus that had been produced by a preceding lytic response, rather than by viral DNA. Recently, diploid human fibroblasts have been transformed by a 59% SV40 DNA fragment, using the calcium technique and the soft agar suspension assay (H. L. Ozer, personal communication). DNA–DNA hybridization experiments demonstrated that these transformed cells contained early viral DNA sequences only.

Cells transformed by SV40 DNA and DNA fragments (the 74 and the 59% fragment) have been shown to contain SV40-specific T antigen by immunofluorescence, complement fixation, and immunoprecipitation (Abrahams et al.[6,22]; P. I. Schrier, unpublished results). BRK cells transformed by intact SV40 DNA yielded infectious virus upon Sendai virus-mediated fusion with BSC-1 cells, but no infectious virus could be rescued from fragment-transformed cells.[22]

BK Virus and Polyoma Virus. The calcium technique has also been used to successfully transform cultured cells with DNA of BK virus.[7] Transforming activity with BK DNA varied between 1.3 and 10 foci/μg DNA. Attempts to transform rat kidney cells with BK DNA using the DEAE–dextran technique were unsuccessful.[7] Transformation of BHK21 cells by polyoma DNA using the hypertonic saline treatment[26] has been reported as early as 1964.[27,28] Relatively high transforming activities, ranging from

[22] P. J. Abrahams, C. Mulder, A. Van de Voorde, S. O. Warnaar, and A. J. van der Eb, *J. Virol.* **16**, 818 (1975).

[23] S. A. Aaronson and G. T. Torado, *Science* **166**, 390 (1969).

[24] S. A. Aaronson, *J. Virol.* **6**, 470 (1970).

[25] S. A. Aaronson and M. A. Martin, *Virology* **42**, 848 (1970).

[26] R. Weil, *Virology* **14**, 46 (1961).

[27] L. Crawford, R. Dulbecco, M. Fried, L. Montagnier, and M. Stoker, *Proc. Natl. Acad. Sci. U.S.A.* **52**, 148 (1964).

[28] P. Bourgaux, D. Bourgaux-Ramoisy, and M. Stoker, *Virology* **25**, 364 (1965).

200 to 1500 colonies/μg polyoma DNA, were obtained when the soft agar suspension assay[29] was used. Unfortunately, no further studies on transformation by polyoma DNA with the hypertonic saline method have been reported, nor have the transformed cells isolated from soft agar medium been further characterized.

Herpes Simplex Virus. With the exception of one report[30] on morphological transformation of rat embryo cells by herpes simplex virus 1 (HSV-1) DNA, most studies on transformation by HSV DNA have been restricted to transfer of the HSV *TK* gene to TK$^-$ cell lines.[8,9,31,32]

Sheared HSV-2 DNA has been used to transform TK$^-$ human cells,[8] while TK$^-$ mouse cells have been transformed with sheared HSV-2 DNA[9,31] or restriction enzyme fragments of DNA from HSV-1[32] and HSV-2.[9] The specific activities obtained with sheared HSV DNA varied from as low as 1–2 colonies/μg[8] to as high as 80–100 transformants/μg DNA.[31] With restriction enzyme fragments the activities again varied from 1–2 transformants/μg genome equivalent[9] up to 20–30.[32] In the latter case,[32] it was found that specific activities varied with the fragment being assayed, a 3.4 kb *Bam*H I fragment of HSV-1 DNA giving about 20-fold higher activity than unfractionated DNA. It is not at all clear, especially in the case of sheared DNA, why the efficiency of the assays for HSV DNA transformation varies so greatly between different laboratories.

Technical Comments on the Calcium Technique

Most of the parameters involved in the calcium technique were examined (and optimized) in studies on the infectivity of adenovirus DNA.[4] One of the first observations was that formation of the calcium phosphate precipitate is essential for activity, and a number of parameters could affect this stage. First, it was found that the final pH should not be too acid; below about pH 6.8, formation of a precipitate was inhibited. Also, at high pH (above about pH 7.4), although a rather heavy precipitate was formed, the resulting DNA calcium phosphate suspension was noninfectious. Furthermore, formation of a calcium phosphate precipitate and biological activity are blocked by the presence of EDTA above a final concentration of about 1 mM. A second parameter affecting the efficiency of the assay is the final DNA concentration. We routinely have used commercial salmon sperm DNA as a carrier at 10 μg/ml in all our assays. How-

[29] J. MacPherson and L. Montagnier, *Virology* **23**, 291 (1964).

[30] N. M. Wilkie, J. B. Clements, J. C. M. MacNab, and J. H. Subak-Sharpe, *Cold Spring Harbor Symp. Quant. Biol.* **39**, 657 (1974).

[31] A. C. Minson, P. Wildly, A. Buchan, and G. Darby, *Cell* **13**, 581 (1978).

[32] M. Wigler, S. Silverstein, L. S. Lee, A. Pellicer, Y. C. Cheng, and R. Axel, *Cell* **11**, 223 (1977).

ever, S. Silverstein and co-workers (personal communication) have recently found that high molecular weight DNA carriers (30×10^6) give approximately two to fivefold greater efficiencies over low molecular weight carriers.

Another (less critical) factor in the efficiency of the technique is the incubation time at 37° following inoculation of the DNA–calcium phosphate suspension into the cultures. We found that at least 3–4 hr were required for DNA uptake and that uptake required the continued presence of calcium ions in excess of that present in growth medium.[4] Thus calcium ions play a dual role in the assay—first in causing the formation of a calcium phosphate precipitate, second in facilitating DNA uptake during the 3–4 hr incubation at 37°.

Modification of the Calcium Technique

A modification of the calcium technique resulting in higher specific infectivities with herpes simplex virus (HSV) DNA has been described by Stowe and Wilkie.[33] The modification consists of treatment of cell monolayers with a dimethyl sulfoxide (DMSO) solution after infection according to the calcium technique. This procedure enhances the infectivity of HSV-1 DNA in BHK cells 100-fold or more. An enhancement of the specific infectivity of Ad2 and Ad5 DNA by approximately tenfold was observed if HeLa cell cultures were treated with DMSO after exposure to the calcium phosphate–DNA mixture.[34]

Recently, preliminary experiments have indicated that posttreatment with DMSO also enhances the frequency of transformation of primary BRK cells by adenovirus and SV40 DNA (A. Houweling, P. J. Abrahams, and R. Dijkema, unpublished results).

The DMSO procedure is as follows[33]: Cells are infected with viral DNA according to the standard calcium technique. After the 4-hr adsorption period at 37°, the medium containing the viral DNA–calcium phosphate precipitate is removed, and the monolayers are washed once with HeBS (pH 7.05, sterilized by autoclaving). Each culture is then treated with 1.5 ml of a 25% solution of DMSO in water for $2\frac{1}{2}$–4 min. (Best results with BRK cells were obtained by a 3-min treatment). After removal of the DMSO solution, the cultures are washed three times with HeBS, standard growth medium is added, and incubation is continued as in the normal assay. For Ad5 DNA, usually a 2 to $2\frac{1}{2}$-fold enhancement of transforming activity is observed, and for SV40 DNA, it is twofold. For both viral DNAs, the effect is noted most clearly at relatively low DNA concentra-

---

[33] N. D. Stowe and N. M. Wilkie, *J. Gen. Virol.* **33**, 447 (1976).
[34] J. Arrand, *J. Gen. Virol.* **41**, 573 (1978).

tions ($< 1 \mu g$ DNA/dish). Substitution of DMSO by a 15% glycerol solution, which was found to give good results for infectivity of adenovirus DNA in certain cell types (J. Williams, personal communication) did not enhance transforming activity of Ad5 and SV40 DNA in primary BRK cells.

Preparation of Specific Viral DNA Fragments for Transformation

Specific fragments of viral DNA can be prepared by digestion with restriction endonucleases. The procedures for enzyme digestion and separation of the DNA fragments by gel electrophoresis are described elsewhere in this volume (Article 40).

Several methods have been used to extract viral DNA from the gel material (cf. Article 46, this volume). From polyacrylamide gels DNA is usually isolated by electrophoresis, e.g., as described by Galibert *et al.*[35,36]

DNA fragments purified by agarose gel electrophoresis can similarly be isolated by electrophoresis, but other extraction procedures are used more commonly. A method that has been applied successfully for the extraction from agarose gels of DNA fragments suitable for transformation studies is described by Mulder *et al.*[37] The method essentially is as follows (slightly modified): agarose gel slices containing DNA fragments are homogenized in electrophoresis buffer (0.04 M Tris-acetate, 0.01 M sodium acetate, 0.001 M EDTA, adjusted to pH 7.8 with acetic acid), using a Dounce homogenizer. The resulting suspension is incubated overnight at $0°$–$4°$. The agarose is removed by centrifugation, e.g., at 35,000 rpm for 20 min in a Spinco SW-50.1, SW-41 Ti, or similar rotor. The agarose pellet is resuspended in electrophoresis buffer, and residual DNA is extracted at $0°$–$4°$ for an additional 4–6 hr. After removal of the agarose by centrifugation, the supernatant liquid is extracted once or twice with phenol, followed by extraction with chloroform–isoamyl alcohol (24 : 1) and the DNA is precipitated with ethanol. The DNA pellet is redissolved in buffer and either precipitated once more with ethanol, or dialyzed against 0.1 M NaCl, 0.01 M Tris-HCl pH8, 0.001 M EDTA to remove residual phenol, etc. Recoveries may vary between 60 and 75%. A disadvantage of this method is that the extracted DNA fragments usually contain a small amount of agarose, which is very difficult to remove. The presence of agarose may exert an inhibitory effect on subsequent enzyme reactions

[35] F. Galibert, J. Sedat, and E. Ziff, *J. Mol. Biol.* **87,** 377 (1974).
[36] P. H. Steenbergh, J. Maat, H. Van Ormondt, and J. S. Sussenbach, *Nucleic Acids Res.* **4,** 4371 (1977).
[37] C. Mulder, J. R. Arrand, H. Delius, W. Keller, U. Pettersson, R. J. Roberts, and P. A. Sharp, *Cold Spring Harbor Symp. Quant. Biol.* **39,** 397 (1974).

with the fragment, but it does not seem to interfere with the transformation assay. This Dounce method may be combined with the "freeze–squeeze" technique, in which the gel slices containing the DNA are frozen in solid CO_2, and squeezed in a garlic-type press. The liquid that is pressed out of the gel contains much of the DNA.[38]

A fourth method consists of dissolving the agarose slices containing DNA in 3–4 volumes of a $5 M$ sodium perchlorate solution at 60° for a few minutes (C. Mulder, G. Hayworth, and P. A. Sharp, unpublished results). After the agarose has been dissolved, the solution is cooled to room temperature and diluted with water to reach a final concentration of 1.5 M sodium perchlorate. The mixture is immediately applied to a hydroxylapatite column, containing 2–3 ml of hydroxylapatite (BioGel HTP; Bio-Rad), e.g., in a 10-ml disposable syringe, and the column is washed with 1.5 M sodium perchlorate. The column is then eluted at 60°, first with 10–15 ml of a 0.14M sodium phosphate buffer pH 6.8, and then with 2 × 4 ml of 0.4M sodium phosphate buffer of pH 6.8. The DNA, which is eluted in the 0.4M phosphate buffer, is dialyzed extensively against 0.1M NaCl, 0.01 M Tris-HCl, pH 8, 0.001 M EDTA to completely remove the phosphate ions (four changes of buffer in 1½–2 days). The DNA is concentrated by ethanol precipitation and is then ready for use. Disadvantages of this method are the elaborate dialysis step to remove the phosphate and the risk of introducing breaks in the DNA in 5 M sodium perchlorate at an elevated temperature. If the heating time in perchlorate is kept to a minimum, however, the DNA fragments extracted in this way will be suitable for transformation studies.

Tumor Induction by Adenovirus and SV40 DNA

Burnett and Harrington[39] reported the induction of tumors in newborn and 21-day-old hamsters after injection of 3 μg of simian adenovirus SA7 DNA. The same group also tested the two molecular halves of SA7 DNA, and observed that both halves contained tumor-inducing capacity.[40] In later work, it was found that the EcoRI fragment with the highest G + C content had tumor inducing capacity, whereas the AT-richest fragment was inactive. (Endo R · EcoRI introduces a single break approximately in the middle of SA7 DNA.)[41]

Subcutaneous injection of 4–5 μg of adenovirus 12 DNA, or an equivalent amount of adenovirus 12 DNA fragment, into newborn Syrian

[38] R. W. J. Thuring, J. P. M. Sanders, and P. Borst, *Anal. Biochem.* **66,** 213 (1975).
[39] J. P. Burnett and J. A. Harrington, *Proc. Natl. Acad. Sci. U.S.A.* **60,** 1023 (1968).
[40] N. Mayne, J. P. Burnett, and L. K. Butler, *Nature (London), New Biol.* **232,** 182 (1971).
[41] J. P. Burnett, N. Mayne, and L. Helton, *Nature (London)* **254,** 158 (1975).

hamsters resulted in the induction of tumors in a small fraction of the animals[42]: 1 out of 31 hamsters injected with intact DNA and 1 out of 20 hamsters injected with the left-terminal EcoRI C fragment. Analysis by reassociation kinetics demonstrated that the tumor induced with intact adenovirus 12 DNA contained most of the viral DNA sequences, and that the EcoRI C fragment-induced tumor contained 15 copies of fragment C and, in addition, approximately one copy of the right-terminal EcoRIA fragment (H. Jochemsen, unpublished results). The presence of fragment A in the tumor cells is likely to be explained by contamination of fragment C by fragment A.

Subcutaneous injection of 1–2 μg of circular SV40 DNA into newborn Syrian hamsters resulted in the development of sarcomas in 11 out of 33 animals.[43] One tumor was grown in tissue culture and was shown by immunofluorescence to contain viral T antigen and to yield infectious virus after Sendai virus-mediated fusion with BSC-1 cells. Addition of calf thymus DNA or poly-L-ornithine as a facilitator had no influence on the frequency of tumor induction. Linear SV40 DNA molecules prepared by digestion with endo R · EcoRI were also capable of inducing tumors in hamsters.

Note: For further modifications to the calcium technique, see F. L. Graham, S. Bacchetti, R. McKinnon, C. Stanners, B. Cordell, and H. M. Goodman (1979). Transformation of mammalian cells with DNA using the calcium technique. *In* "Introduction of Macromolecules into Viable Mammalian Cells" (R. Baserga, C. Croce, and G. Rovera, eds.), in press. Alan Liss, New York.

[42] H. Jochemsen, thesis, Leiden (1979).
[42] C. J. A. Sol and J. Van der Noordaa, *J. Gen. Virol.* **37**, 635 (1977).

[76] Bacteriophage λ Repressor and cro Protein: Interactions with Operator DNA

By ALEXANDER D. JOHNSON, CARL O. PABO, and ROBERT T. SAUER

Bacteriophage λ codes for two repressor proteins. One, the product of the *cI* gene, is known as the phage λ repressor and will herein be referred to simply as repressor. The other, the product of the *cro* gene, is called the cro protein. These proteins act by binding to the phage DNA at two operators, which are several thousand base pairs apart.[1,2] Each operator

[1] M. Ptashne and N. Hopkins, *Proc. Natl. Acad. Sci. U.S.A.* **60**, 1282 (1968).
[2] Y. Takeda, A. Folkmanis, and H. Echols, *J. Biol. Chem.* **252**, 6177 (1977).

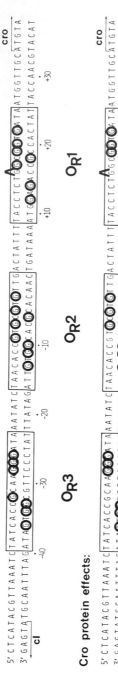

FIG. 1. Sequence of the λ O_R region. O_R1, O_R2, and O_R3 are the λ repressor-binding sites. The transcription starts of the cI and cro genes are indicated by arrows. Guanines which are protected from methylation by repressor or cro protein are circled, and those whose reaction rate is enhanced are indicated by a caret. The top sequence shows the effects of repressor on the methylation of the O_R region, and the bottom sequence indicates the effects of the cro protein. The bases are numbered starting at the HincII cleavage site in O_R. (From Johnson et al.[5])

contains three binding sites for the λ repressor,[3,4] and bases within these same three sites are recognized by the cro protein.[5,6] The DNA sequence of the right operator is shown in Fig. 1. The repressor recognition sequences are boxed. Each repressor binding site consists of 17 base pairs with an axis of partial twofold rotational symmetry through the ninth base pair. The three sites are similar but not identical, and the sites differ in their affinity for repressor. Repressor binds with an order of affinity $O_R1 > O_R2 > O_R3$.[7,8] In contrast, the order of cro protein affinity is $O_R3 > (O_R1, O_R2)$.[5] The different orders of affinity of these two proteins for the three binding sites are responsible for the distinct physiological roles of repressor and cro protein. The mechanism by which repressor and cro protein turn off transcription at O_L and O_R is clear. The recognition sequences (promoters) for RNA polymerase partially overlap the repressor and cro protein binding sites and thus repressor or cro protein bound to the DNA sterically prevents binding of polymerase.[4,5] Repressor also acts as a positive regulator of its own transcription, increasing that transcription where only the strongest binding sites are occupied.[9]

Repressor is a single chain, slightly acidic protein of known sequence, consisting of 236 amino acids.[10] Monomers of repressor are in equilibrium with dimers and higher oligomers.[11] Several lines of biochemical and genetic evidence suggest that the amino terminal section of the molecule is responsible for DNA recognition.[12] The cro protein is a smaller molecule than repressor, containing only 66 amino acids,[13] and solution studies indicate that it exists as a stable dimer.[2,14] The amino acid sequence of cro protein shows no apparent homology with that of repressor,[10,13] nor do the nucleotide sequences of the two genes reveal any obvious relatedness.[15,16]

[3] T. Maniatis, M. Ptashne, K. Backman, D. Kleid, S. Flashman, A. Jeffrey, and R. Maurer, *Cell* **5**, 109 (1975).

[4] M. Ptashne, K. Backman, M. Z. Humayun, A. Jeffrey, R. Maurer, B. Meyer, and R. T. Sauer, *Science* **194**, 156 (1976).

[5] A. Johnson, B. J. Meyer, and M. Ptashne, *Proc. Natl. Acad. Sci. U.S.A.* **75**, 1783 (1978).

[6] A. Johnson, unpublished work.

[7] T. Maniatis and M. Ptashne, *Proc. Natl. Acad. Sci. U.S.A.* **70**, 1531 (1973).

[8] Z. Humayun, A. Jeffrey, and M. Ptashne, *J. Mol. Biol.* **112**, 265 (1977).

[9] B. J. Meyer, D. G. Kleid, and M. Ptashne, *Proc. Natl. Acad. Sci. U.S.A.* **72**, 4785 (1975).

[10] R. T. Sauer and R. Anderegg, *Biochemistry* **17**, 1092 (1978).

[11] P. Chadwick, V. Pirrotta, R. Steinberg, N. Hopkins, and M. Ptashne, *Cold Spring Harbor Symp. Quant. Biol.* **35**, 283 (1970).

[12] R. T. Sauer, C. O. Pabo, B. J. Meyer, M. Ptashne, and K. C. Backman, *Nature (London)* **279**, 396 (1979).

[13] M. W. Hsiang, R. D. Cole, Y. Takeda, and H. Echols, *Nature (London)* **270**, 275 (1977).

[14] A. Folkmanis, Y. Takeda, J. Simuth, G. Gussin, and H. Echols, *Proc. Natl. Acad. Sci. U.S.A.* **73**, 2249 (1976).

[15] T. M. Roberts, H. Shimatake, C. Brady, and M. Rosenberg, *Nature (London)* **270**, 274 (1977).

[16] R. T. Sauer, *Nature (London)* **276**, 301 (1978).

Dimethyl sulfate (DMS) protection studies[5] have revealed that at each repressor binding site, cro protein contacts a subset of the guanines contacted by λ repressor (see below and Fig. 1). A detailed comparison of the interactions of repressor and cro protein with operator DNA may reveal those aspects of DNA–protein interactions which are most important in defining the specificity and strength of binding.

This article discusses some of the biochemical methods used to study the interactions of λ repressor and cro protein with the λ operators. Many of the procedures can be adapted to fit investigations of other repressor–operator systems.

Protein Purifications

Large quantities of λ repressor and λ cro protein can be conveniently isolated from plasmid-containing strains of *E. coli* which overproduce these proteins. These plasmids were constructed using techniques of recombination *in vitro*.[17,18] Bacterial strains carrying plasmid pKB277[17] produce about 1.5% of their soluble protein as repressor. This level is about 150 times that found in a single lysogen. Bacterial strains carrying plasmid pTR214[18] produce 0.5–1.0% of their soluble protein as cro protein. Using the methods described below, gram quantities of repressor and hundred milligram amounts of cro protein can be purified from a kilogram of cells of the appropriate strain.

λ Repressor Purification

Two methods for purifying λ repressor are given below. Method 1 gives excellent recovery of repressor (60–90%) which is highly active as measured by operator binding. In this purification the polyethyleneimine precipitation is the most variable step and most activity losses occur at this stage. Method 2 is easier and faster, but gives somewhat lower recoveries of repressor than method 1. The batch adsorption to the QAE–Sephadex is the most variable step, and in some purifications only 50% of the repressor absorbs to the QAE–Sephadex. A scaled-down version of method 2 has been used in conjunction with gentle lysis procedures emplying lysozyme for isolation of small amounts of ^{35}S-labeled λ repressor. For isolation of ^{35}S-repressor, all buffers are supplemented with bovine serum albumin at a concentration of 1 mg/ml to prevent excessive losses of the small quantities of radioactive protein.

Buffers. Lysis buffer contains 100 mM Tris-HCl (pH 8), 200 mM KCl, 1 mM EDTA, 2 mM CaCl$_2$, 10 mM MgCl$_2$, 0.1 mM dithiothreitol (DTT),

[17] K. Backman and M. Ptashne, *Cell* 13, 65 (1978).
[18] T. M. Roberts, R. Kacich, and M. Ptashne, *Proc. Natl. Acad. Sci. U.S.A.* 76, 760 (1979).

and 5% glycerol. Standard buffer (SB) contains 10 mM Tris-HCl (pH 8), 2 mM CaCl$_2$, 0.1 mM EDTA, 0.1 mM DTT, and 5% (v/v) glycerol. A stock of 1 M potassium phosphate (KP) is made by mixing equal volumes of 1 M KH$_2$PO$_4$ and 1 M K$_2$HPO$_4$. Wash buffer contains 10 mM Tris-HCl (pH 7), 50 mM KCl, 2 mM CaCl$_2$, 0.1 mM EDTA, and 5% dimethyl sulfoxide (DMSO). Assay buffer is wash buffer supplemented with 100 μg/ml bovine serum albumin and 100 μg/ml chick blood DNA. Dilution buffer is standard buffer plus 100 μg/ml bovine serum albumin and 200 mM KCl. All buffers are made using distilled, deionized water, and all pH measurements are made at 20°.

Assay. Repressor can be assayed during purifications by a DNA filter binding assay (see section on filter binding). The ^{32}P-labeled pO_R1 DNA is dissolved at a concentration less than 2 × 10^{-12} M in assay buffer. Aliquots of repressor (1–20 μl) in dilution buffer are added to 250 μl of this assay mix. After 20 min at room temperature (20°–22°) the samples are slowly filtered through 25-mm BA 85 nitrocellulose filters (Schleicher and Schuell) that have been soaked in wash buffer. The filters are washed twice with 300 μl aliquots of wash buffer, and bound DNA is detected by scintillation counting in toluene–liquiflor. Under these conditions, half-maximal DNA binding is observed with approximately 1.1 ng of fully active repressor (see Fig. 2). Other operator-containing DNAs can be used in place of pO_R1 in the filter binding assay. If whole λ DNA is used, half-maximal binding occurs at lower concentrations of repressor since O_L1 has the highest affinity for λ repressor.

Growth and Lysis of Cells. Escherichia coli strain 294/pKB277 is grown to early stationary phase in a 180-liter fermenter in a medium containing 1.5 kg Bacto-tryptone, 0.9 kg Bacto-yeast extract, and 0.9 kg NaCl. The cells are harvested by centrifugation and are resuspended using 1.7 ml lysis buffer per gram wet packed cells. Yields of cells vary from 0.6 to 1.2 kg, and all volumes and amounts in the following purification have been normalized to 1 kg of starting cells.

Cells are lysed by sonication at 4°–8°. The pH is monitored and adjusted to pH 8 by addition of 1 N NaOH when necessary. Sixty milligrams of phenylmethyl sulfonyl flouride are added at the beginning of the lysis procedure. Sonication is monitored by cell counting, and 90% lysis is typically achieved in 2 to 3 hr. The crude lysate is diluted by addition of 8.1 liters SB buffer plus 200 mM KCl, and cellular debris is removed by centrifugation at 13,000 g for 45 min. Repressor remains in the supernatant of this centrifugation and can be stored at 4° for several days or frozen at −20° for several weeks, with no loss of activity. From this point, either of two purification schemes may be followed. All further steps are at 4° unless indicated.

Method 1. The supernatant from the crude lysate centrifugation is made 0.6% in polyethyleneimine,[19] stirred for 10 min, and the precipitate is collected by centrifugation at 6000 g for 15 min. The polyethyleneimine pellet is resuspended in 5 liters of SB plus 200 mM KCl and is recentrifuged. The pellet from this centrifugation is resuspended in 5 liters of SB plus 600 mM KCl, stirred for 10 min, and centrifuged at 6000 g for 15 min. The supernatant containing the repressor is saved.

Repressor in the decanted supernatant is precipitated by addition of 40 g of solid ammonium sulfate per 100 ml of supernatant. The pH is maintained at 8.0 by addition of 1 N NaOH when necessary. The ammonium sulfate suspension is stirred for 30 min, and the precipitate is collected by centrifugation at 13,000 g for 45 min. The ammonium sulfate pellet is resuspended in 500 ml SB buffer and is dialyzed against several changes of SB plus 50 mM KCl.

The dialyzed material is centrifuged to remove any precipitated protein and is loaded onto a 1-liter CM-Sephadex (C-50) column equilibrated in SB buffer plus 50 mM KCl. After loading, the column is washed with 500 ml of SB plus 50 mM KCl. Repressor is then eluted with a linear gradient (8 liters total) from SB buffer plus 50 mM KCl to SB buffer plus 350 mM KCl. Repressor elutes between 150 and 200 mM KCl and is the only major A_{280} peak following the flow through. Fractions containing repressor are located using the DNA filter binding assay or SDS gel electrophoresis[20] and are pooled.

The pooled fractions from the CM-Sephadex column are loaded directly onto a 500 ml hydroxylapatite column (Hypatite C, Clarkson Chemical Co.) equilibrated in 0.1 M KP plus 1.5 mM mercaptoethanol. The column is washed with 500 ml of this buffer, and a 4-liter gradient from 0.1 M KP plus 1.5 mM mercaptoethanol to 1.0 M KP plus 1.5 mM mercaptoethanol is applied. The repressor elutes between 0.5 and 0.75 M KP. Repressor pooled from the hydroxylapatite column is greater than 98% pure by the criterion of SDS gel electrophoresis, and typical preparations yield between 1.5 and 2 g of repressor. Purified repressor is dialyzed into SB buffer plus 200 mM KCl (see Method 2) and is stored frozen at −20° or −70°.

Method 2. Fifty grams of QAE-Sephadex (A-50) is swollen in 10 liters of SB buffer, and this slurry is mixed with an equal volume of the supernatant from the centrifugation of the crude lysate. The slurry is stirred for 1 hr, poured into a glass column, and allowed to drain. The resin is washed with 1.5 liters of SB plus 50 mM KCl, and the repressor is eluted with 3 liters of SB plus 400 mM KCl. Collection of a single 1.5-liter fraction begins as soon as the SB plus 400 mM KCl starts to elute from the

[19] R. R. Burgess and J. J. Jendrisak, *Biochemistry* **14**, 4634 (1975).
[20] U. Laemmli, *Nature (London)* **227**, 680 (1970).

column (at this point a white floculent precipitate is seen in the eluant). This fraction is loaded directly onto a 500-ml hydroxylapatite column which is equilibrated, washed, and developed as in method 1. Pooled repressor fractions at this point are about 95% pure, as determined by gel electrophoresis, and can be used for most purposes. However, the A_{280}/A_{260} ratio is less than that of repressor purified by method 1 (for pure repressor, $A_{280}/A_{260} = 1.81$), and there may be some contamination by oligonucleotides. Further purification of this repressor is achieved by chromatography on CM-Sephadex as in method 1. Since calcium phosphate is relatively insoluble, repressor pooled from the hydroxylapatite column should be dialyzed against SB plus 50 mM KCl without CaCl$_2$, then SB plus 50 mM KCl before CM-Sephadex chromatography. A 200- to 400-ml column is sufficient for 1 to 2 g of repressor.

λ cro Protein Purification

Buffers. Most buffers are adapted from Burgess.[21] Buffer G is 50 mM Tris-HCl, pH 7.9; 2 mM EDTA; 0.1 mM DTT; 1 mM mercaptoethanol; 5% (v/v) glycerol; and 200 mM KCl. Buffer C is 10 mM Tris-HCl, pH 7.9; 0.1 mM EDTA; 0.1 mM DTT; and 5% (v/v) glycerol. A stock solution of buffer C plus 0.2 M KCl plus 100% ammonium sulfate is prepared by saturating buffer C plus 200 mM KCl with ammonium sulfate by stirring overnight at 4°. Solutions containing lower percentages of ammonium sulfate are made by dilution of the 100% stock with Buffer C plus 200 mM KCl.

Assays. The DNA binding assay for cro protein described in detail by Takeda *et al.*[2] can be used to monitor purification. Slab gel SDS electrophoresis (Laemmli,[20] 13.5% acrylamide) can also be used to locate, and assess the purity of, the cro protein. The cro protein migrates near the dye front in this gel system.

Growth and Lysis of Cells. Escherichia coli strain 294/pTR214 is grown to early stationary phase in 180 liters of medium containing 2.8 kg Bacto-tryptone, 1.8 kg NaCl, and 1.8 kg Bacto-yeast extract plus 2 g ampicillin. The following purification scheme is normalized to 1 kg starting cells.

Cells are resuspended in 1.5 liter buffer G and are lysed by sonication (with the addition of phenylmethylsulfonyl fluoride) as described above for the λ repressor purification. All further steps are carried out at 4°. 7.5 liters buffer C and 297 g KCl are added to give a final KCl concentration of 450 mM. Cellular debris is removed by centrifugation at 13.000 g for 45 min, yielding approximately 9.4 liters of supernatant.

Purification. One hundred grams of CM-Sephadex (C-50) are swollen

[21] R. R. Burgess, *J. Biol. Chem.* **244**, 6160 (1969).

overnight in 10 liters buffer C. This slurry is added to an equal volume of the crude lysate supernatant. The mixture is stirred for 2 hr, then vacuum-filtered in batches through a 2-liter sintered glass filter. The resin retained on the filter is scraped into 5 liters of buffer C plus 300 mM KCl. After 1 hr of stirring, the slurry is again filtered, and the retained resin is scraped into a 2-liter chromatography column. Bound protein is eluted with buffer C plus 1 M KCl. The conductivity of the effluent is monitored, and when it begins to rise (after about 2 liters have passed through the column), a single 1.5-liter fraction is collected.

This fraction is dialyzed overnight against buffer C plus 100 mM KCl and loaded onto a 300-ml column of SP-Sephadex (C-25). The column is washed with 500 ml of buffer C plus 100 mM KCl, and is developed with a 2-liter gradient from 100 mM KCl to 1 M KCl in buffer C. The cro protein elutes from the column at a KCl concentration of approximately 300 mM.

Fractions containing cro protein are pooled and precipitated overnight by adding solid ammonium sulfate to 90% saturation (60 g per 100 ml). The precipitated protein is collected by centrifugation at 13,000 g for 45 min. The pellet is then reverse extracted with successively lower concentrations of ammonium sulfate as follows. The 90% pellet is resuspended in 37.5 ml of buffer C plus 200 mM KCl plus 75% ammonium sulfate, stirred for 1 hr, and centrifuged. Both the pellet and the supernatant are saved. The 75% pellet is then resuspended in buffer C plus 200 mM KCl plus 70% ammonium sulfate, stirred for 1 hour, and centrifuged. Following the above procedure, a series of sequential back-extractions are performed at 65, 60, 55, 45, and 40% ammonium sulfate. Supernatant fractions containing cro protein are identified by gel electrophoresis and pooled. These are usually in the range of 45 to 60% ammonium sulfate. At this point, cro protein is the major band seen on SDS gels.

The pooled fractions are brought to 45% ammonium sulfate by addition of buffer C plus 200 mM KCl, loaded onto a 125-ml phenyl-Sepharose (CL-4B) column and rinsed with a column volume of buffer C plus 200 mM KCl plus 45% ammonium sulfate. The column is developed with a 625-ml linear gradient from buffer C plus 200 mM KCl plus 45% ammonium sulfate to 50% (v/v) ethylene glycol. The cro protein elutes late in the gradient, between 6 and 2% ammonium sulfate and comprises the only major A_{280} peak.

Pooled cro fractions are concentrated to 6–8 ml by ammonium sulfate precipitation (>90% saturation) and are loaded onto a 2.5 × 90 cm column of Sephadex G-75 (superfine). The column is run at 5 ml/hr in buffer C plus 200 mM KCl. Again, cro protein is located as the only major A_{280} peak.

This method yields approximately 100 mg of purified cro protein per kilogram of cells which is >95% pure as judged by SDS gel elec-

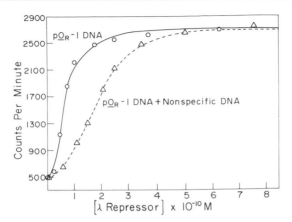

FIG. 2. λ Repressor binding to pO_R1 DNA. Curve 1 (O): nick-translated pO_R1 DNA was dissolved at $2 \times 10^{-12} M$ in assay buffer without the chick blood DNA. Assays were performed at 22° as described (see Repressor Assay section). Repressor concentration refers to repressor monomers. Curve 2 (Δ) was performed identically to curve 1 except that pO_R1 DNA was dissolved in assay buffer including 100 µg/ml chick blood DNA. Each assay tube contained 6700 cpm of pO_R1 DNA.

trophoresis. The overall recovery of cro protein is not known accurately because of the difficulty in determining the cro protein concentration in a crude lysate, either by DNA binding or by gel electrophoresis. We estimate the recovery to be 30%. There are no major losses of cro protein following the CM-Sephadex step.

Filter Binding

Protein–nucleic acid complexes can be trapped by nitrocellulose filtration under conditions where free nucleic acid is not bound.[22,23] Filter binding thus provides a method for measuring the kinetic and equilibrium parameters of protein–DNA interactions,[24,25] and in conjunction with restriction endonuclease cleavage is also useful in identifying and purifying fragments of DNA containing high affinity protein binding sites.

An example of filter binding of ^{32}P-labeled operator DNA by the λ repressor is shown in Fig. 2. Varying amounts of repressor are added to aliquots of a solution containing the labeled DNA which, after equilibration, are filtered through nitrocellulose filters. The filters are then washed,

[22] M. Yarus and P. Berg, *J. Mol. Biol.* **28**, 479 (1967).
[23] A. D. Riggs, S. Bourgeois, R. F. Newby, and M. Cohn, *J. Mol. Biol.* **34**, 365 (1968).
[24] A. D. Riggs, H. Suzuki, and S. Bourgeois, *J. Mol. Biol.* **48**, 67 (1970).
[25] A. D. Riggs, S. Bourgeois, and M. Cohn, *J. Mol. Biol.* **53**, 401 (1970).

and the percentage of bound DNA is determined by scintillation counting. A small percentage of the labeled DNA sticks to the filter in the absence of added repressor (background), and some labeled DNA fails to be bound even at high concentrations of added repressor (see Riggs et al.[24] for discussion). A detailed discussion of the shape of the repressor-operator binding curve is given below.

Sources of Labeled DNA. Phage DNA, plasmid DNA, and restriction fragments can all be used for filter binding experiments. Labeled DNA is routinely prepared by the following methods: (1) *in vivo* labeling of phage DNA,[24] (2) nick translation of purified phage, plasmids, or restriction fragments,[26] and (3) 5' end-labeling of restriction fragments with T4 polynucleotide kinase.[27]

Methods 1 and 2 above result in DNA labeled at high specific activity (10^6–10^7 cpm/μg) which can be used for all the experiments discussed in this section. Occasionally nick-translated DNA has been found unsuitable because of a very high background binding to the nitrocellulose filters or because only a small fraction of the labeled DNA is competent to bind repressor. These problems are rare and have been traced to particular batches of commercial ^{32}P-labeled dXTPs or *E. coli* DNA polymerase I.[28] The use of kinase results in DNA labeled at low specific activities. It is not practical to use this DNA for filter binding experiments requiring low DNA concentrations ($<10^{-12}$ M). Nick translation and kinase have the advantage that stocks of purified DNA ᴄan be quickly labeled when needed.

Phage λ DNA contains multiple repressor binding sites (see above), and thus interpretation of repressor binding to the phage DNA is not straightforward. Two methods are currently employed to isolate DNA molecules bearing a single repressor binding site. (1) λ DNA can be cleaved with the restriction endonuclease *Hinc*II to yield fragments which bear O_L1 only (nominal size, 320 base pairs) and O_R1 only (nominal size, 550 base pairs). Methods for generating and purifying these DNA fragments are described by Maniatis et al.[26,29] (2) A recombinant plasmid (pO_R1) containing a single copy of the repressor binding site O_R1 has been constructed.[6] pO_R1 DNA can be isolated in milligram amounts by the method of Clewell and Helinski,[30] and is a convenient source of DNA bearing a single repressor binding site.

[26] T. Maniatis, A. Jeffrey, and D. G. Kleid, *Proc. Natl. Acad. Sci. U.S.A.* **72,** 1184 (1975).
[27] A. M. Maxam and W. Gilbert, *Proc. Natl. Acad. Sci. U.S.A.* **74,** 560 (1977).
[28] Our unpublished observations.
[29] T. Maniatis, M. Ptashne, and R. Maurer, *Cold Spring Harbor Symp. Quant. Biol.* **38,** 857 (1973).
[30] D. B. Clewell and D. R. Helinski, *Biochemistry* **9,** 4428 (1970).

λ *Repressor–λ Operator Interactions.* Figure 2 shows the binding of λ repressor to DNA that has been labeled by nick translation. This curve is sigmoid because repressor monomers are in concentration-dependent equilibrium with dimers, and the dimer is the form active in binding operator DNA.[11,31,32] Thus at the low repressor concentrations at the beginning of the curve most of the repressor is present as monomers, and monomers do not bind operator DNA strongly. As the repressor concentration is raised, the proportion of dimers increases, giving rise to the sigmoidal increase in binding. Thus, the shape of the binding curve results from two equilibria. One is the association of monomers to form dimers; the other is the binding of dimers to DNA. These equilibria can be represented as follows

$$R_2 \overset{K_1}{\leftrightarrows} 2R \tag{1}$$

$$R_2O \overset{K_2}{\leftrightarrows} R_2 + O \tag{2}$$

The complex equilibria for the binding of repressor monomers to operator can be represented as

$$K_{eq} = K_1K_2 = [R]^2[O]/[R_2O] \tag{3}$$

where the free repressor monomer [R] and free operator [O] concentrations can be related to the total repressor concentration $[R_T]$ and the total operator concentration $[O_T]$ by the following stoichiometry equations

$$[R] = [R_T] - 2[R_2] - 2[R_2O] \tag{4}$$

$$[O] = [O_T] - [R_2O] \tag{5}$$

The value of K_{eq} can be calculated from the binding curve shown in Fig. 2. At the concentration of repressor which gives half-maximal binding of the DNA, the free operator concentration $[O_T - R_2O]$ is equal to the bound operator concentration $[R_2O]$ and thus Eq. (3) reduces to

$$K_{eq} = ([R_T] - 2[R_2O] - 2[R_2])^2 \tag{6}$$

In the experiment of Fig. 2, the operator concentration ($2 \times 10^{-12} M$) is small in comparison with the repressor concentration at which half-maximal DNA binding occurs, and thus the $2R_2O$ term can be ignored. Similarly, from a knowledge of the equilibrium constant for repressor dimers dissociating to free repressor monomers (see below), we can calculate that the concentration of free repressor dimers is very small in comparison with the total repressor concentration, and thus we can make the approximation

---

[31] V. Pirrotta, P. Chadwick, and M. Ptashne, *Nature (London)* **227**, 41 (1970).
[32] R. T. Sauer, Ph.D. Dissertation, Harvard University, Cambridge, Massachusetts (1979).

$$K_{eq} \cong [R_T]^2 \tag{7}$$

Since half-maximal binding occurs at a repressor monomer concentration of $7 \times 10^{-11} M$ ($\pm 1.5 \times 10^{-11} M$), the complex equilibrium dissociation constant K_{eq} is $4.9 \times 10^{-21} M^2/\text{liter}^2$ ($\pm 1.5 \times 10^{-21} M^2/\text{liter}^2$). This represents the value for binding of repressor to the single operator site $O_R 1$.

The monomer–dimer equilibrium has been studied directly using partition chromatography methods,[32] and the equilibrium dissociation constant ($K_1 = [R]^2/R_2$) for this reaction is $1.7 \times 10^{-8} M$ ($\pm 0.4 \times 10^{-8} M$). Since we know the values of K_{eq} and K_1 we can calculate from Eq. (3) the equilibrium dissociation constant ($K_2 = [R_2] [O]/[R_2O]$) for repressor–operator complexes dissociating to free dimers and operators. Under the binding conditions of Fig. 2, K_2 is $2.9 \times 10^{-13} M$ ($\pm 0.9 \times 10^{-13} M$), which corresponds to an interaction energy of -16.7 kcal/mole.

The half-life of the repressor-p$O_R 1$ complex under standard assay conditions is about 10 min, corresponding to a dissociation rate constant of approximately $1.1 \times 10^{-3} \sec^{-1}$. The calculated forward rate constant for repressor dimers binding to $O_R 1$ is thus about 3×10^9 mole/liter sec.

Filter binding experiments can be performed at concentrations of operator DNA higher than that shown in Fig. 2, but then the operator DNA concentration must be known precisely in order to calculate the value of K_{eq} from Eq. (3). At very high DNA concentrations, repressor binding is "stoichiometric," and experiments performed under these conditions can be used to estimate the specific activity of repressor preparations.[11,24] For example, λ repressor purified by method 1 is 50–100% active as determined by binding of p$O_R 1$ DNA at a concentration of $6 \times 10^{-9} M$.

The value of K_{eq} is a function of ionic strength, temperature, and pH. It is also affected by divalent cations.[11,32] In principle, these effects could result either from changes in the repressor monomer–dimer equilibrium or the dimer–operator equilibrium. However, K_1 for the monomer–dimer interaction is reasonably constant from 0° to 20°, from pH 6.5 to 8.0, and in the presence of salt up to 200 mM.[32] Thus the effects on binding probably result from changes in the dimer–operator interaction.

Chadwick et al.[11] have analyzed the binding of λ repressor to phage λ DNA and report that the interaction is strongest at low pH, low ionic strength, and low temperature. For example, the concentration of repressor needed to give half-maximal DNA binding increases by a factor of about 40 in going from pH 6.5 to 8.5. Increasing concentrations of salt weaken the repressor–operator interaction possibly by competing for binding to the ionized phosphates of the DNA. Some of these experiments have now been repeated with DNA containing a single repressor binding site, and the results are qualitatively similar to those reported by Chadwick et al.[11]

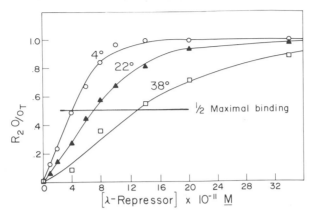

FIG. 3. Binding of λ repressor to pO_R-1 DNA: temperature effects. Nick-translated pO_R-1 DNA was dissolved at 2×10^{-12} M in a buffer containing 10 mM PIPES (pH 6.6), 2 mM CaCl₂, 0.1 mM EDTA, 50 mM KCl, 5% DMSO, and 100 γ/ml BSA. Samples were incubated with repressor for 40 min at 4°, 22°, and 38° and were filtered at these temperatures. Filters were washed twice with the above buffer minus the DNA, BSA, and DMSO. In this experiment, PIPES buffer was used instead of Tris–HCl because the former's pK_a is relatively constant over the temperature range used.

Repressor binds more strongly to operator DNA at 0° than at 20° or 37° (see Fig. 3) indicating that there is a significant enthalpic contribution to the binding energy. Other experiments[11] have shown that the repressor–operator complex is more stable at low temperature, the half-life of the complex being about 25 times longer at 0° than at 20°.

Repressor binds to operator DNA in the absence of divalent cations,[11] but the presence of these ions, especially calcium, stabilizes the binding interaction.[32] One millimolar CaCl₂ reduces by 2.4-fold the concentration of repressor necessary to give half-maximal binding of the HincII "320" restriction fragment (O_L1), when compared with binding in the presence of EDTA. Thus, K_{eq} for the interaction is reduced by approximately sixfold. This entire effect can be accounted for by stabilization of the repressor–operator complex. Figure 4 shows the dissociation of repressor–O_L1 complexes in the presence of varying amounts of Ca²⁺. The half-life of the complexes increases almost sixfold in going from no Ca²⁺ (0.1 mM EDTA) to 1 mM Ca²⁺. Other experiments[32] indicate that the half-life of the complex is unaffected by EDTA or EGTA added in excess after the repressor is bound.

Other divalent cations have similar but less pronounced effects on repressor–operator binding. At concentrations of 1 mM, strontium, barium, and magnesium increase the half-life of the repressor–O_L1 com-

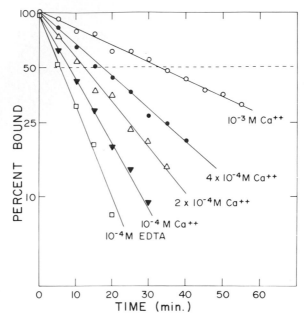

FIG. 4. Dissociation of λ repressor-O_L1 complexes as a function of calcium concentration. A tenfold excess of λ repressor was mixed with nick-translated HincII "320" DNA at a concentration of 1.5×10^{-11} M, in a buffer containing 10 mM Tris–HCl (pH 7), 0.01 mM EDTA, 50 mM KCl, 5% DMSO, 100 γ/ml BSA, and calcium chloride at the indicated concentrations. After 30 min, unlabeled λ DNA was added to a concentration of 7.5×10^{-9} M, and the samples were filtered at the indicated times after this addition.

plex from 6 min (EDTA) to 22, min 17, and 13 min, respectively. Manganese has no effect on the half-life of the complex.

Operator Mutants. A number of single base pair changes lying within O_L1, O_L2, O_R1, and O_R2 reduce the affinity of these specific sites for the λ repressor.[1,29,33] Lambda "virulent" phages bearing these mutated sites were isolated by their ability to grow on cells containing λ repressor and were thus selected for their ability to escape repression. Filter binding of repressor to restriction fragments containing the mutant sites, combined with DNA sequence analysis of the mutations, allows the identification of specific bases in each site which are important in the binding interaction.[29] Moreover, in those restriction fragments which contain more than one intact repressor binding site, comparison of the filter binding of restriction fragments bearing the mutant and wild-type alleles provides a method for deducing the affinity of repressor for each of the sites (see Flashman[33] and Note Added in Proof).

[33] S. Flashman, *Mol. Gen. Genet.* **166,** 61 (1978).

cro Protein–Operator Interaction. A filter binding assay for the cro protein of phage λ has been described.[2,14] The interaction of cro protein with whole λ DNA differs from the repressor–DNA interaction in at least two important respects: (1) the binding curve is hyperbolic rather than sigmoid and (2) the binding of cro protein to operator DNA is much weaker than that of the λ repressor.[2] The cro protein exists in solution as a stable dimer (in the concentration ranges tested) which is presumed to be the DNA-binding form.[2] Although the reported dissociation equilibrium constant for the cro protein–operator interaction (10^{-10} M) was not measured under the same conditions as that for the λ repressor (see above) or to DNA having only a single binding site, it seems likely that the affinity of the cro protein for operator DNA is at least 100-fold less than that of the λ repressor. The dissociation rate constant for the cro protein–λ DNA complex is 2×10^{-2} sec^{-1} at pH 7.3. Because of the short half-life of the DNA–protein complex under these conditions ($\simeq 30$ sec), filter binding with the cro protein is often difficult to perform reproducibly (see Takeda *et al.*[2] for discussion).

Identification of Operator-Containing Restriction Fragments. Once purified repressor protein is available, nitrocellulose filter binding coupled with restriction endonuclease mapping provides a powerful method for identifying those DNA fragments which contain operators. These fragments can then be isolated in larger amounts for functional studies and DNA sequencing.[29,34] The following method was devised for λ repressor, but with minor changes should be applicable to any protein that can trap DNA fragments to nitrocellulose filters.

[32]P-Labeled phage DNA is digested with a restriction endonuclease, phenol extracted, and ethanol precipitated. The restriction digest is resuspended in 250 μl of assay buffer (see discussion of repressor assay above) and a 10- to 100-fold molar excess of λ repressor is added. After 20 min at room temperature, the sample is filtered and washed as described (see discussion of repressor assay above). Operator-containing DNA is eluted from the nitrocellulose filter by vortexing with one to two 200-μl changes of extraction buffer (10 mM Tris-HCl, pH 7.4; 20 mM NaCl; and 0.2% sodium dodecyl sulfate). The DNA is then ethanol precipitated, and the operator fragments are identified by polyacrylamide gel electrophoresis[35] and autoradiography. This method can also be used to estimate the relative affinity of different restriction fragments, by performing the filter binding with limiting quantities of the repressor protein.[29]

[34] W. Gilbert, J. Gralla, J. Majors, and A. Maxam, *in* "Protein-Ligand Interactions" (H. Sund and G. Blauer, eds.), p. 193. de Gruyter, Berlin, 1975.
[35] T. Maniatis, A. Jeffrey, and H. van de Sande, *Biochemistry* **14**, 3787 (1975).

Other Methods for Probing DNA–Protein Interactions

Nuclease Protection. The ability of a bound protein to protect DNA from nuclease digestion has been used to study a number of specific DNA–protein interactions (e.g., *lac*[36] and λ repressors,[7,8] RNA polymerase,[37] and T antigen[38]). The method is particularly useful if the DNA sequence of the general binding region is known.

Maniatis and Ptashne[7] and, more recently Humayun *et al.*[8] have used DNase I digestion to locate the λ repressor binding sites. Varying amounts of λ repressor were mixed with ^{32}P-labeled DNA fragments bearing operators. DNase I was added for a short period of time, and the mixture was then filtered through nitrocellulose to retain DNA–repressor complexes. DNA fragments were eluted from the filter and sized by polyacrylamide electrophoresis. Fragments containing from one to three adjacent repressor binding sites (see Fig. 1) were obtained. The identity of each fragment was deduced by matching its pyrimidine tracts with the known sequence of the operator region.

The smallest fragment generated by nuclease protection of O_R contains the 17 base pairs of O_R1 and additional flanking DNA sequences. This fragment of 25 base pairs is thus larger than the presumptive λ repressor recognition sequence (see Ptashne *et al.*[4] for discussion). The detailed interpretation of nuclease protection experiments is further complicated by the fact that digestion with different nucleases may yield significantly different results, for example, with RNA polymerase promoter protections.[39]

However, in spite of these difficulties, nuclease protection experiments have revealed the approximate size, location, and organization of the λ repressor binding sites. These experiments have been particularly useful in determining the relative affinities of the different binding sites for repressor.

Dimethyl Sulfate Protection. As shown by Gilbert *et al.*[40] the alkalyting agent dimethyl sulfate can be used to locate sites on DNA recognized by proteins and to identify specific points of contact on the DNA. The method relies on the ability of a bound protein to protect from methylation certain purines lying within its recognition site. Fragments of duplex DNA, labeled at one end with ^{32}P, are subjected sequentially to three

[36] W. Gilbert, *Biol. Sys., Ciba Found. Symp., 1972* No. 7 (new ser.), p. 245 (1972).
[37] See W. Gilbert, *in* "RNA Polymerase" (R. Losick and M. Chamberlin, eds.), p. 193, and references therein. Cold Spring Harbor Lab., Cold Spring Harbor, New York, 1976.
[38] R. Tjian, *Cell* 13, 165 (1978).
[39] B. J. Meyer, cited in Ptashne *et al.*[4]
[40] W. Gilbert, A. Maxam, and A. D. Mirzabekov, *Control Ribosome Synth., Proc. Alfred Benzon Symp., 9th,* 1976, p. 139 (1976).

chemical reactions: (1) partial methylation in the presence and absence of bound protein, (2) release of methylated adenine and guanine, and (3) breakage of the DNA backbone at depurinated positions. The products of these reactions are electrophoresed through a denaturing polyacrylamide gel and are autoradiographed to produce a pattern of bands. Each band corresponds to a particular purine in the sequence, and its position on the sizing gel is a measure of the distance from the ^{32}P-labeled end to that purine. The intensity of each band thus reflects the extent of methylation of a specific purine. Bound proteins have been found to suppress and, less frequently, to enhance methylation of particular purines lying in their recognition sites.[5,40-42] These experiments are routinely carried out with λ repressor and λ *cro* protein as follows: DNA fragments are purified from λ DNA or from operator-containing plasmid DNA by electrophoresis on polyacrylamide gels.[26] Restriction fragments are 5' end-labeled with [γ-^{32}P]ATP and T4 polynucleotide kinase,[27] or in some cases restriction fragments with 5' overhangs are 3' end-labeled with [α-^{32}P]dXTPs and DNA polymerase I.[43] Labeled operator containing restriction fragments are cleaved with a second restriction endonuclease as described,[41] and the resulting fragments now labeled at only a single end are isolated by polyacrylamide gel electrophoresis. An end-labeled restriction fragment is added to 50–200 μl of methylation buffer (e.g., 50 mM sodium cacodylate, pH 7.0; 10 mM MgCl$_2$; and 2.5 μg/ml chick blood DNA) at a concentration of approximately 10^{-9} M. Repressor or cro protein is added at a 10- to 1000-fold molar excess over operator DNA. After a 30-min incubation at 0°, dimethyl sulfate is typically added to 50 mM, and the reaction is allowed to proceed at 0° for about 30 min. The reaction is terminated by addition of 0.5 volume of the stop solution described by Gilbert *et al.*[40] A control sample is treated in an identical manner except that no protein is added. Depurination of methylated DNA, strand scission, gel electrophoresis, and autoradiography are carried out according to Maxam and Gilbert.[27]

Figure 1 shows the effects of cro protein and repressor on the methylation of purines in O_R. The methylation patterns at O_L are similar. The following conclusions can be drawn from these experiments: (1) cro and repressor protect guanines but not adenines from methylation, suggesting that both proteins contact DNA primarily in the major groove. This follows from the fact that the sites of methylation probed with dimethyl sulfate, the N^7 ring nitrogen of guanine and the N^3 ring nitrogen of adenine are exposed in the major and minor grooves respectively.[44] (2) The

[41] Z. Humayun, D. Kleid, and M. Ptashne, *Nucleic Acids Res.* **4**, 1595 (1977).

[42] J. Majors, Ph.D. Dissertation, Harvard University, Cambridge, Massachusetts (1977).

[43] E. Schwartz, G. Scherer, G. Hobom, and H. Kössel, *Nature* (*London*) **272**, 410 (1978).

[44] A. Mirzabekov and A. M. Kolchinsky, *Mol. Biol. Rep.* **1**, 385 (1974).

guanine contacts made by the cro protein are a subset of those made by repressor, consistent with its smaller size and weaker binding affinity. In addition to determining some of the particular bases involved in the DNA–protein interaction, dimethyl sulfate protection experiments have revealed the general order of site binding affinity exhibited by repressor and cro protein.[5,41]

NOTE ADDED IN PROOF: Using a modification of the nuclease protection technique, we have recently shown that λ repressor binds cooperatively to the three sites in O_R, whereas cro protein binds noncooperatively [A.D. Johnson, B. J. Meyer, and M. Ptashne, *Proc. Natl. Acad. Sci. U.S.A.* **76** (1979)].

[77] The Isolation and Properties of CAP, the Catabolite Gene Activator

By GEOFFREY ZUBAY

Introduction

Research on CAP, the catabolite gene activator, received its original stimulus from investigations aimed at elucidating the mechanism of catabolite repression. Catabolite repression involves the inhibition of enzyme induction by glucose or its derivatives such as glucose 6-phosphate, fructose, or glycerol. The extent of inhibition is dependent on the bacterial strain, the repressing catabolite, and the growth conditions. Observations on catabolite repression date back to the turn of the century, when Dennert[1] found that the bacterial enzyme "galactozymase" disappeared from cells grown in the presence of glucose.

A turning point in our understanding of catabolite repression was provided by Makman and Sutherland[2] who linked cyclic adenosine 3',5'-monophosphate (cAMP) to the glucose effect. These workers found cAMP in bacteria to the extent of $10^{-4} M$. In the presence of glucose, the intracellular level fell rapidly to about $10^{-7} M$. The supply of cAMP was replenished to normal values upon removal of the glucose. Ullman and Monod[3] noting these results investigated diphasic growth in the presence and absence of cAMP. When one places bacteria in the presence of two sugars, glucose and lactose, maltose, or xylose, a diphasic growth curve ensues. This phenomenon is due to the repressive effect of glucose on the

---

[1] F. Dennert, *Ann. Inst. Pasteur, Paris* **14**, 139 (1900).
[2] S. R. Makman and E. W. Sutherland, *J. Biol. Chem.* **240**, 1309 (1965).
[3] A. Ullman and J. Monod, *FEBS Lett.* **2**, 57 (1968).

METHODS IN ENZYMOLOGY, VOL. 65

enzyme system required for the metabolism of the other sugar. Thus the cell first metabolizes the glucose and then gradually becomes able to use the other sugar as a carbon source. The diphasic growth curve was eliminated in the presence of cAMP. These workers also showed that 5×10^{-3} M cAMP overcame a severe catabolite repression in cells grown on glucose or glycerol media but had no effect on cells grown on a medium containing the nonglucose metabolite, succinate. Independently Perlman and Pastan[4] showed that simultaneous administration of cAMP reversed the repressive effect of glucose on the production of galactosidase in bacteria made permeable by EDTA. Following this, the stimulatory effects of cAMP on genes subject to catabolite repression were demonstrated more directly and convincingly in a cell-free system. This system used DNA containing the β-galactosidase gene, a cell-free extract of *E. coli,* and the low molecular weight components necessary for transcription of the DNA and translation of the resulting messenger RNA (mRNA). In this system it was shown that 10^{-3} M cAMP stimulated β-galactosidase (β-gal) synthesis by as much as 30-fold.[5] It seemed likely that in conjunction with the appropriate mutants such a system would be most useful for detecting and isolating the components responsible for the cAMP effect. With this in mind Schwartz and Beckwith[6] isolated mutants subject to catabolite repression. These mutants fell into two categories, those which could be phenotypically corrected by growing in the presence of 5×10^{-3} M cAMP and those which could not. It was hypothesized that the former mutants were defective in cAMP production and that the latter mutants were defective in the protein(s) with which cAMP interacts to bring about stimulation of β-gal synthesis as well as the synthesis of other proteins subject to catabolite repression. These two types of mutants are now known to originate in the genes *cya* and *crp,* respectively. When the cell-free system was made from extracts of *crp⁻* cells it produced a low level of β-gal which was not stimulated by adding cAMP. The defect could be corrected by adding soluble protein from *crp⁺* cells. Using this effect as an assay, it was possible to fractionate the *crp⁺* cells and isolate a single protein CAP which was responsible for the effect.[7,8] At the present time the properties of CAP are being studied intensively in a number of laboratories. This article is devoted to a discussion of the isolation procedures for CAP and the properties of CAP insofar as they are understood.

[4] R. L. Perlman and I. Pastan, *J. Biol. Chem.* **243,** 5420 (1968).
[5] D. A. Chambers and G. Zubay, *Proc. Natl. Acad. Sci. U.S.A.* **63,** 118 (1969).
[6] D. Schwartz and J. R. Beckwith, *in* "The Lactose Operon" (J. R. Beckwith and D. Zisper, eds.) p. 417. Cold Spring Harbor Lab., Cold Spring Harbor, New York, 1970.
[7] G. Zubay, D. Schwartz, and J. Beckwith, *Proc. Natl. Acad. Sci. U.S.A.* **66,** 104 (1970).
[8] G. Zubay, *in* "The Role of Adenyl Cyclase and Cyclic 3'5'-AMP in Biological Systems" (T. W. Rall, M. Rodbell, and P. Condliffe, eds.), p. 231. Natl. Inst. Health, 1969.

Purification Procedures for CAP

There are from 300 to 1000 copies of CAP per cell, making it somewhat easier to isolate than the usual gene regulatory protein of which there are usually far fewer copies. The high concentration of CAP is a reflection of the large number of genes it affects. As far as is known, the yield of CAP is not very sensitive to growth conditions. In spite of this, to ensure a high quality of starting material, it is recommended that cells be grown by a procedure which gives high yields in late log phase and that cells be stored at $-70°$ for not longer than 1 month before they are used.[9] Several purification procedures are available for the isolation of CAP from E. coli. Each procedure has advantages and disadvantages. The choice of procedure should depend upon the amount of protein desired, the degree of purity and activity required and the amount of work one is willing to do. Gross purity has been decided mainly by analyzing the final produce by SDS–polyacrylamide gel electrophoresis. CAP yields a single protein band on gels with a molecular weight of about 22,000.[10] By this criterion procedures A and D (below) give products which are about 90% pure, whereas procedures B and C give products which are greater than 90% pure. As far as yields are concerned it is not possible to compare procedures A with procedures B and C since procedure A employed a strain with one chromosomal copy of the crp gene, whereas procedures B and C used a strain with an additional copy on an episome. It seems likely that the latter strain produces at least twice the CAP and possibly more than this. Clearly the latter strain would be the best one to use regardless of which procedure one employs. The activity of CAP can be measured in at least three ways: (1) binding of cAMP, (2) cAMP-stimulated binding to DNA, and (3) the ability to stimulate transcription or β-galactosidase synthesis in an in vitro system. Although all procedures yield CAP with these three activities, only in the case of cAMP binding is it clear that the specific activities in the final products are comparable. Given these reservations and the limited information on trace impurities, it is difficult to make judgements as to which procedure to recommend. The original procedure A is recommended as the simplest way of obtaining small amounts of CAP in reasonable purity ($\sim$90%). Procedures B and C probably give a somewhat purer product (greater than 90%) but involve more complex operations. Of the two latter procedures, C is more convenient particularly when working with large amounts. Procedure D claims yields several times higher than procedure C and purity after only 320-fold purification, whereas most procedures require about 1000-fold purification to achieve

[9] G. Zubay, Annu. Rev. Genet. 7, 267 (1973).
[10] A. D. Riggs, G. Reiness, and G. Zubay, Proc. Natl. Acad. Sci. U.S.A. 68, 1222 (1971).

90% of better purity. For these reasons procedure D should be considered even though these claims have not been substantiated.

It is useful to assay the CAP at various stages during purification. The transcription assay in which the DNA-directed synthesis of β-gal is measured is by far the most sensitive and selective assay to use.[7] However, unless one is familiar with the use of the system involved, an easier to set up monitor is the cAMP binding assay. The cAMP binding assay is useless in crude extracts because of the relatively low concentration of CAP and the presence of interfering substances. Once about 50-fold purification has been achieved this assay is both accurate and convenient. Only the cAMP binding assay is reproduced below.

cAMP Binding Assay Method[11]

The binding activity of CAP is measured by incubating the protein sample with cyclic [^{3}H]AMP, precipitating the cyclic [^{3}H]AMP complex with ammonium sulfate and determining the isotope content of the precipitated complex.

Reagents

cAMP, 10 μM and 0.1 M
Cyclic [^{3}H]AMP, 20 Ci/mmole, 1×10^7 cpm/ml
5'-AMP, 0.1 M in 10 mM potassium phosphate, pH 7.7
Casein, 10 mg/ml, 10 mM potassium phosphate buffer, pH 7.7
Saturated solution of $(NH_4)_2SO_4$
NCS-Nuclear Chicago Solvent Liquifluor
Toluene

Procedure. The assay is performed at 1°. To a 10 × 75 mm disposable glass test tube add 10 μl of 10 μM cAMP, 10 μl cyclic [^{3}H]AMP, 10 μl 5' AMP, CAP sample, and H_2O to a final volume of 100 μl. To control tubes add 10 μl of 0.1 M cAMP. cAMP is usually obtained in the acid form, but a 0.1 M solution can be readily made by neutralizing the nucleotide solution with NaOH. Casein, 20 μl, is added to the assay tubes as a carrier protein for measurement of the activity of more highly purified preparations (after phosphocellulose). After a 5-min incubation, add 0.4 ml of cold saturated $(NH_4)_2SO_4$ and mix. The precipitates are collected by centrifugation at 10,000 rpm (12,000 g) for 10 min in an SS-34 rotor in a Sorvall RC2-B centrifuge. Carefully aspirate off the supernatant with a disposable pipette and then wipe the walls of each tube thoroughly with cotton-tipped applicator sticks. Dissolve the pellet in 0.5 ml of NCS, transfer to a count-

[11] I. Pastan, M. Gallo, and W. B. Anderson, this series, Vol. 38, Part C, p. 367.

ing vial containing 10 ml of Liquifluor–toluene scintillation cocktail (42 ml concentrated Liquifluor in 1000 ml of toluene). Wash pipette and tube by washing back and forth four times with scintillation mixture. Sufficient CAP is added to bind 10,000 to 20,000 cpm of cyclic [^{3}H]AMP. One unit of cAMP binding activity is defined as that amount of protein required to bind 1 pmole of cAMP under the conditions of the assay.

Purification Procedure A[10]

About 200 gm of frozen *E. coli* strain 514 homogenized in 700 ml of buffer I (0.01 M Tris-acetate, pH 8.2, 0.01 M Mg(Ac)$_2$, 0.06 M KCl, 6 mM mercaptoethanol) and centrifuged for 30 min at 16,000 g. The sediment containing the bacteria is homogenized and recentrifuged. The final sediment is resuspended in 260 ml of buffer I. The suspension of cells lysed in an Aminco pressure cell at pressures between 4000 and 8000 psi. The lysate is centrifuged for 30 min at 30,000 g in a small Sorvall rotor. The resulting supernatant is centrifuged for 4 hr at 30,000 rpm in a Spinco No. 30 rotor. The resulting supernatant is dialyzed for 16 hr against buffer II [0.01 M K$_2$HPO$_4$–CH$_3$COOH (pH 7.0), 6 mM mercaptoethanol]. This solution is passed over a 2.5 × 15 cm phosphocellulose column (Whatman P11, medium fibrous powder, 7.4 mEq/gm), previously equilibrated with buffer II. After the column is rinsed progressively with 200 ml buffer II and 100 ml of buffer II and 0.4 M KCl, the active fraction, which constitutes about 1% of the protein put on the column, is eluted in buffer II and 0.50 M KCl. The active fraction, detected by ultraviolet absorption, is pooled and dialyzed against buffer III [0.01 M KH$_2$PO$_4$–KOH, pH 7.7, 6 mM mercaptoethanol] overnight. This solution is passed over a 1.4 × 13 cm DEAE–cellulose column previously equilibrated with buffer III. The active fraction, which constitutes about 10% of the protein put on this column, passes through the column with no holdback. The remainder of the protein is retained by the column. Total yield of protein is 300–1000 μg in about 20 ml.

Purification Procedure B[11]

This procedure has been described elsewhere[11] and will not be duplicated here.

Purification Procedure C[12]

Preparation of Chromatographic Materials. Bio-Rex 70 was suspended overnight in an equal volume of 1 M K$_2$HPO$_4$ followed by extensive wash-

[12] E. Eilen, C. Pampeno, and J. S. Krakow, *Biochemistry* 17, 2469 (1978).

ing with deionized water. The Bio-Rex 70 suspension (50%, v/v) was adjusted to pH 7 with HCl. After 15 min, the supernatant was decanted and the resin resuspended in an equal volume of buffer A.

Denatured DNA-Cellulose Preparation. The method of Alberts and Herrick was used.[13] The preparations contained about 0.5 mg of DNA per milliliter of packed volume.

Solutions. All buffers were prepared with deionized water and stored at 4°. Stock solutions of 1 M potassium phosphate, pH 7, and 0.5 M EDTA, pH 7.6, were used to prepare the following solutions: buffer A, 50 mM potassium phosphate (pH 6.8) and 0.1 mM EDTA; buffer B, 10 mM potassium phosphate (pH 7), 0.1 mM EDTA, and 0.1 M KCl; buffer C, 20 mM potassium phosphate (pH 6.5 or pH 8), 1 mM EDTA, and 0.1 M NaCl. A stock solution of 0.2 M Bistrispropane, pH 8, was used to prepare various reaction mixtures.

Lysis and Cell Disruption. Frozen *E. coli* cells (500 g) (KLF 41/JC 1553) were suspended in 1.5 liters of 75 mM Tris-OH and warmed to 20° with stirring. The suspension was rapidly brought to pH 8 with KOH followed by the addition of 50 ml of 0.5 M EDTA, pH 7.6. Lysozyme was added (300 mg in 50 ml of 20 mM Tris-HCl, pH 8) and the mixture stirred for 20 min. After cooling the lysate to 12°, the following additions were made: 50 ml of 1 M Tris-HCl, pH 8, 90 ml of 0.5 M MgSO$_4$, and 50 ml of Brij 58. After stirring for 10 min, the viscous lysate was dispersed in 400-ml batches for 3 min using a Tekmar Super Dispax SD 45 at 7500 rpm. After centrifugation at 12,000 rpm for 20 min in the HA-14 rotor of the Beckman J 21 centrifuge, the supernatant was readjusted to pH 8. All remaining steps were carried out at 4°.

Polyethylenimine Titration. Samples from the supernatant (2 ml) were titrated with 5% polyethylenimine (w/v, pH 8) to determine the amount required for the precipitation of nucleic acids and associated proteins without the loss of CAP as assayed by cAMP binding activity. After centrifugation for 10 min at 10,000 rpm, 20-μl samples of the supernatant were assayed for [^{3}H]cAMP binding. In general, up to 0.1 ml of 5% polyethylenimine per milliliter of supernatant could be added without affecting the cAMP binding activity remaining in solution. The determined volume of polyethylenimine was then added dropwise to the extract with manual stirring over a period of 20–30 min, the supernatant was decanted and adjusted to pH 7 with 1 M acetic acid.

Bio-Rex 70 Chromatography. The polyethylenimine supernatant was batch loaded by stirring with 300 ml of Bio-Rex 70 suspension for 10 min. The beads were then allowed to settle (20 min), and the supernatant was

[13] B. Alberts and G. Herrick, this series, Vol. 21, p. 198.

assayed for cAMP binding. Generally, less than 15% of the activity remained in the supernatant, which was discarded. The Bio-Rex beads were then batch washed four times with 2-liter volumes of buffer A or until the absorbance at 280 nm fell below 0.1. The resin was then poured as a slurry into a glass column to form a bed measuring approximately 3×25 cm. Bound protein was eluted with a linear salt gradient (1 liter total volume) from 0 to 1 M KCl in buffer A. The cAMP binding activity elutes between 0.25 and 0.32 M KCl. The peak fractions were pooled and concentrated by adding ammonium sulfate to 60% saturation at a pH of 6.8–7.0. After centrifugation at 12,000 rpm for 20 min, the precipitate was redissolved in 5–10 ml of buffer B.

Sephacryl S-200 Chromatography. The concentrated fraction from the Bio-Rex 70 step was loaded onto a 2.5×90 cm column of Sephacryl S-200 equilibrated with buffer B. The protein was eluted at 30 ml/hr, and the fractions containing CAP were pooled and adjusted to pH 6.5.

DNA–Cellulose Chromatography. The Sephacryl pool was adjusted to a conductivity of 3 mmho and loaded onto a 1.8×10 cm column of denatured DNA–cellulose previously equilibrated at pH 6.5 with buffer C. The column was washed with 50 ml of buffer C, pH 6.5, followed by the elution of CAP with buffer C at pH 8. The column was stripped with buffer C and 2 M NaCl to remove other proteins that bind DNA but lack cAMP binding activity. The fractions containing the stripped protein were discarded.

Bio-Rex 70 Concentration. The pooled DNA–cellulose fractions were loaded onto a 0.7×14 cm Bio-Rex 70 column equilibrated with buffer A containing 0.1 M NaCl. The protein was eluted in a salt step with buffer A containing 1 M NaCl. The peak fractions were pooled and stored at $-20°$.

The purified CAP, which migrated as a single band on NaDodSO$_4$–polyacrylamide gels, had a specific activity of 6460 units/mg[14] with a 60% recovery of cAMP binding activity (Table I). The procedure can be completed in 5 days.

In preparing for polyethylenimine titrations, small samples (1 ml) should first be titrated with the polycation to ensure that cAMP binding activity remains in the supernatant. Careful attention to pH adjustment of both the reagent and the lysate is required so that CAP remains dissociated from the DNA. It is not unusual for cAMP binding activity to appear depressed during the initial stages of the purification. This may be due to residual polyethylenimine, an inhibitory factor or a cyclic nucleotide phosphodiesterase which may be removed by chromatography on Bio-Rex 70. For this reason, yields have been calculated starting with the first column step (Table I). Chromatography on Bio-Rex 70 is more con-

TABLE I
PURIFICATION OF CAP

Stage of purification	Volume (ml)	Protein (mg)	Activity (units)	Specific activity (units/mg)	Yield (%)	A_{280}/A_{260}
Lysate supernatant	1800	41,400	270,000	6.5	—	0.50
Polyethylenimine supernatant	1300	22,100	163,000	7.4	—	0.72
Bio-Rex 70, I	200	260	352,000	1354	100	0.93
Sephacryl S-200	225	90	276,400	3070	70	1.12
DNA–cellulose	110	35.5	220,000	6300	63	1.70
Bio-Rex 70, II	6.5	32.5	210,000	6460	60	1.78

venient than phosphocellulose[14] when dealing with a large volume, since the Bio-Rex 70 beads can be mixed with the extract and allowed to settle out. Chromatography on DNA–cellulose takes advantage of the cAMP-independent binding of CAP to DNA at pH 6.5 and its low affinity for DNA at pH 8 in the absence of cAMP. This effect, which may be due to the interaction of positively charged groups on CAP with the phosphodiester backbone of DNA, may be overcome by raising the pH in order to lower the positive charge on the protein. When required, the protein may be concentrated on a small Bio-Rex 70 column with minimal losses (Table I).

The yield of CAP represents 60% of the activity in the pooled peak of the Bio-Rex I chromatography step (Table I). This would indicate a nearly 1000-fold purification from the dispersed material or a 765-fold purification of the initial specific activity is calculated on the basis of 352,000 total units.

Purification Procedure D[15]

This is a new procedure which claims to give very high yields of purified protein. It has not been evaluated here.

Physical and Chemical Properties of CAP

CAP migrates on G-100 Sephadex column as a molecule with a molecular weight of about 45,000.[7] After heating in 0.1% SDS, its electrophore-

---

[14] W. B. Anderson, A. B. Schneider, M. Emmer, R. L. Perlman, and I. Pastan, *J. Biol. Chem.* **246**, 5929 (1971).
[15] T. Boone and G. Wilcox, *Biochim. Biophys. Acta* **541**, 528 (1978).

tic migration rate on SDS–polyacrylamide gels indicates a homogeneous species with a molecular weight of 22,000.[10] Since the heating step in SDS should dissociate a protein into its subunits (provided there are no covalent linkages holding the subunits together), it was suggested that CAP is normally a dimer with identical subunits of about 22,000.[10] This was confirmed by equilibrium sedimentation studies of the native protein (estimated MW 44,600) and the reduced alkylated subunits in 6M guanidine hydrochloride (estimated MW 22,300).[16] CAP is a basic protein with an isoelectric point (pI) of 9.2. This high isoelectric point is due to the high content of glutamine and asparagine; the content of lysine and arginine in CAP is not unusually high. Other physical and chemical properties of CAP have been reviewed elsewhere.[16] The biologically significant properties of CAP involve its interaction with cAMP and DNA and its stimulation effect on transcription. These properties are described below.

Interaction of CAP with cAMP and Other Nucleotides

It was originally hypothesized that when cAMP interacts with CAP altering its structure to a form favorable for binding to DNA at the promoter which in turn facilitates polymerase binding at an adjacent site and the initiation of transcription.[17] This model predicts a substantial affinity between cAMP and CAP. Studies[7] involving CAP and [³H]cAMP were done by the dialysis equilibrium technique in a buffer containing 0.01 M Tris-Ac, pH 8.2, 0.01 M Mg(Ac)$_2$, 0.06 M KAc and 1.4 mM dithiothreitol. A Scatchard-type plot indicated a single type of binding site with a dissociation constant, K_D, of 1.7×10^{-5} moles/liter, where K_D is defined by

$$K_D = \frac{[CAP^*][cAMP]}{[CAP^* + cAMP]}$$

Here [CAP*] is the concentration of ligand binding sites. If $n = 1$, then [CAP*] = [CAP]; if $n = 2$, then [CAP*] = 2[CAP]. In view of the dimer structure of CAP it is highly likely that these are two ligand binding sites in the native molecule.

Most recent binding studies[18] done in 40 mM bis(Tris)–propane pH 8.0 gave the same K_D for the cAMP–CAP complex and a K_D for the cGMP–CAP complex of 4.8×10^{-5} M. Binding studies[16] have also been carried out in near saturated (NH$_4$)$_2$SO$_4$ for the cAMP–CAP complex giving a much lower K_D of 10^{-6} mole/liter. Apparently, the ammonium sulfate

---

[16] I. Pastan and S. Adhya, *Bacteriol. Rev.* **40**, 527 (1976).

[17] G. Zubay and D. A. Chambers, *in* "Metabolic Pathways" (D. M. Greenberg, ed.), 3rd ed., Vol. 5, p. 297.

[18] E. R. Eilen, Ph.D. Thesis, City University of New York (1977).

causes an increase in the strength of cAMP binding. All of the binding studies indicate a substantial affinity between cAMP and cGMP for CAP. The fact that salts such as ammonium sulfate can increase the K_D should be taken into account in any quantitative analysis of the significance of this binding. It seems likely that a K_D of 1.7×10^{-5} will be close to that attained under physiological conditions for the cAMP–CAP complex. A number of naturally occurring cAMP derivatives and other cyclic nucleotides have been tested indirectly by determining their effect on cAMP stimulated β-gal synthesis.[19] No nucleotides have been found which can replace cAMP in stimulating β-gal synthesis, but cGMP causes a 70% reduction in the β-gal synthesis when both cyclic nucleotides are present at a concentration of 1 mM. Evidently cGMP binds to the same site as cAMP but does not bring about the conformational change in CAP required for activity. However, Sanders and McGeoch[20] have isolated a mutant containing an altered CAP which is activated by either cAMP or cGMP to stimulate β-gal synthesis. A number of abnormal cyclic nucleotides have also been tested for their ability to bind to CAP.[16]

Physical and chemical studies have been carried out which indicate structural alterations in CAP result from cAMP binding.[16] The chemical studies are particularly interesting as they show that cAMP induces structural alterations in that part of the protein which binds to DNA. Thus native CAP contains two buried and two available cysteine residues.[21] Titration of the two available cysteines with 5,5′-dithiobis(2-nitrobenzoic acid) (DTNB) eliminates cAMP-dependent DNA binding activity which is regenerated by incubating the modified protein with β-mercaptoethanol. The binding of cAMP facilitates a DTNB-mediated disulfide cross-linking of the two available sulfhydryls to produce a 45,000 dalton species on SDS-polyacrylamide gels. Evidently, the binding of cAMP changes the structure of CAP so that the two available cysteine sulfhydryls are brought into close proximity. The two buried disulfides do not react with DTNB unless the structure is partially denatured with 3 M urea. Reaction of the two buried disulfides interferes with cAMP binding.

Limited proteolytic digestion of CAP by chymotrypsin in the presence of cAMP results in the formation of a resistant core molecule with about half the molecular weight.[22] The CAP core retains the dimeric structure and cAMP-binding properties of native CAP, while cAMP-dependent DNA binding is lost. In the absence of cAMP, CAP is relatively resistant to protease digestion.

[19] G. Zubay and D. A. Chambers, *Cold Spring Harbor Symp. Quant. Biol.* **34,** 753 (1969).
[20] R. Sanders and D. McGeoch, *Proc. Natl. Acad. Sci U.S.A.* **70,** 1017 (1973).
[21] E. Eilen and J. S. Krakow, *J. Mol. Biol.* **114,** 47 (1977).
[22] E. Eilen, C. Pampeno, and J. S. Krakow, *Biochemistry* **17,** 2469 (1978).

CAP Binding to RNA Polymerase

All studies aimed at showing an affinity between CAP and RNA polymerase have been negative. For example, when mixed together CAP and polymerase migrate separately through a Sephadex G 100 column (G. Zubay, unpublished observations). Such an experiment would only detect very strong interactions. Clearly more sophisticated techniques should be used before this type of interaction is excluded. The possibility that CAP and polymerase interact directly when attached to the DNA also needs investigation.

CAP Binding to DNA

The primary motivation for investigating the binding of CAP to DNA was to test the hypothesis that CAP, in the presence of cAMP, binds to the *lac* promoter. The carefully contrived conditions under which it was finally possible to demonstrate cAMP-dependent binding to the *lac* promoter bear little resemblance to physiologic conditions (i.e., *in vivo* conditions of ionic strength, CAP concentration). However, the significance of the results seems clear and is supported by genetic correlates which show that mutation altered DNA ineffective in supporting cAMP-stimulated β-gal synthesis *in vivo* is also deficient *in vitro* in the cAMP-stimulated binding of CAP.

The CAP protein has a substantial affinity for many negatively charged species such as DNA, RNA, or phosphocellulose. This nonspecific affinity is higher with lower ionic strength, as would be expected for a reaction between two oppositely charged polyelectrolytes. In seeking conditions under which cAMP-stimulated binding could be demonstrated, it was necessary to use an ionic strength high enough to minimize nonspecific electrostative binding. On the other hand, physiological ionic strengths are too high to see even the specific cAMP induced binding, at least by the commonly used filter binding assay. The final solution was that intermediate ionic strengths between the very low and physiological were used.

The cAMP-Stimulated Binding of CAP to DNA

The affinity of CAP for DNA has been measured by the familiar Millipore filter binding technique.[23] A mixture containing CAP and ^{32}P-labeled DNA was passed through a Millipore filter. DNA was not retained by the filter unless it was bound to protein. It was assumed that all protein,

[23] A. D. Riggs, G. Reiness, and G. Zubay, *Proc. Natl. Acad. Sci U.S.A.* **68**, 1222 (1971).

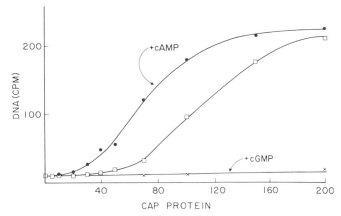

FIG. 1. Bound DNA as a function of CAP.[23] A fixed amount of λ *lac* [^{32}P]DNA (0.05 μg) was mixed with the indicated volume of CAP protein solution in a final volume of 1.3 ml, containing 10 mM KCl, 3 mM Mg(Ac)$_2$, 0.1 mM EDTA, 0.1 mM dithiothreitol, 50 μg/ml bovine serum albumin, 5% dimethyl sulfoxide, and 10 mM Tris HCl, pH 7.4 at 24°. The abscissa units are microliters of a CAP protein solution that contained about 0.4 μg/ml of CAP protein. Each filter received 0.4 ml, containing a total of 260 cpm (□) buffer alone; (○) 3.7 × 10^{-4} M cAMP; (x) 3.7 × 10^{-4} M cGMP. For these studies CAP was purified by procedure A described in the text.

whether free or bound to DNA, was retained by the filter. In this situation, radioactivity retained by the filter served as a measure of the DNA which was complexed to protein. At a fixed level of CAP and DNA about half-maximum binding was achieved at a concentration of 2.5 × 10^{-5} M cAMP. Previous studies in which the K_D for the cAMP–CAP complex showed that about half the cAMP binding sites on CAP should be occupied at 1.7 × 10^{-5} M cAMP. As the concentration of CAP was increased (see Fig. 1), more and more of the DNA was retained until at high levels of CAP most of the DNA was retained in the presence or absence of cAMP. At all concentrations of CAP, added cGMP completely eliminated the retention of DNA in the absence of cAMP. In the presence of cAMP, the cGMP merely reduced the binding. This is in line with the competitive binding of the two cyclic nucleotides to CAP described above. The exponential forms of the binding curve as a function of CAP concentration suggests that more than one CAP molecule is needed to retain DNA on the filter. This result is somewhat misleading, as it suggests that perhaps more than one CAP might be needed to activate the promoter. Arguments are presented below against this idea. It seems likely that the requirement for more than one CAP is a requirement only for filter binding. At a fixed CAP concentration and varying levels of DNA, the binding curve leveled off at about one DNA molecule per 50 CAP molecules suggesting that either

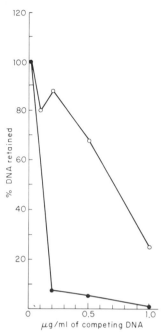

FIG. 2. cAMP stimulated binding of λ *lac* [³²P]DNA in the presence of varying amounts of unlabeled DNA fragments.[25] Competing fragments D_2 and D_{3A} are represented by open and closed circles, respectively. These are *Hin*dII restriction fragments of λp*lac*5 DNA with molecular weights of about 4.1 and 4.5×10^5, respectively. Only D_{3A} contains the *lac* control region. To a fixed amount of λ*lac* [³²P]DNA (0.4 μg/ml) in buffer I containing 4×10^{-4} M cAMP, increasing amounts of unlabeled purified fragments were added as shown. Finally, a constant amount of CAP (0.08 μg/ml) was added and the mixture incubated at 25° for 15 min. The total volume of reaction mixture was 0.5 ml. Aliquots of 0.2 ml were filtered in duplicate and washed with 0.4 ml of buffer II before drying and counting. Data are plotted in terms of percentage DNA retained. A total of 100% DNA retained refers to the quantity bound to the filter in the absence of competitor. This gives 800 cpm and represents about half the labeled DNA present. Buffer I is 10 mM Tris-HCl, pH 7.4; 50 mM KCl; 0.1 mM EDTA; 0.1 mM DTT; 50 μg/ml bovine serum albumin; 10% glycerol. Buffer II is 10 mM Tris-HCl, pH 7.4; 50 mM KCl; 0.1 mM EDTA; 10% glycerol.

much of the CAP was inactive or that more than one CAP per DNA molecule was required for retention of the DNA complex on the filter. It is possible that both of these factors were contributory. When a variety of other DNAs were substituted for λ *lac* DNA in the binding assay essentially the same results were obtained. RNA from various sources competed poorly with DNA. This indicated a preference for DNA in the cAMP-dependent CAP binding reaction but little apparent site specificity. In retrospect the workers in this field had been spoiled by the results of others with *lac* and λ repressor. In both of these cases the binding shows a

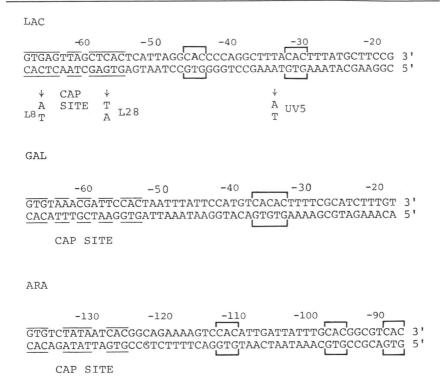

FIG. 3. Selected regions of the *lac*, *gal*, and *ara* promoter–operator regions including presumptive CAP binding sites. Sequence data from Dickson *et al.*[26] for *lac*; Musso *et al.*[28] for *gal* and B. Smith and R. Schleif (unpublished results) for *ara*. Regions showing dyadic symmetry about the center of the presumptive binding sites are overlined and underlined. Numbering of bases is such that +1 is the initiation point for transcription. Regions containing reiteration of the rightmost triplet in binding site are indicated by brackets. It is not clear why half the CAP binding site should be reiterated two or more times but it may have something to do with the dynamic aspects of CAP interaction.

much greater preference for the appropriate promoter DNA so that it was much easier to demonstrate specific site binding.

In the case of CAP, nonspecific binding is a serious factor to contend with, at least under the *in vitro* assay conditions used. Nevertheless, through the use of fractionated small restriction fragments of λ *lac* DNA it has been possible to demonstrate appreciable binding specificity.[24,25] For example, two fragments obtained from a *Hin*dII digest of λ *lac* DNA with near identical molecular weights of about 4.5×10^5 show a tenfold difference in cAMP-induced binding. The fragment with the higher binding, D_{3A}, contains the *lac* promoter. The experiment in Fig. 2

[24] J. Majors, *Nature* (*London*) **256**, 672 (1975).
[25] S. Mitra, G. Zubay, and A. Landy, *Biochem. Biophys. Res. Commun.* **67**, 857 (1975).

illustrates both the preferential affinity of CAP for segments of DNA containing the *lac* promoter as well as the problems associated with nonspecific binding. In this experiment the relative effectiveness of D_{3A} and D_2 has been measured indirectly by using varying amounts of unlabeled fragments in the presence of a fixed amount of ^{32}P-labeled $\lambda plac5$. In this experiment the $\lambda plac5$ is being used as a probe to determine the free CAP available for binding. Less than saturating amounts of CAP are used which leads to the filter retention of about 50% of the $\lambda plac5$ DNA in the absence of competing fragments. It can be seen in Fig. 2 that 0.2 μg of D_{3A} almost completely eliminates retention of $\lambda plac5$ DNA. An equal weight of D_2 causes less than a 10% lowering in the retention of $\lambda plac5$. Evidently the unlabeled D_{3A} fragment, by virtue of its *lac* promoter site, is considerably more effective at binding CAP so that it is unavailable for binding the labeled $\lambda plac5$ DNA. Several other unlabeled restriction fragments known not to contain the *lac* promoter have been examined in parallel experiments and found to be comparable in binding ability to fragment D_2. Finally, Majors[24] has provided the binding experiments with a genetic correlate by showing that *lac* promoter fragments containing mutant defects in the presumptive CAP binding site (see discussion below of the L8 point mutation or the L1 deletion) have lost their extraordinary affinity for CAP. The nucleotide sequence in the *lac* promoter region has been determined, and it is known from a series of genetic deletions that the site necessary for CAP stimulation of the *lac* operon is in the -50 to -80 base pair region.[26] Inspection of the sequence in this region reveals a 14 base segment between -55 and -68 (see Fig. 3) which shows dyadic symmetry for 12 of the 14 base pairs. Gilbert[27] suggested this was the site of CAP interaction from data which indicated that single base pair changes, that block the function of CAP at the *lac* promoter involve GC to AT transitions at positions -66 and -59, respectively (see Fig. 3). This assignment was consistent with the early prediction from the symmetry of CAP and the greater than first power dependence on cAMP concentration for gene activation that the DNA binding site should have dyadic symmetry.[17]

It is noteworthy that the *gal* operon and the *ara* operons which are also stimulated by the cAMP–CAP complex have similarly sized regions of dyadic symmetry (see Fig. 3).[26,28] In the case of these two operons, there is insufficient genetic or biochemical data to verify that these regions are in fact the CAP binding sites of the respective promoters, but it seems likely.

[26] R. C. Dickson, J. Abelson, W. M. Barnes, and W. S. Reznikoff, *Science* **187**, 27 (1975).
[27] W. Gilbert, *in* "RNA Polymerase" (R. Losick and M. Chamberlain, eds.), p. 193. Cold Spring Harbor Lab., Cold Spring Harbor, New York, 1976.
[28] R. E. Musso, R. DiLauro, S. Adhya, and B. deCrombrugghe, *Cell* **12**, 847 (1977).

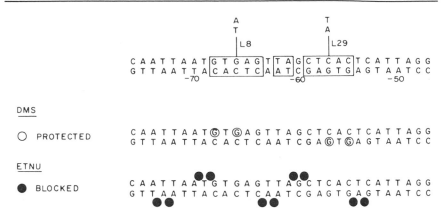

FIG. 4. Presumptive CAP binding region in the *lac* promoter.[30] Boxes indicate regions of dyadic symmetry. Results with dimethyl sulfate (DMS) indicate those bases that are strongly protected from reaction when CAP is bound. Results with ethylnitrosourea indicate those phosphate positions which if ethylated block CAP binding. If the area of interest was wound as a normal DNA double helix, then all the groups implicated in CAP binding would appear on one side of the DNA.

Comparison of the three presumptive binding sites for CAP indicates that only the outer three bases are identical in composition.

Most recently Majors has significantly advanced our understanding of how CAP binds at the *lac* promoter. The techniques used were developed in Gilbert's laboratory[29] for using the alkylating agents, dimethyl sulfate (DMS) and ethylnitrosourea (EtNU) as probes for determining close contacts between proteins and DNA moieties. DMS reacts with the adenine N3 and the guanine N7 positions; EtNU reacts with the phosphates. Both the λ and *lac* repressor interactions have been studied with these probes. Experiments with the DMS probe show that CAP binding protects the two outer guanines on each side of the binding site for methylation; if pre-methylated these groups strongly inhibit CAP binding (see Fig. 4).[30] The pattern of protection demonstrates symmetry and interaction in two adjacent large groves of the DNA. Experiments with EtNU both confirm the importance of the symmetry and show that, like *lac* repressor, CAP interacts with only one side of the DNA helix over the 14 base region covered by the symmetry axis.

Binding of CAP to DNA in the Absence of cAMP

Whereas binding of CAP to DNA can be seen in the absence of cAMP and is greater the lower the ionic strength, such binding may involve nothing more than the attraction between two oppositely charged mac-

---

[29] W. Gilbert, J. Majors, and A. Maxam, *in* "Organization and Expression of Chromosomes" (A. Allfrey *et al.*, eds.), p. 167. Dahlem Workshop Report, Berlin, 1976.

[30] J. Majors, Ph.D. Thesis, Harvard University, Cambridge, Massachusetts (1977).

romolecules. It is likely that at physiological ionic strength *in vivo* this type of binding will be slight and it probably has no biologic influence.

One factor which has been ignored in all binding studies is the tertiary structure of the DNA. There is good reason to believe that chromosomal DNA is highly negatively supercoiled and that the *lac* operon in particular is much more active in transcription in this form. The importance of supercoiling to transcription will be taken up in the next section.

Involvement of CAP in Transcription

Genetic mutants isolated in Beckwith's laboratory have been instrumental in determining the site of action of CAP in the *lac* promoter.[26] This site was localized to a region of about 39 base pairs (bp) by two deletion mutants starting to the left of the operon and stopping at positions -86 and -47 (Fig. 4). The first of these mutants shows completely normal expression; the second shows about 5% of normal expression and this low level of expression is insensitive to *cya* to *crp* mutations which greatly reduce normal expression. The two base replacement mutations L8 and L29 (located at -57 and -66 bp, respectively), which are symmetrically disposed about the 14 bp segment known to bind CAP, both reduce *lac* expression to about 5%. Three base replacement changes at -34, -10, and -9 are present in the so-called *UV5* revertant which shows 50% of normal *lac* expression in *cya* or *crp* mutants. The base changes in *UV5* are believed to influence the binding of RNA polymerase so that prior activation of the promoter by the cAMP–CAP complex is no longer required. In most *in vivo* studies expression of the *lac* operon has been estimated by the amount of β-gal synthesized using a simple colorimetric assay on permeabilized cells. Some experiments have also been done which involve estimating the amount of *lac* mRNA produced.[16] In this case, RNA was measured by hybridization techniques. Such measurements are quantitatively unsatisfactory because of the rapid turnover of mRNA and the concern that an RNA which is not being translated may turnover more rapidly. Such turnover can easily give the impression of less RNA synthesis and lead to an erroneous conclusion with respect to whether control is being exerted at the transcriptional or translational level.

In vitro experiments have led to more definitive evidence about the role of CAP in *lac* expression and the quantitative aspects of the reaction mechanism involved. The complete *in vitro* system is composed of DNA, a crude cell-free extract of *E. coli* (S-30), and all the salts and substrates necessary for RNA and protein synthesis.[9] Complications which might result from the presence of *lac* repressor are overcome by adding antirepressor isopropylthiogalactopyranoside (IPTG) or more simply by using an

S-30 derived from a strain with a deletion of the entire *lac* region including the repressor gene. The use of such a strain also eliminates background *lac* mRNA and β-gal. Usually *lac* expression is measured colorimetrically after synthesis by the hydrolysis of orthonitrophenylgalactoside. This is a highly specific assay for β-gal. Using this system in conjunction with DNA from λd*lac* virus, it was possible to show that cAMP stimulates expression from 8- to 30-fold.[5] The ratio of active to inactive *lac* operon is proportional to the square of the cAMP concentration at low levels, suggesting the involvement of two cAMP molecules in the reaction.[17] The involvement of CAP in the reaction was demonstrated by making the S-30 from a *crp*⁻ strain first isolated in Beckwith's laboratory.[7] Synthesis of β-gal with such an S-30 was reduced to about 5% and showed no enhancement with added cAMP. Purified CAP added to this S-30 system stimulates synthesis but only in the presence of cAMP. The dependence on CAP concentration is linear, suggesting the involvement of one CAP molecule in the reaction. The proposed involvement of two cAMPs and one CAP protein is compatible with the dimeric structures of CAP. Since most dimers have dyadic symmetry, it was predicted from this result that the reaction with DNA would be symmetrical and involve a region of the DNA with dyadic symmetry.[19] The recently obtained results of Majors[30] (discussed above) on the DNA binding of CAP have elegantly verified this prediction. The mutant DNAs L8 and UV5 showed the anticipated behavior in the S-30 system. Thus L8 gave a low level of expression even in the presence of cAMP and CAP, whereas UV5 had a high level of expression which was independent of cAMP and CAP. That cAMP and CAP only affect transcription was demonstrated in two ways in the S-30 system: (1) when *lac* mRNA was used to direct β-gal synthesis instead of *lac* containing DNA the system was insensitive to cAMP,[31] and (2) when DNA was used to direct synthesis, translation can be delayed by leaving out amino acids.[32] If rifampicin or DNAse are added at the same time that amino acids are added, the processes of transcription and translation are temporally segregated. Under these conditions it was shown that cAMP only affected the transcription phase of the reaction.

The S-30 system has also been used to show a stimulation of *lac* transcription by ppGpp and by supercoiling of the DNA. Magic spot, ppGpp, stimulates transcription by two- to three-fold.[16,32] Supercoiling results in about a fivefold stimulation.[33] The physiologic significance of stimulation

[31] G. Reiness and G. Zubay, *Biochem. Biophys. Res. Commun.* **53**, 967 (1973).
[32] G. Reiness, H.-L. Yang, G. Zubay, and M. Caskel, *Proc. Natl. Acad. Sci. U.S.A.* **72**, 288 (1975).
[33] H.-L. Yang, K. Heller, M. Gellert, and G. Zubay, *Proc. Natl. Acad. Sci. U.S.A.* **72**, 3304 (1979).

TABLE II
EFFECT OF CAP PROTEIN AND cAMP ON *lac* TRANSCRIPTION[a]

TABLE II
EFFECT OF CAP PROTEIN AND cAMP ON *lac* TRANSCRIPTION[a]

Template	CAP protein	cAMP	Cpm hybridized to	
			$\lambda plac_L \lambda_L$	$\lambda plac_H \lambda_H$
$\phi 80\ dlac_{III}$	−	−	1	22
$\phi 80\ dlac_{III}$	+	−	189	58
$\phi 80\ dlac_{III}$	+	+	2289	124

[a] [³H]RNA is synthesized in a purified transcriptional system described in detail elsewhere. The reaction mixture was incubated 10 min at 37°, terminated, extracted, and hybridized. [³H]RNA (45,000 cpm) was annealed to separated strands of λplac and λ DNA; the difference is expressed above; cpm annealing to λ separated strands was always less than 0.5% of the [³H]RNA input. Data from Eron and Block.[35]

by magic spot is not clear. This is a compound whose concentration increases appreciably under certain adverse growth conditions, such as amino acid starvation; this has been shown to account for the inhibition of stable RNA synthesis. The physiological significance of the supercoiling seems much clearer, as most of the bacterial chromosome is highly supercoiled.[34] Supercoiling favors the unwinding of duplex DNA which is a prerequisite of initiation. This probably explains the stimulation effect on initiation. This probably explains the stimulation effect on initiation of *lac* mRNA synthesis.

More purified *in vitro* systems have also been used to study transcription from *lac* operon containing DNA. Such systems contain purified RNA polymerase, cAMP, CAP, and those salts and substrates necessary for transcription. When using large DNA templates, such as λd*lac* DNA, only a small fraction of the DNA is the *lac* operon. As a result only a few percent of the RNA synthesized from such a template is *lac* mRNA. In such a situation, the responsiveness of *lac* mRNA synthesis to cAMP and CAP is best seen by assaying the radioactively labeled RNA by selective hybridization. An excellent series of experiments of this type have been reported which use a φ80d*lac* DNA template.[35] Labeled RNA is hybridized to the fractionated strands of λplac DNA. The only regions of homology of these two DNAs are the regions carrying the *lac* operon. Thus the transcript arising from one DNA is hybridized against the other DNA under conditions which eliminate noncomplementary RNA; an accurate estimate of *lac* operon product should be obtained. Such an assay

---

[34] R. Kavenoff and O. Bowen, *Chromosome* **59**, 89 (1976).
[35] L. Eron and R. Block, *Proc. Natl. Acad. Sci. U.S.A.* **68**, 553 (1971).

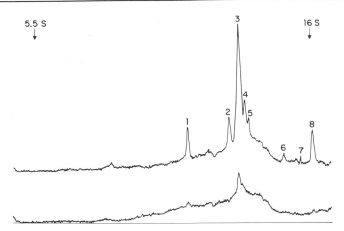

FIG. 5. The transcription profile from D_{3A} in the presence (above) and absence (below) of cAMP and CAP.[37] One microgram of DNA, 2×10^{-4} cAMP in a buffer containing 40 mM Tris-HCl, pH 7.9, 0.1 M KCl, 0.01 M MgCl$_2$, 10^{-4} M EDTA, 10^{-4} M DTT were mixed in the cold. Two micrograms CAP and 1.5 μg of RNA polymerase holoenzyme were added and incubated for a further 2 min period followed by the addition of 8 μg of heparin and then 5 μl of triphosphate solution (ATP, GTP, and CTP at 0.4 mM and [^{3}H]UTP at 0.2 mM). The specific activity of UTP was 2.5 mCi/μmole. Synthesis was done for 20 min at 37° in a final volume of 0.1 ml. RNA was isolated and electrophoresed on a 2% agarose-0.5% polyacrylamide gel, fluorographed, and densitometered by procedures described elsewhere.[38]

shows as strong preference for the synthesis of *lac* mRNA synthesis in the presence of both cAMP and CAP (see Table II).

Purified transcriptional systems have also been used in conjunction with much smaller DNA restriction fragments containing the *lac* promoter and a promoter proximal segment of the *lac* operon. In such cases, the labeled RNA was characterized by gel electrophoresis as one or more homogenous RNAs whose synthesis is initiated at the *lac* promoter in the usual direction. Majors,[36] the first to succeed with this technique used a 200 base pair fragment containing the *lac* operon, obtained by fractionation of a *Hae* III digest of λp*lac* 5 DNA. The main transcript obtained using this DNA as a template in a purified transcriptional system migrated as a single-stranded polynucleotide about 56 bases in length. This is the size of RNA that would result from a *lac* mRNA initiated at the *lac* promoter and terminating in the *lac* operon at one end of the *Hae* III 200 fragment. The synthesis of this RNA was reduced to about 5% if either cAMP or CAP was missing from the synthesis mixture. In another study from our laboratory[37] the template used was a 760 base pair fragment (D_3A) containing the

[36] J. Majors, *Proc. Natl. Acad. Sci. U.S.A.* **72**, 4394 (1975).
[37] S. Mitra, Ph.D. Thesis, Columbia University, New York (1976).

lac operon obtained by fractionation of a *Hin*dII digest of λp*lac*5 DNA. This fragment also contains part of the *lac i* gene, all of the *lac* control region and part of the operator-proximal *lac Z* gene. The main product in this case was an RNA estimated from its electrophoretic migration rate as containing about 460 bases (see peak 3, Fig. 5).[37,38] Again this is about the size of fragment that would be expected from the known distance between the initiation site in the *lac* operon and the *Hin*dII cut in the *lac Z* gene. Here also synthesis is greatly reduced if either cAMP or CAP is left out of the incubation mixture. A number of subsidiary RNA peaks also obtained under optimum conditions have been partially analyzed. The main argument that all of these RNAs are originated from the *lac* promoter is the requirement for cAMP and CAP. A number of other fragments obtained from restriction enzyme digests of λp*lac* which do not carry the *lac* region have been tested and were not stimulated in transcription by cAMP and CAP. The most likely explanation for the complex result with the D_3A fragment is that synthesis does not always stop at the same point. An RNA from the normal initiation point to the end of the D_3A fragment should be about 460 bases in length with a molecular weight of about 150,000. The main product has a molecular weight of about this size. The other products appear to be related to this by having additional RNase-resistant tails added on to the main molecule. Size and preliminary sequence studies indicate that these hairpinned double helix regions include transcripts from both strands of the last third of the D_3A fragment which may be reiterated one or more times to produce a family of different-sized products. It seems likely that these extraordinary RNA molecules are produced by the polymerase turning around on the template, when it reaches near the end of D_3A, and continuing to transcribe from the opposite strand until it reaches the region around base pair 320. At this point either termination takes place or the polymerase can turn around again and make one or more additional passages. Even peak 1 in Fig. 5 has at least one hairpin tail of the type described above. This would give it a molecular weight about 35% larger than peak 3. In spite of this its migration rate is substantially faster than peak 3. This is not unreasonable since this RNA is partially in the duplex form, and double helix RNA migrates much more rapidly than single-stranded RNA on gels containing polyacrylamide.

Although this turning around of the polymerase on the template probably has little to do with the cAMP and CAP reaction specifically, it illustrates the type of complex result which may be obtained *in vitro*. It is not clear if products of this type occur *in vivo*, since the *E. coli* enzyme RNase III would digest double helix RNAs. Despite the type of complica-

[38] H.-L. Yang, G. Zubay, and S. B. Levy, *Proc. Natl. Acad. Sci. U.S.A.* **73**, 1509 (1976).

tions seen here (which may in fact be very informative) the use of the purified transcription systems in conjunction with small restriction fragments is seen as giving most useful results concerning those factors essential for the initiation and elongation of RNA synthesis.

[78] Lactose Operator–Repressor Interaction: Use of Synthetic Oligonucleotides in Determining the Minimal Recognition Sequence of the Lactose Operator

By CHANDER P. BAHL, RAY WU, and SARAN A. NARANG

The sequence of the lactose operator as determined by Gilbert and Maxam[1] contained a 21-nucleotide-long region of twofold rotational symmetry (Fig. 1). A number of O^c mutants have also been sequenced, and all the mutations reside within a stretch of DNA 13-nucleotides long.[2] Thus the sequence of the lactose operator required for specific recognition by the *lac* repressor appears to be between 13 and 21 nucleotides in length. By synthesizing the various sequences in this region of *lac* operator DNA and then studying the binding properties of these synthetic molecules with the *lac* repressor, it is possible to determine the minimal recognition sequence.[3]

The following oligonucleotides were synthesized by the modified triester method of Narang *et al.* ([61], this volume)

> 5' d(A-A-T-T-G-T-G-A-G-C-G-G-A-T-A-A-C-A-A-T-T)
> 5' d(A-A-T-T-G-T-T-A-T-C-C-G-C-T-C-A-C-A-A-T-T)
> 5' d(A-A-T-T-G-T-G-A-G-C-G-G)
> 5' d(A-A-T-T-G-T-T-A-T-C-C-G-C-T-C)
> 5' d(G-A-G-C-G-G-A-T-A)
> 5' d(T-A-T-C-C-G-C-T-C)

Preparation of a 21-Nucleotide-Long Duplex DNA

Two complementary 21-nucleotide-long oligonucleotides, d(^{32}pA-A-T-T-G-T-G-A-G-C-G-G-A-T-A-A-C-A-A-T-T) and d(^{32}pA-A-T-T-G-T-T-A-T-C-C-G-C-T-C-A-C-A-A-T-T) (100 pmoles each), were dissolved in 100 μl of Tris-HCl (pH 7.1), heated to 90° for 1 min and quickly chilled to 0°. The formation of the duplex DNA was carried out by incubating the

[1] W. Gilbert and A. Maxam, *Proc. Natl. Acad. Sci. U.S.A.* **70**, 3581 (1973).
[2] W. Gilbert, J. Gralla, J. Majors, and A. Maxam, *in* "Protein-Ligand Interactions" (H. Sund and G. Blauer, eds.), p. 193. de Gruyter, Berlin 1975.
[3] C. P. Bahl, R. Wu, J. Stawinsky, and S. A. Narang, *Proc. Natl. Acad. Sci. U.S.A.* **74**, 966 (1977).

(A)

5'   A–A–T–T–G–T–G–A–G–C–G–G–A–T–A–A–C–A–A–T–T

3'   T–T–A–A–C–A–C–T–C–G–C–C–T–A–T–T–G–T–T–A–A

      1   3   5   9   13   17   19   21

(B)

5'   T–T–G–T–G–A–G–C–G–G–A–T–A–A–C–A–A

3'   A–A–C–A–C–T–C–G–C–C–T–A–T–T–G–T–T

      3   5   9   13   17

FIG. 1. The sequences in the lactose operator region. (A) The 21-base pair *lac* operator. (B) The 17-base pair *lac* operator (nucleotides 3–19).

mixture at 70° for 30 min, followed by slow cooling (with the water bath turned off) to room temperature in 4–5 hr. The sample was further cooled to 4° in a refrigerator. The duplex DNA was isolated by gel filtration on a Sephadex G-75 column (0.6 × 75 cm) at 4°. The oligonucleotides were eluted with 100 mM NaCl in 10 mM Tris-HCl (pH 7.1). The fractions were counted in a scintillation counter for Cerenkov radiation. DNA first eluted from the column was concentrated to approximately 300 μl, and 3 volumes of cold ethanol were added. After storage at $-20°$ overnight, the sample was centrifuged at 12,000 g for 1 hr. The DNA pellet was dissolved in 200 μl of 10 mM Tris-HCl (pH 7.5) and used for binding experiments.

Binding Assay

The labeled *lac* operator duplex (0.4 pmoles) was dissolved in 100 μl of binding buffer containing 10 mM Tris-HCl (pH 7.5), 10 mM MgCl$_2$, 10 mM KCl, 0.1 mM EDTA, 0.1 mM dithiothreitol, 50 μg/ml of bovine serum albumin and 5% dimethyl sulfoxide. Different volumes of the *lac* repressor (15 μg/ml) were added and the samples were incubated at room temperature for 30 min. Three equal volumes (30 μl) of this incubation mixture were put on 13 mm Millipore membrane filters (HAWP 01300, pore size 45 μm), which were presoaked in washing buffer (binding buffer without the serum albumin and dithiothreitol). The filters were sucked with moderate suction (for approximately 10 sec) and washed once with washing buffer (about 5 sec). Some samples were also incubated in the presence of 1 mM isopropylthiogalactoside (IPTG). The amount of radioactivity retained on the filter in the presence of duplex DNA and repressor was corrected for background radioactivity retained on the filter in the absence of repressor. By plotting the radioactivity retained on filters against the amount of repressor, a typical binding curve (Fig. 2) was

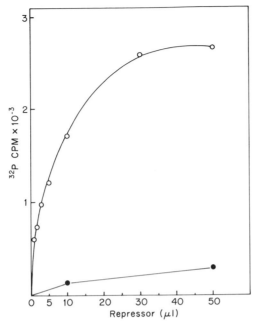

FIG. 2. Binding curve of the synthetic *lac* operator to the *lac* repressor. Each point represents the average amount of radioactive material retained on the filter when 30 µl of sample was filtered. (○) No or (●) 1 mM isopropylthiogalactoside present.

obtained. Binding reached a plateau value at high repressor concentrations.

Synthesis of *lac* Operator Sequences of Varying Length by Repair Synthesis Using AMV Reverse Transcriptase

A mixture of the template strand (100 pmoles) and the primer strand (300 pmoles) in 0.1 M Tris-HCl (pH 8.3) was heated to 90° for 1 min and quickly cooled to 0°. The mixture was incubated at 70° for 30 min and slowly cooled to room temperature and then to 4°. The mixture was adjusted to the following salt concentrations: 50 mM Tris-HCl (pH 8.3), 50 mM KCl, 10 mM dithiothreitol, 10 mM MgCl$_2$, and 5–10 µM each of the desired dNTP's, one of them ^{32}P labeled. After the addition of 10 units of avian myeloblastosis viral reverse transcriptase,[4] the mixture was incubated at 23° for 4–6 hr. The primer extension reaction was terminated by the addition of 5 µl of 0.5 M EDTA and the mixture was loaded onto a

[4] D. L. Kacian and S. Spiegelman, this series, Vol. 29, p. 150.

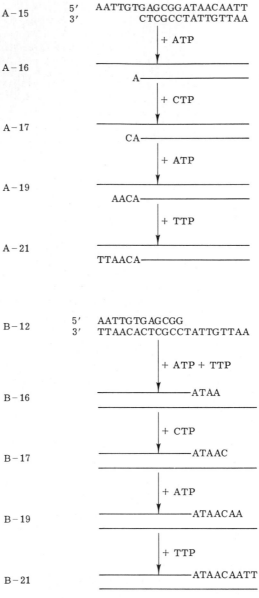

FIG. 3. Schematic representation of the synthesis of *lac* operator molecules of various lengths.

TABLE I

Repressor Binding Efficiency of Different Sizes of Operator Fragments[a]

Length of duplex operator	Binding (%)
16	1.5 ± 0.5
17	12 ± 2
19	40 ± 2
21	40 ± 2

[a] The upper 21-mer single strand was annealed to the 15-mer single strand as shown in Fig. 3, structure A-15. Partial repair synthesis was carried out to produce structures with duplex lengths of 16, 17, 19, and 21. The filter binding efficiency was determined by incubating 1 pmole or the operator sequence (approximately 50,000 cpm) with a tenfold excess of *lac* repressor. Three equal volumes were filtered on three 13-mm filters. The percentage of input counts retained on the filter (% binding) has been corrected for background binding (0.5% of input counts) in the presence of 10^{-3} M isopropyl-thiogalactoside. The 40% plateau value is amost as high as that with the 50,000 base pair long λφ*dlac* DNA which plateaued between 40 and 70% depending on the conditions used.

Sephadex G-50 (fine) column (0.6 × 40 cm) at 4° to remove the dNTPs. The DNA was eluted from the column with a buffer containing 100 mM NaCl and 10 mM Tris-HCl (pH 7.1). The fractions were monitored by counting for Cerenkov radiation, and those containing the duplex DNA were concentrated in a vacuum desiccator to approximately 200 μl. Ethanol (600 μl) was added and the sample stored overnight at −20°. The sample was centrifuged at 12,000 g for 1 hr. The DNA sample was dissolved in 10 mM Tris-HCl (pH 7.1) and used for binding studies or for further extension of the primer strand (Fig. 3). The length of the extended primer was monitored by the mobility of the product on homochromotography in Homo mixture III.[5]

Binding of the Lactose Repressor with Lactose Operator of Varying Lengths

The DNA molecules of varying lengths prepared by repair synthesis were subjected to binding assay with the *lac* repressor as described earlier. The percent binding for varying lengths of the operator sequences is shown in Table I. When the upper 21-mer template and the lower 15-mer primer were extended to duplex lengths of 16, 17, 19 or 21, the 16-mer

[5] R. Wu, E. Jay, and R. Roychoudhury, *Methods Cancer Res.* **12**, 88 (1976).

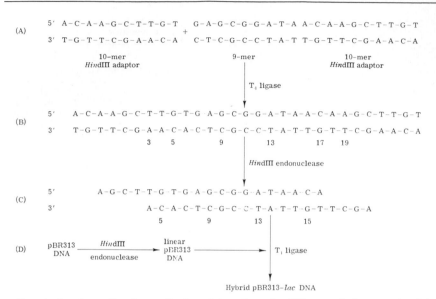

Fig. 4. A scheme for the synthesis and insertion of a 17-base pair *lac* operator into pBR313 plasmid DNA. In structure (B), the underlined segment represents the 17-base pair *lac* operator (nucleotides 3–19). The cells carrying the hybrid pBR313-*lac* DNA produced blue colonies on X-gal plates. Control experiments leaving out T4 ligase in either of the two steps produced no blue colonies.

gave a very low level of *lac* repressor binding (1.5%). The addition of one extra nucleotide to give a 17-mer duplex showed an eight- to tenfold increase in repressor binding, indicating that the G-C pair at position 5 had an important contribution in the specific recognition by the *lac* repressor. Addition of two more A residues, to extend the length of the duplex region to position 3 (Fig. 1A), produced another fourfold increase in repressor binding. This suggests that the two A-T base pairs at position 4 and 3 provided additional contact points with the repressor molecule. Further addition of two nucleotides to make the duplex length of 21 did not increase the binding, indicating that the A-T pairs numbered 1 and 2 were not essential for repressor recognition and binding. Similar results were obtained when lower strand 21-mer was used as a template and duplex DNA of varying lengths were prepared by using 12-mer as primer (Fig. 3B).

From these results, it can be concluded that the A-T pairs at position 1, 2, 20, and 21 in the 21-mer duplex are probably not necessary for repressor recognition. All the essential features of the *lac* operator for maximal repressor binding are included in nucleotides 3–19 (Fig. 1) which constitute a stretch of 17 base pairs.

Synthesis of a Lactose Operator by Blunt-End Ligation of a 9-Mer Duplex
to *Hin*dIII Adaptor Molecules[6]

Two complementary nonanucleotides (9-mer), d(G-A-G-C-G-G-A-T-A) and d(T-A-T-C-C-G-C-T-C), were annealed together to form a duplex structure (Fig. 4A). This 9-mer duplex (12 pmoles) and the synthetic 10-mer *Hin*dIII adaptor molecules (75 pmoles) were joined end-to-end (blunt-end ligation[7]) by incubation at 20° for 6 hr in the presence of 4 units of T4 DNA ligase in 100 μl of a solution containing 20 mM Tris-HCl (pH 7.5), 10 mM dithiothreitol, 10 mM MgCl$_2$, and 35 μM ATP. The solution was heated to 70° for 5 min to inactivate the ligase and then slowly cooled to room temperature. Two volumes of ethanol were added and after 16 hr at $-20°$, the DNA was centrifuged at 10,000 g for 1 hr. Among the reaction products in the DNA pellet was the desired 29-nucleotide duplex DNA containing the 17-nucleotide-long *lac* operator sequence.

Insertion of the 17-Nucleotide-Long Lactose Operator Sequence into pBR
313 Plasmid DNA[8]

The DNA product obtained by the blunt-end ligation of the 9-mer duplex with *Hin*dIII adaptors was dissolved in 50 μl of a solution containing 6.6 mM Tris-HCl (pH 7.5), 6.6 mM MgCl$_2$, and 10 mM dithiothreitol. To this solution was added 1 μg of pBR 313 DNA[9] and 2 units of *Hin*dIII endonuclease (Fig. 4B). The samples were incubated at 37° for 6 hr to produce a 15-nucleotide-long duplex DNA containing 4-nucleotide-long single-stranded ends and pBR 313 DNA with the same single-stranded ends. The samples were then heated at 70° for 5 min, cooled slowly to room temperature, and the DNA precipitated as above with ethanol. The DNA pellet was dissolved in 50 μl of a solution containing 20 mM Tris-HCl (pH 7.5), 10 mM MgCl$_2$, 10 mM dithiothreitol, 35 μM ATP, and 3 units of T4 ligase. The sample was incubated at 12.5° for 24 hr to produce the hybrid pBR 313-*lac* DNA (Fig. 4D) that was used directly for transformation.

[6] C. P. Bahl, K. J. Marians, R. Wu, J. Stawinsky, and S. A. Narang, *Gene* **1**, 81 (1976).
[7] V. Sgaramella, J. H. van de Sande, and H. G. Khorana, *Proc. Natl. Acad. Sci. U.S.A* **67**, 1468 (1970).
[8] C. P. Bahl, R. Wu, and S. Narang, *Gene* **3**, 123 (1978).
[9] R. L. Rodriguez, R. Tait, J. Shine, F. Bolivar, H. Heyneker, M. Betlach, and H. W. Boyer, *Miami Winter Symp.* **13**, 73 (1977).

TABLE II
β-Galactosidase Specific Activity in Cells

Experiment	E. coli strain	Plasmid	Lac operator linked to plasmid	β-Galactosidase specific activity[a] (no IPTG)
1	5346	—	—	0.006
2	5346	pMB9	lac 21	0.30
3	5346	pBR313	—	0.007
4	5346	pBR313	lac 17	0.28
5	MM294	—	—	0.0016
6	MM294	pOP203-1	lac 203[b]	0.55

[a] The β-galactosidase specific activity (Z/B) was expressed as defined by Jobe et al.[11] In experiments 1, 3, and 5 the β-galactosidase assay was carried out at 28° for 1, 2 and 4 hr, respectively, in experiments 2, 4, and 6 it was for 10, 20, and 40 min, respectively. For fully induced culture (with 1 mM IPTG), the assay was carried out for 10, 20, and 40 min and the β-galactosidase activity was approximately 2.2 in all experiments (data not shown).

[b] Escherichia coli strain MM294 and the same strain carrying pOP203-1 plasmid was a gift of F. Fuller, L. Johnsrud, and W. Gilbert.

Transformation and Selection of pBR 313-lac Hybrid Plasmids

Hybrid pBR 313-lac plasmid DNA was mixed together with competent recipient cells[10] (E. coli strain 5346) and incubated at 0° for 30 min. The temperature of the mixture was raised to 42° for 2 min and then dropped to 0° to facilitate uptake of DNA by the cells. Nine volumes of L broth were added and the cells allowed to recover at 37° for 2 hr; ampicillin was added to a concentration of 100 μg/ml. After an additional 30 min at 37°, three different levels of cells were plated on nutrient agar plates containing x-gal (20 μg/ml) and ampicillin (100 μg/ml). The selection of colonies containing lac-pBR 313 hybrid plasmid was based on the overproduction of β-galactosidase that cleaves x-gal to produce a compound with blue color. The plates were incubated at 37° for 24–36 hr, and blue colonies were picked and grown in 1 ml of L broth.[11] An aliquot of the cells was used for the assay of β-galactosidase as described below. Another aliquot of the cells was used to inoculate 100 ml of complete M9 medium for the isolation of DNA after chloroamphenicol amplification.

Determination of β-Galactosidase Activity

The E. coli cells were grown in T broth (Bacto-tryptone, 10 g/liter, NaCl, 8 g/liter). The overnight culture (0.05 ml) was transferred to 2.5 ml of

[10] K. J. Marians, R. Wu, J. Stawinsky, T. Hozumi, and S. A. Narang, Nature (London) 263, 744 (1976).

[11] A. Jobe, J. R. Sadler, and S. Bourgeois, J. Mol. Biol. 85, 231 (1974).

T broth $\pm$ 1 mM isopropylthiogalactoside in a 20-ml test tube and grown at 37° to a density between 0.2–0.3 at 350 nm. The cells were chilled to 0° for 20 min and aliquots (0.02 to 0.20 ml) diluted to 1 ml in Z buffer.[12] The cells were lysed by the addition of 0.1 ml of chloroform and 0.1 ml of 1% sodium dodecylsulfate. The samples were incubated at 28° for 5 min and the β-galactosidase assay initiated with the addition of 0.3 ml of O-nitrophenyl-β-D-galactoside (4 mg/ml). Incubation was at 28° for 10 min to 4 hr. The reaction was stopped by adding 0.5 ml of a 1 M Na$_2$CO$_3$ solution, and the cell debris was pelleted at 6,000 g for 10 min. The absorbance of the clear supernatant solution was read at 420 nm. Units of β-galactosidase specific activity (Z/B) were calculated according to Jobe et al.[11] and are shown in Table II.

The results showed that cells containing a 17-nucleotide-long lac operator (Table II, experiment 4) can produce a high level of β-galactosidase, comparable to that of cells carrying a 203-nucleotide-long natural lac operator region (Table II, experiment 6). Thus, the synthetic 17-nucleotide-long lac operator is functional $in\ vivo$.

[12] J. H. Miller, "Experiment in Molecular Genetics," p. 352. Cold Spring Harbor Lab., Cold Spring Harbor, New York, 1972.

[79] Electron Microscopy of Proteins Bound to DNA

By ROBERT SCHLEIF and JAY HIRSH

Direct observation of proteins bound to DNA is a convenient method to answer a number of questions on mechanisms of gene regulation and nucleic acid metabolism. We have used one of these approaches—binding from a solution onto positively charged hydrophylic carbon grids and staining with uranyl formate—to determine the conditions necessary for RNA polymerase binding to promoters, repressor binding to operators, and formation of induction complexes on the arabinose operon. Also, microscopy was used to determine the positions of the bound proteins or protein complexes on DNA molecules and the conformation of DNA to which repressor or RNA polymerase had bound.[1,2]

Advantages of these techniques are their parsimonious consumption of biological macromolecules and the high information yield of a direct visual method. Additional advantages of the technique are that the mac-

[1] J. Hirsh and R. Schleif, J. Mol. Biol. 108, 471 (1976).

[2] J. Hirsh and R. Schleif, Proc. Natl. Acad. Sci. U.S.A. 73, 1518 (1976).

romolecules are rapidly bound from solution onto the grid. This apparent "snapshot" of molecules as they existed in solution appears not to dissociate weakly associated complexes since there is no surface denaturation during the binding step. Fixing or covalent attachment of proteins to the DNA is unnecessary. The absence of extraneous proteins or of contrast enhancement by shadowing permits high resolution and thus, occasionally, morphological information of proteins may be extracted.

A disadvantage of the electron microscopic approach in contrast to biochemical approaches is the need for expensive and intricate equipment. By contrast to other microscopic approaches, the DNA is not as well spread and it is preferable to use fragments under several thousand base pairs long. The contrast of images is poorer than techniques that use shadowing, and the higher magnifications necessitate using a better functioning microscope and experienced microscopist. Although the preparation of samples for observation requires only a few hours and their observation only another hour, the complete determination of the binding position of a protein on DNA to a precision of ± 20 base pairs requires the equivalent of 2 or 3 day's work. The techniques required are not difficult to learn, and in our experience most biochemists with above average mechanical aptitude have no trouble in applying these methods to their own work.

Equipment Necessary

Our experience is limited to the Phillips EM 300 electron microscope which we have routinely used and an occasional use of the Philips EM 201 electron microscope. All that follows will refer to the EM 300. The scope must be well aligned and relatively clean so that residual astigmation is low. The scope is operated at 60 kV and contains 300-μm condenser apertures and a 25-μm thin gold film objective aperture (Ebtec Corporation, Attleboro, Massachusetts). We replaced the binoculars that were originally provided with the scope with Zeiss 10× binoculars. These give better contrast and illumination and facilitate focusing and astigmating.

The high vacuum evaporator that is used for preparation of the carbon film and activation of the grids does not appear to be critical, but it must be kept clean. We have used a High Vacuum Equipment Corporation Model G-71 evaporator. This is equipped with a liquid nitrogen trap, and Dow Corning No. 704 silicone oil is used in the diffusion pump. The machine attains a vacuum of 2×10^{-5} torr in less than 3 min. This rapid pumping rate minimizes the exposure of the grids to contaminants which could interfere with staining. All interior surfaces of the evaporator are kept as clean as possible. After each session of carbon evaporation the "apparently clean" bell jar is scrubbed with tissue wet with ethanol until no traces of carbon can be picked up by the tissues. The metal base is cleaned as well. Contaminants in the evaporator appear to be responsible for excessively dense staining of the grids.

DNA and Proteins

It is most convenient to use DNA fragments which are between 500 and 2000 base pairs long. These are generally available from restriction enzyme digests of cloned DNA. Usually these have been extracted from agarose or acrylamide gels and therefore are contaminated with gel material. Although such DNA is usable for most enzymological reactions, it is frequently unusable for microscopy as it is covered with the gel material. The agarose or acrylamide are not easily removed by precipitation. In the past we have found passage of the DNA through a DEAE column to be most satisfactory for removing extraneous material. The DNA is diluted into 0.1 M KCl, 0.1 M Tris-HCl (pH 7.4). Whatman DE52 is degassed by resuspending in 0.1 M HCl and aspirating. This is then poured in a Pasteur pipette and flushed with at least 10 column volumes of 0.1 M KCl, 0.1 M Tris-HCl (pH 7.4). The DNA is loaded in any volume. The capacity of the column is about 25 μg DNA per ml of column material. The column is flushed with at least 3 volumes of 0.4 M KCl and 0.1 M Tris-HCl (pH 7.4), and then the DNA is eluted with 2 column volumes of the same buffer whose KCl concentration has been raised to 0.6 M. These steps are all performed at room temperature. Originally yields were greater than 80%, but recently recovery has been much lower, and we have turned to passing the DNA through an agarose A-5m column in 2.5 M KCl and 0.01 M Tris-HCl (pH 7.4). The salt is then removed from the DNA by dialysis, and the DNA is concentrated by ethanol precipitation and resuspension in any convenient storage buffer.

Proteins have also been found to be contaminated by small molecules which bind to the DNA and interfere with staining. These have also been conveniently removed by passing the protein at room temperature through an agarose column in buffer containing 2.5 M KCl. For RNA polymerase agarose A 0.5 M was satisfactory. The KCl is easily removed by dialysis.

Forming the Carbon Layer on the Grids

A typical number of grids to make is 40 to 80 and these last about 2 weeks before the carbon film breaks. The procedure involves forming the Parlodion layer, picking the grids up from the Parlodion, evaporating carbon onto the Parlodion, and dissolving the Parlodion from beneath the carbon layer.

Forming the Parlodion Layer

1. Use a stainless steel tray, 20 × 20 × 1 cm deep. This must be carefully cleaned before use.

2. The Teflon bars, 25 cm long, one square in cross section, the other triangular, must be cleaned before use.

3. Fill the tray with the cleanest available double distilled water[3] until the water is well above the edges, being held there by surface tension.

4. Place the triangular Teflon bar in front on the water surface and supported by the edges of the tray and slide to the rear. This cleans the surface of the water.

5. Place the rectangular bar at the rear, just in front of the beveled bar and draw forward. This leaves the central region wiped twice with the Teflon bars. Watch carefully during these procedures for anything on the surface which is able to slip past the bars at the edges of the tray. If this occurs, more water is needed in the tray.

6. Drop from 2.5 cm above the surface one drop of 5% Parlodion dissolved in amylacetate. Such a solution remains usable for about 6 months. The Parlodion should spread to form an irregular circle about 5 cm in diameter.

7. Wait about 1 to 3 min until edges of the Parlodion ruffle. If the surface is too dirty there will be more than the usual amount of ruffling.

8. Drop 400 mesh grids in a regular pattern, close to each other near the edge, in a flat spot, shiny side down. This takes about 10 min.

Picking Grids Up from the Parlodion

1. Drop a piece of newsprint paper, very carefully, from a height of about ½ inch onto the grids. Wait until the paper has fully wet itself, which may take up to 10 min. The process can be hastened somewhat by dripping water on the back. The grids do not adhere to very thin newsprint or Whatman No. 1 as these papers wet too rapidly.

2. After the paper is fully wet, remove Parlodion from outside the area of the paper and pick the paper up from the corner. All of the grids should remain attached to the paper.

3. Blot dry from the back with filter paper and let dry for about 30 min.

---

[3] We do not know the minimum quality of water adequate for these procedures. The water used in all steps of the descriptions given here is house distilled water that is distributed in tin-lined pipes. This is redistilled in a glass still. At no time is the distilled water or the redistilled water permitted to contact any plastic or any greased fitting or valve.

Evaporating Carbon onto the Parlodion

1. Evacuate the bell jar to about 10^{-4} torr 0.1 μm Hg without liquid nitrogen in the trap.

2. Add liquid nitrogen and pump for about 10 more minutes. The pressure must be no greater than 10^{-5} torr.

3. Pass sufficient current through the carbon rod to heat it to an orange color for about 1 min. This will expel absorbed gases. Allow at least a minute to cool before admitting air. The carbon rods are 2.5 mm in diameter. One rod is flush-ended and the other is sharpened.

4. Place the grids which are still on the newspaper in a petri dish, shiny side up, and weigh down with a few coins and place in the bell jar.

5. After evacuating the bell jar 10^{-5} torr, the carbon is evaporated from a distance of 15 cm for 5 to 10 sec. The carbon rod is heated very bright but should be emitting no sparks. After allowing the carbon rods to cool, the petri plate may be removed from the bell jar. The most satisfactory thickness of carbon is that which leaves barely perceptable shadows of the coins when they are observed in a bright light. Additional carbon may be evaporated if the carbon layer is too thin.

Removing the Parlodion Layer

1. Make a tray about 2×2 cm of Whatman filter paper, being very careful not to touch the surface which will be used.

2. Transfer the grids to the paper, carbon side up, removing them from the newsprint. Note that the paper should be cut to fit in the bottom of a 10 cm diameter jar.

3. Place the paper in the jar without disturbing the grids.

4. Pour isoamyl acetate carefully down the side of the jar until the paper is fully wet. Tightly close the jar and leave 3 hr at room temperature. Remove the paper tray from the jar and place the grids on filter paper in a petri plate until they are needed.

Binding to the Grid and Staining

Suitable samples of protein and DNA are mixed and incubated. We have not extensively explored the suitability of various buffer components, but in several year's work we have found no ingredients incompatible with the technique. Our buffers were confined to those resembling standard physiological conditions in *Escherichia coli*. A typical experiment might use DNA at a concentration of 0.5 μg/ml with 5 μg/ml RNA polymerase. This is incubated at 37° for 10 min in a volume of 20 μl. After the incubation poly(I) is added to a final concentration of 0.3 μg/ml. The droplet is then transferred to a piece of Saran wrap stretched over a petri plate. This and all subsequent steps may be at room temperature. Then an activated grid is touched to the surface and allowed to float on the surface for up to 1 min. A drop of 20 μl will accommodate two grids simultaneously.

The grids must be activated shortly before their use. Typically we activate them about 15 min before we use them. The activation procedure is patterned after that described by Dubochet et al.[4] The grids are placed, shiny side up, on a piece of Whatman filter paper along with several coins. These are placed in the bell jar of the evaporator and the mechanical pump is used to generate a vacuum of 100 μm Hg and then the rough pumping valve is closed. The vapors from 8 to 10 drops of isoamylamine are then admitted to the bell jar. This is conveniently done by opening a glass stopcock to a bottle that had been connected to the air inlet line of the bell jar. During rough pumping the air inlet valve is opened so that the air inlet line is also evacuated. The entrance of the isoamylamine vapors raises the pressure in the bell jar to about 200 μm which is then pumped back down to 100 μm. After the roughing valve has again been closed, the high voltage is turned on for 20 sec. The 10 kV ac voltage is applied to a 15 cm high post in the bell jar surmounted by a 10 cm square of aluminum foil. During the discharge a purple glow is seen in areas of the bell jar. The coins reduce dancing by the grids during the activation.

If the protein concentration is too high, the grid holds stain much too avidly and nothing may be seen. Nonetheless, highly impure protein solutions may be used at high protein concentrations if most of the free protein is removed before binding the DNA and protein to the grids. This is accomplished by passing the sample through a small, 300-μl agarose A-5m column. The ability of such a column to separate macromolecules according to size yields several 20-μl fractions with DNA and a few bound proteins, while later fractions contain the bulk of the free protein. Such a column may be run in a 0.6-ml polyethylene Eppendorf tube using a capillary tube pushed through the end to form the drops. This step is accomplished by slicing most of the end of the tube off with a razor blade first. Running such a column requires about 90 sec, and the fractions are collected directly on the Saran wrap and grids are immediately floated on them.

Staining uses uranyl formate, a stain shown by Brack[5] to stain both DNA and protein. We have used uranyl formate from EM Sciences only. This should be stored as a dry powder in the dark. Within 30 min of the

[4] J. Dubochet, M. Ducommun, M. Zollinger, and E. Kellenberger, J. Ultrastruct. Res. 35, 147 (1971).
[5] C. Brack, Experimentia 29, 768 (1973).

time of staining, 10 mg of small crystals are vigorously mixed with 1 ml of water. Not all the crystals go into solution. Immediately before use the stain solution is centrifuged at 12,000 g for 1 min to pellet all insoluble material. An Eppendorf desk centrifuge is convenient for this step. Different lots and different sources of uranyl formate yield different degrees of contrast, and it is best to try a wide collection of suppliers until a satisfactory lot is found. Different-sized crystals from the same bottle yield different staining properties. Drops of 50 μl of stain are added to the Saran wrap. The grids are transferred from the sample drops to the stain drops. Uniformity in staining is achieved by floating the grid on the surface for a second and removing it, then replacing it for a total of about five times before letting the grid rest on the drop of stain for about 30 sec. The grid is removed from stain and lightly blotted from the edge with filter paper. Only about 80% of the stain which could be removed by blotting should be removed. Excessive removal of stain lowers the contrast of the staining, while leaving too much stain on the grid obscures large regions of the grids. With a little practice more than 95% of all grids prepared by these methods will contain usable areas.

The poly(I) addition reduces the nonspecific binding of RNA polymerase. Heparin may at a concentration of 50 μg/ml also be used for this purpose, but then the sample must be passed over the agarose column before binding to the grids.

Focusing and Astigmating the Phillips EM 300 for High Magnification
    Work

The condition of the scope is crucial for success in high magnification work. In order to have proper illumination at the highest magnification the following qualifications should be met.

1. The gun must be adjusted to have 60 μA of current when the filament is saturated.

2. The condensers C1 and C2 are adjusted to produce a beam of the desired size. Increasing the current through C1, which decreases its focal length, necessitates decreasing the current through C2. This operation has the effect of reducing the diameter of the focused beam. However, it does not increase the luminosity of the beam, since fewer of the electrons are captured by C2 and illuminate the sample. If the beam position were perfectly stable with changes in focus, then the optimum focused beam size would be just the area you observe through the binoculars and all the rest of the specimen would be spared electron damage. However, the prerequisite perfect alignment of the scope is virtually impossible to obtain, and it is necessary to use a larger beam. For working between mag-

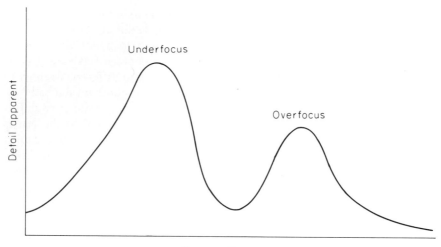

Fig. 1. The apparent change in detail of the background as focus is varied when the astigmatism has been correctly adjusted. There are approximately eight clicks of the finest focus adjustment between the maximum in the underfocus and the maximum in the overfocus positions.

nification taps 13 and 19, we find that C1 set to about 7 and C2 set to whatever will focus the beam to just fill the large viewing screen is optimum.

3. It is essential that the objective aperture and aperture holder be clean. The thin film gold apertures of 20–30 μm must be used in this work.

4. In order that focusing be easy, the focus range control, which is located on the right-hand main panel, should be set to position 4. This adjusts the focus vernier to the finest range, and each click is a change in focus by 125 Å. By comparison, in position No. 1, each click changes the focus by 1000 Å.

5. The astigmatism range control which is located in the left-hand auxiliary panel should be set to position No. 2 or position No. 3. If the scope cannot be astigmated in position 3, then something must be cleaned in the scope, usually the objective aperture or the objective aperture holder. The finest range of this is position 1.

6. The sample holder must be clean and never touched with your hands, only forceps. If the holder is dirty, sonicate with dilute detergent, rinse, sonicate in ethanol, and rinse with ether. Note that small volumes of solvent may be sonicated by placing a beaker containing the solvent in several inches of water in the bath.

7. A grid containing a thin carbon film should be inserted in the scope and scanned in the very low magnification scan made with the objective

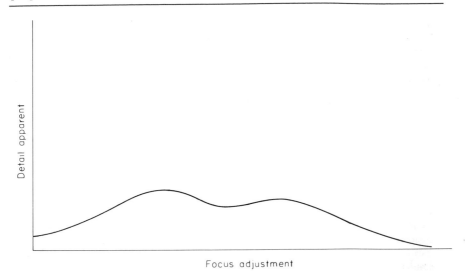

FIG. 2. The apparent change in detail of the background as focus is varied when the astigmatism is far from optimally adjusted.

aperture out to find a position containing a *clean* area. The objective aperture should be reinserted, centered, and the scope set to the magnification position. It should be noted that proper astigmating can be done most easily at the highest magnification. This is most practical at tap 17 or above, but sufficient illumination must be present at these high magnifications. In order to reach approximate focus and high magnification, it is simplest to begin at about tape 13 and reach approximate focus, then switch to tap 15 once again reach approximate focus, and finally go to tap 17 or 19 and reach approximate focus. With magnifiction set to tap 13, move the grid so that the edge of a grid bar crosses the center of the field and adjust the third finest focus adjustment to make a sharp edge. Then move well away from the grid bar and perform the next step. This is conveniently done on some detail near the center of a grid hold such as a crystal of stain or a piece of dirt. At this point it is best to use the second or third finest focus adjustments and scan through the focus range. Observe where the granularity in the image from the crystallites or stain changes from black dots on a white background to white dots on a black background. By scanning back and forth across this region, it is usually straightforward to locate the approximate focus at the position midway between the two "dot extremes." The final focusing is performed by watching the granularity in the background. The amount of detail which can be seen as focus is adjusted near perfect focus, and all astigmatism has been corrected is approximately as shown in Fig. 1.

In the position known as underfocus, a considerable amount of detail is apparent, but this almost vanishes as perfect focus is achieved. This detail is spurious, interferes with proper interpretation of images, and results from wave interference effects. Ordinarily when making initial adjustments considerable astigmatism exists when you reach focus. In this situation, the amount of detail (primarily 10 Å crystalites of carbon) apparent is as shown in Fig. 2.

Thus to begin, you reach tap 19 and try to find the *center* of the focus position. Then adjust the astigmatism controls, first one and then the other, trying to reduce the amount of detail you see in the background. When near properly astigmated, as you adjust either astigmatism control in either direction, an image point appears to smear out. By going first to one side and then to the other, you can find the proper position where no smearing occurs. Having reached as close to proper astigmatism as possible with this focus setting, you can then be considerably more precise in finding the in-focus point. When reaching a better focus point (minimum of detail), it is then possible to readjust the astigmatism controls and make the astigmatism correction more accurate. It is possible to astigmate *only* when the scope is exactly focused, and when you are not observing near a grid bar. When one becomes very skilled at observing these images, the following may be a more efficient means to achieve correct focusing and astigmatism control.

Adjust the focus so that the images of carbon granules are seen as not smeared in either dimension but appear as tiny "×'s". The lengths of the arms of the ×'s are then adjusted to a minimum or zero with the astigmatism controls.

Difficulty in astigmating can arise from being near a grid bar, not having the objective aperture properly centered, or having a dirty objective aperture or sample holder.

Fortunately astigmatism correction does not need to be varied as magnification is varied. Therefore, it is possible to adjust the astigmatism once every 10 or 15 min at the highest magnification and to touch it up only occasionally, and thereafter work at lower magnifications. Figure 3 shows a well formed and astigmated micrograph of RNA polymerase bound to fragments of λ phage DNA.

Photography and Analysis of Data

It is by far the most convenient to use 35-mm roll film to record data. Kodak Release Positive No. 5302 possesses sufficient sensitivity and contrast. This is conveniently purchased in 100-ft rolls and wound from a film bulk loader onto the spools of the 35-mm scope camera. The film is best

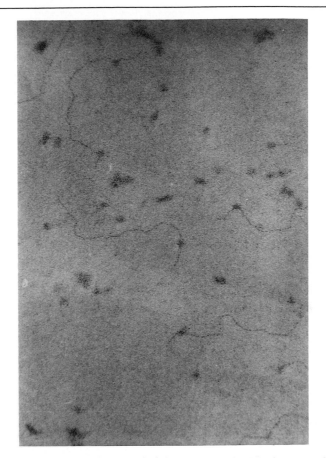

FIG. 3. A typical micrograph. Several of the 970 base pair DNA fragments from phage λ contain an RNA polymerase molecule bound at the promoter 110 base pairs from one end. Additional molecules of RNA polymerase are not bound to the DNA.

exposed so as to yield an optical density of between 1.5 and 2.0 when it is developed. These negatives do not possess sufficient contrast for convenient analysis of their images. They are, therefore, rephotographed onto High Contrast Copy Film, Kodak No. 5069 with a 35-mm camera equipped with bellows, a 50-mm macro lens, and a slide copying attachment. For most work with DNA of 500 to 2000 base pairs, it is convenient to scan grids at a screen magnification of 3.5×10^4 (tap 15) and to record images at a film magnification of 3×10^4 (tap 17) (screen magnification 10^5). The photographic contrast enhancement is done at close to $\times 1$ magnification.

For all the work we have done we have photographed paracrystalline tropomyosin with each roll of film exposed at the scope. Absolute magnification was calculated by assuming a repeat distance in this material of 39.5 nm. With this standard we found a variation of magnification of less than 1% upon removing a grid and reinserting it. Over the period of several years, the magnification of the scope appeared to vary less than 10%. A reasonable compromise magnification standard available to most laboratories would appear to be a segment of sequenced DNA fragment. Reliable use of such a standard would entail micrographs of several DNA molecules in order to attain requisite accuracy (see below).

The most common information to be extracted from micrographs of a protein bound to DNA is the position of the binding. Imperfections in the grids appear to be responsible for the need to measure many molecules in order to determine a binding position with high precision. The standard deviation σ, in the measured lengths of homogeneous DNA molecules whose average length is (L) was found to be $\sigma = 1.4(L)^{\frac{1}{2}}$, where σ and (L) are in the base pairs. Thus for an average length or binding position (L) that has been measured on a population of n DNA molecules, the interval $(L) \pm 2\sigma/n^{\frac{1}{2}}$ has a 95% probability of actually containing the mean length of the population of molecules from which the sample of n was chosen. Typically, measurements on 20 to 100 molecules provide adequate accuracy.

The origin of the differences measured in the lengths of identical molecules has not been determined, but it seems likely that it results from the random foreshortening of the DNA molecules as they drape themselves over the irregular carbon surface. The overall foreshortening in our hands yields an average interbase distance of 2.7 Å in DNA, whereas we would expect it to be 3.4 Å if there were no irregularities.

The final step in the analysis of the rephotographed molecules is tracing them onto paper and measuring them. Many microfilm viewers possess adequate optics, and we have used the Kodak MPE-1 viewer to project the images onto typing paper where they were drawn with pencil. For measurement of hundreds of molecules, the inexpensive Keuffel and Esser 620315 map measurer is appropriate, and if thousands of molecules are to be measured, the Numonics Corp. Electronic Digitizer is convenient.

Author Index

Numbers in parentheses are footnote reference numbers and indicate that an author's work is referred to although his name is not cited in the text.

Subject Index

A

I